System Dynamics

System Dynamics
Modeling, Analysis, Simulation, Design

Ernest O. Doebelin
The Ohio State University
Columbus, Ohio

MARCEL DEKKER, INC.   NEW YORK · BASEL · HONG KONG

Library of Congress Cataloging-in-Publication Data

Doebelin, Ernest O.
   System dynamics: modeling, analysis, simulation, design / Ernest O. Doebelin.
     p. cm.
   Includes bibliographical references and index.
   ISBN 0-8247-0126-7
   1. Systems engineering. 2. Dynamics. I. Title.
TA168.D588   1998
620'.001' 185–dc21   97-44123
                                                                                                                 CIP

An earlier version of this book was published as *System Dynamics: Modeling and Response*, C. E. Merrill, Columbus, Ohio, 1972.

The publisher offers discounts on this book when ordered in bulk quantities. For more information, write to Special Sales/Professional Marketing at the address below.

This book is printed on acid-free paper.

Copyright © 1998 by MARCEL DEKKER, INC. All Rights Reserved.

Neither this book nor any part may be reproduced or transmitted in any form or by any means, electronic or mechanical, including photocopying, microfilming, and recording, or by any information storage and retrieval system, without permission in writing from the publisher.

MARCEL DEKKER, INC.
270 Madison Avenue, New York, New York 10016
http://www.dekker.com

Current printing (last digit):
10 9 8 7 6 5 4 3 2 1

PRINTED IN THE UNITED STATES OF AMERICA

PREFACE

The overall organization and viewpoint of the earlier version* of this book have proven sound over the years and are thus preserved in this edition. Changes have been mainly in the form of expansions that allow the instructor more flexibility in choosing the breadth and depth of treatment desired for the various topics. The main features of the text are:

1. A carefully crafted first chapter that explains how system dynamics fits into an overall curriculum, defines the role it plays within the larger area of engineering design, and classifies the different possible forms of system models and input models
2. In-depth treatment of the elements, to make clear the distinction between math models and real devices, with detailed discussions, illustrations, and design formulas for many practical devices
3. A chapter on energy-conversion devices needed to couple subsystems of different physical types, including induction, servo, and stepper motors, piezoelectric sensors and actuators, positive-displacement hydraulic pumps and motors, electronic amplifiers (smooth, PWM, SCR), and servovalves
4. Comprehensive, but practical and efficient, treatment of differential equation solution methods, including analytical and computer simulation techniques
5. Systematic treatment of all physical forms of first- and second-order systems
6. Extension to systems higher than second order to give a perspective on the general situation
7. A final chapter giving a brief but illuminating comparison of lumped and distributed models, solution methods, and behavior

Since the text is intended to serve not only the needs of an academic course but also those of industrial practice, numerous practical references are footnoted

---

* E. O. Doebelin, *System Dynamics: Modeling and Response*, C. E. Merrill, Columbus, Ohio, 1972.

throughout the text. For example, when discussing the pure/ideal spring element, specific references to automotive industry analytical and experimental studies of the dynamics of real springs are included. The discussion of the elements is in general much more detailed than is found in most system dynamics texts. There are two main reasons for this emphasis. First, I intend the book to be useful beyond the academic course in which it might be used, that is, in the student's later industrial practice. Here one must deal with details of actual hardware and a "spring" must be more than just a number giving its stiffness. Second, a system dynamics course should be an *engineering* course and should begin to make students aware that the simple math models used for springs, resistors, inductors, etc., in basic physics and mechanics courses are only *approximations* to the real devices.

Having now taught system dynamics for 27 years, I am painfully aware that there is not enough time available in the typical course to cover all the "element material" in this text. I do, however, recommend that selected portions of it be assigned for student study and covered in classroom lectures. For mechanical engineering classes, the discussion of the spring element might be a good choice. If this material is given close attention, it will give most students a *general* appreciation of the difference between the commonly used math models and actual physical hardware, for *all* kinds of elements, mechanical, electrical, fluid, and thermal. It will also make them aware that the text has such useful practical content on all the elements, making it a good resource for their future industrial practice.

Another feature of the previous edition that is again prominent in this volume is the early introduction and continual application of *frequency-response* concepts and methods. I have found no better way to define the region of applicability of a particular model than to examine its frequency response—at first analytically and, once hardware is available, experimentally. This viewpoint is presented very early, with the elements, and then extended as the need arises throughout the text. The Fourier spectrum treatment of periodic functions in the earlier edition has been expanded and extended to transients, with a brief but useful discussion of Fourier transform, including use of popular software.

The former edition's strong emphasis on digital simulation, which was unusual at that time, has proven to be correct, and has been expanded in this new treatment. I consider this tool to be much more than an efficient method of getting numerical results. The book uses it as a *teaching and learning* tool, to be integrated throughout the text, rather than being isolated in a separate chapter at the end of the book. This aspect of simulation is most effective when we use a simulation package that employs a graphical user interface. A simulation block diagram resembles the analog computer diagrams of earlier times and shares with them the facility for enhancing the *understanding* of the physical interactions occurring in the real system. When a student first draws, for a mechanical system, a summer whose inputs are the various forces acting on a mass and whose output is the acceleration term in the differential equation, the physical law and its differential equation become "visible" and more easily comprehended. Subsequent integrations to produce velocity and displacement, and the damper and spring forces associated with these motion quantities, further contribute to the understanding of component and system behavior.

While the text mainly uses the SIMULINK software, *all* simulation packages that use graphical interfaces are very similar in application, and this similarity is emphasized in text discussions. I personally would not want to teach system

Preface v

dynamics without student accessibility to a graphical simulation package, but the text certainly can be used in that mode. The simulation diagrams are so closely linked to the system equations that they can be easily understood by readers even if they do not have the benefit of actually running their own simulations. The discussions of simulation *results* are also presented in such a way as to be easily comprehended by readers, without the need to actually run the simulations.

The previous edition used the discussion of mechanical second-order systems as an opportunity to also develop the basic concepts of vibration. This feature has been considerably expanded in this volume since some curricula use part of the system dynamics course as a means of giving all mechanical engineering students a basic introduction to vibration. If the curriculum also includes electives in vibration for those desiring a more comprehensive treatment, the system dynamics course will provide good preparation.

Another area of expansion is that of *electromechanics*. At Ohio State, the two required system dynamics courses in mechanical engineering have recently been reconfigured and renamed "System Dynamics and Vibrations" and "System Dynamics and Electromechanics." To accommodate those who wish to provide examples in the system dynamics course from this important application area, more discussion has been provided. This electromechanics material is presented in such a way that it can easily be left out, if that is desired, without affecting the continuity of other basic topics. The electromechanics emphasis also includes a brief discussion of *mechatronics*, focusing mainly on the dynamic effects of quantization, sampling, and computational delay associated with computer-aided machines and processes. Here, simulation allows these dynamic effects to be understood quickly and easily.

Although the *analytical* foundations of system dynamics will always be rooted in linear, constant-coefficient differential equation theory, we certainly do not want to limit the *design* of practical systems so narrowly. Many system designs can benefit appreciably from the intentional use of nonlinearity, so even beginning students should be made aware of such possibilities. The book uses simulation with many examples of nonlinearity to illustrate the validity of linearized models for small-signal operation, and also the intentional use of specific nonlinearities to achieve improved performance. With simulation so readily available and easily learned these days, even beginning courses in system dynamics should expose students to the positive and negative effects associated with nonlinear operation.

With the increasing curricular emphasis on design, this volume provides several significant design examples. Some of these illustrate the progression from conceptual to substantive to detailed design. In the context of system dynamics, once a design concept has been "invented," one next uses analysis and simulation to find the best values of system-level parameters such as gains, time constants, natural frequencies, and damping ratios. The next level of design chooses numerical values for parameters such as spring constants and damping coefficients. Finally, we are ready to choose a particular form of spring, which then allows the choice of materials and dimensions. Although most courses would not have time for many such comprehensive examples, even one such experience gives most students an improved understanding of the stages encountered in a typical design problem.

The text provides more material than will be covered in the lecture and/or laboratory portion of a typical system dynamics course. I have used a particular teaching method in recent years that I believe to be generally useful: I break up student reading assignments into portions I call "technical browsing" and "study for mastery." In my own engineering career I have found it profitable to develop a "technical browsing" skill, and I believe it is worthwhile for students to cultivate. Practical design engineering requires at least two different kinds of knowledge. The most obvious, and regularly implemented in academic contexts, is the mastery of basic concepts and methods that can be applied to all kinds of situations. The other, less emphasized in most courses, is the simple *awareness* of the existence of certain hardware, reference materials, and engineering techniques. I tell students about these two types of information and try to convince them that they will benefit by devoting some study time in the course to each type. Since it is difficult and probably inappropriate to try to test students on this more nebulous form of information, I don't even try; in fact, I tell them that they won't be tested on this material. Certainly some students will take advantage of my candor and ignore these "browsing" reading assignments, but others will be more conscientious, especially since these readings are practical, mostly devoid of complex equations, and fairly quick to read. I am convinced that this technique of quickly scanning technical material does indeed result in enough "subconscious" retention of useful ideas that it is a worthwhile lifetime pursuit. Much of the "practical" text material on the elements can be treated as technical browsing, since there is certainly not enough time to cover it in class lectures. Depending on the time available and the instructor's own preferences, other text material could also be treated in this way.

For those like myself, who have worked with engineering students for many years, the impression is inescapable that the current crop of students, although quite "computer-competent," is largely unaware of most engineering hardware. While laboratory experiences are perhaps the best way to overcome this deficiency, textbooks that include practical material together with the theoretical are also useful. Since we don't have time to cover much practical material in class, the "technical browsing" approach described above may offer some help in raising students' consciousness of hardware considerations. I have also tried in this book to use for examples, wherever possible, actual industrial hardware that has a specific useful purpose. Thus, when I am presenting some general analysis technique or class of system behavior, I am simultaneously making the reader aware of a practical hardware implementation.

Because system dynamics courses appear at various levels and in various contexts in engineering curricula, it is difficult to recommend a single specific plan for using this text. The previous volume was written for a course at Ohio State that came at the end of the sophomore year. It was the first mechanical engineering course in that curriculum and came immediately after conventional courses in mechanics and differential equations. This volume could certainly be used in this same manner. It could, of course, also be used at any later point in a typical engineering curriculum. If the curriculum includes courses in control, measurement systems, vibration, etc., it would naturally be best if the system dynamics course preceded these, since they can make good use of the background provided by system dynamics.

A perhaps radical approach, but one that I would personally endorse, would be to eliminate the differential equations and electrical engineering service courses often used as prerequisites for system dynamics in a mechanical engineering curriculum. System dynamics, if we make heavy use of simulation, is an ideal place for students

Preface   *vii*

to learn differential equation applications. For non–electrical engineers, drawing on
the preparation of conventional physics courses, the system dynamics treatment of
circuits, electronics, and electromechanics is adequate. This concept is consistent
with current trends to *integrate* typical topics in the curriculum rather than present
them as unrelated fragments in separate courses. This philosophy leads to assigning
more credit hours to system dynamics, using some or all of the hours freed up by
elimination of separate specialty courses. Another desirable result of such integration
is the replacement of several low-credit courses with fewer high-credit courses, allowing students to concentrate their attention. I hope that at least a few schools will give
this idea serious consideration; this text provides sufficient breadth and depth to
implement such a concept.

Depending on the specific prerequisite courses, and the location in the curriculum of the system dynamics course, individual instructors should be able to find in
this text a suitable sequence and selection of topics to meet their needs. A relatively
brief perusal of the table of contents, followed by a more detailed browsing of the
chapter sections, allow an initial selection to be made without much effort. You may
also find the very comprehensive *index* useful in this regard. Naturally, the first
teaching of the course from this text will give a much more reliable reading of
what can be or should be covered in the time available.

Ernest O. Doebelin

CONTENTS

PREFACE   iii

1 INTRODUCTION   1

   1-1   What Is System Dynamics?   1
   1-2   The Input/System/Output Concept   7
   1-3   A Classification of System Inputs   8
   1-4   A Classification of System Models   14
   1-5   System Design   23
       Bibliography   25
       Problems   25

2 SYSTEM ELEMENTS, MECHANICAL   28

   2-1   Introduction   28
   2-2   The Spring Element   29
   2-3   Linearization   43
   2-4   Real Springs   47
   2-5   The Damper (Friction) Element   54
   2-6   Real Dampers   61
   2-7   The Inertia Element   75
   2-8   Referral of Elements Across Motion Transformers   85
   2-9   Mechanical Impedance   90
   2-10 Force and Motion Sources   92
   2-11 Design Examples   99
        Engine Flywheel Example   99
        Accelerometer Transducer Example   104
        Optimum Decelerator Example   107
       Bibliography   116
       Problems   117

Contents

3 SYSTEM ELEMENTS, ELECTRICAL   **123**

3-1	Introduction	123
3-2	The Resistance Element	127
3-3	The Capacitance Element	131
3-4	The Inductance Element	138
3-5	Electrical Impedance and Electromechanical Analogies	148
	Impedance Example	153
3-6	Real Resistors, Capacitors, and Inductors	153
3-7	Current and Voltage Sources	167
3-8	The Operational Amplifier, An Active Circuit "Element"	174
3-9	Modeling and Simulation of Computer-Aided Systems: Mechatronics	185
	Design Example: A Feedback-Type Motion Control System	192
	Bibliography	200
	Problems	201

4 SYSTEM ELEMENTS, FLUID AND THERMAL   **206**

4-1	Introduction	206
4-2	Fluid Flow Resistance and the Fluid Resistance Element	216
	Example: Oscillating Flow	226
4-3	Fluid Compliance and the Fluid Compliance Element	234
	Example: Effective Bulk Modulus	236
4-4	Fluid Inertance	240
	Example: Liquid Inertance	243
4-5	Comparison of Lumped and Distributed Fluid System Models	245
4-6	Fluid Impedance	248
	Example: Use of Differential Equation	252
4-7	Fluid Sources, Pressure and Flow Rate	253
	Example: Real Pressure Source	254
4-8	Thermal Resistance	255
4-9	Thermal Capacitance and Inductance	263
4-10	Thermal Sources, Temperature and Heat Flow	266
	Bibliography	268
	Problems	268

5 BASIC ENERGY CONVERTERS   **272**

5-1	Introduction	272
5-2	Converting Mechanical Energy to Other Forms	272
5-3	Converting Electrical Energy to Other Forms	288
	Example: Induction Motor	298
	Example: Stepping Motor	302
5-4	Converting Fluid Energy to Other Forms	311
5-5	Converting Thermal Energy to Other Forms	313
5-6	Other Significant Energy Conversions	315
5-7	Power Modulators	316
	Example: Motor/Clutch System	330

Contents   **xi**

Bibliography	334
Problems	335

6 SOLUTION METHODS FOR DIFFERENTIAL EQUATIONS   337

6-1 Introduction	337
6-2 Analytical Solution of Linear, Constant-Coefficient Equations:	
The Classical Operator Method	338
Example: Root Finding	340
Example: Complete Solution	342
6-3 Simultaneous Equations	344
6-4 Analytical Solution of Linear, Constant-Coefficient Equations:	
The Laplace Transform Method	350
Linearity Theorem	351
Differentiation Theorem	351
Integration Theorem	351
Example: Simultaneous Equations	354
Laplace Transfer Functions	355
Partial-Fraction Expansion	355
Example: Real Poles	357
Example: Complex Pole Pairs	357
Repeated Roots	358
Example: "Nearly-Repeated" Poles	360
Delay Theorem	363
Example: Discontinuous Input	365
Initial-Value Theorem and Final-Value Theorem	366
Example: Initial Conditions	366
6-5 Simulation Methods	367
Analog Simulation	367
Digital Simulation of Dynamic Systems	370
6-6 Specific Digital Simulation Techniques	385
Generation of Input Signals	385
Side-by-Side Comparisons	387
Event-Controlled Switching	391
6-7 Simulation Software with Automatic Modeling	394
6-8 State-Variable Notation	396
Example: Three-Mass Problem	398
Example: Root Finder Versus Eigenvalues	399
Bibliography	399
Problems	400

7 FIRST-ORDER SYSTEMS   403

7-1 Introduction	403
7-2 Mechanical First-Order Systems	404
Preliminaries to Equation Setup	406
Writing the System Equation	407
The Generic First-Order System and Its Step Response	411

		Experimental Step-Input Testing	417
		Computer Simulation	419
		Design Example: Electric Motor Drive for a Machine Slide	419
		Motion Control by Feedback: An Alternative Design	426
		Optimum Step Response Using a Nonlinear Approach	429
	7-3	Ramp, Sinusoidal, and Impulse Response of First-Order Systems	431
		Ramp Response	431
		Sinusoidal Response (Frequency Response)	433
		Logarithmic Frequency-Response Plotting	437
		Experimental Modeling Using Frequency-Response Testing	441
		Impulse Response of First-Order Systems	444
	7-4	Validation of Linearized Approximations Using Simulation	448
	7-5	Electrical First-Order Systems	450
		General Circuit Laws and Sign Conventions	451
		Practical Examples of Electrical First-Order Systems	452
		Analysis of Passive and Active Low-Pass Filters	454
		Design Example: Low-Pass Filter	457
		Design Example: Approximate Integrator	459
		Design Example: Optical Sensor	462
	7-6	Elementary ac Circuit Analysis and Impedance Methods	466
		ac Circuit Analysis Example	469
	7-7	Fluid First-Order Systems	470
		Basic Laws Useful for Equation Setup	472
		Linearized and Nonlinear Analysis of a Tank/Orifice System	472
		Numerical Example: Nonlinear and Linearized Response of Tank/Orifice System to Step and Sine Inputs	474
		Design Example: An Accumulator Surge-Damping System	478
	7-8	Thermal First-Order Systems	479
		Systems with Several Inputs	483
	7-9	Mixed First-Order Systems	484
		Electromechanical Open-Loop Speed Control	484
		Electromechanical Closed-Loop (Feedback) Speed Control	486
		Hydromechanical Systems: A Hydraulic Dynamometer	489
		Hydromechanical Systems: Open-Loop Hydraulic Speed Control	490
		Thermomechanical Systems: Thermal Expansion Actuators	492
		Thermomechanical Systems: A Simple Friction Brake	496
	7-10	First-Order Systems with "Numerator Dynamics"	498
		Design Example Showing Where System Dynamics Fits in the Overall Design Sequence	503
		Bibliography	512
		Problems	513

8 SECOND-ORDER SYSTEMS AND MECHANICAL VIBRATION FUNDAMENTALS — 521

	8-1	Introduction	521
	8-2	Second-Order Systems Formed from Cascaded First-Order Systems	522
		Cascaded Subsystems: The Loading Effect	523

Contents xiii

		Example: Loading Effect in Two Mechanical First-Order Systems	525
8-3		Mechanical Second-Order Systems	527
		Step Response and Free Vibration of Second-Order Systems	527
		Example: Initial Energy Storage	531
		Example: Design of Package Cushioning for Dropped Packages	535
		Significance of K, ζ, and ω_n	537
		Design Example: High-Speed Scale for Packaging Conveyor	538
8-4		Lab Testing Second-Order Systems Using Step Inputs	539
		Detecting Nonviscous Damping in Transient Testing	542
8-5		Ramp Input Response of Second-Order Systems	544
8-6		Frequency Response of Second-Order Systems	546
8-7		Vibration Isolation and Transmissibility	550
		Design Example: Vibration Isolation of Electric Motor	550
		Force Transmissibility	553
		Motion Transmissibility	555
		Rotating Unbalance	555
		Acceleration to Operating Speed: "Transient Resonance"	558
8-8		Impulse Response of Second-Order Systems	560
8-9		Electrical Second-Order Systems	562
		A Passive Low-Pass Filter	562
		Series Resonant Circuit	568
		ac Power Numerical Example	570
		Band-Pass filters	572
		Notch Filters	573
		Op-Amp Circuits	576
		Design Example: Op-Amp Circuit	578
8-10		Fluid Second-Order Systems	579
		Example: Using Various Checking Methods to Find Errors	581
		Example: Pressure-Measuring System Dynamics	586
8-11		Thermal Second-Order Systems	587
		Improved Tank Heating Model	587
		Accelerated Coffee Cooling	589
8-12		Mixed Second-Order Systems	591
		Hydraulic Material-Testing Machine: Resonance Put to Good Use	591
		dc Motor Control by Field and Armature	595
8-13		Systems with Numerator Dynamics	599
		Automobile Handling Dynamics	599
		Leadlag Dynamic Compensator (Approximate Proportional Plus Derivative Plus Integral Control)	607
		Bibliography	612
		Problems	613
9	**GENERAL LINEAR SYSTEM DYNAMICS**		**621**
9-1		Introduction	621
9-2		System Modeling and Equation Setup	622

9-3	Stability	626
9-4	Generalized Frequency Response	631
9-5	Matrix Frequency Response	640
9-6	Time-Response Simulation	642
9-7	Frequency Spectrum Analysis of Periodic Signals: Fourier Series	644
	Example: Square Wave	646
	Example: Experimental Data	652
	Fourier Series Calculations Using Fast Fourier Transform (FFT) Software	654
	Using Simulation to Compute Complete (Transient and Periodic Steady-State) Response of Linear or Nonlinear Systems to Periodic Inputs	657
9-8	Frequency Content of Transient Signals: Fourier Transform	662
	Example: Rectangular Pulse	663
	Example: Fourier Transform	665
9-9	Experimental Testing Using Spectrum Analyzers	667
9-10	Dead-Time Elements	669
9-11	Another Solution to Some Vibration Problems: The Tuned Vibration Absorber	673
9-12	Improved Vibration Isolation: Self-Leveling Air-Spring Systems	678
9-13	Electromechanical Active Vibration Isolation	683
9-14	An Electropneumatic Transducer Using a Piezoelectric Flapper Actuator	686
9-15	Web-Tension Control Systems	695
	Bibliography	701
	Problems	701

10 DISTRIBUTED-PARAMETER MODELS 707

10-1	Longitudinal Vibrations of a Rod	707
10-2	Lumped-Parameter Approximations for Rod Vibration	713
10-3	Conduction Heat Transfer in an Insulated Bar	719
10-4	Lumped-Parameter Approximation for Heat Transfer in Insulated Bar	724
	Bibliography	726
	Problems	726

APPENDIXES

A	Viscosity of Silicone Damping Fluids	729
B	Units and Conversion Factors	733
C	Thermal System Properties	735

INDEX 737

System Dynamics

1

INTRODUCTION

1-1   WHAT IS SYSTEM DYNAMICS?

Many undergraduate student readers of this book will have completed physics courses which included treatment of the portion of mechanics called *dynamics*. Some will also have followed this with an entire mechanics course on dynamics, as part of the engineering science portion of an engineering curriculum. How does the subject of system dynamics, which first appeared in engineering curricula in the 1960s, relate to these "classical" topics which have been part of engineering programs almost from their beginning in the late 1800s?

Since the words "system" and "dynamics" individually have well-established meanings, we can start from there, but will need to go further to make clear the definition of "system dynamics" as an engineering subject. "Dynamics" refers generally to a situation which changes with time, and a "system" is commonly considered an assemblage of components or elements, so system dynamics must somehow deal with the time-varying behavior of connected components. This statement is true, but it is not "the whole truth," so more must be said.

Let's now summarize the essential features of system dynamics in a few brief statements which will then be elaborated. System dynamics:

1. Deals with *entire* operating machines and processes rather than just isolated components.
2. Treats the dynamic behavior of not just mechanical, but also electrical, fluid, thermal, and "mixed" systems. (Actually, chemical dynamics could also be included, but these are sufficiently unique that their treatment is usually reserved for specialist texts.)
3. Emphasizes the behavioral similarity between systems that differ physically and develops *general* analysis and design tools useful for all kinds of physical systems.
4. Sacrifices detail in component descriptions so as to enable understanding of the behavior of complex systems made from many components.
5. Uses methods which accommodate component descriptions in terms of experimental measurements, when accurate theory is lacking or is not

cost-effective. Develops *universal* lab test methods for characterizing component behavior.

6. Serves as a common unifying foundation for many later courses and practical application areas. Examples: Vibration, measurement systems, control systems, circuit analysis, acoustics, vehicle dynamics.

7. Offers a wide variety of computer software to implement its methods of analysis and design.

Figure 1-1 will be used to elaborate on some of these statements; let's start with statements 1 and 2. The *electropneumatic* (E/P) *transducer* shown in Fig. 1-1 is a generic version of a device offered for sale by a number of firms. It is used in industrial control systems where part of the system is electronic while another part is pneumatic and these two parts must "talk" to each other. A system dynamic study of this device would consider the *entire* transducer as a complete system. A course on the design of machine components would, by contrast, deal only with the design of, say, the cantilever beams used as leaf springs. A course on circuit analysis would deal only with the voltage/current relations in the electric coil. A fluid mechanics course would deal only with the air flow in the nozzle/flapper. System dynamics uses results from all these (and other) fields to deal with the *entire* device.

Before turning to statements 3 to 5 let's use this device to clarify our use of the phrase "entire system." The overall purpose of the E/P transducer is to accept (from some electronic system not shown) an input voltage signal e_{in} in the range 3–15 volts and produce a proportional output air pressure signal p_o in the range 3–15 psig. This output pressure might be used in a pneumatic valve actuator controlling steam flow in a 4-inch-diameter pipe in an industrial heating process (see Fig. 1-2). When we said that system dynamics deals with *entire* machines and processes, the interpretation of "entire" needs to be somewhat flexible. That is, *we* get to choose the "envelope" that defines the portion of the world that we wish to consider for study. If we work for the company that makes E/P transducers, then we consider the device of Fig. 1-1 as our *system*. On the other hand, if we work for the company designing the industrial heating system, the E/P transducer is treated as a *component* which we purchase ready-made and which we assume has certain desirable operating characteristics. We combine these component characteristics with those of the other parts of our large system when we analyze and design it. The methods of system dynamics are used by both the E/P transducer designer and the designer of the larger temperature control system; thus the meaning of "entire system" depends on the context of the application.

To elaborate on statements 3 to 5 it will be useful to give first a word description of device operation. The air supply comes from an air pressure regulator which maintains a constant supply pressure, typically 20 psig. If we would apply an input voltage e_{in}, this would cause a current to flow in the coil. We know from physics that a current-carrying wire in a magnetic field (provided by the permanent magnet shown) feels a force proportional to the current strength, in our case an upward force on the coil. We also know that a cantilever beam (leaf spring) will deflect in proportion to the force applied to it. Thus *voltage causes current which causes force which causes deflection.* Suppose now that we apply such a large voltage that the coil and attached "flapper" move upward so far that the air nozzle is completely sealed off. With the chamber containing pressure p_o now completely sealed, the supply

Introduction

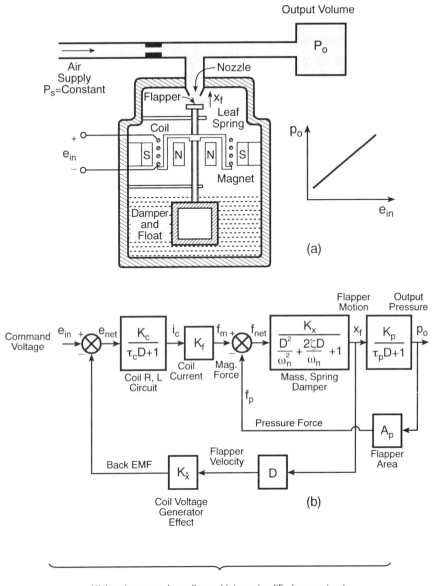

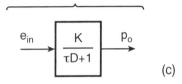

Figure 1-1 Electropneumatic transducer.

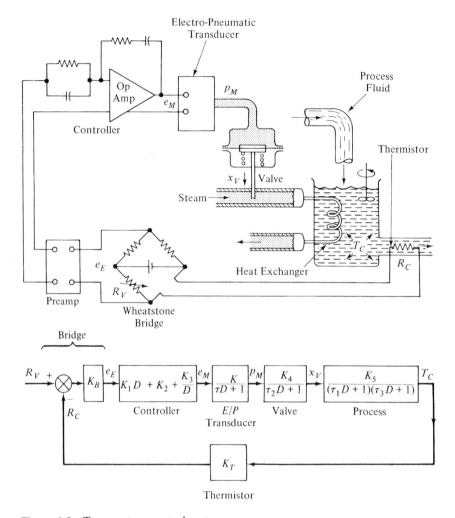

Figure 1-2 Temperature control system.

pressure will force air into this volume until its pressure becomes equal to the supply, in our case 20 psig. If we now consider input voltages *less* than that for which the flapper seals the nozzle, we see that the partially open nozzle is a "leak" and the pressure p_o will go to some pressure less than 20 psig. The lower the voltage and the larger the leak, the lower will be the pressure. Thus we can conveniently control the pressure by changing the input voltage.

While the operating principle just described is not hard to understand, transducer designers have a much harder task. The device must not only work in principle, it must meet certain *performance specifications* if it is to fulfill its practical applications. These specifications usually deal with two main areas: linearity and speed of response. Linearity means that the graph of output pressure versus input voltage (for steady-state conditions) should be very close to a straight line; typical values allow only a 0.5% deviation from perfection. Speed of response means that if we make a sudden change in input voltage, the air pressure must settle to its new

Introduction

steady value within a certain maximum allowable time, 0.2 second being a typical value. There are several electrical, mechanical, and fluid effects in this device that prevent an instantaneous response of the pressure.

Statement 3 mentions an important feature of system dynamics, the organization of diverse physical systems into generic classes which have identical behavior patterns. If we once learn the pattern of a certain class, we never again have to repeat this work when we encounter a new physical example of that class. Two fundamentally significant classes are called *first-order systems* and *second-order systems*. These will be pursued in detail later in this text but we can now start thinking along these lines since the E/P transducer includes examples of each. Since most readers of this book will have some background in simple electrical circuits from a physics course, let's first look at the *R-L* circuit used to model the coil and shown separately in Fig. 1-3a. From the known behavior of resistance and inductance, and from Kirchhoff's voltage loop law we get:

$$Ri_c + L\frac{di_c}{dt} - e_{net} = 0 \tag{1-1}$$

$$\frac{L}{R}\frac{di_c}{dt} + i_c = \frac{1}{R}e_{net} \tag{1-2}$$

$$\tau_c \frac{di_c}{dt} + i_c = K_c e_{net} \tag{1-3}$$

$$(\tau_c D + 1)i_c = (K_c)e_{net} \tag{1-4}$$

The R-L circuit is our first example of the important class of first-order systems (we will be seeing many more later in the text). In Eq. (1-3) we have defined the two standard parameters used to describe *all* first-order systems, the time constant τ_c and the steady-state gain K_c (also called the static sensitivity). The solution of the differential equation when the voltage is a step input is shown in Fig. 1-3. It turns out that speed of response is governed entirely by τ_c; the smaller the value of τ_c the quicker

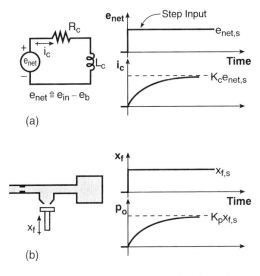

Figure 1-3 Analogous systems: electric and pneumatic.

the response. This is true of *all* first-order systems; once we discover this we never have to repeat the solution process again. The meaning of K_c is also simple: The steady value of the output (current) is just K_c times the value of the input (voltage). Again, this is true of *all* first-order systems. Finally, Eq. (1-4) uses the D-operator ($D = d/dt$) to allow definition of the *operational transfer function*:

$$\text{Operational transfer function} \triangleq \frac{i_c}{e_{\text{net}}}(D) \triangleq \frac{K_c}{\tau_c D + 1} \tag{1-5}$$

(In this text, the symbol $\triangleq$ means "defined to be .") We will have many uses for transfer functions and the concept will be developed in detail later; for the time being it allows us to draw the *block diagram* of Fig. 1-1b.

Note in this diagram that a first-order transfer function is also found between the signal x_f and p_o. This means that the electrical R-L circuit and the pneumatic nozzle-flapper device, two *very different* physical systems, behave in exactly the same way, since their differential equations, when reduced to a standard form, are *identical*. This illustrates statement 3 above. Since the fluid mechanics analysis[1] of the nozzle-flapper device is beyond our scope here we simply ask you to take the result on faith. This analysis shows that at least six parameters are required to describe the behavior. Once we recognize the device as first-order, we see that these six can be "condensed" into only the essential two, τ_p and K_p, as shown in the block diagram. This illustrates statement 4. The *advantage* of describing the overall system with as few essential parameters as possible becomes most apparent when we consider the *design* of such systems. Design generally involves an *optimum* selection of numerical values for all the system parameters. Such optimization studies are much simpler for systems with few parameters, since the possible number of combinations which must be studied grows rapidly with the number of parameters. The block diagram of Fig. 1-1b includes 10 parameters, the absolute minimum. If we had not defined standard parameters such as K's, τ's, etc., there would be many more parameters to deal with at the system design level, making the determination of an optimum design much more difficult and time consuming.

Statement 5, dealing with the use of experimental testing to get numbers for system parameters, can also be illustrated with the E/P transducer. The nozzle-flapper analysis referenced above uses many simplifying assumptions, which cause some inaccuracy in predicting numerical values for K_p and τ_p. Thus these parameters are usually found by experiment as soon as this part of the device has been built. Simple step input tests as shown in Fig. 1-3b are usually sufficient to get the accurate values needed. Note that in complex systems, small errors in each of a large number of parameters can cause large errors in predicting overall system behavior. Thus as components are built or purchased, it is quite common to test them to get the accurate parameter values needed for overall system design.

I hope this brief discussion is giving you a beginning insight into the nature of system dynamics and how it differs from but relates to other topics you may have studied. We next expand our discussion to consider the input/system/output concept, which is a basic viewpoint taken in system dynamics.

---

[1]E. O. Doebelin, *Measurement Systems*, 4th ed., McGraw-Hill, New York, 1990, p. 292.

Introduction   **7**

1-2   THE INPUT/SYSTEM/OUTPUT CONCEPT

In system dynamics, as in other engineering studies, we often use diagrams to aid our own understanding and communicate ideas to others, as we did with Fig. 1-1. Common forms of diagrams include *pictorial, schematic, and block types*. Figure 1-1a is a pictorial diagram, Fig. 1-1b is a block diagram, and Fig. 1-3a is a schematic diagram. Pictorial diagrams are the closest to a "photographic view"; they look like the real object, though simplified. Schematic diagrams use the standard symbols of the technology (such as the familiar R and L symbols in Fig. 1-3a) to *represent* the system, usually for analysis purposes. They don't look like the physical object. Block diagrams can take several useful forms. Sometimes block diagrams contain only word descriptions, but in system dynamics we usually want them to be much more specific, as in Fig. 1-1b). This type of diagram actually contains *all* the information about the system's behavior that is included in the set of equations which describe system operation. That is, the transfer functions shown in the blocks are just a shorthand and graphic way of stating the equations and clearly displaying the relations among the system variables.

These useful diagrams are based on a certain way of thinking about system behavior: the *input/system/output concept*. Any physical device, whether simple or complex, may be considered in this way. Inputs may be thought of as those entities which cause a system to respond with some sort of action or output; that is, there is a cause and effect relation between the inputs and the outputs. In Fig. 1-1b, voltage *causes* current, current *causes* magnetic force, force *causes* spring deflection, deflection *causes* pressure change, etc. The system accepts inputs and responds with outputs. This idea is not limited to system dynamics—you have probably used this kind of reasoning before. System dynamics, however, formalizes the concept and makes it mathematically specific by the use of differential equations and transfer functions.

While the diagrams we have displayed so far don't make this clear, a system need not have only one input and one output. A system can, for example, have three inputs and five outputs, if its physical arrangement and operation dictate this. When this is the case, the definition of transfer function and block diagram must be extended appropriately. In drawing block diagrams and defining transfer functions, we can *choose* the level of detail which we wish to display, according to our needs. A *designer* of the E/P transducer of Fig. 1-1 might prefer the detail of Fig. 1-1b since it makes clear the operation and interconnection of all the components. A *user* of already-designed units (such as the designer of the system of Fig. 1-2), would on the other hand be unconcerned with such detail and prefer the *overall* description of Fig. 1-1c.

You may be wondering how the complex system of Fig. 1-1b could possibly be reduced to the simple single block of Fig. 1-1c. We cannot be totally convincing at this point in the book since the details necessary are yet to be developed. However, the essence of the argument lies in the proper choice of numerical values during the design process. If we *want* the overall behavior to be described by the simple transfer function of Fig. 1-1c, we will choose parameter values which encourage this. The "proof of the pudding" here is a lab test of the final design. If, in the lab test, the overall device behaves nearly as a simple first-order system, this will be an adequate model, no matter how complex the internal details might appear. Another factor allowing such simplification is a restriction on operating conditions. If we know in a specific application (such as that of Fig. 1-2) that the input voltage to the E/P transducer will vary only slowly, then the simplified model may be adequate. Note that if such (justified) simplification of component complexity is *not* actively pursued, then the design of the overall system is complicated by the need to consider a large number of design parameters.

Finally we want to illustrate that a given system can be simultaneously subjected to more than one input. For the E/P transducer of Fig. 1-1, the main input is the voltage command which causes the pressure to change. An *undesired* input which nevertheless must be considered in such precision instruments is the ambient temperature. Temperature, in fact, affects almost every physical phenomenon, so it is an "input" to many engineering systems, whether we like it or not. When temperature changes are extreme and/or system specifications require great precision, this input must be considered in system design. For the E/P transducer, let's pretend that the input voltage is fixed, producing a fixed output pressure. Now let the temperature gradually change. The pressure *should* stay fixed, but it will not, for several reasons. The resistance of the electric coil will change with temperature. So will the strength of the permanent magnet. So will the stiffness of the leaf springs. So will the viscosity of the oil in the damper. So will the density of the air. So will the dimensions of all the mechanical parts. All these changes will affect the air pressure in some way, and the cumulative effect may not be negligible. The E/P designer needs to take all this into account at the design stage. When the device is finally built, the overall temperature effect is easily found by experiment. This "temperature sensitivity" is an important characteristic of such precision instruments and is listed on the specification sheet sent to potential customers.

While you can at this point probably think of many practical machines or processes which could be considered from the input/system/output viewpoint, we offer Fig. 1-4 as some specific examples from the important field of automotive engineering.

We have defined and discussed inputs and we next want to show a useful classification of this aspect of system dynamics.

1-3   A CLASSIFICATION OF SYSTEM INPUTS

Having adopted the input/system/output concept we now want to present useful classifications for inputs and systems. When we define the inputs and the system, the outputs are determined, so there is no need to classify outputs as separate entities.

Recall that by an input we mean some agency which can cause a system to respond. One useful categorization breaks inputs into two broad types: initial energy storage and external driving. Consider a mass suspended from a spring, with the mass stationary at its equilibrium position. We might be interested in the vertical motion of the mass; thus our outputs might be the displacement, velocity, or acceleration of the mass. Note, as we mentioned earlier, that a system can have several outputs, so we might choose to call the displacement the output of interest, or we might choose the velocity or acceleration or some combination of these. How can we get our system to respond? If we grab the mass and pull it further down, when we let go, the mass will surely move even though we are not then exerting any external force on it. This is an example of an input in the form of *initial energy storage*. When we stretch the spring we are giving it potential energy, which it can later give up, causing

Introduction

Engineering Group With Design Responsibility	Inputs of Interest to this Group	Corresponding Outputs
Riding and Handling Qualities Noise Reduction Structural Integrity	Road profile, wind gusts, steering wheel motions, brake application, accelerator application.	Passenger vibration, structural vibration, acoustic noise level, gross vehicle motions, frame and body stresses and deflections.
Environmental Control	Heat and moisture output of occupants, heat output of engine, outside air temperature and humidity, solar radiation flux, air conditioner air flow rates and temperatures.	Vehicle interior comfort level (air temperature, velocity, humidity, odor)

Figure 1-4 Some system examples from automotive engineering.

motion of the mass. In this same system, initial energy storage could take the form of kinetic, rather than potential, energy. Here, we give the mass an initial velocity and then let go.

The above mechanical examples of initial energy storage in potential or kinetic form have counterparts in the other types of system. In electric circuits, we can charge a capacitor or establish an initial current in an inductor. In a thermal system we can heat up a mass. In a fluid system we can pressurize a tank or establish an initial fluid flow. In all these cases, once we "release" the system, no *external* driving agencies act on the system, yet it will respond with changes in its output variables. In general, initial energy storage refers to a situation in which a system is put into a state different from some reference equilibrium state and then "released," free of external driving agencies, to respond in its characteristic way.

The term *external driving* implies that we have conceptually set up an envelope, or boundary, around some assemblage of components, and defined the interior as our *system* and the exterior as the system's *environment*. External driving agencies are physical quantities which pass from the environment, through the envelope (or *interface*) into the system, and cause it to respond. In practical situations there may sometimes be interactions between the environment and the system; however, since we are here dealing with the problem on an introductory level, we often use the concept of an ideal source. Thus in a mechanical system we may assume an external driving force acting on a mass without being explicit as to the exact physical means for providing this force. We simply take the viewpoint that we *choose* to study the response of the system to an ideal force input. Hopefully, there will be some (perhaps many) practical situations which correspond closely to this idealized model. The (unspecified) means for providing such inputs is called an *ideal source*. In mechanical systems, an ideal force source can supply whatever force we choose to assign to it, and is totally unaffected by being coupled to the system it is driving.

In electrical circuits, an ideal voltage source can supply whatever voltage we choose to assign to it, and is totally unaffected by being coupled to some circuit. In physics and mechanics courses which you may have completed, you probably used ideal force and voltage sources in solving problems, but perhaps without realizing that these were *assumptions* rather than real devices. A simple battery makes a good example. In physics problems you just drew the standard symbol for, say, a 6-volt battery and went ahead to solve for the currents in some circuit. You should realize that a *real* "6-volt" battery will *not* supply 6 volts to a circuit! In the real situation, the circuit will draw some current from the battery and the battery's voltage will *drop* to something less than 6 volts, and will in fact *continue* to drop as the battery discharges. Whether this real behavior can be ignored in favor of an ideal "battery" depends on how much current the circuit draws and how accurate the analysis needs to be.

Figure 1-5 shows the classification of inputs which we are developing. Whereas initial energy storage always refers to the state of the system at time = 0, external driving inputs are classified according to how they vary with time, the first broad classification being into deterministic or random time variation. All real-world inputs have at least some element of randomness or unpredictability; thus deterministic models of inputs are always simplifications of reality, although they are quite adequate for many purposes. Note that we use the term "model" since we plan to analyze our systems mathematically, and *any* mathematical description of either actual system inputs or the physical systems themselves *must* be some kind of model (approximation). The real world will always be too complex to describe mathematically with perfect accuracy. Some models are of course very accurate while others mispredict measured values by large percentages, but we should never lose sight of the fact that perfection is a worthy but unreachable goal.

We should also recall that the word "model" has two major meanings in engineering; the physical model and the mathematical model. By a physical model we mean an assemblage of actual hardware, constructed according to appropriate scaling laws, such that it will behave in a manner predictably related to the behavior of the full-scale device or system. Such models need not necessarily reproduce *all* aspects of the full-scale system's behavior; different models may be constructed for evaluating the drag coefficient of a new automobile and its vibration characteristics. While computer analysis of mathematical models continues to reduce the need for scale model studies, they still have an important role to play in engineering.[2]

Returning to Fig. 1-5, deterministic input models are those whose complete time history is explicitly given, as by a mathematical formula or a table of numerical values. They are further subdivided into transient, periodic, and "almost periodic" types. A *transient* input can have any desired shape but exists only for a certain time interval, being constant before the beginning of the interval and after its end. The impact of two cars in a crash is a transient occurrence. *Periodic* input models repeat a

---

[2]E. O. Doebelin, *Engineering Experimentation*, McGraw-Hill, New York, 1995, pp. 323–338.

Introduction

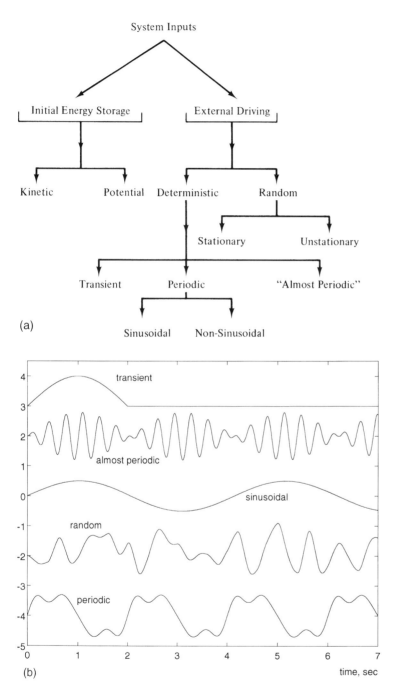

Figure 1-5 Classification of inputs.

certain waveform over and over (ideally forever). They are good models for the "constant-speed" operation of many machines and processes. For example, in an internal combustion engine which is thoroughly warmed up and running at constant speed with a fixed load, every variable in the engine (cylinder pressure, crankshaft torque, exhaust gas flow rate, stress in cylinder head bolts, etc.) will be going through a repetitive cycle. These "cycles" will of course not be *perfectly* repetitive, nor will they "go on forever," but a periodic model can be an excellent approximation.

The simplest periodic input model is a sine wave, for example, a force given by $f = 11.3 \sin(324.8t)$, where the number 11.3 has the units of force, 324.8 is the frequency in radians/sec, and t is the time in seconds. This simple model is of great practical importance, for a variety of reasons. Most directly, for electrical engineers, the vast majority of electrical energy is generated and used as "AC," alternating voltage and current, where the steady-state variables are sinusoidal. For mechanical engineers, most vibration problems are caused by the unbalance of rotating machine parts, which create sinusoidal forces. We shall see later that the response of a system to perfect sine waves (called the system's *frequency response*) tells us how the system will respond to *any* kind of input.

"Almost periodic" input models are a rather special group and receive little attention in an introductory level text such as this. They are continuing functions which are completely predictable but do not exhibit a strict periodicity. One member of this group, the *amplitude-modulated* type, has some important applications and is simply described, so we mention it briefly (more details are available in the literature[3]). An example would be $f = (0.4 \sin 1.52t)(2 \sin 18.67t)$. A graph of such a function (shown in Fig. 1-5) appears to a casual glance to have a repetitive cycle; however, this is disproved on careful examination. Amplitude-modulated stresses, vibrations, and acoustic noises occur unintentionally in gear boxes and rolling-contact bearings. Some electronic measurement equipment uses amplitude modulation by design to achieve certain performance advantages.

Random inputs, in their most general form, could be *exact* representations of real-world physical variables, so they are the most realistic input models. When we try to use them in design calculations, however, we again need to use simplified versions. Random inputs have time histories which cannot be predicted before the input actually occurs, although *statistical* properties of the input can be specified. For example, if an airplane is flown over a certain uniform terrain at a certain altitude and speed, it will encounter air turbulence [vertical gust velocity, $V(t)$] which could be measured and recorded as a function of time, as in Fig. 1-6. (Note that once a random signal has occurred, it is then deterministic; we know its value at every instant of time.) This time history shows no periodicity and is a good example of a random input. While there is no analytical mathematical function available to describe this phenomenon, one *can* compute certain *statistical* properties such as the average value, mean-square value, etc. from the recorded data. If the airplane now turns around and retraces its first flight path through the turbulence, the time history of $V(t)$ will *not* be a duplicate of the first flight; however, the *statistical* properties should be essentially the same (if the weather has not changed, etc.).

Thus when working with random inputs, there is never any hope of predicting a specific time history before it occurs, but statistical predictions can be made and can have practical usefulness. When mathematically studying the response[4] of a

---

[3]Doebelin, *Measurement Systems*, 4th ed., pp. 157–169.

[4]E. O. Doebelin, *System Modeling and Response: Theoretical and Experimental Approaches*, Wiley, New York, 1980, pp. 111–135, 267–282. The most recent printing of this text can be obtained from the author; phone 614-882-2670.

Introduction

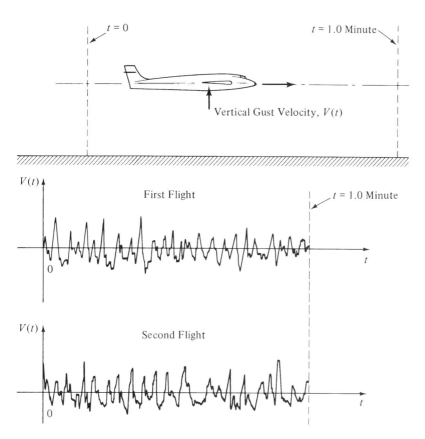

Figure 1-6 Stationary random input.

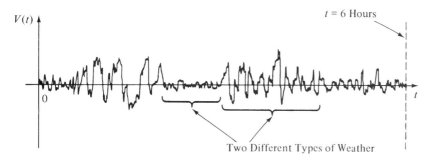

Figure 1-7 Unstationary random input.

system to random inputs it is generally necessary to work with a simplified model of the real random input. The science of probability and statistics provides many such models, and it is necessary to find one which approximates the physical situation with adequate fidelity in each application. The so-called "normal" distribution law (Gaussian probability function) has found wide use in such studies. Returning to the aircraft turbulence example, suppose we consider a complete coast-to-coast flight (Fig. 1-7), which will of course take place at varying altitudes

14 Chapter 1

and will encounter various types of weather. The entire 6-hour record of $V(t)$ will be a random input, but it should be clear that now even the *statistical* properties (mean value, mean square value, etc.) will not be fixed throughout the entire period. Such a random signal is called *unstationary* and can be a realistic representation of an actual physical phenomenon; however, its mathematical treatment is most complex. The more common assumption that statistical properties are time-invariant (*stationary* random signal) allows a more tractable mathematical treatment. In practice, nonstationary signals are often modeled as stationary over a restricted time period. That is, the total time record is treated in *sections* such that each section is approximately stationary, but the statistical properties are different for each section.

1-4 A CLASSIFICATION OF SYSTEM MODELS

Having explored the various types of input models, we now go on to consider models for the system itself. There is not any standard or accepted method of classifying system models, so the upcoming material should be considered as merely one possible (hopefully useful) point of view. We begin by confining our considerations to a macroscopic rather than microscopic scale. That is, we do not concern ourselves with phenomena at the level of molecules, atoms, or subatomic particles/waves, but rather deal with the gross (or so-called "continuum") behavior of matter and energy. At this level, the "fundamental laws of nature" applicable to the study of engineering systems consider matter and energy as being continuously (though not necessarily uniformly) distributed over the space within the system boundaries. Description of the spatial extent of a system in mathematical terms requires setting up a coordinate system such as the familiar XYZ coordinates of Fig. 1-8.

Many system dynamics studies are successfully carried out using rather simple system models, and this text is largely devoted to developing and using such models. Even if we do not use highly complex models, the intelligent use of the simpler models requires that we have some understanding of "what we are missing" when we choose the simpler model over the more complex. Thus we now try to give a qualitative understanding of the more complex models before we go on to concentrate on the simpler ones. The more complex system models referred to are those which consider a "continuous" distribution of matter and energy and thus always lead to partial differential equations. The simpler models are those which concentrate matter and energy into discrete "lumps" and lead to ordinary differential equations. While most of this book uses the simpler models, the last chapter gives a brief, but I feel very useful, introduction to the more complex and realistic partial differential equation models. I hope that your system dynamics course, or self-study, allows you to get into this chapter. It is "capstone" material that puts all the earlier work into better perspective. Let's now look at the more complex models in a qualitative way.

Since, in the most general case, the physical quantities of interest to us in a system vary both with regard to space (location in the system) and time, when we express the natural laws in equation form, we are forced to apply them to an *infinitesimal element* of the system, such as $dx\,dy\,dz$ in Fig. 1-8. Suppose the system of Fig. 1-8 represents a body subjected to various inputs in the form of heat flow rates, and

Introduction

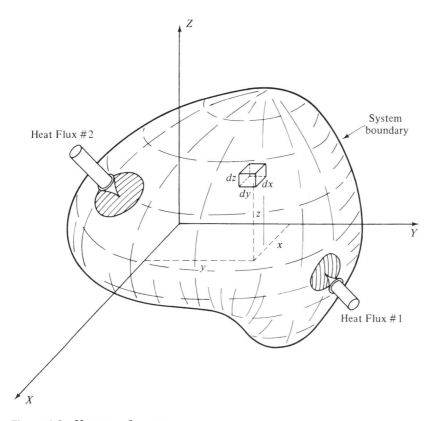

Figure 1-8 Heat transfer system.

the response (output) of interest to us is the temperature of the body. When we say that we wish to find *the* temperature of the body, it should be clear that this temperature varies both with location (x, y, z) in the body and also, at any given location, with time t. If we call the unknown temperature T, we see that T is a function of four independent variables, that is, $T = T(x, y, z, t)$.

Let us consider the body to be solid and with no internal heat-generating mechanisms such as electric heating or chemical reactions. The basic physical laws pertinent to this problem are then the Fourier law of heat conduction and the conservation of energy. The heat conduction law says that the instantaneous rate of heat flow per unit area in a given direction is directly proportional to the rate of change of temperature, with respect to distance, in that same direction. Mathematically,

$$q_s = -k_s \frac{\partial T}{\partial s} \quad \text{W/m}^2 \tag{1-6}$$

where $k_s \triangleq$ thermal conductivity of the material in the s direction

The rate of change of temperature with respect to distance is written as a *partial* derivative since temperature is a function of *several* independent variables. The conservation of energy law, applied to the element $dx\,dy\,dz$, says that, over a time interval dt, the difference between inflow and outflow of heat energy must appear as

16   **Chapter 1**

additional stored energy within the element, manifested as a rise dT of the temperature of the element. Applying this principle to each of the three directions, and assuming that material properties k, c (specific heat), and ρ (density) are constants, leads to the system equation

$$\frac{\partial T}{\partial t} = \frac{k}{\rho c}\left(\frac{\partial^2 T}{\partial x^2} + \frac{\partial^2 T}{\partial y^2} + \frac{\partial^2 T}{\partial z^2}\right) \tag{1-7}$$

We did not provide all the steps in this derivation since we just want to discuss the final equation. Equation (1-7), a partial differential equation, is a relationship among the unknown temperature T, the location coordinates x, y, and z, the time t, and system parameters k, ρ, and c. It is *not*, however, a *complete* description of a heat transfer problem since we have nowhere said anything about the location of the *boundaries* of the system or conditions existing at those boundaries. Also, the state of the system at the initial instant of time ($t = 0$) must be given. For example, it might be that all surfaces of the body (except for the shaded areas over which heat flows 1 and 2 exist) are perfectly insulated. It would also be necessary to state mathematically the nature of the two inputs and the areas over which they act. Finally, the initial temperature distribution would have to be specified as a mathematical function $T(x, y, z, 0) =$ a known function; for example, the simplest case would assume the temperature throughout the body initially uniform, say, at zero degrees [$T(x, y, z, 0) = 0$].

If all the information of the above types were given, then the problem would be completely defined and a solution would theoretically exist, even though it might be impossible to find analytically unless the conditions were quite simple. If a solution *could* be found, this would mean that we would have $T(x, y, z, t)$ as a definite known function such that if someone specifies *any* point x, y, z in the body and *any* time t, we can tell them what the temperature is. That is, given the inputs, (initial temperature distribution and heat flows 1 and 2) and the system model (differential equations and boundary conditions) we can then hope to find the outputs (temperature/time histories at any desired points in the body).

Application of the basic physical laws pertinent to various other types of dynamic phenomena (solid mechanics, fluid mechanics, electromagnetics, etc.) at the macroscopic level will in general lead to problem formulations (models) similar to the one just developed, that is, partial differential equations. This is basically true because we specify *dynamics* problems (thus time is automatically an independent variable) and consider the system variables as quantities which change continuously from point to point in the system, thus giving three spatial independent variables. When the dependent variables (outputs) depend on more than one independent variable, we are bound to get partial differential equations when we apply the physical laws, because these laws generally involve rates of change. These types of models are also called "field" models or *distributed-parameter* models. Most engineers would agree that such models, in their most general form, would behave almost exactly like the real systems at the macroscopic level. Unfortunately, these models can only be analytically solved in a small number of special and simple cases; thus engineers find it necessary and desirable to work with less exact models in many cases, and particularly in system dynamics.

Introduction   **17**

You may be familiar with *finite-element* or other related methods and commercial computer software that "solve" the complex models we just said were unsolvable. These methods are of course extremely important and widely used, but you should recall that these are *approximate, numerical* solutions, not analytical solutions. This means that they always have errors, which may be small or large, and they solve only special cases with specific numerical values of the parameters. Also, such studies usually deal with only *components* of larger systems, rather than the entire system itself. In system dynamics studies, these solution methods are often used as a preliminary step to obtain accurate numerical values for overall component characteristics needed to model the system with lumped parameters.

For example, if a machine member is to be modeled as a spring, but its shape is not a simple coil spring or leaf spring for which formulas are available in handbooks, we may at some point do a finite-element study to obtain a number for the spring stiffness, say, 143 N/m. The finite-element study would also produce a detailed stress and deflection distribution for the entire machine part, and these would be useful in design against failure, but the system engineer at that point would be interested only in the single number for the spring constant. That is all that would be needed to carry out the *system* analysis and design.

We now want to develop the table of Fig. 1-9, which is our classification of system models. This classification is based on the nature of the differential equations used to describe the system. Models toward the top of the table tend to be closer to reality but are mathematically complex and usually analytically unsolvable, although specific numerical cases can be "solved" using finite-element or related methods. The most realistic models are partial differential equation models which allow "complete freedom" in the description of the medium and its properties. In the thermal system problem of Fig. 1-8, we first allow the body to have any shape at all; we don't approximate it with simple geometries such as prisms, cylinders, spheres, etc. We next allow system properties like the thermal conductivity to vary from point to point (nonhomogeneous model) and to be different in different directions (nonisotropic model). For example, in plastic materials reinforced with glass fibers, the thermal conductivity along the fiber length is quite different from that across the fiber length. In a given direction, it could also vary from point to point, if the ratio of fibers to plastic is not uniform over the part.

In setting up equations, the physical laws generally use certain *relationships* between variables. In solid mechanics, we must stipulate the relation between stress and strain. If you have had a course in strength of materials, you may recall that your text assumed "Hooke's law," which says that strain is *proportional* to stress. This of course is an *assumption*; real materials don't behave in exactly this way. The more general case would allow a *nonlinear* relation, and this would be true for all kinds of physical systems, not just solid mechanics. The table columns under "nature of the medium" are intended to summarize these effects. "Continuous" models use an infinitesimal element for setting up the equations while "discrete" models use finite-size "lumps" of the medium. Parameters, such as thermal conductivity, can *vary* from point to point (inhomogeneous medium) and according to direction (anisotropic medium), or can be assumed constant. Relations between variables can be linear or nonlinear. Finally, parameters could also vary with time, randomly or deterministically, or could be assumed constant.

Model Type Number	Nature of the Medium						Time-Variation of System Parameters		
	Continuous (Field Problems)	Discrete (Network Problems)	Space-Variation of Parameters Variable	Space-Variation of Parameters Constant	Nonlinear	Linear	Random	Deterministic, Variable	Deterministic, Constant
1	X		X		X		X		
2	X		X		X			X	
3	X		X		X				X
4	X			X	X		X		
5	X			X	X			X	
6	X			X	X				X
7	X		X			X	X		
8	X		X			X		X	
9	X		X			X			X
10	X			X		X	X		
11	X			X		X		X	
12	X			X		X			X
13		X			X		X		
14		X			X			X	
15		X			X				X
16		X				X	X		
17		X				X		X	
18		X				X			X

Least realistic, easiest to solve → Most realistic, most difficult to solve

Figure 1-9 Classification of system models.

Introduction   **19**

Model "type numbers" 1 through 12 are partial differential equation models, but most of these are analytically unsolvable. Mainly in type 12, linear partial differential equations with constant coefficients, do we find a large number of solvable problems, and then only for simple shapes, such as slabs, circular cylinders, spheres, etc. The classical theories of elasticity, vibrations, acoustics, heat conduction, electromagnetics, fluid flow, etc. are to a large extent based on such equations. Type numbers 13 through 18 are ordinary differential equations, and again, only the linear type with constant coefficients (type 18) allows routine analytical solution.

An understanding of the difference between distributed-parameter and lumped-parameter models is vital to the intelligent formulation and use of the lumped models which this book emphasizes, so we next use an example to develop this comprehension. Our example is from the area of thermal systems and is explained comprehensively in Chap. 10, where distributed models are covered in more detail. Here we give only sufficient explanation to provide a general understanding of the relation between the two types of models. Figure 1-10 shows a simple one-dimensional heat conduction problem which we will model in both ways, so as to see the differences and relationship. The slim metal rod buried in perfect insulation is initially all at 0° Celsius, when at time = 0 the left end ($x = 0$) is suddenly raised to 100°C and left there forever after. We wish to know how temperature varies with time at every point in the rod.

For the distributed-parameter model we choose an infinitesimal element of length dx and apply Fourier's heat conduction law and conservation of energy to get the type 12 model

$$\frac{\partial T}{\partial t} = \frac{k}{\rho c} \frac{\partial^2 T}{\partial x^2} \qquad (1\text{-}8)$$

where we have assumed $k/\rho c$ as constant and neglected temperature variations in the y and z directions (good assumption if rod is "slender"). Details of the solution are shown in Chap. 10; here I simply present and discuss some results, which are graphed in Fig. 1-10 in two ways. If we look at a particular location in the rod, say $x = 5$, we can plot T versus time, giving the right-hand graph (similar graphs for *any* x we choose are also available; several are shown). The form of the solution is such that we can compute T at *any* values of x and t we wish. In the left-hand graph I display the spatial temperature distribution at some selected times. Note that with this distributed-parameter model, we can calculate T for *every* point in time and space.

In Fig. 1-11 we take the very same problem and model it with lumped parameters. Chapter 10 explains details; the essence is as follows. Thermal systems exhibit two fundamental phenomena: resistance to heat flow and energy storage capacity. In our metal rod, the resistance to heat flow is measured by the material's thermal conductivity; the higher k is, the less is the resistance to heat transfer by conduction. Energy storage capacity (called thermal capacitance) per unit volume is the product ρc, which has units of joules/m^3-C°, how many joules of thermal energy it takes to raise the temperature of a cubic meter 1 C°. In the distributed-parameter model the resistance and storage effects are uniformly *distributed* throughout the rod, and separate resistance and capacitance elements need not be calculated and do not appear in the equation. When we decide to use lumped modeling, the first consideration is how many and what size lumps should we use? Once the lumping pattern has

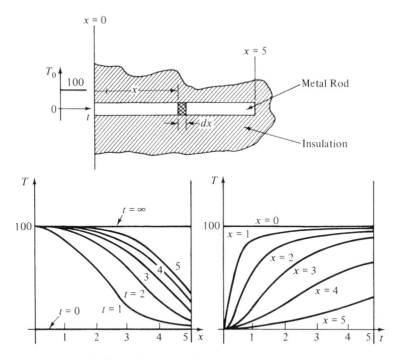

Figure 1-10 Distributed-parameter heat conduction model.

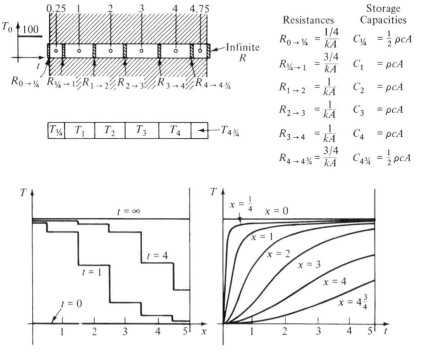

Figure 1-11 Lumped-parameter heat conduction model.

Introduction   **21**

been decided, then the resistance and capacitance effects of each lump can be calculated and entered into conservation of energy equations, one equation for each lump.

In Fig. 1-11 we show six lumps. We intuitively know that our lumped model
will become more accurate (closer to the "exact" distributed model) as we use more
(and thus smaller) lumps since the distributed model uses *infinitesimal* "lumps."
However, it is not at all clear whether we should use 1 lump, 10, or 100. When
the corresponding distributed model *can* be analytically solved, as in our present
example, we have the luxury of being able to compare any lumped results with the
known exact solution. This comparison technique is widely used to verify all kinds of
numerical methods and computer software. One often starts with a small number of
lumps and increases the number until the results don't appear to change much with
further increases. We use a similar approach when reducing the size of the time step
until results "converge," when numerically integrating differential equations in
which time is the independent variable.

In addition to the number of lumps, we also need to consider whether they all
should be the same size. In our present example, two half-size lumps are used at the
ends. This nicety is not necessary, but Chap. 10 explains why it might be useful. Once
the number and sizing of all lumps has (at least tentatively) been decided, we can
compute the thermal resistance and capacitance associated with each lump. Thermal
resistance has units of (heat flow rate in watts)/(lump-to-lump temperature difference
in C°). Thermal capacitance has units of (joules of stored energy)/(lump temperature
rise, C°), and *assumes* a spacewise uniform temperature within each lump. More
details are given in Secs. 4-7 to 4-9 and in Chap. 10; Fig. 1-11 shows the results.
We now express the conservation of energy for each lump in turn, saying for a time
interval dt: (energy into lump through its left resistance − energy out of lump
through its right resistance) = (additional energy stored in that lump during dt).
Implementing this scheme for all six lumps leads to a set of six simultaneous linear
ordinary differential equations with constant coefficients (type 18 model):

$$\frac{3\rho c}{k}\frac{dT_{\frac{1}{4}}}{dt} + 32T_{\frac{1}{4}} - 8T_1 = 2400$$

$$\frac{3\rho c}{k}\frac{dT_1}{dt} + 7T_1 - 4T_{\frac{1}{4}} - 3T_2 = 0$$

$$\frac{\rho c}{k}\frac{dT_2}{dt} + 2T_2 - T_1 - T_3 = 0$$

$$\frac{\rho c}{k}\frac{dT_3}{dt} + 2T_3 - T_2 - T_1 = 0$$

$$\frac{3\rho c}{k}\frac{dT_4}{dt} + 7T_4 - 3T_3 - 4T_{4\frac{3}{4}} = 0$$

$$\frac{3\rho c}{k}\frac{dT_{4\frac{3}{4}}}{dt} + 8T_{4\frac{3}{4}} - 8T_4 = 0 \tag{1-9}$$

This set of six equations, together with the six initial temperatures (all zero),
can be solved to give an explicit solution for each of the temperatures. Figure 1-11
shows plots of typical solutions, using the same ρ, c, and k values as were used for
the distributed-parameter model. When we now choose an x location and plot its
time history, we *cannot* choose any x we please, as we could in the distributed model.
The x values shown are those at the *center* of each lump. When we set up the

equation for each lump, we *assumed* that at any instant of time, that lump had a *uniform* temperature throughout; that's the only way we could express the energy stored in that lump. Thus temperature changes only from lump to lump; within a lump the temperature changes with time smoothly but is *the same* for all x's within that lump. Thus the lumped model sacrifices spatial *resolution* compared with the distributed model. This is true of *all* lumped models, not just this example. For $x = 1, 2, 3$, and 4, we have curves for both the distributed and lumped models, so these can be directly compared. We see that the agreement is visually quite good, so six lumps seems to be adequate for this problem.

The spatial resolution problem becomes more obvious in the other graphs of Fig. 1-11 where we fix time and plot temperature versus x. These "stepped" graphs are clearly crude approximations to the true smooth variation. One could, of course, replace the stepped graphs by "eyeballed" smooth curves drawn through the mid-points of each "flat" section. These smoothed plots then do look quite a bit like the correct distributed results. Such smoothing was in earlier days done "manually" using French curves. Today we can use smoothing software (such as spline functions) to accomplish the same task more efficiently. Similarly, in the graphs against time, we *can* "eyeball" in curves between, say, $x = 1$ and $x = 2$ if we are interested in a point at, say, $x = 1.3$, linearly interpolating between the solution curves. Computer software is again helpful here.

I hope this example will give you some "feel" for the nature of distributed and lumped models in general, not just the thermal system we used. Of course, most distributed models, unlike our example, *cannot* be solved analytically, so we usually won't have the exact solution available and will often choose to "go lumped" right from the start. When we do, we now can see what we are "giving up" when we pursue these simpler models. We get direct results *only* for a finite number of space-wise locations, and the results we *do* get are not perfectly accurate. (Fortunately, *both* these defects can be reduced by using more lumps, at the expense, of course, of solving a *larger* set of simultaneous equations.) What we *gain* is quite significant. The sets of simultaneous ordinary differential equations that *always* result from lumped modeling can only be solved analytically when they are linear with constant coefficients; however, easy-to-use *simulation software* has no trouble at all with the nonlinear and/or variable-coefficient equations associated with more correct models. This allows us to deal realistically with many engineering design and analysis problems. These simulation tools are the "bread-and-butter" solution methods of system dynamics and will be heavily used in this text, as they are in industrial practice.

The examples of Figs. 1-1, 1-2, 1-4, and 1-10 are obviously concerned with "conventional" engineering applications. System dynamics methods are also regularly used to study biological systems such as neuromuscular,[5] blood circulation,[6] temperature regulation,[7] human behavior in aircraft piloting,[8] and human body

---

[5]E. O. Doebelin, *System Dynamics*: *Modeling and Response*, Merrill, Columbus, Ohio, 1972, p. 29.
[6]Ibid., p. 166.
[7]Ibid., p. 32.
[8]Ibid., p. 446. Also, Doebelin, *System Modeling and Response*, pp. 536–556.

Introduction

vibration.[9] Life scientists, medical doctors, and engineers often collaborate on such studies, the language of system dynamics providing a useful means of communication. Large-scale engineering applications include stability studies of interconnected electric power plants scattered across the country and flood-control/irrigation studies of rivers and controlled dams.

The development of a useful model for some machine, process, or phenomenon is actually an *iterative* process; it does not usually proceed in uninterrupted fashion through a precisely ordered set of steps. At various stages we often evaluate what we have at that point and use this new knowledge to return to earlier steps for adjustments and improvements. That is, system dynamic analysis and model building is a *feedback* process, with information from later steps feeding back to modify earlier steps. Figure 1-12 shows a flow chart or block diagram which explains this aspect of modeling.

1-5 SYSTEM DESIGN

While our discussions so far have emphasized the *analysis* of existing systems, it should be pointed out that the major overall function of engineering is the *design* of new products and services which will be useful to society. Of course, a large part of design consists of detailed analysis and evaluation of competing concepts; we should never lose sight of the fact that analysis is only part of the overall process.

Design is also an iterative feedback process and can be diagrammed to clarify its operation (see Fig. 1-13). What role does system dynamics play in design? At the earliest stages, where we are conceiving several competing designs, system dynamics enters mainly by supplying a *point of view*. That is, one who is used to thinking in terms of overall systems has a point of view that may facilitate the generation of a variety of new design concepts. If you are very familiar with available components and have used various combinations of them in the past, it may be easier to now conceive of new arrangements which could accomplish the goals of a proposed new product or service.

At the next stage, where the alternative designs have been roughly formulated and must now be modeled, the systematic approach of system dynamics to modeling can be very helpful. Once models have been stated they can be analyzed and their performance evaluated relative to required specifications. Whether system dynamics analysis methods will be the tool of choice of course depends on the application; if system operation is mainly steady state or slowly varying, system dynamics has little to offer.

While system dynamics provides a basic analysis approach for many important application areas such as vibration, acoustics, measurement systems, control systems, etc., these areas will of course have developed special analysis and design

---

[9]Doebelin, *System Modeling and Response*, pp. 527–536.

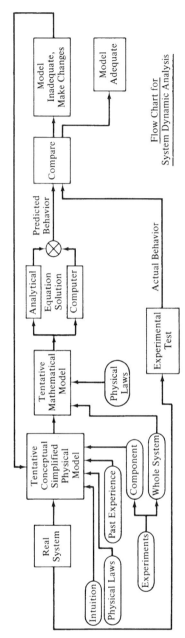

Figure 1-12 The system modeling process.

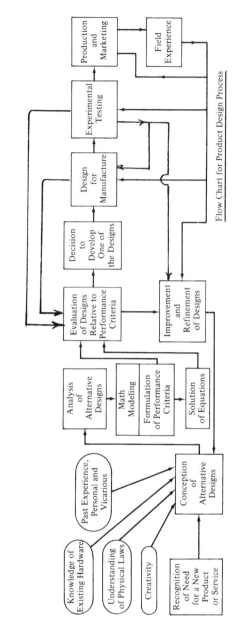

Figure 1-13 The product design process.

Introduction   **25**

tools peculiar to the application. These must be learned by engineers who wish to practice in these areas and then properly combined with the basic tools from system dynamics.

BIBLIOGRAPHY

American Society of Mechanical Engineers, *Journal of Dynamic Systems, Measurement and Control*, published quarterly.

Burton, T. D., Introduction to Dynamic Systems Analysis, McGraw-Hill, New York, 1994.

Cannon, R. H., *Dynamics of Physical Systems*, McGraw-Hill, New York, 1967.

Close, C. M., and D. K. Frederick, *Modeling and Analysis of Dynamic Systems*, 2nd ed., Houghton Mifflin, Boston, 1993.

Cochin, I., and H. J. Plass, Jr., *Analysis and Design of Dynamic Systems*, 2nd ed., Harper Collins, New York, 1990.

Doebelin, E. O., *System Modeling and Response*: *Theoretical and Experimental Approaches*, Wiley, New York, 1980.

Doebelin, E. O., *System Dynamics*: *Modeling and Response*, Merrill, Columbus, Ohio, 1972.

Haberman, C. M., *Engineering Systems Analysis*, Merrill, Columbus, Ohio, 1965.

Martens, H. R., and D. R. Allen, *Introduction to Systems Theory*, Merrill, Columbus, Ohio, 1969.

Ogata, K., *System Dynamics*, 2nd ed., Prentice-Hall, Englewood Cliffs, N.J., 1992.

Reswick, J. B., and C. K. Taft, *Introduction to Dynamic Systems*, Prentice-Hall, Englewood Cliffs, N.J., 1967.

Rosenberg, R. C., and D. C. Karnopp, *Introduction to Physical System Dynamics*, McGraw-Hill, New York, 1983.

Rosenberg, R. C., and D. C. Karnopp, *System Dynamics*: *A Unified Approach*, Wiley, New York, 1975.

Shearer, J. L., A. T. Murphy and H. H. Richardson, *Introduction to System Dynamics*, Addison-Wesley, Reading, Mass., 1967.

Smith, D. L., *Introduction to Dynamic Systems Modeling for Design*, Prentice-Hall, Englewood Cliffs, N. J., 1994.

PROBLEMS

1-1.   Identify possible input and output quantities for the following systems and explain briefly how the inputs affect the outputs. Draw block diagrams in each case, and physical sketches where possible.

    a.   An airplane flying through rough weather

    b.   An airplane touching down for a landing and taxiing to a stop

    c.   An airplane taking off on a rough runway

    d.   An aircraft jet engine

26   **Chapter 1**

    e.   A governor (speed control) for an aircraft jet engine
    f.   An electric generator driven by a steam turbine
    g.   A mechanical pressure gage (include both desired and undesired inputs)
    h.   An engine design group at an automobile manufacturer
    i.   A city government
    j.   The U.S. health-care system
    k.   The human digestive system
    l.   A coal-fired electric power station

1-2.   What might be a source of periodic inputs for:
    a.   A gear with 23 teeth, rotating at constant speed
    b.   A water turbine with seven blades, rotating at constant speed
    c.   A six-cylinder engine, running at constant speed
    d.   A barber's electric clippers, running off 60 cycle AC power
    e.   A railroad bridge carrying a train running at constant speed
    f.   The ringer in a telephone

1-3.   In Fig. 1-1, each block in the block diagram has its own steady-state gain (static sensitivity) K. Describe an experiment you could run to get a numerical value for the K associated with:
    a.   The R-L circuit
    b.   The magnetic force coil
    c.   The nozzle-flapper
    d.   The coil voltage-generator
    e.   The mass/spring/damper system

1-4.   Classify the following inputs according to the scheme of Fig. 1-5 and justify your choice.
    a.   Sonic boom pressure (caused by an airplane) on a house
    b.   Wind on a tall smokestack
    c.   Waves on an ocean liner
    d.   Noise from a riveting machine on the human ear
    e.   Unbalanced tire on a car
    f.   Frictional heating of a truck's brake drums
    g.   Solar heating on a communications satellite
    h.   Pressure inside an automotive airbag
    i.   Electrical signal from a compact disk music recording to the stereo-system amplifier

1-5.   Discuss the relative advantages and disadvantages of lumped-parameter and distributed-parameter modeling methods.

1-6.   In the system of Fig. 1-3a, what feature of the circuit prevents the current from responding instantly? Explain qualitatively why the air pressure in the system of Fig. 1-3b can't respond instantly. If the volume containing pressure p_o were reduced, would the pressure respond more quickly? Why?

1-7.   In the system of Fig. 1-1, for each component or block, only the *desired* input quantity is shown. Unfortunately, all physical hardware is sensitive, to some degree,

Introduction   **27**

to inputs other than that desired. Perhaps the most common such "spurious" input is the ambient temperature: "Almost everything is affected by temperature." In the system of Fig. 1-1, find at least four places where temperature could affect the output of a block, and briefly explain how this occurs.

2

SYSTEM ELEMENTS, MECHANICAL

2-1 INTRODUCTION

Chapters 2 through 4 will introduce the basic building blocks of lumped-parameter modeling, the so-called *system elements*. These might be considered as analogous to the chemical elements of the periodic table; that is, any system encountered in nature can be "built up" from a suitable combination of the system elements, just as the chemical elements are combined to form any of the natural or synthetic materials found in the universe. To develop facility in lumped-parameter modeling of real systems, one must become thoroughly familiar with the basic system elements and their behavior.

The system elements are grouped chapter-wise as mechanical, electrical, fluid, and thermal, mainly as a matter of convenience. In practical applications we sometimes encounter systems which are essentially or entirely, say, mechanical or electrical, but we also find cases where several different forms occur in a single system. The classical area called "shock and vibration," for example, deals almost entirely with mechanical elements, while electric circuit analysis deals with electrical elements. The practical design of automatic control systems for industrial processes and machines, on the other hand, often involves simultaneous consideration of mechanical, electrical, fluid, and thermal elements. In such "mixed" systems we find, in addition, energy conversion devices which couple, say, a mechanical to an electrical element or subsystem. For example, an electric motor (an electromechanical energy converter) can be driven by current coming from an electrical circuit and can then provide torque to drive a mechanical system. Chapter 5 will discuss these energy conversion devices which couple elements of different physical types.

Turning now to the subject of this chapter, the basic mechanical elements are used in modeling those parts of systems involving the motion of solid bodies. Clearly, this encompasses a vast array of practically important devices ranging in size from "micromachines" fabricated by integrated circuit technology to entire rapid-transit trains. Only three elements are required to model the essential features of such systems:

1.   The spring (elastic) element

System Elements, Mechanical   **29**

2. The damper (frictional) element
3. The mass (inertial) element

All these are found in both translational and rotational versions corresponding to the type of motion occurring. In addition to these three *passive* (non-energy-producing) elements, we also consider the *driving inputs* of mechanical systems, i.e., the force and motion sources which cause the elements to respond.

While this chapter is focused on the mechanical elements, it also introduces several *general* concepts and methods of system dynamics that apply to all kinds of elements and systems. That is, in discussing the mechanical elements, we cannot avoid defining and using these general concepts as part of the explanation. Since these concepts are of general applicability, they will be used in the rest of the book, but need not be redefined and explained every time we encounter them. This makes Chap. 2 somewhat longer since it has the dual task of explaining the mechanical elements and introducing these general concepts. These concepts include:

1. Pure and ideal elements versus real devices
2. Transfer functions, operational and sinusoidal
3. Block diagrams
4. Frequency response
5. Linearization of nonlinear physical effects
6. Ideal versus real sources
7. Computer simulation software and methods

2-2   THE SPRING ELEMENT

In a modern technological society even the layperson has a fair understanding of the concept of "springiness"; thus a long discourse on the subject may seem unnecessary for readers who have encountered "springs" in previous physics and mechanics courses. I believe that there are aspects of this subject that introductory physics/mechanics treatments rightfully avoid but that become vital as one moves from an academic environment toward actual engineering practice. This section will expand your understanding of "springs" to include some of these ideas useful in real-world design situations.

In our discussion of *all* the system elements, we use the terms *pure* and *ideal*. A real-world spring, for example, is neither pure nor ideal. That is, if we design or select from available stock a part which we intend to perform the function of a spring alone, we find that this "spring" also has some inertia and friction which are not at all necessary to the function of the system, but which nevertheless exist. The term *pure* thus refers to an "unadulterated" system element (spring, damper, inertia), that is, one which has *only* the named attribute. A pure spring element has *no* inertia or friction and is thus a *math model* (approximation), *not* a real device. Perhaps when you dealt with "springs" in physics and mechanics courses, this distinction was not emphasized or even mentioned. It's now time to embrace the more correct viewpoint. It's especially important in system dynamics because dynamic operation of "springs" sometimes requires that their inertia and/or damping *not* be neglected. The spring problems that you may have encountered earlier were probably "statics" problems and are of course unaffected by inertia, which manifests itself only when acceleration occurs. The concept of "pure" elements carries over into all the other physical types; the "resistor" which you used in circuit analysis was actually a pure resistance element (math model). *Real* resistors always have at least a little inductance and capacitance.

The term *ideal*, as applied to elements, will be defined as meaning *linear*. That is, the pure and ideal elements will be defined by a mathematical relation between the input and output of the element. In an ideal element, this relation will be linear, or straight-line. That is, the output will be perfectly proportional to the input. Linear elements are considered ideal mainly from a *mathematical* viewpoint; they lead to "type 18" models, which are analytically solvable. From a *functional engineering* viewpoint, *nonlinear* behavior may often be preferable, even though it leads to difficult equations. We will shortly be showing some examples where intentional nonlinearity gives significant performance advantages. Computer simulation software now makes analysis of nonlinear systems as quick and easy as linear, so we should not avoid nonlinearity for this reason.

Note that a device can be pure without being ideal (a nonlinear spring with no inertia or friction) and ideal without being pure (a device which exhibits both linear springiness and linear damping). Why do we choose to define and use pure and ideal elements when we *know* that they do not behave like the real devices used in designing systems? A major reason is that once we have defined all three of the pure/ideal mechanical elements, we can use these as building blocks to model real devices more accurately. That is, if our real spring has significant friction, we model it as a *combination* of pure/ideal spring and damper elements. This combination of pure/ideal elements may come quite close in behavior to our real spring. Also, we can approach the more correct *distributed* (partial differential equation) models by representing a machine part with, say, 10 pure/ideal spring elements and 10 pure/ideal mass elements. Let's now start defining the pure and ideal mechanical elements.

The definition of the pure and ideal translational spring element is contained in the input/output relation (see Fig. 2-1a)

$$f = K_s(x_1 - x_2) = K_s x \qquad (2\text{-}1)$$

where

$f \triangleq$ force applied to ends of spring, lb_f or newtons

$x_1 \triangleq$ displacement of one end, inches or meters

$x_2 \triangleq$ displacement of other end, inches or meters

$K_s \triangleq$ spring constant (or spring stiffness), lb_f/in or N/m

$x \triangleq x_1 - x_2$ relative displacement of ends, inch or meter

The origins for the coordinates x_1 and x_2 must be such that the spring is at its "free length" (zero force) condition when $x_1 = x_2$. Throughout this book the symbol $\leftrightarrow$ indicates the assumed positive direction for a displacement, velocity, electric current, fluid flow rate, etc. This symbol does *not* mean that the quantity is going in that direction, but rather that if it *is* going in that direction it will be assigned a positive number. If it happens to be going in the opposite direction, we assign a

System Elements, Mechanical

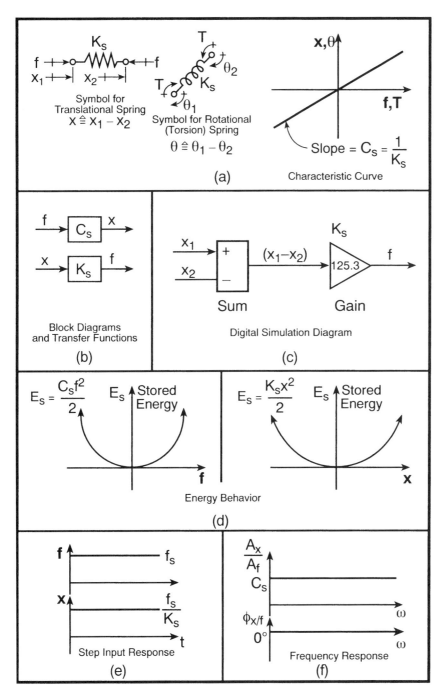

Figure 2-1 The spring element.

negative number. Such *sign conventions* are absolutely essential and must always be chosen *before* we begin to write equations, not after.

Because we must also deal with rotary motion, the rotational version of Eq. (2-1) defines the pure/ideal rotational or torsion spring element.

$$T = K_s(\theta_1 - \theta_2) = K_s\theta \tag{2-2}$$

where

$T \stackrel{\Delta}{=}$ torque (moment) applied to ends of spring, in-lb_f or N-m

$\theta_1 \stackrel{\Delta}{=}$ angular displacement of one end, radians

$\theta_2 \stackrel{\Delta}{=}$ angular displacement of other end, radians

$K_s \stackrel{\Delta}{=}$ spring constant, in-lb_f/radian or N-m/radian

$\theta \stackrel{\Delta}{=} \theta_1 - \theta_2 \stackrel{\Delta}{=}$ relative angular displacement of ends, radian

Note that we use the same letter symbol K_s for both translational and rotary versions, relying on the context to make clear the meaning. The *schematic* symbol shown in Fig. 2-1a is different for the two versions since we usually draw two-dimensional diagrams for translational systems, and three-dimensional for rotary.

The spring constant is also sometimes called the spring stiffness, since a large value of K_s corresponds to a stiff spring. The reciprocal of the stiffness, called the *compliance* C_s, is also sometimes used to describe a spring and is clearly a "softness" parameter; large C_s means a soft spring. Using compliances we have

$$x = C_s f \tag{2-3}$$
$$\theta = C_s T \tag{2-4}$$

Engineers have found *block diagrams* most useful in the design and analysis of all kinds of systems. In a block diagram, rather than showing a "picture" of the system hardware, we show blocks containing mathematical descriptions (transfer functions) of the hardware and connect these blocks with lines which denote the input and output signals for each block. At this stage we consider a transfer function as an output/input ratio (its definition will be expanded and clarified shortly). Since in some practical problems involving springs we know the force or torque and want to find the displacement, while in others we know the displacement and want to find the force or torque, the role of input and output can be reversed, and thus two kinds of transfer function defined.

$$\text{Force-input transfer function} \stackrel{\Delta}{=} \frac{\text{output}}{\text{input}} \stackrel{\Delta}{=} \frac{x}{f} = C_s \quad \frac{\text{in}}{lb_f} \tag{2-5}$$

$$\text{Motion-input transfer function} \stackrel{\Delta}{=} \frac{\text{output}}{\text{input}} \stackrel{\Delta}{=} \frac{f}{x} = K_s \quad \frac{lb_f}{\text{in}} \tag{2-6}$$

and similarly for the rotational case. Using the transfer function concept, one can obtain the output of an element or system by multiplying the input by the transfer function (see Fig. 2-1b).

We have mentioned our intention to use simulation software for many of our system studies. I want to start this right now, even though a simple element hardly requires computer technology for its application. By starting now with the simplest

System Elements, Mechanical 33

examples, we can gradually build up our simulation capability rather than trying to do complex applications "all at once." Digital simulation languages for systems described by ordinary differential equations were first widely used in the 1960s, when they were implemented on large mainframe computers and engineers used them in a batch-processing mode. My 1972 introductory system dynamics text used a language called CSMP[1] (Continuous System Modeling Program), which was widely used around the world in industry and academe. It was still going strong in 1980, when I used it in my more advanced system dynamics book.

The early simulation languages (there were quite a number on the market at any time) might be called "command line" languages, since the user wrote a program line by line, using a library of special simulation statements intermixed with conventional FORTRAN statements. The simulation statements were very powerful, making the writing of CSMP simulations extremely simple and fast. For example, performing a Runge-Kutta numerical integration routine, which would require several pages of FORTRAN code, took only the statement Y = INTGRL(IC,X), where X was the name of the variable to be integrated and IC was the initial condition on Y. Simulation languages on the market today are little different in basic operation from CSMP and its competitors, and are not able to do any simulations that CSMP could not. They are, however, much more accessible to the working engineer since they run in an interactive mode on the engineer's own PC or workstation. Furthermore, while command-line program entry is still used and preferred for some (usually complex) simulations, many languages also provide a graphical user interface ("GUI") that makes program entry even faster and more convenient. We will show both the command-line and GUI techniques, using each where it makes most sense, but after the introductory phase, we'll concentrate mostly on the GUI.

We will use the ACSL[2] and MATLAB/SIMULINK[3] languages for explaining how *all* simulation languages are used in system dynamics. Fortunately, all these languages are very similar and if you learn one of them, it is very easy to pick up any of the others if that should be necessary. Our use of particular languages should not be considered an endorsement. Anyone contemplating purchase of such software should survey what is available and make a reasoned choice based on personal needs, compatibility with existing hardware, cost, etc.

Defining a spring element in command-line mode is of course very quick and simple. Since ACSL is a FORTRAN-based language, and no special operations are needed, we could describe a spring with the statement FORCE = KS*(X1–X2). The spring stiffness KS could be given a numerical value in a separate statement such as CONSTANT KS = 125.3, or we could initially just have written FORCE = 125.3*(X1–X2). Names such as FORCE, KS, X1, and X2 could of course be chosen as we wish, but it is always good practice to choose symbols that remind you of the actual physical quantity.

---

[1]F. H. Speckhart and W. L. Green, *A Guide to Using CSMP*, Prentice-Hall, Englewood Cliffs, N.J., 1976.

[2]Mitchell and Gauthier Associates, Inc., 200 Baker Avenue, Concord, MA 01742-2100, 508-369-5115.

[3]The Math Works Inc., Cochituate Place, 24 Prime Park Way, Natick, MA 01760, 508-653-1415.

If we were developing our simulation using the graphical user interface of SIMULINK, we would start a new file for our simulation and then display on the monitor the menu of icons used in SIMULINK. Here we need just two of these icons: the *summer* and the *gain block*. The needed icons are dragged, using the mouse, from the menu into our "blank screen," to start assembling our simulation diagram. The summer adds and/or subtracts any number of input signals and the gain block multiplies its input signal by a constant of our choice, in our case the spring stiffness KS. Figure 2-1c shows how this would look on the screen. Note that we do *not* type any equations at all; the software "understands" the diagram we are drawing and assembles the corresponding equations for us, "behind the scenes." These simulation diagrams are very similar to the block diagrams that we employ with the transfer function approach, so the two applications reinforce each other.

We will have a "summary" diagram like Fig. 2-1 for each of the elements as we go through Chaps. 2 through 4, and it will always contain the major features of the particular element being defined:

1. The standard letter symbols and schematic symbols for the element
2. The element characteristic curve showing its input/output relation
3. The element block diagram and simulation diagram
4. The energy behavior of the element
5. The element dynamic response: step response and frequency response

Each of the elements we will define has one of two possible energy behaviors: It either stores all the energy supplied to it, or dissipates all of it into heat by some kind of "frictional" effect. For the mechanical elements the spring stores energy as potential (strain) energy, the mass stores energy as kinetic energy, and the damper dissipates energy into heat.

To clarify energy storage in the spring element, consider Fig. 2-2a. If a force is gradually applied (say by your finger) to a spring element and then maintained constant we find that the force has done work in deflecting the spring and this energy is now *stored* in the spring and could be recovered when the spring is allowed to relax. Since the instantaneous power taken from the force source and put into the spring is given by the definition of power as the product of force and velocity, we can write

$$\text{Instantaneous power} \triangleq (\text{instantaneous force}) \times (\text{instantaneous velocity})$$

$$\text{Power} \triangleq P = \left(\frac{f_0 t}{t_1}\right)\left(\frac{C_s f_0}{t_1}\right)$$

$$P = \frac{C_s f_0^2}{t_1^2} t \tag{2-7}$$

Now the total energy put into the spring is given by

$$\text{Stored energy} = \text{work done} = \int (\text{power})\, dt = \int_0^{t_1} \frac{C_s f_0^2}{t_1^2} t\, dt \tag{2-8}$$

where we integrate only to t_1 since the velocity, and thus the power, is zero thereafter. Carrying out the integration gives the stored energy E_s as

$$E_s = \frac{C_s f_0^2}{2} \tag{2-9}$$

System Elements, Mechanical

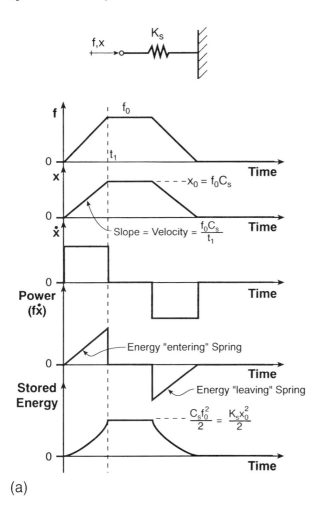

(a)

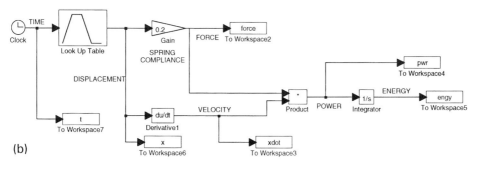

(b)

Figure 2-2 Energy storage and recovery in spring elements.

or alternatively

$$E_s = \frac{K_s x_0^2}{2} \tag{2-10}$$

Actually, this result is independent of the particular time variation of force or motion used in reaching the final force f_0 or position x_0. From the characteristic curve of Fig. 2-1a we can write for an infinitesimal motion dx

$$\text{Work done} = f\, dx = K_s x\, dx \tag{2-11}$$

and thus for a total displacement x_0

$$\text{Work done} = \int_0^{x_0} K_s x\, dx = \frac{K_s x_0^2}{2} \tag{2-12}$$

Figure 2-1d displays these energy relations.

While the above calculations hardly require the services of a computer, we again want to gradually introduce simulation techniques, so we show Fig. 2-2b, a SIMULINK diagram for the above operations. The clock icon is used whenever we need to access the independent variable time. Here we use it in two places: to generate the input displacement as a function of time, and to provide a variable (which we name t) against which to plot any other variables of interest. The "To Workspace" icon is used to record for plotting, tabulation, or further calculation any variable of interest to us.

The "Look Up Table" icon is one of the most useful. It establishes, by means of lists of "x, y" point values, a functional relation between its input quantity and its output quantity. In our case we want the displacement x to follow the time pattern shown in Fig. 2-2a. Since simulation *always* must work with *numerical*, not literal, parameters, we must decide on some numbers to use here. The list of t values entered into the lookup table is given as the "vector" [0 1 2 3 4], where the numbers have the units of seconds of time. The corresponding x vector is [0 5 5 0 0], where these numbers are considered to have the units of displacement, say, inches. The lookup table icon always connects by straight lines the "x, y" point pairs given, so we get the x, t graph that we desire in this case (x goes from 0 to 5 as t goes from 0 to 1, etc.). This icon actually displays a miniature graph of the "x, y" data you enter, if the icon has been "sized" large enough. The size of *all* the icons, as displayed on your screen, can be adjusted by clicking and dragging with the mouse. Don't make them larger than necessary since there is only so much space on your screen for the total diagram. (For diagrams too large for a single screen, SIMULINK provides a way to "compress" portions of a diagram.)

The input to a lookup table need *not* be time; we can use whatever variable we need. Thus if we have lab tested a nonlinear spring and have a table of measured x, f values, we can easily simulate this spring, for which no theoretical formulas may exist, with a lookup table icon. If we want smoother curves than given by the usual straight-line segments, we can use MATLAB's spline function operator to smooth a curve through our data points and then use these new "vectors" in our lookup table lists. For a spring simulation like this, the relative displacement of the two ends of the spring would be calculated somewhere else in our simulation diagram and then "sent" as an input to the lookup table.

System Elements, Mechanical   **37**

The gain icon used here to enter a number (0.2) for the spring compliance has already been discussed, so let's move on to the derivative icon. If you have some background in numerical methods you may recall that numerical differentiation is often difficult to perform accurately since it accentuates any small but fast changes in the quantity to be differentiated. That is, it is a "noise accentuating" process, and is in fact strenuously avoided in using any simulation language. However, sometimes it *can't* be avoided and it may actually work well if the input signal is noise free. Here we use it to get the velocity signal we want from the available displacement signal, and it works almost perfectly in this case.

Next we want to compute power from the available force and velocity signals, and this requires the *product icon*. It can be set up to multiply any number of inputs—here we need only two. Finally we come to the *integrator icon*, which is really the "heart" of any simulation language. That is, the overall problem of all simulation languages is to numerically solve differential equations and this is always accomplished by integrating the highest derivative to get the next lower, integrating that derivative to get the next lower, etc. until we finally get the unknown itself. In our case we only have to integrate the instantaneous power to get the total energy stored in the spring at any instant.

Numerical integration has many subtleties, but *users* of simulation languages need be concerned with only a few of these; the software writers have attended to most of the details and we usually can proceed rather casually. We *do* however always need to select from a menu of available integrators (Euler, Adams, various Runge-Kutta, etc.) one that is suitable for our problem. Usually the SIMULINK default integrator, Runge-Kutta 45, works well, so we generally use it until we notice problems. Starting and stopping times must also be specified; we usually start at $t = 0$ and the stopping time is usually obvious from the physical problem. The RK45 integrator is a *variable* step-size type which tries to optimize both speed and accuracy by varying the computing time step throughout the solution. By choosing the maximum and minimum step sizes as equal, one can force it to be a *fixed* step-size integrator. I usually start this way and only revert to the more efficient variable-step operation when I have my simulation debugged and working well.

To choose a tentative step size to try on your first run, divide the total time by about 1000 if you have nothing else to go by. We want the largest time step which gives acceptable accuracy and sufficiently smooth graphs. If results with a certain time step are considered acceptable, gradually increase the step size until results start to "go bad." This trial-and-error approach will usually define a step size large enough to be fast but small enough to be accurate and "smooth."

Having just shown how a GUI-type simulation would be set up, we now want to simulate the same problem using the command-line approach available in ACSL. An ACSL program might go like this.

```
CONSTANT CS=0.200
X=5*RAMP(0.0)-5*RAMP(1.0)-5*RAMP(2.0)+5*RAMP(3.0)
FORCE=CS*X
XDOT=DERIVT(0.0,X)
POWER=FORCE*XDOT
ENERGY=INTEG(POWER,0.0)
```

We should point out that simulation languages are "nonprocedural." That is, unlike FORTRAN or BASIC, the *sequence* of statements is immaterial, since the language has a sorting algorithm built into it, which sorts the statements into proper order. Thus the above program would run properly even if the sequence shown were "scrambled."

The statement RAMP(0.0) creates a ramp function of slope 1.0, starting at time = 0.0; RAMP(1.0) starts at (is zero until) $t = 1.0$ second, etc. The combination of RAMPs shown above creates the desired x, t pattern of Fig. 2-2. A simple FORTRAN multiplication gets FORCE from X, using the CS value given in the CONSTANT statement. Displacement X is differentiated using command DERIVT to get velocity XDOT; 0.0 is given as XDOT's initial value (it need not be zero). POWER is formed by a simple FORTRAN multiplication and ENERGY is obtained by integrating POWER, the initial value of ENERGY being here taken as zero. Just as in SIMULINK, use of integrators requires that we state starting and stopping times, our choice of integrator type, and maximum and minimum step sizes.

The ACSL "program" shown shows all the computing statements but does not include some "housekeeping" statements that are needed to run the system on any particular computing platform. Also, plotting and tabulating statements are available but not shown since we just want to show the essence of the simulation process.

Parts (e) and (f) of Fig. 2-1 deal with the dynamic response of the pure/ideal spring element and are naturally of particular interest in system dynamics. "Dynamic response" could refer to response to *any* time-varying input, but in defining all the elements we will always discuss only two "standard" inputs, the step input and the sinusoidal input. While other forms of dynamic input will certainly be of interest in particular applications, these two standard inputs are *always* useful. They reveal much about the *general* nature of the dynamic response and are also much used in laboratory testing of both components and complete systems. Frequency response testing (response to sine waves of different frequencies) is so popular that special test equipment is available from many manufacturers around the world. If you are a music aficionado, your stereo system components are all rated in terms of their frequency response, even though music is *not* simple sine waves.

Figure 2-1e shows the response of the pure/ideal spring element to a step input of force. By a *step input* of any variable, we will always mean a situation where the system is "at rest" at time = 0 and we instantly change the input quantity, from wherever it was just before $t = 0$, by a given amount, either positive or negative, and then keep the input constant at this new value "forever." In "generic" graphs, we often jump the input from zero to some positive value; in Fig. 2-1e we jump the force on the spring from zero to a value f_s. Since $x = f/K_s$, the displacement x will *instantly* jump up to a value f_s/K_s. No real spring could do this! The inertia (mass) of a real spring means that a suddenly applied force causes a sudden acceleration ($a = F/m$), but some time must go by before this finite acceleration causes a nonzero velocity, and even more time must go by before this velocity builds up a displacement. Thus the step response of real springs might look more like the graph shown in Fig. 2-3. Whether we can use the simpler pure/ideal model in a particular application depends on a number of factors which we will start explaining before long.

The *sinusoidal input* is perhaps the most important dynamic input since it is the basis of the *frequency-response method* of describing system dynamic response. This method has been successfully used in the design and analysis of all kinds of dynamic

System Elements, Mechanical

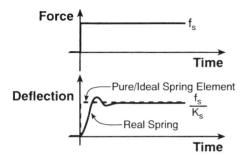

Figure 2-3 Step input response of pure/ideal and real springs.

systems for many years. Before we use it on systems, we will use it on the elements. This introduces the idea in the simplest way and also turns out to be an excellent method of characterizing the deviations of real devices from the theoretical ideal. By frequency response we will always mean how a system responds to an input which is a perfect sine wave. This response generally will change when we change the frequency of the sine wave. We want to "exercise" our systems with sine waves of a wide range of frequencies, ideally from zero to infinity.

For our pure/ideal spring element, the frequency response is very easy to calculate. Let's make the input force be the sine wave $f = f_0 \sin \omega t$, where f_0 is the *amplitude* in lb_f or newtons, and ω is the frequency in radians per second. Recall that while all calculus operations require frequency in rad/sec, engineers often prefer cycles/sec, also called Hertz, and with symbol f. The conversion is of course $f = \omega/2\pi$, one cycle/sec is 2π rad/sec. For a pure/ideal spring element, $x = C_s f_0 \sin \omega t$, giving the graphs of Fig. 2-4. For dynamic systems in general, there will be a phase angle (phase shift) between the input sine wave and the output sine wave. For the spring element it is clear that this phase shift (called angle ϕ in this book) is zero for all values of frequency from 0 to infinity. The *amplitude ratio* A_x/A_f is simply the number C_s, so it also is the same for all values of ω.

For all elements or systems, the *frequency-response graphs* are defined as the graphs of amplitude ratio (output amplitude/input amplitude) and phase shift, each plotted against frequency. It is conventional (and most useful) to always plot these graphs on the *same* sheet of paper, with the amplitude ratio at the top and the phase angle at the bottom. Figure 2-1f shows the frequency-response graphs for the pure/ideal spring element. Note that this element "treats all frequencies the same." This will *not* be true in general. Other elements, and more complex systems, will treat different frequencies differently. This can be either beneficial or disastrous. A radio circuit, when tuned to the frequency of a specific station, greatly magnifies signals of that frequency, compared with those lower or higher. An unbalanced car tire, when driven at a certain speed (frequency) can cause large suspension vibrations.

We earlier mentioned the direct importance of sinusoidal inputs for electrical engineers ("AC power") and mechanical engineers ("unbalanced rotating machine parts"). It turns out, and we will explain in detail later, that *all kinds* of inputs can be expressed numerically in terms of their "frequency content," so the response to sine waves really tells us how a system responds to *any* input, including even random inputs. Also, any real-world device or process will only need to function properly for a certain range of frequencies; outside this range we don't care what happens. For

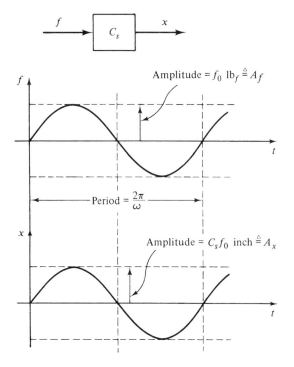

Figure 2-4 Frequency response of spring elements.

example, stereo sound systems need only function "accurately" for frequencies from about 20 to 20,000 Hz since human beings can't hear "sounds" outside this frequency range. If you have ever displayed the output voltage of a microphone on an oscilloscope, and let the "mike" listen to some music, you know that music has *very* complicated waveforms, *not* simple sine waves. Yet the performance of all sound systems *is* given in terms of sinusoidal response. This turns out to be true in general for systems which are nearly linear with constant coefficients: No matter what kind of input our system may in fact be subjected to, we *can* evaluate its performance based on its frequency response. For certain applications we may choose *not* to use frequency response methods, but they are always there if we need them.

Consider an automotive suspension system as a mechanical example. Its major input is the tire vertical force or displacement caused by driving over roads that are not perfectly flat and smooth. For a given road "profile," the frequencies produced will increase as we traverse the profile at higher driving speeds, but the car's top speed is limited, so the frequencies can only go so high. If road "bumps" are closely spaced, this gives higher frequencies for a given speed, but bumps which are *very* closely spaced will be "ignored" by the tires. A great virtue of the pneumatic tire is that its flexibility "envelopes" small, closely spaced bumps, so that they don't cause much vertical force or displacement of the axle. All these known phenomena mean that the frequencies of vertical forcing will be limited to fairly low values, say, between 0 and 30 Hz. This means that whatever models we choose for the springs, shock absorbers (dampers), etc. need only be accurate over this restricted range.

System Elements, Mechanical

We have mentioned that real springs will not behave exactly like the pure/ideal element. One of the best ways to measure this deviation is through frequency response. For the familiar coil spring, used in a compression application, both a theoretical distributed-parameter analysis (type 12 model, includes inertia and springiness but no friction) and frequency-response measurements on real springs are available.[4] The theory gives the frequency response in terms of the *sinusoidal transfer function*, a useful tool which we will discuss in detail when we get to the damper element. This transfer function tells us that the amplitude ratio is given by

$$\frac{A_x}{A_f} = \frac{\tan(\sqrt{m/K_s}\,\omega)}{\omega\sqrt{K_s m}} \qquad (2\text{-}13)$$

where m is the mass of the spring and K_s is our usual spring stiffness. Since $\tan(x)$ approaches x as x goes to zero, we see that the above amplitude ratio starts out, for $\omega = 0$, at $1/K_s = C_s$, just as does the pure/ideal element of Fig. 2-1f. However, as frequency increases, the more correct amplitude ratio of Eq. (2-13) will *not* stay constant at C_s.

One of the seven springs tested had a spring constant of 93 lb$_f$/in and a mass W/g of $(0.509 \text{ lb}_f)/(386 \text{ in/sec}^2)$. Using Eq. 2-13 we get the graph of Fig. 2-5. The four "peaks" shown in the amplitude ratio actually go to infinity since the tangent function does this. At a peak, a tiny force can cause a very large motion, and this

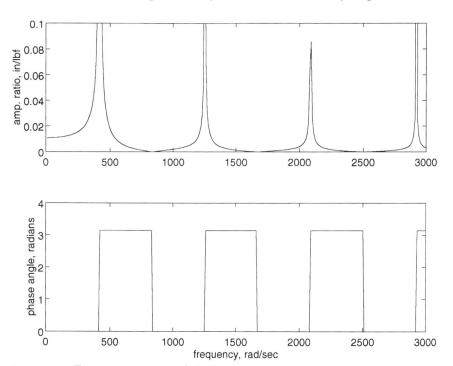

Figure 2-5 Frequency response of distributed-parameter spring model.

---

[4] E. E. Stewart and B. L. Johnson, Transfer Functions for Helical Compression Springs, General Motors Engineering Publication #A-2235, Jan. 18, 1968.

phenomenon is called *resonance*. It actually does occur in real springs, but the peaks are *not* infinite, just very high. This limit on the peaks is due to the *friction* in a real spring, which was *not* included in the distributed-parameter model. When the seven springs were lab tested, the curves matched the theoretical predictions almost perfectly, except, of course, that the peaks were not infinitely high, only several hundred times as large as C_s, the value of the amplitude ratio at zero frequency.

Figure 2-5 also shows that the phase angle of a real spring is *not* zero for all frequencies, but rather "jumps" between zero and pi radians. This was also confirmed in the measurements. To compare the pure/ideal and real behavior more closely we show a "zoomed" version of the amplitude ratio in Fig. 2-6. We see that the pure/ideal model is reasonably accurate up to about 150 rad/sec (24 Hz). Note also that at about 830 rad/sec the amplitude ratio drops theoretically to zero (again this is caused by the tangent function's behavior). This is called an *antiresonance* and physically means that at this frequency it takes a huge force to cause even a small motion. Again the measurements show that we don't get a "perfect zero" here; however, the amplitude ratio does drop to a value several hundred times smaller than C_s.

While the model of Eq. (2-13) is quite accurate for an isolated spring, it is *not* easy to apply it to larger systems, where there may be several springs, masses, and dampers connected in complicated ways. Its main usefulness is thus in *checking* a given spring to see whether we might model it as a pure/ideal element. If we find that it *can't* be treated in this simple way, we will usually formulate a *lumped* model for it,

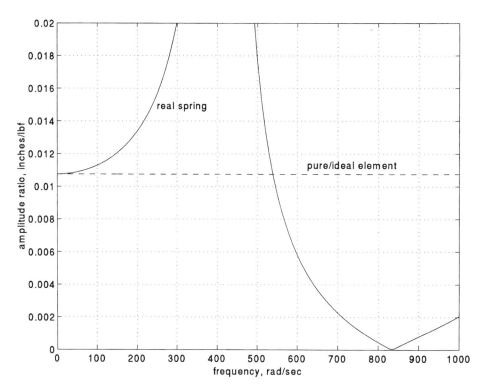

Figure 2-6 Comparison of distributed and lumped spring models.

System Elements, Mechanical

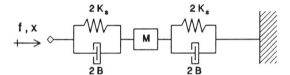

Figure 2-7 A more realistic lumped model for springs.

using pure/ideal spring, mass, and damper elements. Figure 2-7 shows one such model, which we will study later. For the example spring given above, this model would be accurate to about 160 Hz, far beyond the 24-Hz limit of the simple spring element. Furthermore, since it includes friction, the resonant peak will be more correctly predicted than the distributed-parameter model does. Even more important, this model is *easily* included in larger systems, since it uses only ordinary, not partial, differential equations.

2-3 LINEARIZATION

Recall that our definition of "ideal" elements requires that the output/input relation be strictly linear, as in $f = K_s x$ for the spring element. *No* real device can be this perfect; there is always at least a little "curvature" or nonlinearity in real springs, dampers, electrical resistors, etc. Some devices are *intentionally* made quite nonlinear to garner some functional engineering advantage. Nonlinear elements will lead to nonlinear system differential equations, and even though our simulation software can "solve" such problems, we much prefer to deal with approximate linear systems whenever we can. The *general* theory available for such systems is a great aid in design and analysis. An approach often viable is to do the *initial* design studies analytically with linear models, to quickly establish rough values of parameters. We then use simulation methods to check the effects of nonlinearities, perhaps entering them one at a time into the model, gradually building its complexity (and accuracy) until we finish with a quite accurate model.

To "convert" nonlinear models into more tractable linear ones we use the approximation technique called *linearization*. Using a spring as an example, a real force/deflection curve as measured in a lab test must exhibit at least some nonlinearity, as shown in Fig. 2-8. To linearize this behavior we must first choose an *operating point* for the spring. Consider the springs in an automobile suspension system. When the car is sitting still at the curb, its weight will deflect the springs down into an equilibrium position. If we now drive the car over a rough road, the car body vibrations will take place around (above and below) this position. Unless the road is *very* rough, these vibrational displacements of the spring will be small relative to the initial static deflection due to car weight. This situation is ideal for linearization since we want to model the spring behavior for small changes near a given operating point.

Graphically, it is intuitive that, as an approximation, we can replace the actual curve by its *tangent line* at the operating point, and that this will be quite accurate so long as we don't operate the spring too far from the operating point. This is really all there is to the concept of linearization. If our curve is the result of lab data, and no formula for it is known, the fitting of the tangent line is done by eye. If we have a

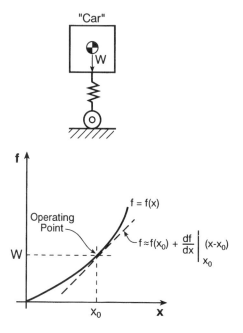

Figure 2-8 Linearization for a nonlinear spring.

formula for the curve, we could use analytic geometry to find the formula of the tangent line, but we prefer another method, which gives exactly the same result.

Recall from calculus the *Taylor series* expansion for any function $y = f(x)$

$$y = f(x_0) + \left[\frac{dy}{dx}\right]_{x_0} (x - x_0) + \left[\frac{d^2y}{dx^2}\right]_{x_0} \frac{(x - x_0)^2}{2!} + \cdots \qquad (2\text{-}14)$$

To get an exact representation of y, the complete series (infinite number of terms) must be used. To get an approximation, the series may be truncated after a finite number of terms. To get a *linear* approximation (which is what we want) we use only the first two terms

$$y \approx f(x_0) + \left[\frac{dy}{dx}\right]_{x_0} (x - x_0) \qquad (2\text{-}15)$$

which will be seen to be precisely the equation of the tangent line at x_0. As an example, consider the nonlinear force/deflection relation

$$f = 2x + 5x^3 \qquad (2\text{-}16)$$

in the neighborhood of the operating point $x = 1.0$. We have

$$f \approx (2 + 5) + (2 + 15x^2)|_{1.0}(x - 1.0) = -10 + 17x \qquad (2\text{-}17)$$

In setting up a linearized differential equation for a system that used this spring we would model the spring force as $-10 + 17x$. At the operating point $x = 1.0$, the *linearized spring constant* would be df/dx at that point, or $17\,\text{N/m}$, if we were using force in newtons and displacement in meters. Such a spring "constant" of

System Elements, Mechanical   **45**

course *changes* when we change the operating point, whereas an ideal spring element has the same stiffness at all points. When our "car" takes on passengers or other loads, and the springs deflect more to take up this load, the spring stiffness will change if the spring is nonlinear.

While it would not be hard to compute a "percent error" between the exact curve and the approximate straight line for any value of *x*, this would *not* tell us how much error to expect in the solution of a differential equation which used this linearization. Such errors can only be found by solving both the nonlinear and linearized equations and then comparing the *solutions*, point by point. Our simulation software makes such comparisons very easy and we will be doing this when we get to the analysis of more complex systems. You may be asking, "If you can easily simulate the nonlinear equation, why bother with linearization?" You need to recall our earlier comments about system *design*, where linear models are much preferable, even though we know that we will later *analyze* our designs (with added nonlinearities) using simulation.

Linearization must sometimes be done where a dependent variable depends on *several* independent variables. Our technique is easily extended to this case using the *multivariable* Taylor series. If $y = f(x_1, x_2, x_3, \ldots)$ we approximate it as

$$
\begin{aligned}
y \approx f(x_{1,0}, x_{2,0}, x_{3,0}, \ldots) + \left[\frac{\partial f}{\partial x_1}\right]_{x_{1,0}, x_{2,0}, x_{3,0}, \ldots} (x_1 - x_{1,0}) \\
+ \left[\frac{\partial f}{\partial x_2}\right]_{x_{1,0}, x_{2,0}, x_{3,0}, \ldots} (x_2 - x_{2,0}) + \cdots
\end{aligned}
\tag{2-18}
$$

When there are only two independent variables [$z = f(x, y)$], we can give a geometrical interpretation of this approximation. The function *z* defines a *surface* and the approximation of Eq. (2-18) represents a *plane*, tangent to the surface at the operating point x_0, y_0. When there are more than two independent variables, the approximation is called a *hyperplane*, but no geometrical interpretation is available. In Eq. (2-18) the partial derivatives (which are all evaluated at the operating point $x_{1,0}, x_{2,0}, x_{3,0}, \ldots$ and are thus *numbers*, not functions) can be thought of as the "sensitivity" of the dependent variable to small changes in that independent variable. If a particular partial derivative is a large number this means that the dependent variable is particularly "sensitive" to changes in this independent variable.

This multivariable linearization provides also a method to develop linear models of complex processes by lab testing. As an example, consider the system of Fig. 2-9, which is used for "levitating" objects with magnetic force. (Figure 2-9 includes a force sensor and gap adjuster not present in actual applications but needed for some lab testing which we shortly discuss.) This principle is used in magnetic bearings for vacuum pumps (lubricated bearings contaminate the vacuum with oil vapor), conveyor systems for moving integrated circuit wafers in clean rooms (conveyors with rolling or sliding bearings contaminate the clean room with tiny wear particles), and high-speed levitated trains (not yet commercialized), used because steel wheels on rails have a restrictive upper speed limit.

While the system of Fig. 2-9 can be analyzed theoretically by an electrical engineer skilled in electromechanics, such analyses are often not highly accurate and experimental testing would be used to get numerical values of parameters for use in analysis of a larger system which included the parts shown here. If mechanical

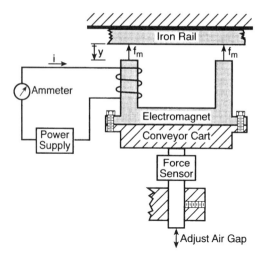

Figure 2-9 Magnetic levitation as example of multivariable linearization.

engineers had purchased these parts to use in a larger system, they would run lab tests to develop the needed model. Most readers of this text are *not* expert in computing magnetic forces; however, it is intuitive that the vertical magnetic force on the levitated object depends on two variables, the air gap y and the coil current i. In fact, most of you would also guess that this force increases as the air gap gets smaller and the current gets larger. Thus it is not unreasonable to assume that magnetic force is some function of y and i; $f_m = f(i, y)$. This function is most likely nonlinear but we can get a linearized model as

$$f_m \approx f(i_0, y_0) + \left[\frac{\partial f}{\partial y}\right]_{i_0, y_0} (y - y_0) + \left[\frac{\partial f}{\partial i}\right]_{i_0, y_0} (i - i_0) \tag{2-19}$$

Let's now consider the application where the object levitated is a clean-room conveyor "cart." To run an experiment which will give us numbers to insert into this model we require an adjustable power supply to set current at desired values, and an ammeter to read the current. We also need an apparatus that will allow us to set desired air gaps and a force sensor to measure the magnetic force exerted on the cart. The desired operating point y_0 for the air gap, say, 0.012 m, would be known from design specifications for the conveyor system so we could adjust the apparatus to create this gap. With the current turned off, the force sensor would read a downward force equal to the weight of the magnet and cart, say, 98 N. As we turn up the current, the magnetic force increases and the force sensor reports a smaller downward force. When it reads zero, we have set the current such that the magnetic force just equals the weight, which is the levitating condition desired. The current at this point is the operating point value i_0, say 2.3 amp, so we have now set our apparatus at the desired operating point.

Next we set the air gap at several values on either side of the operating point and measure the force at each value of y. When we do this, we always check the ammeter, and if the current tries to change from i_0 we readjust the power supply to keep the current constant at all gaps. We then return the gap to y_0 and hold it there

System Elements, Mechanical

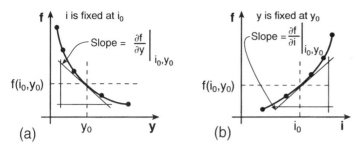

Figure 2-10 Use of experimental testing in multivariable linearization.

as we set various currents above and below i_0, reading the force sensor each time. We now have enough data to plot the graphs of Fig. 2-10. In Fig. 2-10a we measure the slope of the curve at $y = y_0$, say it is -762 N/m. This slope is the value of $\partial f/\partial y$. In Fig. 2-10b we measure the slope at $i = i_0$, getting, say, 67 N/amp, which is $\partial f/\partial i$. Our linearized model for predicting magnetic force from given air gaps and currents is then

$$f_m \approx 98 - 762(y - 0.012) + 67(i - 2.3)$$

This model is linear in the variables i and y and could now be used to set up differential equations for the vertical motion of the cart. The lab technique just explained is not limited to this example but is widely used in all kinds of system dynamics studies.

2-4 REAL SPRINGS

When analyzing an entire system, a spring is usually described by a single number, C_s or its reciprocal K_s, and system design will find an optimum value (or range of values) for this number. The next lower level of design requires that we choose a specific geometrical form, dimensions, and materials which will realize the numerical value of spring constant desired. This section is devoted to reviewing some common (and some uncommon) devices which serve as springs and would be so modeled in a system analysis. We also discuss some further departures of real springs from the theoretical.

In addition to the nonlinearity of the force/deflection curve, real springs also exhibit a noncoincidence of the loading and unloading curves, as in Fig. 2-11. The second law of thermodynamics guarantees that the area under the f vs. x curve (work put into the spring during loading) *must* be greater than that under the unloading f vs. x curve (work recovered from the spring during unloading). That is, it is impossible to recover 100% of the energy put into *any* system. This behavior of real springs indicates the presence of energy-dissipating mechanisms within the spring, whereas pure spring elements have only energy storage and no dissipation. In many springs these energy losses are quite small and it requires expert experimental technique to find a difference between loading and unloading curves; however, such a difference must always exist. (Inexperienced students sometimes get measurements showing *more* energy is recovered than was put in! This lack of expertise is understandable but we hope the lab report has a suitable comment on this impossibility.)

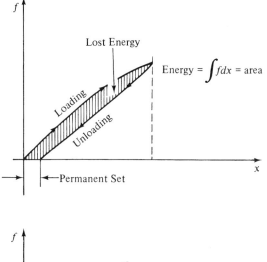

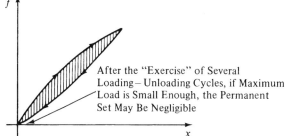

Figure 2-11 Energy losses in real springs.

For precision springs such as are needed in measuring instruments, special alloys such as Iso-Elastic[5] have been developed. This material exhibits a hysteresis (maximum difference between loading and unloading curves) of less than 0.05% of maximum deflection. Where its fragility is not a problem, quartz[6] has also been used as a nearly loss-free spring material. On the other hand, a rubber spring (widely used in shock mounts; see Fig. 2-12d) might have a hysteresis of 3 or 4%. This energy loss might actually be beneficial in such an application, since it would tend to damp out destructive vibrations.

While the coil or helical spring of Fig. 2-12a is perhaps most familiar, a wide variety of different geometrical forms can be and are used for spring functions. In fact, almost any physical object will exhibit springlike behavior in that when you press on it, a deflection nearly proportional to the applied force will occur. Figure 2-12 shows some of the more common forms actually used for springs. The hydraulic spring shown depends on the compressibility of oil for its operating principle and provides high energy storage in a small space.[7] Typical applications are return

---

[5]J. Chatillon & Sons, 7609 Business Park Drive, Greensboro, NC 27409.
[6]Quality Quartz Products, Inc., 8624 East Avenue, Mentor, OH 44060, 216-255-4481. Fused quartz springs measure bacteria weight gain, *Machine Design*, April 12, 1962, p. 33. E. O. Doebelin, Q-flex accelerometer, *Measurement Systems*, 4th ed., McGraw-Hill, 1990, p. 335.
[7]L. L. Johnson, The hydraulic spring, *Machine Design*, May 26, 1960.

System Elements, Mechanical

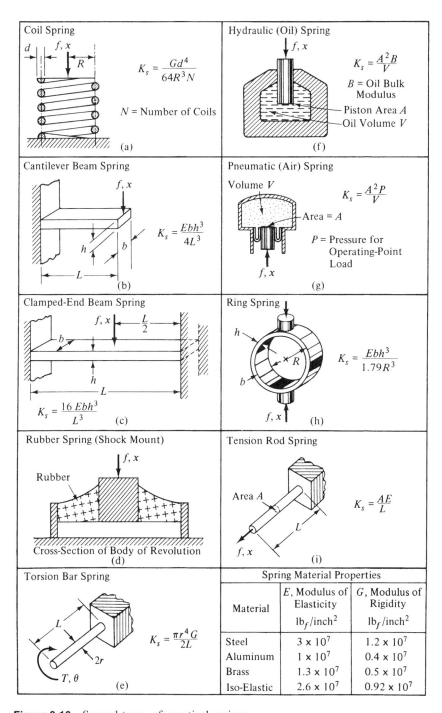

Figure 2-12 Several types of practical springs.

springs ("strippers") in punch-and-die assemblies of punch presses. Air springs (Fig. 2-12g) have many desirable properties for vehicle suspension systems. The most desirable force/deflection curve for this application is nonlinear, as shown in Fig. 2-13.[8] The air spring was actually designed to achieve this desirable characteristic. While a piston-cylinder could be used, the rubber rolling-diaphragm[9] of Fig. 2-12g gives a simpler and better air seal without critical manufacturing tolerances. Air springs can also be "pumped up" to automatically relevel a vehicle when large loads are carried. Since the linearized stiffness of an air spring increases in proportion to the weight supported, and the natural frequency of spring/mass systems depends on the stiffness/mass ratio, air springs used as vibration isolators maintain a constant degree of isolation no matter what weight is supported, a desirable feature.

To demonstrate that almost any object, even of peculiar shape, exhibits spring-like behavior, the "spring" of Fig. 2-14 was constructed of aluminum (pieces screwed together) and experimentally calibrated with dead weights and a micrometer, giving

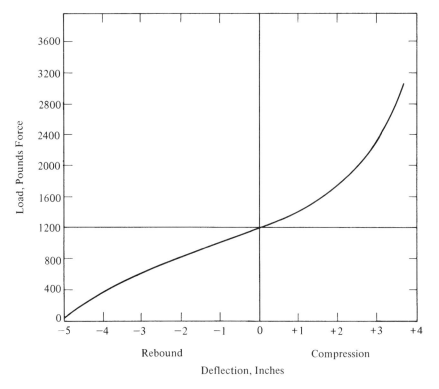

Figure 2-13 Automotive air spring characteristic.

---

[8]V. C. Polhemus and L. J. Kehoe, The development of the General Motors air spring, *General Motors Engineering Journal*, July–Sept, 1957.

[9]Firestone Industrial Products, 701 Congressional Blvd., Carmel IN 46032, 800-888-0650. Goodyear Tire and Rubber Co., Air Spring Department, P.O. Box 185, Greensburg, OH 44232, 800-321-6091.

System Elements, Mechanical

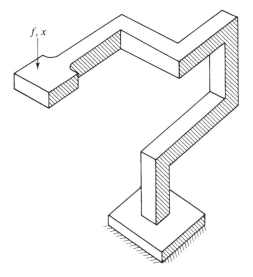

f lb$_f$	x Inch Increasing Load	x Inch Decreasing Load
0	0.0000	0.0002
1	0.0091	0.0094
2	0.0183	0.0186
3	0.0273	0.0276
4	0.0365	0.0366
5	0.0455	0.0456

Figure 2-14 Spring of complex shape.

the results shown. Note that spring-constant formulas for this unique shape would *not* be found in any handbooks and a finite-element analysis would be quite time consuming, whereas a lab test on the existing part took only a few minutes and gives more accurate results than even the finite-element study. Plot the data given to see if this "spring" is linear and, if so, get a number for its spring constant. This lab-test method is quite important since many machine parts, such as the crankshaft of an engine, are not *intended* to be springs, but their unavoidable springiness can cause vibration problems.

We conclude this section with a few more unconventional spring effects. In Fig. 2-15a the horizontal tail ("elevator") of a flying aircraft is deflected an angle θ to the relative wind, to cause the aircraft to pitch upward. If one were to measure the elevator shaft torque due to wind pressure for various values of θ, the graph shown would be obtained. We see that this aerodynamic torque exhibits a springlike behavior and would be modeled as a spring if we were studying the dynamic motion of the elevator as part of an aircraft control-system study. In Fig. 2-15b and c, the action of gravity provides a spring effect in that any motion of the pendulum or liquid column away from their static equilibrium position is accompanied by a restoring force or torque. The buoyancy spring effect in Fig. 2-15d is related to similar phenomena which influence the dynamic behavior of ships. Magnetic and electrostatic "springs" have been used to levitate objects, that is, to support them without physical contact. The magnetic version has been used as a high-speed "frictionless" bearing[10] and to support aircraft models in wind tunnels, while the electrostatic is the basis of sophisticated gyroscopic instruments[11] which exhibit

---

[10] Magnetic Bearings, Inc., 501 First Street, Radford, VA 24141, 703-639-9050.
[11] H. W. Knoebel, The electric vacuum gyro, *Control Engineering*, February 1964, pp. 70–73. Doebelin, *Measurement Systems*, p. 353.

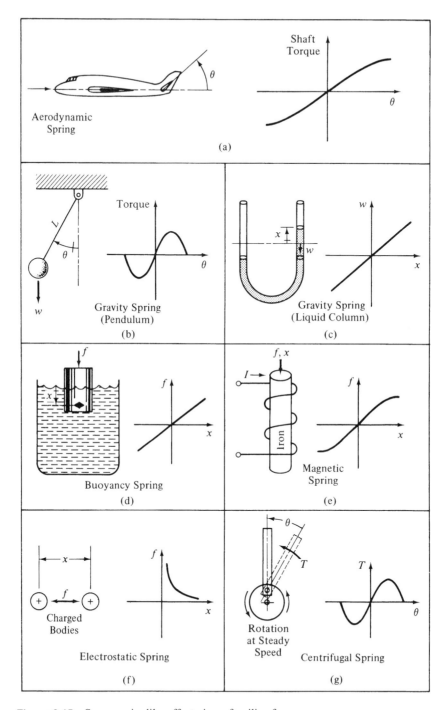

Figure 2-15 Some springlike effects in unfamiliar forms.

System Elements, Mechanical

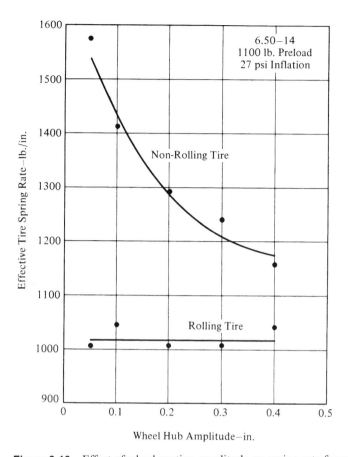

Figure 2-16 Effect of wheel-motion amplitude on spring rate for rolling and nonrolling tires.

extremely low friction effects, since the moving parts are supported electrically rather than in bearings. Centrifugal spring effects, Fig. 2-15g, are somewhat similar to gravity springs in that, in each case, a force "field" creates a preferred position for an object and any displacements away from this position give rise to restoring forces or torques. The pneumatic tires used on vehicles influence the riding and handling qualities of the vehicle and exhibit some interesting spring properties. If the tire is tested when *not* rolling, the force-deflection curve is quite nonlinear, giving a linearized spring constant which varies with amplitude of motion. However, when the same tire is tested[12] while rolling at about 10 mph, the spring constant becomes very nearly independent of amplitude, indicating linear behavior (see Fig. 2-16).

This concludes our discussion of springs, real and ideal. I hope you now have a wider and deeper understanding of these useful devices.

---

[12]Dynamic spring rate performance of rolling tires, General Motors Engineering Publication 3610, 1968.

54   **Chapter 2**

2-5   THE DAMPER (FRICTION) ELEMENT

While a pure spring element stores and returns energy with no loss or dissipation, a pure damper element dissipates *all* of the energy supplied to it. Since energy cannot be destroyed, what we mean by dissipation of energy is that it is converted from mechanical to thermal form (heat) which flows away to the surroundings and is thus no longer available for useful work. Various physical mechanisms, usually associated with some form of friction, can provide this dissipative action. While you have certainly encountered friction forces in earlier physics or mechanics courses, these usually discuss only those forms (called *Coulomb* friction) for which the friction force is proportional to normal force and independent of speed, except perhaps for allowing a different friction coefficient (static friction) when no motion occurs and another (dynamic friction) when motion is occurring. These simple concepts allowed you to work certain practical problems, but if you use these kinds of friction in differential equations, they make the equations nonlinear. Our pure/ideal damper element provides so-called *viscous friction*, which leads to linear differential equations with constant coefficients.

All the mechanical elements are defined in terms of their force/motion relation. When we get to electrical elements we will see that they are defined in terms of their voltage/current relation. For a pure/ideal damper the defining force/motion relation is (see also Fig. 2-17)

$$f = B\left(\frac{dx_1}{dt} - \frac{dx_2}{dt}\right) = B\frac{dx}{dt} \qquad (2\text{-}20)$$

where

$$f \stackrel{\Delta}{=} \text{force applied to ends of damper, lb}_f \text{ or N}$$

$$\frac{dx_1}{dt} \stackrel{\Delta}{=} \text{velocity of one end, in/sec or m/sec}$$

$$\frac{dx_2}{dt} \stackrel{\Delta}{=} \text{velocity of other end, in/sec or m/sec}$$

$$B \stackrel{\Delta}{=} \text{damper coefficient, lb}_f/(\text{in/sec}) \text{ or N}/(\text{m/sec})$$

The damper force is thus seen to be directly proportional to the relative *velocity* of its two ends, whereas the spring force is proportional to the relative *displacement*. Just as in the spring element, the forces on the two ends of the damper are exactly equal and opposite at all times, because both elements have no mass. That is, in Newton's law $\sum F = ma$, if m is zero, the resultant force must be zero at all times, no matter how the damper or spring is moving. This of course is *not* true for *real* springs or dampers.

For rotational systems we have

$$T = B\left(\frac{d\theta_1}{dt} - \frac{d\theta_2}{dt}\right) = B\frac{d\theta}{dt} \qquad (2\text{-}21)$$

System Elements, Mechanical

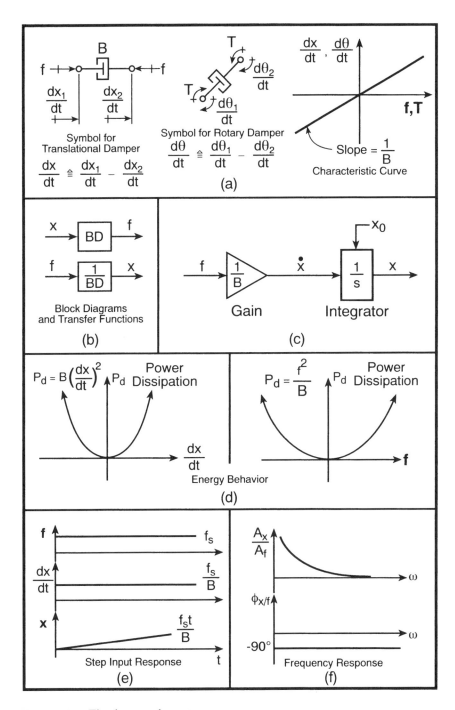

Figure 2-17 The damper element.

56   **Chapter 2**

where

$$T \stackrel{\Delta}{=} \text{torque applied to ends of damper, in-lb}_f \text{ or N-m}$$

$$\frac{d\theta_1}{dt} \stackrel{\Delta}{=} \text{angular velocity of one end, rad/sec}$$

$$\frac{d\theta_2}{dt} \stackrel{\Delta}{=} \text{angular velocity of other end, rad/sec}$$

$$B \stackrel{\Delta}{=} \text{damper coefficient, (in-lb}_f)/(\text{rad/sec) or (N-m)/(rad/sec)}$$

We do *not* "simplify" the units of B to (N-m-sec)/rad since this loses the *physical meaning* easily attached to (N-m)/(rad/sec). That is, what B *means* is: How many N-m of torque does it take to give the damper 1 rad/sec of angular velocity? This meaning is obscured by the "compacted" form (N-m-sec)/rad, so please don't use such forms, even though you may have gotten used to it.

To draw block diagrams, we again need the concept of the transfer fucntion, and it is now necessary to define it more correctly and completely than we did earlier. Usually, our equations will involve derivatives or integrals, not just algebra as was the case for the spring element. Our general definition of transfer function will use the *operator notation*:

$$Dx \stackrel{\Delta}{=} \frac{dx}{dt} \qquad D^2 x \stackrel{\Delta}{=} \frac{d^2 x}{dt^2} \qquad \text{etc.} \qquad D \stackrel{\Delta}{=} \frac{d}{dt}$$

$$\frac{x}{D} \stackrel{\Delta}{=} \int x \, dt \qquad \frac{x}{D^2} \stackrel{\Delta}{=} \int\left[\int x \, dt\right] dt \qquad \text{etc.} \tag{2-22}$$

That is, any quantity found immediately to the right of the *differential operator* D is to be differentiated with respect to time, and the symbol $1/D$ stands for integration with respect to time. (Those readers who know Laplace transforms may want to use s where we use D in transfer functions.)

Applying the operator notation to Eqs. (2-20) and (2-21), we get

$$f = BDx \qquad \text{and} \qquad T = BD\theta \tag{2-23}$$

We define the *operational transfer function* $(f/x)(D)$ by treating these equations as if they were algebraic and forming the output/input ratio

$$\frac{f}{x}(D) \stackrel{\Delta}{=} BD \tag{2-24}$$

This is read "f over x of D is defined to be BD." The notation $(f/x)(D)$ is used since the simpler f/x could be interpreted as an ordinary instantaneous ratio of time-varying quantities f and x, *which the transfer function is not*. That is, the transfer function is *not* $f(t)/x(t)$, but rather a defined symbol which compactly states a differential equation and which allows the drawing of useful block diagrams. So, be sure to always write $(f/x)(D)$, *not* just f/x. Readers who prefer the Laplace transfer functions would write $(f/x)(s) = Bs$.

Sometimes we know the force and want to find the displacement, reversing the roles of input and output, and giving the transfer function

$$\frac{x}{f}(D) \stackrel{\Delta}{=} \frac{1}{BD} = \left(\frac{1}{B}\right)\left(\frac{1}{D}\right) \tag{2-25}$$

System Elements, Mechanical

The meaning of the operator $1/D$ is clarified by writing

$$f\,dt = B\,dx$$

$$\int_0^t f\,dt = B\int_{x_0}^x dx = B(x - x_0)$$

$$x - x_0 = \frac{1}{B}\int_0^t f\,dt \tag{2-26}$$

where x_0 is the *initial value* (value at $t = 0$) of x. This may be compared to Eq. (2-25) rewritten as

$$x = \frac{1}{B}\frac{1}{D}(f) \tag{2-27}$$

If we take the initial value $x_0 = 0$, Eq. (2-26) gives

$$x = \frac{1}{B}\int_0^t f\,dt \tag{2-28}$$

thus the operator $1/D$ indicates *integration* with respect to t, but the constant of integration x_0 is understood, not expressly stated.

In Fig. 2-17c we choose to show the digital simulation diagram for the case where the damper force f is given and we want to find the velocity and displacement caused by this force. We now need an integrator icon, which has the transfer function $1/D$ (SIMULINK uses the Laplace form $1/s$). This icon also provides for setting the initial condition (x at $t = 0$) at whatever value the physical problem requires. To enter this number one double-clicks on the icon and then types in the desired number when a dialog box appears.

To continue our gradual introduction to simulation methods and also to contrast the behavior of spring and damper elements, we now set up the simulation of Fig. 2-18. Here we again use the lookup table to generate a force/time pattern of our choice and then apply this force simultaneously to a spring and a damper. [This technique of running two simulations "side by side" is very useful in design studies where we want to compare the behavior of two (or more) competing designs, to help us decide which is best.] I set the initial displacement at zero, $K_s = 100\,\text{N/m}$, and $B = 100\,\text{N/(m/sec)}$ The force varies as shown in Fig. 2-19, between -1 and $+1$ newtons. The spring displacement xspring obviously will have exactly the same

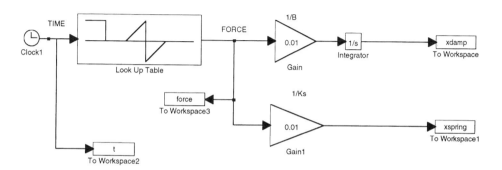

Figure 2-18 Simulation model for comparison of spring and damper behavior.

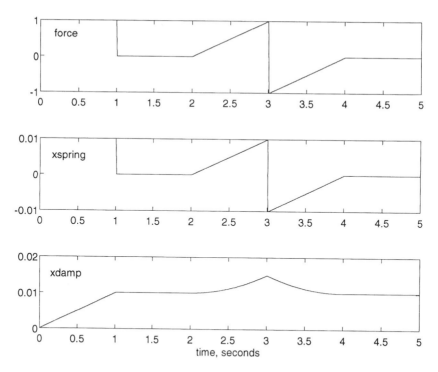

Figure 2-19 Comparison of spring and damper behavior for same force input.

waveform as the force, but varies between $-.01$ and $+.01$ meter. The damper *velocity* (not graphed) of course has the same waveform as the force, but its displacement xdamp looks very different. While the spring suddenly goes from 0 to 0.01 m at $t = 0$, the damper "ramps" up to 0.01, taking 1 second to move this distance. When force drops to zero at $t = 1$, the spring instantly relaxes to zero displacement, while the damper "just sits there" at the displacement it had before the force dropped to zero. When the force ramps up between $t = 2$ and $t = 3$, xdamp follows a parabolic curve (the integral of the ramp).

We have mentioned several times that the damper element dissipates into heat all mechanical energy supplied to it, and we want to now prove this. Using the definition of power as the product of force and velocity,

$$P \triangleq (\text{force})(\text{velocity}) = (f)\left(\frac{dx}{dt}\right) = B\left(\frac{dx}{dt}\right)^2 \quad \frac{\text{N-m}}{\text{sec}} \tag{2-29}$$

Note that any force applied to a damper causes a velocity in the *same* direction. The source which is supplying the force must thus provide power *to* the damper, since when the force on a device and the velocity have the same sign, the power input to the device is positive. With a damper it is *impossible* for the applied force and the resulting velocity to have opposite signs; thus the damper can never supply power to another device — P in Eq. (2-29) is always positive. A spring, however, absorbs power and stores energy as a force is applied to it, but if the force is gradually relaxed back to zero, the external force and the velocity now have *opposite* signs, showing that the

System Elements, Mechanical

spring is *delivering* power (recall Fig. 2-2). For a damper, the total energy dissipated over any time interval is the time integral of the power, $\int P\, dt = $ N-m. A constant force f_0, for example, gives an energy dissipation of $f_0^2 t/B$ N-m for a time interval t.

A step input force f_s instantly (since a pure damper has no inertia) causes a velocity f_s/B which is maintained as long as f_s is maintained (see Fig. 2-17e). This constant velocity produces a displacement which increases linearly with time. This linear increase with time is called a *ramp function*. Thus a step of f causes a step of dx/dt and a ramp of x.

As usual, to study the frequency response, we let $f = f_0 \sin \omega t$. Then

$$x - x_0 = \frac{1}{B}\int_0^t f\, dt = \frac{1}{B}\int_0^t f_0 \sin \omega t\, dt = \frac{f_0}{B\omega}(1 - \cos \omega t) \tag{2-30}$$

These relations are graphed in Fig. 2-20 for an arbitrarily chosen value of x_0. We see that the "sine" wave representing the oscillation of x has a 90° phase lag with respect to the f sine wave, that is, x "starts" at point s, whereas f started at $t = 0$. When a phase angle represents lagging behavior, it is conventionally given a negative sign, thus $\phi_{x/f} = -90°$. Note that this is true for any frequency ω. The amplitude ratio is clearly

$$\frac{A_x}{A_f} = \frac{f_0/B\omega}{f_0} = \frac{1}{B\omega} \tag{2-31}$$

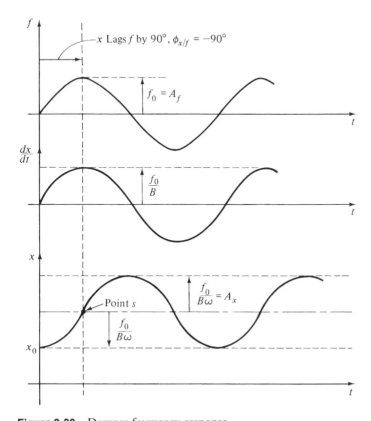

Figure 2-20 Damper frequency response.

giving the frequency response graphs of Fig. 2-17f. Note that a sinusoidal force of any (even small) amplitude can produce a very large displacement (approaching infinity as ω goes to zero) if applied at a low frequency, but a smaller and smaller displacement (approaching zero) as the frequency is raised. Compare this with the behavior of the spring element, which, for a given force, produces exactly the same displacement for every frequency.

The determination of frequency response curves for system elements is relatively quick and simple, but it becomes much more tedious, using our present methods, when complete systems are considered. Fortunately, a shortcut method (which will be derived in a later chapter) called the *sinusoidal transfer function* is available. The sinusoidal transfer function is obtained from our operational transfer function by merely substituting the term $i\omega$ for the D operator wherever it appears; here i is the square root of -1 and ω is the frequency of the sinusoidal input. Applying this general definition to the damper of Eq. (2-25), we get

$$\text{Sinusoidal transfer function} \triangleq \frac{x}{f}(i\omega) \triangleq \frac{1}{i\omega B} \qquad (2\text{-}32)$$

This is read "x over f of $i\omega$ is defined to be 1 over $i\omega B$."

Since sinusoidal transfer functions are usually complex numbers, and we want to be able to compute quickly with them, we now review some basic complex number arithmetic. From Fig. 2-21 recall that complex numbers can be given in rectangular $(a + ib)$ or polar $(M\ /\!\underline{\phi})$ form: $M \triangleq$ square root $(a^2 + b^2)$ and $\phi \triangleq \tan^{-1}(b/a)$. When adding or subtracting complex numbers we usually prefer the rectangular form, whereas the polar form is most convenient for multiplication and division: $(M_1\ /\!\underline{\phi_1})(M_2\ /\!\underline{\phi_2}) = M_1 M_2\ /\!\underline{\phi_1 + \phi_2})$ and $(M_1\ /\!\underline{\phi_1})/(M_2\ /\!\underline{\phi_2}) = (M_1/M_2)\ /\!\underline{\phi_1 - \phi_2}$.

In interpreting sinusoidal transfer functions we will use the polar form $M\ /\!\underline{\phi}$ because it can be shown (in a later chapter) that if we do this, then M will be the

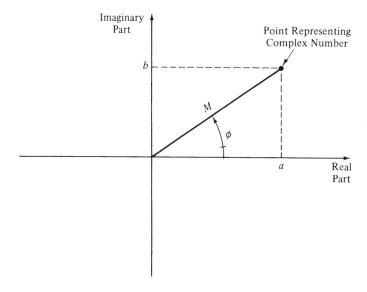

Figure 2-21 Complex number definitions.

System Elements, Mechanical   **61**

amplitude ratio of output over input and ϕ will be the phase shift of the output sine wave with respect to the input sine wave (ϕ is positive if output leads input, negative if output lags input). This holds for *all* sinusoidal transfer functions, not just for a damper. For the damper we get

$$\frac{x}{f}(D) = \frac{1}{BD}$$

$$\frac{x}{f}(i\omega) = \frac{1}{i\omega B} = \left(\frac{1}{B\omega}\right)\left(\frac{1}{i}\right) = \frac{1}{B\omega}(-i) = \frac{1}{B\omega}\ \underline{/-90^\circ} = M\ \underline{/\phi} = \frac{A_x}{A_f}\ \underline{/\phi_{x/f}}$$

$$(2\text{-}33)$$

which we see agrees with our earlier result. We will use this method of calculating frequency response from this point on; there is no easier.

2-6 REAL DAMPERS

Just as with springs, a damper element is sometimes used to model a device designed into a system to accomplish some useful function, and other times for unavoidable "parasitic" effects. Thus automotive shock absorbers (dampers) serve a useful function while the air drag ("friction") force on the car increases gas consumption, but cannot be ignored in modeling the car's forward motion. We will first consider "intentional" dampers. These can take many detailed forms, but remember that to be an energy-dissipating effect, a device must exert a force *opposite* to the velocity. That is, power is always negative when the force and velocity have *opposite* directions. Any device which behaves in this way is some kind of damper.

The classic device is perhaps the viscous (piston/cylinder) damper whose configuration is the basis for the standard damper symbol of Fig. 2-17a. A damper of this type, used to control vibration of the *Ranger* spacecraft's solar panels,[13] is shown in Fig. 2-22. A relative velocity between the cylinder and piston forces the viscous oil through the clearance space h, shearing the fluid and creating a damping force. An analysis in the reference gives

$$B = \frac{6\pi\mu L}{h^3}\left[\left(R_2 - \frac{h}{2}\right)^2 - R_1^2\right]\left(\frac{R_2^2 - R_1^2}{R_2 - h/2} - h\right)\quad \frac{\text{lb}_f}{\text{in/sec}}\qquad (2\text{-}34)$$

where $\mu \triangleq$ fluid viscosity, $\text{lb}_f\text{-sec/in}^2$. Note again the distinction between *system* design and *detail* design. In the system design of the spacecraft vibration controls, the damper is described by the single number B, and system design finds the best numerical value. Once this B value is available, detail design using Eq. (2-34) can commence. There are an infinite number of combinations of L, h, R_1, R_2, and μ that will give the desired B value. Many practical considerations go into the choice of a *single* combination that will actually be used. For example, if h is made too small, dirt particles may cause jamming, temperature changes may cause binding, and manufacturing tolerances may be intolerable.

---

[13]M. Gayman, Development of a point damper for the Ranger solar panels, Jet Propulsion Lab Rept. 32-793, California Institute of Technology, 1965.

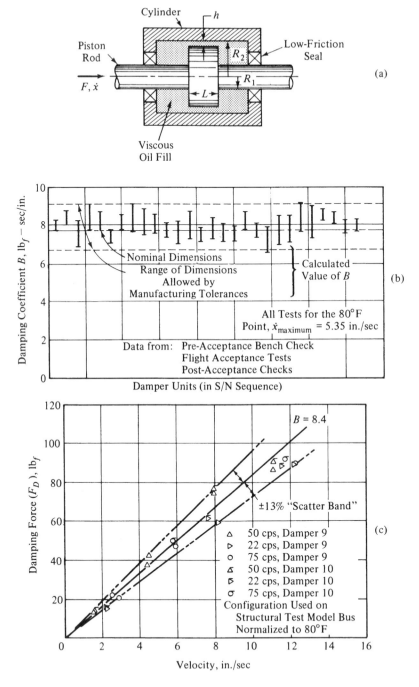

Figure 2-22 Damper used on the Ranger spacecraft.

System Elements, Mechanical

For the application cited, the desired nominal B value was $7.6 \, \text{lb}_f/(\text{in/sec})$. Twenty-eight such dampers were constructed and experimentally tested to determine B (see Fig. 2-22b). The vertical bands indicate the "scatter" of experimental measurements for an individual damper. Note also the possible variation in B due to unavoidable manufacturing tolerances. Nevertheless, the agreement between predicted and actual values is on the average very good. It should be pointed out that, even using silicone oil (which is the least temperature-sensitive liquid suitable for dampers) a temperature change from 70°F to 140°F causes about a 50% decrease in viscosity and thus in B; therefore, experimental values must be temperature-corrected to make a fair comparison with theoretical calculations. Design at the system level for vibration control would actually result in a *range* of acceptable values for B, rather than a single value, because manufacturing tolerances, temperature changes, etc. make the provision of a fixed B value *impossible*. This is true of *every* system, not just this spacecraft. When using simulation at the system level, we often explore the effect of variations from nominal design values for all our system parameters. *Statistical uncertainty analysis*[14] is a useful tool for such studies.

A sinusoidal test method was used to measure B, and the maximum velocity reached was about 12 in/sec. Figure 2-22c shows a fairly linear force/velocity relationship within these limits. If tests have been carried to higher forces and velocities, nonlinear behavior would have been revealed because the fluid flow would change from laminar to turbulent.

Figure 2-23a shows a simple form of damper which is easily analyzed using the basic definition of fluid viscosity given in Fig. 2-23b. Here a flat plate of area A floats on a liquid film of thickness t, pulled by a steady force F which causes a constant plate velocity V. The definition of viscosity is

$$\mu \triangleq \text{fluid viscosity} \triangleq \frac{\text{shearing stress}}{\text{velocity gradient}} = \frac{F/A}{V/t} \qquad \frac{\text{lb}_f\text{-sec}}{\text{in}^2} \qquad (2\text{-}35)$$

The viscosity of fluids is actually measured with an instrument (viscosimeter) based on a rotational version of this scheme in which one measures F, A, V, and t and then calculates the viscosity. Using Eq. (2-35) we get for the damper of Fig. 2-23a

$$F = \frac{2A\mu}{t} V$$

$$B = \frac{F}{V} = \frac{2A\mu}{t} \qquad \frac{\text{lb}_f}{\text{in/sec}} \qquad (2\text{-}36)$$

The rotational versions shown in Fig. 2-24a and b may be similarly analyzed to yield

$$B = \frac{\pi D^3 L \mu}{4t} \qquad \frac{\text{in-lb}_f}{\text{rad/sec}} \qquad (2\text{-}37)$$

$$B = \frac{\pi D_0^{\,4} \mu}{16t} \qquad \frac{\text{in-lb}_f}{\text{rad/sec}} \qquad (2\text{-}38)$$

where the shear area loss due to D_i has been neglected.

---

[14]E. O. Doebelin, *Engineering Experimentation*, McGraw-Hill, New York, 1995, pp. 64–66, 147–152.

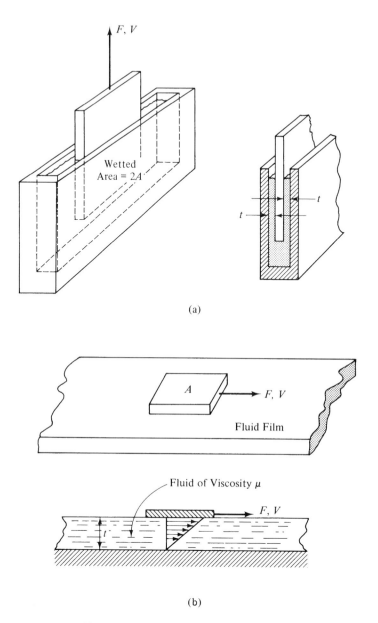

Figure 2-23 Simple shear damper and viscosity definition.

Gases may also be used as the damping fluid. Since their viscosity is much lower, they do not give as large a value of B; however, they are less temperature-dependent. Also, if the gas used is atmospheric air, there is no leakage or sealing problem. Figure 2-25 shows a small damper of this type available commercially.[15] The graphite piston and glass cylinder are fitted to a tolerance of 0.0001 inch and

---

[15]"Airpot," Airpot Corp., 35 Lois Street, Norwalk, CT 06581, 800-848-7681.

System Elements, Mechanical

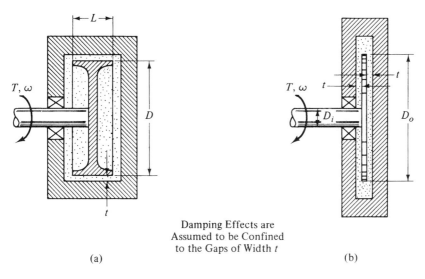

Figure 2-24 Two types of rotary damper.

give practically no rubbing friction since the air forms a thin film between them. The fluid damping action occurs in this air film and also in an adjustable needle valve, which forms a flow restriction between the cylinder and the atmosphere. Flow in the air film is laminar (giving a linear damping relation) while that in the needle valve is more nearly turbulent (giving nonlinear damping) unless the valve is almost shut. If the valve is shut tight we get the strongest damping and, since it is now all due to the air film, it is quite linear. The table of Fig. 2-25 is actual data taken with the valve shut, by applying dead weights to the piston and measuring the resulting steady velocity. Plot this data to check for linearity and also to find B

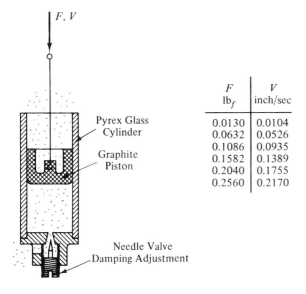

F lb$_f$	V inch/sec
0.0130	0.0104
0.0632	0.0526
0.1086	0.0935
0.1582	0.1389
0.2040	0.1755
0.2560	0.2170

Figure 2-25 A commercial air damper.

for this device. The manufacturer also uses this unique construction to provide a pneumatic actuator which is almost friction-free, and thus ideal for precision motion control.

Just as with springs, linear damping, while "mathematically" ideal, may not always be functionally ideal for particular applications. In the next section, where we study the mass (inertial) element, we will do a design study which shows the required behavior of an *optimum* ("best possible") damper if the application is the deceleration of a moving mass. We will find that such an optimum damper *must* be nonlinear and discover what form this nonlinearity must take.

Certain electrical effects also provide a mechanical damping action which closely approximates that of the pure/ideal damper element. The damping forces available in this way are relatively small but are sometimes sufficient for low-power devices such as measuring instruments. Figure 2-26 shows an *eddy-current* damper.[16] Motion of the conducting cup in the magnetic field generates a voltage

$$E = 10^{-8} B_m \pi D V \quad \text{volts} \tag{2-39}$$

in the cup, where

$B_m \triangleq$ magnetic induction, gauss

$D \triangleq$ cup mean diameter, cm

$V \triangleq$ relative velocity, cm/sec

The resistance of the cup's circular path within the field is

$$R = \frac{\pi D \rho}{bd} \quad \text{ohms} \tag{2-40}$$

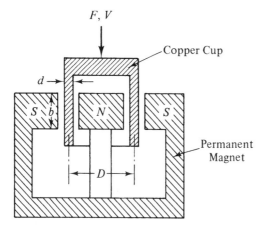

Cross-Section of
Circular Configuration

Figure 2-26 Eddy-current damper.

---

[16]H. K. P. Neubert, *Instrument Transducers*, Oxford University Press, London, 1963, p. 47.

System Elements, Mechanical

where

$\rho \triangleq$ cup resistivity, ohm-cm

$b, d \triangleq$ width and thickness of conducting path, cm

The current in this path is thus $B_m(bdV/10^8\rho)$ and since a current-carrying conductor in a magnetic field experiences a force proportional to the current, we get a force proportional to and opposing the velocity V. The damper coefficient B is found to be

$$B = \frac{B_m^2 \pi Dbd}{10^9 \rho} \quad \frac{\text{dynes}}{\text{cm/sec}} \tag{2-41}$$

The dissipated energy shows up as I^2R heating of the cup. A rotational version is essentially a DC generator with a resistive load. Maximum damping is obtained by just short-circuiting the generator output terminals, giving the minimum total resistance, the generator's internal resistance. Eddy-current damping is relatively insensitive to temperature, as shown in Fig. 2-27.

The use of a porous plug (Fig. 2-28a) as a flow restriction for an air damper has been studied[17] theoretically and experimentally. While the arrangement exhibits some nonlinearity and also a significant air-spring effect, it has been successfully applied in practice. In Fig. 2-28b a capillary tube[18] provides a laminar flow resistance between the ends of a piston/cylinder to give essentially linear damping. Squeeze-film damping,[19] Fig. 2-28c, is quite nonlinear but can provide large forces for small motions. It may employ either gases or liquids. A commercially available[20] rotary damper with adjustable damping is shown in Fig. 2-28d.

Structures used for communications satellites, orbiting precision instruments (Hubble telescope, etc.), and future space stations require damping augmentation

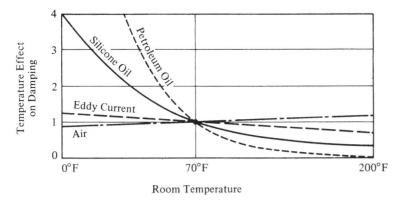

Figure 2-27 Temperature sensitivity of damping methods.

---

[17]R. L. Peskin, and E. Martinez, ASME Papers 65-WA/FE-8 and 65-WA/FE-9, 1965.
[18]H. H. Richardson, Fluid control, components and systems, *Agardograph* 118, December 1968.
[19]E. A. Sommer, Squeeze-film damping, *Machine Design*, May 26, 1966, p. 163.
[20]EFDYN Corp., 7734 East 11th Street, Tulsa, OK 74112-5718, 918-838-1170.

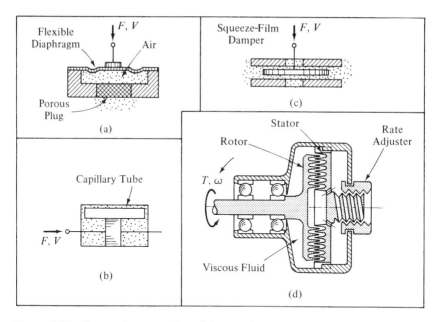

Figure 2-28 Some other examples of damper forms.

since there is no air damping in the vacuum of space and the hysteretic damping of the metal parts is insufficient. Both active damping (feedback control systems, "smart structures") and passive damping are in use. Space platforms for precision instruments can tolerate only the tiniest vibratory motions, so dampers must function for motions the order of 50 nanometers. Figure 2-29 shows a simplified sketch of a viscous damper (Honeywell D-STRUT[21]) designed to work under such conditions. The use of metal bellows, rather than sliding seals, removes all rubbing friction (which would never be "broken loose" for tiny forces and motions) and provides pure viscous damping. The adjusting screw shown (replaced by a stepping-motor drive, for remote electrical adjustment in the actual device) changes the thickness of the oil film in the conical annular gap, giving a 100-to-1 adjustment in B.

Let us now leave the realm of intentionally introduced damping devices and consider briefly the use of the damping element to represent unavoidable "parasitic" energy dissipation effects in mechanical systems. A list of such effects would include

1. Frictional effects in moving parts of machines
2. Fluid drag on vehicles (cars, ships, aircraft, etc.)
3. Windage losses of rotors in machines
4. Hysteresis losses associated with cyclic stress in materials
5. Structural damping due to riveted joints, welds, etc.
6. Air damping of vibrating structural shapes

---

[21]L. P. Davis et al., Adaptable Passive Viscous Damper (An Adaptable D-Strut), SPIE North American Conference, Orlando, Florida, February 1994. Honeywell Satellite Systems Operation, 19019 North 59th Avenue, Glendale, AZ 85308, 602-561-3483.

System Elements, Mechanical

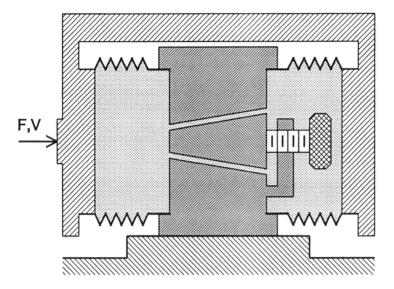

Figure 2-29 Honeywell D-STRUT damper for space structures.

Frictional effects in machines are usually a complex combination of dry rubbing ("Coulomb") friction plus linear and nonlinear fluid friction. The "simple" Coulomb friction considered in introductory physics and mechanics courses is actually quite tricky to accurately simulate in dynamic systems. The main problem here is that, for applied forces less than that required to break loose the static friction, the friction force actually *adjusts* itself to exactly equal the applied force, but no motion occurs. When the applied force just barely exceeds the breakaway friction, the friction force *drops* to the lower dynamic friction level at the instant motion begins. Accurate simulation of this behavior seems to require the versatility of the command-language approach rather than the GUI menu style. An ACSL program for this friction problem is available.[22] We should keep in mind, of course, that real frictional effects are often *not* reliably predictable, so that accurate simulation of some *model* may not correspond closely to the measured behavior on a given day. Coupled shafts are never perfectly aligned and can change their alignment unpredictably from minute to minute, as can thermal expansion and lubrication effects, all of which affect bearing loads and thus friction forces and torques.

As an example of machine friction, consider the hydraulic rotary motor characteristic shown in Fig. 2-30. The friction torque of such a motor is due to rubbing, sliding, and rolling of various parts such as pistons, cylinders, ball bearings, plain bearings, seals, and valve plates. An experimentally measured friction curve for such a motor might appear as in Fig. 2-30a. Just as in springs, the nonlinear characteristic may be linearized for approximate analysis in the neighborhood of an operating point by taking the slope of the curve as a value for B. Dry or Coulomb friction

---

[22] ACSL Reference Manual, Ed. 4.2, p. A-124, Mitchell and Gauthier Associates, 200 Baker Avenue, Concord MA 01742-2100, 508-369-5115.

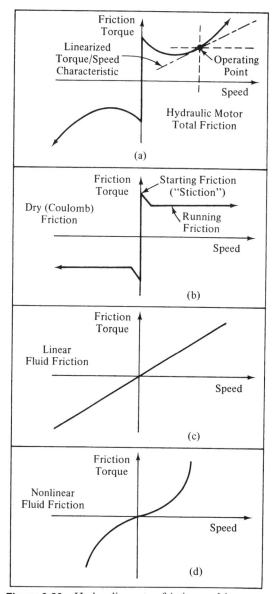

Figure 2-30 Hydraulic motor friction and its components.

is generally assumed independent of velocity except for the difference between static and running friction coefficients (Fig. 2-30b). When the operation of a system involves *large* motions or speed changes rather than small variations about an operating point, linearizing schemes other than the local tangent line may be appropriate (Fig. 2-31). The incentive for linearization is of course the desire to obtain linear differential equations so that rapid and revealing analytical methods may be applied in the early stages of design and analysis. Later, simulation is profitably employed to

System Elements, Mechanical

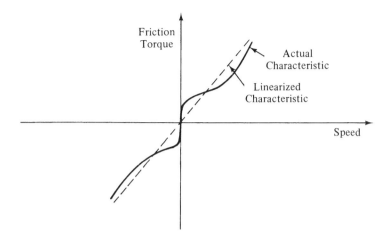

Figure 2-31 Linearization method for large speed ranges.

include nonlinearities, so that we may check whether our earlier linearizations obscured any essential features of system behavior.

The drag force or torque on vehicles or other solid bodies moving in a fluid medium is essentially proportional to velocity for low velocities and becomes proportional to the square of velocity at high speeds. For bodies of simple geometry, theoretical results are available[23] for the low-velocity (viscous) range:

For a sphere of radius r: $\qquad B = \dfrac{F}{V} = 6\pi r\mu \qquad$ (2-42)

For a thin cylinder of radius r and length l: $B = \dfrac{F}{V} = 2\pi l\mu \qquad$ (2-43)

For bodies of complex shape (automobile body, etc.), it is necessary to run experiments to find the force/velocity relation. Even for simple shapes, high velocities cause turbulent flow, and again experiments are needed; however, most such results[24] indicate the drag force to be proportional to the square of velocity, giving a nonlinear damping. The windage torques of rotating electrical machines have a complex nonlinear characteristic.[25]

When structures, such as machine tool frames and aircraft wings, vibrate at resonance, the stresses and deflections are limited only by the damping provided by the surrounding air and the metal hysteresis losses. Spacecraft structures are often damped only by hysteresis, since there is no air damping in the vacuum of space. The treatment[26] of damping due to hysteresis requires a different approach, because one

---

[23]D. G. Stephens and M. A. Scavullo, Investigation of air damping of circular and rectangular plates, a cylinder and a sphere, NASA TND-1865, April 1968.
[24]Ibid.
[25]J. E. Vrancik, Prediction of Windage Power Losses in Alternators, NASA TND-4849, October 1968.
[26]B. J. Lazan, *Damping of materials and members in Structural Mechanics*, Pergamon Press, New York, 1968.

cannot identify an obvious damping "force." Rather, the energy dissipation is occurring at a microscopic level in the metal and is distributed over its volume. The magnitude of the energy loss appears to depend on the local stress raised to some power which, unfortunately, varies over a wide range and must be determined experimentally for each material. Since the stress level also varies over wide ranges and in complicated fashion over the volume of a structure, calculation of total damping is very difficult. Once the structure (or a suitable scale model) has been constructed, however, vibration test measurements allow determination of damping factors associated with each mode of vibration. When the vibration mode of interest is excited, and then allowed to die out freely, the rate of decay of the vibration permits calculation of an equivalent linear damping factor.

We conclude this section with a simulation demonstrating the behavior of the various types of damping discussed. Since damping is often of interest with respect to vibratory motions, our simulation compares the friction forces when the motion has a sinusoidal velocity. Since sine waves are of such general importance, SIMULINK provides an icon (see Fig. 2-32) for generating them; it has adjustable amplitude, frequency, and phase angle. I set this one up to produce a velocity $1.0 \sin 10t$ cm/sec and "sent" this velocity into a gain block set at -5.0 N/(cm/sec), to simulate the pure/ideal damper element. The damping force was named fvisc and graphed at the top of Fig. 2-33, where we see that its waveform is a "negative" sine wave, as expected. This is the *only* type of damping that, when used in a mechanical dynamic system, gives analytically solvable differential equations.

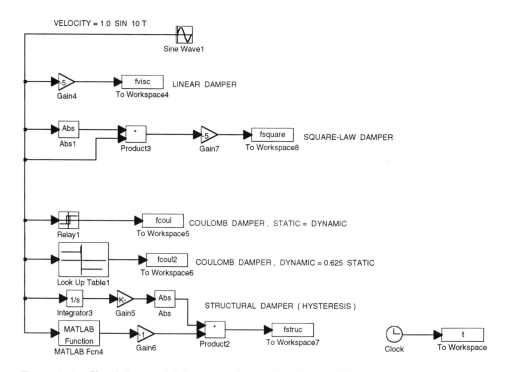

Figure 2-32 Simulation model for comparing various forms of damping.

System Elements, Mechanical

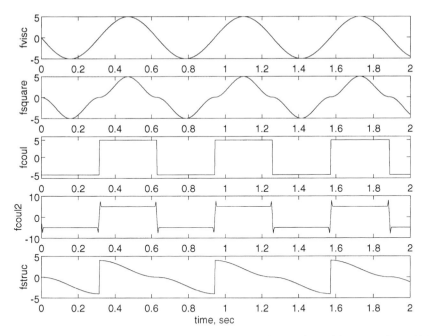

Figure 2-33 Damping force for various forms of damping with same sinusoidal velocity input.

The same velocity is now sent into a "square-law" damper, more representative of those real dampers using fluids with orifices (rather than laminar flow passages), such as automotive shock absorbers. Note that we can't just "square the velocity" since such a term would always be positive and *not* oppose the velocity, as an energy dissipating device must. To get the correct algebraic sign as the velocity changes sign we use an absolute value icon with a product icon, abs $(v) \times (v)$. This is followed by a gain of -5, which makes the peak value of the force the same as for the linear damper, for comparison purposes. Note in Fig. 2-33 that this damping force is *not* a simple sine wave but appears to have some higher-frequency "bumps" in it.

To simulate the simplest model of Coulomb or dry-friction damping, we use SIMULINK's relay icon. When its input signal crosses from negative to positive, its output switches from a given positive value to a given negative value. I have set these output values at $+5$ and -5, to simulate a dry friction force of ± 5 N, always opposing the velocity. No distinction is made between "starting" and "running" friction. Figure 2-33 shows that this form of friction gives a friction force (fcoul) in the form of a square wave, for a sinusoidal velocity.

Recall that we pointed out earlier the difficulty of making an "accurate" simulation of dry friction, allowing a difference between starting and running friction coefficients, and taking into account the "adjustable" nature of the friction force before motion occurs. If we are content with a simulation which at least allows a difference between starting and running friction, a simple approach using a lookup table is possible (see Fig. 2-34). Here, for velocities greater in absolute value than 0.1 cm/sec, the friction force is ± 5 N, just as for the previous example. When velocity drops below 0.1 the friction force gets larger, peaking at ± 8 N for $v = \pm 0.05$. Further drop in velocity, however, causes a *decrease* in friction force, dropping to zero at zero

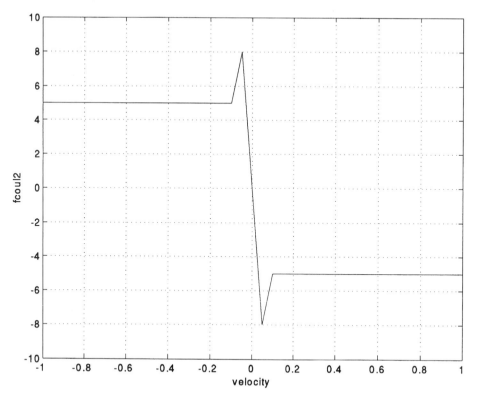

Figure 2-34 One possible model for Coulomb friction with different starting and running values.

velocity. Note that we *could* have set the 0.1 and 0.05 "breakpoints" at any values we please, say, 0.005 and 0.001, to "get closer" to the ideal model of dry friction with starting and running values. However, we *can't* really get what we want with a lookup table because such tables allow *only* single-valued functions. The "true" fricton model here requires a multiple-valued function since the change from starting to running friction occurs precisely at $v = 0$, and for $v = 0$, the starting friction force can actually take on an *infinite* number of values before motion actually commences. It *is* possible to correctly simulate this behavior (recall the ACSL program we referenced [22] earlier), but it *isn't* easy and you *can't* do it with just a lookup table. Our simple approximation fcoul2, however, is closer to reality than was fcoul, so it is useable as long as we understand its limitations, just as with any other model. (What type of damping exists in Fig. 2-34 for $-0.05 < v < 0.05$ cm/sec?)

Our final example shows a model sometimes used to represent structural or hysteresis damping. Recall that we said earlier that this form of energy dissipation occurs on a microscopic level and that a distinct damping *force* cannot be identified. While a force cannot be identified in the real system, nothing prevents us from using a *model* which does include a force, as long as the *effect* of this fictitious force is similar to that of the real damping effect. It has been found that a damping force whose absolute value is proportional to *displacement* gives a good representation of structural damping in many cases. Note that a damping force must always *oppose* the velocity, so we get the simulation diagram for fstruc shown at the bottom of Fig.

System Elements, Mechanical   **75**

2-32. We use an integrator to get displacement x from velocity v, multiply this by a constant of our choice, and then take the absolute value. To get the algebraic sign of the force to always oppose the velocity, we use the SIGN function. This is a MATLAB, not a SIMULINK, function, so we need to use SIMULINK's icon called "MATLAB Function." This icon makes available a wider variety of functions than are in SIMULINK. (I *could* have used the relay icon to get the correct sign, but I want to familiarize you with as many SIMULINK capabilities as I can, so I introduced the Matlab Function icon.)

While the damping forces displayed in Fig. 2-33 are all clearly different in detail, they are *all* legitimate energy-dissipating forces that could be used as models in mechanical system studies. When we include them in larger system models, and solve the differential equations for the unknown motions, whether our chosen friction model is acceptable or not depends on whether these predicted motions agree with lab measurements on the real system. If we work with a certain class of mechanical system, say, automotive suspension systems, we soon learn which models are sufficiently accurate for the kind of equipment we are designing.

2-7 THE INERTIA ELEMENT

A designer rarely inserts a component into a system for the purpose of *adding* inertia; thus the mass or inertia element often represents an undesirable effect which, unfortunately, is unavoidable, since all materials (solid, liquid, or gas) possess the property of mass. There are, of course, *some* applications in which mass itself serves a useful function. Figure 2-35a shows an *accelerometer*,[27] an instrument for measuring acceleration. Every accelerometer *must* contain a mass (called the "proof mass") since the principle of acceleration measurement lies in measuring the *force* required to give the proof mass the acceleration being measured. In Fig. 2-35a the spring element measures the force by deflecting proportionately, while the damper element suppresses spurious vibrations of the proof mass. A displacement transducer[28] converts the spring's deflection into a proportional voltage, since voltage indicating and recording devices are widely used in measurement systems. Note that accelerometer designers and users employ all three of the mechanical elements in their calculations. *Rotary inertia* in the form of *flywheels* is sometimes used as an energy storage device[29] or as a means of smoothing out speed fluctuations in engines or other machines (Fig. 2-35b).

Newton's law defines the behavior of mass elements:

$$\sum \text{forces} = (\text{mass})(\text{acceleration}) \tag{2-44}$$

and refers basically to an idealized "point mass" which occupies infinitesimal space. To apply this law directly to practical situations, the concept of the *rigid body* is introduced in physics and mechanics. For a purely translatory motion (no rotation),

---

[27]Doebelin, *Measurement Systems*, 4th ed., pp. 323–336.

[28]Ibid., pp. 210–308.

[29]D. W. Rabenhorst, Design considerations for a 100-megajoule, 500-megawatt superflywheel, Johns Hopkins App. Physics Lab Rept. TG-1229, 1973. S. Ashley, Flywheels put a new spin on electric vehicles, *Mechanical Engineering*, October 1993, pp. 44–51.

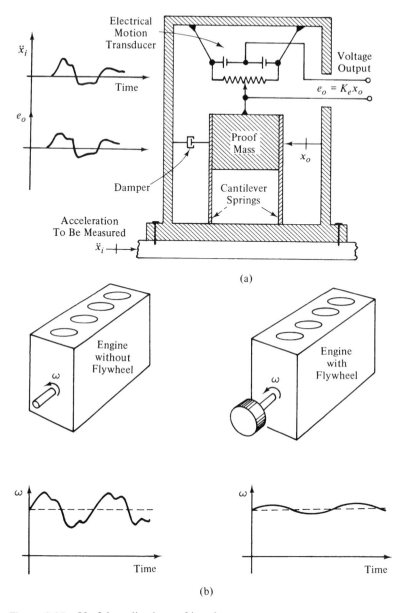

Figure 2-35 Useful applications of inertia.

every point in a rigid body has identical motion and thus Eq. (2-44) applies to such a body (see Fig. 2-36). Real physical bodies can, of course, never display this ideal rigid behavior when being accelerated, since they experience internal elastic deflections which allow relative motion between points in the body. Thus, just as in the case of the spring and damper elements, the pure/ideal inertia element is a model, *not* a real object. Fortunately, in many practical cases, the internal elastic deflections are so small *relative* to the gross motion of the body that the pure/ideal model gives good results in calculations.

System Elements, Mechanical

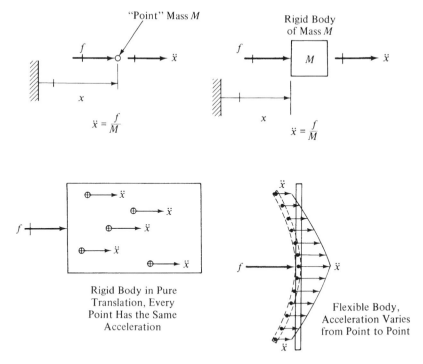

Figure 2-36 Rigid and flexible bodies: definitions and behavior.

For bodies undergoing pure rotational motion about a single fixed axis, we have Newton's law in the rotational form

$$\sum \text{torques} = (\text{moment of inertia})(\text{angular acceleration}) \qquad (2\text{-}45)$$

The concept of moment of inertia J (I is sometimes used instead of J) also considers the rotating body to be perfectly rigid. The "particles" of the body now do *not* all have the same acceleration, but they *do* have accelerations which are intimately related, and in a known way, so that their combined inertia effect (called the mass moment of inertia) can be computed using integral calculus. For a homogeneous right circular cylinder (Fig. 2-37), for example, we may apply the basic (translational) form of Newton's law to the ring-shaped mass element of infinitesimal width dr at radius r. Every particle in this element has exactly the same tangential acceleration $r\alpha$; thus the tangential force to produce this acceleration must be

$$\text{Tangential force} = (\text{mass})(\text{acceleration}) = (2\pi r L\, dr\, \rho)(r\alpha) \qquad (2\text{-}46)$$

The torque associated with this force is simply r times the force and the total torque is obtained by integration:

$$\text{Total torque} = \int_0^R 2\pi \rho L \alpha r^3\, dr = \pi R^2 L \rho \frac{R^2}{2} \alpha = \frac{MR^2}{2} \alpha \qquad (2\text{-}47)$$

Since moment of inertia J is defined by torque $= J\alpha$, we get

$$\text{Torque} = \frac{MR^2}{2} \alpha \stackrel{\Delta}{=} J\alpha$$

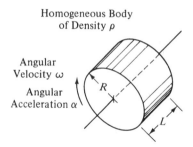

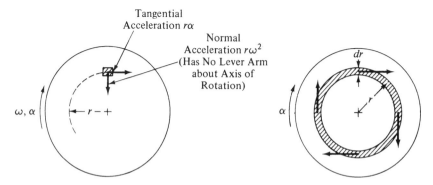

Figure 2-37 Rotational inertia.

$$J \triangleq \frac{MR^2}{2} \quad \text{kg-m}^2 \text{ (SI units)} \qquad (2\text{-}48)$$

$$\text{in-lb}_f\text{-sec}^2 \text{ (British inch units)}$$

$$\text{slug-ft}^2 \text{ (British foot units)}$$

In keeping with the introductory nature of this text, we will treat only those problems where the motion is confined to a plane, rather than the three-dimensional general case, where the dynamics become *much* more complex. We do, however, want to at least mention that to *completely* describe the inertial properties of any rigid body requires specification of its total mass, location of the center of mass, three ("x, y, z") moments of inertia, and three products of inertia. Such detailed information is needed, for example, in the study of vehicle dynamics[30] since vehicles such as aircraft and satellites have complete freedom of three-dimensional motion.

For geometrically simple and homogeneous bodies, J can be calculated with relative ease, and Fig. 2-38 gives a few results useful for the types of problems suitable for this text. For real machine parts, the shapes are usually complex and several materials of different density may be involved, making computation of J difficult and subject to error. However, at the design stage, where the actual part

---

[30]E. O. Doebelin, *System Modeling and Response*, Wiley, New York, 1980, chap. 12.

System Elements, Mechanical

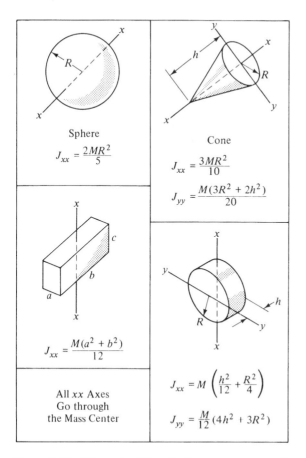

Figure 2-38 Moments of inertia for some common shapes.

exists "only on paper," an estimate of this sort is necessary. Software for automating such calculations for the most general (3-D) case is available.[31] Sometimes these calculations are provided by general-purpose CAD software, as one of the options.

Once a part or device has been constructed, experimental methods of measuring the inertial properties can be used. Apparatus for finding all the inertial properties has been in use for many years, particularly in the vehicle industries. A recent such "machine"[32] measures all the inertial properties of an entire automobile or truck (think how difficult this would be to do from the drawings!). Most such apparatus uses the same principle: let the rigid body become part of a vibrating system, measure the frequency of vibration, and from this, compute the unknown moment of inertia.

---

[31] J. E. Cake, A Fortran code for computing the principle mass moments of inertia of composite bodies, NASA TM X-1754, 1969. Cosmic Program #GSC-13228 (AutoCad to Mass Properties), COSMIC, University of Georgia, Athens, Georgia 30602, 404-542-3265.

[32] G. J. Heydinger et al., The design of a vehicle inertia measurement facility, SAE Paper 950309, 1995.

Let's explain the details for the simplest case, finding the moment of inertia about a known axis of rotation, for a body of arbitrary shape and materials.

Figure 2-39 shows the apparatus with a simple cylindrical object as the test item; it will work, however, with *any* shape and materials. Simple cylinders are often used to *calibrate* or check out the apparatus, since for these we *can* accurately compute J from dimensions and density, and then compare the result to what is measured. The object under test must be mounted in low-friction bearings which constrain it to rotate about the axis for which we desire the moment of inertia. Next, we must provide some kind of torsional spring to exert a torque about the rotation axis. This spring must be calibrated (say, with dead weights) so that we have an accurate value for the torsional spring constant K_s, say, in in-lb_f/rad. We also need an electrical motion transducer to provide a voltage proportional to angle θ, which is recorded as a function of time.

If we now manually deflect the body an angle θ_0 away from its equilibrium position and let go, if the bearings have little friction, the body will oscillate back and forth several cycles before friction causes the motion to stop. The frequency of this torsional vibration can be predicted by solving the differential equation of motion for the apparatus. While differential equation solution will be discussed in detail in a later chapter, many readers will have been exposed to such mathematics before

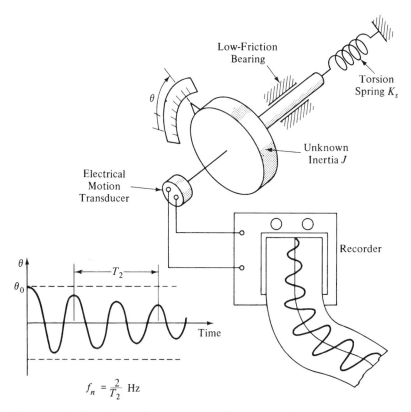

Figure 2-39 Experimental measurement of moment of inertia.

System Elements, Mechanical   **81**

coming to a system dynamics course, so we proceed without much explanation. If your background is lacking here, please just take the results on faith for the moment.

$$\sum \text{torques} = J\alpha = J\,\frac{d^2\theta}{dt^2}$$

$$-K_s\theta = J\,\frac{d^2\theta}{dt^2}$$

$$\frac{J}{K_s}\,\frac{d^2\theta}{dt^2} + \theta = 0 \tag{2-49}$$

The bearing friction torque has been treated as negligible here. Solution of this equation with initial ($t = 0$) conditions $\theta = \theta_0$ and $d\theta/dt = 0$ gives

$$\theta = \theta_0 \cos \omega_n t \tag{2-50}$$

where

$$\omega_n \triangleq \sqrt{\frac{K_s}{J}} \triangleq \text{undamped natural frequency} \qquad \text{rad/sec}$$

Equation (2-50) indicates a sustained oscillation of frequency f_n cycles/sec, where $f_n = \omega_n/2\pi$. Actually, the oscillation will gradually die out due to the bearing friction not being zero. If bearing friction were pure Coulomb friction, it can be shown[33] that the decay "envelope" of the oscillations is a straight line and that friction has *no* effect on the frequency. If the friction is pure viscous (ideal), then the decay envelope is an exponential curve, and the frequency of oscillation *does* depend on the friction but the dependence is usually negligible for the low values of friction in typical apparatus. Figure 2-39 shows the use of 2 cycles of vibration for measurement of the frequency, but it is more accurate to use as many cycles as can be distinctly seen. Once we have measured the frequency f_n we can get J easily as

$$J = \frac{K_s}{4\pi^2 f_n} \tag{2-51}$$

Whereas the linear characteristic curves relating force and displacement for spring elements, and force and velocity for damper elements, are closely approximated only by careful choice of materials, clever design, and limited range of operation, the linear force/acceleration characteristic of the inertia element is for all practical purposes perfectly realized in those (many) cases where the body is "sufficiently" rigid. That is, Newton's law, while being strictly an empirical relation based on experimental measurements, has been found to hold very closely except for relativistic situations in which the velocity of the mass becomes comparable with the speed of light. Thus real inertias may be impure (have some springiness and friction) but are very close to ideal (linear).

Note that the symbol (Fig. 2-40a) for an inertia element has *only one end*, whereas spring and damper elements always have two ends. This is of course because the inertia element is a rigid body and all parts of it have *the same* motion. The operational transfer functions for translational inertia elements with force or dis-

---

[33]W. T. Thomson, *Mechanical Vibrations*, Prentice-Hall, New York, 1948, p. 56.

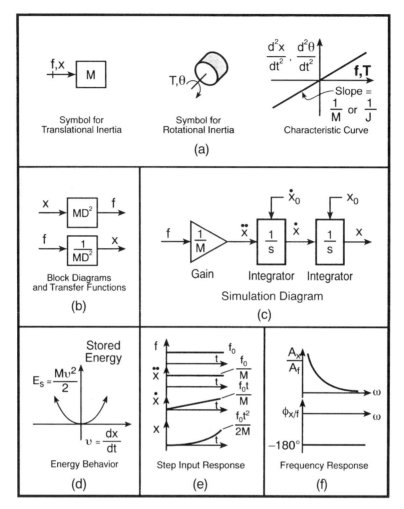

Figure 2-40 The inertia element.

placement inputs, and rotational inertia elements with torque or angular displacement inputs, are

$$\frac{x}{f}(D) = \frac{1}{MD^2} \qquad \frac{\theta}{T}(D) = \frac{1}{JD^2}$$
$$\frac{f}{x}(D) = MD^2 \qquad \frac{T}{\theta}(D) = JD^2$$
(2-52)

with block diagrams as in Fig. 2-40b and simulation diagrams as in Fig. 2-40c. Whereas the spring element stores energy as potential energy of deformation, the inertia element stores it as kinetic energy of motion. A mass M with velocity v has kinetic energy $Mv^2/2$, while a rotary inertia J with angular velocity ω has kinetic energy $J\omega^2/2$ (Fig. 2-40d).

Turning to dynamic response, a step input force of size f_s applied to a mass initially at rest with $x = 0$ causes an *instantaneous* jump in acceleration, a ramp change in velocity, and a parabolic change in position, as we see from Newton's law:

System Elements, Mechanical   **83**

$$f_s = M \frac{d^2x}{dt^2} = M \frac{dv}{dt}$$

$$\int_0^t f_s \, dt = \int_0^v M \, dv$$

$$f_s t = Mv = M \frac{dx}{dt} \tag{2-53}$$

$$\int_0^t f_s t \, dt = \int_0^x M \, dx$$

$$x = \frac{f_s t^2}{2M}$$

The work done by the constant force f_s on the mass M, in moving it the distance x, is $f_s x = f_s^2 t^2 / 2M = Mv^2/2$; thus the mass now has this energy stored and can give it up to another body when it is slowed down by this other body.

The frequency response of the inertia element is easily obtained from the sinusoidal transfer function

$$\frac{x}{f}(i\omega) = \frac{1}{M(i\omega)^2} = -\frac{1}{M\omega^2} = \frac{1}{M\omega^2} \ \angle{-180°} \tag{2-54}$$

and is graphed in Fig. 2-40f. Note that at high frequency, the inertia element becomes *very* difficult to move, since A_x/A_f rapidly approaches zero as frequency increases. Also, the $-180°$ phase angle shows that the displacement is in a direction *opposite* to the applied force.

Since the validity of the ideal and pure inertia element as a model for real inertias rests on the rigidity of the real body, it is instructive to investigate this situation for a body of simple shape. Once again, frequency response provides a perfect way of thinking about this question. The prismatical rod of Fig. 2-41 has its left end driven by a motion input x_i. If the body were perfectly rigid, every particle would have this same motion. While an "exact" analysis requires a distributed-parameter (partial differential equation) treatment, a simple lumped-parameter model reveals the essence of the behavior. The rod is modeled as a single mass M equal to the actual rod mass ρAL and located at the center of mass of the rod. This mass is connected to the left and right ends with massless spring elements having spring constants equal to those of actual pieces of rod of length $L/2$ (see Fig. 2-12i). Applying Newton's law and noting that the spring on the right will exert *no* force on M because it is massless and not attached to anything at its right end, we get

$$(x_i - x_o) \frac{2AE}{L} = \rho AL \frac{d^2x_o}{dt^2} \tag{2-55}$$

$$\frac{\rho L^2}{2E} \frac{d^2x_o}{dt^2} + x_o = x_i \tag{2-56}$$

Using our D-operator notation and defining $\omega_n \triangleq (2E)^{0.5}/(L\rho^{0.5})$

$$\left(\frac{D^2}{\omega_n^2} + 1\right) x_o = x_i \tag{2-57}$$

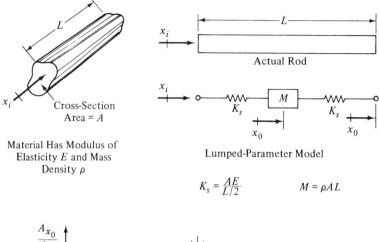

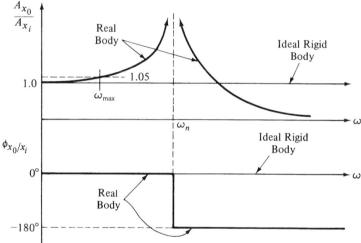

Figure 2-41 Useful frequency range for rigid model of real (flexible) body.

We get the operational transfer function

$$\frac{x_o}{x_i}(D) = \frac{1}{\dfrac{D^2}{\omega_n^2} + 1} \qquad (2\text{-}58)$$

and the sinusoidal transfer function

$$\frac{x_o}{x_i}(i\omega) = \frac{1}{\left(\dfrac{i\omega}{\omega_n}\right)^2 + 1} = \frac{1}{1 - \left(\dfrac{\omega}{\omega_n}\right)^2} \qquad (2\text{-}59)$$

Now for a perfectly rigid body, $(x_o/x_i)(i\omega)$ would be identically equal to 1.0 for *every* frequency, if $x_i = x_{i0} \sin \omega t$. From Eq. (2-59) we see that the real body approaches this as $(\omega/\omega_n) \to 0$, that is, if the forcing frequency ω is small *compared* to ω_n. If we arbitrarily decide that a 5% deviation from perfection is tolerable, we may write

System Elements, Mechanical **85**

$$\frac{x_o}{x_i}(i\omega) = 1.05 = \frac{1}{1 - \left(\frac{\omega_{max}}{\omega_n}\right)^2} \tag{2-60}$$

$$\omega_{max} = 0.218\omega_n = \frac{0.308}{L}\sqrt{\frac{E}{\rho}} \tag{2-61}$$

where ω_{max} is the highest frequency for which the real body behaves "almost" like an ideal rigid body. Notice that as the rod length $L \to 0$, $\omega_{max} \to \infty$, since this corresponds to the real mass approaching a "point" mass, which would behave rigidly for all frequencies. Also, the limiting frequency is higher for "stiffer" materials (larger modulus of elasticity E) and lighter materials (smaller mass density σ). To get some feel for the numbers, consider a steel rod of length 6 inches.

$$\omega_{max} = \frac{0.308}{6}\sqrt{\frac{3 \times 10^7}{0.3/386}} = 10060. \text{ rad/sec} \tag{2-62}$$

$$f_{max} = \frac{\omega_{max}}{2\pi} = 1605. \text{ Hz} = 96200. \text{ cycles/minute} \tag{2-63}$$

We see that this body will act essentially like a rigid body for oscillatory motions up to a frequency of about 96200 cycles/minute. Since no reciprocating and very few rotating machines run at such high speeds, this body could be modeled as a pure inertia in many practical problems. Frequency response is *unmatched* as a technique for defining the useful range of application for all kinds of dynamic systems and we will use it over and over in this way.

We have now presented all three of the mechanical system elements, and also along the way introduced many *general* concepts that will be useful in all our later work (linearization, transfer functions, block diagrams, elementary simulation methods, energy relations, experimental testing, frequency response, and pure/ideal versus real devices).

2-8 REFERRAL OF ELEMENTS ACROSS MOTION TRANSFORMERS

Mechanical systems often include mechanisms such as levers, gears, linkages, cams, chains, and belts (see Fig. 2-42). While the named devices differ considerably in form, they all serve a common basic function, the transformation of the motion of an input member into the kinematically related motion of an output member. While the analysis of systems containing such motion transformers does not require any new elements or methods, it may be simplified in many cases by reducing the actual system to a fictitious but dynamically equivalent one. This is accomplished by a process of "referring" all the elements (masses, springs, dampers) and driving inputs to a *single* location, which could be the input, the output, or some selected "interior" point in the system. We can then write a *single* equation for this equivalent system, rather than having to write *several* (perhaps many) equations for the actual system. It

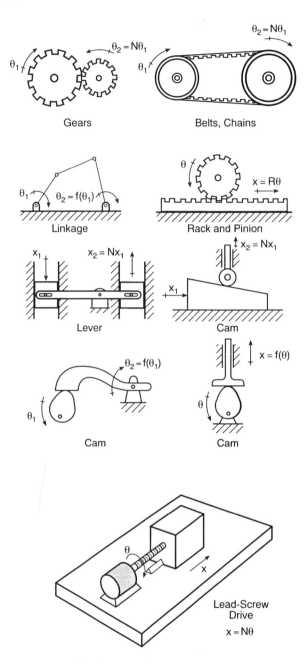

Figure 2-42 Motion transformers.

is not *necessary* to employ such equivalent systems, but it often speeds the work and reduces errors.

We will illustrate the procedure by carrying it through for the lever system of Fig. 2-43. Having done this example, you should be able to apply the method to *any* of the other mechanisms shown in Fig. 2-42. Note that in this example there are three

System Elements, Mechanical

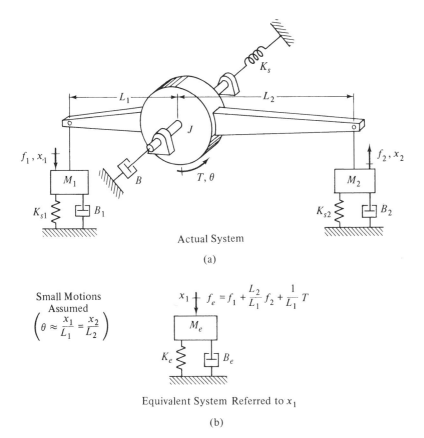

Figure 2-43 Translational equivalent for complex system.

related motions, x_1, x_2, and θ, and that if one specifies any one of these, the other two are immediately known, since they are *kinematically*, rather than dynamically, related. (This statement *assumes* that all the masses are rigid bodies, but we have just shown that this approximation is often valid.) One must first decide which of the three motions is to be retained in our equations and which are to be "eliminated." In a practical application, this choice is usually obvious. Here, let's decide to write our equations in terms of the motion x_1, perhaps because x_1 might be a location where this system couples to another one, so we need to keep this variable in our equations.

All elements and inputs must now be "referred" to the x_1 location and we will define a fictitious equivalent system whose motion will be the same as x_1 but will include *all* the effects in the original system. The rotary part of the real system has only small angular motion, so the x's are taken directly proportional to the rotation angle. We first define a *single equivalent* spring element which will have the same effect as the three actual springs, by (mentally) applying a static force f_1 at location x_1 and writing a torque balance equation as

$$f_1 L_1 = (K_{s1} x_1) L_1 + \left(\frac{L_2}{L_1} x_1 K_{s2}\right) L_2 + \frac{x_1 K_s}{L_1} \tag{2-64}$$

$$f_1 = K_{se} x_1$$

$$K_{se} \triangleq \left(K_{s1} + \left(\frac{L_2}{L_1} \right)^2 K_{s2} + \frac{1}{L_1^2} K_s \right) \quad \text{lb}_f/\text{inch} \tag{2-65}$$

We were able to "isolate" the spring effects by considering a static (stationary) case, where damping and inertia have no influence. *The equivalent spring constant K_{se} refers to a fictitious spring which, if installed at location x_1 would have exactly the same effect as all the springs together in the actual system.*

To find the equivalent damper, "mentally" remove the inertias and springs and again apply a force f_1 at x_1 to get

$$f_1 L_1 = (\dot{x}_1 B_1) L_1 + (\dot{x}_2 B_2) L_2 + B\dot{\theta} = \dot{x}_1 B_1 L_1 + \frac{L_2^2}{L_1} \dot{x}_1 B_2 + \frac{\dot{x}_1}{L_1} B \tag{2-66}$$

$$f_1 = B_e \dot{x}_1$$

$$B_e \triangleq \left(B_1 + \left(\frac{L_2}{L_1} \right)^2 B_2 + \frac{1}{L_1^2} B \right) \quad \frac{\text{in/sec}}{\text{lb}_f} \tag{2-67}$$

Finally, considering only inertias present

$$f_1 L_1 \approx (M_1 L_1^2) \frac{\ddot{x}_1}{L_1} + (M_2 L_2^2) \frac{\ddot{x}_1}{L_1} + (J) \frac{\ddot{x}_1}{L_1} \tag{2-68}$$

$$f_1 \approx M_e \ddot{x}_1$$

$$M_e \triangleq \left(M_1 + \left(\frac{L_2}{L_1} \right)^2 M_2 + \frac{1}{L_1^2} J \right) \quad \frac{\text{lb}_f\text{-sec}^2}{\text{in}} \tag{2-69}$$

While the definitions of equivalent spring and damping constants are approximate due to the assumption of small motions, the equivalent mass just defined has an additional assumption which may be less accurate, so we used the $\approx$ symbol in its definition above. We have treated the masses as *point* masses when we write their moments of inertia as ML^2, and the L's are distances to the mass centers of the M's. More correctly, these moments of inertia should have added to them the moment of inertia of the mass about its own mass center. Whether this correction is important or not depends on numerical values. As an example, suppose M_1 were a sphere of radius 1 inch and L_1 were 10 inches. The correct moment of inertia would then be (using the parallel-axis theorem)

$$J_{M_1} = \tfrac{2}{5} M_1 + M_1(100) = 100.4 M_1 \tag{2-70}$$

which in this case is very close to our approximate value of 100.0. Some applications would, however, have numbers such that the approximation would not be this good, so it is best to check each time.

To refer driving inputs to the x_1 location we note that a torque T is equivalent to a force T/L_1 at the x_1 location, and a force f_2 is equivalent to a force $(L_2/L_1)f_2$. We can now show the dynamically equivalent system as in Fig. 2-43b. If we set up the differential equation of motion for this system and solve for its unknown x_1, we are guaranteed that this solution will be identical to that for x_1 in the actual system. Note that once we have x_1, we can get x_2 and/or θ *immediately* since they are related to x_1 by the simple proportions given in Fig. 2-43; it is *not* necessary to solve any

System Elements, Mechanical   **89**

more differential equations. If we had preferred to *originally* deal with x_2 or θ, we could have set up equivalent systems which referred everything to *those* locations, using the same techniques as we just explained above.

The "rules" [Eqs. (2-65), (2-67), (2-69)] for calculating the equivalent elements without deriving them "from scratch" each time may be summarized as follows:

1. When referring a translational element (spring, damper, or mass) from location A to location B, where A's motion is N times B's, multiply the element's value by N^2. (This is also true for rotational elements coupled by motion transformers such as gears, belts, and chains, although we have not shown it here.)

2. When referring a rotational element to a translational location, multiply the rotational element by $1/R^2$, where the relation between translation x and rotation θ (in radians) is $x = R\theta$. For the reverse procedure (referring a translational element to a rotational location) multiply the translational element by R^2.

3. When referring a force at A to get an equivalent force at B, multiply by N (holds also for torques). Multiply a torque at θ by $1/R$ to refer it to x as a force. A force at x is multiplied by R to refer it as a torque to θ.

These rules apply to *any* mechanism, no matter what its form, so long as the motions at the two locations are *linearly* related. For mechanisms with nonlinear input/output relations (most cams and linkages) these results are good approximations for small motions near an operating point at which the motion-transformation ratio is N.

In addition to making specific numerical calculations, the equivalent system concept also can lead to general guidelines useful in system design. One of these applies to the motion-control systems called *servomechanisms*.[34] Here a motor (electric, hydraulic, pneumatic) drives a load which is to be accurately positioned in response to a command. Since most motors work best at relatively high speed, there is often a rather large motion transformation ratio between the motor and the load; the motor runs at high speed, compared to the load. When this is the case, a "visual" evaluation of the inertia effects in the system can be quite misleading, whereas the equivalent system concept gives us the correct picture. That is, "to the eye," the load inertia appears much larger than that of the motor, so we are led to think that inertia-reduction efforts (to speed up system response) should be concentrated on the load inertia. Actually, in many cases the smaller motor inertia is really critical because its *effective* value involves the square of a "gear ratio."

As an example, consider a case where the inertia on the motor shaft is, say, 10 units while the load inertia is 1000 units. As we look at the two physical objects, it seems that if we want to reduce system inertia we should be working on the load, using light metals and/or drilling lightening holes in it. If the gear ratio is say, 50-to-1, the *effective* motor inertia is $(10)(2500) = 25,000$ units, *25 times* the load inertia. We really should be trying to find low-inertia motors, not redesigning our load!

---

[34]E. O. Doebelin, *Control System Principles and Design*, Wiley, New York, 1985, p. 131.

2-9 MECHANICAL IMPEDANCE

In the theoretical and experimental study of mechanical systems, particularly when trying to predict the behavior of an *assemblage* of subsystems from their calculated or measured individual behavior, the use of so-called impedance methods[35] may have advantages. We cannot pursue this here, but we can introduce some definitions and do some simple examples. *Mechanical impedance* is defined as a transfer function (either operational or sinusoidal) in which force is the numerator term and velocity the denominator. The simplest impedances are those of the spring, damper, and inertia elements themselves.

$$\text{Mechanical impedance of a spring} \triangleq Z_s(D) \triangleq \frac{f}{v}(D) = \frac{K_s}{D} \tag{2-71}$$

$$\text{Mechanical impedance of a damper} \triangleq Z_B(D) \triangleq \frac{f}{v}(D) = B \tag{2-72}$$

$$\text{Mechanical impedance of a mass} \triangleq Z_M(D) \triangleq \frac{f}{v}(D) = MD \tag{2-73}$$

Figure 2-44 shows the frequency response curves for the sinusoidal transfer function versions of these impedances.

When mechanical impedance is determined experimentally, sinusoidal testing is often used. The machine or structure under test is driven by hydraulic or electrodynamic "shakers," and careful measurements of force and velocity are made over wide ranges of frequency to establish the amplitude ratio and phase shift of $(f/v)(i\omega)$ as a function of frequency. Impedance analysis and/or testing can be applied to

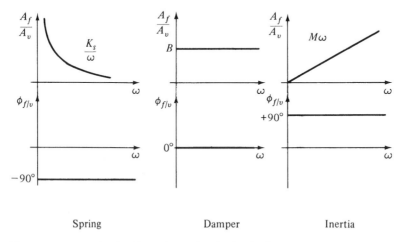

Figure 2-44 Mechanical impedance for the basic elements.

---

[35] R. Plunkett, ed., *Mechanical Impedance Methods for Mechanical Vibrations*, ASME, 1958. Doebelin, *System Modeling and Response*, chap. 7.

System Elements, Mechanical

individual components such as real springs or dampers, or entire structures and machines. One of its most useful types of application is the measurement of impedances of subsystems, which measurements are then used to analytically *predict* the behavior of the complete system formed when the subsystems are connected. For example, an aircraft jet engine can be impedance-tested at the engine manufacturer's plant and an aircraft fuselage impedance-tested at the airframe maker's factory. These measurements can then be analytically combined[36] to predict whether there will be any serious vibration problems when the engine is mounted to the airframe. We can thus discover and correct potential design problems *before* we go to the trouble and expense of actually connecting the engine to the airframe.

Impedance methods also provide "shortcut" analysis techniques for mechanical systems. To illustrate this aspect we quote some useful results without proof. When two elements carry the same force they are said to be connected "in parallel" and their combined impedance follows the same rule as electrical impedances (or resistances) in parallel. That is, the combined impedance is the product of the individual impedances over their sum. For impedances which have the same velocity, we say they are connected in series and their combined impedance is the sum of the individual ones, just as in electrical series circuits. To demonstrate these "rules," consider the system of Fig. 2-45a, where the spring and damper carry the same force, and are thus "in parallel." The impedance $(f/v)(D)$ of the combined system is easily calculated as

$$\frac{f}{v}(D) = \frac{\frac{K_s}{D}B}{(K_s/D) + B} = \frac{K_s B}{BD + K_s} \qquad (2\text{-}74)$$

To get this result *without* using impedance methods is of course possible but takes a little more work.

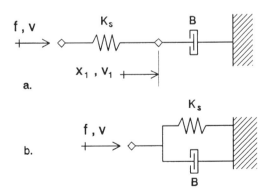

Figure 2-45 Examples for impedance calculations.

---

[36]Doebelin, *System Modeling and Response*, pp. 332–337.

Applied force = f = spring force = damper force

$$f = K_s(x - x_1) = \frac{K_s}{D}(v - v_1) = Bv_1$$

$$v_1 = \frac{K_s}{BD + K_s}v$$

$$f = \frac{K_s}{D}\left(v - \frac{K_s}{BD + K_s}v\right)$$

$$\frac{f}{v}(D) = \frac{K_s B}{BD + K_s} \tag{2-75}$$

The system of Fig. 2-45b has the two elements sharing the same velocity, so they are in series and we quickly get

$$\frac{f}{v}(D) = B + \frac{K_s}{D} = \frac{BD + K_s}{D} \tag{2-76}$$

This result is also easily obtained without using impedance methods.

While sinusoidal test methods are widely used, it turns out that the sinusoidal transfer function can also be found using suitably chosen transient ("pulse testing"), periodic, or random inputs.[37] Pulse testing can be particularly simple in that we strike the structure with an instrumented hammer which both produces the force and measures it. This pulse test data is then mathematically processed to obtain amplitude ratio and phase angle curves of the sinusoidal transfer function. Finally, in reading the literature, you may encounter the term *mobility*. This is nothing but the reciprocal $(v/f)(D)$ of impedance, so no essential new concepts are involved.

2-10 FORCE AND MOTION SOURCES

We have shown the various elements being driven by input forces or motions, but have said little about how these inputs actually arise in practice. Let us first be clear that, to be precise, the ultimate driving agency of any mechanical system is always a *force*, not a motion. This follows from the cause-and-effect relation stated in Newton's laws, i.e., the force *causes* the acceleration, the acceleration does *not* cause the force. Thus, while there are problems in which the concept of a motion input is preferable to a force input, we should not delude ourselves that a motion occurs without a force occurring *first*. Perhaps the proper point of view is simply to ask, at the input of the system, what is *known*, force or motion? If, for example, the motion is known, the fact that this motion is *caused* by some (perhaps unknown) force should not stop us from postulating a problem with a motion input.

It may be appropriate to recall at this point that, at the macroscopic level, the forces available to drive mechanical systems can be put into two classes: forces associated with physical contact between two bodies, and the "mysterious" action-at-a-distance forces, namely, gravitational, magnetic, and electrostatic forces. When

---

[37]Ibid., chap. 6.

System Elements, Mechanical

using Newton's law, Σ forces = (mass)(acceleration), the terms entered into the force summation must arise either from physical contact or else from magnetic, gravitational, or electrostatic origin; *there are no other kinds of forces*. If you have encountered D'Alembert's method of dynamic system analysis, you may at this point be thinking about the so-called *inertia force*, but should recall that this is a *fictitious* force mentally added to convert what is really a dynamics problem into an equivalent statics problem. Inertia force is *not* a real force capable of causing a body initially at rest to move.

Some examples will help to establish a physical feeling for the distinction between force and motion sources. The design of multistory buildings may require consideration of stresses due to both wind and, in some regions of the world, earthquake effects.[38] Figure 2-46 shows an idealized model of a multistory building made up of mass and spring elements. The effect of wind is distributed over the surface of the building, and of course varies with time in a random fashion, but would generally be modeled as a force (or pressure) source. Earthquake-resistant design is a most complex field still based greatly on experience; however, attempts to put it on a rational basis generally consider the structure to be excited by the so-called *ground motion*, that is, a motion, rather than a force, input. The assumption here is that the portion of the earth's crust to which the building is "fastened" is so massive relative to the building that the presence of the building has no effect on this "ground motion," and thus the base of the building is constrained to move with it.

The ground motion caused at a particular location by a particular earthquake is of course impossible to predict, as is the occurrence of the earthquake itself. The engineer must thus rely on measurements of *past* disturbances to aid design. Figure 2-47 shows seismographic recordings of the horizontal component of ground motion, in the north-south direction, of an actual earthquake. Such a record can be used as motion input to a mathematical model of the building to predict stresses and deflections. Since these records are easily obtained on magnetic tape and can thus be reproduced as voltages, one can "play" this input into an analog-to-digital converter to document the curves with closely spaced discrete numerical values. These can then be entered into a lookup table in a digital simulation system (such as the ACSL or SIMULINK we mentioned earlier) and used as input to a simulation model of the building.

To further reduce the need for simplifying assumptions, a *physical* scale model of the building can be constructed, and the base forced to move according to the earthquake ground motion by playing the tape (or digitized versions of it) into an *electrohydraulic shaker*.[39] This is a machine which faithfully reproduces as a motion any electrical voltage-time variation which is applied to its input terminals. Such testing machines allow us to accurately apply predetermined motion inputs of many different forms to systems that we wish to study experimentally. It may also be pertinent to state that the design of such testing equipment relies heavily on system

---

[38]J. A. Blume, Design of Multistory Reinforced Concrete Buildings for Earthquake Motions, Portland Cement Association, 1961.

[39]MTS Systems Corp., Eden Prairie, Minnesota, 612-937-4000. *Hydraulics and Pneumatics*, June 1984, pp. 14–16.

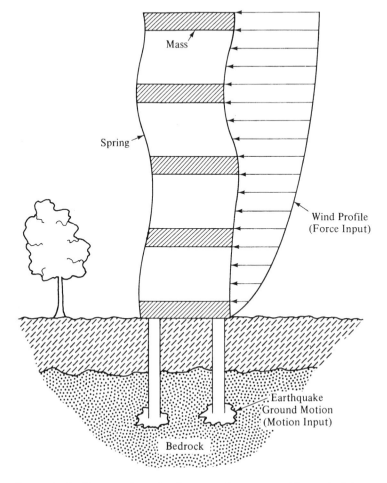

Figure 2-46 Force and motion inputs acting on a multistory building.

dynamic concepts such as those explained in this book. Laboratory simulation testing of this sort has certain advantages over actual field testing and is widely used in many industries[40] to qualify equipment for severe environments.

Many types of complicated machinery (packaging and printing equipment, computer peripherals such as tape and disk drives, machine tools, etc.) are driven by a power source (often an electric motor) which runs at essentially constant speed. This steady rotation may be due to an inherent characteristic of the power source (a synchronous motor, for example), the flywheel action of a large inertia (a punch press, for example), or a feedback control system for speed regulation (steam turbine in a power plant, for example). In any case, the motions of the functional parts of the machine (motions which are generally complex and *not* of uniform velocity) are

---

[40] E. R. Betz, Studying structure dynamics with the Cadillac road simulator, SAE Paper 660101, *SAE Trans.*, vol. 75, 1967.

System Elements, Mechanical

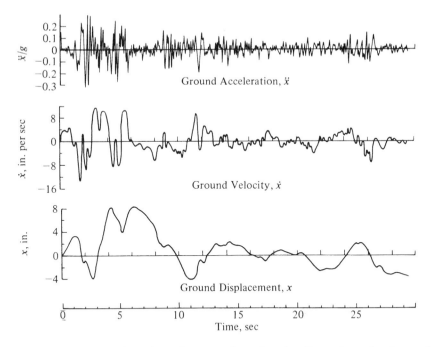

Figure 2-47 Ground motion of an actual earthquake. Ground acceleration, velocity, and displacement, El Centro, California, earthquake of May 18, 1940, N-S component.

usually calculated assuming a known motion (the uniform rotation) at the input of the mechanism; thus we have another example of a motion input.

Inertial navigation systems for submarines must take into account the motion caused by the vehicle being carried along by the earth. That part of the vessel's total motion caused by the earth's motions is a motion input, since the earth's motions are very accurately known, and for all practical purposes, totally unaffected by the presence or absence of a ship. The operation of the gyroscopes and accelerometers used in such navigation systems is also analytically studied by assuming these instruments (which are fastened to the ship and thus are constrained to follow its motions) to be driven by the ship's rotary and translational motions as inputs. Again, the ship is so massive (relative to the instruments) that its motion is unaffected by their presence. Finally, the study of suspension systems for land vehicles generally assumes the vehicle to be driven over a terrain of a certain profile (perhaps random), thus causing the wheels to have a prescribed vertical motion. This motion is the input for a mass/spring/damper mathematical model of the vehicle, and causes motions of the frame and body which can be analytically calculated. If an actual vehicle is available, the electrohydraulic shakers mentioned earlier can be used (one under each wheel) to simulate driving over an actual terrain whose profile has been measured and tape-recorded.

Force inputs that excite vibratory motions often arise from rotational and/or reciprocating unbalance. A rotating rigid body whose mass center does not coincide with its center of rotation is said to be unbalanced and, if rotating at a constant speed ω rad/sec will produce a radial force of magnitude $MR\omega^2$, where M is the mass and R is the distance between the mass center and center of rotation. The vertical component of this force would be $MR\omega^2 \sin \omega t$, an oscillatory force which would act on the bearings, and thereby be transmitted into the machine frame, and possibly cause the machine to vibrate excessively. If the machine is suitably modeled with masses, springs, and dampers, the unbalance force serves as a force input and allows calculation of vibration of other machine parts (see Fig. 2-48).

When force inputs are to be intentionally produced for testing purposes in the laboratory, the *electrodynamic shaker*[41] is often used (see Fig. 2-49). Here, current is passed through a coil suspended in a magnetic field. For a fixed field, a magnetic force directly and instantaneously proportional to the current is produced. A body attached to the coil will, of course, feel this same force. Such shakers are available in sizes from small units one may hold in the hand and producing a maximum force of 1 or 2 pounds, to huge machines weighing 35,000 pounds and producing 25,000 pounds of driving force. The waveform of the force depends on the waveform of the coil current; sinusoidal current gives a sinusoidal force, random current gives a random force. Oscillating forces of many thousand cycles/sec can be achieved. If the test object is free to move, the magnetic force naturally causes a motion of the test object. If we measure this motion, we *could* then consider the input to be a known *motion*, rather than a force input. If we control the shaker current with a feedback system which measures motion, then the shaker becomes even more a *motion* source.

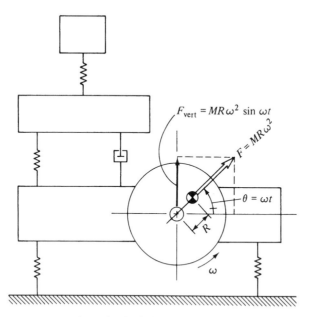

Figure 2-48 A mechanical vibration shaker: rotating unbalance as a force input.

---

[41]C. P. Chapman, Derivation of the Mathematical Transfer Function of an Electrodynamic Vibration Shaker, Jet Propulson Lab Tech. Rept. 32-934, 1966. Doebelin, *System Modeling and Response*, pp. 413–423. Ling Dynamic Systems, Inc., 60 Church Street, Yalesville, CT 06492, 800-468-6537.

System Elements, Mechanical

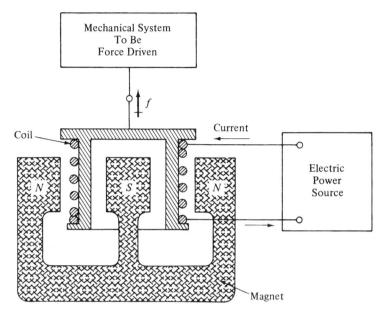

Figure 2-49 Electrodynamic vibration shaker as a force source.

The steering of air, space, and water vehicles is generally accomplished by the manipulation of force or torque inputs. A ship or airplane is maneuvered by deflecting control surfaces (rudders, ailerons, diving planes, etc.) into the relative wind or current. The pressure of the fluid on the control surface produces forces and moments which act on the vehicle to change its attitude and direction of travel. In addition to such *control* inputs, an aircraft is also subject to *disturbing* force inputs in the form of wind gusts, the so-called atmospheric turbulence. For space vehicles, no fluid medium exists to provide control surface pressure; thus designers employ reaction jets, or else swivel the propulsion engine to position the thrust vector so as to cause a turning moment. The Space Shuttle, which operates both in space and in the atmosphere, requires *both* reaction jets and aerodynamic control surfaces as force input devices. Automobiles are maneuvered by force inputs at the tire/road interface, but these are commanded by the driver's motion inputs at the steering wheel. If power steering is used, the driver feels little force at the steering wheel and thus the input is mainly a motion input. When power steering is not provided, the driver definitely feels the tire force reflected back into the steering wheel and the driver input becomes a complex mix of force and motion.

A motion source may sometimes be converted into a satisfactory force source by interposing a "soft" spring between the motion source and the system force input point. Figure 2-50 shows an example of such an arrangement using a Scotch yoke mechanism (a sinusoidal motion source) to produce a sinusoidal force input of selected amplitude and frequency. If motion x_{si} at the input point of the driven system is small compared with the Scotch yoke motion x, then the spring force applied to the system is determined almost entirely by x, whose amplitude and frequency can be set at desired values by adjusting R and ω. The requirement $x_{si} \ll R$ can be satisfied by choosing K_s sufficiently small (a "soft" spring) *relative*

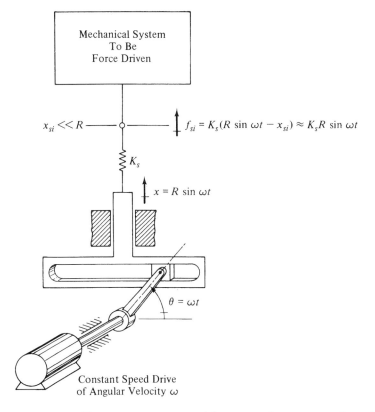

Figure 2-50 Force source constructed from a motion source and a soft spring.

to the "stiffness" of the driven system at its input point. Since small K_s also results in small driving force, this technique may not be useable when large forces are needed.

A little reflection on the above examples of motion and force inputs should suggest to you that, just as in the choice of a suitable model to represent a component or system, the choice of the input form to be applied to the system also requires careful consideration and is subject to some interpretation. Let us conclude our discussion of sources with a brief look at energy considerations. We should first note that a system can be caused to respond only by the source supplying some energy to it. Thus, while we may have chosen to speak of motion and force sources, whichever of these we employ, an interchange of energy must occur between source and system. That is, if we postulate a force source, there will also be an associated *motion* occurring at the force input point. We can compute the instantaneous power being transmitted through this *energy port*, as it is called, as the product of instantaneous force and velocity. If the force applied by the source and the velocity caused by it are in the same direction, power is supplied by the source to the system. If force and velocity are opposed, the system is returning power to the source. The concept of mechanical impedance briefly introduced earlier is of some help here. The transfer function relating force and velocity at the input port of a system is called the *driving-point impedance* Z_{dp}.

System Elements, Mechanical   **99**

$$Z_{dp}(D) \triangleq \frac{f}{v}(D) \qquad (2\text{-}77)$$

$$Z_{dp}(i\omega) \triangleq \frac{f}{v}(i\omega)$$

Now since power $P = fv$ and $v = f/Z_{dp}$

$$P = fv = f\frac{f}{Z_{dp}} = \frac{f^2}{Z_{dp}} \qquad (2\text{-}78)$$

Thus if we apply a force source to a system with a high value of driving-point impedance, not much power will be taken from the source, since the force produces only a small velocity. The extreme case of this would be the application of a force to a perfectly rigid wall (driving-point impedance is infinite since no motion is produced no matter how large a force is applied). In this case the source would not supply *any* energy. The higher the driving-point impedance, the more a real force source behaves like an ideal force source. The lower the driving-point impedance, the more a real motion source behaves like an ideal motion source. More comprehensive studies show that real sources may be described accurately as combinations of ideal sources and an impedance (called the *output impedance*) characteristic of the physical device. A complete description of the situation thus requires knowledge of two impedances: the output impedance of the real source and the driving-point impedance of the driven system. Fortunately, many practical problems do not require numerical consideration of these advanced concepts, but a qualitative understanding gives us a useful perspective.

2-11 DESIGN EXAMPLES

Since we are still early in our development of system dynamics tools, we can't contemplate any extensive design problems; however, certain simple but useful studies are now within our reach and we want to make sure that you don't forget that *design* is what engineers really do.

Engine Flywheel Example.   We have mentioned that inertia in the form of a flywheel is sometimes intentionally added to a machine in order to smooth out speed fluctuations. Added inertia will in general have this effect; however, as is common in most design problems, there is a "downside" or "tradeoff" to this improvement. The increased inertia will slow down the system response when we intentionally change the speed, such as accelerating our car from 40 to 50 mph. That is, the flywheel has the good effect of reducing vibrations due to fluctuating engine speed, but has the bad effect of making the car's response to the throttle more sluggish. Flywheel design must strike a proper compromise between these conflicting goals. System dynamic analysis provides the numerical data needed to make the compromise.

Suppose we have analyzed an engine so that we have a model for its average torque and the torque fluctuation which "rides on top of" this average torque. We also need a model of the load which the engine is driving, perhaps a pump of some sort. The description of the load is often in the form of a torque/speed curve. That is, when the load is running at a particular speed, it requires a certain torque, and this torque is different for different speeds. Such speed/torque curves can be estimated from theory or, if the machine has been built, measured in the lab. Let's assume we have such a curve for our load. At any instant, the combined inertia of engine and load (they are coupled together and run at the same speed) feels the net torque, which is the difference between the engine torque and the load torque. If these two torques are balanced, the engine and load run at a steady speed. If the torques are not equal, the system accelerates or decelerates, according to Newton's law.

We study a situation where the engine is "idling" at about 1000 rpm (104 rad/sec) with average engine and load torques equal at 50 ft-lb$_f$. At time = 1.0 sec, we make a step increase in the throttle, which causes the average engine torque to jump to 70 ft-lb$_f$. The system will now accelerate to a new average speed where the average engine and load torques are again in balance. We wish to study how changes in flywheel inertia (part of the total inertia) affect system dynamic behavior. The SIMULINK simulation diagram of Fig. 2-51 implements Newton's law

$$\sum \text{torques} = \text{engine torque} - \text{load torque} = J\frac{d\omega}{dt}$$

$$\frac{d\omega}{dt} = \frac{1}{J}(\text{engine torque} - \text{load torque}) \tag{2-79}$$

We use the icon Step Fcn1 to create a step change in the average engine torque from 50 to 70 at time = 1.0 sec. The fluctuating component of engine torque goes through a certain pattern which repeats for each complete revolution of the crankshaft. For this example, we take this fluctuation as $(10\sin\theta) + (5\sin 3\theta)$. We need the crankshaft angle θ for this and get it by integrating the speed ω. Having θ, we use the Fcn1 icon to generate the desired function. Engine total torque is obtained by summing the average and fluctuating components. Since we start our simulation at a speed of 104 rad/sec, the initial condition on integrator1 must be set at this value. Integrator2's initial condition is set at 0 to start the crankshaft position at this angle, but we could of course use any angle we wish here.

The load torque is obtained from the speed by implementing the torque/speed curve in Look Up Table1. Since such curves are often given in rpm, we multiply ω by 9.549 to convert rad/sec to rpm. Recall that lookup tables require two "vectors" or lists which give the "x, y" values for the curve being modeled. I want to show a detail here that you can employ in any lookup table you might ever use. Since the SIMULINK lookup table uses linear interpolation between given points, and since the curves we are modeling are usually smooth, it takes a lot of points to get a good fit. The tedium of entering these can usually be avoided by entering only selected critical points which make sure no important "wiggles" are missed, and then using MATLAB's spline function to fit a smooth curve through the given points. The many points generated by the spline operation are then used in the lookup table and you don't have to enter them one at a time. Since SIMULINK is part of MATLAB, the MATLAB command window is always available for such operations. For the present example, the MATLAB statements would be:

```
w=[700 800 900 1000 1100 1200 1300]; list of rpm's
T=[40 42 45 50 56 70 90]; list of corresponding torques
wint=700:10:1300; enlarged list of rpm's
Tint=spline(w,T,wint); torques to go with the new rpm's
save flywhee.m wint Tint
```

System Elements, Mechanical

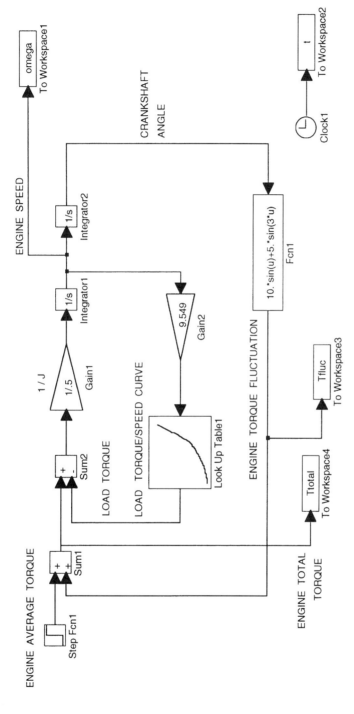

Figure 2-51 Simulation model for engine flywheel design study.

After entering these MATLAB commands, we return to the SIMULINK window and enter wint and Tint as the names of the two lists in the lookup table. We thus get a smooth curve without having to manually enter many points. The save... statement creates a file called flywhee.m which holds the "splined" values wint and Tint. This is done because when you save SIMULINK files, lookup table values obtained from MATLAB are lost. When we return to this simulation at a later time, we must load the file flywhee.m *before* we open the SIMULINK file for Fig. 2-51, or else our lookup table will be "empty."

We can now enter any chosen numerical values for inertia J in the Gain1 block. This J will of course be the *total J* for engine, load, and flywheel. I ran two cases, one with no flywheel ($J = 0.1$ slug-ft^2) and one with a large flywheel which made the total $J = 0.5$. Figure 2-52 shows that, as expected, adding the flywheel reduces the rapid speed fluctuations but significantly slows down the acceleration to the new average speed. With such numerical results available, the designer could make decisions about what size flywheel gave the best compromise. Note how quick and easy it would be to change *any* of the features of this simulation to study their effects. Figure 2-53 shows some more graphical details for this simulation.

As in any other design situation, the *system level* design above, which would find the "best" value for J, is followed by a *detail design* of a real flywheel which would provide the desired J value. You should realize that there are an *infinite*

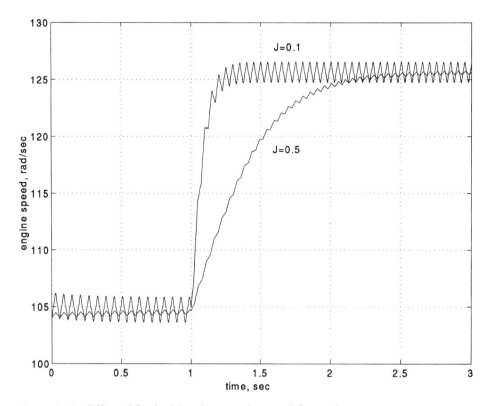

Figure 2-52 Effect of flywheel inertia on engine speed fluctuations.

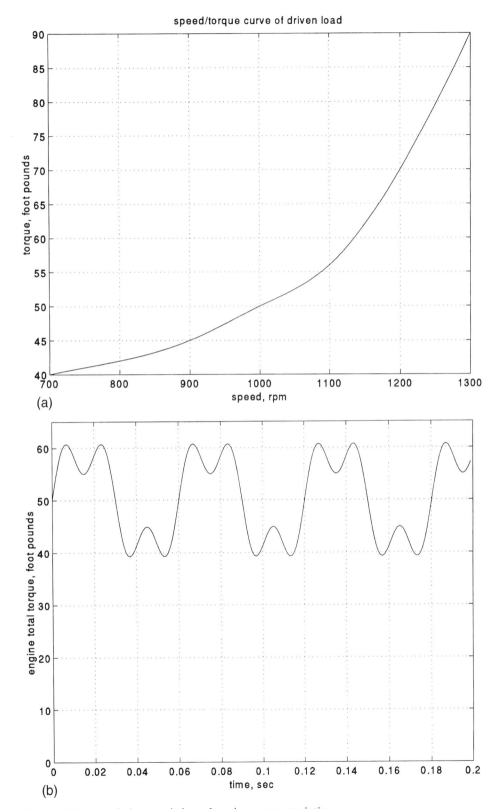

Figure 2-53 Load characteristic and engine torque variation.

103

number of detail designs which would give the same J value, since many different shapes, sizes, materials, and manufacturing processes might be used. Detail design involves many practical and theoretical considerations, some of which are covered in books on the design of machine components, such as flywheels.

Accelerometer Transducer Example. Measuring instruments make good examples for us because system dynamics tools are widely used in both the design and use of this class of devices. Figure 2-35a showed the construction of a typical acceleration transducer or sensor, an instrument for measuring translational acceleration. The engineering design of any device begins with a list of *specifications*; that is, what do we want our device to do? For measuring devices, certain types of specifications are always of interest. For example, commercial accelerometers are always designed for a certain *range* of accelerations. Let's suppose we want an instrument which will measure accelerations in the range $\pm 10g$, that is, $\pm 3860 \text{ in/sec}^2$. (The use of "$g$'s" as a unit of acceleration is common in this industry.)

Size and weight may be important for an accelerometer. If it is to be used to measure the sidewise acceleration of an entire vehicle such as an automobile, it can be relatively large and heavy since there is lots of space available in a car to mount it, and the added mass of the accelerometer will have negligible effect on the operation of the car. On the other hand, if it is to be used to measure the vibration of some sheet metal part of the car, it needs to be small and have a small mass, since its attachment to the sheet metal *will* change the vibration characteristics we are studying and we want this change to be negligible. Let's assume that the intended applications for our device allow a space of about a 2-inch cube and a weight of about 1 pound.

The text near Fig. 2-35 explained the operating principle for steady accelerations, but most accelerometers must also be accurate for certain kinds of *changing* accelerations. The most difficult change for *any* instrument to deal with is a step change in the measured quantity, since this input changes *instantaneously* (in *zero* time) and *no* real measuring device can respond instantly. A more realistic change is called a *terminated ramp*, shown in Fig. 2-54. Here the acceleration changes gradually, more like what can happen in the real world. By adjusting the rise time T_r we can adjust the severity of the change to approximate the worst condition expected in the application. Let's assume that our application requires that the accelerometer respond to terminated ramps with rise times of at least 0.20 second with measurement error never greater than $\pm 5.0\%$ of the final steady acceleration.

Most sensors have a voltage output, as does the accelerometer of Fig. 2-35. Part of accelerometer design is a choice of the displacement transducer that changes

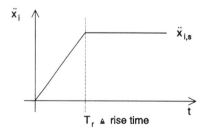

Figure 2-54 Acceleration to be measured by accelerometer.

System Elements, Mechanical   **105**

x_o into a proportional voltage, according to $e_o = K_e x_o$. This choice requires more background than I want to provide here, so let's just assume that if we can get x_o to closely follow the input acceleration, that e_o will be "no problem." (It turns out, in fact, that the main dynamic problems *are* mechanical, not electrical.) The only characteristic we then need for the displacement transducer is its full-scale stroke. This can range from a few microinches to several inches, so we have a wide choice available. Since our allowed space inside the housing is about 1.5 inches, let's arbitrarily assume a stroke of ± 0.2 inch until we run into trouble. That is, for full-scale acceleration of $\pm 10g$ we will get a full-scale spring deflection of ± 0.2 inch.

Recalling now the basic operating principle, that for a steady acceleration A the proof mass M comes to rest *relative to the housing*, at a spring deflection which provides a force MA on the proof mass:

Maximum spring force $= (M)(\text{maximum acceleration})$

$$K_s(0.2) = M(3860)$$

$$\frac{K_s}{M} = 19300. \qquad (2\text{-}80)$$

We now have a "constraint" on the ratio K_s/M. If we later find this constraint intolerable, the *only* way we can change it is to choose a displacement transducer with a different full-scale stroke, since we *must* have a $\pm 10g$ accelerometer. There are of course an *infinite* number of combinations of spring and mass which will satisfy our constraint, so we need to somehow narrow this choice. Since our alloted space (housing outside dimension) is about a 2-inch cube, and maximum weight of the entire transducer is about 1 pound, we have an upper limit on the allowable mass. Steel weighs about 0.3 pounds per cubic inch, so let's try a proof mass of $0.3/386 = 0.000777 \, \text{lb}_f\text{-sec}^2/\text{in}$. This choice makes the spring constant $15.0 \, \text{lb}_f/\text{in}$.

We now have a tentative design which we are sure will meet the requirements of full-scale range, weight and space. Note that damping B has not yet entered in any way, since pure viscous damping exerts *no* force when the proof mass is at rest *relative* to the housing, which *is* the case for a steady acceleration. The purpose of the damper is to suppress spurious vibrations which would occur whenever the acceleration *changed*. Note that coulomb friction is *never* wanted in the mechanical parts of precision instruments since it introduces an unpredictable force. We will later develop analytical tools for quickly finding an optimum value for B in such problems, but for the time being we will use a trial-and-error approach through simulation.

From Newton's law we can see how to set up a simulation diagram.

$$\sum \text{forces on } M = K_s x_o + B\dot{x}_o = M\ddot{x}_m = M(\ddot{x}_i - \ddot{x}_o) \qquad (2\text{-}81)$$

$\ddot{x}_i \triangleq$ absolute accleration to be measured

$\ddot{x}_m \triangleq$ absolute acceleration of proof mass

$\ddot{x}_o \triangleq$ relative acceleration of proof mass and housing

$$M\ddot{x}_o = M\ddot{x}_i - B\dot{x}_o - K_s x_o \qquad (2\text{-}82)$$

From Eq. (2-82) we can understand the simulation diagram shown in Fig. 2-55. This diagram is used to solve for x_o, which is supposed to accurately track the input

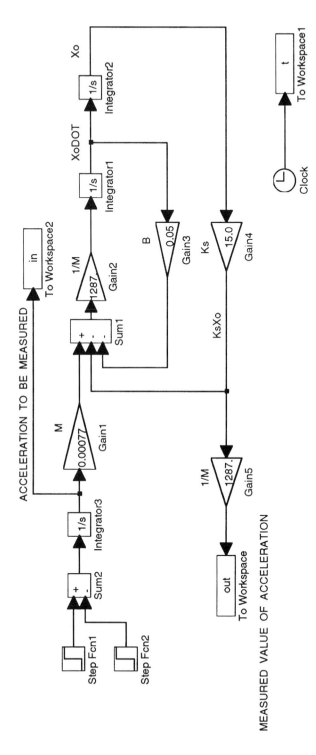

Figure 2-55 Simulation model of accelerometer.

System Elements, Mechanical   **107**

acceleration. Such diagrams are always started with a summer whose output is to be the highest derivative term of the unknown, in our case, $M(d^2x_o/dt^2)$. To force the output to this value, the inputs to the summer must be as shown. These needed inputs are obtained by dividing by M and then integrating, once to get the velocity (needed for the damper force), and once more to get the displacement (needed for the spring force, and also our desired solution). The terminated ramp input is formed using step inputs, integrators, and summers, as shown. Since a *perfect* accelerometer would have $K_s x_o = M(d^2x_i/dt^2)$, we multiply the spring force by $1/M$ to get a quantity ("out" in the diagram) to plot and compare with the actual acceleration ("in" in the diagram).

Figure 2-56a shows the results obtained with $B = 0.0$. During the ramp, we meet the $\pm 5\%$ error criterion but it is exceeded slightly for the constant-acceleration portion. Note that even though we meet the error requirement during the ramp, the accelerometer user might be misled into thinking that the true acceleration had an oscillation in it. This is why accelerometers often have intentional damping. The oscillation in Fig. 2-56a seems to continue "forever"; this is of course due to the complete absence of damping in our model and is unrealistic. Even though we have not designed-in an intentional damper, the real accelerometer *will* have various forms of friction, which would cause the oscillations to decay. Since this "parasitic" damping is often too small, intentional damping may be employed. In our simulation, we can easily try a few values of B and quickly find that $B = 0.05$ gives the good results of Fig. 2-56b. The 0.05 value is *not* a "magic number"; any value in this neighborhood works well. Actually, simulation trial-and-error design is not really necessary for such a simple system. We will later develop *analytical* design tools which predict good damping values accurately.

As usual, the system-level design which has found good values for M, B, and K_s would be followed by detail design of an actual mass, spring, and damper. These detail designs would also involve the final choice of a displacement transducer. All these designs generally *interact* in one way or another. That is, the design of, say, the spring, must consider simultaneously the features of the damper, mass, and transducer, to make sure that everything is compatible. Remember also that the numerical values for M, B, and K_s found at the system-design level may *not* all be physically or economically feasible when we get into the detail design, and the system-level design may have to be adjusted.

Optimum Decelerator Example. When we earlier discussed dampers we noted that the linear ("viscous") damper was ideal mathematically but not necessarily functionally. That is, it gave equations that were analytically solvable, but its functional performance was not always the best that might be achieved. We want to here show that, for a certain class of damping applications, the linear damper is far from optimum and that an optimum damping form can be derived and practically implemented.

The application we consider is one where a mass moving with a certain velocity is to be decelerated to zero velocity in the shortest possible time and distance, under the constraint that the damping force exerted on the mass should never exceed a given value. The limit on the maximum damping force is usually necessary to protect the moving parts from overstress, excessive impact noise, etc. If the maximum damping force is stipulated, it is intuitively clear that our damper should exert this same

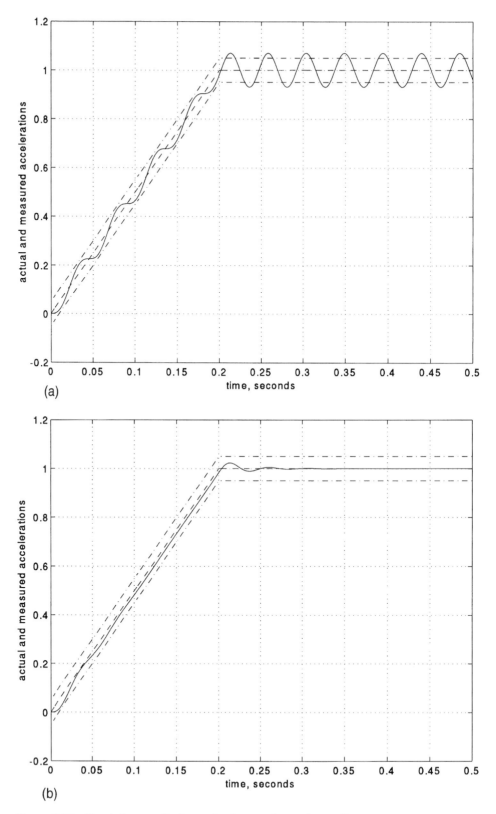

Figure 2-56 Simulation results for evaluating accelerometer performance.

System Elements, Mechanical

force for its entire stroke, if we want to stop the mass in the shortest time. That is, if we are allowed to use a certain maximum force, we should apply this force "all the time." If we do this, there is no *possible* faster way to decelerate the mass.

Having stated the problem, it is immediately clear that our "ideal" linear damper will *not* give us what we want. That is, we *can* choose B so that the maximum allowable force *is* exerted when the mass first strikes the damper at its maximum velocity. However, as the mass slows down, the damper force *decreases* proportional to the velocity, and our "ideal" damper is bound to take longer (and require a longer stroke) than would a damper which *maintained* the maximum force for the entire stroke. What we need is a damper that "increases its B value" as the mass slows down, so that the force stays the same. This will clearly be a nonlinear damper, since a linear damper must have a *constant* B value.

An optimum damper that maintains the force constant as velocity decreased could be constructed in various ways. We will consider a design that corresponds to actual commercial devices. Here the damping is obtained by forcing a damping liquid through a small orifice, as in Fig. 2-57. If this orifice were of a *fixed* size, the damping force, while nonlinear with velocity, would still drop off as velocity decreased—not what we want. Conceptually, we need an "orifice" that *gets smaller* as the stroke proceeds and the velocity drops; however, it has to "get smaller" in exactly the right way, if we want an exactly constant force. We will now derive the needed relation.

If you have not had some fluid mechanics background, we ask you to take on faith a formula from Chap. 4 (fluid and thermal elements) which describes the fluid flow process at the orifice:

$$\text{Volume flow rate through orifice} \triangleq q \ (\text{in}^3/\text{sec}) = C_d A_o \sqrt{\frac{2\Delta p}{\rho}} \qquad (2\text{-}83)$$

$C_d \triangleq$ orifice discharge coefficient, dimensionless

$A_o \triangleq$ orifice area, in^2

$\Delta p \triangleq$ pressure drop across orifice, psi

$\rho \triangleq$ fluid mass density, $\dfrac{lb_f\text{-sec}^2}{\text{in}^4}$

$$q = K_{\text{or}}\sqrt{\Delta p} \qquad (2\text{-}84)$$

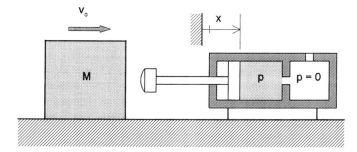

Figure 2-57 Using a damper to decelerate a moving mass.

Chapter 2

where $K_{\text{or}} \triangleq C_d A_o \sqrt{\dfrac{2}{\rho}}$

Since the liquid is treated as incompressible, the rate at which volume is displaced by the advancing piston must equal the volume flow rate through the orifice at every instant. Also, the piston velocity must decrease linearly with time since we insist on a constant decelerating force, which gives a constant deceleration of, say, $a\,\text{in/sec}^2$. The numerical value for "a" is known as soon as we are given the mass M to be decelerated and the maximum allowable decelerating force.

$$q = K_{\text{or}}\sqrt{p - 0} = A_p(v_0 - at) \tag{2-85}$$

$$p = \left(\frac{A_p}{K_{\text{or}}}\right)^2 (v_0 - at)^2$$

Now the pressure p times the piston area A_p is the decelerating force, so this must equal the mass times its acceleration, allowing us to eliminate p from our equations:

$$K_{\text{or}}^2 = \frac{A_p^{\,3}}{Ma}(v_0 - at)^2 \tag{2-86}$$

Now for a motion with constant deceleration a, we know that the displacement x is given by

$$x = v_0 t - \frac{at^2}{2} \tag{2-87}$$

which gives t as

$$t = \frac{v_0 - \sqrt{v_0^2 - 4ax/2}}{a}$$

Substituting for t in Eq. (2-86) finally gives us a formula showing how the orifice must be adjusted to get the constant deceleration that we want.

$$K_{\text{or}}^2 = \frac{A_p^{\,3}}{Ma}(v_0^2 - 2ax) \tag{2-88}$$

We see from Eq. (2-88) that K_{or} starts out ($x = 0$) at a value $A_p^{\,3}v_0^2/Ma$ and ends up at 0.0 (end of stroke, $x = v_0^2/2a$). The definition of K_{or} shows that the only practical way to vary it is by changing the orifice area. While one could surely invent various ways of smoothly changing the orifice area as a function of x (think of the iris diaphragm in a camera, etc.), such mechanisms are needlessly complicated and unreliable. Instead, commercial shock absorbers approximate the smooth area variation defined by Eq. (2.88) by using a sequence of drilled holes, as shown simplified in Fig. 2-58. As the stroke progresses, the total orifice flow area reduces in a stepwise fashion which approximates the ideal smooth curve.

Let's now do a numerical example where the mass weighs $100\,\text{lb}_f$ and has an initial velocity of 5.0 ft/sec. Suppose the maximum allowable damping force is $200\,\text{lb}_f$ and we want to use a piston area of $0.0015\,\text{ft}^2$. The constant deceleration will then be $200/(100/32.2) = 64.4\,\text{ft/sec}^2$, ("2 g's"). I want to do a simulation which compares the performance of four different dampers:

1. A fixed linear (viscous) damper (constant B value)
2. A fixed nonlinear (single-orifice) damper

System Elements, Mechanical

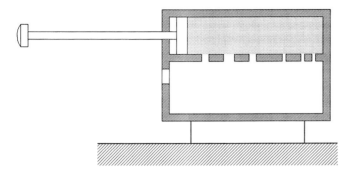

Figure 2-58 Simplified construction of optimum damper.

3. A perfect optimum damper
4. A stepwise approximate optimum damper

A Newton's law for the perfect optimum damper gives

$$-pA_p = M\ddot{x}$$

$$-\left(\frac{A_p\dot{x}}{K_{\text{or}}}\right)^2 A_p = M\ddot{x}$$

$$\left(\frac{-Ma}{v_0^2 - 2ax}\right)\dot{x}^2 = M\ddot{x} \tag{2-89}$$

Figure 2-59 shows simulation diagrams for all four dampers, with the perfect (smooth) optimum at the top. Note that we don't need the usual "tricks" to get the correct algebraic sign on the damping force because the velocity never goes negative in this application. The linear damper needs a B value of $40.0 \text{ lb}_f/(\text{ft/sec})$ to give the desired force of 200 lb_f for the initial velocity of 5.0 ft/sec. Of course its force will drop off as velocity drops.

To set up the stepped approximation to the smooth optimum damper I plotted the function $1/(25. - 128.8x)$ in Fig. 2-60. This is the function which we need to approximate using the "sequence of holes" concept. This function goes from 0.04 to infinity as x goes from 0 to 0.194 ft. The "infinity" is produced by the "adjustable orifice" going *completely* shut at the end of the stroke. In a practical device, one doesn't usually shut *all* the orifices at the end of the stroke, as required by the optimum calculation. Rather, one "finishes" the stroke with a single small orifice open. This will *not* bring the velocity to exactly zero, but will make it slow enough so that when the piston "bottoms out" against a fixed stop; the impact will not be damaging. In approximating the smooth curve we must decide how many holes to use, what size, and where they should be located. This is best done "graphically" by simply drawing in some trial choices and then running the simulation to see whether results are satisfactory or not. In Fig. 2-60 I chose to use five holes. The x spacing and hole size are decided by judging visually whether the stepped curve reasonably fits the smooth one.

The actual numbers for x values and ordinates are "eyeballed" from the graph once the desired straight-line segments have been drawn in by hand. Note that there are short, linear "transition regions" as the moving piston covers up each hole. This is a rough modeling of how the flow area changes as the circular orifices are closed by

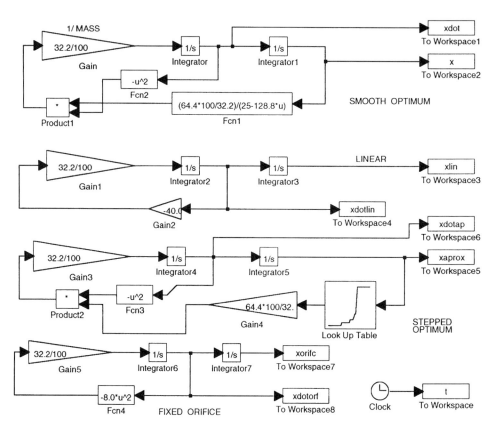

Figure 2-59 Simulation model for optimum damper study.

the moving piston. We could have worked out a *precise* area variation from the geometry of a circle, but this is hardly warranted when there are so many other approximations already.

Once we have numerically defined the stepped curve, we can implement it easily in our simulation with a lookup table. The values are:

```
input=[0 .075 .085 .125 .135 .150 .160 .170 .180 .250]
output=[.04 .04 .10 .10 .16 .16 .27 .27 .80 .80]
```

Our simulation diagram shows a lookup table icon which displays a miniature version of this graph. The rest of the simulation for the stepped optimum damper is similar to that for the smooth optimum.

Our final damper is one with a single orifice, sized to apply the allowed maximum force at the beginning of the stroke. This damping force is easily modeled with the nonlinear function $-8u^2$, as shown in the diagram; $(8)(5^2) = 200 \, lb_f$. Our simulation diagram now includes all four dampers and we can run it to compare their performance. This technique of running several problems "in parallel" when we want to compare various competitive designs is one of the most useful features of simulation.

Since the mass starts with a velocity of 5.0 ft/sec, the first integrator in each of the simulations must have its initial condition set at this value. We then start the

System Elements, Mechanical

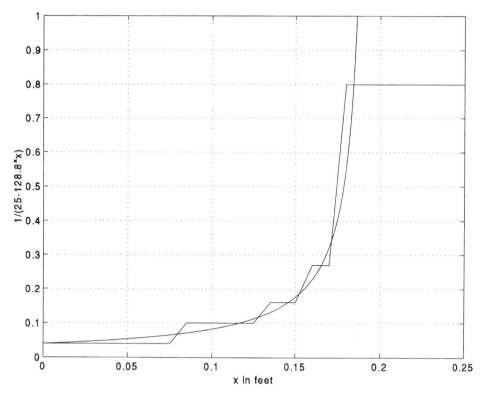

Figure 2-60 Comparison of ideal and approximate damper characteristic.

simulation and graph those quantities of interest to us. Figure 2-61 compares the velocities for all the dampers. The smooth optimum of course takes the velocity precisely to zero at a time of 0.0776 seconds, along a straight-line path; this is the best that can possibly be done. The stepped approximation does a good job, but can't bring the velocity to zero since we intentionally allowed the final orifice to *stay* open at the end of the stroke. The velocity at the end is, however, small enough that it can be brought to zero by impacting a fixed stop. By using more holes, and/or adjusting their location, one could of course bring the velocity curve of the stepped optimum damper closer to the ideal than we have here. This extra effort is of questionable practical value since our models, as usual, neglect some features of the real devices. At this point in the design of a stepped-optimum damper, most engineers would probably turn to lab testing of prototype dampers to get some data on the real behavior. That is, computer simulation is very useful for reducing the amount of lab trial-and-error testing, but we always reach a point where it is cost effective to do some lab work. Such lab work often uncovers neglected aspects of system behavior which we can then include in a more correct simulation.

It is clear from Fig. 2-61 that the "fixed" dampers, either linear or nonlinear (orifice), are far from optimum and will take much longer to reduce the velocity to a small value. They also require much longer strokes, making the damper larger than really necessary. We should also use our simulation to check the damping force since it should not exceed $200\,lb_f$ at any time. Figure 2-59 does not show the forces being sent to the workspace for plotting, but this feature is easily added, giving the graphs

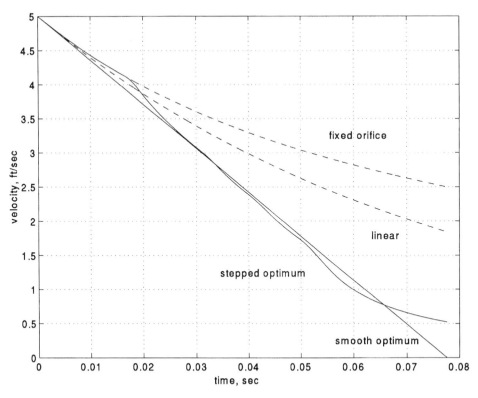

Figure 2-61 Performance comparison of four different dampers.

of Fig. 2-62. The force graph for the stepped optimum damper may be surprising since it shows considerable fluctuation, while our velocity graph was quite smooth. This is explained by noting that the acceleration graph would *also* exhibit these fluctuations, and the velocity graph is the *integral* of acceleration. Integration of *any* time-varying function *always* is a smoothing process, so velocity *can* be smooth with a rather "wild" force graph.

While the average force is near 200 pounds, the instantaneous force clearly violates our 200-pound limit some of the time. If you have access to simulation software, you might try adjusting the stepped approximation to better satisfy this force limit. You will find that unless you use an impractically large number of holes, the only way you can keep the force below 200 is to sacrifice quite a bit of optimum speed and stroke. In fact, lab tests show that the force fluctuation in a real damper of this type is considerably smoothed by the compressibility of real oil and the elasticity of the cylinder walls, seals, etc. Also, the 200-pound limit should be interpreted in terms of its *actual* bad effects in each application, and can often be relaxed. That is, we consider meeting the speed and stroke criteria more important in most cases than the maximum force criterion.

Our sketches of damper construction have been solely for analysis purposes and we now want to show the details of an actual device[42] in Fig. 2-63. If you are

---

[42]Rexroth Crop., 1953 Mercer Road, Lexington, KY 40511, 606-254-8031.

System Elements, Mechanical

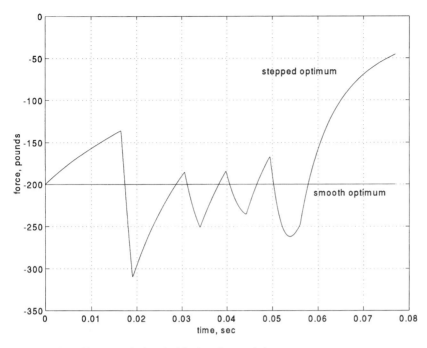

Figure 2-62 Force variation in ideal and actual damper.

relatively inexperienced in engineering design, it is important that you develop a clear understanding of the difference between a *functional concept* (what we have shown in our earlier diagrams) and its *practical implementation* (Fig. 2-63). To satisfy the needs of the marketplace, our simple idea requires considerable embellishment beyond the bare functional concept.

Some features necessary in a practical shock absorber and not provided in the conceptual version of Fig. 2-58 include:

1. A means to *return* the fluid to the cylinder after each stroke
2. A means to *adjust* the overall level of damping to accurately meet the needs of an actual application

The shock absorber consists of a rugged black oxided steel housing, with a full-length male thread on the outside, for mounting and positioning, to accommodate

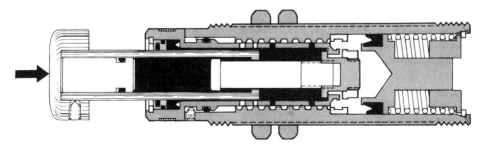

Figure 2-63 Design details of a commercial optimum damper

the motion of the load. A female buttress thread is provided on the inner surface. Inside is a steel cylinder tube with a number of radial holes sized and positioned along its length. The cylinder tube is permanently connected to the knurled adjusting ring. As one rotates the adjusting ring, the tube also rotates so that more or less of its radial holes are blocked by the buttress thread inside the housing. That is, each of the radial holes has more or less of its area covered as the adjusting ring is rotated. This allows the user to "fine-tune" the shock absorber over a range around its nominal value, to more accurately meet the needs of each application.

The piston rod is a hard chrome-plated steel tube with a plug inserted at its outer end. The plug is removed for refilling the shock absorber with oil when that is needed. Protecting the rod end is a case-hardened and tempered rod button. The inner end of the cylinder tube is fitted with a check valve that prevents oil from escaping during the active stroke but allows it to flow freely during the return stroke. Behind the check valve is a spring-loaded accumulator piston. As the chamber fills during the active stroke, the spring-loaded piston retracts, making room for the oil that is being forced out of the main cylinder through the radial holes. When the accumulator piston nears its full stroke, an indicator pin extends out the back end. This pin can be used to actuate a limit switch or other sensor to indicate that the desired full stroke is being achieved or that the oil level is low and needs replenishing.

As the decelerated load is removed, the pressure stored (by the spring) in the accumulator chamber opens the check valve to allow the oil to be returned to the main cylinder in preparation for the next stroke. When a shock absorber is used for cyclic operation in some machine, the absorbed energy gradually heats it up, until the average rate of heat transfer from the damper to its surroundings just matches the average rate of energy absorption, and the temperature stabilizes. This maximum temperature must not exceed an allowable limit. The manufacturer controls this problem by listing a maximum allowable energy dissipation per hour for each model.

BIBLIOGRAPHY

Nakazawa, H., *Principles of Precision Engineering*, Oxford University Press, New York, 1994.

Rothbart, H. A., ed., *Mechanical Design and Systems Handbook*, McGraw-Hill, New York, 1964. (See springs, sec. 33; dampers, sec. 34; mechanisms, secs. 4,5; machine systems, sec. 10; friction, sec. 11; gearing, sec. 32.)

Shigley, J. E., and C. R. Mischke, *Mechanical Engineering Design*, 5th ed., McGraw-Hill, New York, 1989.

Slocum, A. H., *Precision Machine Design*, Prentice-Hall, Englewood Cliffs, N.J., 1992.

Smith, S. T., and D. G. Chetwynd, *Foundations of Ultraprecision Mechanism Design*, Gordon and Breach, Amsterdam, 1994.

Wahl, A. M., *Mechanical Springs*, Penton Publishing, Cleveland, Ohio, 1944.

Young, W. J., *Roark's Formulas for Stress and Strain*, 6th ed., McGraw-Hill, New York, 1989.

System Elements, Mechanical

PROBLEMS

2-1. A nonlinear spring has $f = 100x + 20x^3$, newtons, where x is in centimeters. Find its linearized spring constant for operating points $x = 0, 1, 2$ and 5 cm. For the $x = 2$ operating point, what linearized expression for f would you use?

2-2. Using the graph of Fig. 2-13, estimate the range of linearized spring constant expected over the full range of this air spring.

2-3. A tension rod spring as in Fig. 2-12i is made of steel which can be loaded to a stress of 100,000 psi; the applied stress is f/A. At a load corresponding to the maximum allowed stress, the energy storage is to be 10,000 in-lb$_f$. Find all combinations of A and L which meet this requirement. What is the energy storage per cubic inch of material?

2-4. Derive an expression for the spring constant of the buoyant spring of Fig. 2-15d in terms of dimensions and material properties. The "float" can have any cross-sectional shape, but must be "prismatical." Discuss the effect of a float which is *not* prismatical but, say, spherical.

2-5. Get an expression for the gravity spring torque on the pendulum of Fig. 2-15b and then linearize it for oscillations around $\theta = 0$.

2-6. Derive expressions for the equivalent spring constant of:
    a. The two springs of Fig. P2-1
    b. The two springs of Fig. P2-2

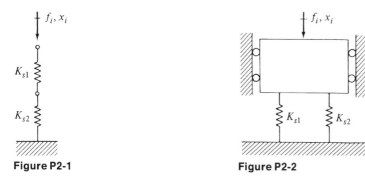

Figure P2-1   Figure P2-2

Discuss the situation when there are *more* than two springs.

2-7. For the damper of Fig. 2-22a, find the viscosity needed to give $B = 10.0$ lb$_f$/(in/sec) if
    $h = 0.05$ inch $R_1 = 0.2$ inch
    $L = 1.0$ inch $R_2 = 1.0$ inch

If this unit is cycled with a displacement $x = 1.0 \sin 10t$ inches (t in seconds), what is the time-average rate of heat generation in watts? Heat transfer calculations show that the damper can transfer heat to its surroundings at the rate of 5.0 watts for each degree of temperature rise above ambient. What will be the steady-state average operating temperature of this damper if ambient temperature is 70°F? Using this temperature, select a suitable fluid from Appendix A. If this damper is inactive and then starts cycling as above, what will be the B value when it first starts cycling?

2-8. Derive Eq. (2-37).

2-9. Derive Eq. (2-38).

2-10. A simplified theory sometimes used for magnetic levitation systems (Fig. 2-9) predicts magnetic force $f_m = Ci^2/y^2$, where C is a known constant for a given system. (Note that this theory predicts *infinite* force for $y = 0$; thus it is used only for nonzero air gaps.) Derive a linearized expression for the magnetic force.

2-11. Repeat problem 2-6, but use dampers rather than springs.

2-12. Using the experimental data of Fig. 2-14, find the spring constant of the structure shown there.

2-13. In Fig. P2-3 a combination of springs from Fig. 2-12b, c, and i is shown. Find the equivalent spring constant for this assemblage. What would this simplify to if the rod connecting the two beams may be considered rigid?

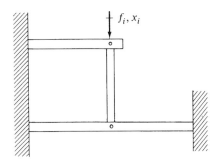

Figure P2-3

2-14. In Fig. 2-12, for all the springs which might be made of metal, what is the effect on K_s of changing from steel to aluminum?

2-15. In Fig. 2-12b, c, e, and i, assume all parameters fixed except L. Sketch a graph showing the variation of spring compliance versus L.

2-16. In Eq. (2-13) discuss the effect of using aluminum versus steel as the spring material, assuming all dimensions are kept the same. You may take aluminum's density and elastic modulus both to be one-third those of steel.

2-17. A rotational damper with one end fixed has a torque given by $(3 - 3t + 62.8 \sin 62.8t)$ applied to the free end, for t going from 0 to 3.0. Find expressions for, and sketch time curves for:
    a. The angular velocity of the free end
    b. The angular displacement of the free end

Draw a simulation diagram for this situation. If you have access to simulation software, choose a B number, run your simulation, and compare results to parts (a) and (b).

System Elements, Mechanical

2-18. A damper with $B = 5\,\text{N}/(\text{cm/sec})$ has one end moving with displacement $(3t + 4\sin 10t)$ cm (t in seconds), while the other end has displacement $1\sin 10t$. Find the force in the damper.

2-19. Figure P2-4 shows the experimentally measured tractive resistance force of an 8000-pound truck. This represents all the power losses except air drag. The air drag force may be estimated as $0.04V^2\,\text{lb}_f$, where V is the vehicle speed in ft/sec. Find a linearized damping coefficient B to represent the total energy dissipation in the neighborhood of speeds of 10, 20, and 30 mph (three different values).

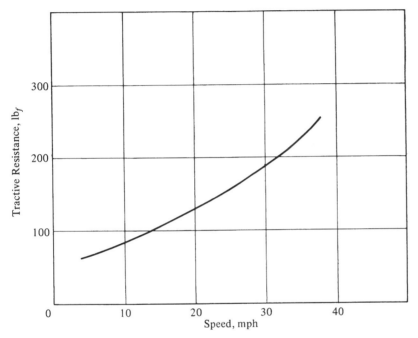

Figure P2-4

2-20. An empirical formula used to estimate the total resistance to motion of a certain type of railway passenger car gives the force as $(130 + 1.5V + 0.034V^2)\,\text{lb}_f$, where V is the car speed in mph. What *kind* of friction is represented by each term in this formula? Sketch a graph of each term versus V for speeds from 0 to 100 mph. Get a linearized damping coefficient B for the total force if V is near 50 mph.

2-21. For a damper with $B = 2.0\,\text{lb}_f/(\text{in/sec})$, calculate and sketch the frequency response curves (Fig. 2-17f). If the amplitude of the sinusoidal driving force is $10.\,\text{lb}_f$, find the frequency at which the displacement amplitude is 0.01 inch.

2-22. Check the correctness of solution Eq. (2-50) by substituting it back into the original differential equation and also checking the initial conditions.

2-23. Obtain the result of Eq. (2-54) without the use of the sinusoidal transfer function.

120   **Chapter 2**

2-24. In Fig. 2-41 the rod acts as an inertia element for sufficiently low frequencies. At low frequencies, how does this same rod behave if its right-hand end is held fast ("built into a wall")?

2-25. In the system of Fig. 2-41, why does the rod cross-sectional area not influence the results?

2-26. For the gears of Fig. 2-42 let $J_1 = J_2 = 0.01\,\text{lb}_f\text{-in-sec}^2$ and let the gear ratio $N = 8$. Find the equivalent inertia of the whole system, referred to the θ_1 shaft. Now find it referred to the θ_2 shaft. If shaft 1 has a damper with $B = 3.0\,\text{in-lb}_f/(\text{rad/sec})$ find the total damping referred to the number 2 shaft.

2-27. In Fig. 2-43

$$L_1 = 1.0\,\text{in} \qquad\qquad L_2 = 3.0\,\text{in}$$
$$f_1 = 3.0\,\text{lb}_f \qquad\qquad f_2 = -2t\,\text{lb}_f$$
$$T = 5\sin 10t\,\text{in-lb}_f \qquad B = 0.0$$
$$W_1 = W_2 = 10\,\text{lb}_f \qquad K_{s1} = K_{s2} = 20.0\,\text{lb}_f/\text{in}$$
$$B_1 = B_2 = 3.0\,\text{lb}_f/(\text{in/sec}) \quad J = 0.25\,\text{lb}_f\text{-in-sec}^2$$
$$K_s = 10.0\,\text{in-lb}_f/\text{rad}$$

Find the equivalent system, referred to x_1.

2-28. Repeat problem 2-27, but refer everything to x_2.

2-29. Repeat problem 2-27, but refer everything to θ.

2-30. Find a mass, and also a spring, that will have the same magnitude of impedance at a frequency of 100. Hz as does a damper with $B = 10\,\text{N}/(\text{m/sec})$ at every frequency.

2-31. Considering a road profile as a motion input to an automobile suspension system, how does the motion input change when the car is driven over the road at various constant speeds? If, at 20 mph the motion input contains frequencies from 0 to 5 Hz, what will the frequency content be at 60 mph? What major frequency would you expect to get when driving at 65 mph over expansion joints spaced every 30 feet?

2-32. In the earthquake record of Fig. 2-47, why are the velocity and displacement traces "smoother" than the acceleration?

2-33. We wish to build a rotating vibration exciter ("shaker") for lab testing, using the principle of Fig. 2-48, but we want *only* vertical oscillating force. "Invent" and explain a modification of the device shown which gives no net horizontal force on the mechanism.

2-34. Design an accelerometer which will fit in a 1-inch cube, measure $\pm 500\,g$'s, and have $\pm 2\%$ accuracy for terminated ramp inputs with rise times longer than 0.01 second. Use analysis and simulation as needed.

2-35. Figure P2-5 shows a viscosimeter (device for measuring fluid viscosity) based on the damper configuration of Fig. 2-24a. Explain in words how this instrument works. Derive a formula showing how viscosity may be calculated from the known dimensions of the device, motor speed, and angular deflection θ. Discuss some reasons why this formula cannot be expected to give perfect accuracy. Such discrepancies between theoretical predictions and actual performance are typical of *all*

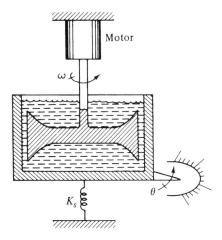

Figure P2-5

real-world devices, but are intolerable in measuring instruments because they often need to be accurate to within 1% or better. (In "ordinary machinery" we use safety factors much larger than 1%, so accuracy is not as critical.) How do instrument designers/users overcome this problem of inadequate theoretical accuracy?

2-36. In the cam of Fig. 2-42, over its specified range of motion, $\theta_2 = 0.5\theta_1 + 0.2\theta_1^2 + 0.05\theta_1^3$. If the θ_1 shaft has a torsion spring with $K_s = 10.$ in-lb$_f$/rad attached to it, refer this spring to the θ_2 shaft for small motions in the neighborhood of $\theta_1 = 0.2$ rad.

2-37. The force of Fig. P2-6 is applied to a pure mass M of 2.5 kg initially at rest at $x = 0$. Calculate the sketch the time history of
  a. Acceleration
  b. Velocity
  c. Displacement
  d. Stored energy

Show a simulation diagram. If you have access to simulation software, run your simulation and compare results to your "hand calculations."

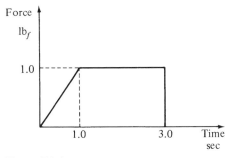

Figure P2-6

2-38. Repeat problem 2-37 substituting a spring $K_s = 1000.$ N/m for the mass. Discuss the difficulties that arise here.

2-39. Repeat problem 2-37 substituting a damper $B = 20.$ N/(m/sec) for the mass. Delete the plot of stored energy and add plots for dissipated power and dissipated energy. Discuss the difficulties that arise here.

2-40. A machine part of complex shape was frequency-response tested with the results as in Fig. P2-7. Over what range of frequencies could it be modeled as a pure spring element? What number would you use for K_s?

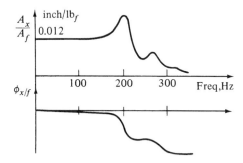

Figure P2-7

2-41. For the "stepped" optimum damper as designed in Fig. 2-60, calculate the diameter of each of the five holes used. Take the discharge coefficient $C_d = 0.61$ and the oil density as 1.55 $lb_f\text{-sec}^2/\text{ft}^4$.

2-42. Suppose we have designed an optimum "smooth" damper according to Eq. (2-88) for a certain mass and initial velocity. In practice, the actual mass may be somewhat different from our design value, and/or its initial velocity may also deviate from that assumed. Use analysis and simulation, as appropriate, to study the effect of these deviations from nominal design values on damper behavior.

3

SYSTEM ELEMENTS, ELECTRICAL

3-1 INTRODUCTION

It is difficult to think of any technological systems of medium or large scale which do not involve electricity in *some* way. If, however, we restrict our consideration to cases in which system functions are performed *largely* by electrical means the following examples might come to mind:

1. Electrical power transmission and distribution
2. Communications (cable, radio, television, microwave, etc.)
3. General-purpose computers
4. Electro-optics (optoelectronics?) (lasers, fiber optics, imaging systems, etc.)

Application areas which, in general, are electromechanical, electrothermal, electro-mechano-acoustic, etc., but which might in specific cases be largely electrical include:

1. Measurement systems
2. Control systems

We described mechanical elements in terms of their force/motion relations. The electrical components or elements of this chapter will be described in terms of their voltage/current relations. To dispel some incorrect but common word usage, note that an "electric" motor is *not* an electrical device but rather an *electromechanical* one, since its description requires specification of not just voltage/current relations but also force/motion relations. The *useful* aspect of an "electric" motor is that it can perform mechanical work. Similarly, a phototransistor (transistor sensitive to light) is an *electro-optical* device, since its description involves voltage, current, and light flux. A loudspeaker would be an electro-mechano-acoustic device while a microphone would be acousto-mechano-electrical. Such "mixed-media" devices are extremely important in many systems, but since they are not strictly electrical we do not include them in this chapter. Some of the more important ones will be treated briefly in Chap. 5, Basic Energy Converters.

We emphasize this careful definition of electrical devices not just for semantic correctness but also to highlight that many devices commonly thought of as "electronic" actually require significant *mechanical* engineering as part of their

123

design. For example, one of the most significant problems in microelectronic devices is overheating, a *mechanical engineering* field. Also, the *manufacture* of almost all "electronic" systems depends critically on many mechanical engineering skills such as vacuum systems, thermal processes, gas flow, robotic wafer handling, etc. Thus many mechanical engineers find employment in the "electronic" industries.

Even with the restriction to strictly electrical devices, the scope is still too large for the purposes of this text and this chapter. Some attempt at classification will help us narrow down to the desired scope. We choose here to classify according to:

1. Network versus field concept
2. Passive versus active device
3. Linear (proportional) versus digital (on-off) device

The *network versus field* classification is essentially that of lumped versus distributed parameters, as discussed in general terms in Chap. 1, and is based on a wavelength/physical size criterion. *If the physical size of a device is small compared to the wavelength associated with signal propagation, the device may be considered lumped and a network model employed.* Wavelengths may be estimated from the known frequency range of a given system and the wave propagation law you encountered in physics courses:

$$\text{Wavelength} = \frac{\text{velocity } V \text{ of wave propagation}}{\text{signal frequency } f} \tag{3-1}$$

The velocity of propagation for electrical waves in free space is 186,000 miles/second. As an example, consider the electrical portion of a high-fidelity music reproduction system. Such systems deal with frequencies in the range 20 to 20,000 Hz, the so-called audio range. The shortest wavelength is associated with the highest frequency, in this case 20,000 cycles/second, so we get (using the simplifying assumption of free-space conditions)

$$\lambda = \frac{186,000 \text{ miles/second}}{20,000 \text{ cycles/second}} = 9.3 \text{ miles/cycle} \tag{3-2}$$

Since a typical resistor or capacitor used in audio circuitry is less than 1 inch long (much less than 9.3 miles), it is clear that audio electrical systems can be (and *are*) treated with the simpler lumped-parameter (network) approach rather than the more general field approach (Maxwell's partial differential equations). It is interesting to note that the wavelength/physical size concept is applicable to any physical system which exhibits wave propagation, such as mechanical vibrating systems and acoustic systems. For acoustic systems in air, the velocity of propagation V is the speed of sound, 1100 ft/sec. Thus to check whether the acoustical portions of a hi-fi system may be treated as lumped or distributed, we calculate the shortest wavelength as $\lambda = 1100/20000 = 0.055 \, \text{ft} = 0.66 \, \text{inch}$. Since a speaker for high frequencies (a "tweeter") may be several inches in diameter and a microphone diaphragm 0.25 to 1.0 inch, we see that the acoustical system is "right on the ragged edge" for validity of a lumped model at the 20,000-Hz frequency. At lower frequencies the lumped model would get better and better. In treating electrical elements we will take strictly the lumped (network) approach and thus eliminate consideration of high-frequency phenomena associated with devices such as radar and microwave antennas, wave-

System Elements, Electrical

guides, etc. This restriction fortunately is not a severe one since many practical systems can be and are treated accurately by the lumped approach.

The distinction between *active and passive devices* is based on energy considerations. From physics you will recall that a resistor, capacitor, or inductor is not a *source* of energy in the sense of a battery or generator. A charged capacitor or current-carrying inductor *does* store energy which can be supplied to another device, but the capacitor did *not* charge itself, nor did the inductor establish its current itself; rather, some energy source was needed for this. A resistor does not even temporarily store energy; it dissipates into heat all the electrical energy supplied to it. Thus the three basic circuit elements—resistance, inductance, and capacitance—are called *passive elements*, since they contain no energy sources.

The basic *active elements* in electric circuits are energy sources such as batteries (electrochemical source), generators (electromechanical source), solar cells (electro-optical source), and thermocouples (thermoelectric source). When these basic sources are suitably combined with the two basic power modulators, the vacuum tube and the transistor, we obtain *active devices* called controlled sources, whose outstanding characteristic is the capability for *power amplification*. These controlled sources will accept as input a low-power voltage or current signal and accurately reproduce it, but at a much higher power level at the output of the device. The vacuum tube (now largely but not completely obsolete) or transistor does not *itself* supply the power difference between input and output; it simply modulates, in a precise and controlled fashion, the power taken from the *basic* source (battery, etc.) and delivered to the output.

These combinations of transistors with their power supplies are generally called active devices and because of their amplification capability are, in a sense, the fundamental base of all electronic systems. The basic principles of electronic amplification are generally introduced in physics courses and, for non-electrical engineers, extended to the point of practical application in a later electronics course of some kind. We will not duplicate this material here but rather emphasize the practical use of what is perhaps the single most useful active linear device, the *operational amplifier*. Through integrated circuit techniques, this device has been reduced in size and cost to the point where it is now treated as an inexpensive circuit element, like a resistor or capacitor. Its function as a basic building block for many different types of useful circuits is enhanced by its ease of application.

That is, careful and expert circuit design is needed to produce the operational amplifier (commonly called "op-amp") itself, but once such a device is available, its further application in the design of instrumentation amplifiers, controllers, filters, and analog computing devices can be successfully accomplished by those with modest electronic expertise, including non-electrical engineers and hobbyists. (Microelectronics of course provides many other linear and digital devices whose use is simple enough to allow nonspecialists to build useful apparatus for many routine applications.) While an op-amp is not strictly an *element*, since it contains resistors, transistors, etc. (which are conventionally considered to be the elements), it is today treated like a component or element and we choose to include it in this chapter.

Since their development in the 1940s, electronic digital computers have had an increasing impact on many aspects of technology and society. We mention these computers here since they are perhaps the most significant applications of *digital electronic devices.* We do not here intend to give a comprehensive description of such devices. Suffice it to say that they essentially perform on-off switching-type functions needed to implement the logic operations required in digital computation. For example, a two-input AND gate produces an "on" output signal only if *both* of the two input signals are simultaneously "on." The "on" and "off" states for both input and output signals can each fall in a wide voltage range and still give correct device operation. For example, *any* voltage between $+2$ and $+5$ volts would represent the "on" state and *any* voltage between 0 and $+0.8$ volts would correspond to the "off" state. Thus the devices are very tolerant of noise voltages and need not be individually very "accurate," even though the overall computer can be extremely accurate.

While high "accuracy" is not needed, these devices must be very small, cheap, and fast, since so many of them are needed to make a computer. In contrasting these digital devices with linear (proportional) devices such as the op-amp, we note that in linear devices the specific waveform of input and output signals is of vital importance, while in digital devices it is simply the presence (logical 1) or absence (logical 0) of a voltage within some wide range that matters; the *precise* value of the signal is of *no* consequence. Just as for op-amps, these digital devices are not really elements in the accepted sense because they contain resistors, transistors, diodes, etc. However, integrated circuit technology produces them in such small sizes and low costs that logic system designers treat them as basic building blocks.

Since a *properly functioning* digital system operates in the realm of arithmetic rather than differential equations, its modeling, analysis, and design do not fit the pattern of linear system dynamics and thus we do not treat digital elements per se. We might mention that when a digital system does *not* function properly, it may be due to dynamic effects which *do* require the use of differential equation models, and system dynamics methods are then again appropriate. However, these dynamic effects are usually characterized by very high frequency ranges and require specialist analysis tools not appropriate to this text. Since digital computers are now very common as *components* of computer-aided machines and processes, we will in this chapter show how, with simulation, we can model those aspects of computer behavior that influence the performance of the overall computer-aided system. These aspects have to do mainly with *sampling, quantization,* and *computational delays.* Fortunately these effects can be treated on a "black box" basis and do not require an understanding of internal computer operations. That is, just as a mechanical engineer will regularly use an oscilloscope in lab work without understanding every detail of its internal operation, just so can we use computers as components of machines and processes which we are designing without becoming experts on computer design.

Having tried to give some picture of those types of basic electrical devices which we will not treat, let's now summarize what will be covered in this chapter:

1. The three basic passive elements: resistance, capacitance, and inductance
2. Energy sources (current and voltage)
3. The active device called the operational amplifier
4. Input/output characteristics of digital computers as used in computer-aided machines and processes

While our treatment will be relatively brief, it will be adequate background for the nonelectrical engineer who works with systems that are partly electrical.

System Elements, Electrical   127

3-2 THE RESISTANCE ELEMENT

The mechanical elements were all defined in terms of their force/motion relations. The electrical elements will all be defined in terms of their voltage/current relations, and we begin with the pure and ideal *resistance element*. The math model (remember this is *not* a real resistor) for this element rigorously obeys Ohm's law, which gives the current/voltage relation as

$$i = \frac{e}{R} \tag{3-3}$$

where

$i \overset{\Delta}{=}$ current through the resistor, amperes

$e \overset{\Delta}{=}$ voltage across the resistor, volts

$R \overset{\Delta}{=}$ resistance of the resistor, ohms

The main features of this element are the strict linearity between e and i, the instantaneous response of i to e (or e to i), and the fact that all the electrical energy supplied is dissipated into heat.

Real resistors are always somewhat "impure" (they exhibit some capacitance and inductance) and nonideal (the i/e characteristic curve is not exactly a straight line). Capacitive and inductive effects make themselves known only when current and voltage are *changing* with time; thus the "impurity" of a resistor will *not* be revealed by a steady-state experiment which establishes the e/i curve by measurements with a voltmeter and ammeter. Such an experiment will, of course, reveal departures from ideal (straight-line) behavior. Many practical resistors are found to be very close to ideal (less than 1% nonlinearity), and this allows us to make resistance measurements with a rather simple instrument, the ohmmeter. In the ohmmeter, a known and fixed current (from a current source) is passed through the unknown resistor, causing a voltage drop across the resistor. This voltage is measured with a voltmeter (either analog or digital) and since (assuming Ohm's law to hold) it is directly proportional to resistance ($e = iR$), the scale of the voltmeter can be marked in *ohms*, rather than in volts, thus giving us a direct reading of resistance.

Such an ohmmeter can, of course, be connected to a *nonlinear* resistance and *will* show a resistance value in ohms. We *can't* then take this resistance value and use it in Eq. (3-3) to predict currents for various values of applied voltage e! The R value obtained is "good" *only* for the one value of current that is equal to the current that the ohmmeter used to measure the resistance. Practical ohmmeters generally have several switch-selectable ranges (300, 3000, 30,000 Ω, etc.) and usually use different currents for the different ranges. A crude check of linearity can be quickly made by measuring the resistor on several ranges. If the indicated R value is significantly different when measured on different ranges, the resistor is nonlinear.

For ideal (linear) resistors we can think of Eq. (3-3) as the *definition* of resistance R in terms of e and i:

$$\text{Resistance } R \overset{\Delta}{=} \frac{e}{i} \quad \text{ohms} \tag{3-4}$$

128   **Chapter 3**

Sometimes the reciprocal of resistance, the *conductance G*, is used:

$$\text{Conductance } G \triangleq \frac{i}{e} \quad \text{siemens (formerly mhos)} \tag{3-5}$$

Note in Fig. 3-1a that a resistor has associated with it only one current i, but two potentials e_1 and e_2 at its terminals. The current is, however, determined by the potential *difference* (voltage) $e_1 - e_2$, which we call simply e. That is, in an ideal resistor, the same current will be caused if we apply $e_1 = 10001$ volts and $e_2 = 10000$ volts, or $e_1 = 1$ volt and $e_2 = 0$ volts. (A *real* resistor, if not carefully designed for high voltage, might be *destroyed* by the first situation.) It is necessary to establish *algebraic sign conventions* for current i and voltage e and Fig. 3-1a shows the accepted form. When e_1 is greater than e_2, e is a positive number. We choose the polarity shown for positive e. If e is negative, the *actual* polarity would be reversed. The fixed $+$ and $-$ signs shown in Fig. 3-1a are thus a *sign convention* for the variable e, *not* an indication of the actual instantaneous polarity, which in dynamics problems, often changes with time.

When we are solving system dynamics problems, the voltage e would be an *unknown*. If the solution for e came out as $e = +6.2$ volts or $e = -3.7$ volts, if we had not *at the beginning of the problem* decided on a sign convention for e, we would not know what actual polarity was meant by either $+6.2$ or -3.7. Since from physics we know that a positive e in Fig. 3-1a causes a current i to flow from left to right, we *must* now take the positive direction for i ($\mapsto$) to the right. Having *chosen* the e sign convention as above, we now have *no choice* in assigning the positive direction for current; it must be as shown, or else Ohm's law would be violated. That is, Ohm's law says that a positive e causes a positive i (R is assumed to be a positive number). Summarizing, we may always choose (from the two possibilities available) either the positive direction for e or i first, but having made that choice, the other sign convention must conform to Ohm's law; that is, a positive voltage *must* cause a positive current.

The block diagrams and transfer functions of Fig. 3-1b should be self-evident from our earlier work with the mechanical elements, as should the simulation diagram of Fig. 3-1c. To determine the energy behavior we recall from physics that the instantaneous electric power supplied to a device is the product of the instantaneous current through the device and the instantaneous voltage across it. If the power has a negative sign it means that the device is *supplying* power rather than using it up. For the resistance element we have

$$\text{Power } P \triangleq ie = i(iR) = i^2 R = e\frac{e}{R} = \frac{e^2}{R} = e^2 G \tag{3-6}$$

We see that the power is always positive, irrespective of the polarity of e or the direction of i, so the resistor always takes power from the source supplying it. Since the resistor cannot return power to the source, all the power supplied is dissipated into heat. That is, the electrical power in watts is also the *heating rate* for the resistor.

If a resistor at room temperature is suddenly connected to a constant-voltage source e, it instantly starts generating heat internally at a rate e^2/R watts $= e^2/R$ N-m/sec $= e^2/R$ joules/sec $= 0.000948$ Btu/sec. [Note that in the SI unit system, *all* forms of power can be (and should be) expressed in watts, and that no conversion

System Elements, Electrical

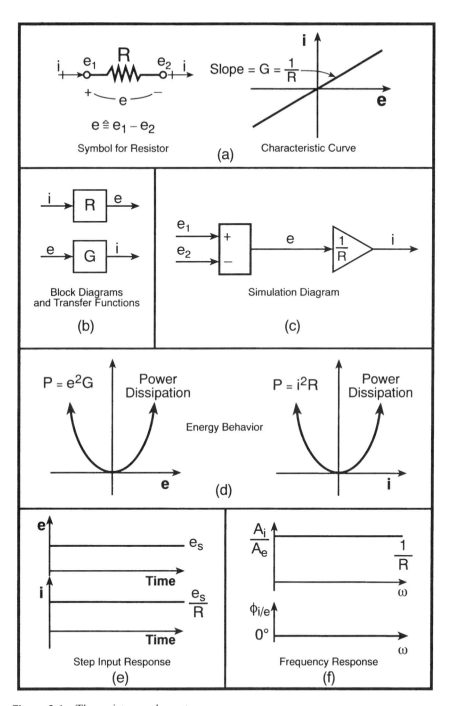

Figure 3-1 The resistance element.

factors between mechanical, electric, and thermal power are needed or desired. The British unit system uses watts, horsepower, and Btu/sec and thus requires inconvenient conversion factors in calculations.] This internal heat generation causes the temperature of the resistor to rise. As soon as the resistor temperature is higher than that of its surroundings, heat transfer by conduction, convection, and radiation causes heat to flow away from the resistor. When the resistor gets hot enough, this heat transfer rate just balances the e^2/R heat generation rate and the resistor achieves an equilibrium temperature somewhere above room temperature. In a real resistor this temperature cannot be allowed to get too high, or else the R value changes excessively or the resistor may actually burn out. Resistors used in electronic equipment are usually rated at 1 or 2 watts or less.

The instantaneous dynamic response of the pure resistance element is shown in the step and frequency-response graphs of Fig. 3-1e and f. Since $i = e/R$ is an algebraic equation, changes in e of any form whatever are *instantly* reflected in proportional changes in i. A random voltage e, for example, will produce a random i of exactly the same shape. The sinusoidal transfer function

$$\frac{i}{e}(i\omega) = \frac{1}{R} \; \angle \underline{0^\circ} \tag{3-7}$$

shows that the amplitude ratio is constant at $1/R$ for all frequencies from zero to infinity and the phase shift between e and i is zero for all frequencies. (In electrical analysis, some authors prefer to use the symbol j for the square root of -1, since i might be confused with current. Since system dynamics treats *all kinds* of systems, not just electrical, we choose to retain i for the square root of -1 since it is universally so used in mathematics.) Real resistors are always impure (contain some inductance and/or capacitance) and this prevents the instantaneous step response, the perfectly flat amplitude ratio, and the zero phase angle. Since practical systems always deal with a limited range of frequencies (*not* zero to infinity), if a real resistor behaves nearly like the pure/ideal model over its necessary range, the fact that it deviates elsewhere is of little consequence.

Resistance elements can be pure without being ideal (linear). Some examples of practically useful nonlinear resistors are vacuum tube diodes (little used today), semiconductor diodes, and the Varistor. Diodes are used for rectification of alternating current in power supplies which convert AC "from the wall plug" into DC, as needed in many electronic circuits and devices. The Varistor[1] is a semiconductor element with a symmetrical e/i relation of approximate fourth-power shape, $i \approx Ke^4$. Its uses include meter overload protection, signal limiting, and low-voltage regulation. These devices are all considered resistive since the current is an instantaneous function of the voltage. However, due to the nonlinearity of the i/e curve, there is some question as to just what the resistance might be as a numerical value.

The most correct interpretation is, of course, that the e/i relation for such devices cannot be defined by a single number; it is given by the e/i *graph*. Sometimes these graphs come from theoretical formulas and sometimes they come from laboratory measurements, and no formula is available. If such a nonlinear

---

[1]Victory Engineering Co., 1-T Victory Road, Springfield, NJ 07081, 201-379-5900.

System Elements, Electrical   **131**

resistor appears in a system we must analyze, our simulation methods (lookup table, or nonlinear function block) have no trouble modeling the e/i behavior, whether there is a formula or not. If one uses an ordinary ohmmeter, as described earlier to measure the "resistance" of such nonlinear devices, the meter *will* produce a reading of so many ohms. What does this reading mean? Recall that ohmmeters force a known and fixed current through the unknown resistance and that this current is different for the various ranges of the instrument. A typical ohmmeter might have ranges and currents as shown in the table.

Range (Ω)	Measuring current (microamperes)
300.	700.
3000.	140.
30000.	20.
300000.	2.
3,000,000.	0.40
30,000,000.	0.04

If, say, we measure the "resistance" of a certain Varistor using the 3000-Ω range and get a reading of 2478 Ω, *all* that this tells us is that when the Varistor current is 140 μA, the Varistor voltage will be $(140 \times 10^{-6})(2478.) = 0.347$ volts. We *can't* use this "R" value to predict *anything* else.

When nonlinear resistors are used with small voltage changes in the neighborhood of a fixed operating point, we can apply the same linearization technique earlier introduced with springs. This procedure allows definition of an *incremental resistance* which is useful for analyzing circuits containing nonlinear devices. This technique is shown in Fig. 3-2c for the Varistor. The incremental resistance will, of course, be different for each operating point.

$$\text{Incremental resistance} \triangleq R_{\text{inc}} \triangleq \frac{de}{di} \tag{3-8}$$

$$e \approx e_0 + R_{\text{inc}}(i - i_0)$$

3-3   THE CAPACITANCE ELEMENT

Two conductors separated by a nonconducting medium (called an insulator or dielectric) form a capacitor whose capacitance is defined by

$$C \triangleq \frac{q}{e} \quad \text{farads} \tag{3-9}$$

where

$q \triangleq$ charge on the capacitor, coulombs

$e \triangleq$ voltage across the capacitor, volts

The process of charging a capacitor consists of removing charge from one conductor and placing an equal amount on the other. The *net* charge of a capacitor is thus

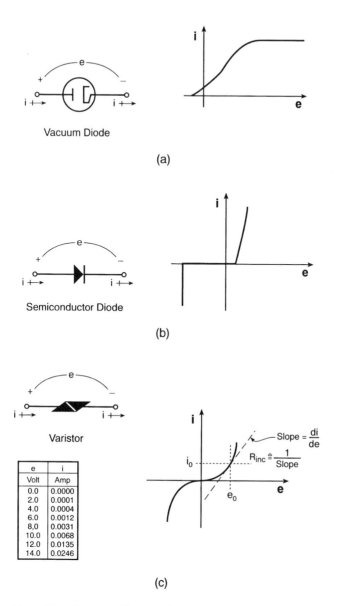

Figure 3-2 Some nonlinear resistances.

always zero and we understand the term "charge on the capacitor" to mean the magnitude of the charge on *either* conductor. Since C is defined to be a positive number, the algebraic sign of charge q is the same as that of voltage e across the capacitor. In the pure and ideal capacitance element, the numerical value of C is absolutely constant for all values of q or e. Real capacitors exhibit some nonlinearity [C as defined by Eq. (3-9) varies with q] and are "contaminated" by the presence of resistance and/or inductance; however the approximation is quite satisfactory in many cases.

System Elements, Electrical **133**

Most labs have various voltage and current measuring instruments but don't often have charge-measuring instruments so we prefer to work with voltage and current, not charge. Since current is defined in terms of charge,

$$i \triangleq \frac{dq}{dt} \tag{3-10}$$

$$e = \frac{1}{C} q$$

$$\frac{de}{dt} = \frac{1}{C} \frac{dq}{dt} = \frac{i}{C}$$

$$i = C \frac{de}{dt} \tag{3-11}$$

Using the D operator we may write

$$i = CDe \tag{3-12}$$

giving the operational transfer function for a voltage input as

$$\frac{i}{e}(D) = CD \tag{3-13}$$

Alternatively,

$$de = \frac{1}{C} i \, dt \tag{3-14}$$

$$\int_{e_0}^{e} de = \frac{1}{C} \int_0^t i \, dt$$

$$e - e_0 = \frac{1}{C} \int_0^t i \, dt \tag{3-15}$$

where e_0 is the voltage which existed across the capacitor at time equal to zero. If e_0 were zero (capacitor initially uncharged), we would have

$$e = \frac{1}{C} \int_0^t i \, dt \tag{3-16}$$

$$\frac{e}{i}(D) = \frac{1}{CD} \tag{3-17}$$

The pure and ideal capacitance element stores in its electric field all the electrical energy supplied to it during a charging process and will give up all of this energy if completely discharged, by, say, connecting it to a resistor. For example, if we apply a constant current i_s to an initially uncharged capacitor, Eq. (3-16) indicates the voltage would rise as a ramp function

$$e = \frac{1}{C} \int_0^t i_s \, dt = \frac{i_s t}{C} \tag{3-18}$$

The instantaneous power into the capacitor is $P = ei = i_s^2 t / C$; thus the total energy supplied up to time t is

134   **Chapter 3**

$$\text{Energy} = \int P \, dt = \int_0^t \frac{i_s^2 t}{C} \, dt = \frac{i_s^2 t^2}{2C} = \frac{Ce^2}{2} = \frac{q^2}{2C} \tag{3-19}$$

Actually, the energy stored by a charged capacitor is $Ce^2/2 = q^2/2C$, *irrespective* of how the final voltage e or charge q was built up; the constant current i_s used above was just an example. This can be shown by recalling from physics that the work done to transfer a charge q through a potential difference e is $e \, dq$. Since for a capacitor $e = q/C$ we have

$$\text{Total energy} = \int_0^q e \, dq = \int_0^q \frac{q}{C} \, dq = \frac{q^2}{2C} = \frac{Ce^2}{2} \tag{3-20}$$

The energy supplied to the capacitor during the charging process is all stored in the capacitor and can be recovered by connecting the charged capacitor to some energy-using device (like a resistor) and letting the capacitor discharge into it. In this process the voltage polarity remains the same as during charging, but now the current is *reversed* (giving it a minus sign) and thus the "power *into* the capacitor" is now negative, which is the same as saying that power is being taken from the capacitor. Recently, "supercapacitors"[2] having C's of about 1 farad in the size of a 1-inch cube (an "ordinary" 1-farad capacitor would need plates with an area of 1 square mile!) became feasible and are used as substitutes for batteries in certain applications.

We should note at this point that when we speak of the "current *through* a capacitor," the current does not really pass through the dielectric material between the metal plates. After all, a dielectric material is one which does *not* conduct electricity. Rather, an equal amount of charge is taken from one plate and supplied to the other by way of the circuit *external* to the capacitor. This flow of charge is, of course, a current, but it does *not* go *through* the capacitor in the same way that it would go through a resistor or inductor. However, as a matter of common usage, we will continue to speak of the current through a capacitor and rely on you to recall the true physical situation. Figure 3-3a shows the standard symbol and sign conventions for current, voltage, and charge. Again, the voltage (and charge) $+$ and $-$ sign convention can be chosen either of two ways, but once this choice is made, the direction of positive current *must* conform to Eq. (3-16); that is, a positive current *must* cause a positive change in e.

The measurement of actual capacitors to obtain a numerical value of C cannot easily be accomplished from the defining Eq. (3-9) as it could for resistors, because instruments for accurately measuring charge q are not widely available in most labs and are difficult to use. Various methods are available; they usually employ application of a sinusoidal voltage of known frequency. For such an AC signal, both current and voltage can be measured and C computed from the sinusoidal transfer function as follows:

$$\frac{i}{e} (i\omega) = i\omega C = \omega C \; \underline{/90^\circ} \tag{3-21}$$

---

[2]Supercapacitors leap battery power in a single bound, *ESD: The Electronic System Design Magazine*, July 1987, pp. 26–28. J. R. Miller and D. A. Evans, Design and Performance of High-Reliability Double-Layer Capacitors, 1990, Evans Co., P.O. Box 4158, East Providence, RI 02914-4158, 401-434-5600.

System Elements, Electrical

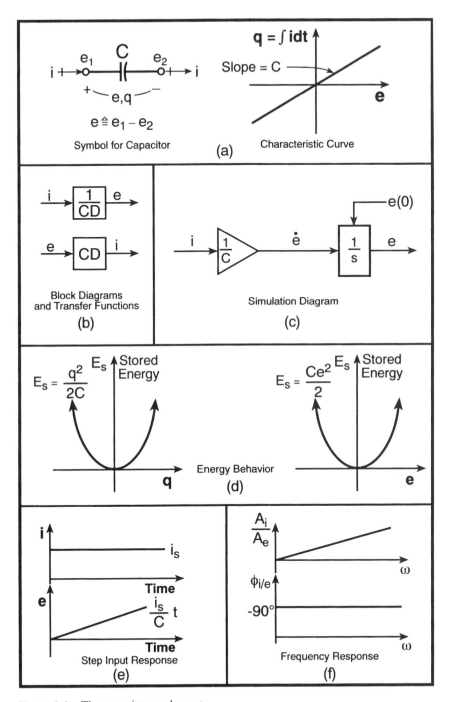

Figure 3-3 The capacitance element.

136   **Chapter 3**

Since the amplitudes I and E of the current and voltage sine waves are easily measured, and frequency ω is known, we compute C from

$$C = \frac{I}{\omega E} \tag{3-22}$$

Actually, commercial capacitance meters generally use an AC bridge method to compare the unknown capacitor with a known standard.

Figure 3-3e shows the voltage response to a step input of current, which presents no mathematical difficulties. If instead we apply a step input voltage, we get

$$i = C\frac{de}{dt} = C\frac{d}{dt} \text{ (step function)} \tag{3-23}$$

Application of the classical definition of the derivative to the step function shows that the derivative is zero everywhere but at the location of the step, where it is *undefined*. That is, the classical definition of the derivative does not allow one to compute (at the location of the discontinuity) a derivative value for *any* function which has instantaneous changes. However, a useful approach, which leads to the definition of a function which may be new to you, is shown in Fig. 3-4. There we approximate the step function as a terminated ramp, a function which *does* have a conventional derivative. By letting the rise time approach zero, we can investigate the meaning of the derivative for step functions.

As we let the ramp get steeper and steeper, we see that the magnitude of de/dt will approach infinity, and its duration will approach zero, but the area under it will always be e_s. The "function" defined by this limiting process is called the *impulse function of strength (area)* e_s. If $e_s = 1.0$ (a *unit* step function), its derivative is a *unit impulse function*, that is, its area is one unit. From Eq. (3-23) we see that a step input voltage produces a capacitor current of infinite magnitude and infinitesimal time duration. Since real physical quantities are limited to finite values, these events cannot, of course, occur in the real world. First, a true (instant rising) step voltage cannot be achieved, and secondly, a real capacitor has parasitic resistance and inductance which limit current and its rate of change. Thus a real capacitor will exhibit a short-lived (but not infinitesimal) and a large (but not infinite) current spike. We will find in later chapters that such a spike (which does have a definite area) may sometimes be treated with good accuracy as a perfect impulse of the same area.

While impulse functions may appear rather exotic and impractical, it turns out that they have many useful applications.[3] Note that they are *not* an electrical phenomenon but rather will appear whenever we try to differentiate discontinuous functions. For example, if we apply a step input of force to a spring, the velocity becomes an impulse function. The nonrigorous approach of Fig. 3-4 would not satisfy a mathematician, but it does produce a correct result. After impulse functions had been invented and used by engineers for several years, mathematicians invented a new branch of mathematics, called the theory of distributions, to put the calculations on a rigorous basis. The Laplace transform method of solving

---

[3]E. O. Doebelin, *System Modeling and Response*, Wiley, New York, 1980, chap. 3 and sec. 6.2.

System Elements, Electrical

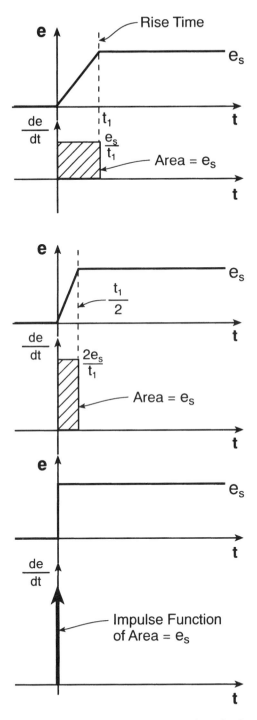

Figure 3-4 Approximate and exact impulse functions.

differential equations, which we will introduce later, also deals correctly with impulse functions.

3-4 THE INDUCTANCE ELEMENT

An electric current (motion of charge) always creates an associated magnetic field. If a coil or other circuit lies within this field, and if the field changes with time, an electromotive force (voltage) is induced in the circuit. The magnitude of the induced voltage is proportional to the rate of change of flux $d\phi/dt$ (webers/sec) linking the circuit, and its polarity is such as to oppose the cause producing it. If no ferromagnetic materials (such as iron) are present, the rate of change of flux is proportional to the rate of change of current di/dt which is producing the magnetic field. The proportionality factor relating the induced emf (voltage) to the rate of change of current is called the *inductance*. The presence of ferromagnetic materials greatly increases the strength of the effects, but also makes them significantly nonlinear, since now the flux produced by the current is not proportional to the current. Thus, iron can be used to get a large value of inductance, but the value will be different for different current levels. For small changes in current about some operating point, one can define an incremental inductance for a linearized analysis using our usual linearizing methods. Large current swings require a nonlinear treatment, and this can be easily studied using simulation methods (lookup table) if the variation of inductance with current has been measured. We will do such a nonlinear simulation shortly.

The pure inductance element has induced voltage e instantaneously related to di/dt, but the relation can be nonlinear. The pure and ideal element has e directly proportional to di/dt ($e = L\,di/dt$); that is, it is linear and free from resistance and capacitance. While it is possible to make resistors and capacitors which are very close to pure and ideal from DC (frequency = 0) to rather high frequencies, a real inductor *always* has considerable resistance. This means that at DC and low frequencies (where di/dt is zero or small), all real inductors behave like *resistors*, not inductors. At high frequencies, all real devices (R, L, and C) exhibit complex behavior involving some combination of all three pure elements. Thus real inductors deviate from the pure/ideal model at *both* low and high frequencies, whereas R and C deviate mainly at high frequencies. One can expect real inductors to nearly follow the pure model only for some intermediate range of frequencies (not including zero). If the inductance value is small enough to be achieved without the use of magnetic material, the behavior may also approximate the ideal (linear).

In applications later in this text we will be using mainly the concept of *self-inductance*; however, we want to at least make you aware of the more general situation which requires also the concept of *mutual inductance*. Self-inductance is a property of a single coil, due to the fact that the magnetic field set up by the coil current links the coil itself. Mutual inductance causes a changing current in one circuit to induce a voltage in *another* circuit. Figure 3-5 shows a configuration illustrating these concepts. Considering voltages induced into circuit A we would have

System Elements, Electrical

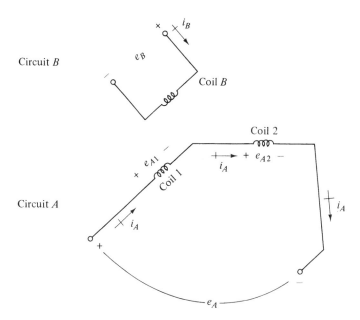

Figure 3-5 Self-inductance and mutual inductance.

$$e_A = e_{A1} + e_{A2}$$
$$= L_1 \frac{di_A}{dt} \pm M_{B/A1} \frac{di_B}{dt} \pm M_{A2/A1} \frac{di_A}{dt} + L_2 \frac{di_A}{dt} \pm M_{B/A2} \frac{di_b}{dt} \pm M_{A1/A2} \frac{di_A}{dt} \tag{3-24}$$

$$e_A = (L_1 + L_2 \pm M_{A2/A1} \pm M_{A1/A2}) \frac{di_A}{dt} + (\pm M_{B/A1} \pm M_{B/A2}) \frac{di_B}{dt} \tag{3-25}$$

where

$$L_1 \triangleq \text{self-inductance of coil 1, henries}$$
$$L_2 \triangleq \text{self-inductance of coil 2, henries}$$
$$M_{B/A1} \triangleq \text{mutual inductance of coils } B \text{ and } A_1, \text{ henries}$$
$$M_{B/A2} \triangleq \text{mutual inductance of coils } B \text{ and } A_2, \text{ henries}$$
$$M_{A2/A1} = M_{A1/A2} \triangleq \text{mutual inductance of coils 1 and 2, henries}$$

(all the e's are in volts, i's in amps, t in seconds)

Note that mutual inductance is symmetrical ($M_{A2/A1} = M_{A1/A2}$). That is, a current changing with a certain di/dt in coil 1 induces the same voltage in coil 2 as would be induced in coil 1 by the same di/dt current change in coil 2. This also holds for *separate* circuits such as A and B; the voltage induced in coil B by di_A/dt in coil 1 would be $\pm M_{A1/B}(di_A/dt)$, where $M_{A1/B} = M_{B/A1}$.

The $\pm$ signs used on the mutual inductance terms in the above equations require explanation. For any *specific* fixed physical orientation of the coils, all signs would be definitely + or definitely −, not the ambiguous ±. That is, the induced voltage in circuit A due to current change in B can either add to or subtract from the self-induced voltage in A. Since a two-dimensional circuit drawing does not show the actual geometry clearly enough, unless additional information is provided we cannot decide whether a given mutual term should be given a + or a − sign. A common convention for providing this polarity information on a drawing (once it has been reasoned out from the physical arrangement or by experimental test) is to place a dot on one end of each coil of a mutual pair. The dots are placed so that the following rule holds:

> If both the assumed positive directions for the two currents are toward the dots (or both away from the dots), then the sign of the M term will be the same as the sign of the L term. Otherwise the M term has the opposite sign of the L term.

The possibility of inductive effects opposing one another is made use of in the manufacture of some wire-wound resistors for high-frequency use, where the parasitic inductance should be minimized. Figure 3-6 shows two such methods in practical use. Note that in each case the currents are directed such that their magnetic fields tend to cancel, thus reducing inductive effects and making the resistor behave more like a pure resistance at high frequencies.

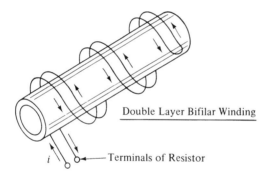

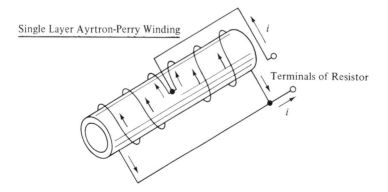

Figure 3-6   Resistors wound to minimize inductance.

System Elements, Electrical

Figure 3-7 displays the behavior of the pure and ideal self-inductance element L. The defining equation is

$$e = L\frac{di}{dt} \tag{3-26}$$

with a consistent set of sign conventions as shown in Fig. 3-7a. If we choose current positive to the right as shown, and if di/dt is positive, we would get a drop in voltage

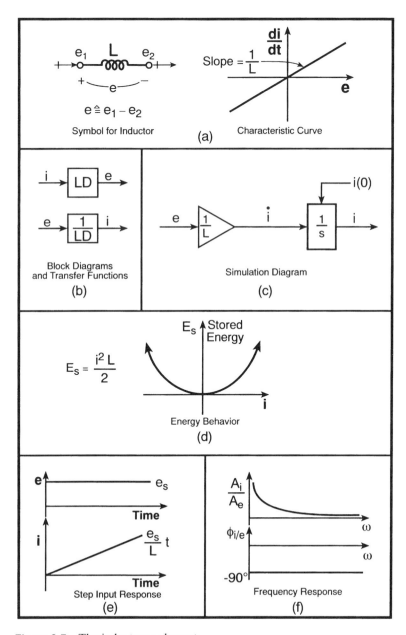

Figure 3-7 The inductance element.

from e_1 to e_2; thus we must take the sign convention on e as shown to conform to Eq. (3-26). That is, L is defined to be a positive number and Eq. (3-26) says that a positive di/dt corresponds to a positive e. If we apply a constant voltage e_s to the inductance element, we cause the current to increase at a constant rate $di/dt = e_s/L$. Since this current and voltage are both positive, the inductor is absorbing energy at a rate $e_s i$. Let us assume that the inductor carried zero current at time zero when the voltage was applied. Then

$$\text{Power} = P = e_s i = e_s \left(\frac{e_s}{L} t \right) \tag{3-27}$$

$$\text{Energy} = \int_0^t P \, dt = \int_0^t \frac{e_s^2}{L} t \, dt = \frac{e_s^2 t^2}{2L} = \frac{i^2 L}{2} \tag{3-28}$$

Thus at an instant t when the current is i, the inductor has received energy in the amount $i^2 L/2$. This energy is actually stored in the magnetic field since we find that if we connect a current-carrying inductor to an energy-using device (a resistor, for example) the inductor will *supply* energy in an amount $i^2 L/2$ as its current decays from i to zero. During this decay process, i (if originally positive) stays positive, but di/dt (and thus e) becomes negative, making power negative and thus showing that the inductor is supplying power to the external circuit. The energy storage $i^2 L/2$ is correct irrespective of how the current i was achieved, as we can see from

$$\text{Power} = ei = L \frac{di}{dt} i \tag{3-29}$$

$$\text{Energy} = \int_0^t iL \frac{di}{dt} \, dt = \int_0^i Li \, di = \frac{i^2 L}{2} \tag{3-30}$$

We have seen that a step input of voltage e_s causes a ramp current $i = e_s t/L$. A step input of current i_s gives rise to a voltage impulse function of size (area) Li_s. The frequency response of the inductance element is obtained from the sinusoidal transfer function as follows:

$$e = L \frac{di}{dt} = L\,Di \qquad \frac{i}{e}(D) = \frac{1}{LD} \tag{3-31}$$

$$\frac{i}{e}(i\omega) = \frac{1}{i\omega L} = \frac{1}{\omega L}(-i) = \frac{1}{\omega L}\ \angle{-90°} = \frac{A_i}{A_e}\ \angle{\phi_{i/e}} \tag{3-32}$$

Note that at very low frequencies ($\omega \to 0$) a small voltage amplitude can produce a very large (approaching ∞ as $\omega \to 0$) current. Thus an inductance is sometimes said to approach a *short circuit* for low frequencies, and under such conditions could be replaced in a circuit diagram with just a piece of "connecting wire." [Recall that in circuit diagrams one shows R's, C's, and L's connected by pieces of perfectly conducting (no voltage drop) "wire."] At high frequencies ($\omega \to \infty$) note that the current produced by any finite voltage approaches zero. Thus we often say that an inductor approaches an *open circuit* at high frequencies, and could thus just be "cut out" of a circuit diagram under such conditions. For a capacitance, just the reverse frequency behavior was observed; the capacitance approaches a short circuit at high frequencies and an open circuit at low frequencies. One can often use these

System Elements, Electrical

simple rules to quickly estimate the behavior of complex circuits at low and high frequency. Just replace the L's and C's by open or short circuits, depending on which frequency range you are interested in. Remember for *real* circuits, however, that real L's *always* become R's for low frequency.

Large inductance values (more than a few tenths of henries) usually require use of magnetic materials, which make the inductor nonlinear. *All* real inductors also have significant resistance. We want to do a simulation example to show how we deal with both of these realities. An actual commercial inductor using magnetic materials to get a very large L value might occupy about a 0.5-inch cube of space, and have $L = 60.$ henries and $R = 5000.$ ohms. Such inductors wind a copper wire coil of many turns around the magnetic material. The 5000-ohm resistance is just the resistance of the copper wire.

The circuit model used for low and intermediate frequencies is shown in Fig. 3-8a. The L value of 60 actually refers to the behavior for currents small enough (less than about 0.0002 amp) that the iron saturation is negligible. To model the nonlinear behavior for larger currents, we need data on the variation of L with current i. Suppose this has been measured, with the results of Fig. 3-8b. Note that the inductance goes toward zero for large currents. These large currents cause the magnetic

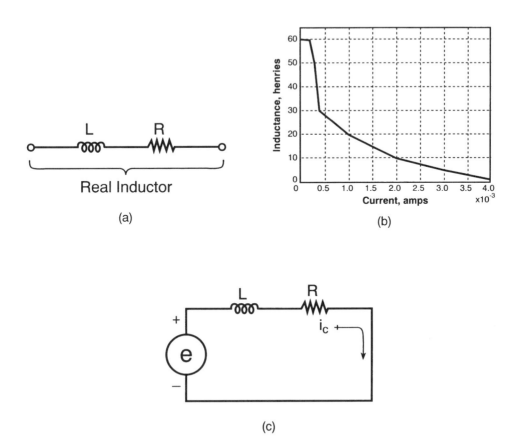

Figure 3-8 Real, nonlinear inductor.

material to *saturate*; that is, increasing the current does not now increase the flux. Since inductance depends on a *changing* flux, L goes to zero as saturation increases. Recall from physics that magnetic materials are thought of as having many tiny magnetic "domains," like tiny bar magnets. Coil current creates a magnetic field which tends to align the domains with the applied magnetic field. When the current is strong enough, all the domains have been exactly aligned with the field and further increases in current can do no more to align the domains; we have reached saturation.

We want to study the behavior of this inductor in the simple circuit of Fig. 3-8c. Remember that the "R" shown is *not* a separate resistor but just the resistance of the inductor itself. We will apply a driving voltage e_i to the inductor and solve for the current i. We could use various forms of driving voltage but a *square wave* is particularly useful and simple. By varying the frequency of this square wave we can study inductor behavior for a range of frequencies. I also want to compare the behavior of this real inductor with that of a pure and ideal 60-henry inductor, and also a linear inductor with $L = 60$. henries and $R = 5000$. ohms. We can simulate all three of these devices in one simulation diagram and run them simultaneously for easy comparison.

To start our simulation diagram, as usual, we must first set up the appropriate equation. Applying Kirchhoff's voltage loop law to our simple circuit we get

$$\sum \text{voltage drops around any complete loop} = \text{zero at every instant of time}$$

$$(3\text{-}33)$$

$$L\frac{di}{dt} + Ri - e_i = 0$$

$$\frac{di}{dt} = \frac{1}{L}(e_i - Ri) \tag{3-34}$$

We can use this equation for all three of our inductors. For the pure/ideal inductor we take $R = 0$ and L constant at 60. For the linear "impure" inductor we take $R = 5000$ and L constant at 60. For the impure and nonlinear inductor we take $R = 5000$ and use a lookup table based on Fig. 3-8b to implement the variable L value. Figure 3-9 shows the simulation diagram for all three inductors. The block labeled Signal Gen. 1 is the SIMULINK icon for a signal generator. This block can be set up to produce periodic waves of sinusoidal, triangular, or square waveform, with an amplitude and frequency of our choice.

We will exercise our simulation with four runs:

1. Low frequency (10 Hz), low amplitude (1 volt)
2. Low frequency, high amplitude (10 volts)
3. High frequency (1000 Hz), low amplitude
4. High frequency, high amplitude

This kind of exploration of ranges of amplitude and frequency is a tool of *general* utility, both in simulation and actual laboratory testing of all kinds of dynamic systems, not just the inductor now under study. It can be done with various waveforms—sinusoidal, square, triangular, etc.—depending on the goals of the particular test.

System Elements, Electrical

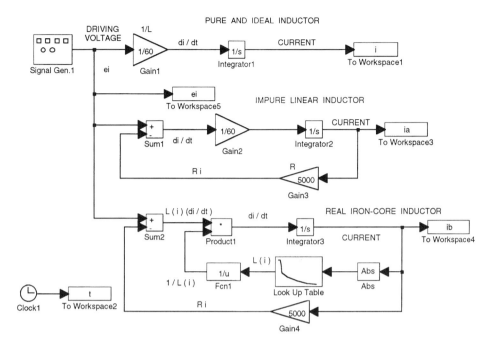

Figure 3-9 Simulation model for comparison of ideal, impure linear, and impure nonlinear inductors.

Figure 3-10a shows the response at low frequency and low amplitude. Note that the pure/ideal response is particularly simple since i is proportional to the integral of e, giving a negative-going ramp function. (The SIMULINK square-wave generator happens to start the square wave of e in the negative portion of its cycle. If you don't like this, it is easy to multiply its signal by -1 in the simulation diagram.) We in fact *chose* to use a square wave because the ideal response is a straight line, allowing easy comparison with the other inductor models. It is clear from the graphs that the pure/ideal inductor model is very poor under these circumstances; the curves for ia and ib are very different from the ideal i. Currents ia and ib are essentially identical (curves overlap) because the current never exceeds 0.0002 amp, which is below the saturation level. In Fig. 3-10b, the 10-volt driving voltage now forces current to get as large as 0.002 amp, and the effects of saturation show up in the difference between ia and ib. Both, however, are still much different from i.

In Fig. 3-11a we are now at 1000 Hz and 1 volt and all three models give very similar response. Even when we increase the amplitude to 10 volts (Fig. 3-11b), we get almost perfect agreement. This last result is less surprising once you note that the current never exceeds 0.00008 amp, well below the saturation level. What has happened is that at 1000 Hz most of the applied voltage appears across the inductor (di/dt is much larger than at 10 Hz) rather than the resistor. (Our graphs show only the first cycle of the square wave. If you have access to simulation software, you might want to run this model for longer times to reveal some more useful information.)

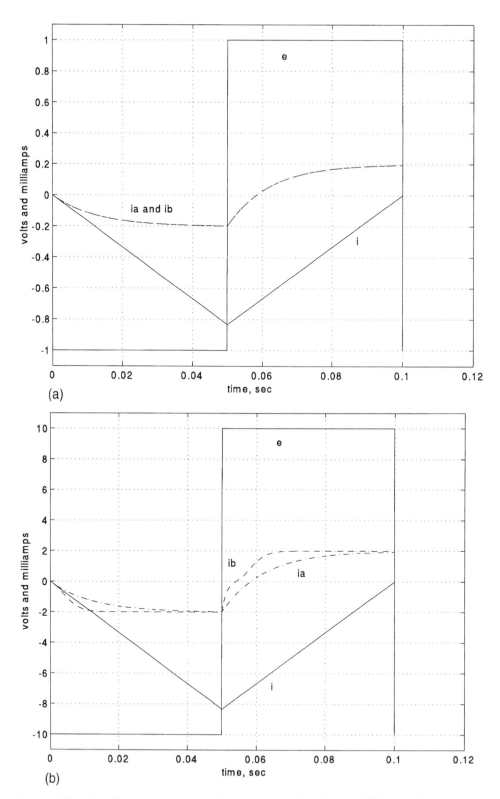

Figure 3-10 Low-frequency response of inductor models at low and high amplitude.

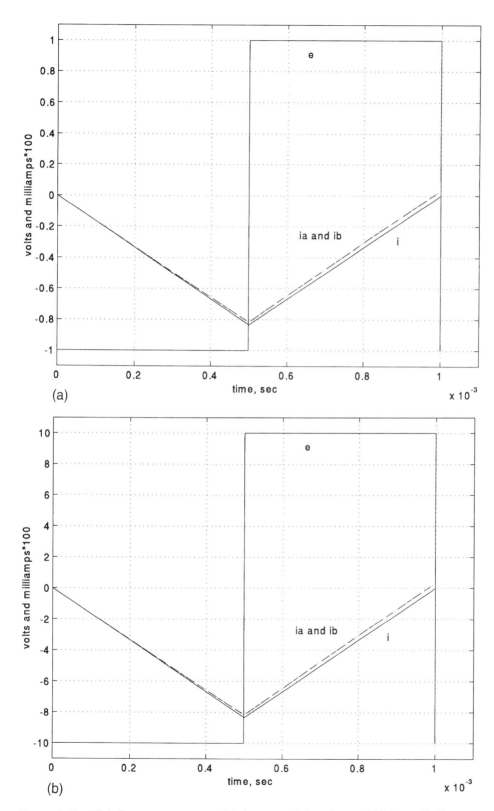

Figure 3-11 High-frequency response of inductor models at low and high amplitude.

148   **Chapter 3**

If we ran our simulation at *very* high frequencies we would find that the pure/ideal model got better and better. This result would *not* be correct because real inductors are *not* accurately modeled as in Fig. 3-8a for such high frequencies. A more complex model and additional numerical data would be needed to treat an inductor under such conditions.

3-5 ELECTRICAL IMPEDANCE AND ELECTROMECHANICAL ANALOGIES

Electrical impedance is a generalization of the simple voltage/current relation called resistance for resistors, so that it can be applied to capacitors, inductors, and in fact entire circuits. Its definition assumes ideal (linear) behavior of the device under study. That is, the current magnitude is directly proportional to the voltage magnitude. As usual, this linear concept is also used for linearized analysis of nonlinear devices for small signals in the neighborhood of an operating point. Electrical impedance is defined as the transfer function relating voltage and current, and as with most transfer functions, is useful in three forms: operational, sinusoidal, and Laplace. These three forms are defined as follows:

$$Z(D) \triangleq \frac{e}{i}(D) \tag{3-35}$$

$$Z(i\omega) \triangleq \frac{e}{i}(i\omega) \tag{3-36}$$

$$Z(s) \triangleq \frac{e}{i}(s) \tag{3-37}$$

Since all these impedances are ratios of voltage over current, their units are *always* taken as ohms. The simplest impedances are those of the pure/ideal elements:

$$Z_R(D) = R \qquad Z_R(i\omega) = R \tag{3-38}$$

$$Z_C(D) = \frac{1}{CD} \qquad Z_C(i\omega) = \frac{1}{i\omega C} \tag{3-39}$$

$$Z_L(D) = LD \qquad Z_L(i\omega) = i\omega L \tag{3-40}$$

The sinusoidal impedances of R, C, and L are graphed in Fig. 3-12a. From the magnitude (amplitude ratio) curves for C and L we can again see the truth of the "simplification rules" given earlier: L acts like a short circuit (impedance approaches zero) at low frequencies and an open circuit (impedance approaches infinity) at high frequencies, while C behaves just the opposite. Recall from Chap. 2 the *mechanical impedances* of damper, spring, and mass:

$$\text{Damper: } Z_B(D) \triangleq \frac{f}{v}(D) = B \qquad Z_B(i\omega) = B \tag{3-41}$$

$$\text{Spring: } Z_s(D) \triangleq \frac{f}{v}(D) = \frac{1}{C_s D} \qquad Z_s(i\omega) = \frac{1}{i\omega C_s} \tag{3-42}$$

System Elements, Electrical

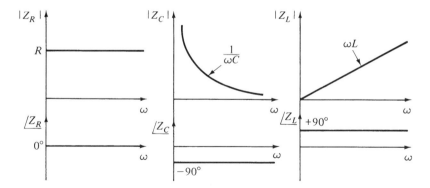

Impedances of Electrical Elements

(a)

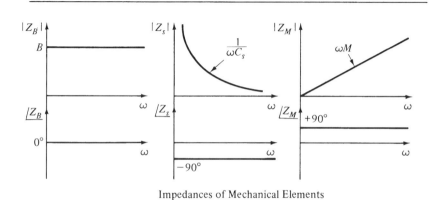

Impedances of Mechanical Elements

(b)

Figure 3-12 Analogous behavior of electrical and mechanical elements: impedance characteristics.

$$\text{Inertia:} \quad Z_M(D) \triangleq \frac{f}{v}(D) = MD \qquad Z_M(i\omega) = i\omega M \qquad (3\text{-}43)$$

These formulas and the graphs of Fig. 3-12b show a striking similarity and establish an *analogy* between electrical and mechanical elements and systems.

A signal, element, or system which exhibits mathematical behavior identical to that of another, but physically different, signal, element, or system is called an analogous quantity or *analog*. Based on the above equations we may thus state that:

Force is a mechanical analog of *voltage*.
Velocity is a mechanical analog of *current*.
A *damper* is a mechanical analog of a *resistor*.
A *spring* is a mechanical analog of a *capacitor*.
A *mass* is a mechanical analog of an *inductor*.

Note that force *causes* velocity, just as voltage *causes* current. A damper dissipates mechanical energy into heat, just as a resistor dissipates electrical energy into heat.

Springs and masses *store* energy in two different ways, just as capacitors and inductors store energy in two different ways. The product $(f)(v)$ represents instantaneous mechanical power, just as $(e)(i)$ represents instantaneous electrical power. We should point out that the above analogy is not the only one that can be established between electrical and mechanical systems,[4] but it is the one I have found the most useful.

Familiarity with analogies may be helpful to an engineer in various ways. For a mechanical engineer working on electromechanical systems, the analogies may make the electrical aspects of the system seem more familiar and understable. Often, already-available mathematical solutions for problems in one field may be directly applied to the analogous problem in the other field, saving the time needed to solve the problem "from scratch." Operating principles of successful hardware in one area may be carried over to "invent" an analogous or similar useful device in another area. It is often possible to reduce an electromechanical system to an all-electric or all-mechanical model using analogies, and electrical engineers sometimes prefer to study systems which are *entirely* mechanical using electrical models. This practice is somewhat a matter of personal preference and may in some cases be the best way to proceed, but I do *not* recommend it as a general rule.

Many years ago, R, C, L electrical analogs of mechanical systems were actually constructed and used in analysis because their assembly and testing was quicker and cheaper than that of the real mechanical system. Later, electronic analog computers were used to "solve" the differential equations of mechanical and other dynamic systems. In these computers the above analogies were *not* used; rather the operations needed to "solve" the equations (integration, summing, coefficient multiplying, etc.) were implemented with op-amp electronics. Both these "analog" approaches are today largely obsolete. We instead "solve" the physical system's equations, as originally derived from the physical laws governing the actual system (not some analog), using convenient simulation software, such as the SIMULINK already used in this text.

My personal preference, and I encourage others to adopt it, is to model systems *directly* rather than to try to force "mixed-media" systems into, say, an all-mechanical or all-electrical form. Real systems, as opposed to overly idealized versions of them, often include significant parasitic and/or nonlinear effects which may not even have a useful analog. When we get to the laboratory testing stage of design, measurements usually relate most closely to the *actual* phenomena, not some assumed analog. At least in medium-sized and larger companies, engineers often work in teams, so the electrical engineer on the team can "educate" the mechanical engineer about some unfamiliar or obscure electrical detail and the mechanical engineer can do likewise for mechanical subtleties. Also, the "boundaries" between the classical disciplines, as seen in the separate academic departments, often are much more blurred in industrial practice, where problems rarely can be neatly compartmentalized into electrical or mechanical specialties. While we may have a degree in a certain specialty, we need to always be willing to continue our education by learning what the job requires, even if it seems to be

---

[4]H. A. Rothbart, ed., *Mechanical Design and Systems Handbook*, McGraw-Hill, New York, 1964, p. 6-33.

System Elements, Electrical

"outside our field." Such broad expertise and flexibility is also much valued in the job market.

Returning to the topic of electrical impedance, we should point out that it can be and is used not just for elements but for complete systems of arbitrary size. In Fig. 3-13a, for example, one can derive a formula for the impedance $(e/i)(D)$ by proper application of physical laws. Sometimes devices are too complex to analyze accurately by theory, or we may wish to check a theory by lab measurements on the actual system. Such measurements are so common that special electronic instruments called *RCL* meters or impedance analyzers are available from several manufacturers.[5] These instruments usually use sinusoidal test signals, so the sinusoidal impedance is what is measured. Figure 3-13b[6] shows measurements made on two inductors of similar nominal value (about 3.5 mH) but different physical construction. Below 0.5 MHz (not shown in the graph), both units behave close to the pure/ideal model, as we can see by the phase angle of about +90° and the amplitude ratio decreasing with frequency. (Of course, at very low frequency, both units would behave like resistors. Also, because of the log-log graph axes the *shapes* of curves will differ from those shown in Fig. 3-12a.)

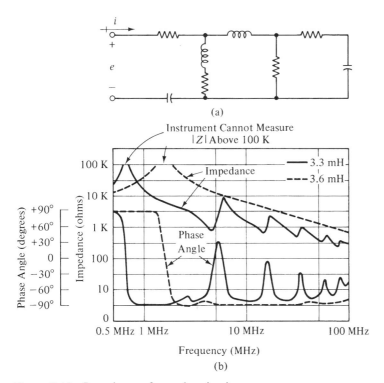

Figure 3-13 Impedance of complex circuits.

---

[5]M. Honda, The Impedance Measurement Handbook, Hewlett-Packard Corp., 1994, tel. 800-452-4844.
[6]Methods of measuring impedance, *Hewlett-Packard Journal*, vol. 18, no. 5, January 1967.

Note that the one unit retains the "perfect" behavior to about 1.5 MHz while the other starts to "go bad" at about 0.5 MHz. One of the main uses of impedance testing is to reveal such differences between supposedly identical components. Beyond the frequency range of inductorlike behavior, the two units show very different responses. Theoretical modeling of such complicated effects would be difficult and inaccurate, while measurements, assuming appropriate instruments are available, are quick and easy. Such impedance measurements are useful not only for the selection and application of components from commercially available stock, but may also be used as quality control checks in the *manufacture* of components. Impedance measurements can be automated to provide 100% testing of manufactured components as they come off the assembly line, or perhaps "spot checks" can be statistically scheduled. Such information may be used not just to reject bad devices but to *diagnose* where in the manufacturing process the fault originates. Instruments for measuring *mechanical* impedance of products are also available[7] and provide similar functions in quality control and process improvement.

While impedance is useful in characterizing the dynamic behavior of components and systems, it also finds application in the solution of routine circuit problems. At any given frequency the sinusoidal impedance of any circuit is $M \angle \phi$ and can be given as a real part R and an imaginary part X:

$$Z = \frac{A_e}{A_i} \angle \phi = M \angle \phi = M\cos\phi + iM\sin\phi = R + iX \qquad (3\text{-}44)$$

where

$R \triangleq$ resistive impedance, ohms

$X \triangleq$ reactive impedance, ohms

To prevent possible misinterpretation, I personally would prefer to here use the symbols Z_R and Z_X instead of R and X, because here R is *not* just the value of some resistor. For example, for the circuit of Fig. 3-14 we have

$$i = i_R + i_L = \frac{e}{Z_R} + \frac{e}{Z_L} = \frac{e}{R} + \frac{e}{LD} = e\left(\frac{1}{R} + \frac{1}{LD}\right) \qquad (3\text{-}45)$$

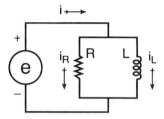

Figure 3-14 Circuit for impedance example.

---

[7]Doebelin, *System Modeling and Response*, chap. 7. R. Plunkett, ed., *Mechanical Impedance Methods for Mechanical Vibrations*, ASME, New York, 1958. Effective Machinery Measurements Using Dynamic Signal Analyzers, Hewlett-Packard AN 243-1, 1996.

System Elements, Electrical   **153**

$$Z(D) = \frac{e}{i}(D) = \frac{RLD}{R + LD}$$

$$Z(i\omega) = \frac{i\omega RL}{R + i\omega L} = \frac{\omega^2 RL^2}{R^2 + \omega^2 L} + i\frac{\omega LR^2}{R^2 + \omega^2 L} \qquad (3\text{-}46)$$

Note in Eq. (3-46) that the real part contains not only resistance elements but also inductance and the frequency, which is why I would prefer to give it a symbol other than "R." However, the "R and X" symbology is well established, so we will live with it, especially since the example has now made clear the true situation.

If X is a positive number, the reactive impedance is "behaving like an inductor" and is called *inductive reactance*; if negative, it is called *capacitive reactance*. Given R and X, one can always compute the magnitude $M = (R^2 + X^2)^{0.5}$ and the phase angle $\phi = tan^{-1}(X/R)$ of the impedance. Since sinusoidal impedance gives the amplitude ratio and phase angle of voltage with respect to current, if the impedance of any circuit (no matter how complex) is known (from either theory or measurement), and either voltage or current is given, we can quickly calculate the other.

Impedance Example. Suppose we apply a 60-Hz sinusoidal voltage of amplitude 120 volts, and zero phase angle to a 0.1-µF capacitor. What will be the sinusoidal current?

$$Z_C(i\omega) = \frac{1}{i\omega C} = \frac{1}{(i)(377)(0.1 \times 10^{-6})} = 26{,}525 \underline{/-90°} \text{ ohms} \qquad (3\text{-}47)$$

$$\text{Current} = \frac{\text{voltage}}{Z} = \frac{120 \underline{/0°}}{26{,}525 \underline{/-90°}} = 0.004524 \underline{/+90°} \text{ amps} \qquad (3\text{-}48)$$

This current could be written as $i = 0.004524 \sin(377t + 90°)$ amps. Suppose next that we have measured the impedance at 60 Hz of a "black box" containing an unknown linear circuit and found it to be $52.4 \underline{/+62°}$ ohms, and that we apply the same voltage to it as above:

$$\text{Current} = \frac{120 \underline{/0°}}{52.4 \underline{/62°}} = 2.290 \underline{/-62°} \text{ amps} \qquad (3\text{-}49)$$

This current could be written as $2.290 \sin(377t - 62°)$ amps. Note that we usually display the phase angle in degrees, but if you are calculating an instantaneous value from $\sin(377t - \phi)$ you *must* use ϕ in *radians* since the frequency 377 is in rad/sec. This example shows how the impedance concept makes sinusoidal response calculations very quick and easy.

3-6 REAL RESISTORS, CAPACITORS, AND INDUCTORS

Just as we saw in mechanical elements, electrical elements are sometimes intentionally designed into a system and other times appear "naturally" (perhaps undesirably) as part of some device. In motion control systems (servomechanisms) which use DC electric motors, for example, the inductance of the motor field must sometimes be

154   **Chapter 3**

included in the model even though it slows the response and is thus undesirable. The inductance is not something that the designer "wired into" the system; it is simply an effect which is present in motor fields that must be taken into account. The cables and wires used to interconnect electrical components are (ideally) perfect conductors devoid of all resistance, capacitance, and inductance; however, at high frequencies they exhibit all three properties and must be so modeled to properly predict system behavior. Transistors are intended as instantaneous power modulators; however, their construction is such that it unavoidably includes parasitic capacitance effects, which again must be modeled for certain types of applications. When circuit elements must be modeled but are unintentional, theory rarely supplies sufficiently accurate numerical values, and measurements, usually using specialized impedance-measuring instruments, become necessary.

Turning now to "intentional" circuit elements, let's begin with resistors. While a pure and ideal resistance element is completely described by a single number R, the practical choice of a resistor involves many complex factors such as:[8]

1. Physical size
2. Wattage rating
3. Stability
4. Shelf life
5. Load life
6. Reliability
7. Frequency range
8. Electrical noise
9. Temperature coefficient of R
10. Voltage coefficient
11. Solderability or weldability
12. Manufacturing tolerance
13. Maximum temperature
14. Shock and vibration tolerance
15. Humidity tolerance
16. Maximum voltage

A number of manufacturing processes[9] are used to fabricate commercial resistors; we here describe only three of these. The most obvious method is that of winding a coil of fine wire on an insulating form, the so-called *wirewound* resistor, Fig. 3-15a. The resistance R of a length L of wire of cross-sectional area A is given by

$$R = \frac{\rho L}{A} \tag{3-50}$$

where $\rho \overset{\Delta}{=}$ material resistivity, ohm-meters. For commercial annealed copper $10^8 \rho = 1.72$, for Constantan (Cu 60, Ni 40) 49.0, for Nichrome 100., and for Manganin (Cu 84, Mn 12, Ni 4) 44.0, all at room temperature. Constantan and Manganin are used when resistance must change little with temperature;

---

[8]G. W. A. Dummer, *Modern Electronic Components*, Sir Isaac Pitman & Sons Ltd., London, 1966, p. 57.
[9]Ibid.

System Elements, Electrical

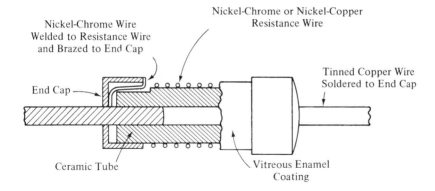

(a) Wirewound Resistor

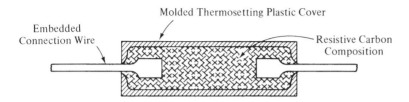

(b) Carbon Composition Resistor

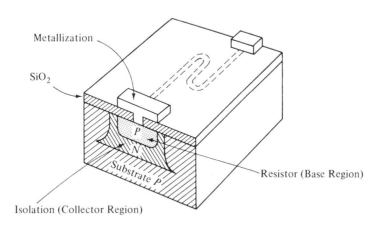

(c) Diffused Microcircuit Resistor

Figure 3-15 Discrete-component and integrated-circuit resistors.

Nichrome is much used to fabricate electrical heaters. The *carbon-composition* resistor (Fig. 3-15b) is molded from a powder made up of carbon black, a resin binder, and a refractory filling. Lead wires may be attached by several methods; Fig. 3-15b, shows wires with enlarged ends molded directly into the carbon rod. In *monolithic integrated circuits*,[10] the tiny resistors needed are produced from semiconductor materials by a diffusion process at the same time that the transistors, diodes, and capacitors are formed (see Fig. 3-15c). An extremely thin and very narrow strip of doped material is diffused into a substrate of oppositely doped silicon. A 100,000-ohm resistor can be put onto a square chip 0.020 inch on a side and 0.006 inch thick.

Resistors vary in size from the microscopic integrated circuit units just mentioned, which can dissipate a few millliwatts of power, to wirewound power resistors rated at 300 watts and measuring 1 by 10 inches. *Stability* of a resistor refers to constancy of the R value under shelf life or working conditions. Precision, hermetically sealed wirewound resistors may have a stability as good as 0.01%. Carbon composition stability is the order of 5% for ordinary conditions, but may be as bad as 25% for severe environments. The *manufacturing tolerance* ranges from 20% for some carbon composition to 0.01% for precision wirewound. *Maximum allowable temperatures* go from 70 to 600°C, while *maximum voltages* go up to about 3000 volts.

The *useful frequency range* of a resistor is determined by the magnitude of the parasitic capacitance and/or inductance effects associated with its construction. Figure 3-16 shows typical behavior for precision metal and carbon film resistors.[11] Note that the model for the resistor includes a parallel capacitance representing an unavoidable parasitic effect. The "resistance" plotted in this figure is actually the resistive impedance of Eq. (3-44). The trend that high-value resistors deviate from the pure resistance model at lower frequencies than low-value resistors is typical of most types of construction. Wirewound resistors may exhibit sufficient inductance to require its inclusion in the device model. Figure 3-17a shows the model employed in such cases. For R values of about 100 ohms or less, the inductance predominates,

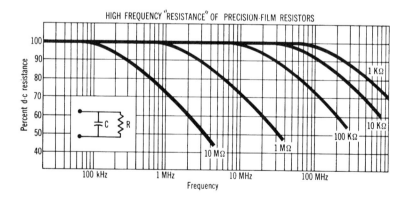

Figure 3-16 High-frequency behavior of film resistors.

---

[10]L. Stern, *Fundamentals of Integrated Circuits*, Hayden Book Co., New York, 1968, p. 73.
[11]Texas Instruments, Brochure 3-66(20M), 1954.

System Elements, Electrical

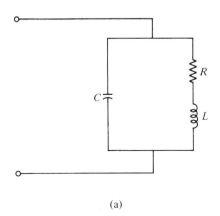

(a)

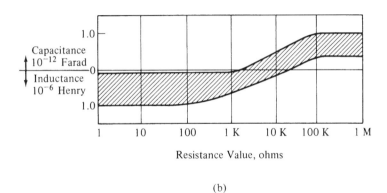

(b)

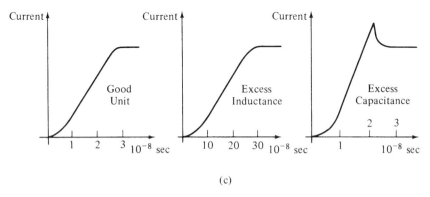

(c)

Figure 3-17 Dynamic behavior of "high-speed" wirewound resistors.

with a gradual shift to capacitive dominance at higher resistance values. Typical ranges of L and C are shown in Fig. 3-17b, and Fig. 3-17c shows step responses for a class of special "high-speed" resistors.[12] Since the resistor mounting method influences the dynamic behavior, tests should be run with mounting methods similar to those expected in the actual application.

Electrical noise refers to random voltages caused internally in the resistor by thermal agitation (Johnson noise) and the so-called current noise. Johnson noise is present in all resistors and contains a very wide range of frequency content. At room temperature a 1-megohm resistor will exhibit about $100\,\mu V$ of noise voltage over a 1-Mhz bandwidth of frequency. Current noise is peculiar to carbon composition resistors and other nonmetallic films, and increases with current through the resistor. It can be as great as several thousand microvolts. The main interest in resistor noise arises when the resistor is used in amplifying circuits for small voltage signals. The presence of noise puts a limit on the smallest signals which can be detected and amplified. The magnitude of noise effects often is much reduced at low temperatures, so some sensitive instruments include *coolers* (liquid nitrogen or helium, thermoelectric, or small mechanical refrigerators) to lower the temperature of sensitive elements.

Temperature coefficient of resistance refers to the fact that all resistance materials change resistance with temperature. For small temperature changes near an operating point, one can linearize the effect and quote a temperature coefficient of resistance as dR/dT. Typical values are: carbon composition $\pm 0.12\%$ per C°, carbon film -0.02 to 0.1%, precision wirewound $+0.002\%$. The variation of resistance with temperature, undesirable in most resistors, is put to *good* use in various temperature *measuring* devices, such as resistance temperature detectors (copper, platinum, and nickel) and thermistors (semiconductors). *Voltage coefficient* is most important for carbon composition resistors and refers to an immediate change in resistance (usually a decrease) following application of a DC voltage. This effect is distinct from the temperature coefficient and can be as much as 0.02% per volt. Wirewound resistors do not show this effect and carbon film have it only to the extent of 0.002% or less.

While resistors themselves have quite good *reliability*, due to the large numbers used in typical electronic equipment they usually are the greatest contributors to failure. Typically, for each transistor, one finds 5 to 10 resistors. The average failure rate for *all* components ranges from 0.0004% per 1000 hours in undersea cable amplifiers (which must be very reliable) to about 2% per 1000 hours in commercial radio and television service. This wide range is due to variations both in component quality and in the severity of the use environment. For resistors used in general-purpose ground-based electronics, average failure rates[13] range from 0.01% per 1000 hours for oxide film types to 0.2% for wirewound. Soldered connections, the second greatest contributor to failure, have a rate of 0.01, whereas transistors range from

---

[12]RCL Electronics, Inc., Catalog 678A.
[13]Dummer, *Modern Electronic Components*, p. 456.

System Elements, Electrical   **159**

0.01 to 0.1. All the failure rates quoted in this section are intended mainly to raise your consciousness about the general problem of failure in engineered systems, and to give a rough idea of typical rates, *not* to supply you with working numbers. Useful working numbers can be obtained only for *specific* devices and manufacturers at the time a design is actually carried out. Even then, they are very dependent on the manufacturer's ongoing *quality control*.

Turning now to capacitors, Fig. 3-18 shows formulas for computing capacitance of some common configurations. Most intentional capacitors are based on the parallel flat plate arrangement, for which

$$C = \frac{\epsilon A}{d} \tag{3-51}$$

where

$\epsilon \triangleq$ permittivity of the dielectric material, F/m

$A \triangleq$ area of plates, meter^2

$d \triangleq$ distance between plates, meters

For a vacuum between the plates, the permittivity is the smallest possible, 8.85×10^{-12}, while for dry air at atmospheric pressure it is 8.85×10^{-12} K, where $K = 1.00059$. The constant K of a dielectric material is called the *dielectric coefficient*, and is simply the ratio of the material's permittivity to that of a vacuum. For a spacing of 1 mm, a 1-farad flat plate air capacitor would be a square 6.5 *miles* on a side, showing that the farad is usually an inconvenient unit of measure. However, as we mentioned earlier, capacitors of the order of 1 farad which are easily held in one hand are commercially available.[14] For most applications the microfarad and picofarad are more convenient, with a 1-μF capacitor (other than the electrolytic type) being considered quite large.

Common dielectrics include impregnated paper ($K = 4$ to 6), glass and mica (4 to 7), polystyrene (2.3), Mylar (2 to 5), and ceramics (6 to 3000). Permittivity for electrolytic capacitors is not generally quoted, because the dielectric film is somewhat indefinite in thickness; however, a K of about 3 appears to be the right order of magnitude. Electrolytic capacitors achieve very large C values in small space because the dielectric film is much thinner ($d \approx 10^{-5}$ cm) than it is practical to make a sheet of paper or plastic film (about 10^{-3} cm minimum). Capacitors may be constructed in several ways, using various materials.[15] One basic method is shown in Fig. 3-19a, where 0.00025-inch aluminum foil (the electrodes) is sandwiched between 0.0005-inch oil-impregnated paper or plastic film (the dielectric), rolled up into a cylinder and provided with terminals and an outer protective case. Metallized paper, a variation on this technique, deposits or sprays a thin metal film onto the paper dielectric to reduce the volume of the finished capacitor.

---

[14]The measurement of electrolytic capacitors, *The Experimenter*, vol. 40, no. 6, June 1966, General Radio Corp. Effective Electrolytic Capacitor Testing, Hewlett-Packard Appl. Note AN-1124-4, 1996.

[15]Dummer, *Modern Electronic Components*, Chaps. 7, 8.

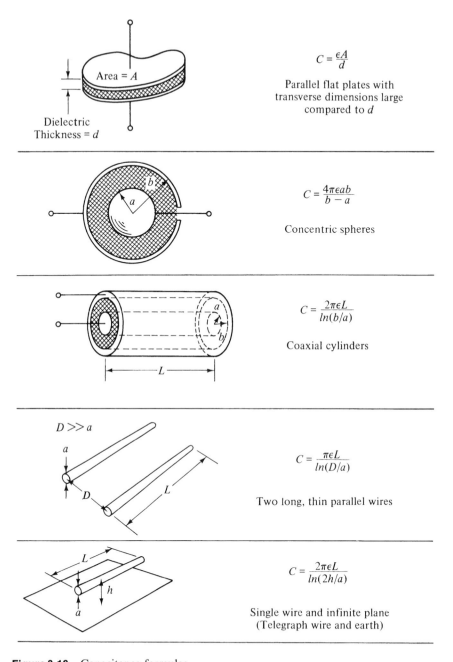

Figure 3-18 Capacitance formulas.

Figure 3-19b shows a tubular ceramic capacitor; the ceramic dielectrics used include steatite, titanium dioxide, and barium titanate. A typical electrolytic capacitor construction is shown in Fig. 3-19c. This type of capacitor has a unique principle of operation which puts it somewhat in a class apart from all other types. The sketch shows sandwiched metal foil and paper separators which appear quite similar

System Elements, Electrical

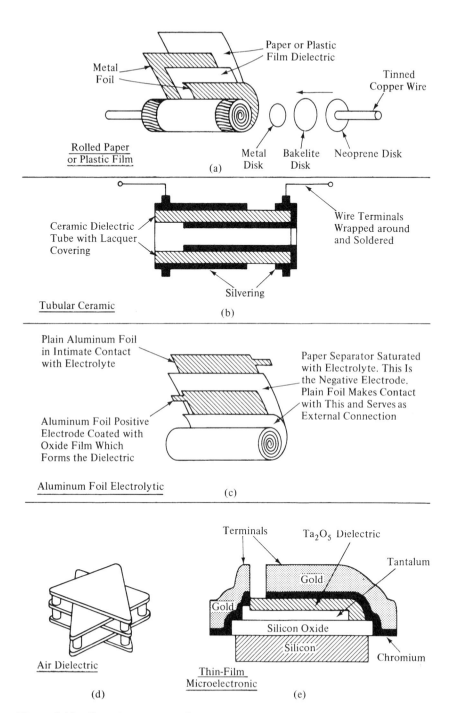

Figure 3-19 Capacitor constructions.

to a "conventional" capacitor; however, the paper is *not* the dielectric in this case. The paper is saturated with an electrolyte paste of glycol and ammonium tetraborate. One of the 0.002-inch aluminum foils has had a very thin layer of aluminum oxide (the dielectric) formed on it by applying a constant voltage while it was immersed in an ammonium borate solution. This foil will be the anode, or positive terminal, of the capacitor, and *must never have a negative voltage applied to it.* That is, the electrolytic capacitor is a *polarized* device and is not usable where currents and voltages actually reverse, whereas all other types of capacitors are completely "reversible." (Actually, reversible electrolytics can be and are made by preforming the dielectric film on *both* of the aluminum foils and making suitable connections; however, most electrolytics are of the polarized type and must be used with this in mind.)

Air-dielectric capacitors, Fig. 3-19d, are used mainly as laboratory standards and are made in parallel-plate and concentric cylinder configurations. They are extremely accurate and stable, varying in the order of 0.01 to 0.04% over several years and having a temperature coefficient of capacitance of less than 10 ppm (parts per million) per °C. Diffused capacitors for monolithic integrated circuits are produced by techniques similar to those explained for the resistor of Fig. 3-15c.[16] Thin-film capacitors are tiny discrete components used in microelectronic circuits when capacitors of higher quality than can be produced by the diffusion process are needed. A chip 0.0024 inch thick and 0.032 inch square accommodates a 3000-pF, ±5% capacitor. Tiny "beam leads" 0.006 inch long, 0.004 inch wide, and 0.0004 inch thick are used to attach the capacitors to other circuit elements. Figure 3-19e shows a cross section of a typical[17] unit using tantalum oxide as the dielectric.

While a pure capacitor stores and can then release all the energy supplied to it, real capacitors exhibit losses for various reasons. The *power factor* of a capacitor is defined as the ratio of energy wasted per cycle of AC voltage, divided by the energy stored per cycle. It may vary with frequency, and values measured at 1000 Hz range from 0.00001 for precision air-dielectric types through 0.0005 for polystyrene film to 0.05 for some electrolytics. This wasted energy shows up as heat in a real capacitor, whereas a pure capacitor experiences no temperature rise whatever. An equivalent circuit sometimes used to model a real capacitor is shown in Fig. 3-20a. The above-mentioned power losses are due to the resistive elements in this model. The presence of the parallel resistance element R_p, called the *leakage resistance*, in the model shows that a real capacitor will allow a (*very* small) DC current to flow. For a given type of capacitor, the leakage resistance is very nearly inversely proportional to the capacitance; that is, CR_p is constant. A charged capacitor will not hold its voltage indefinitely; it will "leak off" through R_p, taking about $5CR_p$ seconds to discharge completely. The time for this decay varies from several days for Teflon and polystyrene types to a few seconds for some electrolytics. For polystyrene, for example, the product CR_p at room temperature is about 10^6 seconds, where R_p is in megohms

---

[16]V. Lehmann et al., A new capacitor technology based on porous silicon, *Solid State Technology*, November 1995, pp. 99–102.

[17]W. E. Wesolowski, and M. Tierman, Beam-leaded, thin-film capacitors, *Electronic Capabilities*, Winter 1969–1970, p. 34.

System Elements, Electrical

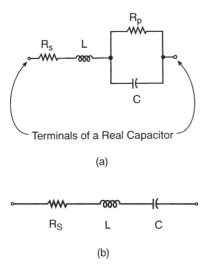

Figure 3-20 Models for real capacitors.

and C in microfarads. A 1-μF unit would thus be expected to have an R_p of about 10^{12} ohms. While the existence of R_p is necessary to explain the observed "leakage" of real capacitors, it often has little effect on dynamic behavior and the simpler model of Fig. 3-20b is found adequate for many real capacitors.

Dielectric absorbtion refers to the reappearance of a voltage after a charged capacitor has been discharged by short-circuiting and is then open-circuited. As an example, for a 200-volt initial charge applied for 1 minute and a 2-second short-circuit discharge, the voltage reappearing after 1 minute is 0 for an air capacitor, 0.02% (40 mV) for polystyrene, and 2.0% for oil-impregnated paper. Together with other effects, the above-mentioned deviations from perfection place frequency restrictions on the application of various capacitor types as shown in Fig. 3-21. The *temperature coefficient* of capacitance in ppm per °C varies from +10 for precision air capacitors through about ±200 for paper and plastic films to about +1500 for many electrolytics. Failure rates in percent per 1000 hours are in the range of 0.005 for polystyrene to 0.5 for aluminum foil electrolytics.

Commercial inductors are not generally available in as wide a selection of sizes and types as are resistors and capacitors. In fact many circuit synthesis methods endeavor to achieve the required performance by the use of only R and C because of practical difficulties associated with the construction and use of L (recall our earlier comments on "L becoming R at low frequency" and the nonlinearity of the iron-core inductances needed to get large L values). We shall see later that the use of active (operational amplifier) network methods allows one to achieve, with R and C alone, circuit behavior that would otherwise require L.

The theoretical calculation[18] of the inductance of configurations of most practical shapes (see Fig. 3-22) can be carried out with good accuracy so long as no

---

[18]F. W. Grover, *Inductance Calculations*, D. Van Nostrand, New York, 1946.

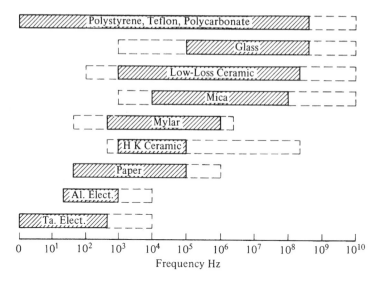

Figure 3-21 Useful frequency range for various capacitor types.

magnetic materials are used. When magnetic materials are used, L values computed from the air-core formulas of Fig. 3-22 may be multiplied by the relative permeability for the magnetic material to estimate L values for small-signal operation near an operating point. For large-signal operation, L varies significantly with current, and simulation methods such as we used in Fig. 3-9 are helpful. Magnetic materials can greatly increase the L value since relative permeabilities can be very large: about 5200 for iron and 100,000 for special alloys such as Permalloy.[19]

For real inductors the *quality factor Q* plays a role similar to that of the power factor in capacitors; that is, it gives an indication of the energy losses due to the presence of resistance. Since most inductors are coils of many turns of wire, they must of necessity have considerable resistance. At low and intermediate frequencies a real inductor may be modeled as in Fig. 3-23a. For such a model, Q is defined by

$$Q \triangleq \frac{\omega L}{R} \tag{3-52}$$

where $\omega \triangleq$ frequency of AC voltage, rad/sec, and is proportional to the ratio of stored energy to dissipated energy. We see that for a pure inductor ($R \equiv 0$) the quality factor would be infinity irrespective of the frequency; thus high Q indicates a more nearly pure inductor. Note that this trend is *opposite* to that for power factor in capacitors; there one wants a *small* power factor. For real inductors, $R \neq 0$ and Q approaches zero for low frequencies. If the model of Fig. 3-23a held for all frequencies, Q would approach infinity as ω approached infinity; however, at intermediate frequencies core-loss effects appear which change the model to that of Fig. 3-23b. Core losses are energy losses due to eddy currents and hysteresis in the core of

---

[19]G. Rizzoni, *Principles and Applications of Electrical Engineering*, 2nd ed., Irwin, Chicago, 1996, p. 804.

System Elements, Electrical

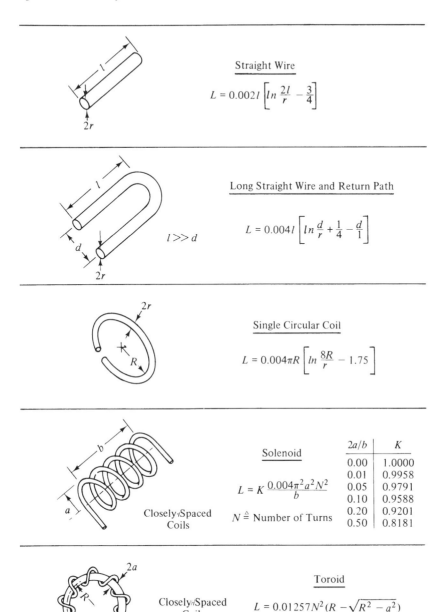

Figure 3-22 Inductance formulas (all lengths in cm, inductance in μH).

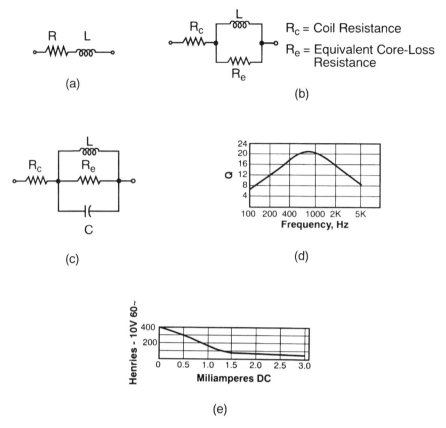

Figure 3-23 Models and behavior of real inductors.

magnetic material and are represented by an equivalent shunt resistance. For this new circuit Q is given by

$$Q \triangleq \frac{\omega L}{R_c + \frac{R_c + R_e}{R_e^2}(\omega L)^2} \tag{3-53}$$

which approaches zero for both very low and very high frequencies. At high frequencies the parasitic capacitance must be taken into account, as in Fig. 3-23c.

Figure 3-23d shows a measured Q curve for a commercial[20] inductor made as a magnetic core toroid. A very large inductance (60 henries) is achieved in a space of about a 0.5-inch cube with a weight of 0.2 ounce. The DC resistance is 5160 ohms. Because magnetic materials are used, the inductance varies with current; to keep this variation below 5% the DC current must be limited to 0.2 mA or less. At a given

---

[20]Miniductor ML-10, United Transformer Co., New York.

System Elements, Electrical   **167**

frequency the inductance may be quite constant over a fairly wide range of AC voltage amplitudes so long as the instantaneous AC current remains at or below the DC limiting value just mentioned. A 2-henry unit with DC resistance of 130 ohms and DC current limit of 8 mA shows only a 3% change in inductance for a 400-Hz voltage of 0.1 to 35 volts amplitude. A typical curve showing inductance variation (measured at 60 Hz, 10 volts AC) with DC current is given in Fig. 3-23e. The leveling off of L at high currents is due to the saturation of the magnetic core, which reduces the permeability.

Lest the large L values (60 and 2 henries) and low-frequency ranges just quoted mislead the reader, we should indicate that many practical inductors of millihenry and microhenry value work in high kilohertz and megahertz frequency ranges. A 0.15-μH inductor is about 0.1 inch in diameter and 0.4 inches long, with a DC resistance of 0.02 ohms and a Q of about 65 at 25 MHz, the frequency at which L was measured. If we take the simple definition of Q as $\omega L/R$ in this example, we predict a Q of 1180, much higher than the value actually measured. One explanation of this discrepancy lies in the so-called *skin effect*, which makes the resistance at high frequency much higher than its DC value. At DC and low frequencies, the current is uniformly distributed over the cross section of a conductor, while at high frequency it is "crowded" toward the surface, making the effective cross section smaller and thus raising the resistance. This is caused by self-induced emf's set up by variations in the internal flux in the conductor. In our numerical example, it appears that at 25 MHz the effective resistance is about $1180/65 = 18$ times the DC value. Part of this resistance could also be caused by the core losses mentioned earlier. The skin effect is not peculiar to inductors; it occurs in all conductors at high frequency and provides an explanation of several otherwise puzzling observed phenomena, so add it to your "catalog" of useful electrical effects.

3-7   CURRENT AND VOLTAGE SOURCES

The energy sources which drive electrical systems may be conveniently classified as voltage sources or current sources. An *ideal voltage source* supplies the intended voltage to the circuit no matter how much current (and thus power) this might require. An *ideal current source* supplies the intended current to the circuit no matter how much voltage (and thus power) this might require. For example, an ideal 10-volt source applied to a 10-ohm resistor produces a 1-amp current and 10 watts of power. The same 10-volt source applied to a 0.001-ohm resistor produces 10,000 amps and 100,000 watts of power. No *real* voltage source, such as a battery or electronic power supply, can produce *unlimited* power in this ideal manner; however, over *restricted* ranges, they may approach the ideal. Furthermore, the behavior of real sources may be modeled by a combination of an ideal source and a passive element, or elements, such as a resistor. Just as in mechanical systems, where the force source is fundamental because of the cause-and-effect relation given by Newton's law (a force *causes* a motion, not the reverse), we might say that voltage sources are fundamental since it is the electromotive force (voltage) which *causes* a current to flow. Nevertheless, in practice, use of a current source as a model may be quite correct in some systems, since real devices which behave very much like current sources are available.

Batteries[21] of various types are widely used as electrical power sources, particularly when portability of the equipment (such as a laptop computer) is essential. Experimentally measured performance curves of a rechargeable lead-acid unit[22] are shown in Fig. 3-24a. By plotting voltage versus current for an early time (say, 1 second) when the battery is still fully charged, we get Fig. 3-24b. We there see that this battery is not an ideal voltage source, since voltage drops off as more current is drawn. However, the curve is very nearly a straight line of slope -0.1 volts/amp, suggesting that we can model this real battery as an ideal 12.6-volt source in series with a 0.1-ohm resistor, as in Fig. 3-24c. The (fictitious) 0.1-ohm resistor is called the *internal impedance* of the real source. An ideal voltage source has zero internal impedance; thus real voltage sources approach perfection as their internal impedance approaches zero.

Recently, "supercapacitors"[23] have replaced batteries in certain applications. Capacitors have always been known as energy-storage devices but rarely were used for that specific purpose as battery replacements. Supercapacitors have very high C values in small space, require no maintenance, and have a higher cycle life, simpler charging circuitry, and lower cost than batteries. A major application area is as backup power for computer memories.

Worldwide, the vast majority of electric power is produced by the rotating mechanical-to-electrical power converters called *generators*, usually of the AC type ("alternators"). The mechanical power to drive the generators often comes from steam turbines, whose ultimate power source is heat from a burning fuel or a nuclear reaction. Generators are basically voltage sources which again exhibit an internal impedance, so that their terminal voltage drops off as more current is drawn from them. Since electrical equipment in homes and factories is not usually connected *directly* to the output terminals of a generator, we might be more interested in modeling the source represented by, say, a 120-volt, 60-Hz "wall plug." We will not pursue this analytically, but you might want to think about how you would set up and run an experiment to gather the information needed to formulate such a model.

Direct-current generators are somewhat simpler to model than alternators and Fig. 3-25 shows a 5000-watt unit which is driven by a 7.5-hp AC induction motor running at a speed of 3450 rpm. The output voltage of the generator is controlled by adjusting the DC voltage applied to its field circuit. This changes the current in a coil with an iron core, which produces the magnetic field of the generator. When the field

---

[21]Batteries, *Machine Design*, April 11, 1963, p. 189. Nickel-Cadmium Battery Application Engineering Handbook, 2nd Ed., General Electric Publ. GET-3148A, 1975. Mathematical storage-battery models, *NASA Tech Brief*, vol. 8, no. 3, Item 27, 1984. Battery Testing, Appl. Note 372-2, Hewlett-Packard Corp. 1988. J. B. Bates et al., Rechargeable thin-film lithium batteries, *Solid State Technology*, July 1993, pp. 59–64. D. Maliniak, Intelligence invades the battery pack, *Electronic Design*, January 9, 1995, pp. 153–159.

[22]Delco Energette, Brochure DR-9647, Delco-Remy Division.

[23]Supercapacitors leap battery power sources in a single bound, *ESD*, July 1987, pp. 26–28. S. Ashley, Surging ahead with ultracapacitors, *Mechanical Engineering*, February 1995, pp. 76–79. The Evans Capattery, Evans Co., P.O. Box 4158, East Providence, RI 02914-4158, 401-434-5600.

System Elements, Electrical

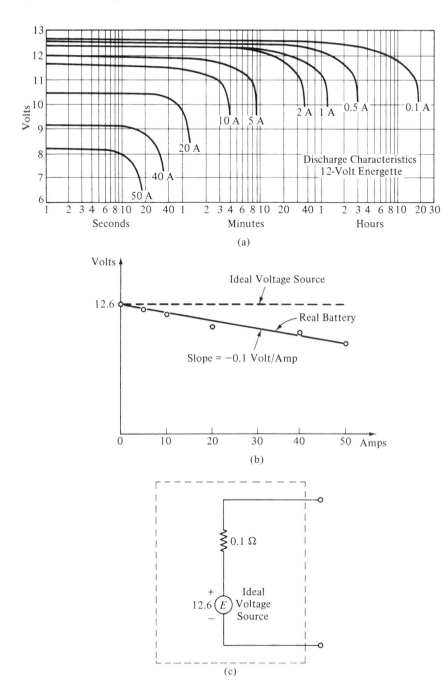

Figure 3-24 Modeling a real battery.

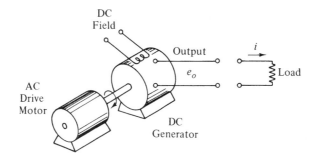

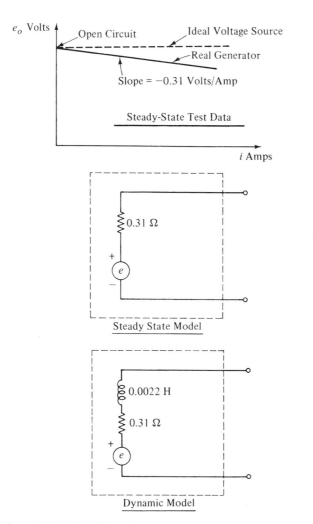

Figure 3-25 Modeling of a DC generator power source.

System Elements, Electrical   **171**

voltage (and thus the magnetic field itself) is set at zero, the generator output voltage is zero, even though it is running at full (3450 rpm) speed. By adjusting the field voltage we can control the generator output voltage proportionally, the highest attainable voltage in this unit being 100 volts. In fact, by *reversing* the field voltage polarity, the coil current and magnetic field are also reversed, which also reverses the polarity of the generator ouput voltage. Thus, by manipulating the field voltage, we can control the generator output in the working range of ±100 volts. The machine is also designed to supply as much as ±50 amps of current.

If the field voltage is fixed and the generator is not connected to some electrical load (output terminals open-circuit), the generator will produce a definite output voltage. If we now connect a load, current will be drawn from the generator and its terminal voltage will drop below the open-circuit value, just as we saw earlier for the battery. This drop is caused by the internal impedance of the generator armature circuit (the rotating part of the generator). In the machine being considered, this was measured as 0.31 ohm. Thus, if we have, say, 10 volts open-circuit voltage and then connect a load which draws 1 amp, the voltage will drop to 9.69 volts. Dynamic tests reveal the armature to have 2.2 mH of inductance; thus if the load is changed *suddenly* (say, by switching a resistor across the output terminals) the generator behaves as a source with dynamic characteristics. Figure 3-25 depicts some of these concepts.

Electronically regulated DC power supplies, rather than batteries, are used as power sources for much electrical equipment. These power supplies take 110-volt/60-Hz AC power from the "wall plug" and provide regulated DC voltage or current at their output. Many such regulators use feedback control principles and provide performance closely approaching the ideal voltage or current source, at least within their designed operating range. Details of operation of such supplies are beyond the scope of this text, but may be found in manufacturer's catalogs.[24] Operating specifications of a typical 15-volt supply are:

> Current range: 0–1.2 amp
> Voltage change for 0–1.2 amp current change: <0.05%
> Voltage change for 105- to 125-volt power line change: <0.05%
> Temperature effect on voltage: <0.05% per Celsius degree
> Time drift: <0.05% in 8 hours

The internal impedance is quoted as 0.008 ohm for DC to 100 Hz, 0.02 ohm for 100 to 1000 Hz, and 0.1 ohm plus 1 µH of inductance for 1 to 100 KHz. Since all such power supplies produce DC by rectifying and filtering the input AC power, there is always a little AC "riding on top" of the DC, since no filter is perfect. For the unit just described, this ripple is <0.0005 volt, which is <0.003% of the 15-volt DC. From these specifications we can see that this device comes very close to the ideal voltage source, *as long as we don't exceed its current rating.*

While voltage sources are by far the most common, some applications require current sources. Voltage sources which use feedback control principles can usually be

---

[24]Kepco Inc., 131-38 Sanford Avenue, Flushing, NY 11352, tel: 718-461-7000, catalog 146-1255 or current equivalent.

converted to current sources by connecting a current-sensing resistor[25] in series with the load, and feeding back the voltage across this resistor to the regulator cirucit. This circuit now tries to keep the voltage across the current-sensing resistor constant, which is the same as keeping the load current constant; thus we now have a current source. Specifications for a unit of this type designed to provide 0 to 0.5 amp at 0 to 40 volts are:

> Current change for full-rated (0–40 volt) voltage change is $<0.005\%$ or $2\,\mu A$, whichever is greater.
> Current change for AC input voltage change from 105 to 125 volts is $<0.0005\%$ or $0.2\,\mu A$, whichever is greater.
> 8-hr time drift is $<0.02\%$ or $2\,\mu A$, whichever is greater.
> Temperature effect is $<0.01\%$ per $C°$.
> Ripple is $<0.02\%$.

The output impedance is 800,000 ohms plus $0.1\,\mu F$ of shunt (parallel) capacitance. Note that a current source has a very *high* internal impedance—just the opposite of a voltage source. This 800,000 ohms does *not* correspond to an actual 800,000-ohm resistor in the power supply circuitry; rather it represents the ratio of the load voltage change necessary to cause a current deviation from the desired value. A current source is loaded by connecting it to a device (load) which requires a certain voltage to force the set current through it.

For example, if we set the current control at 0.5 amp in the unit above and short-circuit its output (attach a load of "zero" resistance), the current drawn will be 0.5 amp and the load voltage will be zero. If we now attach an 80-ohm resistance load, the current should ideally stay precisely at 0.5 amp and the load voltage should go to 40 volts. The 800,000-ohm internal impedance is the *ratio* of the voltage change (0–40 volts) to the current change from the ideal value. In this case the current change is $40/800,000 = 0.00005$ amp, which is 0.01% of the set current of 0.5 amp, and $\pm 0.005\%$ if we take a ± 20-volt variation around 20 volts. This 0.005% corresponds to the specification on current regulation given earlier. Figure 3-26 shows how a real current source with a known internal impedance can be modeled as an ideal current source in parallel with this impedance. In Fig. 3-26a the load is a short circuit, so no current goes through the internal resistance, the voltage e_{ab} is zero, and all the current goes from the source into the load. By adding a 80.008-ohm load in 3-26b, we cause e_{ab} to be 40 volts and the source current i now splits between the load and the internal resistance such that $i_{IR} = 0.00005$ and $i_L = 0.49995$. The current deviation from the set value is thus 0.00005, in agreement with our earlier calculation. Note that if the internal resistance R_{IR} were *infinite* (an open circuit), then *all* the source current goes to the load irrespective of the load resistance and we have a perfect current source. Thus a real current source approaches perfection as its internal impedance approaches infinity.

The practical voltage and current sources mentioned so far have provided either constant (DC) or sinusoidally varying (AC) voltages or currents. In analyzing

---

[25]M. Martich and G. Wedeking, Precision low-ohmic resistors provide accurate current sensing, *PCIM*, June 1992, pp. 15–19.

System Elements, Electrical

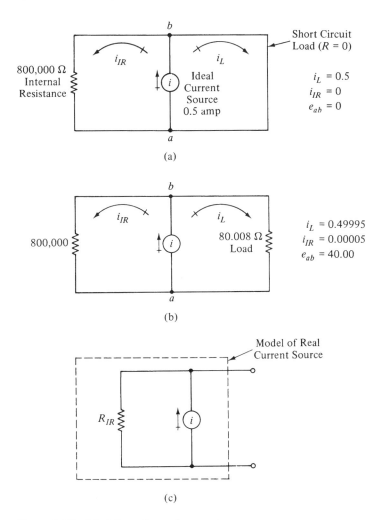

Figure 3-26 Modeling of a real current source.

electrical systems we can, of course, assign any time variation we wish to our sources; that is, $e(t)$ and $i(t)$ may have any form whatever, to suit the physical situation. In experimental work, versatile electronic *signal* generators (usually voltage sources) are available which provide a wide variety of useful forms of $e(t)$. In Fig. 1-5b we displayed various classes (transient, amplitude-modulated, periodic, random) of input signals that might be applied to *all kinds* of systems, not just electrical. Since *electrical* signal generators are usually more convenient and versatile than any other kind, when we require in lab work a certain form of driving input for a system, we very often generate this signal first as a voltage signal. We then apply this signal to the appropriate form of *transducer* to convert the electrical signal to the physical form we really want. For example, when we want rapidly time-varying forces or motions, we often use the electrodynamic vibration shaker of Fig. 2-49 to *transduce* an electrical signal of the desired waveform into its mechanical counterpart. Similar transducers are available for converting electrical commands into proportional temperatures, pressures, flow rates, etc. These devices are often in the form of feedback control systems in which the electrical signal plays the role of system command and the desired physical output is the controlled variable.

A typical[26] electronic signal generator of this type provides single pulses, bursts of several pulses, and periodic waves of sine, square, triangular, ramp, $\sin x/x$, exponential rise or fall, and cardiac (ECG) waveforms. The frequency of the periodic waves can be accurately varied from $100\,\mu$Hz to 15 MHz and the pulses can have corresponding ranges of time duration. The basic waveforms can also be amplitude modulated, frequency modulated, burst modulated, and FSK modulated. Frequency sweeps (frequency of the periodic waveform changes smoothly from a low value to a high value at a selectable rate) are also available. Most versatile of all, one can program *arbitrary* waveforms with as many as 16,000 selectable points to match a desired shape, and then repeat this waveform periodically at rates from $100\,\mu$Hz to 5 MHz. This digital synthesis capability has a resolution of 12 bits (1 part in 4096). The amplitudes of all the waveforms are adjustable up to a maximum of 10 volts peak to peak. A Gaussian random signal of frequency bandwidth 10 MHz is also available.

Signal generators such as that just described can only provide small current and power. The maximum current is often about 0.10 amp; thus at 5 volts peak, the power supplied to the load would be only 0.5 watt. Also the generator internal impedance is often 50 ohms, so voltage drops off as more current is drawn. When we want a certain waveform at a higher power level, we must send the generator's output to the input of a *power amplifier*. Such amplifiers are available in various models which can supply from 10 watts to many kilowatts of power to a load. An electrodynamic vibration shaker system of such a type can produce forces as large as 90,000 pounds.

When random signals with more "adjustability" are needed, special random function generators are available.[27] The referenced unit provides both Gaussian and binary random signals of adjustable amplitude and frequency content. The frequency bandwidth of the signal can be set anywhere from 0.00015 to 50,000 Hz, a useful range for many physical systems which we may want to study. Random test signals are often preferred since they quickly "exercise" the system being tested, over a wide frequency range.[28]

3-8 THE OPERATIONAL AMPLIFIER, AN ACTIVE CIRCUIT "ELEMENT"

As mentioned earlier, the operational amplifier is not strictly an element in the usual sense, because it *contains* elements, such as resistors and transistors. We include it in this chapter because of its great utility and the fact that it may be physically as small

---

[26]Hewlett-Packard 33120A Function Generator/Arbitrary Waveform Generator, User's Guide, Hewlett-Packard Corp., 1994, tel. 800-452-4844.

[27]Hewlett-Packard Model 3722A.

[28]Doebelin, *System Modeling and Response*, pp. 267–282.

System Elements, Electrical

as a resistor or capacitor, and some models cost less than a dollar. Also, like R, L, and C it *is* a basic building block used in a wide variety of practical circuits. In fact it is probably the most widely used integrated circuit device. While the broad range of types, performance, and cost prevents one from showing a "typical" op-amp, Fig. 3-27 will give some idea of the size and internal circuitry of an integrated circuit model. Although this circuitry looks complex, and requires special electronics expertise to *design*, once "experts" have designed the op-amp, those with modest electronics background (including even "hobbyists" without engineering degrees) can *use* the op-amp to build instrumentation amplifiers, filters, controllers, analog signal processing devices, etc. A properly designed op-amp allows us to use certain simpli-

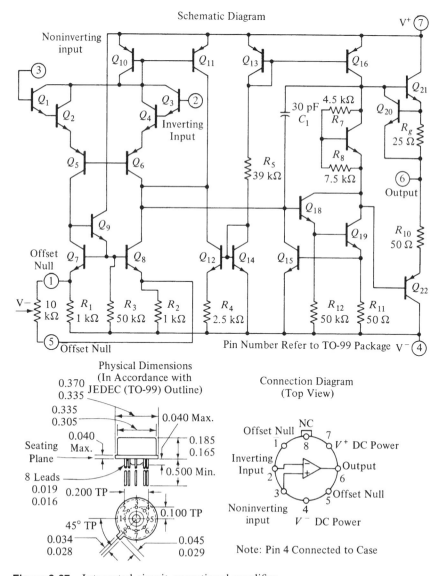

Figure 3-27 Integrated-circuit operational amplifier.

fying assumptions when analyzing a circuit which uses op-amps. We accept these assumptions "on faith" and then find that they make op-amp circuit analysis quite simple. The simplified models which we will use are explained in many books and manufacturers' literature.[29]

In Fig. 3-28a the input voltages e_{i1} and e_{i2} are applied to the amplifier input terminals and produce an amplified signal $A(e_{i2} - e_{i1})$, where A is the gain (amplification, volts/volt) of the amplifier. The amplifier of course also requires connections to appropriate DC power supply voltages, such as ± 15 volts, but from here on we will not explicitly show these. The configuration shown is called a *differential-input* amplifier; if e_{i2} is connected to ground (allowing input only at e_{i1}), it is called *single-ended*. Both configurations have useful applications; we will concentrate mainly on the single-ended. The impedances Z_i and Z_o are, respectively, the input and output impedances of the amplifier. To get the simplest model, but one which is still useful for many practical applications, we make the following assumptions:

1. The op-amp's gain A is infinite.
2. Z_i is infinite; thus *no current* is drawn at the input terminals.

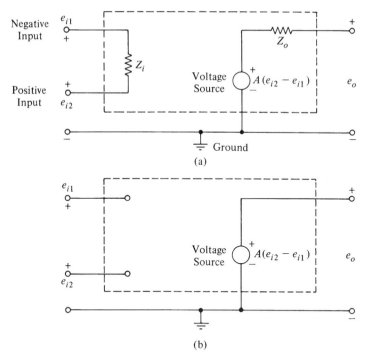

Figure 3-28 Models of op-amp input and output.

---

[29]W. G. Jung, *I-C Op-Amp Cookbook*, 2nd ed., Howard W. Sams & Co., Inc., Indianapolis, IN 46268, 1980. R. G. Irvine, *Operational Amplifier Characteristics and Applications*, 3rd ed., Prentice-Hall, Englewood Cliffs, N.J., 1994. Linear Products Data Book, Burr-Brown, 6730 South Tucson Boulevard, Tucson, AZ 85706. 1992 Amplifier Applications Guide, Analog Devices, One Technology Way, P.O. Box 9106, Norwood, MA 02062-9106.

System Elements, Electrical   **177**

3. Z_o is zero; thus $e_o = A(e_{i2} - e_{i1})$.
4. The time response is instantaneous.
5. The output voltage has a definite design range, such as $\pm$ 10 volts. Proper operation is possible only for output voltages within these limits.

Users of op-amps rely on op-amp *designers* to provide products which will fulfill these requirements with sufficient accuracy for practical use. Certain applications require certain of the assumptions to be fulfilled particularly closely. Op-amp designers recognize these needs and provide particular op-amp designs to meet them. Meeting such special needs may raise the cost and/or require that some *other* assumptions be met *less* stringently. Figure 3-28b shows the simplified model corresponding to the above list of assumptions. (We will later explore briefly the effects of a real amplifier deviating from this ideal behavior; however, our emphasis is on the simpler model. It facilitates understanding of many practical applications and gives results close to actual measured behavior in most cases.)

The op-amp was invented in the 1940s and was originally used mainly to build the general-purpose analog computers that dominated system dynamics simulation until the 1960s when digital simulation started to take over these tasks. Today, *general-purpose* simulation is largely done digitally (using software) rather than with analog hardware. However, analog signal processing and *special-purpose* simulation still use the op-amp analog methods, so we still want you to be familiar with these system dynamics tools. The three basic elements of an analog computer for solving differential equations are:

1. The coefficient multiplier
2. The integrator
3. The summer

The operations performed by these devices are still needed today in many measurement, data processing, and control applications, even though general-purpose analog computers are largely obsolete. Figure 3-29 shows all three of these devices, which we now analyze.

Rarely used "by itself," the op-amp is usually combined with passive elements, mainly resistors and capacitors. Figure 3-29a shows the *coefficient multiplier*. The function of this device is to accept a time-varying or constant input voltage e_1 and produce an output voltage e_o which is Ke_1, where K is an adjustable constant of our choice. (The name coefficient multiplier comes from the analog computer application, where a signal representing, say, the velocity v of a mass, needed to be multiplied by a damper coefficient B, to simulate the damping force Bv in a differential equation being solved to find the motion of the mass.) In many, but not all, op-amp applications, the terminal where e_{i2} is applied (called the positive input) is grounded, making this voltage exactly zero at all times. We also connect ordinary resistors R_i (called the *input resistor*) and R_{fb} (called the *feedback resistor*) as shown.

To analyze op-amp circuits we use the two basic electrical circuit "laws" (Kirchhoff's voltage loop and current node laws) together with the op-amp simplifying assumptions listed earlier. You will find that these assumptions make op-amp circuit analysis *easier* than "ordinary" circuit analysis! We wish to find a relation between e_o and e_1. At the point called the *summing junction*, since the current into the

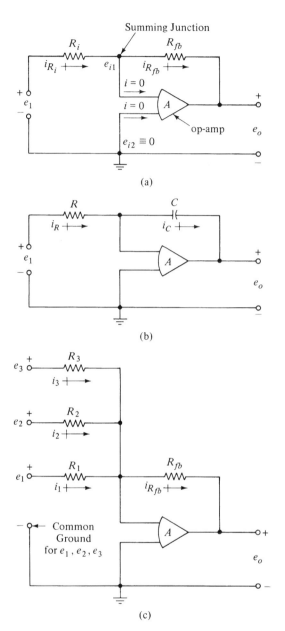

Figure 3-29 Coefficient-multiplier, integrator, and summer.

amplifier is assumed zero, the current node law says that $i_{R_i} = i_{R_{fb}}$. Since the voltage across R_i is $e_1 - e_{i1}$ and the voltage across R_{fb} is $e_{i1} - e_o$ we have

$$i_{R_i} = \frac{e_1 - e_{i1}}{R_i} = i_{R_{fb}} = \frac{e_{i1} - e_o}{R_{fb}} \quad (3\text{-}54)$$

Now the voltage e_{i1} is related to e_o by $e_o = -Ae_{i1}$, where A is the amplifier gain; thus $e_{i1} = -e_o/A$. Then Eq. (3-54) becomes

System Elements, Electrical   **179**

$$\frac{e_1 + (e_o/A)}{R_i} = \frac{(-e_o/A) - e_o}{R_{\text{fb}}}$$

(3-55)

and if $A = \infty$,

$$e_o = -\frac{R_{\text{fb}}}{R_i} e_1$$

(3-56)

Of course A cannot be infinite, but it can be, say, 10^6 volts/volt and then the terms neglected in Eq. (3-55) become very small compared to the terms retained, making Eq. (3-56) a very good approximation. For example, suppose we choose R_{fb} to be 100,000 ohms and R_i to be 1000 ohms, to construct a coefficient multiplier that multiplies by 100. If our op-amp has a design output range of ± 10 volts, then the design input range would be ± 0.1 volt and the largest value of the neglected term e_o/A is 0.00001 volt, much smaller than the full-scale value of e_1 and thus legitimately neglected relative to it.

While the device just analyzed is called a coefficient multiplier when it is used in an analog computer, when used in an instrumentation context it would simply be an amplifier, in our example case an amplifier with a gain of 100 volts/volt. It might be used to amplify voltage from a strain-gage pressure sensor (which might supply 0.100 volt for full-scale pressure) to the 10-volt range for entry into an analog-to-digital converter, which *requires* a 10-volt full-scale input signal. You might at this point be thinking, "Why don't you just use the op-amp *directly* for your amplifier since it has more than enough gain?" The problem here is that the op-amp gain can be relied upon to be "very large" but *cannot* be relied upon to be an accurate stable value. That is, the gain A is "guaranteed" only to be, say, in the *range* 1 to 5 million V/V. One op-amp that you might buy would have 1.6 and another of the *same* type might have 3.8. Also, any one op-amp might have 1.6 on Tuesday and 2.8 on Wednesday. Such uncertainty in gain is totally intolerable in precision applications. Note, however, that as long as A is *larger* than the value used in a design study, the approximation used to get to Eq. (3-56) *will* be valid. Note that the gain of 100 that we achieved in our example depends for its accuracy and stability on the values of two fixed resistors, and *not* on the value of A, *so long as A is "large enough."* This is one of the main reasons why op-amps are so valuable as basic circuit building blocks. Using them, we can construct circuits whose performance depends mainly on passive components such as R and C, which can be selected to have accurate and stable values. The op-amp gain can "wander around" without causing any problem, so long as it always is "large enough."

The op-amp gain A is sometimes called the *open-loop gain* while the e_o/e_1 ratio in Eq. (3-56) is called the *closed-loop gain*. We see of course that the closed-loop gain is *negative*; the output voltage has a polarity opposite that of the input. Note that this is *not* a consequence of the minus sign in $e_{i1} = -e_o/A$. We would get the same result in Eq. (3-56) whether A was positive or negative. Thus the minus sign in Eq. (3-56) is due to the circuit *configuration* used and is unavoidable. In, say, a measurement system using such an amplifier, the minus sign is not an insurmountable problem since we *know* it is there; we can easily interpret our measured voltage correctly. However, it is easy to get an amplifier with *positive* gain if we must. One simply builds a *second* amplifier with a gain of -1 (use equal-value input and feedback resistors) and send the output of the first amplifier into the input of the second.

180   **Chapter 3**

The two amplifiers together would of course then have (for our earlier example numbers) a closed-loop gain of $+100$ V/V. The amplifier with the gain of -1 can be used *anywhere* where a sign change is needed and is given its own name, the *inverter*. The combination of two amplifiers is simply accomplished in practice since one can purchase integrated-circuit chips with *several* op-amps on a single chip. This is done both to simplify the wiring and also to get the two amplifiers to be better *matched* in their electrical and temperature characteristics. This matching of characteristics is one of the great advantages of integrated-circuit methods as compared with "soldering together" separate discrete components.

From Eq. (3-55) and the fact that A may be treated as infinite, we see that the voltage e_{i1}, which is called the summing junction voltage, can *always* be treated as zero in those op-amp circuits where the positive input is grounded. From now on we won't bother to even write terms like e_o/A since we now know that they can be treated as zero. Because the summing junction voltage is "practically zero," the summing junction is often called a *virtual ground*, since its voltage is for all practical purposes zero, the same as the *true* ground, whose voltage is *exactly* zero. When op-amp circuits *don't* ground the positive input (differential input), the *difference* $(e_{i2} - e_{i1})$ is taken as being practically zero. Such circuits force the relation $e_{i1} \equiv e_{i2}$, which has many practical applications.

Using the assumptions just employed above, Fig. 3-29b may be quickly analyzed by writing

$$i_R = \frac{e_1 - 0}{R} = i_C = C\,\frac{d(0 - e_o)}{dt} = -CDe_o \tag{3-57}$$

$$e_o = -\frac{1}{RCD}\,e_1 = -\frac{1}{RC}\int e_1\,dt \tag{3-58}$$

We see that this configuration provides the analog computing operation of integration with respect to time. If we take, say, $R = 10^6$ ohms and $C = 0.5\,\mu\text{F}$, we have $e_o = -2.0\int e_1\,dt$, so the output voltage is -2.0 times the integral of the input voltage. (The minus sign can of course again be removed by use of an inverter.) Integration is the fundamental operation needed to solve differential equations and the integrator is thus the "heart" of electronic analog computers, just as *numerical* integration software is the heart of digital simulation languages such as the SIMULINK we used earlier. Electronic analog integration continues to be widely used in measurement and signal processing, so the circuit of Fig. 3-29b is far from obsolete. For example, if we use an accelerometer (which is a mechanical device but produces an electrical output signal) to measure an unknown acceleration, we can easily also get a *velocity* signal by connecting the output voltage of the accelerometer to the input of an integrator.

In Fig. 3-29c we show a device for adding voltages, the summer. (Recall from earlier SIMULINK examples that a summer is also needed there, but it is done in software, not op-amp hardware.) When solving differential equations, summers are used to "add up" the various terms in the equation. Using our usual assumptions we can write

$$i_1 + i_2 + i_3 = i_{R_{\text{fb}}} \tag{3-59}$$

System Elements, Electrical

$$\frac{e_1 - 0}{R_1} + \frac{e_2 - 0}{R_2} + \frac{e_3 - 0}{R_3} = \frac{0 - e_o}{R_{fb}} \tag{3-60}$$

$$e_o = -\left(\frac{R_{fb}}{R_1} e_1 + \frac{R_{fb}}{R_2} e_2 + \frac{R_{fb}}{R_3} e_3\right) \tag{3-61}$$

If we take $R_{fb} = R_1 = R_2 = R_3$ we get

$$e_o = -(e_1 + e_2 + e_3) \tag{3-62}$$

showing how the three input voltages are summed. By use of inverters we can both add and subtract voltages. Also, we can *combine* the summing and coefficient multiplier operations in a "summer" by choosing proper values for the resistors, as can be seen from Eq. (3-61).

While the coefficient multiplier, integrator, and summer are fundamental to solving differential equations, op-amp circuits have many other uses. Figure 3-30 shows an arrangement which allows easy design of many devices useful in the electronic portions of dynamic measurement, control, and signal processing systems. These include high-pass filters, low-pass filters, band-pass filters, band-reject filters, lead controllers, lag controllers, lead-lag controllers, approximate integrators and differentiators. In the figure, the impedances Z_i and Z_{fb} represent arbitrary impedances, that is, any combination of R, C, and L exhibiting two terminals, such as Fig. 3-13a. (Actually, L is rarely used, for reasons explained earlier.) From the definition of impedance,

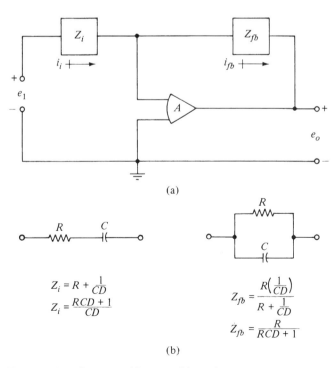

Figure 3-30 Op-amp with general impedances.

$$i_i = \frac{e_1 - 0}{Z_i(D)} = i_{\text{fb}} = \frac{0 - e_o}{Z_{\text{fb}}(D)} \tag{3-63}$$

$$\frac{e_o}{e_1}(D) = -\frac{Z_{\text{fb}}(D)}{Z_i(D)} \tag{3-64}$$

If the impedances are as in Fig. 3-30d, for example,

$$\frac{e_o}{e_1}(D) = \frac{-R_2 C_1 D}{(R_1 C_1 D + 1)(R_2 C_2 D + 1)} \tag{3-65}$$

This transfer function represents a band-pass filter, a device for rejecting signals above and below a selected band of frequencies, allowing only those signals with frequency in the pass band to get through.

We conclude our brief treatment of op-amps with a quick look at the deviations of real op-amps from the ideal assumptions used in op-amp circuit design. Many applications don't require these refinements, but they may be necessary when specifications become stringent. Then an op-amp with better-than-normal behavior with respect to some aspect of performance may be needed. Consider first the effect of noninfinite gain A on the circuit shown in Fig. 3-29a. Equation (3-55) may be manipulated to give

$$e_o = -\frac{R_{\text{fb}}}{R_i} e_1 \left(1 + \frac{1}{A} + \frac{R_{\text{fb}}}{AR_i}\right) \tag{3-66}$$

The open-loop gain A may be in the range of 10^4 to 10^8, while R_{fb}/R_i rarely exceeds 10^3; thus, the error upper limit is from about 10^{-5} (0.001%) to 0.1 (10%). The *meaning* of this error is that if one selects precision resistors for R_i and R_{fb} so as to get a precise e_o/e_1 ratio, and if the gain A is too low, the ratio will be inaccurate. Of course, if A is known and *fixed*, we could select the resistors to compensate for the error due to low A. However, as we noted earlier, A may drift in random fashion due to temperature, age, etc., reducing the effectiveness of the compensation.

The next errors we consider are those due to *offset voltage* and *bias current*. Offset voltage refers to the fact that if e_1 in Fig. 3-29a is made zero by grounding it, e_o will *not* be exactly zero, due to imperfections in the amplifier. This offset can be trimmed out at a given instant, but temperature drift will cause it to reappear. Figure 3-31a shows a model for computing error due to this effect. Analysis gives

$$e_o = e_{\text{os}} \left(1 + \frac{R_{\text{fb}}}{R_i}\right) \tag{3-67}$$

Since e_o should be zero, this is the error voltage. The best values of offset voltage e_{os} are the order of $30\,\mu\text{V}$ over a temperature range of -25 to $+85°\text{C}$, with a temperature coefficient of about $0.2\,\mu\text{V}/\text{C}°$. Inexpensive op-amps may have $10\,\text{mV}$ and $15\,\mu\text{V}/\text{C}°$. Taking an e_o full-scale voltage of 10 volts, for $R_{\text{fb}}/R_i = 1000$, the maximum error as a percentage of full scale could range from 0.3 to 100% if no attempt were made to trim the error. The "trimming" (also called nulling) referred to at several places in our op-amp discussions usually employs some simple additional circuitry with *adjustable* resistors, and is explained in detail in the references.[30]

---

[30]Jung, *I-C Op-Amp Cookbook*, pp. 125–134.

System Elements, Electrical

Bias current is the small current that flows in the amplifier input leads, even when no input voltage is applied. The model in Fig. 3-31b leads to the result

$$e_o = -i_{b1} R_{fb} \qquad (3\text{-}68)$$

Values of i_{b1} can be as small as 75 fA (1 fA = 10^{-15} amp) at 25°C. Over the range 0 to 70°C, the offset current of this unit would never exceed ±4 pA. For op-amps not optimized for low bias current, these numbers would be considerably larger.

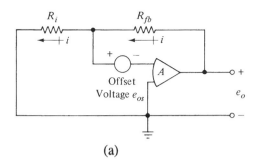

(a)

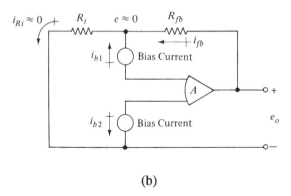

(b)

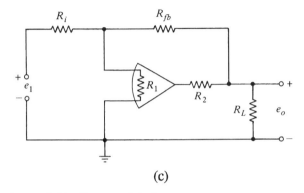

(c)

Figure 3-31 Error models for op-amps.

The next deviation of real amplifiers considered here is the noninfinite input impedance and nonzero output impedance. Figure 3-31c shows a model for these effects; note that a load resistance R_L is also shown, to represent the input resistance of any device which would be connected to the op-amp output. Analysis shows that the effect of noninfinite input resistance is equivalent to a loss of open-loop gain A, the effective value being given by

$$A_{\text{eff}} = \frac{A}{1 + \dfrac{R_i R_{\text{fb}}}{R_1 (R_i + R_{\text{fb}})}} \tag{3-69}$$

A similar effect is produced by nonzero output impedance:

$$A_{\text{eff}} = \frac{A}{1 + \dfrac{R_2}{R_{\text{fb}}} + \dfrac{R_2}{R_L}} \tag{3-70}$$

Input resistances are in the range of 10^6 to 10^{13} ohms while output resistances are the order of 100 ohms.

The speed of response of op-amps is specified in several different ways. One method considers the *closed-loop* frequency response when the op-amp is connected as a coefficient multiplier, that is, $(e_o/e_i)(i\omega)$ in Fig. 3-29a. The fastest op-amps will have this frequency response flat to about 500 MHz when the coefficient is set at 1.0 (input resistance and feedback resistance equal). For a coefficient (closed-loop gain) of 20, the flat range of amplitude ratio drops to about 80 MHz. Another method uses the *settling time* after a step input is applied to the coefficient multiplier. Times to settle within 1, 0.1, and 0.01% of the final value may be quoted. For the unit just described, and for a closed-loop gain of 2, the 1% settling time is about 6 nanoseconds. Manufacturer's data books, such as those listed in the chapter bibliography, give full specifications for a wide selection of op-amps and also contain much useful application information, including recommended circuits for many different applications. A phone call to application engineers at the manufacturer will also produce advice tailored to your specific needs.

Most op-amps are themselves able to supply only limited electrical *power* at their output terminals. Typically the output voltage is limited to about ± 10 volts and the maximum current rarely exceeds 0.05 amp. If higher power is needed to drive loads such as motors or loudspeakers, the op-amp output can be connected to the input of a separate *power amplifier*. Another approach integrates the op-amp and power amp into a single device, called a *power op-amp*. These devices provide the usual versatility of the op-amp in building circuits with useful dynamic behavior but also provide an output capable of, say, 50 volts and 10 amps. Because of the high power level, special consideration must be given to design factors such as cooling.[31]

---

[31]W. W. Olschewski, Designing with Power Op-Amps, Apex Microtechnology Corp., 5980 North Shannon Road, Tucson, AZ 85741, 800-421-1865.

System Elements, Electrical   **185**

3-9   MODELING AND SIMULATION OF COMPUTER-AIDED SYSTEMS: MECHATRONICS

Mechatronics is a term of relatively recent origin and usually is defined as a combination of mechanics, electronics, and computer technology, often in the context of a feedback control system. Automotive engine control systems are a good example. Here a multitude of sensors measure various temperatures, pressures, flow rates, rotary speeds, and chemical composition and send this information to a microcomputer. The computer integrates all this data with preprogrammed engine models and control laws and sends commands to various valves, actuators, fuel injectors, and ignition systems so as to manage the engine's operation for an optimum combination of acceleration, fuel economy, and pollution emissions.

In a mechatronic system, computer technology often allows changes in design philosophy which lead to better performance at lower cost. Coordinate measuring machines[32] are used in quality control labs and on the factory floor to quickly and accurately measure the critical dimensions of manufactured parts. These instruments usually provide accurately controlled motion along three independent (x, y, z) axes to position a sensitive probe at locations on the part that need to be measured. The "classical" design philosophy for precision gaging was to manufacture the gaging devices to the highest possible mechanical accuracy, leading to accurate but expensive equipment.

Today, many designers are using computer technology to relax the accuracy requirements on individual parts of the measuring machine, resulting in significant cost savings. Ths is possible because many of the mechanical "errors" in the machine are "systematic" (reproducible) rather than being random and unpredictable. For example, the pitch of a lead screw may not be perfectly uniform over its entire length, but the deviations from perfection of a given screw are nearly constant from day to day. In a coordinate measuring machine, we can allow such reproducible errors in all the parts and then correct for their total accumulated effect by *calibrating* each machine against an accurate standard such as a laser interferometer.[33] This calibration is done at the factory before each machine is shipped, and the measured errors are recorded in computer software which is also shipped with the machine. When the customer puts the machine into service, the computer corrects each measurement as it is taken, using the calibration data that it has "memorized." Since the mechanical errors are largely reproducible, the original factory calibration may "last" several years before a recalibration is needed. This philosophy of relaxing manufacturing tolerances on individual parts of a product and then correcting them all in "one fell swoop" by a final calibration step is a powerful tool of the designer and is being used more and more. The computer, of course, is vital since it "memorizes" the calibration data and then commands the system actuators to make the needed corrections.

In engine control systems, coordinate measuring machines, and all other computer-aided machines and processes, we always require *interfacing* between

---

[32]E. O. Doebelin, *Measurement Systems*, 4th ed., McGraw-Hill, New York, 1990, pp. 355–364.
[33]Ibid., pp. 271–277.

the largely analog world of sensors and actuators and the digital world of the computer. Figure 3-32 shows the general configuration of such systems, most of which use the feedback principle. The desired behavior of our machine or process is entered into the computer in the form of programs and numerical values. The actual behavior is measured by sensors "attached" to the machine or process. The computer compares desired and actual and if they differ, decides (using a control law we have given it) how to manage the actuators so as to bring actual closer to desired.

The interfacing operation and the internal workings of the computer itself introduce certain effects which must be modeled in the system dynamic studies needed for design and analysis of the overall system. In this section we will show how the essential features of these effects may be modeled and conveniently simulated. The main effects we consider may be described as

1. Sampling
2. Quantization
3. Computational delay

Simulation software, such as ACSL and SIMULINK, provides convenient means for including these effects in our overall system simulation.

Sampling refers to the fact that most sensors, being analog devices, provide information about the measured quantity as a smoothly varying physical signal (often a voltage). The digital computer, however, can only compute with one specific number at a time; it will *not* accept as input a smoothly varying voltage. Thus the *interface* between the sensor and computer must include a *sampling device*, which takes discrete "samples" of the smoothly varying voltage, usually at fixed time intervals. The time interval between samples must be sufficiently short or else we will lose important information present in the analog signal. Sampling requirements are usually stated in *frequency response* terms, showing again the wide utility of methods based on sine waves. A famous theorem, Shannon's sampling theorem, states that we must take at least two samples per cycle of the *highest* frequency present in our analog signal.

At this point in our study, we have shown you how to deal directly with sine waves of any frequency, but we have not shown that *all* waveforms (transients, periodic signals, random signals, etc.) can be described in terms of their frequency

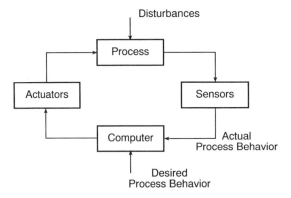

Figure 3-32 Functional block diagram for feedback control systems.

content. That is, a transient of a particular waveform, which lasts, say, 0.015 seconds, *can* be said to contain only frequencies below, say, 300. Hz. In fact, *every* physical signal will contain frequencies only up to some limiting value, and *nothing* beyond this. Thus it makes sense to talk about the highest frequency present in *any* signal, not just sine waves. The truth of this assertion will be mathematically demonstrated later in this text; we ask you to take it on faith for now.

Shannon's theorem requires only 2 samples per cycle, but in practice we find that most working systems use 7 to 10 samples per cycle. When less than 2 samples per cycle are used, really disastrous results occur, the most common being the phenomenon called *aliasing*. Here the sampled waveform shows frequencies *lower* than any truly present in the analog signal, giving completely misleading results. We can easily demonstrate this aliasing effect in Fig. 3-33, where a 10. Hz sine wave is sampled every 0.105 second, and the sampled points are then plotted as a dashed curve. To our surprise, we see what appears to be a nice sine wave of about 0.5-Hz frequency, something *totally absent* in the original analog signal! The aliasing effect in this example is quite obvious, but it can be rather subtle (but still very damaging) when we sample signals of more complex waveform. To avoid such aliasing effects, we need to sample rapidly enough to have 7 to 10 samples per cycle of the *highest* frequency present in our analog signal.

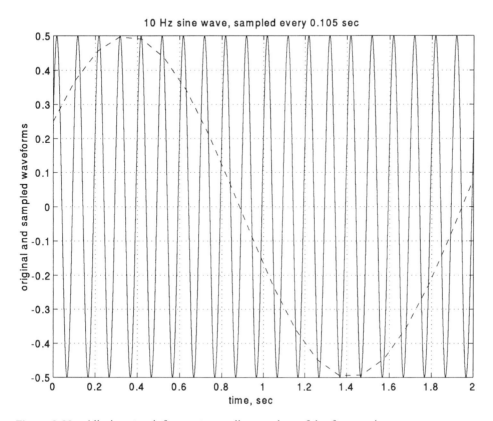

Figure 3-33 Aliasing: too infrequent sampling produces false frequencies.

In practical systems, the analog signal is not only sampled but also "held," using electronics called a *sample-and-hold amplifier*.[34] That is, the digital computer can not deal with the sampled value, which theoretically exists only for one value of time. It requires a *finite* time interval to do whatever calculations we require with the sampled value; thus sampled values are always *held constant* until the next sample is taken. In Fig. 3-34 we show a SIMULINK diagram which we will use to illustrate all of the features (sampling, quantization, computational delay) of computer-aided systems mentioned earlier. Other simulation languages, such as ACSL, provide similar functions. The icon labeled "zero order hold 1" in our diagram provides the sample-and-hold function. By double-clicking on it, one can choose the sampling interval. In our example, we choose this at 0.005 second, so the analog waveform would be sampled every 0.005 second and then held constant until the next sample is taken.

After sampling and holding, the signal is next *quantized* in an analog-to-digital converter[35] characterized by its bit value. An n-bit converter divides the total voltage range of the sampled signal into 2^n equal subranges. Each subrange takes up $1/2^n$ of the total range. For example, if the total range is 0–10 volts, a 2-bit A/D converter splits this into 4 subranges: 0–2.5, 2.5–5, 5–7.5, and 7.5–10 volts. An analog voltage which falls into the first subrange is quantized as 0 volts, one in the second range becomes 2.5 volts, the third range gives 5.0 volts, and the last range becomes 7.5 volts. For example, an analog voltage of 3.672 volts will enter the computer as the number 2.500. Clearly, converters with small n are very inaccurate, so we usually use n in the range of about 8 to 16, which gives, respectively, resolutions of 0.195 to 0.00153% of the full range. Our brief description here gives enough information for our current purpose, but you should consult the reference for more practical details when you need to select an actual A/D converter.

In Fig. 3-34 the SIMULINK icon for either an A/D or D/A converter is called *quantizer* and we apply a 5-bit A/D converter to the sampled-and-held signal produced by the zero-order hold. To set up a quantizer, just double-click on its icon and you can enter the resolution corresponding to the number of bits you want. In Fig. 3-34 I used much "coarser" quantization than is usually practical, so that the effects of quantization would be very obvious in graphical displays. Figure 3-35 shows the analog signal (a 10. Hz sine wave), the sampled-and-held signal ("zhold"), and the quantized signal ("adout") which is the number that actually enters the computer. The "inaccuracy" of a 5-bit converter is clearly shown here.

Our final feature of computer operation occurs in the computer itself rather than in the interfacing hardware (sample-hold amplifier, A/D converter). The computer itself also quantizes the data but most computers use more bits than do typical A/D converters, so the accuracy of the numbers is determined by the A/D resolution, *not* the number of bits the computer uses. When data enters the computer (either a single signal or several signals), we generally want some calculations done with this data; that is, we are using some software (programs) to process the data. These calculations take a finite amount of time, depending on how complex they are.

---

[34]Ibid., pp. 826–834.
[35]Ibid.

System Elements, Electrical

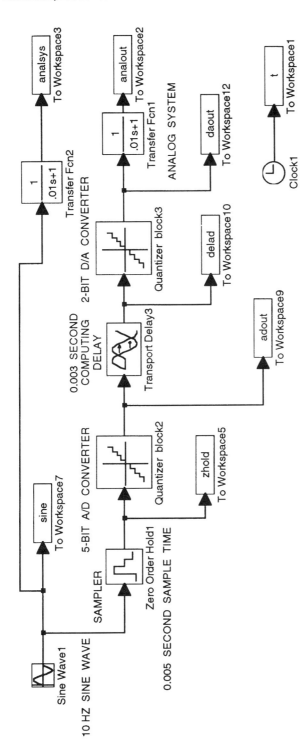

Figure 3-34 Simulation model for demonstrating sampling, quantization, and computational delay.

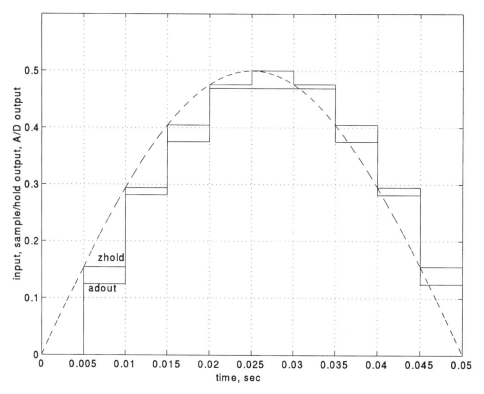

Figure 3-35 Behavior of 5-bit A/D converter.

When the calculations are finished, we can output the digital result through a digital-to-analog converter and send this to actuators (usually *analog* devices, like motors) which manipulate process variables according to our wishes. The computational delay inherent in these operations may have important effects on the behavior of the overall computer-aided system, so it must usually be included in our system models.

All simulation languages provide for the modeling of "pure time delays," also called *transport lags* or *dead times*. They regularly occur in systems which *don't* involve digital computers. For example, pneumatic pressure signals in pipelines travel at the speed of sound, about 1100 ft/sec. A pressure signal entered into a 110-ft long pipeline arrives at the far end 0.10 second later. Radio signals to and from a lunar robot vehicle travel at about 186,000 miles/sec. A human "driver" observing the lunar terrain through a TV camera on the vehicle will receive the TV picture on the earth about 1.3 seconds after the "action" happens on the moon. There will be another 1.3-second delay before steering commands from earth are received by the moon vehicle's steering system. These delays make the remote steering task difficult and must be taken into account in the design of such systems.

In Fig. 3-34 a computational delay of 0.003 is modeled with the icon called transport delay 3. As usual, we can set the desired delay by double-clicking on this icon. The delay is clearly shown in Fig. 3-36. In this example we are not doing any

System Elements, Electrical

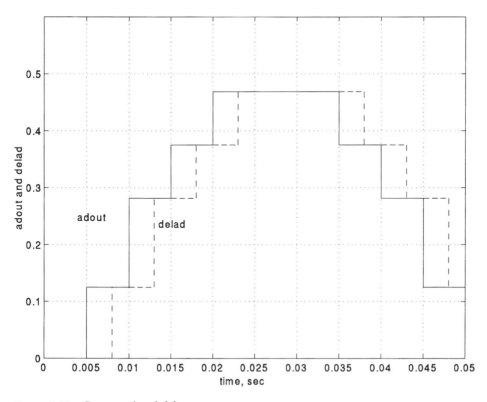

Figure 3-36 Computational delay.

calculations with the signal "adout" but are simply delaying it as if calculations *were* being done. Once calculations are complete, the computer needs to send these results out to actuators used to regulate our machine or process. Since these actuators are usually analog devices, the computer's digital output must be converted back to an actual voltage, in a D/A converter. This converter also has a bit value which determines the resolution of the voltage signal it sends out. In our example I made this resolution *much* coarser than one usually would, so that the effect would be obvious in graphs. The 2-bit D/A converter divides the ±0.5 volt range into only four subranges.

In Fig. 3-34 the D/A output, which is a sequence of step changes between the four values available to a 2-bit converter, is sent into an analog system described by a first-order differential equation. This analog system might represent some kind of actuator. The original analog signal (a 10-Hz sine wave of 0.5-volt amplitude) is also sent *directly* into an identical system. Thus we can compare the behavior of the computer-aided system with a pure analog system in Fig. 3-37. Note first that the pure analog system starts to respond *immediately* just after $t = 0$, while the digital system, due to sampling and computational delay, has *no* response at all until $t = 0.008$. When it *does* respond, its response is "jerky" compared with the smooth behavior of the pure analog system. Due to coarse quantization, its response is also *smaller* than it should be. Lest you now believe that computer-aided systems are "no good," we hasten to remind you that *all* of these bad features can usually be vastly

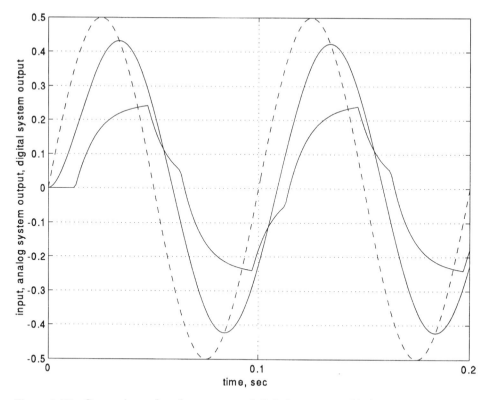

Figure 3-37 Comparison of analog system and digital computer-aided system.

improved by using faster computers (less delay), more frequent sampling, and finer quantization. We gave a "horrid" example only to make the basic phenomena more obvious to readers who may never have seen them before.

With respect to the aliasing phenomenon caused by too-coarse sampling, we sometimes include in our systems *anti-aliasing filters*. These are low-pass filters which are placed just before the sampling device. Their purpose is to filter out any frequency content higher than the highest frequency of interest to us. Suppose that the highest frequency of interest were, say, 100 Hz, and we therefore selected a sampling rate of 1000 samples/second (10 points per cycle). Suppose also that our analog signal contained some frequencies (perhaps electrical noise) that were higher than 100 Hz. If these high frequencies got to the sampler, they would cause aliasing because our sampling rate is adequate only up to 100 Hz. A low-pass anti-aliasing filter could be used to allow frequency content below 100 Hz to pass through to the sampler unaffected, but filter out the offending higher frequencies before they got to the sampler and caused aliasing.

Design Example: A Feedback-Type Motion Control System. To illustrate the practical application of the simulation tools just described, we will now study the operation of a computer-aided motion-control system of the feedback type. The task of such a machine is to accurately and quickly position a translatory load in response to commands from a computer or a human operator. Such motion-control systems

System Elements, Electrical **193**

have myriad applications in manufacturing processes, such as integrated circuit processing, micromachine fabrication, assembling circuit boards, precision measuring machines, etc.

When a load mass is to be moved in translation, we must first select the type of actuator to provide the necessary force. Possibilities include pneumatic cylinders, hydraulic cylinders, rotary electric motors with rack-and-pinion or lead screw motion converters, or translatory electric motors. When the needed force is modest and the stroke is not very large, engineers often will use a *voice-coil actuator*[36] for the task. This device is very similar to a loudspeaker, except it is designed for a longer stroke and positions a load mass rather than the speaker's paper cone. A cylindrical shell called the armature is wound with many turns of copper wire, forming a coil, which is firmly fastened to the armature with epoxy cement. The armature is placed in the field of a permanent magnet. When current is passed through the coil, a magnetic force proportional to the current is produced. By attaching the armature to the mass which is to be positioned, the magnetic force is applied to the mass, causing it to move.

In a feedback-type motion-control system we measure the mass's displacement with a suitable sensor, creating a voltage proportional to displacement. This voltage is sampled and digitized so that we can send information about the mass's actual position to our computer. There the actual position is compared with the desired position. The desired position might vary with time in a specific way. We could write a program which told the computer how the desired position varied with time. At each sampling instant the computer would compare the digital value of the desired position with the digitized value of the measured position. If there is a difference between actual and desired positions, the computer sends a control signal, through a D/A converter to a power amplifier which sends current to the voice coil. The magnitude and direction of the current are such as to drive the load towards the desired position. When desired and actual positions are the same, no magnetic force is needed and thus no current is produced.

Figure 3-38 shows a pictorial and block diagram of such a system. The load is modeled as a pure/ideal mass element and a pure/ideal damper (viscous) element. The damper represents the shearing of an oil film which separates the moving mass from its guideway; we neglect any coulomb ("dry rubbing") friction. The sensor is assumed to measure the displacement perfectly and instantly. The power amplifier is assumed to produce a current instantly proportional to the displacement error signal provided by the computer. In electromechanical systems, the analog electronic portions of the system, while not really instantaneous in response, are usually so much faster than the portions with moving parts that the assumption of no lag is a reasonable one.

The design of all types of feedback control systems, including the motion control system just described, rests on a well-developed theory and set of routine design tools.[37] Academic courses on this subject usually come *after* a system

---

[36]Voice-Coil Actuators: An Applications Guide, BEI Motion Systems Co., 150 Vallecitos de Oro, San Marcos, CA 92069, 619-744-5671, 1996.

[37]E. O Doebelin, *Control System Principles and Design*, Wiley, New York, 1985.

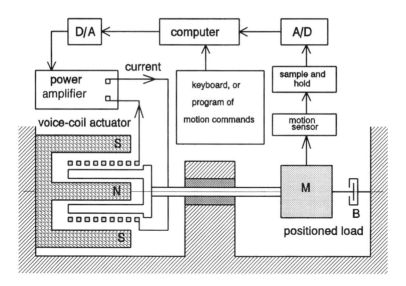

Figure 3-38 Computer-aided motion-control systems.

dynamics course. While study of this background theory certainly speeds design and protects against foolish mistakes, it is surprising how much we can do right now, using simulation methods. Our upcoming study is not intended to supplant courses in control systems but rather serves a motivational purpose since it uses available tools to analyze and design a practical machine.

We first need to get some numerical values for system parameters. The mass of the object to be moved would be known at the beginning of the problem, but the total moving mass includes also the mass of the coil member in the voice-coil actuator. These actuators come in various sizes, so we need to choose a particular one, at least tentatively. (Once analysis of a tentative design starts, we may uncover facts that require changes in our initial choices of parameter values.) Let's assume that the *total* moving mass weighs $0.5\,lb_f$, making its mass $0.5/386 = 0.001295\,lb_f\text{-sec}^2/\text{inch}$. Note that while the British unit of mass, when we are using length in feet, has its own name, the *slug*, the mass unit when inches are being used has never been given its own name. Since the damper in our system represents the shearing of a lubricating oil film on the guideway bearings, we would probably estimate a numerical value for B by making some force/velocity measurements for these bearings. Let's assume that the force/velocity curve plotted from this data is reasonably linear with a B value of $0.259\,lb_f/(\text{in/sec})$.

A catalog from the voice-coil actuator manufacturer will give us an estimate of the constant relating magnetic force to coil current for the actuator we have selected. Catalog values are *always* estimates since they are typical or average values for the maker's production. Once we have the actuator in our lab, it is wise to test it to get a more accurate value of its force constant. We just apply different known coil currents, measure the resulting force, and plot a force versus current graph to check for linearity and get a number for the force constant. Let's assume the catalog estimate is all we have for now and that it is $0.50\,lb_f/\text{amp}$. The coil current is supplied by a power amplifier which gets its input voltage from the output of the D/A converter.

System Elements, Electrical

Digital-to-analog converters typically produce full-scale voltages of, say, ± 5 volts, but can provide only about 10 milliamps of current, not enough to drive a motor, so an external power amplifier is needed.

We here use the type called a *transconductance* amplifier; that is, it accepts a voltage input, but produces a proportional *current* at its output. Thus its amplification factor is not the usual volts/volt, but rather amps/volt. Such amplifiers are used to speed up the response of circuits containing inductance, such as the coil of our motor. If we used a "voltage" amplifier, the coil current would be an unknown in a Kirchhoff voltage-loop equation containing a resistive drop, an inductive drop, and a back emf (generator effect present in every motor). This makes analysis more difficult (though certainly possible) and also results in a *slower* response of current to voltage, caused by the inductance. The transconductance amplifier *suppresses* these undesired effects by *enforcing* the commanded current. This may appear somewhat "magical," but really can be implemented, since the transconductance amplifier is *itself* a feedback device which measures its current and if it is not what is commanded, supplies a sufficiently high *voltage* to "make the current behave."

While the moving mass is more or less fixed because the load to be driven is given, and the damper is also fixed by the bearings being used, the gain of an amplifier is readily adjustable. This is fortunate because there *must* be several adjustments possible in any practical design if we are to meet given performance specifications. If everything is given, there is nothing left to design! At this point we can write a Newton's law for the moving mass:

Magnetic force + damper force = mass × acceleration

$$K_{mf}i - B\frac{dx}{dt} = M\frac{d^2x}{dt^2}$$
$$e_{D/A}K_{pa}K_{mf} - B\frac{dx}{dt} = M\frac{d^2x}{dt^2} \tag{3-71}$$

Here $e_{D/A}$ is the voltage output of the D/A converter, $K_{mf} = 0.5\,\text{lb}_f/\text{amp}$, and K_{pa} is the power amp gain, which we now choose as 2.0 amps/volt, as a trial value within the capability of our selected amplifier. Motion sensors are available in a wide range of sensitivities and provide another component whose steady-state gain is easily adjustable. Let's use 1.0 volt/inch for a trial value.

The simplest mode of smooth control is called *proportional control*. Here the controller (we will use a digital computer) simply multiplies the error between desired and actual position by a constant number. If this number is too low, the system will be slow and inaccurate. If this number is too high, the system will be unstable and possibly destroy itself with wild oscillation! With a digital computer, it is of course very easy to set this number at any value we wish. It is now perhaps time to display a simulation diagram for our system; see Fig. 3-39. We enter the commanded position digitally from our keyboard, or by means of a computer program written to move the load mass according to our needs. In Fig. 3-39 we enter a simple step command of 0.0823 inch.

Our sensor, an analog device, has its output digitized with a 7-bit A/D converter, giving a resolution of 0.0078125 inch. I have intentionally made this quantization coarser than normally used, to illustrate a certain behavior. Once the

196 Chapter 3

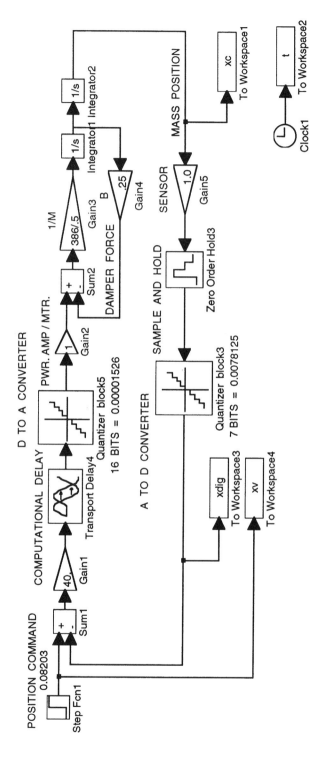

Figure 3-39 Simulation diagram for computer-aided motion-control system.

System Elements, Electrical   **197**

displacement information is digitized, we can compare the actual and desired positions by a subtraction in the computer, producing a digital error signal. For our proportional control law, the computer multiplies this error by 40.0, our trial value. We next model the computational delay with a dead-time, which we can set at zero or any other value of our choice. The computer itself carries many digits, so it does not contribute significantly to quantization error. I chose the D/A quantization at 16 bits, which gives a resolution of 0.00001526 volt for the ±5 volt range of the D/A, so there will be negligible quantization error here also.

In Fig. 3-39 the power amp and motor have been combined into a single block with a gain of (2.0 amps/volt) (0.5 lb$_f$/amp) = 1.0 lb$_f$/volt. The Newton's law equation is implemented in our usual way with a summer, gain blocks for $1/M$ and B, and two integrators. For our first test, I set the computational delay at zero. Delays of *any* kind degrade feedback system performance, so we start with the best possible situation first. As mentioned earlier, the sensor quantization was deliberately set too coarse, to illustrate an important general feature of digital systems. The commanded position is 0.08203 inch, whereas the only values available from the A/D are increments of size 0.0078125. Thus the tenth increment would be 0.078125 and the eleventh would be 0.0859375. That is, the system is "looking for" 0.08203 but "can't find it"; thus it will "hunt" or oscillate between the two values it *can* find. This hunting is usually unacceptable but *must* occur when quantized (rather than smoothly varying) signals are used. Of course one usually can (and does) set the sensor quantization resolution fine enough to make the hunting unnoticeable. (Real effects such as Coulomb friction in the load bearings may also stop the hunting.)

Figure 3-40 shows this behavior for the system parameters set as in Fig. 3-39. Note that the load displacement xc is a smooth curve but its quantized measurement xdig follows it inaccurately in a stepwise fashion. Also, xc never "settles down," but goes into a continuous oscillation, called a *limit cycle*. The average value of this xc oscillation is also *not* the desired value of 0.08203, but is offset from it by a *steady-state error*. In Fig. 3-41 I have made the A/D resolution equal to the D/A (0.00001526) and all these problems disappear. Actually, xc is *still* hunting around the desired value but now the amplitude is so small that it can't be seen on the graph.

All the above results were obtained with a computational delay of zero, which may be unrealistic. Actually, the computational delay is important only *relative* to analog "delays" present in the rest of the system. Thus if the computer is controlling a very slow analog system, the computer's delay may be relatively so small that it could be neglected in analysis. Temperature control systems for buildings are a good example; one can't change such temperatures quickly. In our present example, the mass/damper system does not respond instantly when force is applied; it takes some time for force to cause a displacement. We shall see in later chapters that the mass/damper system is a *first-order* dynamic system and has a *time constant* given by $M/B = 0.005$ second. While this is *not* the same as a dead time of 0.005 second, it *is* a dynamic lagging effect.

To show the bad effect of excessive computational delay in computer-aided feedback systems I next set this delay at 0.008 second, a little larger than the 0.005-second analog lag. I am still using the "good" resolution on both the A/D and the D/A. Figure 3-42 shows the (disastrous!) result. The system has become *absolutely unstable*, with oscillation of ever-increasing amplitude. Such oscillations may cause mechanical or electrical failures if left unchecked. What may also happen is that as

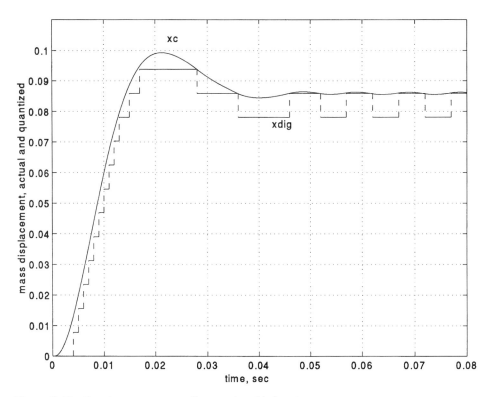

Figure 3-40 Step input response of computer-aided system.

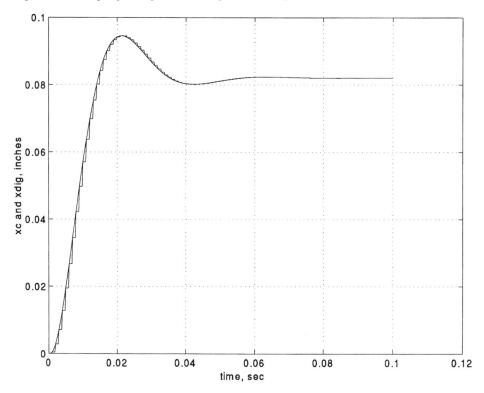

Figure 3-41 Response improved with finer quantization.

System Elements, Electrical

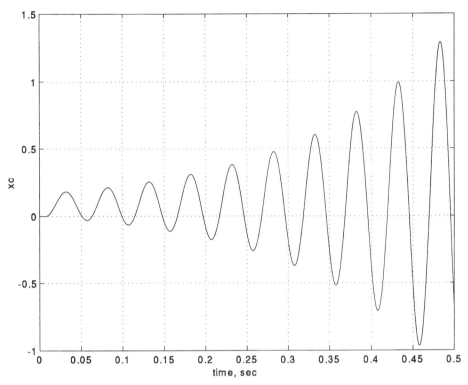

Figure 3-42 System instability caused by excessive computational delay.

the oscillation amplitude increases, various system components which were assumed to behave linearly now go nonlinear, usually by "saturating." For example, any real amplifier has a *limit* on its output current; it can't produce more than a certain number of amps. When this limit is reached, the oscillations will no longer build up, they will continue at a *fixed* amplitude, but this "limit cycling" is of course also unacceptable. Since SIMULINK provides a saturation module, we could easily include this realistic nonlinear feature in any simulation that might need it.

While excessive computational lag will always make a feedback system unstable, if such a lag is unavoidable we can still get a stable system by compromising some other aspect of system performance. In our present example, we used a gain of 40 in the proportional controller; this gave a certain speed of response when the computational lag was taken as zero. In general, reduction of controller gain in feedback systems usually improves stability, but at the expense of speed and accuracy. To show this, we reduce this gain to 10 but retain the 0.008-second delay that caused instability in Fig. 3-42. This design change results in a stable system but one which is somewhat slower to respond (see Fig. 3-43).

Our motion control example could be expanded in many useful ways, using our simulation tools, but we choose to conclude it at this point. Those readers who may later go on to a complete course in automatic control will there learn many useful analysis and design methods which enable one to wisely choose initial configurations

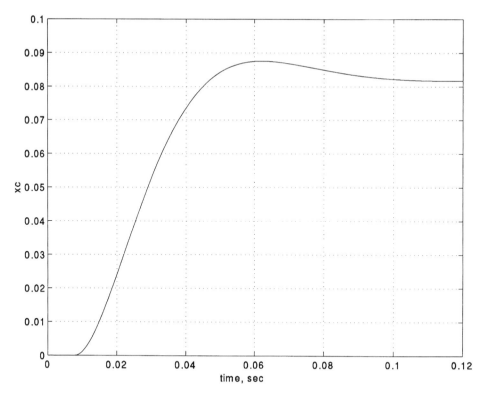

Figure 3-43 Stability recovered by system gain reduction, but with loss of speed.

for control systems and also make reasonable choices of starting values for parameters. Such choices allow simulation studies to start fairly close to the final design, thus saving much engineering time and computer expense.

BIBLIOGRAPHY

1992 Amplifier Applications Guide, Analog Devices, One Technology Way, P.O. Box 9106, Norwood, Mass., 02062-9106, 1992.

Horowitz, P., and W. Hill, *The Art of Electronics*, 2nd ed., Cambridge University Press, New York, 1989.

Irvine, R. G., *Operational Amplifier Characteristics and Applications*, 3rd ed., Prentice-Hall, Englewood Cliffs, N.J., 1994.

Jung, W. G., *I-C Op-Amp Cookbook*, 3rd ed., Howard W. Sams & Co., Indianapolis, Ind., 46268, 1986.

Linear Design Seminar, Analog Devices, One Technology Way, P.O. Box 9106, Norwood, Mass., 02062-9106, 1994.

Rizzoni, G., *Principles and Applications of Electrical Engineering*, 2nd Ed., Irwin, Homewood, Ill., 1996.

System Elements, Electrical

PROBLEMS

3-1. Calculate the wavelengths of a television signal of frequency 80 MHz and a radar signal of frequency 2000 MHz. Estimate the size at which a field-type treatment of components would become advisable.

3-2. Figure P3-1 shows a test set up for measuring static current/voltage relationships such as that of Fig. 3-2c. Plot this data and check to see how well it conforms to the relation $i = Ke^4$. Also find the linearized incremental resistance for an operating point of 10 volts. Plot the power dissipated versus both i and e.

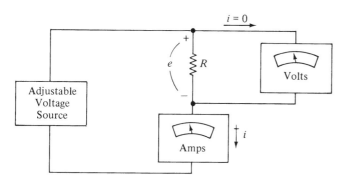

Figure P3-1

3-3. In the ohmmeter of Fig. P3-2, the current is fixed at 0.0001 amp. If we wish to measure resistances from 1 to 10,000 ohms, what full-scale voltage ranges are needed in the voltmeter? A more rugged voltmeter (higher voltage range) could be used if the current were increased. Is there any disadvantage to this? When an ohmmeter is used to measure a nonlinear resistance, what does the reading mean?

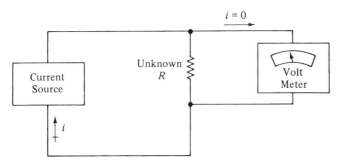

Figure P3-2

3-4. A particular 0.25-watt resistor can dissipate heat to its surroundings at the rate of 1.75×10^{-6} Btu/sec for each degree Fahrenheit of temperature rise above ambient. When the temperature rises, the resistance changes by 0.05% for each degree F. How much does the resistance change for a load change from 0 to 0.25 watt?

3-5. What is the maximum steady voltage which may be applied to a 1000-ohm, 0.5-watt resistor? If the voltage is a pulse waveform as in Fig. P3-3, what is the

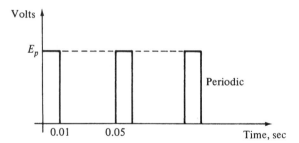

Figure P3-3

maximum allowable? (*Hint*: Use the average power as the criterion.) This technique is sometimes used to increase the output signal of strain-gage transducers (devices for measuring force, pressure, or acceleration). The strain gages are resistors which can only stand so much heating, but their output signal depends on the *peak* value of the transducer excitation voltage. By making this excitation similar to that in Fig. P3-3, the output voltage can be increased without causing overheating.

3-6. A 1000-ohm resistor has a voltage $e = -3 + 4t + 5e^{-t} + 6\sin 10t$ across it, where e is in volts and t is in seconds. Using available digital simulation, find and plot versus t the current, the instantaneous power dissipation, and also the total energy dissipated, for t from 0 to 1 second.

3-7. In xy plotters (which you may have used in a lab), the time-sweeps are generated by applying a ramp voltage to the servo drive system of one of the axes. The ramp voltages are made by sending a constant current to a capacitor. If the servo system sensitivity is 0.10 inches/volt and a current of 0.00001 amp is used, what capacitance is needed to give a sweep speed of 1 inch/sec?

3-8. Find and sketch the current needed to produce the voltage of Fig. P3-4 across a 1-μF capacitor. Using available digital simulation, find and graph versus time the current, instantaneous power, and stored energy of this capacitor. Assume the capacitor was initially ($t = 0$) charged to 5.0 volts.

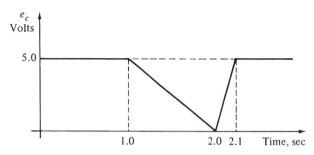

Figure P3-4

3-9. Find the current through a 1-μF capacitor if the voltage across it is $e = 1\sin t + 1\sin 10t + 1\sin 1000t$ volts, t in seconds.

3-10. Find and sketch the voltage across a 1-μF capacitor if the current is as shown in Fig. P3-5.

System Elements, Electrical

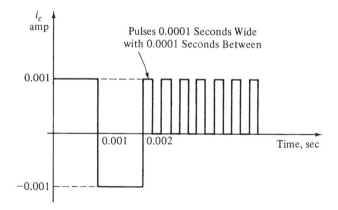

Figure P3-5

3-11. In Fig. P3-6 the mutual inductance between coils 1 and 3 and between coils 1 and 2 is negligible, while that between 2 and 3 is 0.01 henry. Find e_1 if $i_1 = -3 + 1000t$ amps, and $i_2 = 10 - 3000t$ amps, t in seconds.

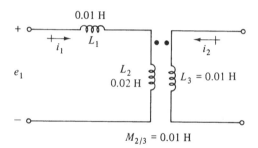

Figure P3-6

3-12. For a 0.001-henry inductance with current as in Fig. P3-7, find and sketch the voltage.

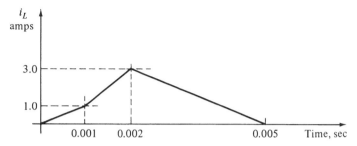

Figure P3-7

3-13. Find the current through a 1-henry inductor if the voltage across it is $e = 1 \sin t + 1 \sin 10t + 1 \sin 1000t$ volts, t in seconds.

3-14. It is desired to use a 50,000-ohm film resistor of the type shown in Fig. 3-16. What is the highest frequency at which it behaves like a pure resistor?

3-15. What would be an appropriate circuit model and range of numerical values for a 500,000-ohm resistor of the type shown in Fig. 3-17b. How would the model change if the resistance were 100 ohms?

3-16. Estimate the capacitance of 10 feet of coaxial cable with

$$a = 0.01 \text{ inch} \qquad b = 0.03 \text{ inch}$$

The cable's insulation has a dielectric coefficient $K = 5.0$. When we use piezoelectric sensors[38] to measure force, pressure, or acceleration, we must know the capacitance of the cable which connects the sensor to the voltage-measuring instrument.

3-17. For a capacitor constructed as in Fig. 3-19a, with electrodes 0.00025 inch thick, and Mylar dielectric 0.0005 inch thick, estimate the total area needed for a 1-μF unit. If it is 2 inches long, estimate the diameter.

3-18. In Fig. 3-20, what are the values of R_s, R_p, and L in a pure capacitor?

3-19. For a solenoid as in Fig. 3-22, with $a = 1$ inch and $b = 4$ inches, find the number of turns needed to get $L = 0.25$ henry. If the wire used has a diameter of 0.01 inch and a resistance of 100 ohms/1000 feet, what will be the resistance of the "inductor"? Using Eq. (3-52), compute Q for $\omega = 100,000$ rad/sec. If, instead of an air core, a magnetic core with relative permeability of 1000 is used, what would L be?

3-20. For the inductor of Fig. 3-23d ($L = 60$ henries, $R = 5160$ ohms) use Eq. (3-52) to estimate Q and compare with the experimental result. Comment on the discrepancy at high frequency. Could R_e of Fig. 3-23b be estimated from the data given and the formula for Q given in the text? Explain how.

3-21. What voltage is produced by an open-circuited ideal current source? What current is produced by a short-circuited ideal voltage source? Discuss.

3-22. How would you set up and run an experiment to determine a model for the electrical source represented by the 110-volt, 60-Hz wall plug in your laboratory?

3-23. In Fig. P3-8, the voltage source provides $10 \sin \omega t$ volts (t in seconds). Calculate and sketch the current amplitude versus frequency for R, L, and C separately. Can we add these amplitudes at any frequency to get the amplitude of the *total* current supplied by the voltage source? Explain.

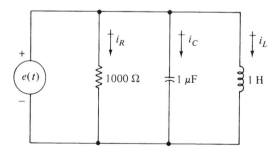

Figure P3-8

---

[38] E. O. Doebelin, Measurement Systems, 4th ed., McGraw-Hill, New York, 1990, pp. 261–268.

System Elements, Electrical   205

3-24. A complex circuit's impedance was measured at 60 Hz and found to be $2827 \,\underline{/-38°}$ ohms. If we apply a voltage $53.4 \sin 377t$, what will be the current? If the voltage were $53.4 \sin(377t + 10°)$ volts, what would the current be?

3-25. If $R_{fb} = 1.0$ megaohm, design an op-amp summing circuit to implement: $e_o = -4.2e_1 - 6.7e_2 - 10.0e_3$.

3-26. We are measuring acceleration with an accelerometer which provides a voltage output of $2.5 \sin 1000t$ when the acceleration is given by $3860 \sin 1000t$, inch/sec^2. Design an op-amp integrator which will accept the accelerometer output voltage as its inplut. We want 1 volt coming out of the integrator to represent 1 inch/sec of velocity.

3-27. Using only R's and C's, design op-amp circuits to produce the following transfer functions for $(e_o/e_1)(D)$:

   a. $5D/(10D + 1)$
   b. $(0.01D + 1)/(0.001D + 1)$
   c. $-1000/D$
   d. $(0.001D + 1)/(0.01D + 1)$

3-28. The motion-control system of Fig. 3-38 can be modified in several ways to accommodate various practical applications. Using available digital simulation, investigate system behavior for the following situations.

   a.   Suppose the load being positioned includes a linear spring between the mass and the frame. Try various spring constants and adjust the amplifier gain if needed to achieve system stability.
   b.   Sometimes the load mass is subjected to an external force which tries to push the load away from the commanded position. Model this force as a step input and see how the system responds to it.
   c.   The spring effect of part (a) and the disturbing force of part (b) each produce a steady-state error in the position. That is, the load does *not* come to rest at the commanded position. Both these error effects can sometimes be alleviated by use of a more sophisticated controller. Instead of the proportional controller, try a proportional-plus-integral type. Model this by sending the position error signal to a gain block and also to an integral block followed by a gain, and then sum these two signals in a summer. Since integral control can reduce stability, you may have to adjust the gains to get a stable system.

4

SYSTEM ELEMENTS, FLUID AND THERMAL

4-1 INTRODUCTION

While the lumped-parameter approach usually employed in system dynamics has been successfully applied to the analysis and design of many different types of physical systems, some areas have used the approach more than others. Many electrical systems are originally conceived by their designers thinking in terms of putting together a combination of R, L, C, op-amps, and other integrated-circuit modules to achieve a new device with some useful function. Workers in these areas find a wide selection of such components available to implement their circuit concepts. Mechanical systems on the other hand, are rarely initially conceived in terms of some sort of connection of K_s, B, and M. Rather, designers draw upon their knowledge of basic mechanisms (cams, gears, linkages, etc.), various power sources (hydraulics, pneumatics, electric motors, etc.), sensing instruments, and control schemes to create a system which will, at least nominally, perform the desired functions. To check the details of performance of this proposed system, a dynamic model must often be formulated and analyzed, and at this point, lumped-parameter system analysis may be very useful.

Fluid and thermal systems follow a somewhat similar pattern in that system dynamics may receive relatively light conscious emphasis during the early conceptual phases. Furthermore, due to the generally less-well-defined shapes of bodies of fluid (as compared to solid bodies) and the fact that heat flow rarely is confined to such simple and obvious paths as is current in an electric circuit, these types of systems may appear to be less well suited to the lumped-parameter viewpoint. Alternative approaches, such as those of computational fluid dynamics, however, may be better suited to studies of components rather than complex systems. System dynamics can preserve the identity of individual components while comprehending the entire system, and thus often gives insights into needed design changes at both the component and system levels. We thus recommend the consideration of system dynamics methods for thermal and fluid machines and processes, even though they initially seem less well suited to these more amorphous systems.

206

System Elements, Fluid and Thermal

Fluid and thermal systems are generally less familiar than mechanical and electrical ones to engineering students at the sophomore level, since mass/spring oscillators and simple electrical circuits are often introduced in beginning physics courses. As a prelude to our detailed discussion of the fluid and thermal elements, we thus want to show and explain some practical examples of actual hardware which has been successfully studied using system dynamics tools. Figure 4-1 shows the propellant feed system for a liquid-fueled rocket engine.[1] Such rocket engines may exhibit undesired unstable oscillatory behavior of several types; the one called *chugging* refers to relatively low-frequency (100–500 Hz) combustion-chamber pressure oscillation. A dynamic model of this system is needed to allow analytical prediction of those sets of system parameters which result in stable operation and those which will be unstable. Also, it is desired to investigate the feasibility of adding oscillation-suppression devices to unstable systems to stabilize them. The model used in this study had to take into account the inertia, springiness (fluid compliance), at the turbopump inlet due to cavitation, and at the combustion chamber injector dome,

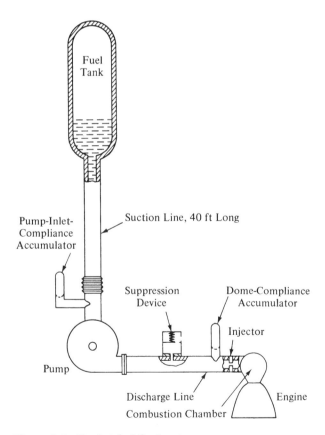

Figure 4-1 Rocket fuel feed system.

---

[1] D. J. Wood and R. G. Dorsch, Effect of Propellant-Feed-System Coupling and Hydraulic Parameters on Analysis of Chugging, NASA TN-D-3896, 1967.

the combustion chamber pressure/flow dynamics, and the pump pressure/flow relation. We will see shortly that fluid systems can be modeled in terms of three basic elements (inertia, elasticity, and friction) just as we found for mechanical and electrical systems.

In Fig. 4-2 we see a fluid flow model of the cooling water systemm of the Scattergood steam power plant of the city of Los Angeles, California.[2] Seawater is drawn into the system from the ocean, then proceeds through various valves and chambers to the steam condensers where it cools and condenses the steam and is itself thereby heated, and finally returns to the ocean which is used as a heat sink for the power plant. The model shown is intended for analysis of dynamic response of fluid flows and pressures only; the thermal aspects of the processes are essentially "decoupled" from the fluid mechanics in such a system. A thermal analysis (which would require its own model) might also be quite important in such a system, not only to ensure proper operation of the steam plant, but also to protect the ocean environment from thermal pollution.

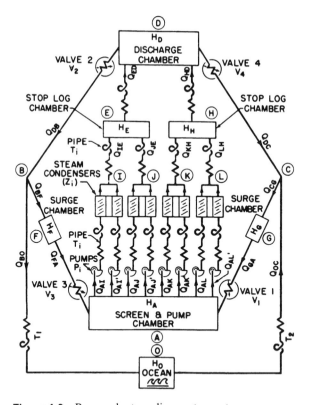

Figure 4-2 Power plant cooling water system.

---

[2]A. Reisman, On a Systematic Approach to the Analysis and Synthesis of Complex Systems Involving the Unsteady Flow of Fluids, ASME Paper 64-WA-FE-36, 1964.

System Elements, Fluid and Thermal

Large aircraft such as the DC-10 of Fig. 4-3,[3] use hydraulic power for many functions, such as control-surface positioning, wheel steering and braking (including antiskid systems), and landing-gear actuation. A modern jet transport may have as much as 2000 hp of hydraulics on board. Since system reliability is so important to passenger safety, the use of redundant ("backup") systems, as in Fig. 4-3, is common. If engine and/or hydraulic failures occur in one of the systems, a backup system takes over to allow continued safe flight. Hydraulic power is generated by engine-driven pumps and is distributed to motors, cylinders, and valves by metal tubing. The motors and cylinders convert the fluid power back into mechanical power used to move the ailerons, rudder, etc. Some of these hydraulic systems are part of the autopilot system, which is a feedback system, thus dynamic behavior is very important to the stability and control of the entire aircraft. Modeling of such systems requires mechanical elements, fluid elements, and fluid/mechanical transducers (energy converters) such as pumps and motors. We might also mention that the motion of the aircraft itself is usually modeled using standard system dyamics tools such as transfer functions.

Figure 4-4[4] shows the overall layout and transmission details of a German city bus which uses computer-controlled hydraulics to achieve energy savings of about 25% compared to conventional buses. A conventional diesel engine is the power source, but clever use of hydraulics results in good performance and significant energy savings. It has long been known that automotive efficiency could be improved if continuously variable (rather than conventional stepwise) transmissions were used to connect the engine to the drive wheels. The discrete choices of the conventional first, second, and third gears preclude operation of the engine at maximum efficiency. By using variable-displacement hydraulic pump/motors, the "gear" ratio can be continuously adjusted to allow maximum efficiency at all stages of acceleration.

Further efficiency improvements are obtained by using the hydraulic motors as pumps during deceleration when the bus needs to stop. That is, instead of wasting the bus's kinetic energy with the usual friction brakes, this kinetic energy is used to drive the pumps, producing fluid power which is temporarily stored in hydraulic accumulators, where nitrogen gas is compressed by the flow of high-pressure oil. The shaft torque to drive the pumps acts as a braking torque on the bus axle. The stored energy of the compressed nitrogen is later used to drive oil through the hydraulic motors to help the bus accelerate during the next start/stop cycle. Coordination of the engine and transmission is managed by a microcomputer.

The human body utilizes many flow processes in its operation. Perhaps the most obvious of these are the respiratory system, involving the flow of air, and the circulatory system, providing a vital flow of blood to all parts of the organism. Engineers are involved in the modeling of such systems, both because they have the fluid mechanics expertise which many medical people lack, and also because they are designing machines to aid doctors in their work, including artificial organs

---

[3]The DC-10 and Its Hydraulic System, Vickers Aerospace Fluid Power Conference, 1968.

[4]Hydro Bus with Stepless Transmission and Hydraulic Energy Recovery, RIQ International Edition, 4/1989, Mannesman Rexroth.

HYDRAULIC POWER SYSTEM AND PIPING

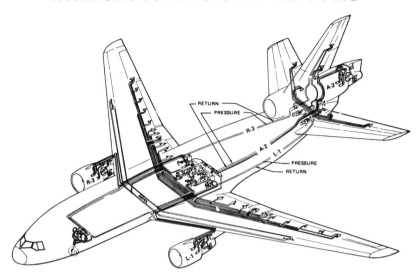

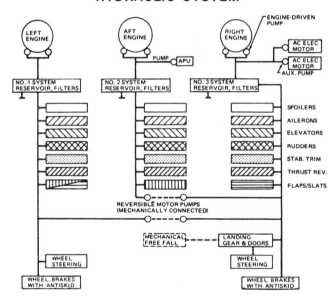

Figure 4-3 Aircraft hydraulic system.

System Elements, Fluid and Thermal

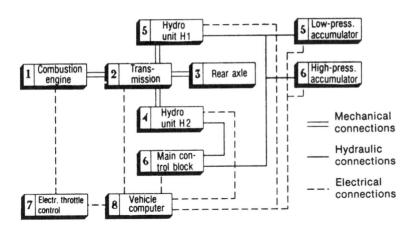

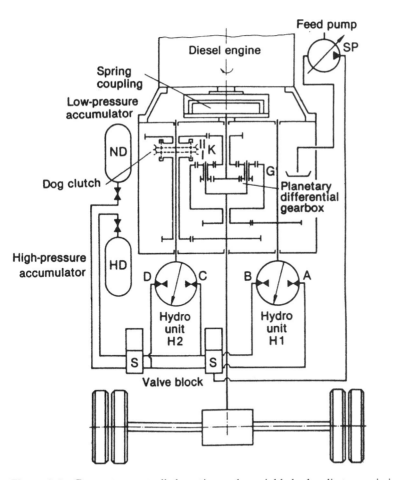

Figure 4-4 Computer-controlled continuously variable hydraulic transmission for a city bus.

212   **Chapter 4**

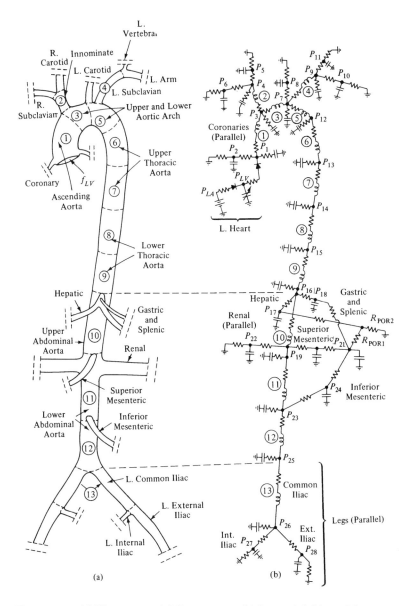

Figure 4-5 (a) Human arterial flow system, (b) lumped fluid model.

such as heart valves and even complete hearts. Figure 4-5[5] shows a rather comprehensive model of the human systemic arterial system based on lumped-parameter modeling methods. The diagram of Fig. 4-5b looks like an electric circuit, but only because the symbols for electrical resistance, capacitance, and inductance were used

---

[5]M. F. Snyder, V. C. Rideout, and R. J. Hillestad, Computer modeling of the human systemic arterial tree, *J. Biomechanics*, vol. 1, 1968, pp. 341–353.

System Elements, Fluid and Thermal 213

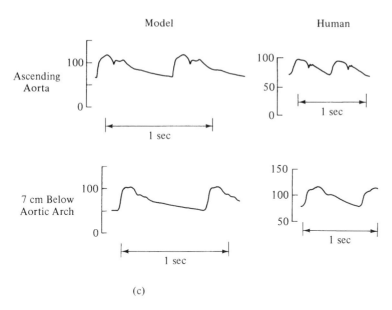

(c)

Figure 4-5 (Continued). (c) Pressures in mmHg versus time.

to represent fluid resistance (friction), compliance (springiness), and inertance (inertia).

Actually, the fluid mechanics equations were directly simulated on an electronic analog computer, a popular tool of system dynamics in the 1950–1970 period; an intermediate electrical model was not employed. The fluid model and equations would not change today, but we would now employ digital simulation software rather than analog computer hardware as our analysis tool. The ultimate test of any mathematical model is to see whether it can duplicate behavior measured in the actual system. If it does this faithfully, it can be a very useful tool, since "experiments" can be run on the model much more easily, quickly, and safely than on a human subject. Figure 4-5c shows some pressure-time records indicating this model to be quite realistic.

Polymerization processes in the chemical industry require careful temperature control, and are often carried out in jacketed kettles[6] as in Fig. 4-6. The kettle is loaded with monomer, catalyst, and water, and is initially heated to start the reaction. A constant desired temperature is maintained by proportioning the flow rates of steam and cold water to the kettle jacket. This temperature control becomes particularly difficult when the reaction becomes exothermic (producing heat rather than requiring heat addition), because this situation is inherently unstable; more heat produces higher temperature, which causes a faster reaction, which produces more heat, etc. The control system stabilizes this process by providing just the right amount of cooling to maintain an essentially constant temperature, as desired.

---

[6]G. L. Rock and Lee White, Dynamic analysis of jacketed kettles, *ISA Journal*, March–April, 1961.

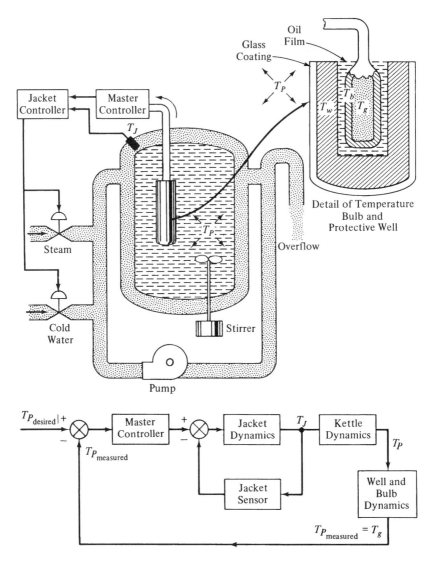

Figure 4-6 Chemical process temperature control system.

A careful dynamic analysis of this thermal system is necessary to achieve proper operation. The control loop on jacket temperature T_J can usually be made "tight" enough so that T_J follows its command very rapidly, and thus these dynamics are neglected in analyzing the rather slow overall system. In *designing* the T_J loop for this behavior, however, one would have to consider its thermal dynamics. These involve heat flows due to mass flows of water and steam, heat transfer through the walls of the jacket, and heat storage in the jacket fluid. Similarly, for the process vessel itself, there is heat transfer from the jacket, and endothermic or exothermic reaction heat, frictional heat due to the stirring device, and heat storage in the process fluid. Additional important thermal dynamics are found in the temperature measuring device and its protective well. Heat flows from

System Elements, Fluid and Thermal   **215**

the process fluid through the thermal resistance of the protective well, and is stored in the metal wall when its temperature rises. This heat flow continues through the oil film separating well and thermometer bulb, and into the thermometer bulb wall, where part of it is stored when the bulb wall temperature rises. Since the thermometer[7] is one using the pressure of the gas in the bulb as its operating principle, heat must flow into this volume of gas, raising its temperature and thus its pressure. We will shortly show that modeling of thermal systems requires only *two* basic elements, thermal resistance and thermal capacitance, rather than the three needed in mechanical, electrical, and fluid modeling. This is basically so because no thermal phenomenon obeying the mathematical relationship which would correspond to mechanical mass, electrical inductance, or fluid inertance has been discovered.

Our final thermal system example is from the field of satellite temperature control. While human journeys to the moon are stupendous technical achievements, perhaps the most significant *economic and social* accomplishments of the space programs have been the various unmanned satellite programs. These have revolutionized communications, weather forecasting, military operations, and navigation. Whatever the specific mission, equipment carried aboard satellites must often be kept within fairly narrow temperature ranges to ensure adequate performance and long life. Both "passive" and "active" control schemes are in use. In a passive temperature control scheme, no separate heating or cooling components are used; the vehicle's thermal properties are cleverly adjusted so that temperature stays within the desired range "naturally," for the anticipated thermal environment. When passive methods cannot meet requirements, active systems using feedback controlled heating and/or cooling may be employed. To design either active or passive systems, a thermal model of the vehicle is needed.

Figure 4-7 shows a thermal model of a cylindrical spacecraft with four solar-cell paddles, such as the Advanced Orbiting Solar Observatory (AOSO).[8] The cylindrical body is divided into four sections, each of which is assumed at a uniform but time-varying temperature. Each section receives and transmits various heat fluxes from other sections, the sun, the earth, internal heat generated by instruments, etc. The difference between incoming and outgoing heat flux must be stored in the section, causing its temperature to rise and fall. Since many of the heat fluxes are by radiation (which depends on the fourth power of temperature) the problem has significant nonlinearity, making the model equations impossible to solve analytically. At the historical time of the reference, the analog computer was the tool of choice for such problems. Today, the thermal system modeling and equations remain unchanged but we use digital simulation for the analysis. Figure 4-8 shows some results indicating how adjustment of thermal properties can significantly modify the temperature cycles of the various parts of the satellite.

---

[7]E. O. Doebelin, *Measurement Systems*, 4th ed., McGraw-Hill, New York, 1990, pp. 623–625.

[8]F. J. Cepollina, Use of Analog Computation in Predicting Dynamic Temperature Excursions of Orbiting Spacecraft, NASA TM-X-55432, 1966.

216 Chapter 4

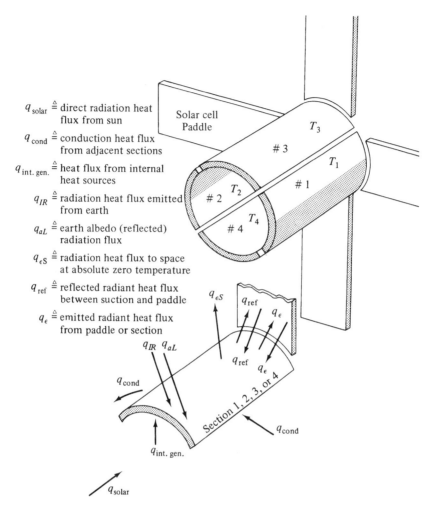

Figure 4-7 Thermal model of satellite.

4-2 FLUID FLOW RESISTANCE AND THE FLUID RESISTANCE ELEMENT

Having shown several examples of fluid and thermal systems of varying complexity, we now want to define and discuss the three basic elements used in lumped-parameter modeling of the fluid portions of dynamic systems. We begin with the element that, like mechanical friction and electrical resistance, performs the energy-dissipating function. The dissipation of fluid energy into heat occurs in all fluid devices to some extent. A straight length of pipe, tubing, or hose, pipe fittings such as tees and elbows, a partially open valve, an orifice, leakage paths in fluid machines—all of these exhibit fluid resistance whose effect may need to be modeled in a system analysis. Since pipes are present in almost any fluid system and they are perhaps the simplest example, we begin our study of fluid resistance with them.

System Elements, Fluid and Thermal

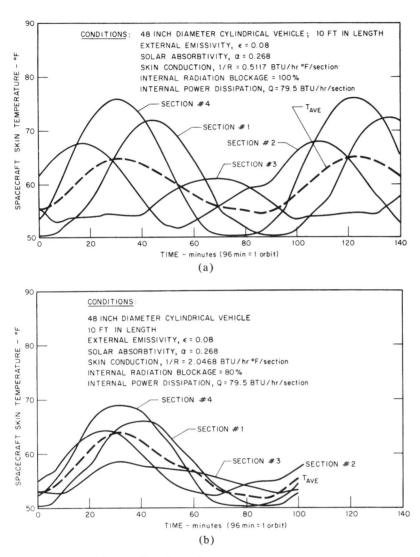

Figure 4-8 Orbiting satellite thermal response curves.

Consider the flow of a fluid (recall that the word fluid includes liquids, gases, and vapors) in a constant-area, rigid-walled conduit as in Fig. 4-9. The variables of primary interest are the average fluid pressure p (lb_f/in^2, or pascal) and the volume flow rate q ($inch^3/sec$, or m^3/sec). The average flow velocity v (inch/sec or m/sec) is defined as q/A, where A is the conduit cross-sectional area (in^2 or m^2). Note that the product of p and q has the dimensions ($inch$-lb_f)/sec and is in fact the *fluid power*, just as ($f \times v$) is mechanical power and ($e \times i$) electrical power. While the actual fluid pressure and velocity vary from point to point over the flow cross section in a real fluid flow, we will assume a so-called *one-dimensional* flow model in which the velocity and pressure are *uniform* over the area. Thus, the average velocity and average pressure correspond numerically with the values at *any* point in the cross

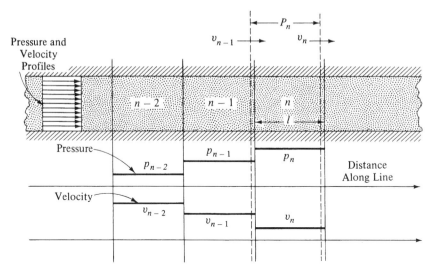

Figure 4-9 Lumped model of fluid pipeline.

section. Note that these average quantities are *spacewise*, not *timewise*, averages. This one-dimensional model has been found to give good results in many (but not all) applications, and is in fact emphasized in many basic fluid mechanics textbooks. In a lumped-parameter dynamic analysis the pipeline is broken into segments, as shown in Fig. 4-9. Within each segment or lump, the pressure and velocity may vary arbitrarily with time, but are assumed uniform over the volume of the lump. By considering the behavior of one typical lump (the nth) we are led to definitions of all three basic fluid elements, even though the focus of this section is on only one, the fluid resistance.

In describing the behavior of a single lump, we will apply conservation of mass and also Newton's law. During a time interval dt, the difference between mass flow into and out of a lump must equal the additional mass stored in that lump. Also, the force (pressure times area) difference between left and right ends of a lump must equal the lump mass times its acceleration. In applying conservation of mass, we will need to use the fluid property called the *bulk modulus*, which is a measure of the fluid's compressibility. As with most material properties, the numerical value of bulk modulus is found by experiment. In this case we compress a fluid sample of volume V and measure the volume change ΔV caused by a pressure change ΔP:

$$B \triangleq -\frac{\Delta P}{\Delta V/V} \quad \text{psi} \tag{4-1}$$

This experiment can be run under two different constraints. If we maintain a constant temperature during the compression, we get the *isothermal* bulk modulus. If we prevent any heat from being transferred, we get the *adiabatic* bulk modulus. For liquids, there is not much difference between the two; for gases the difference is considerable. For small gas pressure changes in the neighborhood of an operating point absolute pressure p_0, the isothermal bulk modulus is given by p_0, while the adiabatic is kp_0, where k is the ratio of specific heats. In system analysis, we use the

System Elements, Fluid and Thermal

isothermal bulk modulus when the process is slow enough to allow the available heat transfer processes to maintain a roughly constant absolute temperature. For rapid processes, there is not enough time for much heat transfer to take place, and we use the adiabatic modulus. For liquids, we often assume the bulk modulus to be independent of pressure, a typical value for hydraulic oil being about 250,000 psi.

The topic of fluid friction is a complex one, which is treated at length in courses on fluid mechanics. Since one of the features of system dynamics is the regular use of experimentally defined coefficients, we here take a phenomenological viewpoint and state that experiments show that when a fluid is forced through a pipe at a constant flow rate, a pressure difference related to that flow rate must be exerted to maintain the flow (see Fig. 4-10). One would intuitively expect that it would take larger pressure differences to cause larger flow rates and this behavior is what is observed. In general the relation between pressure drop and flow rate is nonlinear; however, some situations give a nearly linear effect. We thus will define the pure and ideal fluid friction or fluid resistance element by the linear relation:

$$\text{Fluid resistance} \triangleq R_f \triangleq \frac{p_1 - p_2}{q} \quad \frac{\text{psi}}{\text{in}^3/\text{sec}} \tag{4-2}$$

I recommend that you display the units as shown, rather than "simplifying" them to psi-sec/in^3, which obscures the *physical* meaning of R_f. (This comment on dimensions applies to *many* situations, not just this one.) For nonlinear fluid resistances, we can define linearized values in the neighborhood of an operating point in our usual way.

Returning now to Fig. 4-9, consider conservation of mass as applied to the nth lump over an infinitesimal time interval dt. Mass enters the lump from the left at a rate $Av_{n-1}\rho$ and leaves at the right at a rate $Av_n\rho$, where ρ is the fluid mass density. We now assume that pressure and temperature changes are small enough to treat the

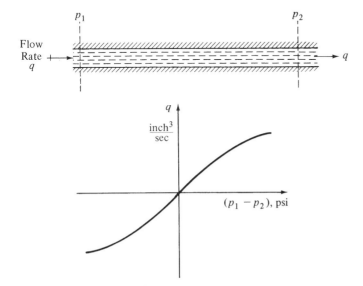

Figure 4-10 Experimental determination of fluid flow resistance.

density as a constant, corresponding to a constant operating-point pressure and temperature. For constant density, conservation of mass is the same as conservation of volume. Using our definition of bulk modulus we can then write:

$$(Av_{n-1} - Av_n) \, dt = dV = \frac{V}{B} \, dp_n = \frac{Al}{B} \, dp_n$$

$$(q_{n-1} - q_n) \, dt = \frac{Al}{B} \, dp_n \tag{4-3}$$

$$p_n = \frac{1}{C_f} \int (q_{n-1} - q_n) \, dt \tag{4-4}$$

$$C_f \triangleq \frac{Al}{B} \triangleq \text{fluid compliance (fluid capacitance)} \tag{4-5}$$

If we take the net volume flow rate $(q_{n-1} - q_n)$ as analogous to electric current, and the pressure p_n as analogous to voltage drop in Eq. (4-4), we see the analogy to electrical capacitance.

Newton's law for the nth lump gives us

$$Ap_{n-1} - Ap_n - AR_f q_n = \rho Al \frac{dv_n}{dt} = \rho l \frac{dq_n}{dt} \tag{4-6}$$

$$(p_{n-1} - p_n) - R_f q_n = \frac{\rho l}{A} \frac{dq_n}{dt} \tag{4-7}$$

Since Eq. (4-7) contains both the resistance (friction) and inertance (inertia) effects, we consider each (in turn) negligible, to separate them. If the fluid had zero density (no mass) we would get

$$(p_{n-1} - p_n) = R_f q_n \tag{4-8}$$

while if the resistance (friction) were zero, we would have

$$(p_{n-1} - p_n) = \frac{\rho l}{A} \frac{dq_n}{dt} = I_f \frac{dq_n}{dt} \tag{4-9}$$

$$I_f \triangleq \frac{\rho l}{A} \triangleq \text{fluid inertance} \tag{4-10}$$

Taking $(p_{n-1} - p_n)$ as analogous to voltage drop, and q_n analogous to current, we again see the analogy to electrical resistance and inductance. Since we have earlier established electrical/mechanical analogies, the fluid elements clearly have mechanical analogs also.

We will return to the fluid compliance and inertance elements in more detail in later sections, since they occur in more general contexts, not just in pipelines. They were briefly introduced in this section to allow a clearer discussion of resistance as a separate element and also to illustrate that the three elements are always present whenever a body of fluid exists. Whether all three will be *included in a specific system model* depends on the application and the judgment of the modeler, just as in mechanical and electrical systems. We now want to give a more general definition of fluid resistance and give an organized display of its features, as we did with all the mechanical and electrical elements.

System Elements, Fluid and Thermal

When a (one-dimensional) fluid flow is steady (velocity and pressure at any given point not changing with time), the inertance and compliance cannot manifest themselves (even though they exist), and only the resistance effect remains. We can thus experimentally determine fluid resistance by steady-flow measurements of volume flow rate and pressure drop, as in Fig. 4-11, or, if we attempt to calculate fluid resistance from theory, we must analyze a steady-flow situation and find the

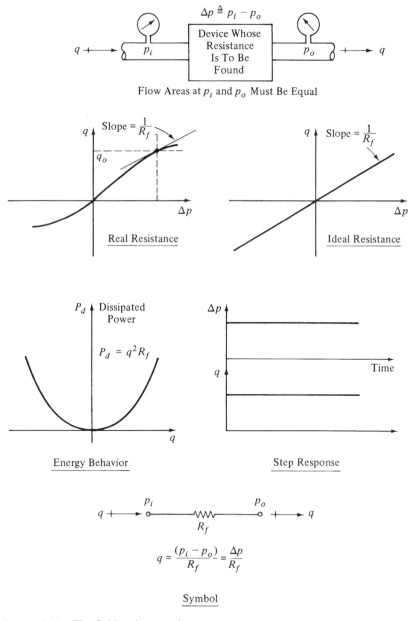

Figure 4-11 The fluid resistance element.

relation between pressure drop and volume flow rate. If a nonlinear resistance operates near a steady flow q_0, we can define a linearized resistance, good for small flow and pressure excursions from that operating point, by the slope of the curve, as in Fig. 4-11. Note that Eq. (4-8), which defines fluid resistance, is an *algebraic*, not differential, equation; thus the pure fluid resistance element exhibits an instantaneous response of q to an applied Δp, or Δp to an applied q. The frequency response is thus, of course, flat from zero to infinite frequency with zero phase angle.

Just as in electrical resistors, a fluid resistor dissipates into heat all the fluid power supplied to it. (This heat warms the fluid and any surrounding piping or machinery.) The concept of "fluid power" is widely used in hydraulics (liquid flows) and also applies to gas flows if the density does not change much. For these situations, we define fluid power at a flow cross section as the rate at which work is done by the pressure force at that cross section. For our assumed one-dimensional flow at velocity v, the pressure force is just pA, so the work done in time dt by this force would be $(pA)\,dx = pA(v\,dt)$, which makes the power equal to $pAv = pq = (\text{lb}_f/\text{in}^2)(\text{in}^3/\text{sec}) = (\text{in-lb}_f)/\text{sec}$. At the entrance (station 1) to a fluid resistance element, the fluid upstream is doing work on the resistance element at a rate $p_1 q_1$, while at the exit (station 2), the resistance element is doing work on the downstream fluid at a rate $p_2 q_2$. The resistance element thus receives (and dissipates into heat) the net power. For incompressible flow, the volume flow rate is the same at both stations, so the power dissipated into heat is just $q(p_1 - p_2) = q\,\Delta p = q^2 R_f = \Delta p^2/R_f$. Note in Fig. 4-11 that when we measure fluid resistance as shown there, the gages for reading the pressures must be located at points where the flow areas are equal. If this is not done, the pressure difference measured would include a component caused by the velocity change associated with an area change. This change in "pressure energy" is *not* dissipated into heat but rather has been *converted* into kinetic energy of velocity; thus it should *not* be included in measurements intended to characterize energy dissipation.

While we can (and often *must*) determine flow resistance values by experimental steady-flow calibration as in Fig. 4-11, it is of course also desirable to be able to calculate from theory, before a device has been built, what its resistance will be. For certain simple configurations and flow conditions, this can be done with fairly good accuracy. Flow conditions may be categorized in several useful ways, one of which is whether the flow is laminar or turbulent. *Laminar flow* occurs at relatively low velocities, and is characterized by an orderly and mathematically tractable motion of the fluid governed by viscosity effects rather than inertia. *Turbulent flow* occurs at higher velocities, where inertia effects outweigh those of viscosity. While one can still speak of an average velocity at a cross section, the individual fluid "particles" have random transverse velocity components superimposed on their gross forward motion, making a detailed mathematical analysis which documents the motion of each particle, effectively impossible. While certain aspects of turbulent flow are subject to analysis yielding useful results, frictional resistance effects generally require experimental study, the results of which may, however, often be generalized.

It has been found for steady flows that one can predict whether laminar or turbulent flow will occur, by calculating a dimensionless parameter called the *Reynolds number* N_R, which is effectively a ratio of inertial to viscous forces. If N_R

System Elements, Fluid and Thermal   **223**

is low enough, laminar flow will occur. As N_R increases, a transition region is encountered in which a clear-cut distinction between laminar and turbulent operation is not possible. (In designing fluid systems, one usually tries to avoid this region since operation there may be unpredictable.) Above this transition region, turbulent flow definitely occurs. In practical engineering systems, turbulent flow is more common than laminar. For flow inside a smooth, straight pipe of circular cross section, N_R is defined by

$$\text{Reynolds number} \triangleq N_R \triangleq \frac{\rho D v}{\mu} \tag{4-11}$$

where

$$\rho \triangleq \text{fluid mass density, lb}_f\text{-sec}^2/\text{in}^4$$

$$D \triangleq \text{pipe inside diameter, inch}$$

$$v \triangleq \text{fluid average velocity, in/sec}$$

$$\mu \triangleq \text{fluid viscosity, lb}_f\text{-sec/in}^2$$

This relation holds for both liquids and gases. If $N_R < 2000$ the flow will be laminar; if $N_R > 4000$ the flow will be turbulent, unless extreme care is taken to prevent disturbances which initiate turbulence. In most applications, $N_R > 4000$ essentially guarantees turbulent flow. In the transition region $2000 < N_R < 4000$, flow conditions are not reliably predictable, so systems should be designed to avoid this region, if at all possible. For flow "devices" other than smooth, straight pipes, the transition from laminar to turbulent flow again depends on a Reynolds number but its definition varies from device to device and you should consult fluid mechanics texts to find the proper formula for a particular application.

Laminar flow conditions produce the most nearly linear flow resistances, and these can also be calculated with good accuracy from theory, for passages of simple geometrical shape. The most common case is a long, thin flow passage called a *capillary tube*. For a circular cross section, theoretical analysis gives the result (ideally for incompressible fluids, but usable for gases also, as long as density changes are small)

$$\text{Volume flow rate} = q = \frac{\pi D^4}{128\mu L}\Delta p \tag{4-12}$$

where

$$D \triangleq \text{pipe diameter, inch}$$

$$\mu \triangleq \text{fluid viscosity, lb}_f\text{-sec/in}^2$$

$$L \triangleq \text{pipe length, inch}$$

$$\Delta p \triangleq \text{pressure drop, psi}$$

The fluid resistance is thus

$$R_f \triangleq \frac{\Delta p}{q} = \frac{128\mu L}{\pi D^4} \quad \frac{\text{psi}}{\text{inch}^3/\text{sec}} \tag{4-13}$$

Note that increasing the length increases the resistance in direct proportion, while decreasing the diameter by, say, 2 to 1 will cause a 16 to 1 increase in resistance. Since the usual manufacturing tolerances on small-diameter tubing are not negligible, if one purchases tubing specified, as say, 0.042 inch inside diameter, the *actual* diameter can be several thousandths of an inch smaller or larger. This variation causes a corresponding uncertainty in design calculations for fluid resistance and must be taken into account since the fourth power effect on D magnifies any changes in this dimension. Also, some fluids, such as ordinary tap water, will gradually deposit solids inside the tubing, again changing dimensions from those assumed in design. Finally, while Eq. (4-12) allows *any* numerical value of Δp to be inserted, if this Δp causes a flow rate with $N_R > 2000$, we do not have laminar flow and the formula is *not* applicable.

The theory leading to Eq. (4-12) actually assumes the length L is a section of an infinitely long pipe, so that end effects can be neglected. In real capillary tubes, these end effects cause nonlinearity and may not be negligible. A formula for estimating such end effects is[9]

$$R_f = \frac{128\mu L}{\pi D^4}\left(1 + 0.0434\,\frac{D}{L}\,N_R\right) \tag{4-14}$$

If $0.0434 D N_R/L$ is negligible relative to 1.0, then the capillary will be nearly linear. If not, the nonlinearity can be made obvious by substituting for N_R to get

$$R_f = \frac{128\mu L}{\pi D^4}\left(1 + \frac{0.1736}{\pi}\,\frac{\rho}{L\mu}\,q\right) \tag{4-15}$$

which shows R_f to depend on flow rate q, making it nonlinear. Figure 4-12 shows theoretical laminar flow resistances for some other shapes of flow passages. In estimating N_R for such shapes, to check for laminar flow, use the so-called *hydraulic diameter* D_h for D in the N_R formula:

$$D_h \triangleq \frac{4(\text{cross-section area})}{\text{perimeter}} \tag{4-16}$$

For turbulent flow of liquids (or gases with small density change), the pressure/flow relation has been found by experiment to be well fitted by the empirical formula[10]

$$\Delta p = \frac{0.242 L\mu^{0.25}\rho^{0.75}}{D^{4.75}}\,q^{1.75} \tag{4-17}$$

Note that the fluid mass density ρ is now present, and also that q is raised to the 1.75 rather than 1.0 power; thus the $\Delta p/q$ relation is nonlinear. It is still, however, an algebraic relation, so that the time response is still instantaneous. When digital simulation such as our SIMULINK, is used for fluid system studies, this nonlinear relation is easily modeled. If a linear model is desired for studies of small flow excursions about a steady-flow operating point q_0, we may linearize in the usual way and define an incremental linearized resistance as $d(\Delta p)/dq$ as follows.

---

[9]H. E. Merritt, *Hydraulic Control Systems*, Wiley, New York, 1967, p. 33.
[10]Ibid., p. 39.

System Elements, Fluid and Thermal

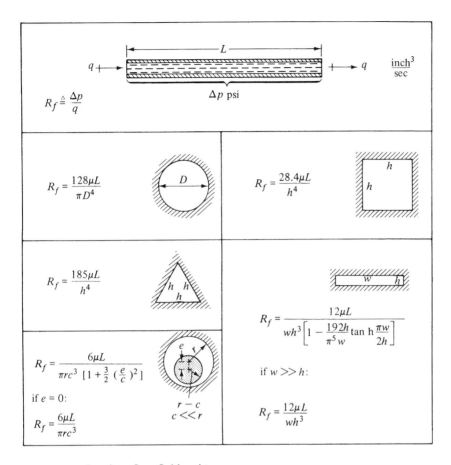

Figure 4-12 Laminar-flow fluid resistances.

$$\Delta p = \frac{0.242 L \mu^{0.25} \rho^{0.75}}{D^{4.75}} q^{1.75} = K_q q^{1.75} \tag{4-18}$$

$$\left[\frac{d(\Delta p)}{dq}\right]_{q=q_0} = 1.75 K_q q_0^{0.75} = R_f \tag{4-19}$$

$$K_q \triangleq \frac{0.242 L \mu^{0.25} \rho^{0.75}}{D^{4.75}} \tag{4-20}$$

For noncircular pipes, use the hydraulic diameter [Eq. (4-16)] in place of D for estimating pressure drops and flow resistances.

When we say that the response of q to an applied Δp is instantaneous for the laminar and turbulent pipe-flow resistance elements discussed above, it is important to remember that the bodies of fluid within the pipe length L also have compliance and inertance. Thus if one applies a sudden Δp to the fluid at rest, it does *not* suddenly achieve a flow rate q, since friction is *not* the only effect present. A simple frequency-response study may be helpful in illustrating this point.

EXAMPLE: OSCILLATING FLOW

In Fig. 4-13, an air pressure controller allows us to apply a sinusoidal pressure difference $\Delta p = p_1 - p_2 = A_p \sin \omega t$ to the ends of a pipe of length L, connected between two large shallow tanks. The fluid in the pipe will be considered incompressible (compliance is zero), but will exhibit friction and inertance. The difference between the applied Δp and the frictional pressure drop $R_f q$ is available to accelerate the inertance. Let's assume that the amplitude of sinusoidal flow is low enough that laminar flow formulas for steady flow can be used, as an approximation, for this unsteady flow. (This approximation will be discussed shortly.) While, for one-dimensional flow, the fluid inertance treats the "slug" of fluid as a rigid body with all particles having the same velocity, steady laminar flow has a parabolic velocity profile, which leads to the effective mass being $\frac{4}{3}$ of the actual mass $\rho A L$. This makes the laminar flow inertance $4\rho L/3A$. Newton's law for the pipe of Fig. 4-13 thus gives

$$A(\Delta p - R_f q) = A\left(\frac{4\rho L}{3A}\frac{dq}{dt}\right) \tag{4-21}$$

$$\left(\frac{4\rho L}{3AR_f}D + 1\right) = \frac{1}{R_f}\Delta p \tag{4-22}$$

$$\frac{q}{\Delta p}(D) = \frac{1/R_f}{\frac{4\rho L}{3AR_f}D + 1} \tag{4-23}$$

$$\frac{q}{\Delta p}(i\omega) = \frac{1/R_f}{\frac{4\rho L}{3AR_f}i\omega + 1} \tag{4-24}$$

In Eq. (4-24), note that, for a given pipe and fluid, if frequency ω is sufficiently low,

$$\frac{q}{\Delta p}(i\omega) \approx \frac{1}{R_f} \tag{4-25}$$

that is, the pipe/fluid is essentially a resistance element for low frequency excitation.

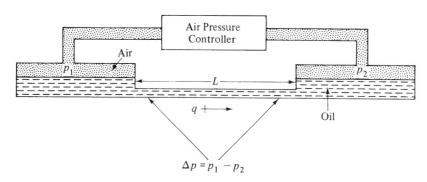

Figure 4-13 Fluid system for studying oscillating flow.

System Elements, Fluid and Thermal

To get some actual numbers, let's take $D = 0.05$ inch and $L = 20.0$ inch. For a typical oil at room temperature, $\rho = 7.95 \times 10^{-5}$ lb$_f$-sec^2/in^4 and $\mu = 4.0 \times 10^{-6}$ lb$_f$-sec/in^2, giving

$$R_f = \frac{128\mu l}{\pi D^4} = \frac{128 \times 4 \times 10^{-6} \times 20}{3.14 \times 625 \times 10^{-8}} = 522 \; \frac{\text{psi}}{\text{inch}^3/\text{sec}} \tag{4-26}$$

$$\frac{4\rho L}{3AR_f} = \frac{4 \times 7.95 \times 10^{-5} \times 20}{3 \times 1.96 \times 10^{-3} \times 522} = 2.06 \times 10^{-3} \text{ sec} \tag{4-27}$$

If we want the imaginary part to be small relative to 1, say, 0.1, then frequency ω must be less than 64.4 rad/sec = 10.3 Hz. Thus the given tube and fluid behaves essentially like a fluid resistance so long as the frequency content of the input pressure is less than about 10 cycles/sec. For higher frequencies, the inertance becomes significant and should be taken into account.

In all the above calculations and experimental measurements of flow resistances, the approach has been to use formulas relating flow rate and pressure drop for *steady* flows as if they held for general (unsteady) flows. This approach is widely used, and usually of sufficient accuracy; however, it should be recognized as an approximation. For mechanical engineering students, the usual undergraduate course in fluid mechanics concentrates mostly on steady flow and such texts rarely have much to say about fluid friction for unsteady flow. If we search in the technical literature for papers rather than textbooks, we find that such questions *have* been studied, using partial-differential equation models to derive a transfer function relating Δp to q. This function will in general have the form

$$\frac{\Delta p}{q}(i\omega) = R_f + i\omega I_f \tag{4-28}$$

where the real part R_f is the fluid resistance, and the imaginary part (divided by ω) is the fluid inertance I_f. For our analysis,

$$\frac{\Delta p}{q}(i\omega) = \frac{128\mu L}{\pi D^4} + i\omega \frac{4\rho L}{3A} \tag{4-29}$$

For laminar incompressible flow in circular tubes, the distributed-parameter model can be solved, giving a complicated solution in terms of Bessel functions.[11] For low frequencies ($\omega < 32\mu/\rho D^2$) the exact expression becomes our simple result in Eq. (4-29). For high frequencies ($\omega > 7200\mu/\rho D^2$) the result is

$$\frac{\Delta p}{q}(i\omega) = \frac{8L}{\pi D^3}\sqrt{2\rho\mu\omega} + i\omega \frac{\rho L}{A} \tag{4-30}$$

---

[11]C. K. Stedman, Alternating Flow of Fluid in Tubes, Statham Instrument Notes, #30, January 1956. G. B. Thurston, Periodic flow through circular tubes, *J. Acous. Soc. Am.* 24: 653–656, 1952.

Note that at high frequencies the inertance corresponds to the "physical" mass ρAL, just as for simple one-dimensional flow. The reference gives plots showing a smooth monotonic transition between low and high frequency regions; thus the inertance is always bracketed between $4\rho L/3A$ and $\rho L/A$. Perhaps the most interesting feature of the resistance is that it is frequency-dependent, increasing with $\omega^{0.5}$.

We now leave pipes, tubing, and hoses (where fluid resistance is distributed over a considerable length L) and consider *orifices* (where resistance is concentrated in a short distance). In Figure 4-14a, a flowing liquid discharges through a sharp-edge orifice into atmospheric air. Note that the liquid pressure p drops from the upstream value p_u to p_{at} over a very short distance. In Figure 4-14b we see the more common situation, where an orifice is inserted in a pipe. Downstream of the orifice the liquid flow spreads out so that it again fills the pipe. The pressure drop across an orifice is basically due to a conversion of energy from the form of "fluid power"

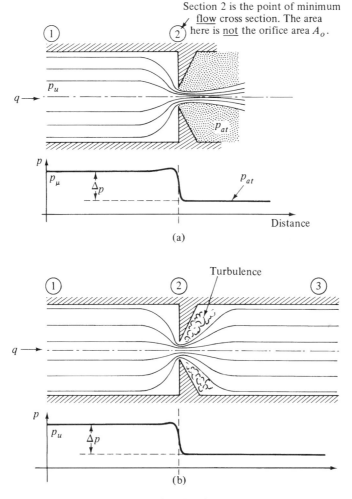

Figure 4-14 Characteristics of orifice flow.

System Elements, Fluid and Thermal 229

(pressure times volume flow rate) that we defined earlier, to the power of kinetic energy. Fluid mechanics texts show from conservation of energy that for level flow of a frictionless incompressible fluid:

Pressure/flow power + kinetic energy power = a constant   (4-31)

Considering any two locations 1 and 2, we may write

$$p_1 q + (\text{kinetic energy per unit time})_1 = p_2 q + (\text{kinetic energy per unit time})_2$$

(4-32)

For a section of liquid of length L, the kinetic energy is $\rho A L v^2 / 2$. This kinetic energy passes a cross section in a time interval AL/q, so the kinetic energy crossing the boundary per unit time is $\rho v^2 q / 2$. Our power equation is then

$$p_1 q + \frac{\rho v_1^2 q}{2} = p_2 q + \frac{\rho v_2^2 q}{2} \tag{4-33}$$

$$p_1 - p_2 = \Delta p = \frac{\rho}{2} \left(v_2^2 - v_1^2 \right) \tag{4-34}$$

showing that if v_2 is to be larger than v_1, we must expect a pressure drop Δp. Since $q = A_1 v_1 = A_2 v_2$,

$$\frac{2\Delta p}{\rho} = \left(\frac{q}{A_2} \right)^2 - \left(\frac{q}{A_1} \right)^2 = \frac{1 - (A_2/A_1)^2}{A_2^2} q^2 \tag{4-35}$$

$$q = \frac{A_2}{\sqrt{1 - (A_2/A_1)^2}} \sqrt{\frac{2\Delta p}{\rho}} \tag{4-36}$$

This is the basic pressure/flow relation for an orifice, which we see to be nonlinear. Note that fluid viscosity is not present, but density is. The analysis applied between stations 1 and 2 can also be applied between stations 1 and 3, or 2 and 3. Since the area at 3 is the same as at 1, we would find that there is a pressure *rise* from 2 to 3 that exactly equals the drop from 1 to 2. That is, the pressure at 3 is the same as at 1; we "recover" all of the pressure that we "lost" between 1 and 2. This result is a consequence of conservation of energy and the lack of an energy dissipating effect in the flow model used.

Experiments on real orifices show that the pressure at station 3 does *not* come back up as frictionless theory predicts; there is a "permanent" pressure loss. This is due to turbulence in the flow and the viscosity of real fluids, and cannot be accurately predicted from theory. Another problem is that the areas which should be used in our formulas are not the areas of metal parts such as pipes and orifices, but rather the areas of the fluid stream, which are difficult to accurately predict. For these reasons, the practical formula used to predict the pressure/flow relation for orifices in pipelines uses experimental data, defines the pressure drop as that between stations 1 and 3, and uses only the cross-sectional area of the hole in the metal orifice. This formula takes the form

$$q = C_d A_o \sqrt{\frac{2 \Delta p}{\rho}} \tag{4-37}$$

where

$A_o \triangleq$ orifice cross-sectional area

$C_d \triangleq$ experimental discharge coefficient

$\Delta p \triangleq$ pipeline permanent pressure drop (station 1 to station 3)

Numerical values of C_d depend mainly on Reynolds number and the area ratio $A_{\text{pipe}}/A_{\text{orifice}}$ and may be found in fluid mechanics texts and handbooks. Often this area ratio is greater than 25 and the Reynolds number is greater than 10,000. Then the value of C_d will be close to 0.61. When high accuracy is required, it would be best to experimentally calibrate the actual orifice using the actual fluid. The data would then be curve-fitted using the square root function of Eq. (4-37) to get the best value of C_d. If for some reason (unlikely) the data is not well fitted by a square root function, and if the data is sufficiently reproducible, other functions might be used. If the model is to be used in the digital simulation of a fluid system, one could also use the experimental data directly in a lookup table, rather than trying to fit an analytical function to it.

While sharp-edge orifices have more predictable characteristics and are less viscosity (and thus temperature) sensitive, manufacturing costs are lower for simple drilled holes, giving an orifice in the form of a short length of pipe (see Fig. 4-15). Discharge coefficients may be estimated from the formulas[12]

$$C_d = \frac{1}{\sqrt{1.5 + 13.74 \frac{L}{DN_R}}} \qquad \frac{DN_R}{L} > 50 \qquad (4\text{-}38)$$

$$C_d = \frac{1}{\sqrt{2.28 + 64 \frac{L}{DN_R}}} \qquad \frac{DN_R}{L} < 50 \qquad (4\text{-}39)$$

where the Reynolds number is calculated from

$$N_R \triangleq \frac{\rho D q}{A_o \mu} \qquad (4\text{-}40)$$

Sharp-edge orifices are the most nearly pure fluid resistances since the space over which the pressure drop occurs is very small. Since a small volume of fluid is

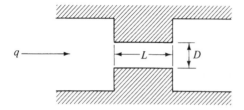

Figure 4-15 Short-tube orifice.

---

[12]Merritt, *Hydraulic Control Systems*, p. 42.

System Elements, Fluid and Thermal   **231**

involved, the compliance and inertance will be correspondingly small and usually negligible. If these effects can not be neglected, an analysis in the literature[13] provides some guidance.

For small flow changes around a steady-flow operating point, the orifice resistance can be linearized in our usual way.

$$q = C_d A_o \frac{\sqrt{2\,\Delta p}}{\rho}$$

$$\left[\frac{dq}{d\,\Delta p}\right]_{\Delta p = \Delta p_0} = \frac{1}{R_f} = \frac{C_d A_o}{2\sqrt{\Delta p_0}} \sqrt{\frac{2}{\rho}} \tag{4-41}$$

$$R_f = \frac{\sqrt{2\Delta p_0 \rho}}{C_d A_o} \tag{4-42}$$

If an R_f value is wanted for operation about a *zero* flow condition, $\Delta p_0 = 0$ and Eq. (4-42) gives $R_f = 0$. Actually, as Δp is decreased, the orifice flow becomes laminar and the flow resistance becomes linear and *not* equal to zero. A short-tube orifice (Fig. 4-15) can be treated like a capillary tube if DN_R/L is less than about 2; Eq. (4-39) actually becomes

$$C_d \approx \sqrt{\frac{DN_R}{64L}}$$

for small DN_R/L, and if this is substituted into Eq. (4-37), we get Eq. (4-12), the capillary tube formula. For sharp-edge orifices, a theoretical laminar flow result is available[14]

$$q = \frac{\pi D^3}{50.4\mu} \Delta p \tag{4-43}$$

for estimating flow resistance near zero flow. The transition from laminar to turbulent flow occurs at about $N_R = 9.3$. Above 9.3, Eq. (4-37) with $C_d = 0.61$ can be used.

Figure 4-16 shows some pressure/flow curves actually measured for water flowing in capillary tubes and an orifice in a student system dynamics lab at Ohio State University. (These fluid resistances are later used in dynamics experiments on complete fluid systems.) Note that the capillary tube of length 5.5 inches is really quite nonlinear even at $N_R = 1000$. To get the same flow resistance with better linearity, three tubes of the same diameter but three times as long were connected in parallel. This design change has two good effects as predicted by Eq. (4-14). First, D/L will now be one-third of what it was before. Also, for a given total flow rate, each tube now carries only one-third the flow, and thus has one-third the Reynolds number. The calibration data show that the predictions of Eq. (4-14) are indeed borne out, and linearity is now much better. The orifice shown is in the form of a

---

[13]Doebelin, *Measurement Systems*, p. 571.
[14]Merritt, *Hydraulic Control Systems*, p. 44.

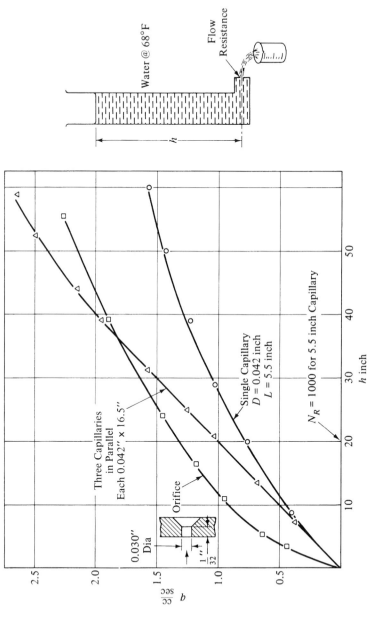

Figure 4-16 Experimental curves for various flow resistances.

System Elements, Fluid and Thermal

short tube, with $D/L \approx 1$; thus Eqs. (4-38) and (4-39) are applicable for C_d calculations. A curve fit of the data shows good correlation with the square-root relation predicted by Eq. (4-37).

The fluid resistances discussed so far are all intended to be essentially constant in numerical value. Many fluid systems require *adjustable* resistances, and often these take the form of some kind of *valve*. Valves are of many different types and perform various functions in fluid systems. We are here considering, not the "on-off" type of valve, which is either completely open or completely shut, but rather those valves used to smoothly *modulate* the flow rate. Such valves are used in many manual or automatic fluid control systems. The vast majority of valves have a square-root type of pressure/flow relation, such as Eq. (4-37), and usually require experimental calibration if we want a reasonably accurate flow model. The flow modulation is achieved by somehow varying the "orifice" area A_o. A complete calibration would thus give a *family* of flow rate versus pressure drop curves, one for each flow area.

An interesting example of a flow control valve using *micromachine technology* is shown in Fig. 4-17.[15] Microelectromechanical systems (MEMS) use manufacturing techniques similar to those used for microcircuits to produce sensors and actuators on a microscopic scale. Most of the commercial devices using this technology so far have been sensors, such as pressure transducers and accelerometers.[16] The valve shown is one of the earliest MEMS devices which provides an actuator function and is available as an off-the-shelf product. Microvalves require some electrical method of producing a force which causes a motion of valve parts that modulates the flow area. Most micromachines use piezoelectric or capacitance methods for producing a force from an electrical signal. These methods were found inadequate for the microvalve. It turns out that thermal actuation has many advantages for these small-scale devices.

The thermal principle used is the same as in ordinary "large-scale" bimetallic thermometers. That is, two materials with different coefficients of thermal expansion

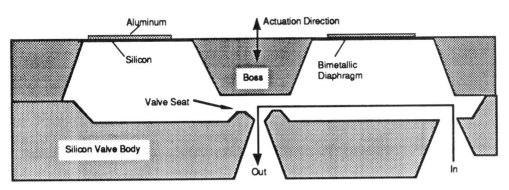

Figure 4-17 Flow-control valve using micromachine technology.

---

[15] Understanding Microvalve Technology, IC Sensors Inc., 1701 McCarthy Blvd., Milpitas, CA, 95035-7416, 408-432-1800.
[16] Doebelin, *Measurement Systems*, pp. 335, 740.

are bonded together. When the temperature changes, the differential thermal expansion causes the "sandwich" to deform, giving the desired force and motion. For micro devices, the two materials must also be compatible with the microscale manufacturing processes. The valve shown uses a bimetallic circular diaphragm of silicon and aluminum, the basic structure being silicon, with a thin layer of aluminum deposited on it. A diffused electrical resistor (about 40 ohms) is provided under the aluminum layer. Application of voltage to this resistor causes heating, thermal expansion, and thus opening of the valve, which is normally closed. About 4 volts is needed to completely open the valve; smaller voltages cause partial opening, as desired. In a typical valve, for maximum opening, the temperature rise is about 50 C°, if motion is restrained, about 4300 dynes of force is developed, and if motion is allowed, the deflection is about 27 μm. A complete ON/OFF/ON cycle can be accomplished in about 0.050 second. This microvalve is available as a separate item and is also part of a gas flow controller which uses a micro pressure sensor, thus employing both microsensing and microactuation.

4-3 FLUID COMPLIANCE AND THE FLUID COMPLIANCE ELEMENT

We already have seen that a fluid *itself*, whether a liquid or a gas, exhibits compliance due to its compressibility. Certain *devices* may also introduce compliance into a fluid system, even if the fluid were absolutely incompressible. Metal tubing and (in particular) rubber hoses will expand when fluid pressure increases, allowing an increase in volume of liquid stored. Accumulators use spring-loaded cylinders or rubber air bags to provide intentionally large amounts of compliance. A simple open tank exhibits compliance, since an increase in volume of contained liquid results in a pressure increase due to gravity. In general, the compliance of a device is found by forcing into it a quantity of fluid and noting the corresponding rise in pressure. For liquids, the input quantity is a volume of fluid V, and the ideal compliance is defined by

$$\text{Fluid compliance} \triangleq C_f \triangleq \frac{V}{P} \quad \frac{\text{inch}^3}{\text{psi}} \tag{4-44}$$

or, in terms of volume flow rate q,

$$C_f \triangleq \frac{\int q \, dt}{p} \tag{4-45}$$

For nonlinear compliances, the actual p-V curve can be implemented in a computer lookup table, or, if linearized analysis is desired, the local slope can be used to define an incremental compliance. A standardized symbol for fluid compliance has not been agreed upon, although some writers use the electrical capacitance symbol with one end of the "capacitor" always connected to ground. However, since it is preferable to work with the fluid equations directly, rather than to use analogies, the symbol of Fig. 4-18 is suggested as consistent with fluid circuit diagrams.

In Eq. (4-5) we calculated the compliance of a section of hydraulic line, due to the bulk modulus of the liquid itself, as Al/B. Additional contributions to compliance which may be significant here are due to entrained air bubbles and the flexibility of the tubing. Suppose air (or other gas) bubbles take up x percent of the total

System Elements, Fluid and Thermal

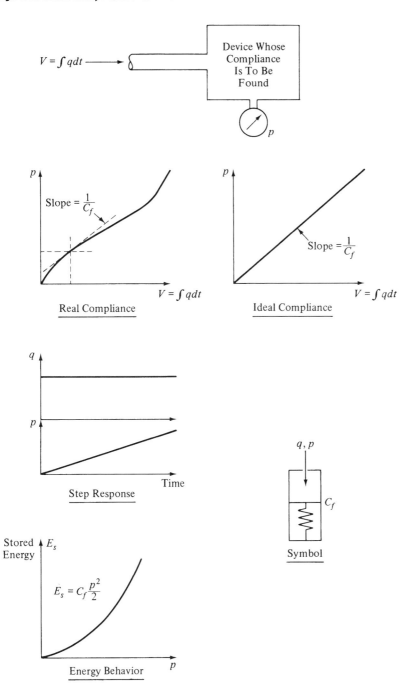

Figure 4-18 The fluid compliance element.

volume V. A pressure increase of Δp will cause a change in the liquid volume of $\Delta p(1 - x)V/B_l$ and of gas volume $\Delta p x V/B_g$. The fluid volume change is thus

$$\Delta V_{\text{fluid}} = \left[\frac{(1-x)V}{B_l} + \frac{xV}{B_g}\right]\Delta p \tag{4-46}$$

For thin-walled tubing the tubing volume change is given by

$$\Delta V_{\text{tubing}} = \frac{VD}{tE}\Delta p \tag{4-47}$$

where

$t \triangleq$ tube wall thickness

$D \triangleq$ tube mean diameter

$E \triangleq$ tube material elastic modulus

The total volume change is

$$\Delta V_{\text{total}} = \left[\frac{D}{tE} + \frac{1-x}{B_l} + \frac{x}{B_g}\right]V\,\Delta p \tag{4-48}$$

and the total compliance is

$$C_f = \left[\frac{D}{tE} = \frac{1-x}{B_l} + \frac{x}{B_g}\right]V \quad \frac{\text{inch}^3}{\text{psi}} \tag{4-49}$$

EXAMPLE: EFFECTIVE BULK MODULUS

Suppose we have a steel pipe with $D/t = 10$, 1% air bubbles, $B_l = 250{,}000$ psi, and $B_g = 500$ psi. The B_g value of 500 psi assumes the system operating point pressure is 500 psia and slow (isothermal) pressure changes. [For rapid (adiabatic) changes B_g would be $1.4p = 700$ psi.] The relative importance of the three terms in Eq. (4-49) is:

Steel tube: $\dfrac{D}{tE} = \dfrac{10}{3 \times 10^7} = 0.33 \times 10^{-6}$

Oil: $\dfrac{1-x}{B_l} = \dfrac{0.99}{2.5 \times 10^5} = 4.0 \times 10^{-6}$

Air bubbles: $\dfrac{x}{B_g} = \dfrac{0.01}{500} = 20.0 \times 10^{-6}$

It is clear from this example that even a small amount of entrained air can severely reduce the "stiffness" of hydraulic fluid. This results in slower response speed ("lower natural frequency") in hydraulic motion control systems using piston/cylinder or rotary motors. A practical problem is that the percent of entrained air in an operating hydraulic system is nearly impossible to determine accurately and varies from minute to minute. Hydraulic system designers thus usually use a working value of bulk modulus based on operating experience. This number varies with the

System Elements, Fluid and Thermal   **237**

application and from designer to designer but typically might be about 75,000 psi. The concept of an effective or equivalent bulk modulus which combines all three compliance effects into one number can be defined from Eq. (4-48).

$$\frac{1}{\text{Effective bulk modulus}} \triangleq \frac{1}{B_e} \triangleq \frac{\Delta V/V}{\Delta p} = \left[\frac{D}{tE} + \frac{1-x}{B_l} + \frac{x}{B_g} \right] \qquad (4\text{-}50)$$

For our example numbers, $B_e = 41{,}200$ psi, much lower than the 250,000 of the liquid itself.

Figure 4-19 shows some devices called *accumulators*, which are designed intentionally to exhibit fluid compliance. In Fig. 4-19a, we have a simple spring-loaded piston and cylinder; Fig. 4-19b shows a flexible metal bellows; while in Fig. 4-19c the compliant element is a nitrogen-filled rubber bag. The devices of Fig. 4-19a and c can be designed to store large amounts of fluid energy and are widely used in hydraulic power systems for short-term power supplies, pulsation smoothing, and to reduce pump size in systems with intermittent flow requirements. Since metal bellows are somewhat limited in volume change, they are more often found used as dynamic elements in low-power devices such as instruments. Due to their complex shape, the compliance of metal bellows is difficult to calculate but it can be measured easily once a bellows has been constructed. For the spring loaded piston

$$C_f = \frac{\Delta V}{\Delta p} = \frac{A\,\Delta x}{K_s\,\Delta x/A} = \frac{A^2}{K_s} \quad \frac{\text{inch}^3}{\text{psi}} \qquad (4\text{-}51)$$

where

$$A \triangleq \text{piston area, inch}^2$$

$$K_s \triangleq \text{spring constant, lb}_f/\text{inch}$$

In Fig. 4-19c, the rubber bag is initially pressurized (through a "tire valve" not shown) with nitrogen, so that the bag completely fills the steel pressure vessel. Then hydraulic oil is forced in, compressing the gas more and partially filling the vessel with liquid. If the accumulator is now connected to a hydraulic system so that liquid can flow in and out, we can analyze the behavior as follows, beginning with the perfect gas law:

Perfect gas law: $pV = MRT$ (4-52)

When the rubber bag is initially charged with gas, a definite mass M of gas is put into it. Since the bag is then sealed, this M is constant. For slow pressure changes the temperature T is assumed to stay nearly constant (isothermal process). The gas constant R also stays nearly constant for the pressure and temperature ranges with which accumulators work; thus MRT will be taken as constant. At any instant t, we can write

$$V_{\text{gas}} = \frac{MRT}{p_{\text{gas}}} = V_{\text{gas,op}} - \int_0^t q_{\text{liq}}\, dt \qquad (4\text{-}53)$$

where $V_{\text{gas,op}}$ is the gas volume at time $t = 0$, corresponding to an operating point (see Fig. 4-19d). As long as there is any liquid in the vessel, $p_{\text{liq}} \equiv p_{\text{gas}}$ since the

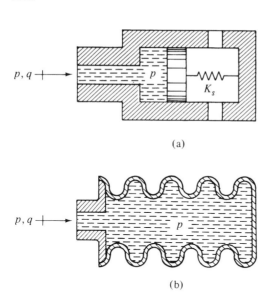
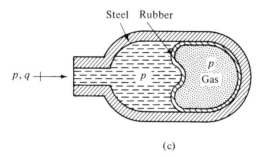
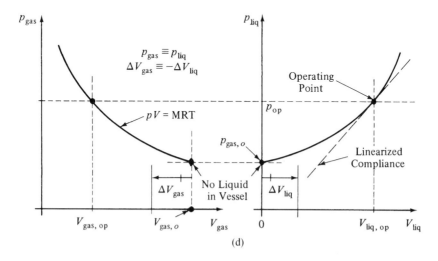

Figure 4-19 Accumulators [liquid compliances, (a)–(c)] and gas-bag accumulator characteristics (d).

rubber bag is flexible enough that it cannot support a pressure difference across it. We thus have

$$p_{\text{liq}} = \frac{MRT}{V_{\text{gas,op}} - \int_0^t q_{\text{liq}}\,dt} = \frac{MRT}{V_{\text{gas},0} - V_{\text{liq}}} \tag{4-54}$$

This is the pressure/flow relation defining the compliance effect of the accumulator, and is seen to be nonlinear in the p_{liq} versus V_{liq} relation. This nonlinearity generally prevents analytical study but is easily modeled with simulation. As usual, we can avoid simulation if we can justify the approximation of a linearized, small-signal analysis near a chosen operating point.

$$C_f \triangleq \frac{dV_{\text{liq}}}{dp_{\text{liq}}} = \frac{(V_{\text{gas},0} - V_{\text{liq,op}})^2}{MRT} = \frac{V_{\text{gas,op}}^2}{MRT} = \frac{MRT}{p_{\text{op}}^2} \tag{4-55}$$

In Fig. 4-20a a vertical cylindrical tank of cross-sectional area A is supplied with a volume flow rate q; the pressure at the tank inlet is p, liquid height is h. The vertical motion of the liquid in such tanks is usually slow enough that the velocity and acceleration have negligible effects on the pressure p and it is simply given by $p = \gamma h$, where γ is the specific weight of the liquid in lb_f/in^3. That is, thinking of the tank as a large-diameter vertical pipe, R_f (velocity effect) and I_f (acceleration effect) are negligible relative to the height effect. If we add a volume V of liquid to the tank, the level h goes up an amount V/A and the pressure rises an amount $V\gamma/A$. The tank compliance is thus

$$C_f = \frac{\text{volume change}}{\text{pressure change}} = \frac{V}{V\gamma/A} = \frac{A}{\gamma} \quad \frac{\text{inch}^3}{\text{psi}} \tag{4-56}$$

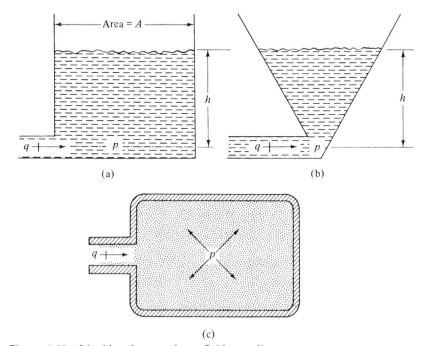

Figure 4-20 Liquid and gas tanks as fluid compliances.

For noncylindrical tanks, such as in Fig. 4-20b, the compliance effect is nonlinear, but can be linearized in the usual way if desired.

In Fig. 4-20c a rigid tank of volume V contains a gas at pressure p. For slow (isothermal) pressure changes in which the fluid density is nearly constant, we may write $pV = MRT$. If we force a mass $dM = \rho\, dV$ of gas into the tank we cause a pressure change dp given by

$$dp = \frac{RT}{V}\, dM = \frac{RT}{V}\, \rho\, dV = \frac{RT}{V}\, dV\, \frac{p}{RT} \tag{4-57}$$

$$C_f \triangleq \left[\frac{dV}{dp}\right]_{p=p_0} = \frac{V}{p_0}\ \frac{\text{inch}^3}{\text{psi}} \tag{4-58}$$

This is a linearized compliance useful for small changes near an operating point p_0. For rapid (adiabatic) but still small pressure changes, analysis shows the compliance is $C_f = V/(kp_0)$, where k is the ratio of specific heats (1.4 for air, for example). In dealing with the compression and expansion of gases in this chapter we have emphasized the ideal processes called isothermal and adiabatic. Real processes actually follow a *polytropic* relation ($p/\rho^n = \text{constant}$), which is "somewhere between" the isothermal and adiabatic. The polytropic exponent n varies with the application and must be determined by experiment, but in general lies between 1.0 (isothermal) and k (adiabatic). A theoretical and experimental study[17] which used accurate nonlinear resistance and capacitance relations showed perceptible, but small, differences between adiabatic, isothermal, and polytropic models. The experiments were, however, limited to small pressure changes near atmospheric pressure. One would expect more significant differences between the models for large pressure changes.

4-4 FLUID INERTANCE

While devices for introducing resistance (capillaries, orifices) and compliance (tanks, accumulators) are often intentionally designed into fluid systems, the inertia effect is more often than not a parasitic one; thus inertance "devices" are relatively unknown as commercial components. In terms of analytical treatment, the inertance of *pipes* is perhaps most commonly encountered and we shall emphasize it. Any flowing fluid has stored kinetic energy because of its density (mass) and velocity. The inertance of a finite-size lump of fluid represents a summing up of this kinetic energy over the volume of the lump. The simplest assumption possible here is that of one-dimensional flow (Fig. 4-21a), where all the fluid particles have identical velocities at any instant of time. Since every fluid particle has the same velocity, a lump of fluid can be treated as if it were a rigid body of mass $M = \rho AL$. A pressure drop Δp across a pure inertance element will cause a fluid acceleration according to Newton's law:

---

[17]J. Dagan and C. K. Kwok, Study of Pneumatic Capacitors, ASME paper 71-WA/Flcs-3, 1971.

System Elements, Fluid and Thermal 241

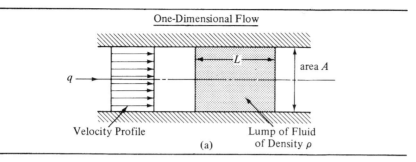

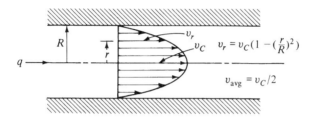

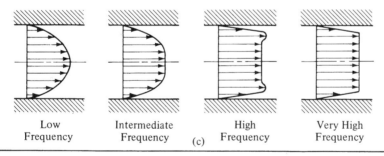

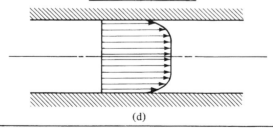

Figure 4-21 Velocity profiles for various flow conditions.

$$A \Delta p = \rho A L \frac{dv}{dt} = \rho A L \left(\frac{1}{A} \frac{dq}{dt} \right) \qquad (4\text{-}59)$$

$$\Delta p = \frac{\rho L}{A} \frac{dq}{dt} \triangleq I_f \frac{dq}{dt} \qquad (4\text{-}60)$$

where

$$I_f \triangleq \frac{\rho L}{A} \qquad (4\text{-}61)$$

Equation (4-60) is of course analogous to $e = L(di/dt)$ for inductance in electrical systems and $f = M(dv.dt)$ for mass in mechanical systems. The electrical inductance symbol (Fig. 4-22) is widely used for fluid inertance, and since it causes no conceptual difficulties in setting up fluid equations, we will also adopt it.

Theoretical analysis of steady laminar flow shows the velocity profile to be parabolic, as in Fig. 4-21b. By computing[18] the kinetic energy of a unit length of fluid with this velocity profile, and equating it to that of a uniform-velocity mass with the same *average* velocity, we find that the uniform-velocity mass must be $\frac{4}{3}$ the actual mass. This result is also a close approximation for *unsteady* flows if the velocity time variation is not too rapid (signal frequency content restricted to low frequencies). As frequency content extends to higher frequencies (Fig. 4-21c), the velocity profile becomes more "square" and the correct mass approaches the "physical mass" $\rho A L$. The inertance for laminar flow is thus always between $\frac{4}{3} \rho L/A$ and $\rho L/A$, the midpoint $\frac{7}{6} \rho L/A$ occurring at about $\omega = 50\mu/(R^2 \rho)$.[19]

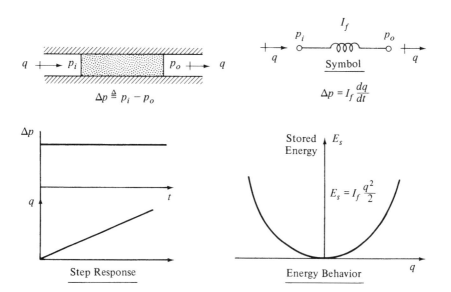

Figure 4-22 The fluid inertance element.

---

[18] Doebelin, *Measurement Systems*, 4th ed., p. 477.
[19] Stedman, Alternating Flow of Fluid in Tubes [11].

System Elements, Fluid and Thermal   **243**

EXAMPLE: LIQUID INERTANCE

For a typical hydraulic oil ($\mu = 2 \times 10^{-6}$ lb$_f$-sec/in^2, $\rho = 0.8 \times 10^{-4}$ lb$_f$-sec^2/in^4) flowing in a tube of radius $R = 0.05$ inch, the frequency ω is about 500 rad/sec, for example. Thus, well below 80 Hz we use $\frac{4}{3}$ the physical mass, near 80 Hz we use $\frac{7}{6}$, and well above 80 Hz we use the physical mass itself.

In steady turbulent flow the velocity profile cannot be calculated from theory, but experiments have shown that for rough or smooth pipes the measured profiles are quite accurately fitted by equations of the form

$$\frac{v_r}{v_c} = \left(1 - \frac{r}{R}\right)^{1/n} \tag{4-62}$$

The parameter n varies over the range 4 to 5 for rough pipes and from 6 (Reynolds number 4000) to 10 (Reynolds number 3×10^6) for smooth pipes. In all cases the profile is quite square and calculations of kinetic energy show that the effective mass is nearly equal to the "physical" mass ρAL. For turbulent flow it is thus quite reasonable to use the value of inertance corresponding to one-dimensional flow, that is, ρLA.

While measurements of fluid resistance can be made without having compliance and inertance effects present (in *steady* flow, dp/dt and dq/dt are zero), the separate "steady-state" measurement of fluid inertance would require a nonzero dq/dt with a zero q and a zero dp/dt. These conditions are generally not realizable in practice and thus steady-state measurements of fluid inertance are not commonly made. For this reason, a "characteristic curve" relating Δp and dq/dt for inertance [corresponding to the $q/\Delta p$ curve for resistance (Fig. 4-11) and the $p/\int q\, dt$ curve for compliance (Fig. 4-18)] is not given in Fig. 4-22. For the pure and ideal inertance element, the Δp versus dq/dt curve would, of course, be a straight line. Since real resistance and compliance generally exhibit some (sometimes considerable) nonlinearity in their characteristic curves, one wonders whether real inertance behaves similarly. Unfortunately, the above-mentioned lack of an experimental method for getting the required characteristic curve for inertance prevents a straightforward comparison.

If we appeal to theory we see that, for a fluid "particle," the inertance effect is nothing more nor less than the force/acceleration characteristic of Newton's law, which experiments have proven to be extremely linear, except when velocities approach the velocity of light. Since fluid systems operate with velocities which are entirely negligible relative to the speed of light, the inertance effect for a particle should be essentially linear. In practical problems however, we are concerned with the inertance effect of the *total* flow in a pipe or machine, which clearly involves a complicated summing-up of the effects of myriad "particles," the motions of which are influenced not just by inertia but also by resistance, for example. The basic concept of linearity requires that a change in the level of the input quantity produces a strictly *proportional* change in the level of the output quantity. For fluid inertance, the input would be an applied Δp and the output would be the rate of change of flow, dq/dt. Suppose we apply to a "lump" of fluid in a pipe a sinusoidal pressure difference $\Delta p = \Delta p_0 \sin \omega t$ of fixed frequency ω. If the amplitude Δp_0 is sufficiently small, we would expect the peak flow rate to also be small enough to allow laminar flow at all times, and if ω is small enough, the inertance will be very nearly that of steady laminar flow, $4\rho L/3A$. If input Δp_0 is now increased (keeping ω fixed), it is intuitively clear that this larger accelerating force will cause larger peak flow rates, and at some point the flow will become turbulent rather than laminar. While little is known about the velocity profiles of unsteady turbulent flow, it would be unreasonable to expect them to be identical with those of laminar flow; thus the inertance must change as Δp_0 is increased, showing a nonlinear inertance behavior. Thus while the fundamental inertia law for a particle is linear, the gross inertance effect in a real fluid may be nonlinear. Fortunately this nonlinearity is not excessive, since it is most likely bounded by the inertance values $4\rho L/3A$ and $\rho L/A$ in all practical situations, a 33% range at worst.

One other type of application where fluid inertia effects may be important and where some useful results are available is the motion of solid bodies immersed in a fluid. When a solid moves through a fluid, we know intuitively that the mass being moved includes not just that of the solid body, but also some nebulous amount of fluid which "goes along for the ride." For example, if we measure the natural frequency of vibration of a spring/mass system in air and then in water, the frequency in water will be lower, since the effective mass moved (solid plus liquid) is now larger. The difficulty of course is to know how much fluid mass is to be added to the solid's mass, since each particle of the moving fluid has a different and complex motion. This problem has been addressed both theoretically and experimentally and we now simply quote some useful results[20] without any attempt at derivation.

For a circular disk of diameter D, moving perpendicular to the plane of the disk, the added fluid mass (called the *hydrodynamic mass*) is equal to that of a fluid disk of the same diameter and of thickness $0.424D$. This is for a disk that is not close to any other solid object. If the disk is "close" to a parallel wall (nominally a distance g from the wall), the added mass is that of a fluid disk of diameter D and thickness $D^2/(32g)$. Note that as the "gap" g approaches zero, the hydrodynamic mass goes to infinity! While the assumptions of the analysis are such that we should not let $g \to 0$, the *trend* is correct and can be intuitively understood. When g is small (relative to D), as the disk approaches the wall at some velocity, the fluid (assumed incompressible) must escape to the sides through a narrow passage. Conservation of volume requires that this sidewise velocity will be much higher than the disk's velocity toward the wall. This high velocity fluid has a lot of kinetic energy, which must be equaled by that of the (fictitious) hydrodynamic mass, which moves only at disk velocity. Thus the hydrodynamic mass must be very large if it is to have the same kinetic energy as the fluid in the narrow passage between disk and wall. We can also note that the "disk/wall" formula does *not* give the correct result when g becomes large. That is, for large g, we should get the same result as from the "disk away from other objects" formula, which we don't (the disk/wall formula predicts a hydrodynamic mass of *zero* as g gets large). Thus the disk/wall formula should be used only for an intermediate range of g values, neither too small nor too large.

---

[20]R. J. Fritz, The effect of liquids on the dynamic motions of immersed solids, *ASME J. Engineering for Industry*, February 1972, pp. 167–173.

System Elements, Fluid and Thermal

For a square plate of side length L, not close to any other object, the hydrodynamic mass is that of a fluid "square" of side length L and thickness $0.375L$. For a sphere of radius R, not close to any other object, the hydrodynamic mass is one-half that of a fluid sphere of radius R. The reference gives many other results, some derivations, and also discusses resistance and compliance effects.

4-5 COMPARISON OF LUMPED AND DISTRIBUTED FLUID SYSTEM MODELS

While this text emphasizes lumped-parameter modeling, we need to keep reminding ourselves that such models are always approximations to the more-correct distributed models, which use partial differential equations. Our practical use of lumped models is less likely to get into difficulty if we at least have some familiarity with the behavior of the distributed models. In the case of fluid systems, a good overview is available from my more advanced system dynamics text.[21] We will look at one specific example which, however, gives some useful general insight.

Figure 4-23 shows a pipeline terminated in a sharp-edged orifice. An actual system of this sort has been experimentally tested, in addition to being modeled with both distributed and lumped methods. We can thus compare the two types of analytical models with the actual measured behavior. The distributed model results in partial-differential equations of the classical type called *wave equations*. Wave equations, whether applied to systems of electromagnetic, mechanical, or fluid type always lead to a relation between wavelength λ (ft/cycle), frequency f (cycles/sec), and velocity of propagation c (ft/sec):

$$\lambda = \frac{c}{f} \tag{4-63}$$

Propagation velocity is the speed with which a disturbance propagates through the medium; a familiar example is the speed of sound in air (about 1100 ft/sec). For pipelines, the reference shows that the propagation velocity c is given by $(B_e/\rho)^{0.5}$, where B_e is the effective bulk modulus of Eq. (4-50) and ρ is the fluid mass density. The pipeline was a 68-ft length of stainless steel tubing, 1-inch O.D. and $\frac{1}{16}$ inch wall thickness. The fluid was JP-4 jet fuel ("kerosene") with density of 7.26×10^{-5} lb_f-sec^2/in^4 and bulk modulus of 173,600 psi. The apparatus was a carefully built and operated lab setup rather than a typical industrial system, so measures

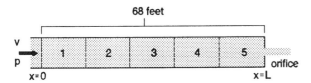

Figure 4-23 Five-lump model of liquid pipeline with orifice.

---

[21] E. O. Doebelin, *System Modeling and Response: Theoretical and Experimental Approaches*, Wiley, New York, 1980, pp. 363–408.

were taken to minimize air bubbles and the calculated B_e assumed no air in the liquid, giving a propagation velocity of 3900. ft/sec.

While the distributed-parameter model is more accurate, it cannot usually be solved for the time response, but frequency-response calculations are not difficult. Thus, as we have seen several times before, the comparison of different models for a system is often best done in terms of the response to sine waves. When our fluid system is driven at steady state by sine wave inputs (pressures or flow rates), theory shows that the response (output pressure or flow rate) at any location in the system is also sinusoidal in time, with the same frequency as the input. What is perhaps *less* obvious is that, at any instant of time, the variation of pressure or flow rate with *distance* along the pipeline is also sinusoidal, with a wavelength λ given by Eq. (4-63).

At this point we need to recall that any practical dynamic system will experience excitation (input signals) whose maximum frequency will be limited to a definite value. Engineers experienced in a particular application generally have a good feel for the highest frequency excitation that might reasonably be expected. Let's suppose that in our pipeline application, excitation with frequency content higher than 30 Hz is unlikely. (We can do our calculations for *any* assumed highest frequency.) Using Eq. (4-63), we find the wavelength corresponding to 30 Hz is about 130 ft/cycle. What does this mean? It means that if our system is operating in a sinusoidal steady state at 30 hz, the spacewise variation of both pressure and flow rate will go through one complete cycle in a distance of 130 ft. At this point we need to mention that the sinusoidal fluctuations are actually small changes superimposed on a steady operating-point pressure or flow rate. For example, the steady flow velocity might be 3.0 ft/sec with a sinusoidal variation of ±0.3 ft/sec "riding on top." Thus the total flow velocity would never actually reverse, but rather oscillate between 2.7 and 3.3 ft/sec.

In Fig. 4-9, where we used *lumped* modeling, the variation of pressure and velocity along the length of the pipeline was *assumed* a *stepwise* one. Within a given lump there was *no* variation, but pressure and velocity *did* change when we went to a neighboring lump. All distributed models allow a *smooth*, not stepwise, variation, and this is of course more correct. For sinusoidal excitation of our pipeline model, this smooth variation is found to be sinusoidal. It is clear that as a lumped model uses more and smaller lumps, the stepwise variation more nearly approximates the true smooth variation. The question is, how many lumps are needed to get "accurate" results with a lumped model? Experience with many kinds of systems shows that if we choose 10 lumps per wavelength of the *highest* frequency, we usually get good results. That is, a stepwise variation is an acceptable approximation to a sine wave if there are 10 steps per wavelength. Note that if we decide that our lumped model needs to be "good" for excitations of *higher* frequency than we originally planned, the lumps *must* get smaller and there must thus be more of them.

For our 30-Hz maximum frequency, the wavelength was about 130 feet, so each lump should be about 13 feet long. (If we used 60 Hz this length would be about 6.5 feet.) Since the pipeline is 68 feet long, we model it with five lumps, each 13.6 feet long. This slight deviation from 13.0 feet will have only minor effects. The reference chooses to solve the following problem. Suppose we apply as input a sinusoidal variation in flow velocity at the location $x = 0.0$ in Fig. 4-23 and inquire as to what the pressure fluctuation (output) at that same location would be. This is essentially asking for the sinusoidal transfer function relating these two quantities. We

System Elements, Fluid and Thermal

have given earlier all the numerical values needed except those describing the orifice located at $x = L$. To get an accurate relation between pressure and velocity for the orifice, its steady-flow calibration curve was measured and linearized at the experiment operating point. This relation between v and p at the orifice location was found to be $v_L = 0.495 p_L$, where v is in inch/sec and p is in psi. The orifice was the only fluid resistance included in either the lumped or distributed models; the fluid itself was treated as ideal (zero viscosity).

Using the numerical values given above and the lumped element values derived from Fig. 4-9, the reference computed the desired transfer function as:

$$\left[\frac{p}{v}(i\omega)\right]_{x=0} = \frac{5.74[(i\omega)^5 + 854(i\omega)^4 + 8.28 \times 10^5 (i\omega)^3 + 1.42 \times 10^8 (i\omega)^2 + 3.43 \times 10^{10}(i\omega) + 1.17 \times 10^{12}]}{(i\omega)^5 + 2421(i\omega)^4 + 8.28 \times 10^5 (i\omega)^3 + 4.02 \times 10^8 (i\omega)^2 + 3.43 \times 10^{10}(i\omega) + 3.33 \times 10^{12}} \tag{4-64}$$

Using the same physical parameter values but the distributed-parameter method, this transfer function was:

$$\left[\frac{p}{v}(i\omega)\right]_{x=0} = \frac{\cosh(0.0174 i\omega) + 1.68 \sinh(0.0174 i\omega)}{0.495[\cosh(0.0174 i\omega) + 0.595 \sinh(0.0174 i\omega)]} \tag{4-65}$$

Figure 4-24 shows the amplitude ratio and phase angle curves for both lumped and distributed models, together with the values actually measured in the lab test. The

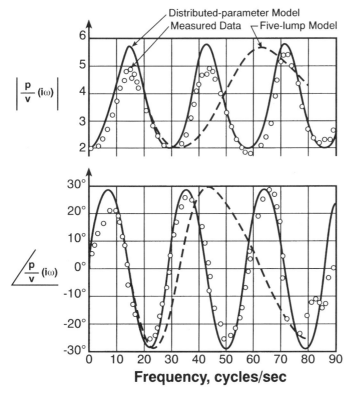

Figure 4-24 Comparison with measured results: lumped and distributed models for a liquid pipeline.

distributed model predicts an infinite number of resonant peaks, three of which are shown. The measured peak frequencies agree very well with this prediction, but the peak amplitudes are slightly too high. This is due to the neglect of viscosity; the reference shows another (more complex) model which includes viscosity and predicts the peak heights accurately. The five-lump model agrees almost perfectly with the distributed model up to about 30 Hz, which was the maximum frequency for which the lumped model was designed. Peak height here could also be improved by including viscosity in the model. Beyond 30 Hz the lumped model becomes progressively less accurate. If we needed accuracy at higher frequency, we need to use more and smaller lumps, as explained above.

Although our discussion has involved only one specific example, the basic concepts will carry over to all fluid systems. Once we have a formula for the velocity of propagation and choose the highest frequency of interest, we can pick a size and number of lumps which will give good accuracy up to that frequency, using the "10 lumps per wavelength" guideline. Higher operating frequencies require more and smaller lumps. If you are asking yourself, "Why use lumped models when distributed are available?", remember that the lumped models can be solved easily for the *time* response to any form of input and they also allow easy simulation with standard software such as our SIMULINK. The distributed models offer none of these important features, even though they are more accurate representations of the physical facts.

4-6 FLUID IMPEDANCE

Most fluid systems do not really *require* the separation of pressure/flow relations into their resistive, compliant, and inertial components; this separation is mainly one of analytical convenience. For complex fluid systems where experimental measurements may be a necessity, the measurement of *overall* pressure/flow characteristics has become a useful tool. The term *fluid impedance* is directly analogous to mechanical and electrical impedance discussed earlier, and is defined as the transfer function relating pressure drop (or pressure), as output, to flow rate as input, that is,

$$\text{Fluid impedance} \triangleq \frac{\Delta p}{q}(D) \quad \frac{\text{psi}}{\text{inch}^3/\text{sec}} \tag{4-66}$$

For the individual fluid elements we have

$$\text{Fluid resistance:} \quad \frac{\Delta p}{q}(D) = R_f \qquad \frac{\Delta p}{q}(i\omega) = R_f \tag{4-67}$$

$$\text{Fluid compliance:} \quad \frac{\Delta p}{q}(D) = \frac{1}{C_f D} \qquad \frac{\Delta p}{q}(i\omega) = \frac{1}{\omega C_f} \, \underline{/-90^\circ} \tag{4-68}$$

$$\text{Fluid inertance:} \quad \frac{\Delta p}{q}(D) = I_f D \qquad \frac{\Delta p}{q}(i\omega) = \omega I_f \, \underline{/+90^\circ} \tag{4-69}$$

and Fig. 4-25 shows the frequency response curves.

System Elements, Fluid and Thermal

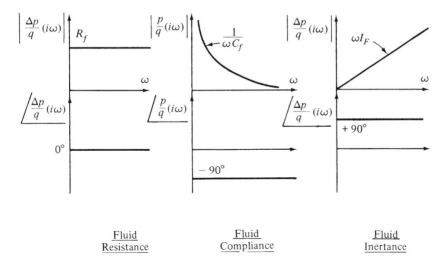

Figure 4-25 Fluid impedance of the three basic elements.

A good example of a complex fluid system[22] in which impedance measurements have been found useful is shown in Fig. 4-26. The system under study is a boiler using as the working fluid Freon (a refrigeration fluid widely used at the time of the reference but now outlawed to protect the earth's ozone layer). Since in a boiler the working fluid enters as a liquid and leaves as a gas (vapor), the flow situation (called two-phase flow) is quite complex and difficult to analyze. Accurate dynamic performance data on the boiler is needed to properly design the larger thermal power system of which it will be a part, so an experimental fluid impedance study was run. To minimize the effect of nonlinearities, the impedance is taken for small perturbations around an equilibrium operating point. By running several such tests at different operating points, one can explore the degree of nonlinearity. (If all operating points gave exactly the same impedance curves, the system would be perfectly linear.)

The test procedure is to set the flow-control servovalve at some fixed position and establish steady flow. Then the valve is oscillated sinusoidally about the original position, causing a small oscillation of flow rate and, thereby, pressure. (For example, the data of Fig. 4-27 had an equilibrium flow rate of 445 lb_m/hr and a pressure of 25.5 psia. The flow rate oscillation was set at an amplitude of about 40 lb_m/hr; the resulting pressure oscillation amplitude ranged from 0.04 to 0.28 psi.) The flow rate oscillation was measured with a turbine flow meter and the pressure with a piezoelectric pressure tranducer.[23] Measurements were made at 30 frequencies between 0.05 and 4.0 Hz, giving the sinuosidal transfer function of the impedance $(P'_I/W'_F)(i\omega)$ as shown in Fig. 4-27. (Note that the flow rate is given as a mass

---

[22]E. A. Krejsa, J. H. Goodykoontz, and G. H. Stevens, Frequency Response of Forced-Flow Single Tube Boiler, NASA TN-D-4039 (June 1967).

[23]Doebelin, *Measurement Systems*, 4th ed., pp. 458, 577.

250   Chapter 4

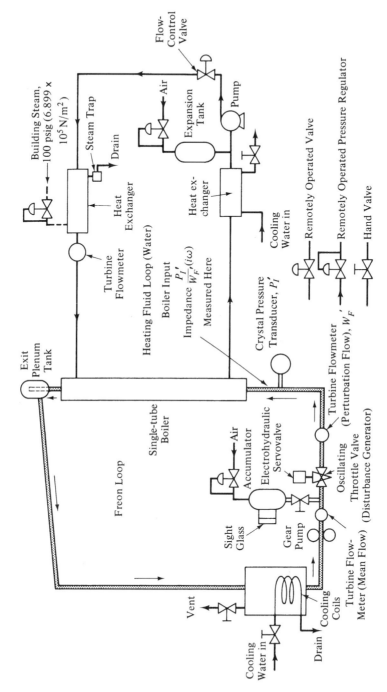

Figure 4-26 Complex flow system (Freon boiler) with measurement instrumentation.

System Elements, Fluid and Thermal

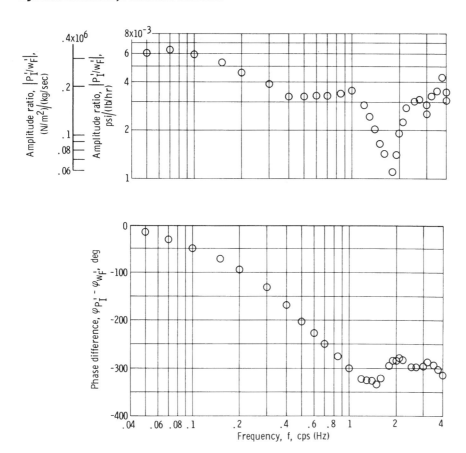

Figure 4-27 Measured fluid impedance of a Freon boiler.

flow rate rather than our usual volume flow rate, but it can be easily converted if we know the density.) A theoretical model of this system has also been derived and comparison made[24] with measured behavior. Good agreement was achieved under some, but not all, operating conditions, showing again the importance of experimental testing in validating theoretical models or revealing their defects so that improvements can be made.

If a fluid impedance is known as an operational transfer function $(\Delta p/q)(D)$, it should be clear that one can then calculate the response to any given input by solving the corresponding differential equation.

---

[24]E. J. Kresja, Model for Frequency Response of a Forced Flow, Hollow, Single Tube Boiler, NASA TM-X-1528, March, 1968.

252   **Chapter 4**

EXAMPLE: USE OF DIFFERENTIAL EQUATION

Suppose that a system has the transfer function

$$\frac{\Delta p}{q}(D) = \frac{10.0}{5D+1} \quad \frac{\text{psi}}{\text{in}^3/\text{sec}} \tag{4-70}$$

and that we apply an input flow rate $q = 1 + 2t$. The system differential equation would then be

$$(5D+1)\,\Delta p = 10q = 10 + 20t \tag{4-71}$$

$$5\frac{d\,\Delta p}{dt} + \Delta p = 10 + 20t \tag{4-72}$$

To solve this equation we need to know one initial condition; let's suppose that $\Delta p = 0$ at $t = 0$. The solution then turns out to be

$$\Delta p = 90(e^{-0.2t} - 1) + 20t \tag{4-73}$$

If a fluid impedance is measured by the frequency response technique, we then do *not* have a transfer function in equation form, as we did in the above example; we have only the curves, as in Fig. 4-27. The response to *sinusoidal* inputs is of course easily calculated from such curves. Suppose, however, that we want to find the response to an input which is *not* a sine wave but rather has a time-variation of arbitrary shape. Two methods of solving such problems are available. In the first method, we curve-fit the measured frequency-response curves with analytical functions, trying different forms and numerical values until an acceptable fit is achieved. Software to expedite such curve-fitting is available, for instance, the module called INVFREQS in the MATLAB SIGNAL PROCESSING TOOLBOX. Once we have a formula for the transfer function, this is the same as having the system differential equation, and we can then solve this for *any* form of input, either analytically or with simulation if that is necessary or desirable. In the second approach,[25] we use the measured frequency-response curves directly, without any curve fitting. To do this, we must compute the *Fourier transform* of the desired time-varying input signal to get its representation in the frequency domain. This operation "converts" the time function into its corresponding frequency function, which will be a complex number which varies with frequency. This complex number is multiplied, one frequency at a time, with the complex number which is the measured sinusoidal transfer function, giving a new complex number, whose magnitude and phase can be graphed versus frequency. This new complex number is the frequency representation of the *output* of our system. The final step is to use the *inverse Fourier transform* to convert this function back into the time domain, to give our system output as a specific, plottable, function of time. All these operations are again available in standard computer software, such as MATLAB.

---

[25] E. O. Doebelin, *Measurement Systems*, 4th Ed., pp. 147–157.

System Elements, Fluid and Thermal

All the above discussion of course applies to *any* linear, time-invariant, dynamic system, not just fluid systems. That is, if we can measure the frequency response, we can get the response to *any* form of input, not just sine waves.

4-7 FLUID SOURCES, PRESSURE AND FLOW RATE

An *ideal pressure source* produces a specified pressure at some point in a fluid system, no matter what flow rate might be required to maintain this pressure. Similarly, an *ideal flow source* produces a specified flow rate, irrespective of the pressure required to produce this flow. In fluid systems, the most common source of fluid power is a pump or compressor of some sort. A *positive-displacement* liquid pump draws in, and then expels, a fixed amount of liquid for each revolution of the pump shaft. When driven at constant speed, such a pump closely approximates an ideal constant-flow source over a considerable pressure range. Its main departure from ideal behavior is a decrease in flow rate, due to leakage through clearance spaces, as load pressure increases. This leakage flow is proportional to pressure; thus one can represent a *real* pump as a parallel combination of an ideal flow source and a linear (and large) flow resistance R_{fl} as in Fig. 4-28. If the inlet flow impedance of the load is low relative to

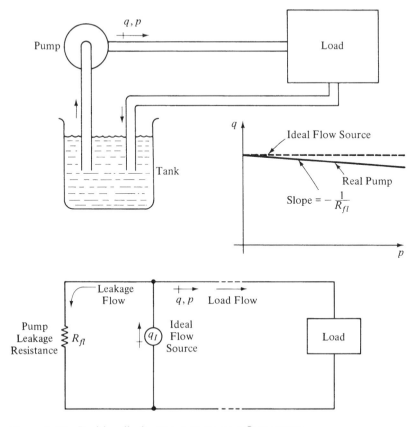

Figure 4-28 Positive-displacement pump as a flow source.

254   **Chapter 4**

R_{fl}, most of the flow goes into the load rather than the pump leakage path, and the pump acts nearly as an ideal flow source.

When we need to manipulate a flow rate as function of time, several possibilities exist. A fixed-displacement pump may be driven at a time-varying speed. A versatile and accurate version of such a concept uses an electric motor to drive the pump shaft, a flow sensor to measure the flow, and a feedback controller which compares desired flow with actual (measured) and commands the motor to change speed so as to keep actual flow close to desired at all times. Since most flow sensors have an electrical output, the desired flow rate may also be entered as a proportional electrical signal, with an amplifier to boost the flow error signal sufficiently to drive the motor. Since electronic signal generators can provide desired-flow commands of any waveform, such a feedback system can provide a versatile flow source. Instead of a fixed-displacement pump, we could use a variable-displacement pump. Here the pump shaft speed is constant, but pump output per revolution can be varied, while the pump is running. A stroking mechanism allows flow rate to be varied smoothly and quickly from full flow in one direction, through zero flow, to full flow in the reverse direction. The stroking mechanism could be driven directly or we could again use a feedback scheme as we did above with the fixed-displacement pump.

By combining a positive-displacement pump with a relief valve, one can achieve a practical constant-pressure source. This real source will not have the perfect characteristic of an ideal pressure source, but can be modeled as a combination of an ideal source with a flow resistance. A relief valve is a spring-loaded valve which remains shut until the set pressure is reached. At this point it opens partially, adjusting its opening so that the pump flow splits between the demand of the load and the necessary return flow to the tank. To achieve this partial opening against the spring, the pressure must change slightly; thus we do not get an exactly constant pressure (see Fig. 4-29). This real source can be modeled as a "series" combination of an ideal pressure source with a small flow resistance. (The pump leakage effect is usually small enough to neglect.)

EXAMPLE: REAL PRESSURE SOURCE

Measured data for a p versus q graph (Fig. 4-29) for a real pressure source follow essentially a straight line for $0 < q < q_p$. The pressure at $q = 0$ is 1000 psi, and just below q_p (which is 10 gal/min) the pressure is 950 psi. How would you model this source? The equivalent flow resistance of this system would be

$$R_{fv} = -\frac{\Delta p}{\Delta q} = \frac{50}{10} = 5.0 \, \frac{\text{psi}}{\text{(gal/min)}}$$

The source model would thus be an ideal 1000 psi pressure source in series with a flow resistance of 5.0 psi/(gal/min). If this source were connected to a pure resistance load with $R_{fL} = 200$ psi/(gal/min), what would be the load flow rate and pressure drop? The circuit total resistance is 205, so the flow rate is 1000/205 = 4.88 gpm and the pressure is $1000 - (4.88)(5) = 976$ psi.

System Elements, Fluid and Thermal 255

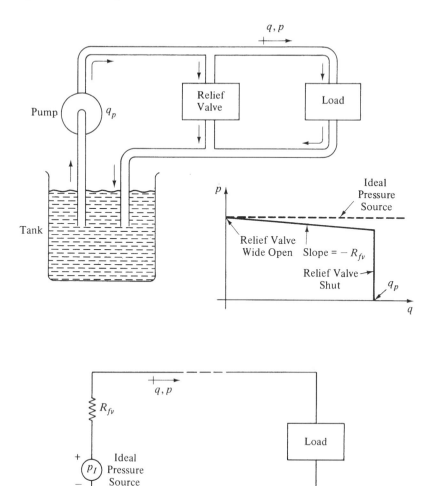

Figure 4-29 Positive-displacement pump with a relief valve: a pressure source.

The above two examples do not, of course, exhaust the possibilities with regard to power sources in fluid systems, but they should give some idea of how real sources may be modeled in terms of ideal sources and passive elements. Other fluid power sources encountered in practice include centrifugal pumps, accumulators (used for short-term power supplies), elevated tanks or reservoirs (gravity is the energy "source"), the human heart (a complex pump), etc.

4-8 THERMAL RESISTANCE

We begin our study of thermal elements with *thermal resistance*. Whenever two objects (or two portions of the same object) have different temperatures, there is a tendency for heat to be transferred from the hot region to the cold region, in an attempt to equalize the temperatures. For a given temperature difference, the rate of heat transfer varies, depending on the thermal resistance of the path between the hot and cold regions. The nature and magnitude of the thermal resistance depend on the modes of heat transfer involved: conduction, convection, and radiation.

In Fig. 4-30 the two bodies at temperature T_1 and T_2, respectively, are connected by a solid rod of constant cross-sectional area A and length L. (To consider *only* the conduction mode of heat transfer, we assume the surface of the rod to be perfectly insulated.) The rod is made of a material with *thermal conductivity k*. Fourier's law of heat conduction may be written in the following form as a means of defining thermal resistance.

$$\text{Heat transfer rate} \triangleq q = \frac{kA}{L}(T_1 - T_2) = \frac{kA}{L}\Delta T \quad \frac{\text{Btu}}{\text{sec}} \text{ or watts} \quad (4\text{-}74)$$

where

$A \triangleq$ cross-sectional area, inch^2 or m^2

$L \triangleq$ length, inch or m

$k \triangleq$ thermal conductivity, (Btu/sec)/[in^2-(F°/in)] or (watts)/[m^2-(C°/m)]

$T_1, T_2 \triangleq$ temperatures, °F or °C

Thermal conductivity is a material property which is found by experiments based on Eq. (4-74). That is, q, A, L, and ΔT are all measured for a steady-state situation and k is then calculated from Eq. (4-74). Ideally, k is a constant, but in reality it may vary with temperature, position in the body, and direction of heat flow.

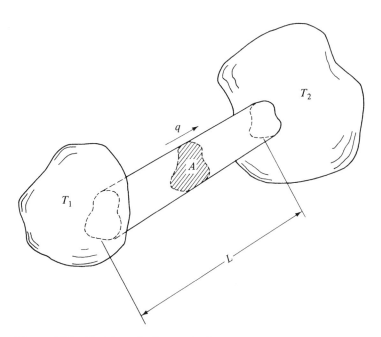

Figure 4-30 Heat transfer by conduction.

System Elements, Fluid and Thermal

When T_1 and T_2 are widely separated and k varies greatly (a good example being cryogenic systems) the method of *thermal conductivity integrals* allows a more convenient calculation of the heat flow. For small dynamic changes in either of the two widely separated temperatures a linearized approach is feasible, but large changes require simulation, using a lookup table for the thermal conductivity integral. If the cross-sectional area changes along the heat flow path, a *geometry factor* can take this into account. Both the geometry factor and the technique of thermal conductivity integrals are discussed in the literature.[26] Printed-circuit boards are a good example of anisotropic (direction-sensitive) behavior of thermal conductivity. Here the material is in the form of a "sandwich" with layers of high-conductivity copper and low-conductivity epoxy-fiberglass. Thermal conductivity of the composite sandwich in a direction perpendicular to the plane of the board may be only 0.05 times that for the parallel direction.[27]

Having pointed out some exceptions, we now concentrate on the simple linear case where k is assumed to be constant. Equation (4-74) indicates an *instantaneous* relation between ΔT and q. This would be true if the rod had no thermal capacitance (heat storage capability). Since a real rod will have thermal capacitance, we must think of Eq. (4-74) as defining *only* the resistive component of rod dynamics; its thermal capacitance will be taken into account separately. We can now define the pure and ideal thermal resistance for conduction heat transfer as follows.

$$q = \frac{\Delta T}{R_t}$$

$$R_t \triangleq \frac{\Delta T}{q} = \frac{L}{kA} \quad \frac{F^\circ}{Btu/sec} \text{ or } \frac{C^\circ}{watt} \tag{4-75}$$

The analogy to electrical resistance is clear if we think of ΔT as the driving force (voltage) and heat flux q as the current.

When heat flow occurs through the interface where two solid bodies share a common surface, the phenomenon of *contact resistance* is observed. If the contact surface were perfectly smooth, the contact resistance would be zero and the temperatures of the two bodies would be identical at the contact surface. Real objects always have some surface roughness, which causes essentially a step change in temperature across the interface. This effect can be modeled with a thermal contact resistance, which depends mainly on the roughness of the surfaces, and the contact pressure, for any two given materials. Contact resistances must be obtained by lab testing and are quite unreliable and difficult to predict, but may be critical in some calculations. Some typical values[28] are available in the literature, but must be applied with large safety factors, due to their uncertainty. For example, aluminum-to-aluminum joints may have resistance values ranging from 8.3×10^{-5} to $45. \times 10^{-5}\, C^\circ/watt$, for an area of $1.0\,m^2$. The two aluminum pieces themselves,

---

[26] Cryogenic Heat Flow Calculations, Lake Shore Cryotronics, 64 E. Walnut St., Westerville, OH, 43081-2399, 614-891-1392. R. L. Garwin, *Rev. Sci. Instrum.* 27 (1956), 826.

[27] J. E. Graebner, Thermal conductivity of printed wiring boards, *Electronics Cooling*, vol. 1, no. 2, October 1995.

[28] A. F. Mills, *Basic Heat and Mass Transfer*, Irwin, Chicago, 1995, p. 57.

taking 5 mm as a typical thickness, would have a total resistance of about $5. \times 10^{-5}$ for the same 1.0-m² area, showing clearly the large error caused by ignoring contact resistance. Note also that even though the resistance of the aluminum itself can be predicted fairly accurately, the contact resistance, and thus the overall resistance, will be quite uncertain and also may change unpredictably if contact pressure changes.

Many practical situations involve heat flow through fluid/solid interfaces by *convection*. Here the heat flows by conduction through a thin layer of fluid (called the *boundary layer*) which adheres to the solid wall. At the interface between the boundary layer and the main body of the fluid, the heat is carried away into the main stream by the constantly moving fluid particles. This overall process is called convection heat transfer, and is illustrated in Fig. 4-31. Experiments have shown that this process may be described by the equation

$$q = hA(T_1 - T_2) = hA\,\Delta T \tag{4-76}$$

where $h \triangleq$ film coefficient of heat transfer $\dfrac{\text{Btu/sec}}{\text{in}^2\text{-F}^\circ}$ or $\dfrac{\text{watts}}{\text{m}^2\text{-C}^\circ}$

The film coefficient h depends on the geometry of the solid bodies, the nature of the fluid flow, and the fluid properties. It must be found by experiment, but for many configurations the experimental results have been generalized so that h may be predicted with fair accuracy from calculations. Also, h varies somewhat with temperature, so Eq. (4-76) is really a linearized version of reality, but the accuracy is often adequate. The pure and ideal thermal resistance associated with convection is thus

$$R_t \triangleq \frac{\Delta T}{q} = \frac{1}{hA} \quad \frac{\text{F}^\circ}{\text{Btu/sec}} \text{ or } \frac{\text{C}^\circ}{\text{watt}} \tag{4-77}$$

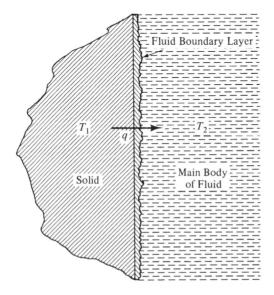

Figure 4-31 Heat transfer by convection.

System Elements, Fluid and Thermal

Often conduction and convection are combined, and we can define an overall heat transfer coefficient and thereby an overall thermal resistance. Figure 4-32 shows a cross section of an automobile "radiator" (convector might be a better name) in which heat flows from the hot internal coolant fluid, through the metal wall, and into the air forced over the radiator by a fan. Since the same heat flux q goes through all three of the resistances we may write

$$q\left(\frac{1}{h_W A}\right) + q\left(\frac{L}{kA}\right) + q\left(\frac{1}{h_A A}\right) = T_W - T_A \tag{4-78}$$

$$q = \frac{T_W - T_A}{\frac{1}{h_W A} + \frac{L}{kA} + \frac{1}{h_A A}} \triangleq \frac{\Delta T}{R_t} \triangleq UA\Delta T \tag{4-79}$$

$$\text{Overall resistance} \triangleq R_t \triangleq \frac{1}{h_W A} + \frac{L}{kA} + \frac{1}{h_A A} \tag{4-80}$$

where U is called the *overall coefficient of heat transfer*. We see that the overall resistance is just the sum of the individual resistances, as we might expect from the electrical analogy.

The final mode of heat transfer that we consider is *radiation*. Here, two bodies can exchange thermal energy with no physical contact whatever (see Fig. 4-33). This mode often contributes a relatively small portion of the total heat transfer unless the temperatures are quite high. However, if other modes are inhibited, then radiation can be important even at low temperature. For example, the heat transfer at the outer surface of an orbiting satellite must be entirely due to radiation since it is exposed only to the vacuum of space, defeating any conduction or convection.

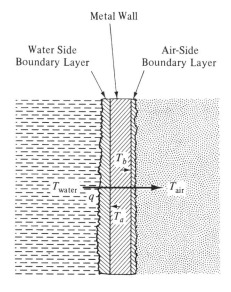

Figure 4-32 Combined conduction/convection: overall heat transfer.

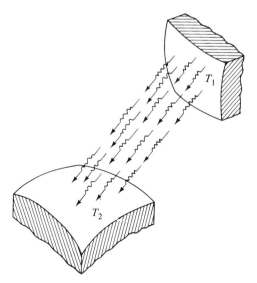

Figure 4-33 Heat transfer by radiation.

The rate of radiation heat transfer depends on a surface property of each body called the *emissivity*, geometrical factors involving the portion of emitted radiation from one body that actually strikes the other, the surface areas involved, and the absolute temperatures of the two bodies. Here again we need to warn that emissivity values are usually more uncertain than conductivities or convection coefficients, so highly accurate calculations should not be expected. Note that with conduction and convection we could use either "ordinary" (°F or °C) temperatures or absolute (°R or K), since the subtraction involved in ΔT gives the same value. For radiation, we *must* use absolute temperatures.

For a given configuration and materials, the defining equation takes the form

$$q = C(T_1^{\,4} - T_2^{\,4}) \tag{4-81}$$

where C includes all the effects other than temperature. This mode of heat transfer is clearly nonlinear, but can be linearized for approximate analyses as long as the temperatures do not vary greatly from defined operating points $T_{1,o}$ and $T_{2,o}$.

$$q \approx C(T_{1,o}^4 - T_{2,o}^4) + \left[\frac{\partial q}{\partial T_1}\right]_{T_{1,o}, T_{2,o}} (T_1 - T_{1,o}) + \left[\frac{\partial q}{\partial T_2}\right]_{T_{1,o}, T_{2,o}} (T_2 - T_{2,o}) \tag{4-82}$$

$$q \approx -3CT_{1,o}^4 + 3CT_{2,o}^4 + (4CT_{1,o}^3)T_1 - (4CT_{2,o}^3)T_2 \tag{4-83}$$

While Eq. (4-83) is linear in T_1 and T_2, it does *not* allow definition of a thermal resistance unless $T_{1,o} = T_{2,o} = T$, that is, the operating point must be one of zero heat transfer. For this case,

$$q \approx 4CT^3(T_1 - T_2) \tag{4-84}$$

$$R_t \approx \frac{\Delta T}{q} = \frac{1}{4CT^3} \tag{4-85}$$

System Elements, Fluid and Thermal   261

The Taylor series linearization of Eq. (4-82) does not lead to a general resistance expression but another approach provides a useful result. Equation (4-81) can be exactly factored as

$$q = C(T_1 + T_2)(T_1^2 + T_2^2)(T_1 - T_2) \tag{4-86}$$

which gives

$$R_t = \frac{\Delta T}{q} = \frac{1}{C(T_1 + T_2)(T_1^2 + T_2^2)} \tag{4-87}$$

This resistance varies with T_1 and T_2 and is thus nonlinear; however a linear approximation near a given operating point is

$$R_t \approx \frac{1}{C(T_{1,o} + T_{2,o})(T_{1,o}^2 + T_{2,o}^2)} \tag{4-88}$$

Figure 4-34 shows the symbol used for thermal resistance in system diagrams; it is identical with that for electrical resistance. The temperature difference $T_1 - T_2$ is a driving potential for heat flux q, just as a voltage difference $e_1 - e_2$ is a driving potential for current i. Since the equation relating q to ΔT is an algebraic one, the response of q to ΔT is instantaneous. When energy behavior is considered, the thermal/electrical analogy breaks down, since the heat flux q is *already* power, whereas the analogous current i is not. Also, all the heat flux entering the thermal resistance at one end leaves at the other end, and none is lost or dissipated; whereas the electrical energy supplied to a resistor is all converted into heat, and is thus lost to the electrical system. Appendix C lists some numerical values relating to conduction, convection, and radiation resistance, to give some idea of orders of magnitude, ranges, and relative sizes.

In discussing thermal resistance we have concentrated on the basic modes of heat transfer. In thermal system analysis and design, the *overall* thermal resistance of

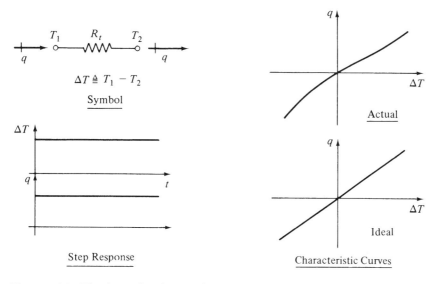

Figure 4-34   The thermal resistance element (see table in Appendix C).

262   **Chapter 4**

hardware components, obtained from lab testing, is also widely used, particularly in electronic and electromechanical applications. Since only the overall resistance is desired and can be directly measured, no attempt at a separate analysis of the various modes (conduction, convection, radiation) is attempted. Since geometry and materials would often make accurate analysis difficult or impossible, lab testing of actual equipment is appropriate. As electronic equipment is miniaturized more and more, the heat generated is concentrated in smaller volumes. This tends to raise component temperatures, and the junctions of electronic devices can withstand only moderate temperatures,[29] typically 115°C for microelectronics (as low as 65°C for some military applications), to as high as 180°C for power electronics. Heat flows from the device junction outward to the ambient air, usually through three specific thermal resistances connected in series: the device resistance from junction to case, a contact resistance between the device case and the heat sink, and the heat sink resistance from sink metal temperature to ambient cooling air.

For a specific component, say a transistor, its resistance from junction to case is considered fixed and beyond the control of the cooling system designer. The contact resistance between the device and the heat sink can be quite important and is often reduced by use of special thermal greases or sheet materials placed in the joint and/or by maintaining good clamping pressure. A typical 0.002-inch-thick layer of such grease may have a thermal resistance of 0.2 C°/watt for a 1-inch square area. Heat sinks are of many different types, air-cooled finned aluminum extrusions being the most common. For natural convection (no fan), thermal resistances (for 1 inch³ of heat sink volume) range from 30 to 50 C°/watt. With forced convection air flow of 5 m/sec, the resistance drops to 3 to 5 C°/watt. All these numerical values are from the S. Lee reference.

When air-cooled heat sinks can not provide the needed cooling, liquid-to-air heat exchangers used with liquid-cooled plates may provide a solution.[30] Here the electronic equipment is mounted on a metal plate which has internal passages for a flow of cooling water. The water is continuously cooled by passing it through a heat exchanger which uses a fan-induced flow of cooling air. These more complicated devices are also usually described in terms of the thermal resistance concept. A typical liquid-cooled plate has a thermal resistance of 0.2 to 0.5 C°/watt. Using this together with the electronic device thermal resistance, one can calculate the needed performance of the liquid-to-air heat exchanger. This performance is described in terms of watts/C°, which we see is just the reciprocal of thermal resistance. *Heat pipes*[31] are sealed tubes with a phase-change liquid and a wick structure inside. They transfer heat along the length of the tube, similar to a solid rod, except that their thermal resistance is much less than that of typical metal rods. For example, a 0.25-inch-diameter, 12-inch long heat pipe with a 20 F° temperature difference might conduct 189 Btu/hr while a copper rod of identical dimensions would conduct only 4.3 Btu/hr. Heat pipes are used in a variety of applications ranging from electronics cooling in notebook computers to mold temperature con-

---

[29]S. Lee, How to select a heat sink, *Electronics Cooling*, vol. 1, no. 1, June 1995, pp. 10–14.
[30]Thermacore Inc., 780 Eden Road, Lancaster, PA 17601, 717-569-6551.
[31]Thermacore Inc., Common Questions About Heat Pipes, 1995.

System Elements, Fluid and Thermal   **263**

trol in injection molding machines. Application of heat pipes also uses the concept of thermal resistance.

While the accuracy and speed of motion-control systems which use electric motors depend on electromechanical parameters such as inertia, friction, and magnetic torque, the choice of the *size* of a particular motor is largely based on thermal considerations related to the temperature limits of the electrical insulation used in the motor's construction. Since materials and geometry again (just as in electronic devices) make accurate thermal analysis difficult, lab testing is used to thermally characterize electric motors, using the concept of thermal resistance. A brochure[32] describing a line of brushless servomotors includes a complete set of specifications, including thermal parameters. For each motor, two thermal resistances have been measured. One corresponds to a "worst case" situation where the motor is bolted to a mounting flange which is a poor thermal conductor or is already hot from other heat sources in the machine. In this case the motor's heat must all be dissipated to the surrounding air, giving a relatively large thermal resistance (2.6 C°/watt for a 0.15-kW motor and a 0.16 for a 20-kW motor). If the motor is mounted to a "cool" flange, the thermal resistance is lower: 0.96 and 0.112, respectively, for the small and large motor. Later in this text, when we study electromechanical system dynamics, we will use thermal parameters such as these in our system simulations.

4-9 THERMAL CAPACITANCE AND INDUCTANCE

When heat flows into a body of solid, liquid, or gas, this thermal energy may show up in various forms such as mechanical work or changes in kinetic energy of a flowing fluid. If we restrict ourselves to bodies of material for which the addition of thermal energy does *not* cause significant mechanical work or kinetic energy changes, the added energy shows up as stored internal energy and manifests itself as a rise in temperature of the body. For a pure and ideal thermal capacitance, the rise in temperature is directly proportional to the total quantity of heat energy transferred into the body, giving the following definition:

$$T - T_0 = \frac{1}{C_t} \int_0^t q \, dt \tag{4-89}$$

where

$T \overset{\Delta}{=}$ temperature of body at time t

$T_0 \overset{\Delta}{=}$ temperature of body at time $= 0$

$C_t \overset{\Delta}{=}$ thermal capacitance, Btu/F° or J/C°

Since we refer to *the* temperature of the body, we are assuming that, at any instant, the body's temperature is *uniform* throughout its volume. For fluid bodies, this ideal situation is closely approached if the fluid is thoroughly and continuously mixed. For solid bodies, uniform temperature requires a material with infinite ther-

---

[32]Brushless Servomotors, Vickers Inc., 5435 Corporate Drive, Suite 350, P.O. Box 302, Troy, Michigan 48007-0302, 313-641-0145, 1989.

mal conductivity k, since then, for any heat flow rate q through the body, the temperature difference $\Delta T = -q\,\Delta x/(kA)$ would be zero. No real material has inifnite k; thus there is always some nonuniformity of temperature in a body during transient temperature changes. Many practical problems involve solid bodies immersed in fluids, and for this situation a useful criterion for judging the validity of the uniform-temperature assumption is found in the Biot number[33] N_B (see Fig. 4-35):

$$N_B \triangleq \frac{hL}{k} \qquad (4\text{-}90)$$

where

$h \triangleq$ film coefficient at surface

$L \triangleq$ volume/(surface area)

$k \triangleq$ thermal conductivity of solid body

For bodies whose shape approximates a plate, sphere, or cylinder, if $N_B < 0.1$, the error in assuming the solid to have a uniform temperature is less than about 5%. For example, a 1-inch-diameter spherical steel ball being heated in stagnant air has $h \approx 2$, $k \approx 35$, and $L = \frac{1}{6}$, making $N_B \approx 0.0095$; thus this ball's internal temperature may safely be assumed uniform in a thermal system analysis. The thermal capacitance of the boundary layer fluid film is generally negligible, since the film is very thin; thus convective resistances are very nearly pure resistances.

When the Biot number is less than 0.1, the assumption of uniform temperature is acceptable, except for the "early times" of a step change in fluid temperature. For such early times, the temperature change of the solid body is localized in a thin

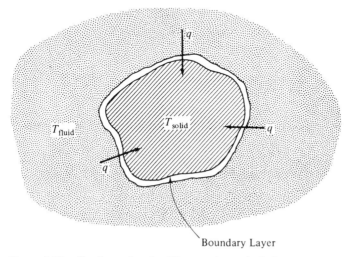

Figure 4-35 Configuration for Biot number calculations.

---

[33] Mills, *Basic Heat and Mass Transfer*, p. 913.

System Elements, Fluid and Thermal

"skin" near the fluid/solid interface.[34] As time goes by this skin grows until it extends through the entire body, whereupon the uniform temperature assumption becomes acceptable. The division between "early" times and "later" times is not precise but can be estimated from another dimensionless group, the Fourier number. Diffusion of heat through a solid body is governed by its *thermal diffusivity* α, where $\alpha \triangleq k/(\rho c)$. A large value of α means rapid diffusion of heat. The Fourier number N_F is defined as $\alpha t/L^2$, where t is time and L is the thickness of a plate, or the radius of a cylinder or sphere. A conservative requirement on the Fourier number is that it be greater than about 10 for the uniform temperature assumption to be accurate. Since steel has $\alpha \approx 0.1 \, \text{cm}^2/\text{sec}$, the steel sphere used as an example above could be treated as uniform temperature for times longer than 161 seconds. Unfortunately, most dynamic applications of thermal systems involve continuously changing temperatures, rather than simple step changes, so this use of the Fourier number must be somewhat qualitative. If the spatial variation of temperature, rather than an "average" temperature, in the solid body must be predicted, we should use several (or many) lumps of thermal capacitance, rather than just one, in our model. Sometimes we must begin our modeling with several lumps and let these results tell us if we can simplify the model to fewer, or just one, lump of thermal capacitance.

The calculation of numerical values of thermal capacitances is relatively straightforward, since the temperature rise of a body when heat is added is given by

$$\text{Heat added} = \int q \, dt = \text{mass} \times \text{specific heat} \times \text{temperature rise} \qquad (4\text{-}91)$$

Thus,

$$C_t \triangleq \frac{\text{heat added}}{\text{temperature rise}} = \text{mass} \times \text{specific heat} = Mc \qquad (4\text{-}92)$$

The specific heat c of real materials varies somewhat with temperature; however in many cases it is sufficiently accurate to use a constant value which is the average for the range of temperature covered. This keeps the behavior linear, if we are trying to keep our system differential equations linear with constant coefficients to allow analytical solution. When c undergoes large changes and we insist on greater accuracy, the variation of c with temperature can be included in simulation models through curve fits or lookup tables. For fluids (particularly gases) the specific heat is often measured for two different situations: constant volume and constant pressure. Since these values are quite different (for air at $32°F$, $c_p = 0.240$ and $c_v = 0.171$), one must be careful to use the value which corresponds most closely to the actual application. When heat is added to or taken away from a material which is changing phase (melting or freezing, vaporizing or condensing) the thermal capacitance is essentially *infinite*, since one can add heat without causing *any* temperature rise.

---

[34]A. Bejan, *Heat Transfer*, Wiley, New York, 1993, p. 144.

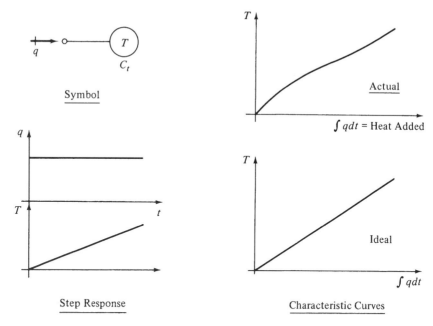

Figure 4-36 The thermal capacitance element (see table in Appendix C).

In Eq. (4-89) we may define $T_0 \equiv 0$ if we wish, giving the transfer function

$$\frac{T}{q}(D) = \frac{1}{C_t D} \qquad (4\text{-}93)$$

and the step response of Fig. 4-36. A standard symbol for thermal capacitance has not be defined; that given in Fig. 4-36 is suggested as a simple and reasonable one.

By analogy to electrical systems, a *thermal inductance* L_t would have a q/T characteristic given by $T_1 - T_2 = L_t(dq/dt)$. No physical effect following this relation has yet been discovered, thus thermal inductance is not necessary for the description of thermal system behavior and is not defined or used. Mechanical, electrical, and fluid systems each require three different elements for their description, two of which are energy storage elements which store energy in two different ways. Thermal systems require only two elements, and only one of these stores energy. This energy storage feature allows mechanical, electrical, and fluid systems to display *natural ("free") oscillations*, where energy is traded back and forth between the two types of storage elements. Thermal systems cannot and do not display such free oscillations, though they can, of course, be put into *forced* oscillation by external periodic driving agencies. These behaviors will be discovered mathematically when we later solve system differential equations. They of course are also observed in laboratory testing.

4-10 THERMAL SOURCES, TEMPERATURE AND HEAT FLOW

The ideal temperature source maintains a prescribed temperature (either constant or time-varying) irrespective of how much heat flow it must provide, while an ideal

System Elements, Fluid and Thermal

heat-flow source produces a prescribed (constant or time-varying) heat flow irrespective of the temperature required. Constant-temperature sources may often be quite well approximated by utilizing materials undergoing phase change. A well-stirred bath of ice and water remains very nearly at 32°F, even if heat flows are entering or leaving it; similarly for water boiling at atmospheric pressure and 212°F. The melting points of various metals and salts are similarly used to establish desired constant temperatures at various levels. A large vessel of liquid, even if *not* changing phase, will maintain a nearly constant temperature for short time intervals as long as the heat flows in or out are not too large. When a specific time-varying temperature is required, a liquid bath (or air flow) with a feedback temperature-control system may be necessary (Fig. 4-37).[35] Here the bath temperature is measured and converted to a proportional voltage, which is then compared with a command voltage representing the desired (constant or time-varying) temperature. If desired and actual temperatures are not equal, the controller modulates the power to the electric heater so as to provide more or less heat, as needed. Many other schemes for providing controlled temperature or heat flows may be found in the literature.[36]

Perhaps the most convenient heat flow source for many applications is electrical resistance heating. A constant or time-varying voltage $e(t)$ applied to a resistance heating coil produces an electrical heat generation rate of $e^2(t)/R$ if inductance is negligible. Suppose such a coil is immersed in a fluid bath to act as a heat-flow source. The electrically generated heat goes partly into heating the coil itself, and the rest of the heat flows away to the fluid as intended. If the thermal capacitance of the

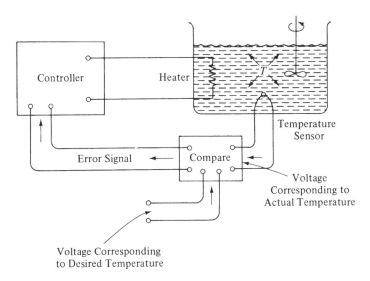

Figure 4-37 Feedback control system for temperature source.

---

[35] E. O. Doebelin, *Control System Principles and Design*, Wiley, New York, 1985, pp. 250, 278, 441.

[36] E. O. Doebelin, *Engineering Experimentation*, McGraw-Hill, New York, 1995, chap. 6 (Apparatus Design and Construction).

268   **Chapter 4**

coil metal (relative to that of the fluid bath), and the convective resistance between coil and fluid are both sufficiently small, then most of the electrically generated heat flows into the fluid and we have a good approximation to a heat-flow source. Radiant heat flux may also be usable as a heat-flow source. The sun, for example, provides about 400 (Btu/hour)/ft^2 at high altitudes, and as much as 300 at the earth's surface on a clear day. These heat flows are of course quite unaffected by the presence or absence of an object to receive them, and thus represent nearly ideal heat-flow sources. Radiant heat lamps are available when a controllable source of this type is needed. Mechanical shutters can turn radiant beams "on and off" very quickly if step changes in heat flow are needed.

BIBLIOGRAPHY

Andersen, B. W., *The Analysis and Design of Pneumatic Systems*, Wiley, New York, 1967.
Doebelin, E. O., *System Modeling and Response: Theoretical and Experimental Approaches*, Wiley, New York, 1980.
Fox, R. W., and A. T. McDonald, *Introduction to Fluid Mechanics*, Wiley, New York, 1985.
Merritt, H. E., *Hydraulic Control Systems*, Wiley, New York, 1967.
Mills, A. F., *Basic Heat and Mass Transfer*, Irwin, Chicago, 1995.

PROBLEMS

4-1.   In Fig. 4-12, compare the fluid resistance R_f for circular, square, and triangular "pipes" of the same area and length. Speculate on why circular pipes are so common.

4-2.   For water flowing in a 0.25-inch-diameter smooth pipe of length 10 feet, what is the fluid resistance for laminar flow? What is the largest pressure drop which will give laminar flow and what is the flow rate for this condition? If the pressure drop is made 10 times this value, will the flow rate also increase by 10 times? What *will* the flow rate now be? Use viscosity of 2.09×10^{-5} lb$_f$-sec/ft^2 and density 1.93 lb$_f$-sec^2/ft^4.

4-3.   For the pipe of problem 4-2, calculate and plot versus flow rate the incremental linearized flow resistance for turbulent flow. For a pressure drop of 300 psi, how much does this R_f change for a $\pm 10\%$ change in pressure drop?

4-4.   For a laminar pipe flow, if viscosity changes by 100%, how much does R_f change? Compare this with the change in R_f caused by a similar viscosity change, but in turbulent flow. If we desire to minimize the effects of viscosity changes (perhaps caused by temperature changes), should we design for laminar or turbulent flow?

4-5.   In Eq. (4-14) if N_R is 500, what is the maximum allowable D/L to give 5% nonlinearity? Using this D/L value and $\mu = 1.4 \times 10^{-7}$ lb$_f$-sec/in^2, design a capillary tube to give $R_f = 100$ psi/(in^3/sec). What is the maximum flow rate this tube can pass with laminar flow? Use density 9.31×10^{-5} lb$_f$-sec^2/in^4.

4-6.   In Eq. (4-24) reduce the term $4\rho L/3AR_f$ to its simplest form in terms of basic constants. Now state the requirements for negligible inertia effect in terms of ρ, D, μ, and ω. Why does length L have no effect on this? If ω_{max} is the frequency at which

System Elements, Fluid and Thermal   **269**

the imaginary term in Eq. (4-24) is 0.1, plot a curve of $D^2\rho/\mu$ versus ω_{max}. What is the usefulness of such a curve in designing flow resistances?

4-7. Using Eq. (4-37), plot q versus Δp for water at 70°F flowing through an orifice of 0.1 inch diameter, taking $C_d = 0.60$. What is the linearized resistance in the neighborhood of $\Delta p = 100$ psi? How much does it change for a $\pm 10\%$ change in Δp?

4-8. For a single capillary in Fig. 4-16, compute a theoretical pressure/flow curve and compare with the actual measured behavior, using:

    a. Equation (4-12)
    b. Equation (4-15)

4-9. Repeat problem 4-8 for the three capillaries in parallel in Fig. 4-16.

4-10. Using the appropriate formulas from the text, plot a theoretical pressure/flow curve for the orifice of Fig. 4-16, and compare with the measured result.

4-11. Using Eq. (4-37), fit an empirical curve to the orifice data of Fig. 4-16. That is, find K in $q = K(\Delta p)^{0.5}$. How well does this curve fit the data? For the K you found, what is the corresponding value of C_d?

4-12. What D/t ratio is required in steel tubing for the tubing compliance to equal the compliance of oil with bulk modulus of 200,000 psi? What would D/t be for aluminum tubing?

4-13. "Rubber" hose is often a complex composite material with layers of rubber, fabric, and woven metal reinforcement. This makes theoretical calculation of compliance a practical impossibility. Explain how you would set up experiments to find the compliance, remembering that it probably will be nonlinear.

4-14. Design a piston-type accumulator as in Fig. 4-19a to supply a hydraulic load which consumes 0.2 horsepower for a 1-minute period. Assume the load can use all the stored energy and design for a maximum pressure of 3000 psi. Find all combinations of A and K_s which meet these requirements. If space limits A to 5 in.^{2}, find K_s and the total stroke.

4-15. Repeat problem 4-14 for the case where the load can only use the energy when the pressure is between 2000 and 3000 psi.

4-16. Find an expression for the linearized compliance of the conical tank in Fig. 4-20b.

4-17. Repeat problem 4-16 for a spherical tank.

4-18. Using Eq. (4-62), compute the kinetic energy, effective mass, and inertance for any value of n. Evaluate the effective mass for $n = 4$ and for $n = 10$.

4-19. For the system whose fluid impedance is given in Fig. 4-27, find the flow rate if:

    a. Pressure $= 1.0 \sin 0.1\pi t$, psi, t in seconds.
    b. Pressure $= 1.0 \sin \pi t$.
    c. Pressure $= 1.0 \sin 3.6\pi t$.
    d. Pressure $= 1.0(\sin 0.1\pi t + \sin \pi t + \sin 3.6\pi t)$.

4-20. In the system of Fig. 4-28, $q_I = 30$ in.^{3}/sec, $R_{fl} = 1000$ psi/(in.^{3}/sec), and the load is a fluid resistance of 10 psi/(in.^{3}/sec).

    a. What pressure will this system run at?
    b. What flow is supplied to the load?
    c. What fluid power is supplied by the pump?

d. Suggest a reasonable definition of efficiency for such a system and compute its numerical value.

e. Plot a q versus p curve for this pump.

4-21. For the system of Fig. 4-29, $p_I = 1000$ psi, $R_{fv} = 1$ psi/(in^3/sec), and the pump flow is 30 in^3/sec.

a. Plot a p versus q curve for this system.

b. What is the percent pressure change over the full range of the power supply?

c. Suggest a reasonable definition of efficiency for such a system and then compute its numerical value at $q = 0$, 15, and 30 in^3/sec. What happens to the wasted portion of the energy?

d. Pump leakage was neglected in the model of Fig. 4-29. Explain how the behavior changes if this leakage is *not* neglected and sketch the effect on the p versus q graph.

4-22. The brass rod of Fig. P4-1 carries a steady heat flux of 50 Btu/hr. Compute the thermal resistance of each section of the rod, and then its total resistance. If the left end is at 400°F, calculate and plot the variation in temperature from left to right. Compute the thermal capacitance of each section of the rod, and also its total capacitance. Suggest a way to use these capacitance values to compute the total stored energy when the rod is at a non-uniform temperature as in this example. If we arbitrarily call the bar's energy content zero when it is all at 0°F, compute the stored energy when it is heated as above.

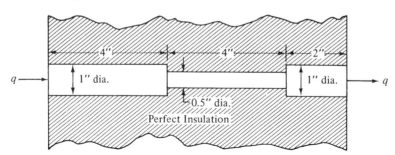

Figure P4-1

4-23. In Fig. 4-32, let $h_W = 500$ Btu/(hr-ft^2-°F), $h_A = 20$ Btu/(hr-ft^2-°F), and let the metal wall be 0.030-inch-thick brass. Compute the total thermal resistance per square foot of area. Which component of resistance dominates the total? How much error is caused by entirely neglecting the others?

4-24. If $h = 2.0$ Btu/(hr-ft^2-°F), what is the size of the largest steel cube for which the internal temperature may be assumed uniform? Repeat for a silver cube. How do these results change if $h = 200$?

4-25. Based on the data of Appendix C, what material should we use for thermal energy storage if we wish to use the minimum space? Which materials would be "second-best" and "third-best"?

4-26. A 500-ohm electric resistance heater is embedded in a brass sphere of 5 inch diameter. The surface of the sphere is perfectly insulated. If the temperature is 70°F

System Elements, Fluid and Thermal 271

at time zero when 100 volts is applied to the heater, calculate and plot versus time, the temperature rise of the sphere. Make and state any assumptions needed to solve this problem. If the 100-volt source is turned on and off in a repetitive cycle (2 minutes on, 3 minutes off), again calculate and plot temperature versus time.

4-27.   In problems similar to 4-26, we would like to assume uniform temperature at any time, to allow direct and simple use of the concept of thermal capacitance. For situations like that of Fig. 4-35 we are able to use the Biot number to make a proper judgement, but the conditions of problem 4-26 are *not* those associated with the Biot number. While a simple and widely applicable guide such as the Biot number is not available for such problems, there are some rough approximations which can indicate whether the desired assumption is reasonable or not. The idea is as follows.

The internal heat source produces a definite and known heat flow at its surface, and this heat must then propagate out through the sphere to heat the rest of the material. For heat to flow through the sphere material by conduction, there must of course be a temperature difference from point-to-point in the sphere. This immediately tells us that the temperature *cannot* possibly be uniform, but the question is, how *non*uniform is it? We can estimate such temperature differences by assuming a *steady-state* (rather than the true transient) condition, and computing how large a temperature difference would need to exist in order to produce the known heat flow rate by conduction in the object under study.

For a hollow spherical shell of inner radius r_i and outer radius r_o, it can be shown[37] that the thermal resistance is given by $R_t = (1/r_i - 1/r_o)/4\pi k$. This result allows us to calculate the temperature difference from inside to outside, for any given steady heat flow. In problem 4-26, no geometric details of the buried heater were given; it probably would *not* be spherical. To use the given theoretical result outside its assumption of a spherical cavity we could, as a further approximation, replace the actual heater with a spherical heater of the same volume. Let's assume that our heater has a volume equal to that of a 0.5-inch-diameter sphere. Compute the temperature difference, from inside to outside, corresponding to the known heat flow from the heater. Discuss this result with respect to the calculations of problem 4-26. How would your confidence in the uniform-temperature assumption be affected by:

    a.   A heater with a much higher heating rate

    b.   A sphere material which was a good thermal insulator

[For *cylindrical* shells of length L, the reference shows (page 59) that the thermal resistance is $(\log_e(r_o/r_i))/2\pi kL$.]

---

[37] A. F. Mills, *Basic Heat and Mass Transfer*, Irwin, Chicago, 1995, p. 64.

5

BASIC ENERGY CONVERTERS

5-1 INTRODUCTION

Chapters 2 to 4 introduced the basic elements of mechanical, electrical, fluid, and thermal systems. Practical machines and processes often include hardware which involves several of these fields, since the different functions in a process may each be best accomplished in a particular way. If different parts of a process operate with different forms of energy but all must work together, it means that devices called *energy converters* must be available to couple the diverse parts. (The word *transducer* is sometimes used in place of "energy converter.") Also, systems often exhibit coupling between different forms of energy, which was not intentionally designed into the system but must nevertheless be accounted for in analysis. This chapter will give a brief introduction to some of these devices and effects which accept energy input in one form and produce an output in another form. In most cases just a word description or diagram will be given, but in a few instances the analytical models are sufficiently simple and practically useful that they will be stated and explained in some detail. Our main purpose here is not to develop a broad theory of energy conversion, but rather to introduce the reader to some basic devices which we will use in later chapters to model complete systems involving more than one form of energy. Some of these, such as electrical and hydraulic drives for machinery, are of tremendous economic and technical importance.

5-2 CONVERTING MECHANICAL ENERGY TO OTHER FORMS

We define mechanical power as the product of torque and angular velocity for a rotating shaft, or force and translational velocity for a translating shaft. Figure 5-1 summarizes some of the methods of converting power available in mechanical form to other forms.

The most important mechanical-to-electrical energy conversion process is undoubtedly that associated with the rotating electrical machines called generators

272

Basic Energy Converters

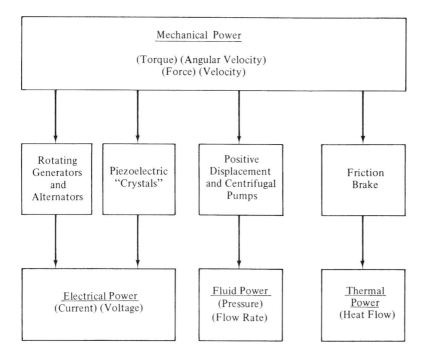

Figure 5-1 Converting mechanical power to other forms.

(dc) or alternators (ac). In these machines the mechanical power is applied to the rotating member (rotor) and the electrical power is taken off windings on either the stationary member (stator) or the rotor. For the dc machine the relationships are relatively simple, and we will state them without detailed derivation. The generator basically produces an output voltage by causing "wires" to cut through a magnetic field. The stronger the field and the higher the wires' velocity, the larger the voltage generated. In small generators the field may be produced by permanent magnets, which require no electrical power supply to maintain the field; the field is there "forever" (assuming the magnets are not damaged by excessive temperature or other effects). For larger machines, permanent magnets are not practical and the magnetic field is provided by a coil of wire wrapped around an iron core. Such "wound" fields require a continuous supply of dc current to the coil, which has resistance and thus wastes power continuously.

In Fig. 5-2 we show a wound-field generator and the field circuit which represents the coil, which is the nonrotating part. Field current i_f is produced by the field voltage e_f, which can be constant or variable, depending on how the generator is controlled. If we have a PM (permanent magnet) field, then there is no field circuit, but we still have the necessary field, though it is now, of course, not electrically controllable. The field circuit shown models the coil as resistance and inductance in series. Since the coil uses an iron core, the inductance will be somewhat nonlinear.

The rotating part, called the armature, carries the "wires" which cut the magnetic field as the armature rotates. A commutator and brushes (not shown) perform a switching function which maintains the desired polarity of the dc output voltage as the armature rotates; it also allows power to flow from the rotating parts to the

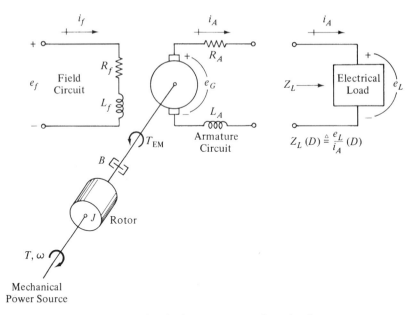

Figure 5-2 Mechanical-to-electrical energy conversion: the dc generator.

stationary output terminals. The armature circuit includes the passive elements of resistance and inductance and a voltage source, the generated voltage e_G. The armature resistance and inductance are actually located in the rotating member, but are conventionally shown external, as in Fig. 5-2. The armature *circuit* resistance R_A is the sum of the armature's resistance and that of the brushes. These can be combined into a single resistance since they both carry the same current i_A. In a small PM generator these two resistances might each be about 1 ohm. While field inductance (assuming a wound field) is usually nonnegligible, armature inductance often can be neglected. When we later describe dc *motors*, we will find that all the above features will also be present, since, at least in principal, the dc generator and dc motor are *the same* machine. In the generator, we apply mechanical power to the shaft and take off electrical power at the armature. In the motor, we apply electrical power to the armature (and perhaps the field) and take off mechanical power from the rotating shaft. While a dc motor (generator) *can* be run as a generator (motor), there are some subtle operational requirements that make the detail design of the two machines somewhat different, to optimize the behavior of each for its intended function.

In Fig. 5-2 we see some relatively simple circuits, which we can analyze using the usual basic laws and element behavior, and a simple rotating mechanical system, also easily analyzed with Newton's law. There are, however, two electromechanical effects which may be unfamiliar and which must be included in the overall system model. These are the *magnetic torque effect* and the *generated voltage effect*. When we apply mechanical power to the shaft, say by turning it with our hands, we would feel an opposing magnetic torque, which must be included in our Newton's law statement. Also, when the generator rotor rotates, a voltage e_G is produced, which must be included in the Kirchhoff's law statement for the armature circuit. Both these effects are usually explained in a first course in electrical physics, and a *designer*

Basic Energy Converters   **275**

of generators would need to understand these effects in terms of basic material properties and geometry. *System engineers* usually are users, *not* designers of generators and motors, and their understanding of generator operation need not be as detailed.

Both theoretical analysis and lab testing show that the magnetic torque, which resists our attempts to turn the generator shaft, is proportional to the product of field strength and armature current. Field strength is fixed for a PM machine and is proportional to field current for a wound field, so long as the iron in the field has not saturated. Thus we can write:

Electromagnetic torque $= K_{Twf} i_f i_A$ Wound-field machine   (5-1)

Electromagnetic torque $= K_{Tpm} i_A$ PM-field machine   (5-2)

These are instantaneous relations; the torque follows the currents instantly. A designer of generators would use physical analysis to obtain formulas for estimating K_{Twf} or K_{Tpm} from material properties, dimensions, and basic laws. A system engineer would purchase an existing generator from the manufacturer, whose catalog would provide numerical values for these constants. These would very likely *not* be theoretical values but rather the results of lab tests. Furthermore, they would be average values, since each machine, though intended to be identical, will differ somewhat.

We should note that a system engineer, while not usually competent to theoretically predict the needed constants, can easily run experiments and thus get numbers specific to the actual machine to be used. This is of course preferable to using average values. Figure 5-3 shows the results of typical experiments of this kind, for a wound-field machine. Since Eq. (5-1) shows that the torque is independent of speed, we can measure the torque for the simplest condition, zero speed. We do this by applying a force measuring device such as a spring scale to the end of a lever arm fastened to the generator shaft, thus restraining the shaft from rotating. We then use electrical power supplies to provide field current and armature current. Note that since we are *applying* electrical power to the armature, we are actually using the generator as a motor, but Eq. (5-1) holds in either case. That is, the torque constant is the same number, whether the machine is used as a generator or a motor. This type of test would be called a "stalled-torque" test, but the results apply at all speeds, not just the zero speed we use for convenient measurement. Since machine cooling is much poorer at zero speed than when rotating, be careful to take the measurements quickly enough to prevent overheating.

In Fig. 5-3a we set the armature current at a fixed value and vary the field current. Torque increases in proportion to field current until the iron in the field saturates (the magnetic field no longer increases), whereupon further increase in field current gives no increase in torque. We repeat this test with several different armature currents, getting the family of curves shown. We also fix the field current and then vary the armature current, as in Fig. 5-3b. Note now that increasing armature current does not produce a saturation effect; torque continues to increase with armature current, even for the case where the field is saturated. While we can get arbitrarily large torques by increasing armature current, these cannot be used in practice because the machine will overheat and damage the electrical insulation.

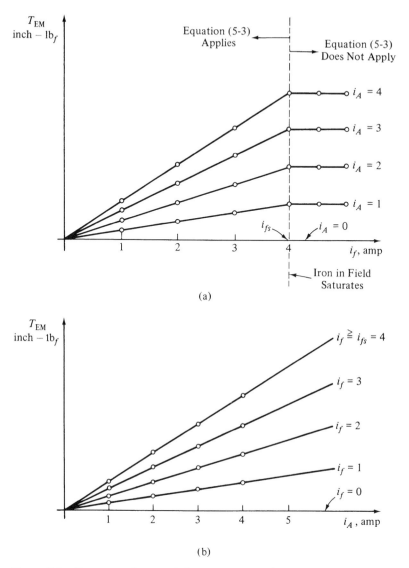

Figure 5-3 Generator characteristic curves: magnetic torque.

If the machine behaved *exactly* as given by Eq. (5-1), only *one* test run would be needed to get a number for K_{Twf}, since we would have numbers for one value each of i_f, i_A, and torque. Our more extensive testing is usually justified since it reveals any subtle departures from linearity, and also gross nonlinearities such as the saturation effect. By using *all* the measured data points (except those in saturation) and a statistical regression program, we can get a single value for the torque constant which is more reliable than that obtained from only one run. The lab testing needed for a PM machine is even simpler and should be obvious at this point.

We are now in a position to write the Newton's law for the rotating parts in Fig. 5-2.

Basic Energy Converters 277

$$\sum \text{torques} = J\alpha$$

$$T_a - B\omega - K_{\text{Twf}}i_f i_A = J\frac{d\omega}{dt} \tag{5-3}$$

where

$T_a \triangleq$ input torque applied by mechanical power source, N-m

$B \triangleq$ viscous friction coefficient of rotor bearings and windage, N-m/(rad/sec)

$\omega \triangleq$ rotor speed, rad/sec

$J \triangleq$ moment of inertia of everything that rotates with the rotor shaft, kg-m^2

Note that, even if torque T_a were given, we could not solve Eq. (5-3) because it has three unknowns. To get the additional two needed equations, we need to analyze the field circuit and armature circuit. The field circuit is a simple series RL circuit and is easily solved for the field current once the input field voltage is given. In the armature circuit we would need to define a *specific* electrical load, rather than the generic one shown. Once this is done, we still need a description of the generated voltage e_G. Here, theory and lab testing show that

$$e_G = K_{\text{Ewf}}i_f\omega \qquad \text{Wound-field machine}$$

$$e_G = K_{\text{Epm}}\omega \qquad \text{PM-field machine} \tag{5-4}$$

For an existing machine, the needed constants can again be found by lab testing, as shown in Fig. 5-4 for a wound-field machine. The generator is run at various speeds and with various field currents, and the voltage e_G is measured with a high-resistance voltmeter, so as to keep armature current i_A essentially zero. (Recall that the resistance R_A is *internal* to the generator and would cause an incorrect voltage reading if current were allowed to flow through it.) That is, the voltage e_G is the *open-circuit* output voltage of the generator.

We would now have all the numerical parameter values needed to write and solve the set of three simultaneous differential equations in the three unknowns i_f, i_A, and ω, assuming that the inputs e_f and T_a were also given as functions of time. Later in this chapter we will see that our discussion of generator systems will be directly applicable to dc motor drive systems, widely used in all kinds of machines and processes, and much more common than generator systems. In this chapter we have used *rotary* machines as our discussion examples, but *translatory* dc motors are also common and can be modeled in exactly the same way, using appropriate force and voltage constants, as we have already seen in the system of Fig. 3-38. In later chapters we will study the design of several types of electric-motor drives used in motion-control systems.

Figure 5-1 shows one other class of mechanical-to-electrical energy conversion device, the piezoelectric "crystal." (The dictionary gives the preferred pronunciation as pie-eezo-electric but the secondary pronunciation pee-ayzo-electric is also common.) Just as in the generator/motor systems we have discussed above, the piezoelectric effect is also "reversible." We can apply mechanical input energy and convert it into electrical output energy or apply electrical input energy and convert it into

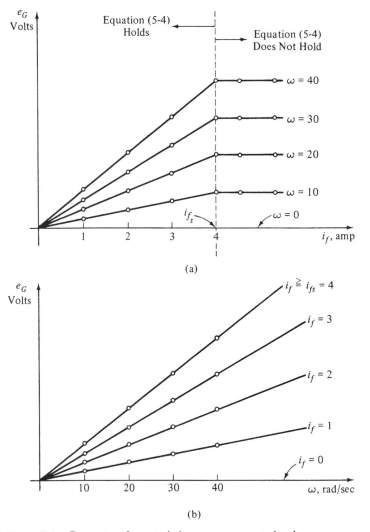

Figure 5-4 Generator characteristic curves: generated voltage.

mechanical output energy. We should first make clear this effect is used only at relatively low power levels, whereas electromagnetic motors and generators are used at milliwatt to megawatt levels. The mechanical-to-electrical applications of piezoelectricity are mainly in the field of sensors (measuring instruments), in particular, force, pressure, and acceleration measurement.[1] These are actually all *force* measurements since the pressure is converted to a force using an elastic diaphragm, and the accelerometer uses a lumped mass to generate a force proportional to acceleration (recall Fig. 2-35a).

---

[1] E. O. Doebelin, *Measurement Systems*, 4th ed., McGraw-Hill, New York, 1990, pp. 261–268, 325–331, 404–407, 458–460.

Basic Energy Converters

Figure 5-5 shows the simplest configuration of a piezoelectric element: a "crystal" sandwiched between metal electrodes and subject to direct tension or compression input forces. In a piezoelectric sensor, the force f produces an output voltage e which is a measure of the force, pressure, or acceleration, which may be varying rapidly with time. The voltage can be recorded on a variety of voltage-measuring instruments or further processed in a digital data acquisition system. Materials which exhibit the piezoelectric effect include natural and synthetic crystals (quartz and Rochelle salt are examples), and synthetic polycrystalline ferroelectric ceramics (barium titanate, lead zirconate titanate, etc.) which can be made piezoelectric by suitable processing. Some flexible polymer films also produce this effect.

We use the word "crystal" somewhat loosely in our discussions to cover all piezoelectric materials. Since all these materials are good electrical insulators (dielectrics), when we apply the electrode plates as in Fig. 5-5, we create a capacitor. This capacitor is now quite unusual, however, since it is also a *generator* of electrical charge whenever the crystal is deflected by the application of mechanical force. Since a charged capacitor has a proportional voltage, we can see the basis of the sensor applications. We show only the direct tension/compression mode of stressing, but the piezoelectric effect can be implemented for many other types of deformation: shear, torsion, bending, twisting, etc. This makes the effect quite versatile. For small forces, for example, a bending mode will be much more sensitive than direct compression.

While the detailed analysis of piezoelectric devices can become very complicated, some of the observed overall behavior can be described in simple terms. Again we take the system engineer's viewpoint and couch our description in terms of

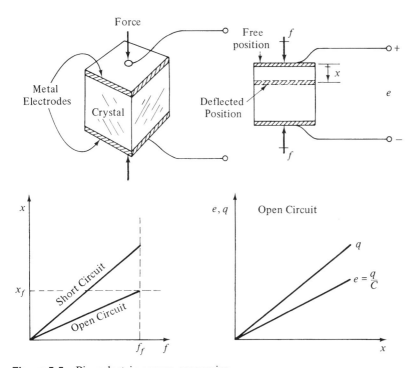

Figure 5-5 Piezoelectric energy conversion.

parameters measurable for an existing device. If we gradually apply a force to the crystal of Fig. 5-5 with its electrical terminals open-circuit, the crystal deflects (like a spring) in proportion to the applied force, and simultaneously a charge q and voltage $e = q/C$ appear on the capacitor, the amount of charge being directly proportional to the deflection. The mechanical work done by the applied force is $f_f x_f/2$ and a portion of this work has been converted into electrical energy, since the charged capacitor has stored energy $q^2/2C$. If the deflection is reversed from compression to tension, the polarity of the charge and voltage also reverses. The fraction of the mechanical work which can be converted into electrical energy varies from material to material. A material constant k, called the *electromechanical coupling coefficient*, is defined by

$$k^2 \triangleq \frac{\text{electrical energy}}{\text{input mechanical energy}} \tag{5-5}$$

and varies from about 0.1 (1% energy conversion) for quartz, through 0.5 to 0.7 for synthetic ceramics, to 0.9 (81% energy conversion) for Rochelle salt.

Due to the coupling between mechanical and electrical effects, some unusual behavior is observed. If we apply a force with the electrical terminals short-circuited (so that the capacitor cannot be charged), the crystal will be observed to be a "softer" spring (more compliant) than if the terminals were open-circuit. The relation between the spring constants K_{oc} and K_{sc} (N/m or lb_f/in) for the two conditions is given by

$$\frac{K_{oc}}{K_{sc}} \triangleq \frac{K_{\text{open circuit}}}{K_{\text{short circuit}}} = \frac{1}{1 - k^2} \tag{5-6}$$

This relation can be derived from Eq. (5-5) by the following reasoning. With the terminals open circuit, apply a force f which causes a deflection x. The total energy put into the crystal by the mechanical power source is $x^2 K_{oc}/2$, the electrical energy produced is $k^2(x^2 K_{oc}/2)$ and the stored mechanical energy is given by $x^2 K_{oc}(1 - k^2)/2$. If we now hold the crystal fast so that x cannot change (and thus no mechanical work can be done), and then short-circuit the terminals, the electrical energy will be dissipated, and the force will relax to the value associated with the deflection x and the short-circuit spring constant K_{sc}. We may then equate the stored mechanical energy to that associated with the short-circuit spring constant

$$\frac{x_f^2 K_{oc}(1 - k^2)}{2} = \frac{x_f^2 K_{sc}}{2} \tag{5-7}$$

$$\frac{K_{oc}}{K_{sc}} = \frac{1}{1 - k^2} \tag{5-8}$$

Figure 5-6 illustrates this process.

If a piezoelectric generator is connected to an electrical load with input impedance Z_L, as in Fig. 5-7, and then a driving force $f(t)$ is applied, the following equations hold.

$$\sum \text{forces} = \text{mass} \times \text{acceleration} \tag{5-9}$$

$$-B\dot{x} - K_s x - C_1 e + f(t) = M\ddot{x} \tag{5-10}$$

Basic Energy Converters

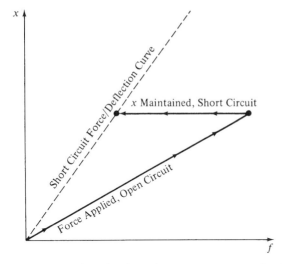

Figure 5-6 Piezoelectric spring constants: open-circuit and short-circuit.

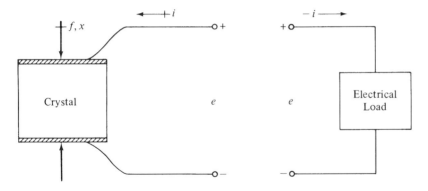

Figure 5-7 Piezoelectric crystal with electric load.

where

$B \triangleq$ effective damping coefficient of crystal

$M \triangleq$ effective inertia (mass) of crystal

$C_1 e \triangleq$ piezoelectric force, proportional to crystal voltage

The spring constant K_s is defined by applying a static force f with the terminals short-circuited so that $e \equiv 0$. Since $d^2x/dt^2 = dx/dt = 0$ for a static force, Eq. (5-10) gives $K_s = f/x$, which can be measured in a lab test by applying dead weights and measuring the deflection. The piezoelectric force constant C_1 is found by clamping the crystal rigidly so that $x = dx/dt = d^2x/dt^2 \equiv 0$, applying a known voltage e, and measuring the force produced by the crystal pushing against the clamp. Equation (5-10) then gives us $C_1 = f/e$. The electric circuit equations are

$$C \frac{de}{dt} - C_q \frac{dx}{dt} = i \qquad (5\text{-}11)$$

282   **Chapter 5**

for the crystal and

$$\frac{e}{i}(D) = -Z_L \tag{5-12}$$

for the electrical load.

Think of Eq. (5-11) as a superposition of an "ordinary" capacitor with the usual voltage/current relation and a current generator with current proportional to "crystal velocity." (Crystal charge generation is proportional to deflection, so current dq/dt is proportional to velocity.) In Eq. (5-11), if we clamp the crystal so that $dx/dt \equiv 0$, then we find that $C = i/(de/dt)$, showing that C is just the capacitance of the crystal, which can be measured with an ordinary capacitance-measuring instrument, but it *must* be done with the motion constrained. (If we measure C with the crystal free to deflect, we will get a *different* number for C.) The constant C_q is defined by leaving the terminals open-circuit, so that $i \equiv 0$, and then applying a known deflection x which will produce a voltage e which can be measured, giving $C_q = C(de/dt)/(dx/dt) = Ce/x$. Finally, note that in Eq. (5-12) the minus sign is necessary since the definition of positive current in the crystal is opposite to that which would be used to define the load impedance Z_L.

We have defined the constants C_1, K_s, C, and C_q in terms of measurements performed on a device already built, since this gives the most accurate values and also allows a system engineer (who is *not* a piezo specialist) to work with the devices. Device *designers* (who *are* piezo specialists) have methods for estimating all these constants before a device is constructed, from dimensions and fundamental material properties. These material properties must themselves be measured experimentally, but this need be done *only once* for each material, not for each new device. Since a crystal usually does not exhibit an obvious lumped mass and damper, the numerical values of M and B are neither theoretically accessible nor directly measurable. This difficulty can be circumvented by dividing through Eq. (5-10) by the spring constant K_s, giving

$$\frac{M}{K_s}\ddot{x} + \frac{B}{K_s}\dot{x} + x + \frac{C_1 e}{K_s} = \frac{f(t)}{K_s} \tag{5-13}$$

which we rewrite as

$$\frac{\ddot{x}}{\omega_n^2} + \frac{2\zeta}{\omega_n}\dot{x} + x + \frac{C_1 e}{K_s} = \frac{f(t)}{K_s} \tag{5-14}$$

where

$\omega_n \triangleq (K_s/M)^{0.5} \triangleq$ undamped natural frequency of crystal vibration (short circuit), rad/sec

$\zeta \triangleq B/(2(K_s M)^{0.5}) \triangleq$ crystal damping ratio, dimensionless

Methods for calculating and/or measuring ω_n and ζ *directly* (without knowing M or B separately) are available (see Chapter 8) and relatively convenient. In later chapters we will study complete piezoelectric systems in more detail.

Returning to Fig. 5-1, let's now consider some methods for converting mechanical power to fluid power, with emphasis on equipment designed for use with liquids, since the relations for compressible fluids are more complex and beyond our intended scope. Two main types of pumps account for the majority of such energy

Basic Energy Converters

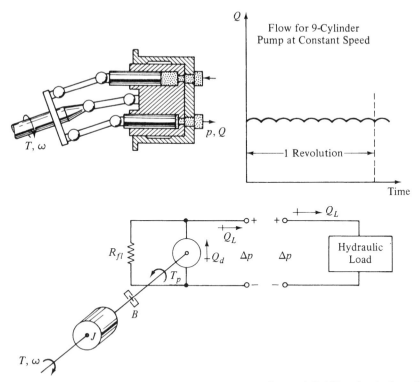

Figure 5-8 Positive-displacement pumps: construction and fluid/mechanical model.

conversion: positive-displacement (Fig. 5-8) and centrifugal (Fig. 5-9). While positive-displacement pumps take a variety of forms (piston, vane, gear, etc.), their overall characteristics are basically similar, and a general model adequate for system dynamic analysis can be formulated. Figure 5-8 shows a multiple-piston pump with a rotary mechanical input, which we shall use as a concrete example in developing the general model. As the input shaft is rotated, the individual pistons are sequentially forced in and out of their cylinders, drawing fluid from the input port and expelling it at the output. Valves (not shown) are properly sequenced with rotation so that each cylinder is alternately exposed to the inlet port and then the discharge. The outflows from each cylinder are summed at the discharge port, so that, while each individual cylinder flow rate is pulsating, the total pump flow rate is relatively smooth. Intuitively one would guess that smoothness would increase with the number of cylinders; commercial pumps with seven or nine cylinders are not unusual. While hydraulic motors will be discussed later in this chapter, it might be well to mention now that, just as in the dc motor/generator, the hydraulic pump and motor are in essence the *same* machine. That is, in Fig. 5-8, if we force fluid under pressure through the machine, we will develop a mechanical torque at the shaft, which is the function of a motor.

A fundamental parameter of a positive displacement pump is D_p, its displacement of fluid per radian of shaft rotation, that is, how many cubic inches or cubic centimeters of fluid are passed through the pump when the shaft turns through 1 radian. It is easily found for an existing pump by measuring its flow rate at constant speed, with no back pressure (so that leakage is negligible). The torque T_p

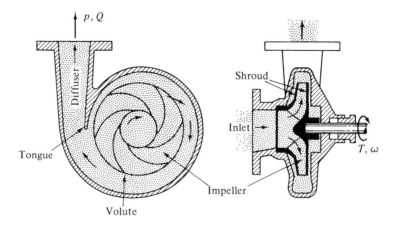

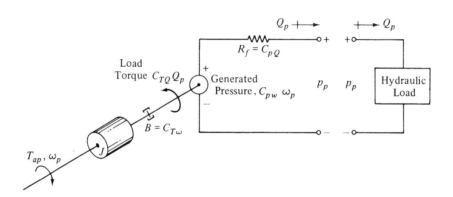

Figure 5-9 Centrifugal pumps: construction and small-signal fluid/mechanical model.

required to drive the pump is directly related to D_p and the pressure rise Δp created by the pump, as revealed by the following analysis. Assuming perfect energy conversion (no losses due to friction, etc.), all the mechanical energy put into the pump shows up as fluid energy. In turning the shaft through a small angle $d\theta$, a torque T_p does mechanical work $T_p\, d\theta$. At the same time, a volume of fluid $D_p\, d\theta$ has been forced through the pump against the pressure difference Δp. The mechanical power would be $T_p\, d\theta/dt$, and the fluid power (flow rate times pressure drop) would be $D_p(d\theta/dt)\,\Delta p$; thus, for no losses:

$$T_p \frac{d\theta}{dt} = D_p\, \Delta p\, \frac{d\theta}{dt} \tag{5-15}$$

$$T_p = D_p\, \Delta p \tag{5-16}$$

The instantaneous torque felt by the drive shaft is thus given by Eq. (5-16) in terms of the instantaneous pressure difference.

We can now state the equations describing system behavior. Using Newton's law,

Basic Energy Converters   **285**

$$T - B\omega - D_p \, \Delta p = J\dot{\omega} \tag{5-17}$$

where

$B \triangleq$ viscous damping of pump moving parts

$J \triangleq$ moment of inertia of pump moving parts

Since most pumps have both rotating and reciprocating parts which all contribute to the total inertia J, numerical values for J are not easily calculated, but pump manufacturers usually supply this number. As usual in real machines, the friction (damping) will not be perfectly viscous, will be difficult to estimate accurately from theory, and will change due to effects such as temperature, shaft alignment, bearing wear, etc. You should thus treat the number B as rather uncertain. For the fluid circuit we treat the liquid as incompressible, thus conservation of volume gives

$$\omega D_p - \frac{\Delta p}{R_{fl}} = Q_L \tag{5-18}$$

$$\text{Fluid impedance of load} \triangleq Z_L(D) \triangleq \frac{\Delta p}{Q_L}(D) \tag{5-19}$$

The pump leakage resistance R_{fl} can be found by lab testing of an existing pump. As soon as the hydraulic load is specified in detail, Eq. (5-19) can be made specific. For example, if the load were simply a flow resistance R_L, we would have $R_L = \Delta p / Q_L$.

Turning now to centrifugal pumps (Fig. 5-9), we first note that while a positive displacement pump is basically a flow source, delivers a flow rate proportional to speed and independent (except for the small leakage) of pressure, and will stall if the flow is shut off, a centrifugal pump is more like a pressure source. If the flow is shut off, it continues to run and develop pressure (though it will overheat under such conditions). In applications, positive displacement pumps are often used in fluid-power systems for actuating machinery. The same fluid is circulated back to the pump when discharged from the hydraulic load, such as a hydraulic motor, and is used over and over again. Centrifugal pumps are not generally used in this way; rather they are employed to *move fluids* from one place to another in chemical processing plants, refineries, power plants, sewage treatment plants, municipal water supplies, hydroelectric pumped-storage systems, etc. In these applications, the fluid is not usually recirculated through the pump; the processes are "once through" processes.

While turbomachinery theory may be applied to the study of pumps to predict characteristics, system analyses often rely on the use of measured characteristic curves. Figure 5-10 shows a test setup used to determine the two families of curves needed to describe a pump. These curves are a graphical presentation of the functional relations

$$p = p(Q, \omega)$$
$$T_p = T_p(Q, \omega) \tag{5-21}$$

While mathematical formulas for these relations are not available, computer simulation can work directly from the experimental curves using table lookup or curve-fitting techniques. Using such an approach, the system equations would be

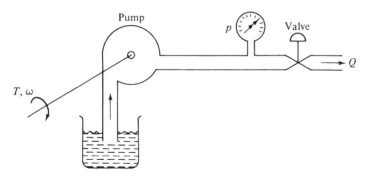

Figure 5-10 Centrifugal pump lab test setup.

$$T_a - T_p(Q, \omega) = J\dot{\omega} \tag{5-22}$$

$$p = p(Q, \omega) \tag{5-23}$$

$$Z_L(D) = \frac{p}{Q}(D) \tag{5-24}$$

An analysis suitable for studies of small perturbations from an operating point can be carried out by linearizing the nonlinear functions $p(Q, \omega)$ and $T_p(Q, \omega)$, using our usual Taylor series approximation. The resulting linear equations allow a more general evaluation of the effect of system parameters on response than do numerical computer simulation studies, and are particularly useful in the early stages of system analysis and design. We assume steady-state operation with constant values T_{p0}, Q_0, ω_0, p_0 of torque, flow rate, speed, and pressure when at $t = 0$ a small change T_{ap} is made in the driving torque T_a. The nonlinear functions may be approximated as

$$T_p(Q, \omega) \approx T_p(Q_0, \omega_0) + \left[\frac{\partial T_p}{\partial Q}\right]_{\omega_0, Q_0} (Q - Q_0) + \left[\frac{\partial T_p}{\partial \omega}\right]_{\omega_0, Q_0} (\omega - \omega_0) \tag{5-25}$$

$$T_p(Q, \omega) \approx T_p(Q_0, \omega_0) + \left[\frac{\partial T_p}{\partial Q}\right]_{\omega_0, Q_0} Q_p + \left[\frac{\partial T_p}{\partial \omega}\right]_{\omega_0, Q_0} \omega_p \tag{5-26}$$

where

$$Q_p \triangleq Q - Q_0 \triangleq \text{perturbation in } Q$$
$$\omega_p \triangleq \omega - \omega_0 \triangleq \text{perturbation in } \omega$$

Similarly,

$$p(Q, \omega) \approx p(Q_0, \omega_0) + \left[\frac{\partial p}{\partial Q}\right]_{\omega_0, Q_0} Q_p + \left[\frac{\partial p}{\partial \omega}\right]_{w_0, Q_0} \omega_p \tag{5-27}$$

The numerical values of the partial derivatives in Eqs. (5-26) and (5-27) are found from the experimental curves as in Fig. 5-11. We can then write the system equations as

$$(T_{p0} + T_{ap}) - (T_{p0} + C_{TQ}Q_p + C_{T\omega}\omega_p) = J\dot{\omega}_p \tag{5-28}$$

$$T_{ap} - C_{TQ}Q_p - C_{T\omega}\omega_p = J\dot{\omega}_p \tag{5-29}$$

Basic Energy Converters

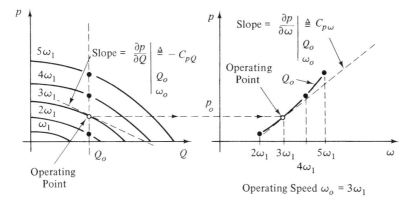

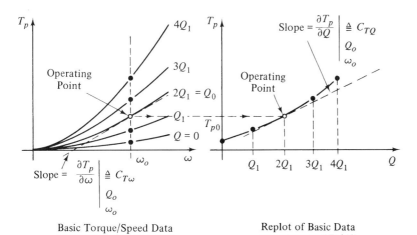

Figure 5-11 Definition of pump-model parameters from measured curves.

$$p - p_0 \triangleq p_p = -C_{pQ}Q_p + C_{p\omega}\omega_p \tag{5-30}$$

$$Z_L(D) = \frac{p_p}{Q_p}(D) \tag{5-31}$$

These equations may be interpreted so as to yield the model of Fig. 5-9. The term $-C_{TQ}Q_p$ represents a load torque presented by the pump, while $-C_{T\omega}\omega_p$ has the form of a viscous damping torque. In Eq. (5-30), $C_{p\omega}\omega_p$ is the generated pressure, while $-C_{pQ}Q_p$ represents a pressure drop due to flow resistance.

The final conversion process considered in Fig. 5-1 is that of mechanical power into thermal power. Perhaps the most common instance of this is found in various frictional processes. Generally, friction is considered an undesirable parasitic effect; however, certain practical devices, notably clutches and brakes, rely on it for their principle of operation. The heat generated by the friction is usually an undesirable but unavoidable byproduct, and must be taken into account in the design of fric-

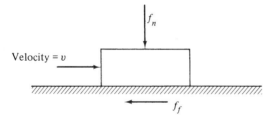

Figure 5-12 Rubbing friction as heat source.

tional devices. A good example is found in disk brakes for aircraft wheels. These brakes must absorb and dissipate into heat a portion of the large amount of mechanical kinetic energy possessed by the moving aircraft to bring it to a controlled stop. In doing this, the temperature of the brake lining and other parts must be kept low enough to prevent damage or excessive fading of the brakes. We have thus an example of a mechanical/thermal system in which the heat flow into the brakes is provided by the conversion of mechanical energy into heat by the friction process.

The simplest model of solid (rather than fluid) friction assumes the friction force directly proportional to the normal force, and independent of rubbing speed, temperature, or any other influences. Thus in Fig. 5-12,

$$f_f = \mu f_n \tag{5-32}$$

where

$f_f \triangleq$ friction force, lb_f or N

$f_n \triangleq$ normal force, lb_f or N

$\mu \triangleq$ friction coefficient, assumed constant

At any instant when the rubbing velocity is v in/sec, the mechanical friction power would be $\mu f_n v$, in-lb_f/sec. Since 1 in-lb_f/sec is equal to 0.000107 Btu/sec, the frictional dissipation represents a heat flow source for which

$$\text{Heat-flow rate} = q = 0.000107 \mu f_n v \quad \frac{\text{Btu}}{\text{sec}} \tag{5-33}$$

In more complex models the friction coefficient may be taken as some function of rubbing speed, temperature, and other pertinent factors, as revealed by experimental testing. When these complex friction models are embedded in an overall dynamic system model, computer simulation methods will likely be necessary.

5-3 CONVERTING ELECTRICAL ENERGY TO OTHER FORMS

Figure 5-13 lists some devices for converting electrical power to mechanical, fluid, and thermal form. Beginning with electromechanical conversion, the electric motor in its various forms is obviously of overriding importance. You need only to begin to count the motors in any building you happen to be in, to appreciate the widespread usefulness of this device. Motors based on the interaction of magnetic fields take many different forms; dc varieties include permanent-magnet field, wound field separately excited, shunt, series, and compound, while induction and synchronous

Basic Energy Converters   **289**

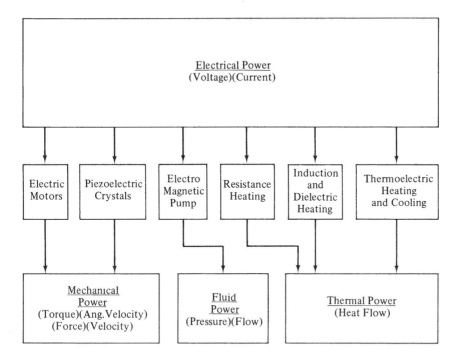

Figure 5-13   Converting electrical power to other forms.

are common ac types. Stepper motors, which are commanded by electrical pulse trains and produce a discrete step (often 1.8°) of mechanical rotation for each command pulse, are widely used in motion control systems. While rotary motors most often come to mind, the operating principles of most types can be adapted to a translational configuration.

While the general subject of electric motors is vast and requires entire books and academic courses to develop design competence, some specific types are simple enough and so widely used that we want to present sufficient coverage to allow our usual system-engineering-oriented description. These descriptions will be used in later chapters where motors arise naturally as part of dynamic motion control systems. Our focus will mainly be on the separately excited dc motor as shown in Fig. 5-14. As we noted earlier for dc generators, the field may be provided by permanent magnets, rather than the wound field shown. This simplifies the construction and analysis of the motor, but makes its control less versatile since the field cannot now be manipulated for control purposes. Fortunately, armature control of motors with fixed fields has been developed to the point where field control is rarely needed. In earlier years, PM-field motors were limited to rather small sizes, and high-power motors were of necessity wound field types. Today PM motors are available up to about 50 hp and are in fact the most widely used type for motion control systems. Our Fig. 5-14 relates directly to a brush-type rotary motor, but it turns out that the model and equations developed can be used for both rotary, translational, brush-type, and brushless DC motors, either wound field or PM field. These categories include a large percentage of the applications in dynamic motion-control systems, so our time will be well spent in getting familiar with this model.

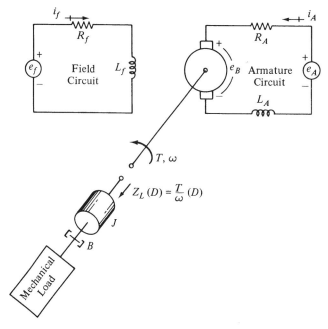

Figure 5-14 Model for dc motor and load.

As mentioned in Sec. 5-2, the torque constants K_{Twf} and K_{Tpm} and the voltage constants K_{Ewf} and K_{Epm} are identical for motors and generators, and the lab test methods described earlier for generators are used also for motors. In reality, the torque and voltage constants are *not* independent parameters that can be individually picked when designing a system. They are actually *proportional*, so if we choose one, the other is fixed. In fact, if we use SI units [volts/(rad/sec) and N-m/amp], the two constants are numerically *equal*. Motor catalogs often obscure this equality by using "peculiar" units such as oz-in/amp and volts/1000 rpm; however, if you convert the units, you will see the equality. If a wound field is used, the field current is easily found from its simple RL circuit; if a PM field is used then the field is fixed and no field circuit need be analyzed. For the armature circuit, Kirchhoff's voltage loop law gives

$$i_A R_A + L_A \frac{di_A}{dt} + K_{Ewf} i_f \omega \text{ (or } K_{Epm}\omega) - e_A = 0 \tag{5-34}$$

Here the voltage e_B produced by the generator action of the motor is called the *back emf*, and is given by the same formula as the generated voltage of a generator. In a motion-control system the armature drive voltage e_A would come from the output of an amplifier. Armature inductance is often negligible. When it isn't, we may want to use a transconductance amplifier, rather than the usual voltage amplifier. A transconductance amplifier has a *current* output which faithfully follows the command of the amplifier input voltage. Since armature current is now related only to amplifier input voltage, there is no need to write a Kirchhoff law to solve for it. This also means that the back emf $K_{Ewf} i_f \omega$ (or $K_{Epm}\omega$) and armature inductance now have no effect on system dynamics. This speeds up the system response. On the other hand,

Basic Energy Converters

when pulse-width-modulation (PWM) amplifiers are used to improve efficiency in high-power systems, a certain amount of armature inductance is *necessary*, to smooth out the current waveform. If the motor's armature inductance is not enough to meet this requirement, we may have to *add* an intentional inductor to the armature circuit.

The Newton's law for the rotating parts is relatively simple since the magnetic torque produced by the motor is given by the same formula we showed earlier for the torque needed to drive a generator:

$$K_{Twf} i_f i_A \text{ or } (K_{Tpm} i_A) = T = Z_L(D)\omega \tag{5-35}$$

If, for example the load is just more inertia and friction:

$$Z_L(D) = J_{total} D + B_{total}$$

$$K_{Twf} i_f i_A \text{ or } (K_{Tpm} i_A) = (J_{total} D + B_{total})\omega \tag{5-36}$$

where J_{total} is the sum of motor and load inertias and B_{total} is the sum of motor and load viscous friction.

The models just developed hold for both brush and brushless dc motors, but we want to give a little physical description of the two types since many readers may be unfamiliar with the brushless version. Figure 5-15 shows simplified radial cross sections of these two motors (both are PM types), and also an ac induction motor which we will shortly discuss. The brush-type dc motor is the "classical" dc motor which is covered in most electrical physics courses, sometimes even in high school. The PM field is stationary, fastened to the machine stator, and creates a radial magnetic field. The rotating armature ("rotor") is made of iron, slotted to carry the winding of copper wires. Brushes and a commutator (not shown) connect the rotor winding to the external (stationary) armature supply e_A, which drives current through the windings. The current-carrying conductors are in a transverse magnetic field and thus feel a tangential magnetic force which tends to rotate the armature. As the rotor turns, conductors which were under a north (south) pole find themselves now under a south (north) pole, which causes the magnetic torque to reverse, giving

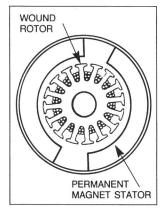
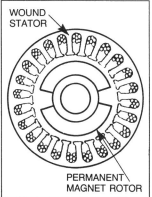
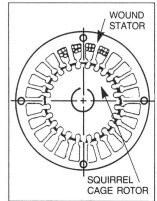

Figure 5-15 Brush-type dc, brushless dc, and squirrel-cage ac induction motors.

an oscillatory motion rather than the desired continuous rotation. The brushes and commutator perform a switching function which reverses the connections of the external armature voltage to the rotating armature, just at the point where the magnetic field reverses. This keeps the magnetic torque always in the same direction, giving continuous rotation. The magnetic torque on the rotor is also felt (in a reversed sense) by the field magnets, but they of course do not move, being fixed to the stator, which is stationary.

The brushless motor is a more recent invention and is actually more correctly called a type of synchronous ac motor; however, its equations and dynamic model, when properly interpreted, are identical to the brush-type model we just described above. In Fig. 5-15 we see that now the permanent magnets are fastened to the rotor and rotate with it, while the windings are found in the stationary stator. The windings are again called the armature, but they now do not rotate. If we applied to the windings a fixed polarity armature voltage we would again produce an oscillatory rotation. Since the brushless motor, by definition, does *not* have either brushes or a mechanical commutator, the oscillation problem must be solved in another way. The method used is to provide angle sensors on the rotor which send rotor position information (voltages) to an electronic "commutation" circuit, which manipulates the armature voltage so as to always produce a torque in the same direction, as the rotor turns.

In a brushless motor "the winding" is actually *several* windings, which are mechanically "phased" with respect to each other. The electronic commutation circuit manipulates the drive voltage to each phase so as to create a rotating magnetic field which "drags" the permanent magnet rotor along with it, similar to a conventional ac synchronous or induction motor. While the ac motors use a field that rotates at a fixed frequency (usually the 60-Hz power line frequency), the "brushless dc" machine rotates the field so that it always is oriented perpendicular to the permanent magnet field of the rotor (this is why rotor position sensors are needed). The perpendicular position is that which gives the maximum magnetic torque.

Practical brushless machines must have at least two phased windings and we now do an analysis of this simplest system. The two windings A and B are geometrically located in the stator so that the magnetic torque of each is related to rotor position angle θ as follows.

$$T_a = i_A K_{T\text{pm}} \sin \frac{P}{2} \theta \tag{5-37}$$

$$T_B = i_B K_{T\text{pm}} \cos \frac{P}{2} \theta \tag{5-38}$$

Here P is the number of permanent magnet poles in the rotor. The machine of Fig. 5-15 has $P = 2$; other even values of P are possible. Using the rotor position information (angle θ) supplied by the sensors, the electronic commutation circuits adjust the currents in the two phases such that

$$i_A = i \sin \frac{P}{2} \theta \tag{5-39}$$

$$i_B = i \cos \frac{P}{2} \theta \tag{5-40}$$

Basic Energy Converters
293

The total torque T of the motor is then given by

$$T = T_A + T_B = iK_{Tpm}\left(\sin^2 \frac{P}{2}\theta + \cos^2 \frac{P}{2}\theta\right) = iK_{Tpm} \tag{5-41}$$

We see that motor torque is directly proportional to armature current i, just as in the brush-type motor. The brushless motor back emf (generator action) can also be shown to behave the same way as in the brush-type motor. Thus the models for the brush and brushless motors are essentially the same, and catalogs[2] quote numerical values for the torque and back emf constants, to be used in Newton and Kirchhoff law equations just as we have done. [The perfectly smooth (ripple-free) torque predicted by Eq. (5-41) is usually not realized in commercial motors because considerations of cost and complexity dictate the use of nonideal rotor position sensors and commutation circuits. Our simple model using torque and back emf constants *is*, however, still used in modeling motion control systems, since the accuracy is adequate.] We should also note (see fig. 5-16) that a PM brush-type motor has only two wires connecting the motor (armature) to the power supply, whereas a brushless motor requires several more since the rotor position sensors need a few wires to send information to the commutation circuit and the amplifier output must supply power to *several* (phased) windings. The *overall* operation, however, can still be modeled in the same simple way as for a brush-type motor.

The choice between a brush or brushless motor for a specific application is a design decision requiring careful consideration, the details of which are beyond the scope of this text. (Motor manufacturers who offer *both* motor types are a good source of such information.) We do, however want to give at least a few of the major differences. The absence of the rubbing mechanical contact of brushes and commutator means that brushless motors will enjoy a longer life with less maintenance and will also produce fewer contaminating wear particles, critical in applications like clean-room manufacturing processes. Brush arcing (sparking) limits the highest speeds of brush-type motors and also produces high-frequency electrical interference ("static") which may affect sensitive electronic circuitry. The major heat source in brushless motors is in the motionless stator, where cooling is more easily applied. Both types of motors can be used in vacuum environments, but the brushless motor adapts more conveniently. The main disadvantage of brushless motors is the higher cost and complexity.

The squirrel-cage ac induction motor shown in Fig. 5-15 has traditionally been used mainly for constant-speed applications and some speed-control systems, but not for position control. Recent developments in control systems ("flux-vector control," "field-oriented control") are allowing this simple and reliable motor to compete with dc systems for both speed- and position-control applications,[3] but these systems are not easily modeled, so we limit our presentation here to just making you aware of their existence. The three-phase squirrel-cage induction motor has long

---

[2]Brushless Motors: An Application Guide, BEI Motion Systems Co., 804-A Ranchero Drive, San Marcos, CA 92069, 619-744-5671, 1996.

[3]D. Y. Ohm, Field oriented control of induction motors, *Motion*, March/April 1991, pp. 3–14. W. Leonhard, *Control of Electrical Drives*, Springer, New York, 1985. Indramat Division, Rexroth Corp., 255 Mittal Drive, Wood Dale, IL 60191, 708-860-1010.

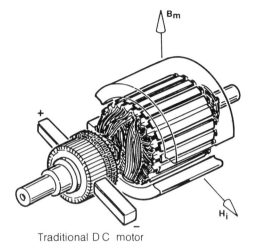

Traditional D C motor

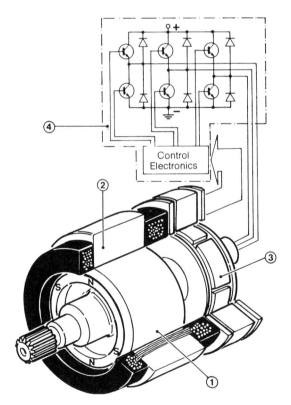

"Brushless" D C motor

Figure 5-16 Brush-type dc and brushless dc motor construction.

Basic Energy Converters   **295**

been, and will continue to be, the "workhorse" for applications which are nominally constant-speed. Two aspects of these motors' operation have dynamic (unsteady-speed) behaviors which can often be simply analyzed, so we want to briefly cover these.

While an application may be nominally constant-speed, the motor and load must always be started from rest and accelerated up to operating speed. The details of this acceleration are sometimes of engineering interest and can often be easily calculated. A detailed dynamic analysis[4] of polyphase induction motors leads to a complicated set of nonlinear differential equations, including both electrical and mechanical dynamics, and requiring simulation software for solution. In many applications, the electrical transients are much faster than the mechanical ones and can be neglected with little error when calculating the motor/load motion. That is, as long as our interest is not in the instantaneous values of voltages and currents, but only in the motion, a simple approximate calculation often suffices. (We should warn however that the ignored electrical transients are accompanied by an oscillating magnetic torque. If the frequency of this torque is near the natural frequency of any mechanical spring/mass loads, destructive vibrations *can* occur.[5]) Assuming that our motor drive system does *not* have any vibration problems, we are able to use the motor's steady-state speed/torque curve to predict the motor/load acceleration with good accuracy. Motor speed/torque curves can be estimated from theory and are also available as measured data from motor manufacturers.

In a three-phase induction motor the stator carries a winding of three phases which are powered from the 60-Hz power line in such a way as to produce a rotating magnetic field. Note that the stator is mechanically stationary; the field rotates "electrically" because of the clever arrangement of the windings. (This rotating-field concept was invented and patented in the United States about 1880 by Nikola Tesla, an immigrant from Yugoslavia. Tesla, a contemporary of Thomas Edison, championed ac power while Edison favored dc.) For a specific 60-Hz motor, this stator field might rotate at a speed (called the *synchronous speed*) of 1800 rpm (other typical synchronous speeds might be 900, 1200, and 3600 rpm). The rotor also has a "winding" in the form of a "squirrel cage" of axial conductors connected to two end rings, but there are no brushes or commutator to connect it to the stationary parts of the motor. The only mechanical contact between rotor and stator is at the rotor bearings; no brushes or slip rings are used.

Voltages are induced in the squirrel-cage conductors because the rotating magnetic field produced by the stator cuts through the conductors. These voltages cause currents to flow internally in the rotor and the interaction of these currents with the rotating field creates the motor torque. While the stator currents will always have the power line frequency (say 60 Hz), the frequency of the rotor currents *changes*, depending on the rotor speed, because it depends on the *relative* motion of the

---

[4]S. A. Nasar and L. E. Unnewehr, *Electromechanics and Electric Machines*, Wiley, New York, 1979, pp. 419–425.

[5]T. Iwatsubo, Y. Yamamoto, and R. Kawai, Startup torsional vibration of rotating machine driven by synchronous motor, in *Dynamics of Multibody Systems*, G. Bianchi and W. Schiehlen, eds., Springer, Berlin Heidelberg, 1986.

rotor and the rotating field of the stator. A completely unloaded motor (not possible in the real world because of bearing friction and rotor air drag) would run in steady state at synchronous speed, that is, the same speed as the rotating field of the stator. A real motor runs at a lower speed, depending on its mechanical load. *Motor slip S is* defined by

$$S \triangleq \frac{\text{synchronous speed} - \text{actual speed}}{\text{synchronous speed}} \tag{5-42}$$

The frequency of the rotor currents at any steady motor speed is given by the product of the slip, and the stator (fixed) frequency. For example, if the actual speed were 1746 rpm and the synchronous speed 1800, the slip would be 0.03 (3%) and the frequency of the rotor current would be $0.03 \times 60 = 1.8\,\text{Hz}$.

The key to the simple calculation of motor/load acceleration (assuming the drive has no vibration problems) is the motor steady-state speed/torque curve. While measured speed/torque curves are more accurate than theoretical predictions, a theoretical formula is certainly useful for a number of purposes and is available in several references.[6]

$$T_m = \frac{N_{ph} N_{pp} M^2 V_s{}^2 R_r \omega_s / S}{[R_r R_s / S - \omega_s{}^2 (L_r L_s - M^2)]^2 + \omega_s{}^2 (L_s R_r / S + R_s L_r)^2} \tag{5-43}$$

The terms in this formula are now defined and numerical values given for an example motor quoted by Nasar and Unnewehr.[7]

$$S \triangleq \text{motor slip}$$

$$N_{pp} \triangleq \text{number of motor pole pairs} = 2$$

$$V_s \triangleq \text{RMS stator voltage, 180 volts}$$

$$N_{ph} \triangleq \text{number of motor phases} = 3$$

$$T_m \triangleq \text{motor magnetic torque, N-m}$$

$$k \triangleq \text{maximum coupling coefficient}$$

$$\triangleq L_{rs}^2 / L_r L_s \triangleq M^2 / L_r L_s \text{ no units}$$

$$\omega_s \triangleq \text{synchronous speed} = 188.5\,\text{rad/sec}$$

$$L_s \triangleq \text{inductance of each of three stator windings} = 0.02039\,\text{hy}$$

$$L_r \triangleq \text{inductance of each of three ``fictitious'' coils used to model the motor rotor} = 0.02039\,\text{hy}$$

$$L_{rs} \triangleq \text{magnitude of the mutual inductance between pairs of stator/rotor}$$
coils. These *vary* as the cosine of the rotor position angle; however, the formula uses only the magnitude (peak value), which is 0.0200 hy for our motor.

$$R_r \triangleq \text{resistance of each of three rotor ``coils''} = 0.083\,\text{ohm}$$

$$R_s \triangleq \text{resistance of each of three stator windings} = 0.063\,\text{ohm}$$

---

[6]S. Seely, *Electromechanical Energy Conversion*, McGraw-Hill, New York, 1962, p. 213.
[7]Ibid., p. 422.

Basic Energy Converters

Substitution of the numbers for the example motor (a 30-hp unit) gives for Eq. (5-43)

$$T_m = \frac{13.72S}{0.02134S^2 + 0.006889} \quad \text{N-m} \tag{5-44}$$

This relation is easily computed and graphed to give the motor theoretical speed torque curve, Fig. 5-17. In Fig. 5-17a we see that the motor torque starts out at 486 N-m when the motor is at rest, builds up to a peak of about 560 as the motor accelerates to about 800 rpm, and then drops off to zero at the synchronous speed of 1800 rpm. The speed/torque data can of course also be plotted as in Fig. 5-17b, and in fact this form is more common.

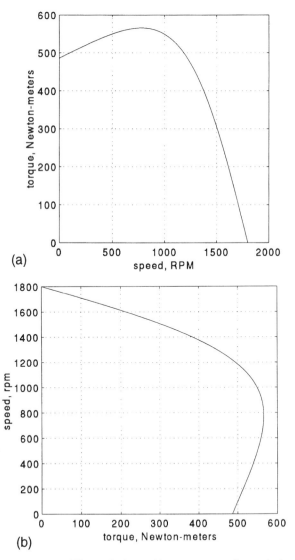

Figure 5-17 Theoretical speed/torque curves for ac induction motor.

The speed at which the motor will actually run depends on the speed/torque curve of the mechanical load which it is driving. The speed will rise until the motor torque just equals the total load torque; then the torques are balanced and Newton's law says that acceleration stops. If the motor is not connected to *any* mechanical load, it will speed up to nearly synchronous speed, since its bearing friction and rotor air drag torque are relatively small. Note in Fig. 5-17a that for speeds below the peak of the curve (about 800 rpm), a reduction of speed results in a reduction of torque, which itself causes a *further* loss of torque, an *unstable* situation. Thus, if the load's speed/torque curve intersects the motor's curve in this region, this operating point will *not* be a stable one and cannot be used. Actually, most induction motors are designed to operate well beyond the peak, in regions of relatively low slip (3 to 5%), though designs for special applications may have slip as low as 1% and as high as 13%. Note also that a "30-hp" motor (as in our example) does *not* necessarily develop 30 hp. It is designed to operate "best" at 30 hp but will of course supply only the power needed to drive its attached load at the speed corresponding to the intersection of the motor and load speed/torque curves. For our example motor, 30 hp occurs at about 1680 rpm. If desired, we could easily get a graph of horsepower versus speed, since power is just the product of torque and speed, and Eq. (5-44) relates torque to speed. Such a graph shows horsepower peaking at about 83, with speed 130 rad/sec.

EXAMPLE: INDUCTION MOTOR

We now want to use the speed/torque curve of our example motor to study the acceleration from rest, of a system comprising our motor and a mechanical load consisting of inertia and viscous friction. The motor's inertia is 0.06 kg-m^2 and the geared load contributes 0.24 (referred to the motor shaft), making the total inertia 0.30 kg-m^2. When the total inertia is large compared with the motor inertia, the acceleration is slowed, favoring our assumption of neglecting the electrical transients. The driven machine is a "paddle-wheel" type of mixer, and measurements have shown that it resists motion with a viscous type of torque, with a coefficient (referred to motor shaft) of 0.6 N-m/(rad/sec). All other frictional torques, including the motor bearing friction are assumed negligible relative to the mixer viscous torque. After the mixer has reached steady speed, an additional load torque on the motor occurs when a clutch engages an auxiliary load. This load (called "pulse load torque" in the simulation diagram) exerts a constant torque of −50 N-m from 0.3 to 0.5 seconds, and then is removed.

The system differential equation is given by Newton's law as

$$\sum \text{torques} = J\alpha$$

$$T_m - B\omega + \text{pulse load torque} = J\frac{d\omega}{dt} \tag{5-45}$$

The SIMULINK simulation diagram of Fig. 5-18 should be largely self-explanatory. Step functions 1 and 2 produce the pulse load torque which jumps to −50 at time = 0.3 second and then drops to zero at 0.5 second. Motor/load speed in radians/second is used in the Fcn block to compute slip, and then slip is used in

Basic Energy Converters

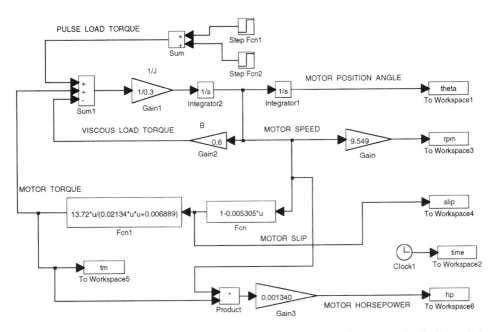

Figure 5-18 Simulation diagram for induction motor accelerating a mechanical inertia/friction load.

the Fcn1 block to compute motor torque from Eq. (5-44). Instantaneous motor horsepower is computed from instantaneous speed and torque.

Figure 5-19 shows the results of this simulation. Note that the vertical scale shows rpm directly, but the other curves have been scaled so that they are all clearly visible. As in any simulation, we can also get tables of values if the curves do not provide sufficient readability. We see that slip starts at 1.0 (100%) and quickly drops to a low value typical of steady-state operation near the design point of the motor. Horsepower rises quickly from zero and peaks at over 80 hp before dropping to a steady-state value near 25 hp. When the pulse load comes on at 0.3 second, horsepower rises to about 35 and later returns to its previous value when the pulse load disappears. Speed rises smoothly from zero to about 1700 rpm in about 0.2 second, and this occurs as the motor rotates through about 25 radians or 4 revolutions. We see again the power of simulation as a tool for understanding the operation of complex systems and providing useful design information. Recall that more accurate results would be obtained if we used an *experimental* motor speed/torque curve rather than the theoretical one of our example, and that our lookup-table simulation capability makes this easily possible.

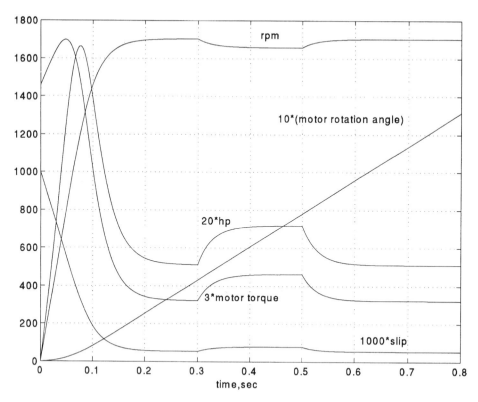

Figure 5-19 Simulation results for motor/load acceleration.

We next want to present a useful model for the dynamic behavior of the popular *stepping motor*. This class of motors also has several versions and corresponding analysis models,[8] and we choose one of the simpler descriptions[9] for our presentation. Figure 5-20 shows a simple permanent-magnet stepping motor. The rotor is a permanent magnet with a single pair of poles. The stator has four windings (electromagnets) which can be independently energized by connected them to electrical power supplies (voltage or current sources). By reversing the voltage polarity or current direction of the electromagnets, we can create either north or south poles at the ends of the electromagnets facing the rotor. By properly energizing all four electromagnets, we can create a stator magnetic field which will attract the rotor into specific angular positions. By switching the windings in the proper sequence, we can

---

[8] A. Leenhouts, *The Art and Practice of Step Motor Control*, Intertec International Inc., 1987; A. Leenhouts, *Step Motor System Design Handbook*, Litchfield Engineering Co., 1991; C. Raskin, Stepper motion . . . What's it all about? *Motion Control*, January 1993, pp. 17–22; C. K. Taft and R. G. Gauthier, Stepping motor selection for point-to-point positioning systems, *PCIM*, October 1985, pp. 22–33; P. G. Krause and O. Wasynczuk, *Electromechanical Motion Devices*, McGraw-Hill, New York, 1989, chap. 8.

[9] D. J. Robinson, Dynamic Analysis of Permanent-Magnet Stepping Motors, NASA TN D-5094, March 1969.

Basic Energy Converters

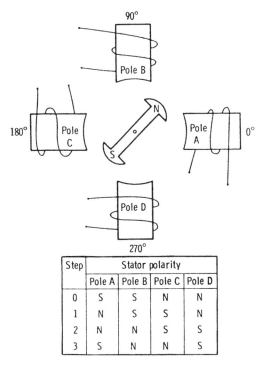

Figure 5-20 Diagram of a simple permanent-magnet stepping motor and table of stepping sequence.

cause the rotor to move, in a stepwise fashion, in either a clockwise or counterclockwise direction.

For the condition called "step 0" in the table, the rotor would be attracted to the 45° position shown in Fig. 5-20. For the "step 1" condition, the rotor would move CCW to a 135° position, while "step 3" would cause a CW motion to the 315° position. Similarly, step 2 provides the 225° position. We see that the motor can be caused to rotate, in 90° steps, in either the CW or CCW direction, by energizing the windings in the proper sequence. By making this sequence occur slowly or rapidly, we can control the average speed of the motor's rotation. Electronic circuits are available that will provide, on command, the sequence (and speed of repetition) that we desire, allowing versatile control of motor motion.

Note also that when the motor is at rest in a certain position, if we try to twist the rotor away from this position, we will feel a "springlike" magnetic torque which tries to keep the rotor in its preferred position. If we twist the rotor away from this equilibrium position and then let go, the rotor will perform a damped oscillatory motion, at a natural frequency determined by the magnetic "spring constant" (N-m/rad) and the inertia J of the motor rotor. When we switch the windings to cause the motor to move from one step to the next, we can expect this motion to also exhibit some oscillation before friction causes the rotor to settle into its new location.

While the motor of Fig. 5-20 has four steps per revolution, it is not difficult to construct motors with steps of almost any desired size. One such variation, the *variable-reluctance* stepping motor, has a rotor which is not a permanent magnet

302   **Chapter 5**

but rather is made of soft iron with many radial teeth, somewhat like a gear. The stator still has a number of electromagnet windings. When the windings are energized, the iron teeth are attracted in such a way as to create a large number of preferred angular positions. By using the proper number of windings and rotor teeth, a motor with the desired number of steps per revolution can be designed. A very common arrangement gives steps of size 1.8°, that is, 200 steps per revolution.

Since Newton's law determines the rotor's motion, we need to have an expression for the motor magnetic torque if we want to analyze the motor's response to commands. The Robinson reference [9] shows that a reasonable approximation for the motor torque is

$$\text{Motor torque} = K_T i \sin[N_{s,90}(\theta_c - \theta_m)] \qquad (5\text{-}46)$$

where

$K_T \overset{\Delta}{=}$ motor torque constant, N-m/amp

$N_{s,90} \overset{\Delta}{=}$ number of steps in 90° of motor rotation

$i \overset{\Delta}{=}$ winding current, amp

$\theta_c \overset{\Delta}{=}$ commanded motor angle, rad

$\theta_m \overset{\Delta}{=}$ actual motor angle, rad

Note that when the commanded and actual motor angles are equal, the magnetic torque is zero; we are at one of the equilibrium positions. If we try to twist the rotor away from such a position, we feel an opposing "magnetic spring" torque that varies sinusoidally with the angular deviation from the equilibrium position. For small motions away from the equilibrium point, the sinusoidal torque variation is nearly linear, allowing definition and lab measurement of a linearized magnetic spring constant, K_{ms} N-m/rad. The natural frequency ω_n of the oscillations discussed above can be estimated from $(K_{ms}/J)^{0.5}$, where J is the moment of inertia (kg-m^2) of everything that rotates with the motor rotor. If J is not known, its value can be estimated from the given formula by measuring the natural frequency and the spring constant.

The motor torque equation shows that torque is proportional to winding current i, just as in the dc motors we studied earlier. If the windings are supplied from a current source, then when we switch from one step position to the next, we assume that the new currents appear instantly, with no dynamic lag. If voltage sources are used, then the currents must be solved for using a Kirchhoff voltage loop equation with resistive, inductive, and back emf voltage-drop terms, again similar to a dc motor armature circuit. The currents will now *lag* the applied voltage, slowing the motion response of the motor.

EXAMPLE: STEPPING MOTOR

We want to study the response of a 200-step-per-revolution stepping motor to position commands. The total inertia is 0.002 kg-m^2, viscous damping is 0.5 N-m/(rad/sec), and maximum motor torque is 5.0 N-m. We also want to investigate the effect of intermittent disturbing load torques that might act on the load. Newton's law for the rotating inertia gives

Basic Energy Converters

$$T_m - B\omega + T_L = J\frac{d\omega}{dt} \tag{5-47}$$

and the simulation diagram of Fig. 5-21. We are here assuming the simplest model, a current source, for the winding excitation, to avoid the additional dynamic effects associated with voltage sources. The disturbing torque is produced as a rectangular pulse by the two step functions and the summer. We can make this torque pulse occur at any time and be of any size and duration. Nonlinear block Fcn1 is used to produce the sinusoidal motor torque, with Gain3 being used to set its maximum value. We can easily make the position command take any form we like; the diagram shows a sinusoidal variation with time, for which we can set the amplitude and frequency at will. Since the step motor must be commanded in discrete steps of the correct size, we sent the "true" command angle (the ideal motion) into a quantizer set at 0.0314159, the radian equivalent of the 1.8° steps.

If we set the command amplitude and frequency low enough, the motor should have little trouble following the command without losing any steps. Note that this motion control system is "open-loop", *not* a closed-loop or feedback system. That is, the load motion is *not* measured, so if the motor does not respond perfectly to every commanded step, the system is unaware of this and an error occurs. Many practical step-motor systems *are* open-loop, so they must be carefully designed and used so that steps are not lost. If we command steps too rapidly, *any* open-loop system will lose steps, so we always test our systems to find their limitations and then operate them within those limitations. Steps will also be lost if disturbing torques are too large. At the design stage, we do the testing by simulation; in development, we lab-test the actual system.

Figure 5-22a shows the behavior for a command amplitude of 0.1 radians at a frequency of 1 Hz. We see that each step command is faithfully followed. Of course the "steppy" motion deviates considerably from the ideal sine wave command, but

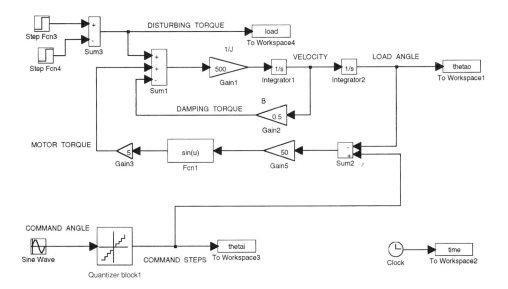

Figure 5-21 Simulation diagram for stepping motor.

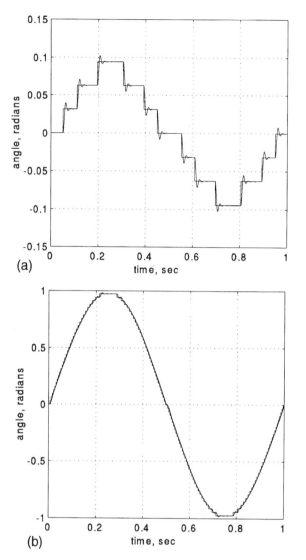

Figure 5-22 Step motor response for (a) low-amplitude sinusoidal command, (b) high-amplitude sinusoidal command.

this is obviously due to the finite *resolution* of one step, 0.0314 radian. If this is inadequate for the application, a number of design changes are possible. We could put a 100-to-1 gear train between the motor and load, so that the load moves 1/100 of the motor motion. We might try to find a motor with more than 200 steps per revolution. We might try *microstepping*,[10] a technique where the winding currents are adjusted in small increments to create many new equilibrium points between the original 200 of the basic motor. When a rotary motor is used to make a translational

---

[10] K. McCarthy, Getting more out of microstepping, *Motion Control*, March 1991, pp. 12–16.

Basic Energy Converters

motion, we often use a precision lead screw for the motion conversion. Note that a 200-step-per-revolution motor coupled to a lead screw with a pitch of 5 rotations per inch gives a translational resolution of 0.001 inch, adequate for many positioning applications.

In Figure 5-22b we have left the frequency at 1 Hz but increased the amplitude to 1.0 radian. Now the motor motion visually duplicates the command quite nicely, though of course the resolution of 0.0314 radian is still present. In Fig. 5-23a we raise the command frequency to 98 rad/sec and find that the motor now cannot keep up, even at the low amplitude of 0.1 radian. In Fig. 5-23b we command at 1-Hz fre-

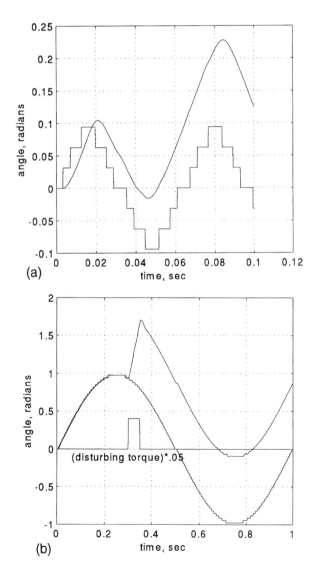

Figure 5-23 Loss of synchronism in step motor (a) due to excessive command speed, (b) due to excessive load torque.

quency and 1-radian amplitude, but apply a pulse disturbing torque of 8 N-m that lasts from 0.30 to 0.35 second. This torque "bumps" the motor out of synchronism with the command pulses, causing a large position error. Stepping motors can be and are used in closed-loop configurations,[11] which may be useful in overcoming some of the limitations of the open-loop systems we have here described.

Our simple step-motor example has revealed most of the essential features of this type of drive. However, step-motor technology is quite diverse and our short section does not pretend to tell the whole story. A vast and comprehensive literature, which we have lightly sampled in our references, is available for readers needing additional details. Don't overlook the manufacturer's catalog/handbooks, which often give excellent practical advice on applications, such as the "hybrid" step motor, a combination of the PM and variable-reluctance principles.[12]

While the "magnetic" motors we have emphasized in Fig. 5-13 for electrical-to-mechanical energy conversion and motion control certainly dominate the market in an economic sense, the other conversion process (piezoelectric) shown there has recently assumed importance in some narrow, but technologically significant areas. When (1972) the first edition of this text was produced, the use of piezoelectric actuators (electrical input, mechanical output) was quite restricted and limited to a few special-purpose applications. Today there are many applications for such devices and several manufacturers[13] stock a wide range of general-purpose motion-producing devices and associated electronic amplifiers and motion sensors. Since the motions produced are the elastic deflections of the "crystals," they are limited in most cases to less than 100 μm.

Control of motion at these low levels is vital to a number of processes which can be gathered under the general titles of *microtechnology* and *nanotechnology*—the art, science, and engineering of mechanical and electrical devices at the scale of micrometers and nanometers. The Polytec PI catalog and handbook referenced gives a good overview of how piezoelectric positioning technology is applied in various areas:

PRECISION MECHANICS AND MECHANICAL ENGINEERING

Adjusting tools   Correcting wear
Controlling injection nozzles   Micropumps
Linear drives   Piezo hammers
Extrusion tools   Microengraving systems
Active vibration isolation

---

[11] R. Gfrorer, W. Siefert, T. Baur, Closed-loop control improves 5-phase step motor performance, *PCIM*, March 1989, pp. 52–56.

[12] Engineering Reference and Application Solutions (pp. A1–A80), Compumotor Division, Parker Hannifin, 5500 Business Park Drive, Rohnert Park, CA 94928, 800-358-9070.

[13] Products for Micropositioning, Polytec PI, Inc., 23 Midstate Drive, Suite 104, Auburn, MA 01501, 508-832-3456.

OPTICS AND MEASURING TECHNOLOGY

Mirror positioning	Holography
Interferometry	Laser tuning
Fiber optic positioning	Fast mirror scanning
Adaptive and active optics	Image stabilization
Autofocus	Stimulation of vibrations

MEDICINE

Micromanipulation	Cell penetration
Microdosing devices	Audiophysiological stimulation
Shockwave generation	

MICROELECTRONICS

Wafer and mask positioning	Microlithography
Inspection systems	

In Fig. 5-5 we showed a basic piezoelectric device with one layer of piezo material sandwiched between metal electrodes, and we developed the system equations for the (mechanical input)/(electrical output) mode of operation. Just as we were able to use the dc generator equations for the dc motor application, we can also use the piezo equations for either mode of operation. That is, in Fig. 5-5 we now *apply* a drive voltage at the e terminals and this will *produce* a piezoelectric force and motion x. In Eq. (5-10) or (5-14) the $f(t)$ term now plays the role of a *disturbing* force, and may or may not be present in a particular application. The piezoelectric force $C_1 e$ is now not a "resisting" force but rather the driving force which causes the desired output motion. We thus change the sign convention for this force so that a positive voltage causes a positive motion, and move it to the right side of the equation since it now plays the role of an *input*.

$$\frac{\ddot{x}}{\omega_n^2} + \frac{2\zeta}{\omega_n}\,\dot{x} + x = \frac{C_1}{K_s}\,e + \frac{1}{K_s}\,f(t) \qquad (5\text{-}48)$$

The driving voltage e is usually the output voltage of an amplifier, so we need to consider the modeling of the amplifier output circuit. Recall that the "passive element" aspect of the piezo device is that of a capacitor, so the amplifier is essentially driving a pure capacitance load. The piezo force is proportional to the capacitor voltage e, but this voltage can only be changed by an amplifier current charging or discharging the capacitor. If we use a conventional "voltage amplifier," then a step change in amplifier input voltage should ideally produce a step change in output voltage; however, for a capacitor load, this requires an *infinite* current! We have sometimes suggested the use of "current" (also called transconductance) amplifiers to nullify electrical (inductive) lags, but this artifice won't work here because *no* amplifier can provide the needed infinite current.

To get a more correct picture of system dynamics we need to model the amplifier output circuit with the capacitor present and include this in our overall system description. In Fig. 5-24a we show a voltage amplifier, whose own output resistance is R_{ao}, connected to the piezo capacitance C. Circuit analysis gives us Eq. (5-49), which shows that a sudden change in amplifier input voltage causes a *gradual* (exponential) rise in voltage across the piezo element, and thus a gradual rise in

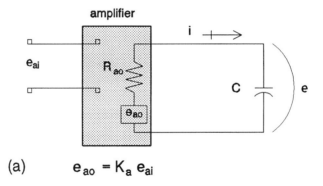

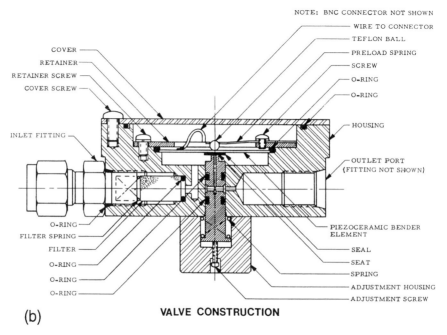

Figure 5-24 (a) Piezo crystal as capacitive load on amplifier; (b) piezoelectrically actuated flow control valve.

piezo force, delaying also the motion x. In addition to this linear delaying effect there is also the possibility of having amplifier *saturation*, a nonlinear, but real,

$$e_{ai}K_a = i_C R_{ao} + \frac{1}{CD} i_C$$
$$e = \frac{1}{CD} i_C$$
$$\frac{e}{e_{ai}}(D) = \frac{K_a}{R_{ao}CD + 1} \qquad (5\text{-}49)$$

aspect of every amplifier. That is, if we make a *large* and rapid change in amplifier input, output current will saturate and capacitor charging will be retarded in a

Basic Energy Converters   **309**

nonlinear way. If we know the amplifier's saturation characteristics, this nonlinear effect can be easily included in a simulation model.

We have now presented all the descriptions needed to model a complete piezoelectric actuation system. Before leaving the subject we want to familiarize you with some of the actual hardware. While the piezoelectric effect is bidirectional, the materials used are weak in tension, so actuators are usually designed to move the attached load mass by expanding the piezo element, putting it in compression. A typical large actuator is a cylinder about 2.5 cm in diameter and 18 cm in length. To give its maximum expansion of 180 μm requires 1000 volts. Since most general-purpose amplifiers do not provide such high voltages, the piezo manufacturers often provide the special amplifiers needed. The quoted actuator can safely exert 4500 newtons of compressive force and 500 newtons tension. Its capacitance is 3 μF and its mechanical natural frequency is about 2000 Hz. No intentional damping is provided and the parasitic frictional effects are small, so the resonance at 2000 Hz has quite a high peak. We usually drive such actuators with signals whose frequency content is far below (maybe 10% of) the natural frequency, so the low damping is not a problem. Of course if the actuator is attached to a load which has appreciable mass, the system natural frequency will be less than the 2000 Hz of the "bare" actuator. The device stiffness is about 40 N/μm, so we can estimate the system natural frequency as soon as we know the load mass.

The piezoelectric force is basically related to the electrostatic force on a charged particle in an electric field. The piezoelectric materials have an unsymmetrical internal charge distribution, so when they are exposed to an external electric field, an internal force is generated. The external electric field is simply the voltage across the terminals divided by the thickness of the piezo element. For a typical material, a field strength of about 1000 V/mm gives the maximum practical expansion. Thus a piezo "wafer" 1.0 mm thick might require about 1000 volts. A single wafer at maximum expansion would only give a few micrometers of motion, so "stacked" actuators are necessary to get the large (180-μm) expansion quoted in our above example. Here, a large number of wafers are stacked, with the needed electrodes sandwiched between them, and a single applied voltage excites them all in parallel. To allow use of low-voltage amplifiers, some stacked actuators use very thin wafers. This works because the deflection depends on the electric field (V/mm), not the applied voltage itself. Thus a 0.1-mm-thick wafer can give maximum expansion with only 100 volts.

The motion provided by piezoelectric actuators exhibits some hysteresis. In applications where this might not be acceptable, it is possible to use the actuator in a feedback system. This requires a motion sensor; strain gages, LVDTs, or capacitance sensors are in use here. The most sensitive and accurate systems use capacitance sensors to achieve resolutions of 0.1 nm.[14] Piezoelectric positioners are used in "atomic force microscopes," which allow measurement of many physical effects at the scale of molecular and atomic dimensions. In addition to the basic unidirectional actuators described above, many multiaxis stages are available "off the shelf" for translational and rotational motions. Piezoelectric actuation is also used for ink-

---

[14]P. D. Atherton, Moving and measuring to better than a nanometer, *Motion*, March/April 1993, pp. 2–10; Queensgate Instruments, 516-623-9725.

pumping purposes in some ink-jet printers, and valve actuation in fluid flow control valves (Fig. 5-24b). The valve shown[15] uses the piezo material in a bending mode, requires 100 volts to actuate, and has a response time of about 0.002 second. Some active vibration isolation systems use piezo actuators to rapidly move masses to create counterforces which reduce vibration in structures and machines.

In Fig. 5-13, the direct conversion of electrical power to fluid power can be accomplished by the *electromagnetic pump* of Fig. 5-25, but this mode of energy conversion is of limited practical importance. Only fluids of high electrical conductivity, such as certain liquid metals, can be pumped efficiently in this way. Without going into detail, the principle is the same as that of the electric motor, in that a current-carrying conductor in a magnetic field feels a force, except here the "conductor" is the fluid being pumped. If the flow is blocked, the force is still felt; thus the pump can produce pressure at zero flow. When flow is allowed, a "back emf," which opposes the applied voltage, develops, just as in a motor.

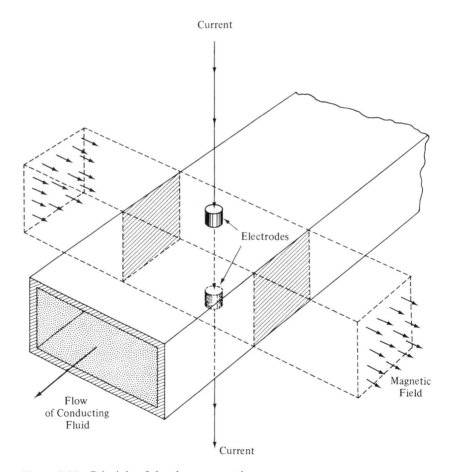

Figure 5-25 Principle of the electromagnetic pump.

---

[15]Maxtek, Inc., Torrance, CA.

Basic Energy Converters   311

Electrical heating processes of various types are of considerable commercial importance. For the simplest form, resistance heating, the conversion of electrical to thermal power follows the simple relation

$$\text{Rate of heat generation} = i^2 R \quad \text{watts}$$
$$= 0.000949\, i^2 R \quad \text{Btu/sec} \tag{5-50}$$

While resistance heating can be accomplished using either ac or dc, induction and dielectric heating basically require ac. In induction heating, a coil carrying ac power induces eddy currents into the piece being heated and these currents cause $i^2 R$ (resistive) heat generation within the piece, frequencies in the range 480 Hz to 450 KHz being employed. The workpiece must be a fairly good electrical conductor. At high frequencies the "skin effect" crowds the current near the surface, allowing concentration of the heating for surface heating processes such as case-hardening of metal parts. In dielectric heating (2 to 40 MHz) the workpiece is a fairly good electrical insulator, and the heating effect is uniformly distributed over its volume. The heating effect is produced by the dielectric loss coefficient of the material being heated. Thermoelectric heaters and coolers employ circuits of two properly chosen dissimilar materials to convert electric power directly to heat flow. They are mainly used, at low power levels, to cool electronic devices. Laser beams are widely used in manufacturing processes such as cutting or welding, where heat generation certainly is involved. The laser is electrically controlled, but the beam itself might be thought of as "optical" power, which is converted to "thermal" power by complex processes occurring at the point where the beam impinges on the workpiece material. More details on all the heating/cooling processes briefly described here can be found in the literature.[16]

5-4 CONVERTING FLUID ENERGY TO OTHER FORMS

Positive-displacement machines (cylinders for translation and motors for rotation) and turbines are widely used to convert fluid power to mechanical form. While there are differences in some details of construction, these devices are essentially the same as the corresponding pumps (positive-displacement and centrifugal) already discussed in Sec. 5-2, except that fluid energy is now the input, and mechanical energy the output. Our treatment can thus be fairly brief, since the models are so similar. The torque developed by a positive-displacement motor is given by

$$T_m = D_m\, \Delta p \quad \text{inch-lb}_f \tag{5-51}$$

where

$D_m \overset{\Delta}{=}$ motor displacement, in^3/rad

$\Delta p \overset{\Delta}{=}$ instantaneous pressure drop across motor, psi

---

[16]E. O. Doebelin, *Engineering Experimentation*, McGraw-Hill, New York, 1995, pp. 387–402.

Newton's law for the motor and attached load is thus easily written; the motor contributes its own inertia J and damping B to the total load. The motor flow rate in terms of motor speed ω and pressure drop is given by

$$\text{Motor flow rate} = D_m \omega + \frac{\Delta p}{R_{\text{fl}}} \tag{5-52}$$

where R_{fl} is the motor leakage flow resistance. As soon as the pressure/flow characteristics of the fluid power source driving the motor are known, they may be combined with those just given for the motor to get an overall fluid system equation. Some rotary motors allow their displacement to be *varied* while the machine is running, as a means of controlling the motion. Such variable-displacement motors will have an additional equation relating D_m to the action of the control mechanism.

Turbines and centrifugal pumps have much in common; both fall in the category called *turbomachines*, the pump converting mechanical power to fluid power, the turbine accomplishing the reverse conversion. Water turbines, such as are used to drive generators in hydroelectric plants, may be modeled in a fashion quite similar to that used for centrifugal pumps, using experimental characteristic curves analogous to those of Fig. 5-11. Gas and steam turbines are somwhat more complex, since both thermal and fluid aspects must be considered; however, the linearization of families of characteristic curves is still a useful technique.

The electromagnetic flowmeter[17] of Fig. 5-26 is really a measuring instrument, rather than a power-generating device; however, it does convert fluid power into an

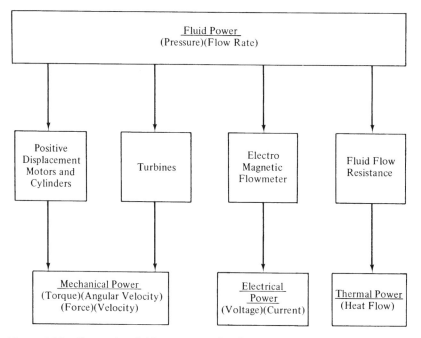

Figure 5-26 Converting fluid power to other forms.

---

[17]Doebelin, *Measurement Systems*, 4th Ed., pp. 583–588.

Basic Energy Converters

electrical signal. The principle is identical to that of the pump of Fig. 5-25, except now the flow is the input, which produces a voltage output at the electrodes, in direct proportion to the flow rate. Both the voltage produced and the resulting current flowing into the external voltage-measuring circuitry are very small, and thus the power output is negligible; however, this type of flowmeter has considerable practical importance because of certain advantages it has over other methods of measuring flow rates.

Fluid flow resistance (friction) as a means of generating heat is usually an undesirable parasitic effect. If frictional pressure drops are not kept small enough in fluid power systems, the working fluid may heat up to the point where it deteriorates, or critical components such as seals fail. To prevent this, heat exchangers with cooling water may be added to maintain fluid temperature low enough to ensure long equipment life. Since frictional pressure drop times volume flow rate has the dimensions of mechanical power (in-lb$_f$/sec), one can convert to heat flow using familiar conversion factors.

Sometimes conversion of fluid power to heat is intentionally used as a means of control, as in the *servovalve*, used to control the motion of fluid actuators (cylinders or motors). The servovalve is supplied by a pressure source, and meters flow to the actuator by partially opening a flow port between the source and the actuator. As fluid flows from the source, through the valve port, and then into the actuator, the product of valve pressure drop and flow rate is fluid power which is entirely converted into heat by the flow resistance of the valve port, which is essentially a variable-size orifice. The total fluid power expended is the sum of that wasted into heat across the valve port and that usefully converted into mechanical power in the actuator. In effect, the servovalve controls fluid power by controlling the fraction that is converted into heat. When the valve is barely open, most of the power is wasted and only a little goes to the actuator. When the valve is wide open, less power is wasted and more usefully converted. This scheme may seem wasteful (the overall efficiency rarely exceeds 30%), but servovalve control is often chosen for applications requiring maximum speed. We might point out that linear transistor amplifiers (and their vacuum-tube predecessors) operate in exactly the same way, modulating power by wasting a controlled portion into heat. Such electronic amplifiers also have equally poor efficiencies.

5-5 CONVERTING THERMAL ENERGY TO OTHER FORMS

The direct conversion of thermal energy to mechanical energy may be accomplished through the phenomenon of thermal expansion (see Fig. 5-27). Any object subjected to heat addition will experience a temperature rise, and the accompanying expansion may be caused to do mechanical work by letting it push a load. This process is not widely used to generate mechanical power, but does serve a useful function in some temperature-measuring and control devices such as bimetal thermometers, and gas, liquid, and vapor pressure thermometers.[18]

---

[18]Ibid., pp. 619–625.

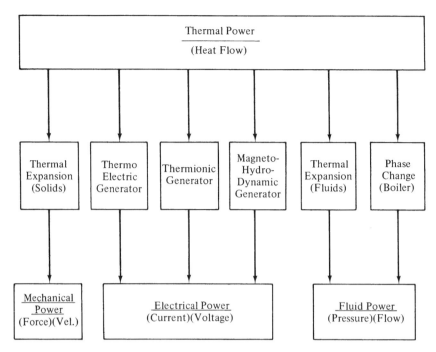

Figure 5-27 Converting thermal power to other forms.

General-purpose thermal actuators based on this principle are not common, but at least one company[19] is presently marketing such devices. The expansion medium is a special polymer which pushes a piston as it expands. A travel of 0.15 inches and a maximum force of 200 pounds are possible, but the total work done must always be less than 7.5 in-lb$_f$. A heating rate of 150 watts gives a piston speed of 0.08 in/sec. Full expansion takes a few seconds but retraction depends on natural cooling, which is slower. Also, a spring or other force is required to actually return the piston to its unexpanded position.

Since electrical power is of such great importance, the direct generation from thermal energy has received considerable attention and several classes of useful devices are in various stages of practical development. While some of these are now in actual service, it is only fair to say that their total contribution to the world's electrical generating capacity is exceedingly small. They do, however, find application in certain specialized situations where other methods are at a disadvantage. The thermoelectric effect, in which the application of heat to an electric circuit made up of two properly chosen dissimilar materials results in a current flow, was known for many years and is still usefully employed as a temperature-measuring instrument, the thermocouple. The effect has been practically employed to generate small quantities of electric power in specialized applications. Thermionic generating devices employ

---

[19]TCAM Technologies, Inc., 33800 Curtis Blvd., Suite 114, Eastlake, OH 44095, 216-942-2727.

Basic Energy Converters   **315**

the same sort of principle as used at the cathode of an electronic vacuum tube—the "boiling off" of electrons from a suitable material by the application of heat. Again, only small amounts of power are produced in this way and applications are very limited. Magnetohydrodynamic (MHD) methods of power generation have the potential for large-scale power production but have not yet been brought to practical realization, even after many years of research. The reader interested in learning more about these processes of direct energy conversion of heat to electricity will find a number of texts devoted entirely to this subject.[20]

Thermal expansion of fluids allows conversion of thermal power to fluid power. Such energy conversion is limited to low power levels, such as in measuring instruments, when liquids are involved. In the case of gases and also when phase change from liquid to gas occurs, large amounts of power can be converted. Actually, the addition of heat to a gas does not result in a straightforward conversion to fluid power (flow rate times pressure); several forms of energy are present and must be properly taken into account using principles of thermodynamics and fluid mechanics. Internal combustion engines (gasoline, diesel, natural gas, etc.) and gas turbines are good examples of such energy conversion. The combustion process converts the chemical energy of the fuel into heat, which in turn is converted to fluid pressure and flow, and ultimately into mechanical power available at the shaft. In a steam turbine, the combustion occurs external to the turbine in a furance, where the heat is applied to a boiler. The boiler accepts a heat input and uses it to vaporize the water, increasing its temperature and pressure. We thus have a conversion of thermal energy into fluid energy which the turbine can turn into mechanical shaft power. Again, several forms of energy (not just fluid energy in the form of pressure times flow) are present and must be properly accounted for.

Conversion of heat directly to fluid power is used in the bubble-jet type of ink-jet printer.[21] Here tiny resistance heaters vaporize the ink, causing pressure to rise and expel a tiny drop of ink toward the paper. The process requires 0.04 mJ of energy per drop, while an "ordinary" thermal printer uses 3.4 mJ per dot. This large reduction in energy allows a battery-operated portable printer to print several hundred pages per battery charge. The *Hewlett-Packard Journal* issue referenced is devoted entirely to the invention, research, design, development, and manufacture of this unique device. I found it fascinating reading and highly recommend it to you.

5-6   OTHER SIGNIFICANT ENERGY CONVERSIONS

Since earlier chapters concentrated on mechanical, electrical, fluid, and thermal forms of energy, the present chapter emphasizes the interactions and couplings among these. While we do not intend this brief treatment to be comprehensive in the field of energy conversion, certain processes not falling into the above categories are of sufficient interest that we wish to at least mention them (Fig. 5-28). The importance of chemical and nuclear fuels can hardly be overemphasized since they

---

[20]S. Angrist, *Direct Energy Conversion*, Allyn and Bacon, Boston, 1965.
[21]*Hewlett-Packard Journal*, May 1985.

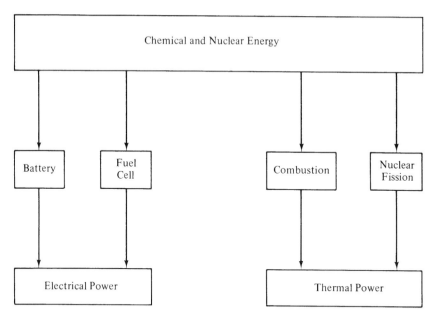

Figure 5-28 Conversion of chemical and nuclear energy to other forms.

are the fundamental source of practically all of our useful power. Conversion of this stored energy into directly usable form involves combustion of chemical fuels (coal, oil, gas, etc.) and fission of nuclear fuels to produce thermal energy. System studies of overall power plants must thus take into account the dynamic behavior of these combustion or nuclear processes.[22] Direct conversion of chemical energy to electricity is accomplished by various forms of batteries and the fuel cell. Dynamic behavior of batteries has not received (or apparently needed) much consideration; however, fuel cell dynamics[23] are of considerable importance since control systems are needed to obtain proper cell performance.

5-7 POWER MODULATORS

This chapter is intended mainly to familarize you with some basic energy conversion devices which we will want to use in designing and analyzing dynamic systems in later chapters. To conclude the chapter, this section will discuss some important

---

[22]R. Dolezal, *Process Dynamics: Automatic Control of Steam Generation Plant*, Elsevier, New York, 1970; M. A. Schultz, *Control of Nuclear Reactors and Power Plants*, McGraw-Hill, New York, 1955.

[23]P. R. Prokopius, Internal Voltage Control of Hydrogen-Oxygen Fuel Cells–Feasibility Study, NASA TN-D-7956, April 1975; P. R. Prokopius and R. W. Easter, Mathematical Model of Water Transport in Bacon and Alkaline Matrix-Type Hydrogen-Oxygen Fuel Cells, NASA TN D-6609, March 1972.

Basic Energy Converters

power modulators. These are not necessarily energy *converters,* but it is not unreasonable to include them in this chapter since they are hardware of great practical importance that you need to be aware of and that we will soon use as our study turns more to consideration of complete systems. By power modulator we here mean some device which takes a more or less steady source of energy, draws power from it in a controllable way, and applies this modulated power to some kind of "actuator" in a dynamic system.

Perhaps the most common and important power modulators are the various kinds of electronic amplifiers which show up in so many dynamic systems, even those that are mainly mechanical, fluid, or thermal. We want to describe three main types of amplifiers and show how to simulate their behavior in a larger system of which they are a part. As usual, we take the system engineer's viewpoint and give only enough detail to allow basic understanding, definition of measurable parameters, and simulation. We definitely do *not* explore the design of the amplifiers themselves, a task certainly reserved to electrical engineering specialists.

We of course have earlier discussed in some detail *operational amplifiers.* These are mainly used in low-power applications in measurement and signal processing, and our earlier treatment is adequate for those purposes. In this chapter we instead deal with amplifiers of higher power level, as needed to drive the *actuators* in our system. In fact, the output of operational amplifiers is often used as the input to these power-amplifying stages. Of the many types of power amplifiers, we will restrict ourselves to three:

1. The "smooth" or linear transistor amplifier
2. The pulse-width-modulation (PWM) or switching amplifier
3. The silicon-controlled-rectifier (SCR) amplifier

The "smooth" or linear power amplifier (sometimes called Class AB) is the simplest to use and model since its output voltage is smoothly proportional to its input voltage. When it is part of a dynamic system which has also mechanical, and/ or fluid and thermal portions, the amplifier is usually so much faster than the other parts that it is modeled as *instantaneous* (transfer function is just K) with little error. Of course it is not *really* instantaneous, and if you use it in a system which is *totally* electrical, its dynamics might not be negligible. Also, as we saw in Fig. 5-24a, certain kinds of electrical loads may affect the output/input voltage relation. Usually, however, this ratio is just taken as a simple constant K, from zero frequency up to several thousand hertz, often well beyond the highest response frequency for the mechanical, fluid, or thermal parts of the system. Be sure, of course, to ask the amplifier supplier for the maximum output current limit. If your load tries to draw more current than this, the amplifier output voltage is no longer proportional to the input voltage; rather it saturates. This does not usually damage the amplifier, but unless you have included the saturation effect in your simulation model, your simulated results will be incorrect, often in a nonconservative direction.

While the smooth amplifier just described is ideal in terms of simplicity of analysis and most performance attributes, it suffers from a severe economic penalty. To implement this type of amplifier, the electrical engineer uses the transistors as smooth modulation devices rather than switching devices. It turns out that this mode of operation is extremely energy inefficient (efficiency is 20 to 60%); most of the power taken from the amplifier dc power supply is *wasted* into heat, which also creates temperature problems for the transistors. For low-power applications, say less than 1000 watts, the poor efficiency can be tolerated because the cost of the wasted power is not large (50% of 1000 watts is 500 watts; 50% of 20000 watts is 10000 watts). Also, if a high-power load can not tolerate the "roughness" of the PWM or SCR amplifier types, we may elect to use a smooth amplifier even if the cost of wasted power is large. An example might be a large electrodynamic vibration shaker. Here, large power is involved, but in vibration testing, purity of the force waveform is often essential; the "jerkiness" of PWM and SCR waveforms might be unacceptable.

To get high energy efficiency from transistors, they must be used as "switches" rather than smooth modulators, and the PWM amplifier does exactly that to achieve 75 to 90% efficiency. The prices to be paid for this advantage are reduced bandwidth (frequency response) and some degree of jerkiness in the output voltage waveform. A major application is in motion control using electric motors. Fortunately here, motor winding inductance has a smoothing effect on the current; a jerky voltage produces a less jerky current. Motor torque is directly proportional to current, so torque will be as jerky as current; however, motor *velocity* will be smoothed by inertia and damping, and motor *position* will be smoothed even more. Thus, in most cases, a PWM amplifier will be used for power ranges from about 500 watts to 20 kilowatts (27 hp) to gain the energy efficiency. The switching frequency of PWM amplifiers is usually in the range of 5000 to 20000 Hz. A rough rule of thumb is that the amplifier effective bandwidth (flat frequency response) is about 10% of the switching frequency. Flat amplifier response of 500 to 2000 Hz is usually ample for systems with mechanical moving parts; they generally can't move that fast, so the amplifier doesn't limit system performance. The higher switching frequencies are harder to implement electronically but have the advantage of reduced acoustic noise; 20000 Hz is beyond the range of human hearing.

How does a PWM amplifier actually work? Sometimes experienced system designers will just ignore the details and model the PWM amplifier with a simple K factor, just like a smooth amplifier, especially if system bandwidth is low ("slow" systems). This is almost necessary if we try to solve the equations analytically. If we use simulation, a more correct model is not hard to implement. In Fig. 5-29 we compare the response of a smooth amplifier to that of a PWM for the same sinusoidal input signal. A PWM amplifier always includes a triangle-wave oscillator which sets the basic frequency of the pulse train, in our example, 5000 Hz. The amplifier input signal is summed with this triangle wave. This shifts the triangle wave up or down, depending on whether the input signal is positive or negative. The shifted triangle wave is positive for one portion of each cycle and negative for the remaining portion. By sending this shifted waveform into a switch, the switch output (which is the amplifier output) becomes a rectangular wave which spends part of each cycle at the positive power supply voltage (+30 volts) and the rest at the negative voltage (−30 volts). The output transistors are thus used as switches, which is their most efficient mode of operation. When the amplifier input signal is zero, the output spends exactly half of each cycle at +30 volts and half at −30 volts, giving an average value of zero. When the amplifier input signal goes positive, the output has a positive average value, and when it goes negative, the output has a negative average value. A motor, which has inductance and inertia, tends to respond to the average value, since the fluctuating portion of the signal is too fast for it to follow. In the

Basic Energy Converters

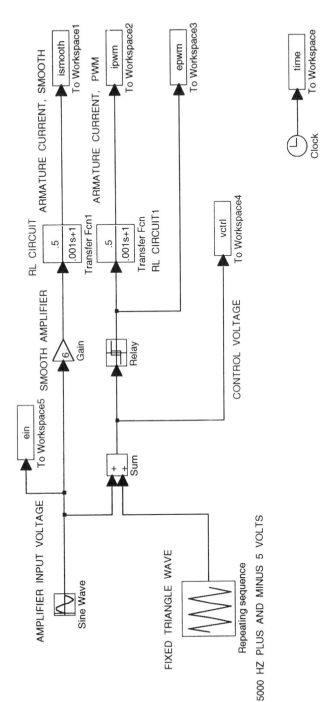

Figure 5-29 Simulation comparison of "smooth" and PWM amplifiers.

figure, we send the amplifier output to an *RL* circuit (like a motor armature circuit) to show this smoothing effect. If we wanted to model a complete motion control system, we would see even more smoothing due to motor/load inertia and damping.

Figure 5-30a shows the PWM control voltage vctrl and the PWM amplifier output voltage epwm when the amplifier input voltage is a sine wave of 5 volts amplitude and 250 Hz frequency. The amplifier output is extremely "jerky," jumping back and forth between +30 volts and −30 volts. However, if we apply this voltage to

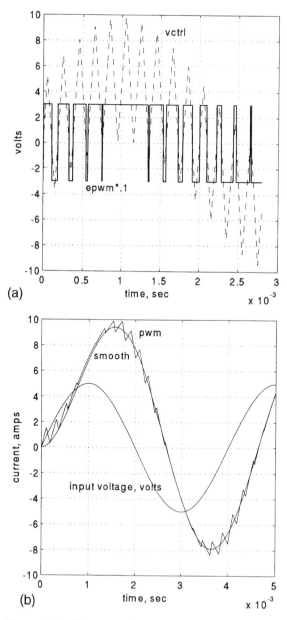

Figure 5-30 Response of "smooth" and PWM amplifiers to a sinusoidal input.

Basic Energy Converters   **321**

a series RL circuit, such as might appear in the armature of a motor, the current becomes much smoother. The RL circuit voltage-to-current transfer function shown has $R = 2$ ohms and $L = 0.002$ hy. Figure 5-30b compares the current of the PWM amplifier with that of a "smooth" amplifier and we see that the PWM current, while not perfectly smooth, is much less jerky than the PWM voltage which causes it. We also show the amplifier input sine wave, which reveals the lag (phase shift) between input voltage and output current. This lag is due to the inductance. Thus a larger inductance has the good effect of smoothing the current more, but the bad effect of causing more system delay, which can cause trouble in some motion control systems. If motor armature inductance is too low, and we add more, we need to strike a compromise between these two conflicting design goals.

Let's turn now to the third and last amplifier type, the SCR. Both the smooth and PWM amplifier types require for their operation dc power supplies. That is, we usually use the ac power line as our ultimate power source, so if a device needs dc, we have to "make it" from ac. In fact, these dc supplies represent a large fraction of the cost, size, and weight of smooth and PWM amplifiers. The SCR amplifier does not need or use such dc supplies; it works directly from the ac power line. Its efficiency is even better than that of PWM types: 85 to 95%. It can be used in low or moderate power systems but it becomes the method of choice when the power level gets much above 30 hp, where its cost becomes signficantly less than the PWM. It can be used up to thousands of kilowatts, for both motion control and other applications such as electric heating. Its main drawback is a slower response speed, but this is often acceptable since the high-power rotating machines will tend to themselves be slow.

There are several versions of SCR amplifiers; single-phase, polyphase, half-wave, full-wave, etc. We again do not go into the electronic details but rather show a simulation model useful for overall system studies. All SCR amplifiers share the same basic concept. By "turning off" the sinusoidal ac voltage waveform for a fraction of each cycle, we control the average voltage, current, and power supplied to the load. If we want a positive average value we use only the positive part of each cycle; if we want a negative average value, we use the negative part. For heating (rather than motion) control, both positive and negative voltages give equal heating, so an SCR amplifier for heating control can be fairly simple. In fact, you have probably used a *light dimmer* in your house or apartment to smoothly control the level of illumination; this is actually a simple SCR-type of control. Since the SCR device uses a 60-Hz basic waveform (rather than the high frequencies typical of PWM systems), its speed of response (bandwidth) will be some fraction of 60 Hz, much slower than the PWM. Again, a rough rule of thumb is a bandwidth of about 10% of the power line frequency, say about 6 Hz. Another possible advantage of SCR systems in motor control is the capability of *returning* power to the ac line when a motor is slowed down ("dynamic braking"). Other amplifier types may have some difficulty in efficiently handling this situation.

Figure 5-31 shows a simulation diagram for a single-phase, full wave, SCR amplifier. This diagram will behave essentially the same as the real electronics, but it does *not* correspond to a device-by-device replica of the actual hardware. We leave an explanation of this simulation to the reader; I have included enough detail on the diagram that you should be able to puzzle it out. As in our earlier PWM simulation, we use a low-frequency sine wave as our input signal (low now means less than 60 Hz). Figure 5-32 shows the amplifier voltage output, which consists of "pieces"

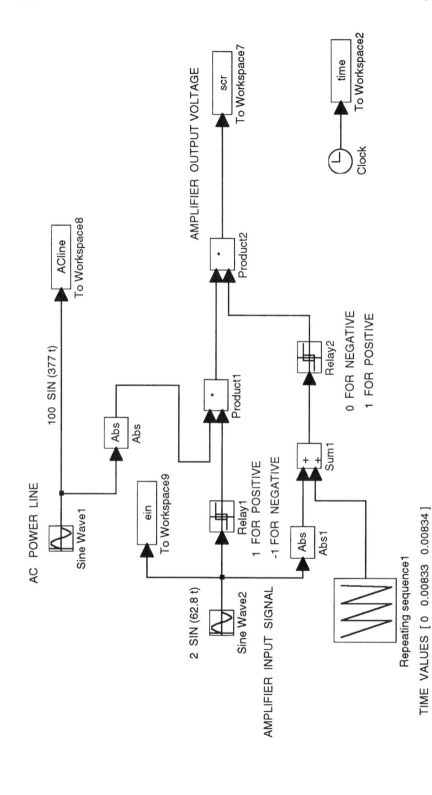

Figure 5-31 Simulation of SCR (silicon-controlled-rectifier) amplifier.

Basic Energy Converters

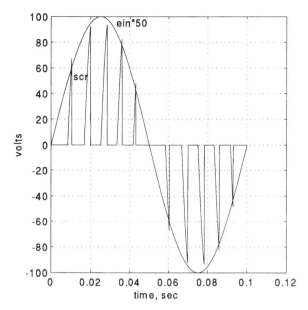

Figure 5-32 SCR amplifier response to sinusoidal input.

of the 60-Hz power line voltage, the length of the pieces depending on the magnitude of the input voltage and the polarity depending on the polarity of the input voltage. We could have applied the "jerky" output voltage to an *RL* circuit (as we did in the PWM example) and again seen a smoothing effect. Other variations of the SCR amplifier can be simulated with "tricks" similar to those in Fig. 5-31.

In hydraulic systems, the *servovalve* plays a power-modulating role similar to that of the smooth electronic amplifier in electromechanical systems. Just as an electronic amplifier has a dc power supply, the servovalve has a constant-pressure power source as we discussed in Fig. 4-29 and as we show again in Fig. 5-33. The essential features of a servovalve are a moveable spool and stationary flow ports, which the spool covers or uncovers when it is moved (displacement x_{sv}). The servovalve is basically a flow control valve, directing flow from the high-pressure source (1000 psig in our example) to the actuator, and from the actuator to the low-pressure return line to the system reservoir. The return pressure must be above atmospheric to urge the fluid into the atmospheric tank, but the low flow resistance of the return line guarantees that this pressure will be close to 0 psig, and it is usually modeled as such, relative to the supply and cylinder pressures, which are much higher.

In the diagram, the servovalve is shown in its "null" position, where the spool exactly covers the ports, preventing any flows to and from the cylinder actuator, so the actuator is at rest. This rest position could be anywhere within the stroke limits of the actuator. If the servovalve spool is now moved to the right, high-pressure oil will enter the left side of the actuator, and drive it to the right, while the right side of the actuator, now open to the return line, exhausts its fluid back to the tank. If the spool displacement x_{sv} is small, the actuator velocity dx_{act}/dt will also be small. If the valve is wide open (full spool stroke is quite small, typically 0.050 inch for a medium-size valve), we get maximum actuator velocity. By properly positioning the spool we

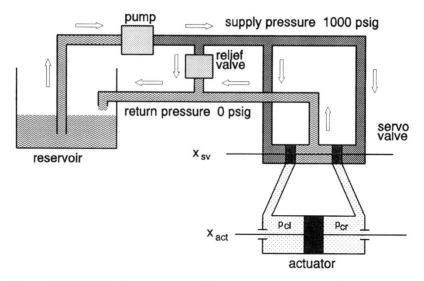

Figure 5-33 Hydraulic servovalve with constant-pressure supply, driving a translational actuator.

can thus smoothly control the actuator speed and position. Note that moving the spool to the left of the null position simply reverses the actuator motion. Also, whenever the actuator gets to the desired location, we can stop it at that point by moving the spool to the null position.

The force required to position the valve spool is quite small, often less than a pound, while the actuator force can be thousands of pounds. The mechanical power (force × velocity) input at the spool is thus much less than the actuator output power, so it is legitimate to call the servovalve a *power amplifier*. Of course, just as in electronic amplifiers, we do not violate conservation of energy. The output power is taken from the power supply, not the spool input power. Many motion control systems are *electro*hydraulic, since we want to use computers and electrical motion sensors, so the positioning of the valve spool is often done with a small electromagnetic actuator. For "large" valves (wide-open flow rate exceeds about 5 gpm), these magnetic actuators are no longer capable of accurately moving the large spool, and *multistage* valves are used. Here the magnetic actuator positions a low-power first-stage device, and a miniaturized hydromechanical servo system positions the main spool. The largest valves (about 1000 gpm) use as many as three stages. Figure 5-34[24] shows a two-stage valve available in seven models, ranging from 1.3 to 31.5 gpm (at 1000-psi supply pressure). The magnetic actuator for the first stage requires about 0.050 amp full scale, and valve frequency response is flat to about 10 Hz.

In later chapters we will study complete systems using servovalves. Here we just want to note that the flow processes at the valve ports are treated as orifices of variable area, with the flow area proportional to the spool displacement x_{sv}. At an

---

[24]Ultra Hydraulics, P.O. Box 30809, Columbus, OH 43230-0809, 614-759-9000.

Basic Energy Converters

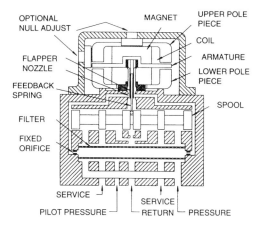

Figure 5-34 Construction details of hydraulic servovalve.

instant when the spool is to the right of the null position, the flows through the two valve ports would be

$$\text{Flow rate into left chamber of cylinder} = Kx_{sv}\sqrt{1000 - p_{cl}}$$
$$\text{Flow rate out of right cylinder chamber} = Kx_{sv}\sqrt{p_{cr} - 0} \quad (5\text{-}53)$$

The cylinder chambers are treated as fluid compliances, taking into account the fluid bulk modulus and the elasticity of chamber walls, tubing or hoses. When a rotary actuator is used in place of a cylinder, leakage across the motor may need to be modeled as fluid resistance. Various nonlinear effects are present and can be linearized for analytical models or treated directly with simulation if greater accuracy is needed.[25]

Just as in the smooth electronic amplifier, servovalve systems have low energy efficiency. In fact, in the system of Fig. 5-33, when the servovalve is in the null position, the output power is zero, while the pump is working at full flow against full pressure and thus exerting full system power, giving *zero* efficiency. Even when the servovalve is active, much power is being wasted in the pressure drop across the valve ports. In high-power systems, this poor efficiency may be economically unacceptable, and other approaches should be explored. Various versions of *pump control*, rather than valve control, can improve efficiency. The basic idea is to *generate* the fluid power only when it is needed. The most direct version of pump control uses a variable-displacement pump driven at constant speed by an electric motor, usually an ac (induction) motor. Such pumps (see Fig. 5-35[26]) have a stroking mechanism which can smoothly adjust the pump's displacement, while the pump is running, between full flow in one direction and full flow in the opposite direction. The pump output flow is sent to a translational or rotary actuator just as in a valve-controlled system (Fig. 5-35 shows a fixed-displacement rotary motor).

---

[25] Doebelin, *Control System Principles and Design*, Wiley, New York, 1985, pp. 95–113.
[26] Sunstrand Corp., Bulletin 9779, Ames, Iowa, 1982.

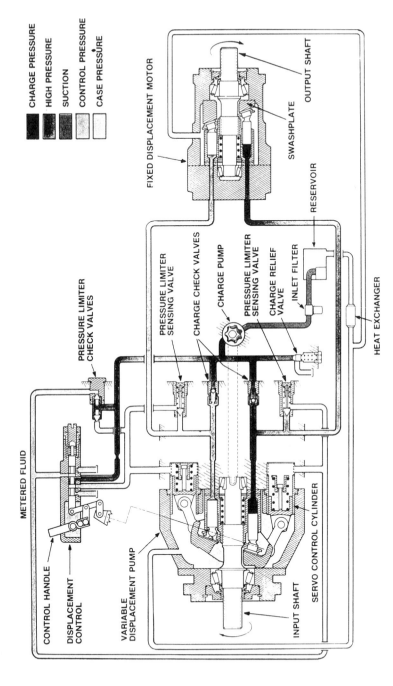

Figure 5-35 Variable-displacement pump driving fixed-displacement rotary motor ("hydrostatic transmission").

Basic Energy Converters

When we require no change in the load position, the pump goes to its neutral (zero displacement) stroke. Even though the electric motor and pump are running at full speed, little power is used since the pump is *not* pumping. Only when we require load motion does the pump stroke mechanism move away from its neutral position and cause pump flow in the desired direction and at the desired rate. While the servovalve "stroking" mechanism requires little power, the pump stroking mechanism involves considerable force and inertia. Usually a magnetic actuator is inadequate, and in fact a small valve-controlled servosystem is often used to stroke the pump. This subsystem is quite low power, so its inefficiency does not have a major effect on overall system efficiency. In Fig. 5-35 the pump displacement is controlled "by hand" with the control handle shown. Motion of this handle requires little human force since a small hydromechanical servo system uses hydraulic pressure in the servo control cylinders to actually stroke the pump.

Ideally the pump's two flow ports would simply be connected to the motor's two flow ports to form a complete system, but certain practical considerations result in the more complex arrangement shown in Fig. 5-35. A small, fixed-displacement pump ("charge pump") is needed to continuously make up system leakage. This pump also provides "supercharging" to prevent the main pump from pulling a vacuum at its inlet port when a large inertial load initially keeps the motor from moving. If low absolute pressure (vacuum) were allowed to occur, the fluid would exhibit *cavitation*, a destructive and noisy effect which limits system life and performance. Various check ("one way flow") and relief valves are also needed to isolate the charge system from the main system and to limit the maximum pressures that might be caused by trying to quickly decelerate a fast-rotating inertia load.

Our final power modulators of this section are *friction brakes and clutches*. When a motion control system requires rapid cycling of a load, servo-controlled motors or open-loop stepping motors are widely used. However, when high power levels and rapid cycling rates are required, modulating power at the motor may not be practical, and clutch/brake systems may be more suitable. Here a nominally constant-speed motor (ac induction motors are usually used) running "unloaded" (near synchronous speed) is connected, upon command, to the stationary load by a friction clutch. The clutch must of course initially slip, but can exert a large accelerating torque on the load, rapidly bringing it up to motor speed. The motor will of course slow down somewhat during this process. When the two shafts achieve the same speed, the clutch stops slipping since the static coefficient of friction is larger than the dynamic (slipping) coefficient. At this point the motor and load accelerate together to the final steady speed, which is dictated by the intersection of the motor speed/torque curve and the "load" speed/torque curve, which includes windage and bearing friction of the motor.

Sometimes loads must also be *de*celerated, and then a friction *brake* is needed. When both actions are necessary, a brake/clutch system is used. If the load must alternately be driven in both directions, then two clutches and one brake will do the job. Figure 5-36[27] shows all four versions of such systems. Whereas "dry" friction clutches and brakes are certainly in use, the referenced manufacturer uses lubricated

---

[27]Force Control Industries, Inc., 3660 Dixie Hwy., Fairfield, OH 45014, 513-868-0900.

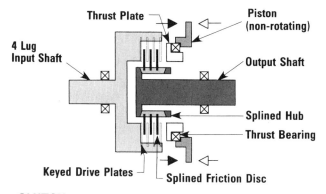

CLUTCH

A clutch consists of multiple rotating steel drive plates keyed to the input shaft and alternating friction discs splined to the hub of the output shaft. Pressure acting on the non-rotating piston exerts clamping pressure on the clutch stack through a thrust bearing and a rotating thrust plate. The clutch is engaged and torque is transmitted from the input to the output shaft.

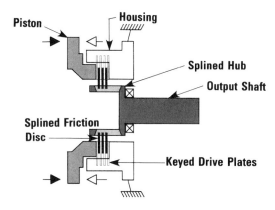

BRAKE

A brake consists of multiple non-rotating steel drive plates keyed to the housing and alternating friction discs splined to the hub of the output shaft. Pressure acting on the non-rotating piston exerts clamping pressure on the brake stack. The brake is engaged and torque is transmitted from the output shaft to the housing.

Figure 5-36 Friction clutches and brakes for motion-control systems.

Basic Energy Converters

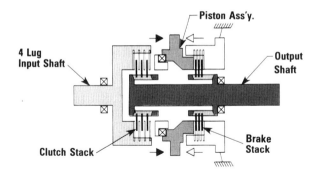

CLUTCH/BRAKE

A clutch/brake is a combination of both a clutch stack and a brake stack operating about a common output shaft. As a centrally located piston assembly is shifted to exert clamping pressure on the clutch stack torque is transmitted from the input shaft to the output shaft. When shifted away from the clutch stack to the brake stack the clutch is automatically released, and braking torque is transmitted to the output shaft. The single-centrally located piston prevents clutch and brake overlap.

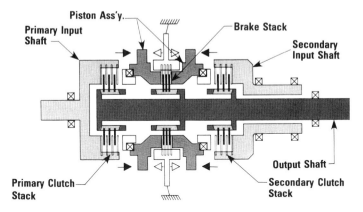

DUAL CLUTCH/SINGLE BRAKE

A dual clutch/brake consists of two clutches and a brake operating about a common output shaft. Two separate pistons are used to exert clamping pressure on either the primary clutch, secondary clutch, or the brake. The primary and secondary input shafts may be driven in a variety of ways to select different speeds or directions as desired.

Figure 5-36 (*Continued*)

330 Chapter 5

friction plates to get smoother and more accurate action, longer wear, and better heat dissipation. The "stacks" of friction plates are usually pressed together by a pneumatic piston, giving the normal force which causes the friction torque. Most systems apply a fixed air pressure to the piston, but the friction torque can also be smoothly varied by applying a varying control air pressure, perhaps using an electropneumatic transducer like that of Fig. 1-1. Even when a fixed pressure is intended, some pneumatic dynamics may need to be modeled since the command signal to the clutch or brake goes to an on-off type of solenoid air valve. The valve takes some time to fully open and then there is further delay for the flow to build up the piston pressure.

EXAMPLE: MOTOR/CLUTCH SYSTEM

We want to model and simulate a simple clutch system which uses the same induction motor we studied earlier (Fig. 5-17), to accelerate a load of inertia and viscous friction. Since there are two rotating shafts, two Newton's laws can be written. The only part of this system which presents some new difficulties is the friction torque of the clutch. This torque acts on both the motor shaft and the load shaft; it tends to slow down the motor and speed up the load. Even though the friction plates are lubricated, the manufacturer states that we should treat the friction torque as essentially constant while any slipping is taking place, just as in simple dry friction. This aspect of the friction model is easily treated. The difficulty arises when the two shafts achieve the same speed and thereafter move together. Now the friction torque is *not* the simple constant value we assume during slipping. Rather, the friction torque "adjusts itself" to provide exactly the amount of torque needed to keep the two shafts moving at the same speed. This torque is *not* constant or known "ahead of time," because it is a manifestation of the well-known behavior of "dry" friction forces *before* any slippage occurs; the friction force can be *any* value less than the maximum available, and adjusts itself to just balance all the other forces.

This "simple" friction model can actually be quite difficult to accurately simulate, as we noted earlier in Sec. 2-6. We will here use a "trick" that seems to work well for this type of application. In the SIMULINK system simulation diagram of Fig. 5-37, the Newton's laws for the two shafts should be easily recognized and related to Fig. 5-18. The inertia on the motor shaft is now that of the motor *alone*, $0.06 \, \text{kg-m}^2$, while that of the load is taken as 0.5. Viscous damping coefficients of motor and load shafts are, respectively, 0.05 and 0.10 N-m/(rad/sec). The motor speed/torque curve is the same as in Fig. 5-18. Initially (before actuating the clutch) the motor is running at the steady speed (1792 rpm) dictated by its speed torque curve and the viscous friction, so the initial condition for integrator2 is set at the rad/sec equivalent of this speed. Integrator3, whose output is the load speed, has its initial condition set at 0 rad/sec. When we want to accelerate the load we electrically activate a solenoid air valve, which opens wide, letting air flow from a fixed supply (say, 100 psi) into the clutch cylinder. The clutch piston air pressure takes some time to build up to 100 psi, so the clutch friction torque also lags. We model all these dynamics between clutch input command (the step function shown) and clutch torque, with a gradual (actually exponential) rise that takes about 0.030 second. The transfer function $1/(0.01s + 1)$ implements this behavior [see Fig. 1-3b and Eq. (1-5) for review of the "first-order behavior"].

Basic Energy Converters

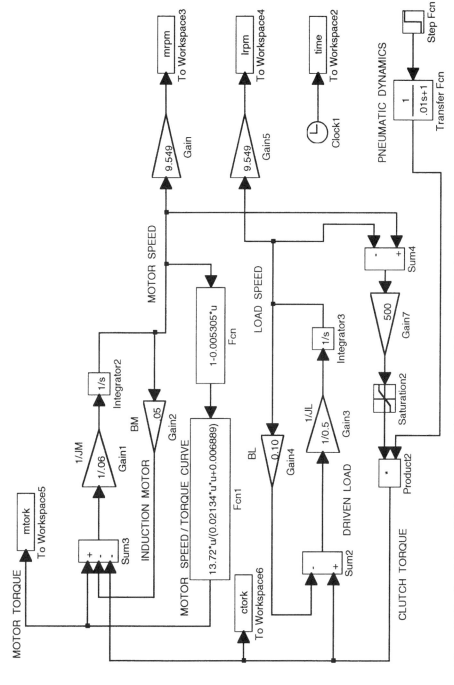

Figure 5-37 Simulation of induction motor/clutch system driving mechanical load.

The clutch friction behavior when changing from the "slipping" condition to the "locked-up" condition is approximated as follows. Even though the friction is physically of the dry (Coulomb) type, we use a viscous model as part of our scheme. That is, in Fig. 5-37, we multiply the speed difference (slipping velocity) between motor and load by the constant 500, just as for a viscous damper between these two shafts. By itself, this would of course be incorrect, since it would make the friction torque proportional to slip velocity, whereas the true dry friction torque is *constant* for all slip velocities other than zero. To correct this error, we send the "viscous" torque signal into a *saturation* effect, available in our simulation software. A saturation effect makes its output proportional to its input, but only until a certain point is reached. When the input goes beyond this point, the output is given a *fixed* value of our choice.

In our example we choose this fixed value to be the dry friction torque value for the slipping condition (700 N-m). Thus, whenever the difference between motor and load speed is large enough, the friction torque is fixed, but when the load speed comes "close" to the motor speed, the friction torque "self-adjusts" to keep the motor and load speeds nearly equal. The number 500 used in Gain7 is not a "magic value"; it must simply be "large enough." That is, in the actual system, when "lockup" occurs, the motor and load speeds are *exactly* the same, since the static friction coefficient is larger than that for slipping. In our approximation, the two speeds *never* become exactly the same, but by making the number 500 very large, they will be very nearly the same. In an actual simulation we can easily adjust this number until the two speeds for the locked-up condition are as close as we wish.

Figure 5-38a shows results using the numerical values of Fig. 5-37. We should note that the numbers chosen give a motor speed drop that is larger than usually used in practice, so as to give visually convenient curves. Motor speed of course starts at its initial steady value of 1792 rpm, with the load at rest. Clutch friction torque rises smoothly to the saturation value (700 N-m) in about 0.03 second, as dictated by the valve opening and air pressure buildup dynamics. (The motor torque and clutch torque curves are plotted as *twice* the true numerical values, again for visual clarity.) Motor speed drops and load speed rises until they become "equal" at about 0.06 second, whereafter the two curves rise together toward a final equilibrium state where the motor torque just balances the two viscous friction torques of motor and load. This final speed will be less than 1792 rpm since the motor now has to balance *both* viscous torques, not just that of the motor.

One of the advantages of clutch-type systems is that we can use clutch torques *larger* than the maximum torque of our motor, thus accelerating the load more rapidly. Such torques are possible since we are using the kinetic energy available in the rapidly rotating motor inertia to create them. In Fig. 5-38a the peak motor torque shown there is actually the maximum this motor can produce (see Fig. 5-17a), while the peak clutch torque is clearly larger. In Fig. 5-38b, all the numbers are the same except we have reduced the clutch friction torque from 700 to 300 N-m. Motor speed now does not drop nearly so much, yet the load accelerates to final speed in about the same time. Using this type of simulation, the design engineer can quickly explore many different variations to find the best combination of system parameters. We can easily add additional features such as computing the instantaneous heating rate, total heat energy, motor and load position angles, etc.

Basic Energy Converters

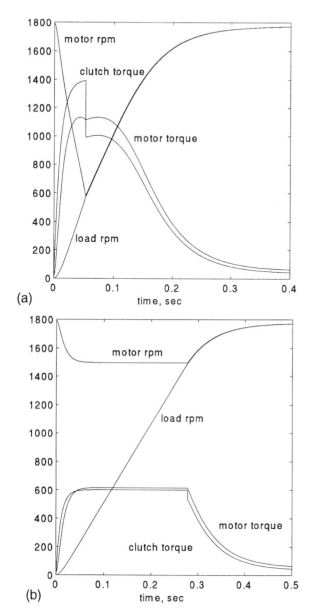

Figure 5-38 Simulation results for motor/clutch/load system.

While clutch/brake systems are often used open-loop, certain applications benefit from closed-loop (feedback) operation and such equipment is available.[28] These applications involve the positioning of a load in a *repetitive* cycle. Examples include conveyors which start and stop at fixed locations, cutoff systems which shear a

---

[28]CLPC Series II Closed-Loop Position Control, Force Control Industries, Inc., [27].

continuous web of material into like-size sheets, and packaging machines which repeat motions over and over. Feedback control of such clutch/brake systems is quite different from the conventional configuration. To position the load at a desired location, the clutch is first engaged, accelerating the load and moving it toward the desired "home" position. Before this position is reached, the clutch is disengaged and the brake engaged, hopefully bringing the load to rest at the desired home position. The instantaneous position of the load is measured with an incremental encoder,[29] a device which produces distinct electrical pulses every so many degrees of motion. For example, the encoder might produce 1000 pulses/revolution. System electronics include also a pulse counter, a keyboard or thumbwheel switches for entering the desired motion as a certain number of pulses, and a comparator to compare the desired pulse count with the actual.

As the load moves toward the desired home position, the encoder pulse count is monitored, and when it reaches a certain value, the brake is engaged. The load will now eventually stop at some location, hopefully close to that desired. *No matter where it stops, the system accepts this position and makes no effort to correct it!* This behavior is very different from the conventional servo system, where, if the position is not correct, the actuator tries to correct it, moving in reverse if necessary, before finally settling into a steady state. To make clutch/brake systems of this type successful, they have to be initially "tuned," by trial and error, so that the correct instant of brake application is found. Also, the system performance specifications will always allow some ± tolerance on the final position. That is, if we command, say, 2346 counts, we will accept a position within ±5 counts of this value. Once such a system has been tuned, we rely on the *repeatability* of all system components and environmental factors to maintain accuracy.

If we did no more than thus far described, it would be incorrect to call this a feedback system; the actual load position is measured but not really *used*. Actually, an important additional operation is included. For every "move," the system records the error and computes a running average of, say, the last five moves. If this average error exceeds that allowed, on the *next* move, the instant of brake engagement is *adjusted* in the direction and amount to bring the system "back on target." Such a scheme will work quite nicely so long as any "drifts" in the system occur rather slowly. The fact that such systems are successfully marketed by several manufacturers testifies to the fact that the scheme is indeed workable.

BIBLIOGRAPHY

Andersen, B. W., *The Analysis and Design of Pneumatic Systems*, Wiley, New York, 1967.
Angrist, S., *Direct Energy Conversion*, Allyn and Bacon, Boston, 1965.
Burwen, R., Stop!, *Motion Control*, March/April 1996, pp. 12–18. An excellent discussion of amplifier/motor dynamic braking.
Jones, D. H., Choosing the right servo amplifier, *Control Engineering*, January 1973, pp. 40–43.

---

[29]Doebelin, *Measurement Systems*, 4th Ed., pp. 300–306.

Basic Energy Converters   **335**

Krause, P. C., and O. Wasynczuk, *Electromechanical Motion Devices*, McGraw-Hill, New York, 1989.

Leonhard, W., *Control of Electrical Drives*, Springer-Verlag, New York, 1985.

Merritt, H. E., *Hydraulic Control Systems*, Wiley, New York, 1967.

Nasar, S. C., and L. E. Unnewehr, *Electromechanics and Electric Machines*, Wiley, New York, 1979.

Shepherd, D. G., *Principles of Turbomachinery*, Macmillan, New York, 1961.

Tal, J., Amplifier operation modes and their effect on servo systems, *Motion*, September/October 1992, pp. 20–23.

White, D. C., and H. H. Woodson, *Electromechanical Energy Conversion*, Wiley, New York, 1959.

PROBLEMS

5-1. In the generator of Fig. 5-2, assume no mechanical or electrical losses ($B \equiv 0$, $R_A \equiv 0$). By equating input mechanical power to output electrical power, show that the generator torque constant and generated voltage constant are numerically equal, when expressed in SI units. (This holds also for motors.)

5-2. Draw and explain a simulation diagram for solving Eq. (5-3) for ω, assuming T_a, i_f, and i_A were given inputs.

5-3. A short-circuited piezoelectric sensor deflects $10\,\mu\text{m}$ when a force of 1000 newtons is applied. When the sensor is clamped and 10 volts are applied, the clamp feels 100 newtons force. Find numerical values for the constants K_s and C_1 in Eq. (5-10).

5-4. The sensor of problem 5-3 has a small steel ball dropped on it, causing it to vibrate at its natural frequency, which is measured to be 50,000 Hz. What is the effective mass of this sensor?

5-5. The pump of Fig. 5-8 produces 10 gpm when driven at 1750 rpm with no back pressure. Find its displacement in in^3/rad. With 1000-psi back pressure, at the same speed, the flow rate drops to 9.5 gpm. Find the pump's leakage flow resistance in psi/(in^3/sec).

5-6. Draw a simulation diagram to solve for ω in Eq. (5-17), assuming T and Δp were given inputs.

5-7. In Fig. 5-11, if the pump were a *perfectly* linear device, sketch the shape of all the curves shown. Explain.

5-8. Assuming e_A and i_f given, draw a simulation diagram to solve for ω from Eqs. (5-34) and (5-36).

5-9. For the motor of Eq. 5-44, compute and plot a curve of horsepower versus speed.

5-10. A centrifugal pump and associated piping system has a speed/torque curve given by

$$\text{torque (N-m)} = 0.0001333\,(\text{rpm})^2$$

If this pump system is driven by the motor of Fig. 5-17a, at what speed will the pump/motor system run?

336   **Chapter 5**

5-11. Modify (and explain) the simulation diagram of Fig. 5-21 for the case where the windings are driven by a voltage source. The winding circuit has R, L, and a back emf proportional to motor velocity, all in series.

5-12. Stepping motors used for constant-velocity applications are often brought up to speed *gradually*, so as to not lose any steps. One such method commands a constant acceleration until the desired steady speed is reached. At this point we switch to a constant-velocity command. We want to accelerate a 200 steps/rev motor to 10 rev/sec in 2 seconds. Modify Fig. 5-21 to implement this scheme.

5-13. Explain how the curves of Fig. 5-17 would be affected if the amplitude of the motor sinusoidal voltage was changed.

5-14. Using any simulation software available to you, modify the diagram of Fig. 5-18 to replace the pulse load torque with a sinusoidal one given by $25\sin(31.4t)$, N-m, t in seconds. Run your simulation and prepare a graph similar to Fig. 5-19.

5-15. For the amplifier of Fig. 5-24a, compute and graph the sinusoidal transfer function $(e/e_{ai})(i\omega)$, both amplitude ratio and phase angle, if the amplifier output resistance is 100 ohms and the piezo element capacitance is 5 µF. If we require the amplitude ratio to be within 5% of flat, what is the highest frequency we can use? If we want to extend this range of frequencies and not use a different piezo element, what should we look for in purchasing a new amplifier?

5-16. The piezoelectric actuator referenced in the text has a natural frequency of 2000 Hz and a spring stiffness of 4×10^7 N/m. Compute its effect mass in kilograms. If we attach this actuator to a 5-kg load mass, what will now be the natural frequency?

5-17. Using simulation software available to you, reproduce the simulation of Fig. 5-29 and explore the following variations.

   a. Increase the circuit inductance by 5 times. [This makes the RL circuit transfer function $0.5/(0.005s + 1)$.]
   b. Make the change of part (a) and also make the triangle wave frequency 1000 Hz.

5-18. Carefully explain the operation of the simulation of Fig. 5-31, using sketches of the waveforms of all the signals.

5-19. Using simulation software available to you, reproduce the simulation of Fig. 5-31 and add a series RL circuit driven by the amplifier output voltage. Let the resistance be 5 ohms and explore the effect of various inductance values.

5-20. Modify the simulation diagram of Fig. 5-37 to include a brake system which can be actuated after the load is up to speed and the clutch is disengaged. This simulation must also compute the load position, not just the speed.

5-21. Modify the simulation diagram of Fig. 5-37 to include calculation of the instantaneous frictional heating rate.

6

SOLUTION METHODS FOR DIFFERENTIAL EQUATIONS

6-1 INTRODUCTION

Application of lumped-parameter models to dynamic analysis of physical systems leads to a system description in terms of ordinary differential equations. Except for very simple systems, the description will be a set of several (perhaps many) *simultaneous* differential equations, as many equations as there are system unknowns. These equations can be solved directly, to find system behavior, by three general methods:

1. Analytically
2. Using an analog computer (now largely obsolete)
3. Using digital simulation (numerical analysis)

Analytical solutions are largely limited to linear equations with constant coefficients, while approximate computer methods (analog or digital) handle all types of equations, including nonlinear and linear with time-varying coefficients.

We prefer analytical methods since they produce *formulas* for the solution, whereas computer methods produce only tables of numbers, or graphs. Formula solutions show directly how system parameters affect system response. This information is most useful for design purposes, where we are trying to find the best combination of system parameters. Computer solutions can provide similar information but only with repeated trials, using different combinations of parameters. When there are many parameters, such "search" methods can become expensive. A common approach is to first *linearize* our equations, to allow analytical solution, even though we get approximate results. Using the formula solutions to help us arrive at an optimum set of parameters, we "rough out" a trial design. Using these trial parameter values, we go to digital simulation, including now some or all of the nonlinear or time-varying-coefficient effects. In a complex system we may add one nonlinearity at a time, so that we appreciate its effect without the confusing presence of others. Because the superposition theorem of linear differential equations does not apply here, we finally need to include *all* the significant nonlinearities. That is, the effect of one nonlinearity may be *changed* by the simultaneous presence of another.

337

We often thus use a hierarchy of system models, starting with overly simple ones which give a quick but rough understanding through analytical solutions. The model is gradually embellished with more and more realistic features, trying to build our understanding in comprehensible stages. Sometimes we will have to also do some lab testing of part or all of the actual hardware to verify a model or get a numerical value for a parameter. This lab testing may be required at several stages of the development. As computer simulation has improved in comprehensiveness and accuracy, the need for lab testing has become less, and it may be pushed later and later in the design cycle. It is doubtful, however, that we will ever be able to completely forgo this "proof of the pudding." How much lab testing to use, and when in the cycle, is a vital decision that requires experience in theory, simulation, and experimentation. Too much reliance on computer predictions can lead to technical and economic disasters.

We will begin with analytical solutions, showing two alternative methods, the classical operator method and the Laplace transform method. After this, simulation methods of various types will be discussed and illustrated. Some simulation software, such as the SIMULINK which we have already used several times, requires that the engineer physically analyze the system and actually write out the describing differential equations. Other software completely avoids these steps and requires only that the user "draw a picture" of the system. The software then itself generates the describing equations and solves them numerically. While this latter method can be very efficient in the hands of an experienced engineer who has previously done a lot of equation writing, it can be dangerous for beginners (such as students) who lack the experience needed to evaluate the validity of the solutions provided. Our emphasis in this text is thus on software which requires you to write the equations yourself, since I feel that this is a necessary prelude to moving on to the more "automated" types of software. We will, however, give enough description of the more advanced methods that you can appreciate their operation.

6-2 ANALYTICAL SOLUTION OF LINEAR, CONSTANT-COEFFICIENT EQUATIONS: THE CLASSICAL OPERATOR METHOD

While certain nonlinear and variable-coefficient linear differential equations have closed-form analytical solutions, the majority of these equations which arise in engineering practice have no such analytical solution and yield only to approximative computer methods. Only for linear equations with constant coefficients do general solution techniques exist which "always work." We will now briefly review the method usually called the *classical operator method*. Later we will also show the *Laplace transform method*, an alternative approach which solves the same class of equations. While the two methods both solve the same equations, it is useful to know both since each has certain advantages and drawbacks.

If we look at the physical laws used to generate lumped-parameter models of mechanical, electrical, fluid, and thermal systems we see that the highest derivatives involved are the second derivatives. Newton's law, for example, always has the second derivative of displacement on the right-hand side, and the forces on the

Solution Methods for Differential Equations   **339**

left-hand side generally involve derivatives no higher than the first. Our general solution method, however, deals with an equation having the nth derivative as its highest, where n can be any value. The reason we must deal with the higher derivatives, even though they do not appear in the original physical laws, is that we usually have as our system model a *set of several simultaneous* equations. The solution methods *combine* these several equations into a single system equation, and in this process, higher derivatives appear and must be dealt with.

The general form of equation which we treat is thus of the form

$$a_n \frac{d^n q_o}{dt^n} + a_{n-1} \frac{d^{n-1} q_o}{dt^{n-1}} + \cdots + a_1 \frac{dq_o}{dt} + a_0 q_o = b_m \frac{d^m q_i}{dt^m} + \cdots + b_1 \frac{dq_i}{dt} + b_0 q_i$$

$$(6\text{-}1)$$

where

$$a\text{'s and } b\text{'s} \overset{\Delta}{=} \text{physical parameters, assumed constant}$$
$$q_o \overset{\Delta}{=} \text{output quantity of a system, the unknown}$$
$$q_i \overset{\Delta}{=} \text{input quantity of system}$$
$$t \overset{\Delta}{=} \text{time}$$

To actually solve such an equation, the input quantity must be given as a known function of time, whereupon the differentiations on the right-hand side of Eq. (6-1) can be carried out, making the entire right-hand side itself a known function of time, call it $f(t)$. The classical operator method of solution is a "three-step" process:

1. Find the complementary solution, called q_{oc}.
2. Find the particular solution, called q_{op}.
3. Add q_{oc} to q_{op} to get the total solution: $q_o = q_{oc} + q_{op}$. Now apply the initial conditions to find the constants of integration, and thereby the final and complete solution.

Steps 1 and 2 can be done in either sequence; step 3, of course, must be done last. We will shortly show a method for finding the complementary solution which *always* works. For the particular solution, *no* method which always works exists, and several methods which sometimes work are known. For some $f(t)$'s, *no* solution method is known. Fortunately, one method, called the method of undetermined coefficients, works for most cases of practical interest.

To find the complementary solution we first write our equation in operator notation:

$$(a_n D^n + a_{n-1} D^{n-1} + \cdots + a_1 D + a_0)q_o = f(t) \tag{6-2}$$

The *system characteristic equation* is defined by setting the terms inside the parentheses equal to zero:

$$a_n D^n + a_{n-1} D^{n-1} + \cdots + a_1 D + a_0 = 0 \tag{6-3}$$

We now treat this equation as if it were an algebraic equation in the unknown D, even though D is really an operator, not a number. (When we later show the more rigorous Laplace transform method, the characteristic equation *really will be* an algebraic equation.) We must now solve the algebraic characteristic equation for its roots, of which there will be n. For example, if we had

340   **Chapter 6**

$$3.2D^2 + 8.5D + 1.6 = 0 \qquad (6\text{-}4)$$

we would quickly find from the quadratic formula the two roots -0.20 and -2.5. In general we have n roots; let's call them $s_1, s_2, \ldots, s_n$. While "everybody" knows the formula for finding the two roots of quadratic equations, you may not be aware that formulas for getting the exact roots of cubic and quartic equations are also available. However these formulas are so complicated that we rarely use them, even when they are available in math software such as MATHCAD. When n is greater than 4, it has long ago been proven that formulas for the roots are not possible. This means that it is *impossible* to carry out the solution of the differential equation *in letter form* when n is greater than 4. In fact, because the known formulas are so complicated, we generally don't try to get letter-form solutions for cubic or quartic equations.

Thus, for n greater than 2, we almost always will have to work with *numerical* values of system parameters rather than letter values. This is undesirable because in design studies, we would like to have explicit formulas which show how each system parameter affects system behavior. Since the values of the characteristic equation roots *must* be found to proceed with the solution, we often need to insert specific numerical parameter values and then use a numerical *root finder* to get approximate values for the roots. When using such a root finder you should always remember that it is an *algorithm*, not a precise formula like the quadratic formula, and may give incorrect values. Fortunately, good root finders usually give accurate values, and we can always check them by substituting the answer back into the equation.

Since we have been using SIMULINK for our simulations, we will also use the root finder that is in MATLAB, since SIMULINK is part of MATLAB. Root finders are of course available in much other mathematical software.

EXAMPLE: ROOT FINDING

Suppose we have the system characteristic equation

$$0.005D^6 + 0.45D^5 + 10.75D^4 + 105.5D^3 + 893.0D^2 + 4700.0D + 800.0 = 0$$
$$(6\text{-}5)$$

This equation is for a system of three moving masses (three second-order equations lead to one sixth-order equation). All the masses, springs, and dampers in this system *had* to be given numerical values. It is *impossible* to solve such a problem with letter values. Once we assign numerical values, the root finder makes our job quick and "painless." For the MATLAB root finder we would just write

```
C=[.005  .45  10.75  105.5  893.3  4700.0  800.0]
S=roots(C)
```

and the six roots appear "immediately";

```
-21.023  -8.8661  -0.17598  -58.686
-0.62448+i9.0956  -0.62448-i9.0956
```

In any problem, once we have the roots (either in letter form or number form), we *immediately* write down the complementary solution, using a set of rules proven

Solution Methods for Differential Equations
341

in courses on differential equations, but which we here just accept "on faith." If, in a math course, you were taken through some intermediate steps at this point, I urge you to now ignore these needlessly time-consuming operations. That is, as engineers, when a mathematical process is nothing but a routine (dare I say cookbook!), we treat it as such and spend our time on more creative aspects of our work.

We now quote, without proof, the rules needed to write out the complementary solution directly from the list of roots. Four cases are possible and must be treated. We can get real roots, and they can be single roots or repeated any number of times. We need a rule for each situation. It can be proven that if the system parameters (spring constants, resistances, etc.) are all real (not complex) numbers, then if any complex roots appear, they *always* appear in pairs of the form $a \pm ib$. Such root pairs can be repeated or unrepeated, again requiring two rules to cover these two cases. The needed rules are:

1. For any unrepeated real root s, the solution is Ce^{st}, where C is a constant of integration which cannot be found as a number until step 3 of the solution process.

2. For any unrepeated complex root pairs of form $a \pm ib$ the solution is $Ce^{at} \sin(bt + \phi)$, where C and ϕ are constants of integration to be found later.

3. For any repeated root, for example, -1.5, -1.5, the solution is $C_1 e^{-1.5t} + C_2 t e^{-1.5t}$. For a "triple repeat," say, -1.5, -1.5, -1.5, the solution would be $C_1 e^{-1.5t} + C_2 t e^{-1.5t} + C_3 t^2 e^{-1.5t}$. These two examples establish a pattern which you can follow for roots repeated *any* number of times.

4. For any repeated complex root pair, say, $-3 \pm i4$, $-3 \pm i4$, the solution would be $C_1 e^{-3t} \sin(4t + \phi_1) + C_2 t e^{-3t} \sin(4t + \phi_2)$. If the root pair should be repeated more times, use the same pattern shown in the rule for repeated real roots.

These four rules cover *all* the possibilities; no other rules are needed. As soon as you have a list of the roots for any characteristic equation, write out the complete complementary solution *immediately*; *no* intermediate steps are needed. For our example above, we thus can write out the complementary solution as:

$$q_{oc} = C_1 e^{-21.023t} + C_2 e^{-8.666t} + C_3 e^{-0.17598t} + C_4 e^{-58.686t}$$
$$+ C_5 e^{-0.62448t} \sin(9.0956t + \phi) \tag{6-5a}$$

Since the complementary solution procedure "throws away" the right-hand side $f(t)$ of the differential equation, this right-hand side, often called the *forcing function* of the system, must somewhere be taken into account. The particular solution fills this need. Since the particular solution will depend on the nature of $f(t)$, a mathematician can always concoct a sufficiently "pathological" $f(t)$ to thwart any solution method that we might come up with. This is the reason that no *general* method which always works is available. Fortunately several methods are available which work for most $f(t)$'s that engineers encounter, and of these, one method meets most of our needs. This method is usually called the *method of undetermined coefficients*, and we now present it as the only method explained in this book.

When we are dealing with a method which sometimes works and sometimes doesn't, it is helpful to begin with a test to see whether it will work or not, so we

342   **Chapter 6**

don't waste our time. Such a test is available and goes as follows. We must first differentiate, with respect to the independent variable (time), the $f(t)$ on the right-hand side of our equation, over and over again. Our particular solution method will work if one of two things happens:

1. Successive time derivatives eventually all become zero. For example, if the right-hand side were $6t^2 + 5t$, after three derivatives have been taken, all subsequent ones are clearly zero.
2. Successive time derivatives repeat themselves. For example, differentiating $5 \sin(10t)$ will never give anything except $\sin(10t)$ and $\cos(10t)$ functions (the numbers "in front of" the functions don't matter).

If neither case 1 nor 2 occurs (successive derivatives keep producing *new* kinds of functions), then this method will not work.

　　If we apply this test and find that the method works, we proceed as follows. The particular solution is written out as a sum of terms, each term corresponding to one of the different forms of time function occurring in the right-hand side itself and all its successive derivatives. Each of these terms is multiplied by an "undetermined coefficient," whose value we can *now* find, we *don't* have to wait until the third step as we did with the constants in the complementary solution. To find the values of these coefficients, and thus the particular solution, substitute the assumed particular solution into the differential equation and require it to be an identity. That is, gather terms of like form on the left- and right-hand sides and require that their coefficients be identical. This procedure will always generate as many algebraic equations as there are coefficients. This set of equations can then be solved to find all the coefficients, which of course gives us the complete particular solution. Let's now do an example which illustrates the entire three-step procedure.

EXAMPLE: COMPLETE SOLUTION

Suppose we have a differential equation and initial conditions

$$\frac{d^2x}{dt^2} + 3\frac{dx}{dt} + 2x = 4e^{-5t} \qquad \text{at } t = 0^+, \frac{dx}{dt} = 0, \; x = 2.0 \qquad (6\text{-}6)$$

The symbol $t = 0^+$ refers to a time an infinitesimal amount after the time $t = 0$, and is the time at which "initial" conditions *must* be evaluated when using the classical operator method of solution. This subtlety may not have been emphasized in your math course, but it is vital in certain types of problems, so be sure to take note of it. We generally take $t = 0$ as the beginning of any inputs to our systems, so $t = 0^+$ means a time just *after* the input has been applied. In many systems, the distinction between times just before and times just after the input is applied have no consequence (the conditions are the same either way), but in some systems this is *not* true and we *must* define "initial" as $t = 0^+$ to avoid incorrect answers. When we later show you the Laplace transform method, there "initial" always means just *before* the input is applied. Both solution methods give exactly the same answers, but the word "initial" must be interpreted correctly for each.

　　We can always get either the complementary or particular solution first; let's here get the complementary, which means we need the roots of the system characteristic equation

Solution Methods for Differential Equations   **343**

$$D^2 + 3D + 2 = 0 \tag{6-7}$$

The quadratic formula quickly gets us the roots $s_1 = -2$, $s_2 = -1$ and the complementary solution $x_c = C_1 e^{-2t} + C_2 e^{-t}$. Repeated differentiation of the forcing function $4e^{-5t}$ gives only terms of the form Ae^{-5t}, so the method of undetermined coefficients will work. The solution $x_p = Ae^{-5t}$ is substituted into Eq. (6-6) to give

$$25Ae^{-5t} - 15Ae^{-5t} + 2Ae^{-5t} \equiv 4e^{-5t} \tag{6-8}$$

$$12Ae^{-5t} \equiv 4e^{-5t}$$

$$A = \tfrac{1}{3}$$

The complete solution is thus

$$x = x_c + x_p = C_1 e^{-2t} + C_2 e^{-t} + \tfrac{1}{3} e^{-5t} \tag{6-9}$$

To find C_1 and C_2 we apply the initial conditions.

$$x(0^+) = 2 = C_1 + C_2 + \tfrac{1}{3} \tag{6-10}$$

$$\dot{x}(0^+) = -2C_1 - C_2 - \tfrac{5}{3} \tag{6-11}$$

$$C_1 = -\tfrac{10}{3} \qquad C_2 = 5$$

The complete specific solution for the given initial conditions is thus

$$x = -\tfrac{10}{3} e^{-2t} + 5e^{-t} + \tfrac{1}{3} e^{-5t} \tag{6-12}$$

One nice feature of the topic of differential equations is that *there is never any reason to accept an incorrect answer*. If the solution, such as Eq. (6-12), is substituted into the original equation (6-6) and makes it an identity, and if it satisfies the initial conditions, then it *must* be the one and only correct solution.

The above simple routines will enable you to solve any ordinary linear differential equation with constant coefficients irrespective of its order (the order of its highest derivative), as long as $f(t)$ can be handled by the method of undetermined coefficients. Two special cases which occur rarely, but should be mentioned, require a slightly modified procedure. If a term in x_p has the *same* functional form as one in x_c, the term in x_p should be multiplied by the lowest power of t which will make it different from all the x_c terms associated with the root which produced the x_c term. For example, if the right-hand side of Eq. (6-6) had been $4e^{-t}$, then x_p would have had the form Ae^{-t}, the same as $C_2 e^{-t}$ in x_c. We should thus modify x_p to be Ate^{-t} before finding A. If in addition the left-hand side of Eq. (6-6) had been $D^2 + 2D + 1$, with roots $s_1 = s_2 = -1$ and $x_c = C_1 e^{-t} + C_2 t e^{-t}$, then x_p would have to be modified to $At^2 e^{-t}$. The second special case arises if the characteristic equation has the form

$$D^m(a_n D^{n-m} + a_{n-1}D^{n-m-1} + \cdots + a_{m+1}D + a_m) = 0 \qquad a_m \neq 0 \tag{6-13}$$

When writing x_p for such a situation we must include, in addition to the usual terms, terms in the first, second, . . . , mth integrals of $f(t)$.

We should at this point mention the important *principle of superposition*, which applies only to linear differential equations. If the driving function $f(t)$ in Eq. (6-2) is

344   **Chapter 6**

composed of a sum of terms, $f_1(t)$, $f_2(t)$, etc., this principle allows us to find the particular solution x_p for each term of the driving function *separately*, and then get the total x_p by simply adding all the individual solutions. In addition to its direct mathematical utility in getting equation solutions, this principle also has two important general consequences relative to the behavior of linear systems. The first might be called the "amplitude insensitivity" of linear systems. To explain this we first need to note that, except for systems which have some *zero* roots for their characteristic equation, and *unstable* systems (some roots with positive real parts) which are rarely useful, the complementary solution generally goes to zero as time passes. This is because the negative exponentials (Ce^{-at}) we usually see here clearly go to zero. We thus usually will call the complementary solution a "transient" solution (dies away) and the particular solution the "steady-state" solution (remains after transient disappears).

By amplitude insensitivity, we mean that if we have found the steady-state solution for a driving function, say $4e^{-5t}$, if we scale up this driving function to $8e^{-5t}$ or scale it down to $2e^{-5t}$, the steady-state response (particular solution) will similarly scale up or down. "Nothing new" is thus found out about the response of linear systems by changing the *size* of the driving inputs; as long as the *form* of the input remains the same, the steady-state responses are directly proportional. This follows from the superposition principle by noting that $8e^{-5t}$ can be written as $(4e^{-5t} + 4e^{-5t})$; thus the x_p for $8e^{-5t}$ is just twice that for $4e^{-5t}$. Such statements *cannot* be made for nonlinear systems; the response to an input of doubled size may be *entirely different in form* from the response to the original input. This fact is used in lab testing to define ranges of inputs for which a system is essentially linear (remember *no* real system can be exactly linear). We apply inputs of a given form, say step inputs, and increase the size. When the system response to a "large" input is no longer nearly proportional to that for a "small" input, we have found the amplitude limits of "linearity" for that system. (This statement applies to nonlinearities in the form of smooth curves. Some nonlinearities, such as a dead space or lost motion in a mechanism, effect *small* motions more than large ones, and the system's linear range may then be limited also on the low side.)

The second general consequence of superposition is that if we know how a system responds to each of two different inputs when they are separately applied, then there will be no "surprises" when they are *simultaneously* applied. That is, the behavior for the combined inputs is just the sum of the responses to the individual inputs. Again, nonlinear systems do not behave so simply; the response to a combination of inputs may show features found in *none* of the individual responses, so knowledge of such individual responses is no guarantee against "surprises" when they occur combined. In nonlinear systems which go unstable, for instance, the system may be stable for each input applied separately, but unstable when they are applied together.

6-3 SIMULTANEOUS EQUATIONS

As we mentioned earlier, high-order differential equations do not arise directly from physical laws but rather are the result of mathematical processes applied to sets of simultaneous equations of low order. A physical system need not be very complex

Solution Methods for Differential Equations

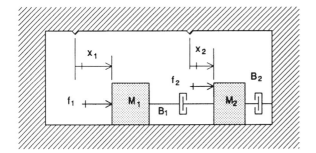

Figure 6-1 Two-mass system for simultaneous-equation example.

for its description to require several simultaneous equations, thus we definitely need to know how to deal with them. Let's use the mechanical system of Fig. 6-1 to develop the general solution technique for any set of simultaneous equations. We assume that the two input forces $f_1(t)$ and $f_2(t)$ will be given as specific functions of time and that we wish to solve for the two velocities $v_1(t)$ and $v_2(t)$. When a mechanical system has any number of moving masses, we simply write a Newton's law for each one, thereby generating a set of simultaneous equations, one for each mass.

$$f_1 - B_1(v_1 - v_2) = M_1 \frac{dv_1}{dt}$$
$$f_2 + B_1(v_1 - v_2) - B_2 v_2 = M_2 \frac{dv_2}{dt} \tag{6-14}$$

Neither of these equations can be solved by itself since each contains *both* unknowns; however, the pair can be solved simultaneously. We now develop a general method which can solve simultaneous sets with any number of equations and unknowns. Using the classical operator approach (we will shortly show the Laplace transform method as an alternative), the first step is to rewrite the equations using our *D*-operator notation. It can be shown, but we do not here prove, that, once written in operator form, the equations may be treated as a set of simultaneous *algebraic* equations, and any method you learned for solving sets of linear algebraic equations can now be applied to the operator equations. Such methods include substitution and elimination, but the most systematic is the use of determinants. Determinants are almost necessary when dealing with more than three equations in three unknowns, and also can be nicely computerized. Let's now rewrite our equations using *D* operators and also use a systematic row-and-column arrangement to prepare for determinant operations.

$$(M_1 D + B_1)v_1 \quad + (-B_1)v_2 = f_1$$
$$+(-B_1)v_1 \quad + (M_2 D + (B_1 + B_2))v_2 = f_2 \tag{6-15}$$

Note that we have forced all the unknowns into columns on the left-hand side and moved all the inputs (driving functions) to the right-hand side. Each equation has its own row, but the order of these rows is not material. That is, we could have made our first row the second and vice versa, and the solution process would not change. For a simple set of equations, substitution and elimination may be quicker than determinants and you should try this approach and compare your results to those we now obtain by determinants. Using the methods of linear algebra (Cramer's rule and determinants) we can reduce the set of two equations in two unknowns to single equations, each involving only one unknown; we are therefore ready for the solution procedure reviewed in Sec. 6-2.

$$v_1 = \frac{\begin{vmatrix} f_1 & -B_1 \\ f_2 & M_2D + (B_1 + B_2) \end{vmatrix}}{\begin{vmatrix} M_1D + B_1 & -B_1 \\ -B_1 & M_2D + (B_1 + B_2) \end{vmatrix}} \tag{6-16}$$

$$v_2 = \frac{\begin{vmatrix} M_1D + B_1 & f_1 \\ -B_1 & f_2 \end{vmatrix}}{\begin{vmatrix} M_1D + B_1 & -B_1 \\ -B_1 & M_2D + (B_1 + B_2) \end{vmatrix}} \tag{6-17}$$

Recall that each of the unknowns is equal to the ratio of two determinants. The denominator determinant is formed from the n by n array of coefficients on the left-hand side and need be computed only once; it is the same for all the unknowns. The numerator determinant is different for each unknown and is formed by replacing the column of coefficients associated with that unknown by the column which is the right-hand side.

At this point you need to recall how to expand a determinant. One popular method uses the known procedure for a 2 by 2 determinant plus a rule (called expansion by minors) for reducing an n by n determinant to a combination of 2 by 2's. We should also point out that if you have a symbolic processor as part of your math software, this software will expand determinants *in letter form*, which is our present problem. For example, MATHCAD has a subset of MAPLE, which provides this determinant expansion capability. Proceeding "by hand" and recalling the rules for expanding 2 by 2 determinants, that is,

$$\begin{vmatrix} a & b \\ c & d \end{vmatrix} = ad - bc$$

we get

$$v_1 = \frac{(M_2D + (B_1 + B_2))f_1 + (B_1)f_2}{M_1M_2D^2 + (M_1(B_1 + B_2) + B_1M_2)D + B_1B_2} \tag{6-18}$$

$$v_2 = \frac{(B_1)f_1 + (M_1D + B_1)f_2}{M_1M_2D^2 + (M_1(B_1 + B_2) + B_1M_2) + B_1B_2} \tag{6-19}$$

We now "cross multiply" to get the single differential equations for each unknown.

$$(M_1M_2D^2 + (M_1(B_1 + B_2) + B_1M_2)D + B_1B_2)v_1 = (M_2D + (b_1 + b_2))f_1$$
$$+ (B_1)f_2 \tag{6-20}$$

$$(M_1M_2D^2 + (M_1(B_1 + B_2) + B_1M_2)D + B_1B_2)v_2 = (B_1)f_1 + (M_1D + B_1)f_2 \tag{6-21}$$

If forces f_1 and f_2 were now given as explicit functions of time, and if all needed initial conditions were known, we could get complete solutions for the unknown velocities.

While the quadratic formula would allow us to proceed in letter form, we leave this as an exercise and instead choose some specific numerical values. Let's assume that at time $= 0$ both masses have zero displacement and zero velocity. Let force f_1

Solution Methods for Differential Equations 347

be zero and let force f_2 be a step input of 1.0 pound that occurs just an instant after $t = 0$. Take both masses to be 1.0 slug and the damper coefficients to each be $1.0\,lb_f/$ (ft/sec). Suppose we are only interested in the velocity of mass 1. Equation (6-20) then becomes

$$(D^2 + 3D + 1)v_1 = 1.0 \tag{6-22}$$

$$\text{Roots} = -0.382 \text{ and } -2.618$$

$$v_{1c} = C_1 e^{-0.382t} + C_2 e^{-2.618t}$$

$$v_{1p} = 1.0$$

$$v_1 = C_1 e^{-0.382t} + C_2 e^{-2.618t} + 1.0 \tag{6-23}$$

We need two initial conditions to find the numerical values for the constants of integration. Recalling that "initial" means just an instant *after* the step force is applied, we need to know the velocity and acceleration of mass 1 at this time. We see that the two initial conditions given earlier are *not* the ones we now need. The velocity and displacement were known just *before* the input was applied and now we see that we need the velocity and acceleration just *after* the force is applied. The reason we need an initial value of acceleration is that our solution procedure has taken two first-order equations and reduced them to a single *second*-order equation. Since in general the needed initial conditions are on the unknown itself and all its derivatives up to the $(n-1)$st, we need an initial velocity and acceleration. If we had combined three second-order equations into one sixth-order, we would need initial values for derivatives up to the fifth. This need for initial values of high-order derivatives is one of the drawbacks of the classical operator solution method relative to the Laplace transform, which does not have this feature. These initial values can always be found, but it is extra work.

We now find the needed values just after the force is applied. If the velocity of mass 1 was given as zero *before* the force is applied to mass 2, we now show that this velocity will *still* be zero just after the force is applied. If we consider mass 2, its acceleration *will* suddenly jump up when the 1-pound force is applied to it; the acceleration just before and just after the input is applied *are* different. This acceleration is, however, finite (actually $A = F/M = 1/1 = 1.0\,ft/sec^2$), and a finite acceleration cannot cause a finite change in velocity over the infinitesimal time from $t = 0$ to $t = 0^+$. This means that the velocity of mass 2 is *still* zero at $t = 0^+$, and thus the force of damper 1 on mass 1 will be zero at this time. Since this damper force is the *only* force on mass 1, the acceleration of mass 1 at $t = 0^+$ must be zero. Our two "initial" ($t = 0^+$) conditions are thus both zero. Using these in Eq. (6-23) we find

$$v_1 = -1.171 e^{-0.382t} + 0.171 e^{-2.618t} + 1.0 \tag{6-24}$$

We could similarly solve for velocity v_2, but note that there both initial conditions will *not* be zero.

When a system involves several inputs and several outputs, we can define and use several transfer functions. In Eq. (6-18) the superposition principle lets us set $f_2 = 0$ and get the transfer function

$$\frac{v_1}{f_1}(D) = \frac{M_2 D + B_1 + B_2}{M_1 M_2 D^2 + (M_1 B_1 + M_1 B_2 + M_2 B_1)D + B_1 B_2} \quad (6\text{-}25)$$

or, alternatively, set $f_1 = 0$ to get

$$\frac{v_1}{f_2}(D) = \frac{B_1}{M_1 M_2 D^2 + (M_1 B_1 + M_1 B_2 + M_2 B_1)D + B_1 B_2} \quad (6\text{-}26)$$

We could get two more transfer functions from Eq. (6-19) in similar fashion. Superposition has allowed us to break up the total system response into its two parts, one due to force 1 and the other due to force 2. When we draw our usual block diagrams, superposition requires that we now "put these two pieces together again," as in Fig. 6-2.

The fact that $(v_1/f_2)(D) \equiv (v_2/f_1)(D)$ is not a coincidence, but rather an example of a general relation called the *reciprocity theorem of linear systems*. Its physical interpretation is usually amazing when we are first made aware of it. It says that if we push on mass 1 with a time-varying force of any kind whatever, thus causing mass 2 to have a definite motion, and if we then apply the same force to mass 2, mass 1 will have the same motions as did mass 2! (This holds for systems with any number of masses. If we push on mass 37 and observe mass 95, we get the same motion as when we push on mass 95 and observe mass 37.) While perhaps physically surprising, the mathematical interpretation is quite clear. In Eqs. (6-15), the coefficient array on the left side may be interpreted as a matrix, and systems which obey the reciprocity theorem have a special kind of matrix called a *symmetric matrix*, that is, $a_{ij} = a_{ji}$, in our case, $-B_1 = -B_1$. The reciprocity theorem is not just a mathematical oddity, but rather has useful applications. When we analyze many physical systems we make assumptions that lead to symmetric matrices. To check whether the real system behaves closely to our model predictions, we will often run tests in which we apply force at, say, location 12 and observe motion at location 35, and then reverse the locations of forcing and motion measuring. If the two measured motions nearly agree, it is a verification of our model. Reciprocity is of course not limited to mechanical systems but applies to any model which leads to symmetric matrices.

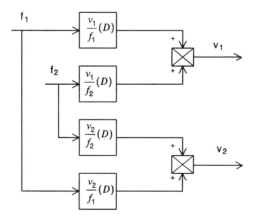

Figure 6-2 Block diagram showing superposition in system with two inputs and two outputs.

Solution Methods for Differential Equations 349

We have used an example which requires only the ability to expand 2 by 2 determinants, since most readers of this book will already have this skill. For those who may not recall how to deal with the general case (n by n), we now give a quick review. Remember, of course, that if you have symbolic processor capability in your math software, you may be able to avoid this work. The method of *expansion by minors* is a systematic way of expanding large determinants, but actually a 3 by 3 example should be sufficient to explain the general method. In Fig. 6-3 we can choose any row or any column for our expansion; if some rows and columns have zeros, you may save some work by choosing one of these. Having chosen a particular row or column, we work our way entirely through that row or column, as follows. First we must attach a plus or minus sign to every cell in the array. This is done by starting at the upper left with a plus sign and then alternating plus and minus signs, going to the right in the first row, dropping down to the second row and going right to left, etc. Figure 6-3 shows the general pattern, using a 5 by 5 example.

Returning now to the row or column of your choice, cross out (mentally or actually) the row and column of the first cell. This will leave a 2 by 2 determinant. The first term in our expansion is the product of this determinant and the element in the chosen cell, with the algebraic sign of that cell. In Fig. 6-3 I have chosen to use the first row for my expansion. You then go to the next cell in the row and again cross out the row and column of that cell, leaving a different 2 by 2 determinant. Proceed in this fashion until you come to the end of your chosen row or column to complete the expansion of your determinant. Be sure you understand the example of Fig. 6-3 and can extend it to any size determinant. I would suggest you try a 4 by 4 right now, unless you are already expert in such operations. Note that a 4 by 4 will first produce 3 by 3's, which then must be reduced to 2 by 2's. Clearly this procedure can get quite tedious and error-prone, so symbolic processor software which does it for us is quite welcome if we have to do this very often. Note that even when we

$$\begin{vmatrix} a_{11} & a_{12} & a_{13} \\ a_{21} & a_{22} & a_{23} \\ a_{31} & a_{32} & a_{33} \end{vmatrix} = +a_{11}(a_{22}a_{33}-a_{23}a_{32})$$

$$-a_{12}(a_{21}a_{33}-a_{23}a_{31})$$

$$+a_{13}(a_{21}a_{32}-a_{22}a_{31})$$

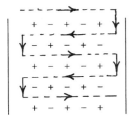

rule for signs

Figure 6-3   Rules for expanding determinants by minors.

350   **Chapter 6**

choose numerical values rather than letters (M's, B's, etc.) for system parameters, we *still* need a symbolic processor because the D operators are letters, *not* numerical values. When determinants have *only* numerical elements, then the expansion does not require a symbolic processor and goes much faster.

Be sure to note that the denominator determinant is *always* the same, no matter which unknown you might be solving for, so it need be expanded only once. This also means that the left-hand side of each differential equation for a single unknown will also be exactly the same. Thus a system has *only one* characteristic equation and only one set of roots needed for the complementary solution.

This completes our treatment of the classical operator method of solution for linear differential equations with constant coefficients, and simultaneous sets of such equations. The methods we have shown will handle equations of any order and simultaneous sets with any number of equations and unknowns.

6-4   ANALYTICAL SOLUTION OF LINEAR, CONSTANT-COEFFICIENT EQUATIONS: THE LAPLACE TRANSFORM METHOD

The Laplace transform method does not solve any equations that could not be solved by the classical operator method, however each method has its own features, so it is useful to know both. As usual we do not provide any proofs but simply show you how to use the method as efficiently as possible. Some features of the Laplace transform method are:

1. Separate steps to find the complementary solution, particular solution, and constants of integration are not used. The complete solution, including initial conditions, is obtained at once.

2. There is never any question about *which* initial conditions are needed; the solution process automatically introduces the correct ones. For sets of simultaneous equations, the "natural" initial conditions (those physically known) are all that are needed, whereas the classical operator method requires that we mathematically derive some *additional* initial conditions. Also, initial conditions in the classical method are evaluated at $t = 0^+$, a time just *after* the input is applied. For some kinds of systems and inputs, these $t = 0^+$ conditions are *not* the same as those *before* the input is applied, so extra work is required to find them. The transform method uses conditions *before* the input is applied; these are generally physically known and are often zero, simplifying the work.

3. For inputs that cannot be described by a single formula for their entire course, but must be defined over segments of time, the classical method requires piecewise solution with tedious matching of final conditions of one piece and initial conditions of the next. The Laplace transform method handles such discontinuous inputs very neatly.

All the theorems and techniques of the Laplace transform derive from the fundamental definition for the direct Laplace transform $F(s)$ of the time function $f(t)$.

Solution Methods for Differential Equations 351

$$\text{Laplace transform of } f(t) \overset{\Delta}{=} \mathscr{L}[f(t)] \overset{\Delta}{=} F(s) \overset{\Delta}{=} \int_0^\infty f(t)e^{-st}\,dt \qquad t>0 \quad (6\text{-}27)$$

$s \overset{\Delta}{=}$ a complex variable $\overset{\Delta}{=} \sigma + i\omega$

The integral of Eq. (6-27) cannot be evaluated for all $f(t)$'s but when it can, it establishes a unique pair of functions, $f(t)$ in the time domain and $F(s)$ in the s domain. It is conventional to use capital letters for s functions and lowercase for t functions. Since comprehensive tables of such Laplace transforms have been published, it is rarely necessary for a transform user to actually work out Eq. (6-27). Table 6-1 is a brief table adequate for the purposes of this book and most practical applications. When we use Laplace transforms to solve differential equations, we must transform entire equations, not just isolated $f(t)$ functions, so several theorems necessary for this will now be stated, without proof.

Linearity Theorem

$$\mathscr{L}[a_1 f_1(t) + a_2 f_2(t)] = \mathscr{L}[a_1 f_1(t)] + \mathscr{L}[a_2 f_2(t)] = a_1 F_1(s) + a_2 F_2(s) \qquad (6\text{-}28)$$

This theorem says we may transform an entire equation by adding the transforms of the individual terms. Also, the transform of a constant (a_1, a_2) times $f(t)$ is just the constant times the transform of $f(t)$.

Differentiation Theorem

$$\mathscr{L}\left[\frac{df}{dt}\right] = sF(s) - f(0) \qquad (6\text{-}29)$$

$$\mathscr{L}\left[\frac{d^2 f}{dt^2}\right] = s^2 F(s) - sf(0) - \frac{df}{dt}(0) \qquad (6\text{-}30)$$

$$\mathscr{L}\left[\frac{d^n f}{dt^n}\right] = s^n F(s) - s^{n-1}f(0) - s^{n-2}\frac{df}{dt}(0) - \cdots - \frac{d^{n-1}f}{dt^{n-1}}(0) \qquad (6\text{-}31)$$

This theorem allows one to transform a derivative of any order, and automatically inserts the necessary initial conditions into the solution process. That is, $f(0)$, $(df/dt)(0)$, and the like are the initial values of $f(t)$ and its derivatives, evaluated numerically at a time instant just *before* the driving input is applied.

Integration Theorem

$$\mathscr{L}\left[\int f(t)\,dt\right] = \frac{F(s)}{s} + \frac{f^{(-1)}(0)}{s} \qquad (6\text{-}32)$$

where $f^{(-1)}(0)$ is the initial value of $\int f(t)\,dt$. For example, if $f(t)$ were the *velocity* in a mechanical motion problem, $f^{(-1)}(0)$ would be the numerical value of the *displacement* just *before* the system input was applied. While this chapter title says *differential* equations, it would be more correct to say *integrodifferential* equations, since our models sometimes have both derivatives and integrals in them. For example, a Kirchhoff voltage-loop law for a circuit with resistors, inductors, and capacitors will have terms such as $(1/C)\int i\,dt$ in it, in addition to derivative terms for the inductors. This is an example of why we need a transform for integrals. If higher-order integrals appear:

Table 6-1

	$F(s)$	$f(t)$
1	s	$\delta'(t)$, first derivative of unit impulse
2	1	$\delta(t)$, unit impulse
3	$1/s$	1, unit step, $u(t)$
4	$1/s^2$	t
5	$\dfrac{1}{s^n}$	$\dfrac{1}{(n-1)!}\,t^{n-1}$
6	$\dfrac{1}{s+a}$	e^{-at}
7	$\dfrac{1}{s(s+a)}$	$\dfrac{1-e^{-at}}{a}$
8	$\dfrac{1}{s^2(s+a)}$	$\dfrac{e^{-at}+at-1}{a^2}$
9	$\dfrac{s+a_0}{s^2(s+a)}$	$\dfrac{a_0-a}{a^2}e^{-at}+\dfrac{a_0}{a}t+\dfrac{a-a_0}{a^2}$
10	$\dfrac{s^2+a_1 s+a_0}{s^2(s+a)}$	$\dfrac{a^2-a_1 a+a_0}{a^2}e^{-at}+\dfrac{a_0}{a}t+\dfrac{a_1 a-a_0}{a^2}$
11	$\dfrac{1}{(s+a)(s+b)}$	$\dfrac{e^{-at}-e^{-bt}}{b-a}$
12	$\dfrac{s+c}{(s+a)(s+b)}$	$\dfrac{(c-a)e^{-at}-(c-b)e^{-bt}}{b-a}$
13	$\dfrac{1}{s(s+a)(s+b)}$	$\dfrac{1}{ab}+\dfrac{be^{-at}-ae^{-bt}}{ab(a-b)}$
14	$\dfrac{s+c}{s(s+a)(s+b)}$	$\dfrac{c}{ab}+\dfrac{c-a}{a(a-b)}e^{-at}+\dfrac{c-b}{b(b-a)}e^{-bt}$
15	$\dfrac{s^2+a_1 s+a_0}{s(s+a)(s+b)}$	$\dfrac{a_0}{ab}+\dfrac{a^2-a_1 a+a_0}{a(a-b)}e^{-at}$ $-\dfrac{b^2-a_1 b+a_0}{b(b-a)}e^{-bt}$
16	$\dfrac{1}{s^2(s+a)(s+b)}$	$\dfrac{t}{ab}-\dfrac{a+b}{(ab)^2}+\dfrac{1}{a-b}\left(\dfrac{e^{-bt}}{b^2}-\dfrac{e^{-at}}{a^2}\right)$
17	$\dfrac{s+a_0}{s^2(s+a)(s+b)}$	$\dfrac{a_0-a}{a^2(b-a)}e^{-at}+\dfrac{a_0-b}{b^2(a-b)}e^{-bt}$ $+\dfrac{a_0}{ab}t+\dfrac{ab-a_0(a+b)}{(ab)^2}$
18	$\dfrac{s^2+a_1 s+a_0}{s^2(s+a)(s+b)}$	$\dfrac{a^2-a_1 a+a_0}{a^2(b-a)}e^{-at}$ $+\dfrac{b^2-a_1 b+a_0}{b^2(a-b)}e^{-bt}$ $+\dfrac{a_0}{ab}t+\dfrac{a_1 ab-a_0(a+b)}{(ab)^2}$
19	$\dfrac{1}{(s+a)(s+b)(s+c)}$	$\dfrac{e^{-at}}{(b-a)(c-a)}+\dfrac{e^{-bt}}{(a-b)(c-b)}$ $+\dfrac{e^{-ct}}{(a-c)(b-c)}$

	$F(s)$	$f(t)$
20	$\dfrac{s + a_0}{(s + a)(s + b)(s + c)}$	$\dfrac{a_0 - a}{(b - a)(c - a)} e^{-at}$
		$+ \dfrac{a_0 - b}{(a - b)(c - b)} e^{-bt}$
		$+ \dfrac{a_0 - c}{(a - c)(b - c)} e^{-ct}$
21	$\dfrac{s^2 + a_1 s + a_0}{s(s + a)(s + b)(s + c)}$	$\dfrac{a^2 - a_1 a + a_0}{(b - a)(c - a)} e^{-at} + \dfrac{a_0}{abc}$
		$+ \dfrac{b^2 - a_1 b + a_0}{(a - b)(c - b)} e^{-bt}$
		$+ \dfrac{c^2 - a_1 c + a_0}{(a - c)(b - c)} e^{-ct}$
22	$\dfrac{1}{s^2 + a^2}$	$\dfrac{\sin at}{a}$
23	$\dfrac{s}{s^2 + a^2}$	$\cos at$
24	$\dfrac{1}{(s + a)^2 + b^2}$	$\dfrac{1}{b} e^{-at} \sin bt$
25	$\dfrac{s + a_0}{(s + a)^2 + b^2}$	$\dfrac{1}{b} [(a_0 - a)^2 + b^2]^{1/2} e^{-at} \sin(bt + \phi)$
		$\phi \triangleq \tan^{-1} \dfrac{b}{a_0 - a}$
26	$\dfrac{1}{s[(s + a)^2 + b^2]}$	$\dfrac{1}{b_0^2} + \dfrac{1}{bb_0} e^{-at} \sin(bt - \phi)$
		$\phi \triangleq \tan^{-1} \dfrac{b}{-a}, \; b_0 \triangleq \sqrt{a^2 + b^2}$
27	$\dfrac{s + a_0}{s[(s + a)^2 + b^2]}$	$\dfrac{a_0}{a^2 + b^2} + \dfrac{1}{b\sqrt{a^2 + b^2}}$
		$\times [(a_0 - a)^2 + b^2]^{1/2} e^{-at} \sin(bt + \phi)$
		$\phi \triangleq \tan^{-1} \dfrac{b}{a_0 - a} - \tan^{-1} \dfrac{b}{-a}$
28	$\dfrac{1}{s^2[(s + a)^2 + b^2]}$	$\dfrac{1}{a^2 + b^2}\left[t - \dfrac{2a}{a^2 + b^2} + \dfrac{1}{b} e^{-at} \sin(bt - \phi) \right]$
		$\phi \triangleq 2 \tan^{-1} \dfrac{b}{-a}$
29	$\dfrac{1}{(s + c)[(s + a)^2 + b^2]}$	$\dfrac{e^{-ct}}{(c - a)^2 + b^2}$
		$+ \dfrac{1}{b[(c - a)^2 + b^2]^{1/2}} e^{-at} \sin(bt - \phi)$
		$\phi \triangleq \tan^{-1} \dfrac{b}{c - a}$
30	$\dfrac{1}{s(s + c)[(s + a)^2 + b^2]}$	$\dfrac{1}{c\sqrt{a^2 + b^2}} - \dfrac{1}{c[(a - c)^2 + b^2]} e^{-ct}$
		$+ \dfrac{1}{b\sqrt{a^2 + b^2}\,[(c - a)^2 + b^2]^{1/2}} e^{-at}$
		$\sin(bt - \phi)$
		$\phi \triangleq \tan^{-1} \dfrac{b}{-a} + \tan^{-1} \dfrac{b}{c - a}$

354   **Chapter 6**

$$\mathcal{L}[f^{(-n)}(t)] = \frac{F(s)}{s^n} + \sum_{k=1}^{n} \frac{f^{(-k)}(0)}{s^{n-k-1}} \tag{6-33}$$

where

$$f^{(-n)}(t) \triangleq \int \cdots \int f(t)(dt)^n \quad \text{and} \quad f^{(-0)}(t) \triangleq f(t) \tag{6-34}$$

We will shortly give a few more useful theorems but we now actually have enough, together with the transform table, to solve many differential equations. Let's first repeat our earlier example of Fig. 6-1, but now use the Laplace transform method.

EXAMPLE: SIMULTANEOUS EQUATIONS

When using the Laplace transform method, we *always* start with the "raw" simultaneous equations, just as they come from the physical laws. *Do not* do any combining of equations or *D*-operator manipulations to reduce the set of simultaneous equations to single equations in each unknown! Such operations would not lead to wrong answers, but they are needless work which also *defeats* some advantages of this method. We would thus start with Eqs. (6-14) and apply our theorems to each of these equations. We begin by Laplace-transforming both sides of each equation. This is clearly valid since performing the same operation (no matter what it is) on *both* sides of any equation is always correct.

$$F_1(s) - B_1 V_1(s) + B_1 V_2(s) = M_1[s V_1(s) - v_1(0)] \tag{6-35}$$

$$F_2(s) + B_1 V_1(s) - B_1 V_2(s) - B_2 V_2(s) = M_2[s V_2(s) - v_2(0)] \tag{6-36}$$

Note that only two initial conditions are needed and these are the velocities just *before* the inputs are applied. These velocities would be known in any practical problem and would often be zero. Equations (6-35) and (6-36) *really are* algebraic equations, whereas in the *D*-operator method we had to say "treat them *as if* they were algebraic." Let's now insert the same numerical values used in our earlier example.

$$0 - V_1 + V_2 = sV_1 - 0 \tag{6-37}$$

$$\frac{1}{s} + V_1 - V_2 - V_2 = sV_2 - 0 \tag{6-38}$$

We are here giving up the $V(s)$ notation for the simpler V since the upper case *implies* a function of s. The step-input force $f_2(t)$ of size 1.0 transforms into $1/s$, according to entry 3 of our table. (Note that, by the linearity theorem, a step input of *any* size, say C, transforms into C/s. This kind of proportionality of course holds for *all* the table entries.) Equations (6-37) and (6-38) can easily be solved for both velocities, using determinants or substitution and elimination. Solution for V_1 gives

$$V_1 = \frac{1}{s(s^2 + 3s + 1)} = \frac{1}{s(s + 0.382)(s + 2.618)} \tag{6-39}$$

This s function appears in the table as entry 13, so we simply copy the solution for $v(t)$ from the table:

$$v_1(t) = 1.0 + 0.171e^{-2.618t} - 1.171e^{-0.382t} \tag{6-40}$$

Solution Methods for Differential Equations 355

This of course agrees with our earlier solution by the *D*-operator method [Eq. (6-24)]. Note that root finding is still needed.

Laplace Transfer Functions. We have defined and used since early in the book the concepts of operational and sinusoidal transfer functions. Suppose we Laplace-transform a set of simultaneous differential equations, *take all the initial conditions to be zero*, and then reduce the set down to single algebraic equations in single unknowns. If we now form the ratio of any output quantity to any input quantity, this ratio will be a function of *s* and is defined as the *Laplace transfer function* relating that pair of output/input variables. If we apply this idea to Eqs. (6-35) and (6-36) we see that the pair of equations looks exactly like Eqs. (6-15), except that *D*'s are replaced by *s*'s and the variables are written uppercase rather than lowercase. When we use determinants or other algebraic means to reduce the set of equations to single equations, we will get results like Eqs. (6-18) and (6-19), except we again have *s*'s instead *D*'s.

We can obtain Laplace transfer functions "from scratch" by transforming the physical equations with zero initial conditions, using algebra to reduce the equation set to single equations, and then forming whatever output/input ratios we want. If we had previously obtained *D*-operator or sinusoidal ($i\omega$) transfer functions, all we need do is replace every *D* or $i\omega$ by an *s*. For example, from Eq. (6-18),

$$\frac{V_1}{F_1}(s) = \frac{M_2 s + B_1 + B_2}{M_1 M_2 s^2 + (M_1 B_1 + M_1 B_2 + M_2 B_1)s + B_1 B_2} \tag{6-41}$$

Thus, if we have any one form of a transfer function (*D*, $i\omega$, or *s*) we can quickly get any of the others by simple substitution.

Partial-Fraction Expansion. Although a comprehensive transform table will allow inverse transformation [$F(s)$ to $f(t)$] of many practical problems, we will occasionally need the partial-fraction expansion method to handle general cases. We assume that the function $F(s)$ to be inverse-transformed is a ratio of polynomials, since this is the form we usually encounter in solving ordinary differential equations.

$$F(s) = \frac{N(s)}{D(s)} = k \frac{s^p + B_{p-1}s^{p-1} + \cdots + B_1 s + B_0}{s^n + A_{n-1}s^{n-1} + \cdots + A_1 s + A_0} \tag{6-42}$$

Using a root-finder if necessary, we can factor both the numerator and denominator to give the form

$$F(s) = k \frac{(s - z_1)(s - z_2) \cdots (s - z_p)}{(s - p_1)(s - p_2) \cdots (s - p_n)} \tag{6-43}$$

The numerical values p_i are called the *poles* of the function $F(s)$ and the values z_i are called its *zeros*. Note that when *s* takes on the value of a zero, the function $F(s)$ becomes equal to zero. When *s* takes on the value of a pole, $F(s)$ become infinite. All *s* functions, including specific interpretations such as Laplace transfer functions, have poles and zeros. Except for the multiplying factor called *k*, giving the poles and zeros of an *s* function completely specifies that function.

Usually, $n > p$ (proper fraction); if not (improper fraction), divide $D(s)$ into $N(s)$ to get

$$F(s) = \cdots + L_2 s^2 + L_1 s + L_0 + \frac{N_1(s)}{D(s)} \tag{6-44}$$

The term $N_1(s)/D(s)$ will now be a *proper* fraction, which is the reason for doing the division. That is, before we can do our partial-fraction expansion, we need a *proper* fraction. Usually this occurs "naturally," but sometimes we need to do the division. The "L terms" will inverse transform into impulse functions of time (see Fig. 3-4), and derivatives of impulse functions. If our $F(s)$ function represents some real physical variable, such as a force or a motion, an impulse function (or one of its derivatives) means that this variable goes to infinity, which is impossible. Thus we rarely will be dealing with improper fractions, in fact if you get one, it would be wise to examine your previous analysis for possible mistakes. We do, however, include the case of improper fractions for mathematical completeness and also because impulse functions *do* sometimes appear in correct physical analyses as intermediate steps, where their "infiniteness" does not violate any physical laws.

From algebra we know that we can write

$$F(s) = \frac{N(s) \text{ or } N_1(s)}{(s - p_1)(s - p_2) \cdots (s - p_n)}$$

$$\frac{N(s) \text{ or } N_1(s)}{D(s)} = \frac{K_1}{(s - p_1)} + \frac{K_2}{(s - p_2)} + \cdots + \frac{K_k}{(s - p_k)} + \cdots + \frac{K_n}{(s - p_n)} \tag{6-45}$$

where we have assumed there are no repeated roots (they will be dealt with shortly), and the K's are real or complex numbers but do not involve s. This expression shows that complicated $F(s)$'s (high-degree polynomials) which do *not* appear in any table can be broken down into a sum of very simple s functions which *are* in even a simple table. Note also that our method generally requires *root-finding* to factor the denominator and allow its separation into the partial fractions. We must now show a method for getting numerical values of all the K's, since we can't inverse transform until we have these numbers. Our method initially assumes that there are no repeated roots, but we will extend it to include this case shortly.

The poles in our expression can be either real or complex numbers. Let's first treat the case of any real pole, call it p_k. To find K_k we multiply Eq. (6-45) by $(s - p_k)$, to make K_k "stand alone."

$$\frac{[N(s) \text{ or } N_1(s)](s - p_k)}{(s - p_1) \cdots (s - p_k) \cdots (s - p_n)} = \frac{K_1(s - p_k)}{(s - p_1)} + \cdots + K_k + \cdots + \frac{K_n(s - p_k)}{(s - p_n)} \tag{6-46}$$

Since this equation is true for any value of s, choose $s = p_k$ to "wipe out" all the right-side terms except K_k, giving

$$K_k = \left[\frac{N(s) \text{ or } N_1(s)}{(s - p_1) \cdots (\quad) \cdots (s - p_n)} \right]_{s = p_k} \tag{6-47}$$

Let's now do a simple example to illustrate the general method.

Solution Methods for Differential Equations 357

EXAMPLE: REAL POLES

Our earlier example, Eq. (6-39), appears directly in the table, but we *can* use the partial-fraction expansion method if we wish.

$$V_1(s) = \frac{1}{s(s^2 + 3s + 1)} = \frac{K_1}{(s+0)} + \frac{K_2}{(s+0.382)} + \frac{K_3}{(s+2.618)} \tag{6-48}$$

$$K_1 = \left[\frac{1}{s^2 + 3s + 1} \right]_{s=0} = 1.0$$

$$K_2 = \left[\frac{1}{s(s + 2.618)} \right]_{s=-0.382} = -1.171$$

$$K_3 = \left[\frac{1}{s(s + 0.382)} \right]_{s=-2.618} = 0.171$$

Using table entries 3 and 6, we see that

$$v_1(t) = 1.0 + 0.171e^{-2.618t} - 1.171e^{-0.382t} \tag{6-49}$$

which agrees with our earlier result.

When some of the poles are complex, we can still use the same approach to find numerical values for the K's associated with those poles. Because complex poles always appear as pairs $(a \pm ib)$, an efficient method deals with pairs of poles, rather than individual terms. We show a method which requires that you find the K only for the term which has the negative imaginary part. Let's do a more comprehensive example which will develop the general technique for complex pole pairs.

EXAMPLE: COMPLEX POLE PAIRS

$$F(s) = \frac{1}{s(s + 1)(s + 2)(s + 1 + i1)(s + 1 - i1)(s + 2 + i2)(s + 2 - i2)}$$
$$= \frac{K_1}{s} + \frac{K_2}{s + 1} + \frac{K_3}{s + 2} + \frac{K_4}{s + 1 - i1} + \frac{K_5}{s + 1 + i1} + \frac{K_6}{s + 2 - i2} + \frac{K_7}{s + 2 + i2} \tag{6-50}$$

$$K_1 = \frac{1}{(1)(2)(2)(8)} = 0.0313 \qquad K_2 = \frac{1}{(-1)(1)(1)(5)} = -0.20$$

$$K_3 = \frac{1}{(-2)(-1)(2)(4)} = 0.0625 \qquad K_4 = \frac{1}{(-1 + i)(i)(1 + i)(i2)(1 + i3)(1 - i)}$$
$$= 0.056 \; \underline{/-26.5°}$$

$$K_6 = \frac{1}{(-2 + i2)(-1 + i2)(i2)(-1 + i3)(-1 + i1)(i4)} = 0.0041 \; \underline{/-8.2°}$$

358   **Chapter 6**

First note that we compute only K_4 and K_6, the terms with the negative imaginary parts; however, the transform given below takes into account *all* the terms. This is possible because it can be shown that K_5 and K_7 are complex conjugates (same real part, negative imaginary part) of K_4 and K_6; thus there is never any need to compute them separately. We now give a rule which inverse transforms any pair of terms with denominators of the form $(s + a + ib)$, $(s + a - ib)$ once we have found the K, given as $K \underline{/\phi}$, associated with the negative imaginary part.

$$f(t) = 2Ke^{-at} \sin(bt + \alpha)$$

$$\alpha \overset{\Delta}{=} \phi + 90° \tag{6-51}$$

Using this rule we can now write out our $f(t)$:

$$f(t) = 0.0313 - 0.2e^{-t} + 0.0625e^{-2t} + 0.112e^{-t} \sin(t + 63.5°)$$
$$+ 0.0082e^{-2t} \sin(2t + 81.8°) \tag{6-52}$$

We have now shown how to use the partial-fraction expansion to inverse-transform terms associated with any number of unrepeated real poles and any number of unrepeated complex pole pairs. While repeated poles of $F(s)$ or repeated roots of the characteristic equation are not common since they require a special relationship among the system parameters, they do occur, so we need to know how to deal with them.

Repeated Roots. We need to first make clear some distinctions between "pure" mathematics and practical engineering. If we want to "manufacture" a polynomial which will have *perfect* repeated roots or factors, we simply start with the desired factors and multiply them out to get the polynomial. For example, $(s + 1)(s + 1) = s^2 + 2s + 1$, and we *know* that the root $s = -1$ is perfectly repeated. In practical engineering work, the polynomial arises when we combine simultaneous physical equations which have coefficients that represent physical parameters such as spring constants, masses, inductances, etc. Since the physical parameters are never known precisely, and even if they were so known *momentarily*, slight changes in temperature or other environmental factors would cause them to change, the polynomial coefficients in the equations of real systems are never known precisely and "drift around." Since the polynomial coefficients have this inherent uncertainty, and perfectly repeated roots require a *perfect* relation among the coefficients, we can't really get repeated roots in such situations. Furthermore, for equations of fifth- and higher-order (*not* unusual), the procedure for finding the roots is itself approximate, so even if we gave a root finder a "manufactured" polynomial with perfectly repeated roots, the roots it found might be "close," but *won't* be perfectly repeated.

Even though perfect repeated roots are not to be expected in practical problems, we can give a method for getting the partial-fraction expansion for such perfect situations. Let's show a simple example to outline the method.

$$F(s) = \frac{1}{s(s + 1)(s + 1)} = \frac{K_1}{s} + \frac{K_2}{(s + 1)^2} + \frac{K_3}{s + 1} \tag{6-53}$$

Solution Methods for Differential Equations   **359**

First note that when a root appears twice, the partial-fraction expansion has *two* terms. One has the usual form we have seen, but the other has the denominator squared. If the root appeared *three* times, there would be three terms, one with $1/(s+1)$, one with $1/(s+1)^2$, and one with $1/(s+1)^3$, each with its own K. This is a general pattern, you can extend it to any number of real or complex roots, repeated any number of times. Let's now try to find the K's, using our standard method.

$$K_1 = \frac{1}{(1)(1)} = 1$$

$$\frac{(s+1)^2}{s(s+1)(s+1)} = \frac{K_1(s+1)^2}{s} + K_2 + \frac{K_3(s+1)^2}{s+1} \tag{6-54}$$

If we now set $s = -1$ as usual, we get $K_2 = -1$ with no difficulty. However, if we now proceed to K_3:

$$\frac{s+1}{s(s+1)(s+1)} = \frac{K_1(s+1)}{s} + \frac{K_2(s+1)}{(s+1)^2} + K_3 \tag{6-55}$$

we get

$$\infty = \infty + K_3$$

when we set $s = -1$, frustrating our attempt to find K_3.

To resolve this difficulty, return to Eq. (6-54) and note that if we *differentiate* this equation with respect to s we can cause K_3 to "stand alone."

$$\frac{d}{ds}\left[\frac{1}{s}\right] = \frac{d}{ds}\left[\frac{K_1(s+1)^2}{s} + K_2 + K_3(s+1)\right] \tag{6-56}$$

Carrying out the differentiation and setting $s = -1$:

$$\frac{-1}{s^2} = -1 = \left[\frac{K_1[(s)2(s+1) - (s+1)^2]}{s^2}\right]_{s=-1} + 0 + K_3 = K_3 \tag{6-57}$$

Observe that the differentiation of the right-hand side of Eq. (6-56) is (except for the term involving K_3) a "waste of time" since in the *next* step (insertion of $s = -1$) all these terms will always go to zero. This differentiation scheme allows one to find the K's for any number of repeated real roots, each repeated any number of times; however, for roots appearing three or more times, *repeated* differentiation will be necessary to cause the desired K to stand alone. Finally, the inverse transforms of terms such as $K/(s+a)^n$ will give time functions such as $Kt^{n-1}e^{-at}/(n-1)!$, as we might expect from our experience with the classical solution method.

The treatment of *repeated complex root pairs* follows essentially the same pattern as for repeated real roots; the partial-fraction expansion contains terms involving *powers* of the factors, and *differentiation* with respect to s is necessary to find the K's other than the one associated with the highest-power term.

$$F(s) = \frac{N(s)}{(s^2 + 2\zeta\omega_n s + \omega_n{}^2)^n D_1(s)}$$

$$= \frac{K_1}{(s+a-ib)^n} + \frac{K_1'}{(s+a+ib)^n}$$

$$+ \frac{K_2}{(s+a-ib)^{n-1}} + \frac{K_2'}{(s+a+ib)^{n-1}}$$

$$+ \cdots$$

$$+ \frac{K_n}{(s+a-ib)} + \frac{K_n'}{(s+a+ib)} \tag{6-58}$$

We find the K's (they will be complex numbers) as shown earlier for repeated real roots. The time function takes the form

$$f(t) = 2|K_1| \frac{t^{n-1}}{(n-1)!} e^{-at} \sin(bt + \underline{/K_1} + 90°)$$

$$+ 2|K_2| \frac{t^{n-2}}{(n-2)!} e^{-at} \sin(bt + \underline{/K_2} + 90°)$$

$$+ \cdots$$

$$+ 2|K_n| e^{-at} \sin(bt + \underline{/K_n} + 90°) \tag{6-59}$$

If you have MATLAB, it will do most of the work of partial-fraction expansions for you, including the handling of improper fractions and repeated roots. Let's do such an example which will also bring out some useful additional information on repeated roots in real physical systems. A SIMULINK simulation will also be helpful here.

EXAMPLE: "NEARLY-REPEATED" POLES

Figure 6-4 shows simulations of two systems. One has a perfectly repeated pair of complex roots and the other has roots which are "not quite" repeated. We show each of them in two forms: one as factored into two quadratic terms and the other multiplied out into a fourth-degree polynomial. These two forms are of course equivalent. If we send an input (say a step function) into each of them, the responses will be identical. The responses of the "perfectly repeated" and "unrepeated" systems will of course be different, but we would guess that they shouldn't be *much* different since the physical parameters of both must be quite close to each other. We are, perhaps, a little concerned about this because the analytical solutions for repeated and unrepeated roots *are* quite different. The repeated root system has a multiplying factor of t in its solution which the unrepeated does not. Multiplying by t makes a function tend toward infinity as time goes by, although an e^{-at} term can overpower this effect.

Let's start our study by asking MATLAB to "do" the partial-fraction expansion for us. We need to enter the numerator and denominator polynomials of our $F(s)$ function into MATLAB, calling the numerator b and the denominator a. Let's first do the system with the unrepeated quartic polynomial transfer function. We enter into MATLAB

```
b=[1]; a=[1 .401 2.0502 .403 1.01 0]
```

Solution Methods for Differential Equations 361

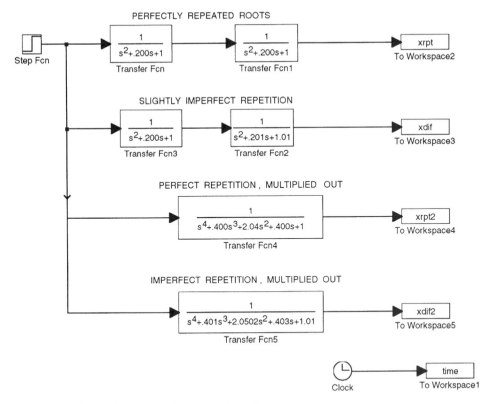

Figure 6-4 Simulation study of "repeated root" systems.

The denominator polynomial is actually fifth degree because our input is a step function of size 1.0. This makes the $F(s)$ function for the system output equal to $1/s$ times the transfer function, raises the degree to 5, and makes the last term zero, rather than 1.01. The MATLAB statement that does partial fraction expansions is called RESIDUE and is used as follows.

```
[r,p,k]=residue(b,a)
```

The above statement is *always* written exactly as shown, assuming you have defined a and b as we did. Once you enter this statement, results for r, p, and k will be displayed. The vector p will be a list of the roots of the denominator polynomial, that is, the poles of $F(s)$. The vector r will be a list of the partial-fraction constants, which we have called the K's, and which mathematicians call the *residues*. The vector k will be empty unless you have an improper fraction. If you *do* have an improper fraction (rare) the vector k will list the "L-terms" in our Eq. (6-44). For the above a and b values we get

```
p=  -.1005 +.9999i
    -.1005 -.9999i
    -.1000 +.9950i
    -.1000 -.9950i
     0
```

362   **Chapter 6**

We see that the two root pairs are close, but not identical, as expected. For r we get

```
r=  48.9999  -9.9755i
    48.9999  +9.9755i
   -49.4949 +10.0504i
   -49.4949 -10.0504i
    0.9901   -0.0000i
```

The vector k is empty since we had a proper fraction. We could at this point use the r values to compute three K's as in Eqs. (6-50) and then get $f(t)$ as in Eqs. (6-51) and (6-52). Let's defer these calculations and instead let MATLAB work on the perfectly repeated system.

```
b=[1];   a=[1. 400 2.04 .400 1 0];
[r,p,k]=residue(b,a)
```

MATLAB returns

```
p=  -.1000  +.9950i
    -.1000  -.9950i
    -.1000  +.9950i
    -.1000  -.9950i
     0
```

These appear to be perfectly repeated root pairs, but if you ask MATLAB for more digits ("format long") you will see that they are *not* perfect. If we look at the residues we see

```
r=   .6156E7  -1.8963E7i
     .6156E7  +1.8963E7i
    -.6156E7  +1.8963E7i
    -.6156E7  -1.8963E7i
    0.0000E7  -0.0000E7i
```

These residues (and therefore our K values) are huge! In fact the MATLAB manual warns us when describing the residue operation that the algorithm may be unreliable when dealing with repeated or "close to repeated" roots. For our earlier residue calculation with the "unrepeated" roots, it turns out those calculations are also wrong. Thus, whenever our root finder comes up with roots that are close to each other, the roots themselves may be OK, but the residue results are suspect.

Does this mean that it is impossible to do accurate response calculations in such situations? Fortunately the answer is usually no, there *are* ways to get valid results. If we run the SIMULINK simulation, this does *not* involve root finding or residue calculating, but rather uses numerical integration to find the system response. These methods are usually reliable and the presence of repeated or near-repeated roots has no bad effects. If we insist on getting "analytical" solutions, that may also be possible. MATLAB provides another routine, called RESI2, for dealing with repeated poles. Since, in practical problems there is no such thing as *perfectly* repeated poles, it really deals with poles that are "nearly" repeated. Let's try it on *both* of our example systems.

You start by using RESIDUE first, exactly as we did above. When you see that you are getting some near-repeated poles in the list called p, identify the p's with the

Solution Methods for Differential Equations 363

positive imaginary parts as follows. When MATLAB list the p's, it calls the first one p(1), the second one p(2), the third one p(3), and the fourth one p(4). When you use RESI2 you have to identify them this way. Our examples, either the "perfect" repeated roots or the "not quite perfect," both have one pair of complex poles, "repeated" once, so in our Eqs. (6-58) and (6-59), we need to compute K_1 and K_2. Both these constants are associated with *one* pole, the one called p(1), because it has the positive imaginary part. Note that when the *pole* has a positive imaginary part, the *term*, such as that of K_4 in Eq. (6-50), has a *negative* imaginary part. To get K_1, invoke RESI2 as follows

```
resi2(b,a,p(1),2,1)
```

Here the ,2,1 means that the pole is of multiplicity 2 and that we want to compute K_1. MATLAB returns K_1 as $-0.50000 + 0.07563i$, which we convert to $0.5057 \underline{/171.4°}$. To get K_2,

```
resi2(b,a,p(1),2,2)
```

gives $0.02525 + 0.2513i$ which we convert to $0.2525 \underline{/84.3°}$. These results are for the "perfect" system. If we change a and b to the numbers for the "not quite perfect" system we get the results $K_1 = -0.4835 + 0.0737i$ and $K_2 = 0.0249 + 0.2457i$. Note that these values are different from, but close to, the values for the "perfect" system. This is reassuring; when we make small changes in the polynomial coefficients, we expect small changes in the system response.

We still cannot be sure that these calculations are accurate, so we compare the analytical solution with that given by SIMULINK simulation. Using Eq. (6-59) we get

$$f(t) = 0.505te^{-0.1t}\sin(0.995t + 3.042) + 1.011e^{-0.1t}\sin(0.0995t + 4.562) + 1.0$$
$$(6\text{-}60)$$

for the "perfect" system's response. Figure 6-5 shows simulation results for both the perfect and not-quite-perfect systems, simulated as two cascaded second-order systems and also as single fourth-order systems. All four response curves lie essentially on top of each other. The fourth-order models should of course respond *exactly* like their respective cascaded second-order models. The not-quite-perfect system has coefficients so close to those of the perfect system that its response, though not *exactly* the same, is so close we can't see the difference on the graph. When we graph Eq. (6-60) (or a similar equation obtained for the "not-quite-perfect system") we again duplicate the curve of Fig. 6-5. We see thus that "repeated" roots require extra care, but we can get analytical solutions if that is desired. If we are satisfied with simulation (numerical integration of the differential equations), then no special efforts are needed.

Delay Theorem. This theorem has two main uses. One is to model that aspect of physical system response called *dead time* (also called transport lag, transport delay, or discrete delay). In a system that has dead time, when we apply an input, *nothing* happens at the output until a finite time later. In systems involving wave propagation, such as a fluid pressure pulse traveling down a pipeline at a finite speed, or

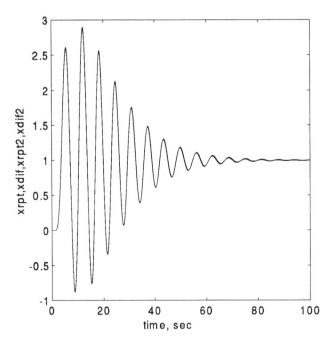

Figure 6-5 Results for "repeated root" systems (four curves "lie on top of each other").

radio wave propagation from earth to the moon (again at a finite speed), when the sending and receiving stations are separated by a certain distance, there will be *no* output until enough time has passed for the wave to travel that distance. We also encounter dead time in computer-aided systems, where the digital processor takes a finite length of time to process data (recall the system of Fig. 3-38).

The other use of this theorem involves response calculations for systems whose input is "discontinuous" with time. That is, the input cannot be described by a *single* mathematical function for its entire extent, but rather must be broken into segments. The classical operator method treats such cases only with difficulty, whereas the Laplace transform method uses the delay theorem to ease the solution. To prepare for the delay theorem we first define the delayed unit step function as follows.

$$\text{Unit step function} \triangleq u(t) \triangleq 1.0 \quad t > 0$$
$$0 \quad t < 0$$
$$\text{Delayed unit step function} \triangleq u(t - a) \triangleq 1.0 \quad t > a$$
$$0 \quad t < a \quad (6\text{-}61)$$

Figure 6-6a shows these definitions while Fig. 6-6b shows the behavior of the dead-time dynamic element. The delay theorem states that

$$\mathscr{L}[f(t - a)u(t - a)] = e^{-as}F(s) \quad (6\text{-}62)$$

Note that multiplying $f(t - a)$ by $u(t - a)$ "turns off" (multiplies by zero) the f function for all $t < a$. From Eq. (6-62) and Fig. 6-6b it is clear that the Laplace transfer function of a dead-time element is given by

Solution Methods for Differential Equations

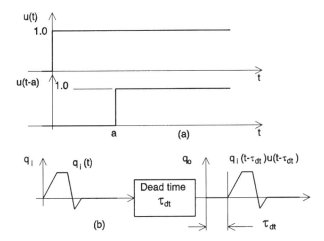

Figure 6-6 Definition of dead-time element.

$$\frac{Q_o}{Q_i}(s) = e^{-\tau_{dt}s} \tag{6-63}$$

To illustrate the use of the delay theorem for discontinuous inputs, let's do the following example.

EXAMPLE: DISCONTINUOUS INPUT

Figure 6-7 shows a simple circuit with an input voltage that is discontinuous. We first write the input in a form that facilitates the application of the delay theorem. That is, we want it to employ forms like $f(t - a)u(t - a)$. For the given waveform, one version might be

$$e_i(t) = tu(t) - (t-1)u(t-1) - u(t-1)u(t-1)$$

Using the delay theorem:

$$E_i(s) = \frac{1}{s^2} - \frac{e^{-s}}{s^2} - \frac{e^{-s}}{s} \tag{6-64}$$

Let's assume all initial conditions are zero, giving

$$E_o(s) = \frac{E_o}{E_i}(s)E_i(s) = \left[\frac{1/RC}{s + 1/RC}\right]\left[\frac{1}{s^2} - \frac{e^{-s}}{s^2} - \frac{e^{-s}}{s}\right] \tag{6-65}$$

Let $RC = 1.0$:

$$E_o(s) = \frac{1}{s^2(s+1)} - \frac{e^{-s}}{s^2(s+1)} - \frac{e^{-s}}{s(s+1)} \tag{6-66}$$

Using table entires 7 and 8 and the delay theorem:

$$e_o(t) = (e^{-t} + t - 1) - u(t-1)[e^{-(t-1)} + (t-1) - 1] - u(t-1)[1 - e^{-(t-1)}] \tag{6-67}$$

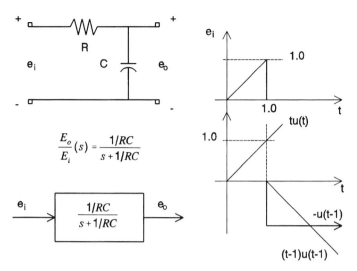

Figure 6-7 System with discontinuous input.

In graphing $e_o(t)$, note that terms multiplied by $u(t-1)$ contribute *nothing* to the graph for $t < 1$.

Initial-Value Theorem and Final-Value Theorem. Our last two theorems are used when we have an $F(s)$ and want only to find the value of $f(t)$ at either $t = 0$ or $t = \infty$, without bothering to get a complete inverse transform. The final-value theorem states

$$\lim_{t \to \infty} f(t) = \lim_{s \to 0} sF(s) \tag{6-68}$$

This theorem is useful *only* when the system's output $f(t)$ approaches a *constant* value as t goes to infinity. It is not of great utility; its results can generally be obtained as quickly by classical methods, because the particular ("steady state") solution in such cases is obvious by inspection. The initial-value theorem

$$\lim_{t \to 0^+} f(t) = \lim_{s \to \infty} sF(s) \tag{6-69}$$

is occasionally useful for finding the value of $f(t)$ just *after* ($t = 0^+$) the input has been applied. In getting the $F(s)$ needed in Eq. (6-69), our usual Laplace definition of initial conditions as those *before* the input is applied must be used.

EXAMPLE: INITIAL CONDITIONS

For the system of Eq. (6-19), using the numerical values of that example, the velocity of mass 2 for a unit step input of force 2 can be written in Laplace form. Recall that when we used the classical operator method on this system, we had to figure out the accelerations at $t = 0^+$ by physical reasoning. With the initial-value theorem, we can get this information from the Laplace equations *without* any additional physical reasoning. Note below that if we have a result for $V(s)$, acceleration $A(s)$ is just $sV(s)$.

Solution Methods for Differential Equations 367

$$V_2 = \frac{s+1}{s(s^2 + 3s + 1)}$$

Acceleration of mass $2 = A_2 = sV_2 = \dfrac{s+1}{s^2 + 3s + 1}$

$$a_2(0^+) = \lim_{s \to \infty} sA_2(s) = +1.0 \tag{6-70}$$

This result of course agrees with physical reasoning. For practice, you might want to use this method to find $a_1(0^+)$ for this example; we earlier found by physical reasoning that this was 0.

6-5 SIMULATION METHODS

We have used some simple simulation methods earlier in the book, so the idea is by now not entirely strange to you. In this section we take a more organized and comprehensive viewpoint, rather than just showing specific examples. While *analog* simulation is, as a general-purpose analysis tool, today largely obsolete, it still finds occasional application in special-purpose contexts, such as built-in equipment in machines or processes. For this reason and also for historical interest, we begin with a quick overview of this technology.

Analog Simulation.   The heart of electronic analog simulation is comprised of the op-amp integrators, summers, and coefficient multipliers we discussed in Sec. 3-8. If we have a sufficient number of these devices, we can "solve" any simultaneous set of linear differential equations with constant coefficients. By adding to these tools the *variable multiplier and divider* and the *arbitrary function generator* of Fig. 6-8, our capability extends to time-varying coefficients and nonlinear equations. Figures 6-9, 6-10, and 6-11 show, respectively, the analog simulation diagrams for the equations

$$A\frac{d^2e}{dt^2} = F\sin\omega t - B\frac{de}{dt} - Ce \tag{6-71}$$

$$(1 - At)M\ddot{x} = F - B\dot{x}^3 - K\sin x \tag{6-72}$$

$$\begin{aligned} \dot{x}_1 + 2x_1 - 2\dot{x}_2 + 3x_2 &= 4 \\ 2\dot{x} + x_1 + \dot{x}_2 - x_2 &= 2t \end{aligned} \tag{6-73}$$

These analog simulation diagrams look almost exactly like the digital simulation diagrams used with a graphical interface software product such as the SIMULINK we have been using. One obvious difference is the *sign changing* that is inherent in the op-amp devices but of course missing in the digital version.

　　While general-purpose analog computers with hundreds of computing elements were highly successful engineering tools of the 1950–1980 era, they were much less convenient to use than the digital computers and simulation software that superseded them. Being analog hardware devices, they were subject to limited accuracy and required constant "tuning up" to maintain accuracy in the face of hardware

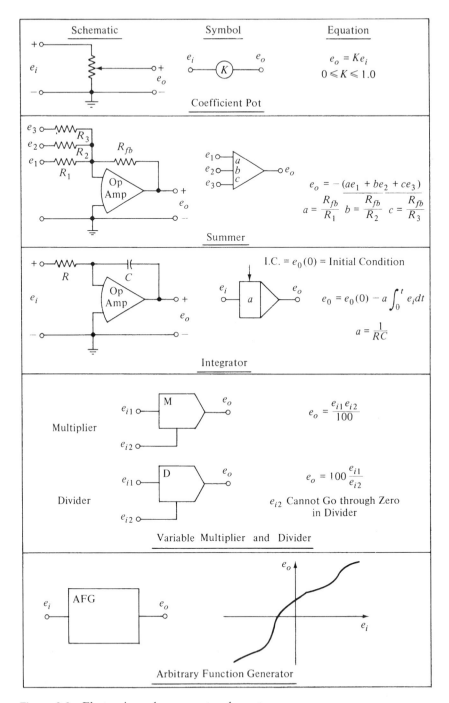

Figure 6-8 Electronic analog computer elements.

Solution Methods for Differential Equations

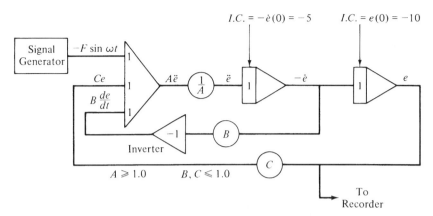

Figure 6-9 Analog computer solution of linear differential equation with constant coefficients.

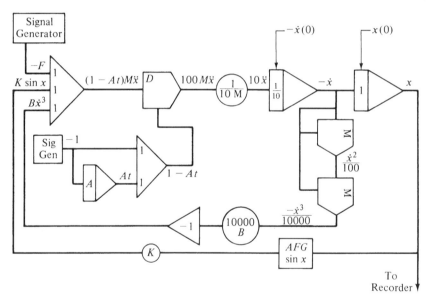

Figure 6-10 Analog computer solution of nonlinear differential equation with time-varying coefficient.

drift. They also required *time scaling* and *magnitude scaling* of the equations to suit the voltage limits and operation speed of the particular machine.[1] Fast and inexpensive personal computers and convenient simulation software have overcome these analog drawbacks and made digital simulation of dynamic systems a pleasurable and cost-effective tool available to the individual engineer on an as-needed basis.

---

[1] E.O. Doebelin, *System Dynamics*, Merrill, Columbus, Ohio, 1972, pp. 251–261.

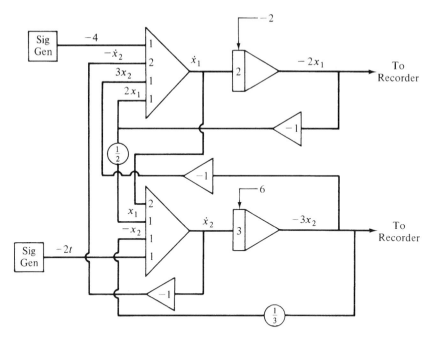

Figure 6-11 Analog computer solution of simultaneous differential equations.

Digital Simulation of Dynamic Systems. Since the term "digital simulation" could reasonably be applied to *any* software application that dealt with physical systems, we want to narrow our scope to those products whose focus is on the numerical solution of the ordinary differential equations which describe lumped-parameter models of dynamic systems. This type of software first appeared around 1960, and one of the most popular versions at that time was called CSMP (Continuous System Modeling Program), an IBM program product. At that time we were using mainframe computers, batch processing, punched-card input, and output on large line printers. When the computer was "busy," we sometimes waited hours to see the results of a run. In terms of problem capability, however, CSMP and similar software of that time could handle anything that we do today. In fact, the underlying numerical mathematics and algorithms have not changed much over the years. We used CSMP for all kinds of course work, research, and in several books that I wrote (*System Dynamics*, 1972; *Measurement Systems*, 3rd ed., 1983; *Control System Principles and Design*, 1985).

When CSMP was no longer supported by a software company we gradually switched over to products such as ACSL (Advanced Continuous Simulation Language) and MATLAB/SIMULINK, which were available for personal computers and provided graphical user interfaces. Quite a number of competitive software products of this class are on the market and a potential user needs to carefully survey them to choose that most suited to local needs and hardware. Explanations in this book are mainly in terms of MATLAB/SIMULINK since this software happens to be currently available to me and my students. Fortunately, all such software shares many similarities, so if you learn one of them, it is very quick and easy to pick up any of the others.

Solution Methods for Differential Equations

While we often prefer a graphical user interface (GUI), a basic understanding of the workings of such software is best obtained by looking at a command-line version first. Also, when simulations get more complicated, a command-line approach may actually be preferred (or necessary) over a GUI. We will use the command-line version of ACSL[2] (it also has a GUI version) as a vehicle for our explanations. Let's use the mechanical system of Fig. 6-12 as a physical example. To see how nonlinearities and time-varying coefficients are handled, we include there a cubic spring, a square-law damper, and a damper whose coefficient changes with time (perhaps due to heating effects). Using Newton's law on each mass in turn we get a set of three simultaneous nonlinear differential equations, one having a time-varying coefficient.

$$f_1 - 5x_1 - 8(x_1 - x_2)^3 - 1(\dot{x}_1 - \dot{x}_2)|(\dot{x}_1 - \dot{x}_2)| = 0.5\ddot{x}_1 \tag{6-74}$$

$$8(x_1 - x_2)^3 + 1(\dot{x}_1 - \dot{x}_2)|(\dot{x}_1 - \dot{x}_2)| - 2(x_2 - x_3) = 1.0\ddot{x}_2 \tag{6-75}$$

$$f_3 + 2(x_2 - x_3) - 40(1.0 - 0.0025t)\dot{x}_3 = 2.0\ddot{x}_3 \tag{6-76}$$

These equations would be solved by running the ACSL program shown below.

```
X1DOT2=-4.0*STEP(0.0)-10.*X1-16.*(X1-X2)**3-2.*(X1DOT-X2DOT)*ABS(X1DOT-X2DOT)
X1DOT=INTEG(X1DOT2,0.0)
X1=INTEG(X1DOT,0.0)
X2DOT2=8.0*(X1-X2)**3+1.0*(X1DOT-X2DOT)*ABS(X1DOT-X2DOT)-2.0*(X2-X3)
X2DOT=INTEG(X2DOT2,0.0)
X2=INTEG(X2DOT,0.0)
X3DOT2=2.5*SIN(1.0*T)+1.0*(X2-X3)-20.*(1.0-.0025*T)*X3DOT
X3DOT=INTEG(X3DOT2,0.0)
X3=INTEG(X3DOT,0.0)
```

The pattern of these equations, which is used for *all* simulations, is to write a statement (ACSL uses the FORTRAN language) for the highest derivative of each of the unknowns, and then numerically integrate to get the lower derivatives and finally the unknown itself. For each variable we make up convenient names such as X1DOT2

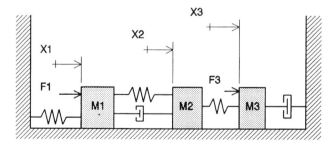

Figure 6-12 Mechanical system with nonlinearities and time-varying parameters for simulation example.

---

[2]MGA Software, 200 Baker Ave., Concord, MA 01742-2100, 800-647-2275.

for d^2x_1/dt^2. The INTEG statement causes the numerical integration of the named derivative, starting from the initial value given. Thus, INTEG(X1DOT2,0.0) gets us X1DOT from X1DOT2, and INTEG(X1DOT,0.0) gets us X1 from X1DOT. Some nonlinearities, such as absolute value (ABS), powers (**3), etc., are already available in standard FORTRAN. Many special operations such as saturation, dead space, quantizers, transfer functions, etc. are provided by special ACSL statements.

In our example, the spring between masses 1 and 2 has a cubic nonlinear relation while the damper has a "square-law" nonlinearity. Note that we can't just square (X1DOT-X2DOT), since this would not reverse the damper force when the relative velocity reversed. Using the absolute value as shown *does* give both the proper magnitude and algebraic sign for this damper force. The force f_1 has been taken as a step input at time 0, of size -2.0 pounds. The ACSL statement STEP(0.0) implements a step function of size 1.0 at time $= 0$; you can make steps "happen" at any time TS of your choice by using STEP(TS). The force f_3 has been made a sine wave of amplitude 5.0 pounds and frequency 1.0 rad/sec. The damper attached to mass 3 has a coefficient which varies with time as $40(1 - .0025t)$, which is written with standard FORTRAN operations.

Statements which specify how long the problem is to run, what kind of integrator to use, the size of the time step, which variables are to be graphed or tabulated, the format of tables and graphs, etc. allow us to conveniently control our simulation. While writing and running a command-line program like our example would certainly not be difficult or time consuming, if we have available GUI-type software, it would probably be preferred by most users for such simple problems. Since the MATLAB/SIMULINK software of this type is available to me, I have used it earlier in this text and will now explain its use in more general terms. Learning to use this specific software will really give you good preparation for using *any* dynamic-system simulation software.

I will be using the PC Windows version of SIMULINK but the other versions are nearly identical in usage. Of course, software is always accompanied by manuals explaining its use, but my experience has been that it is a great aid to beginners to provide a condensed and reorganized "manual" which extracts the essentials from the usual overwhelming detail. When you first log on to SIMULINK you see a main menu with seven items, as in Fig. 6-13. Double-clicking on each of these brings forth a more detailed menu of that particular item, from which you actually choose the items you need to "build" your simulation diagram.

Before going to your computer you should have arranged your system equations so that the highest-derivative term of each unknown appears all by itself on the left-hand side. While we could force the coefficient of this term to be 1.0, as we did with our command-line example, with GUI-type software I prefer to leave these terms just as they come from the physical equation. In our example this means

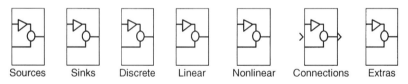

Figure 6-13 SIMULINK's main menu.

Solution Methods for Differential Equations

that the left-hand sides are all of the form $(M)(d^2x/dt^2)$. If you always do this, then the *first* block that you "draw" in your simulation diagram will be a *summer* whose output is this highest-derivative term and whose inputs are all the terms on the right-hand side of that equation, in our example, all the forces acting on that mass. A summer can be configured to have any number of inputs, and each input can have whatever sign ($\pm$) that you need for your equation. The output of the summer is the algebraic sum of its inputs. Thus the summer is actually a graphical statement of the equation. When we go to the menu to select items, we will thus need to get as many summers as there are equations in our set. (Actually, if you get only *one* summer, you can *copy* it as many times as you want by clicking on it with the right mouse button and dragging the icon.)

The signal coming out of the summer will have a coefficient (in our example a mass) multiplying the highest derivative. To get the highest derivative all by itself, so we can integrate it, we next multiply by the reciprocal of the coefficient ($1/M$ in our case). To multiply a signal by a constant we use a *gain* block. We will need as many gain blocks as we have multiplications by constants, but you can again get only one, and later copy it as many times as you need to. Once you have the highest derivative "standing alone," you can integrate it successively until you have all the lower derivatives and the unknown itself. This stepwise process is shown in Fig. 6-14 and in fact is used at the start of *all* simulation problems. In step 1 of Fig. 6-14 we simply *assume* that we will be able to lay our hands on the inputs to the summer which has the highest derivative term as its output. These inputs are either known forcing functions (like the step and sinewave forces in our example) or else somehow depend on the unknown and its derivatives. Since our integration of the highest derivatives always produces all the other derivatives and the unknown, we *will* be able to provide all the inputs which we *assumed* available in step 1.

By inspecting your original set of equations, you should be able to see *all* the mathematical operations needed in your simulation diagram. In our example we

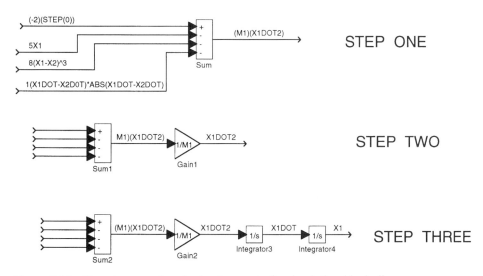

Figure 6-14 Stepwise procedure in development of a simulation block diagram.

need summers, gain blocks, integrators, multipliers, absolute value blocks, and certain known functions of time (sine waves, step functions, etc.). We also need ways to record and graph our results. All these and more are available on the various menus, so we now give some details on how to access the needed items. When the main SIMULINK menu of Fig. 6-13 is in view, you will also see a menu bar of word items: FILE, EDIT, OPTIONS, SIMULATION, STYLE. Click on FILE and then select NEW to open a new file. When you do this a new window called UNTITLED appears; this "empty" window is where you will "build" your simulation block diagram. Before going any further it is probably a good idea to *save* your new file, so click on FILE and then SAVE AS, giving the file a name of your choice. Then return to your block diagram window to start building the diagram. The window should now have a title given by the filename you just created; it won't be called UNTITLED any more. As you progress in building the diagram, it is a good idea to SAVE the file whenever you have made a few changes, so that if you "crash," you won't lose much work.

When building any simulation diagram, there are three major types of blocks that are needed: sources, systems, and readout (graphing) devices. Let's first discuss the SOURCE type of blocks. These are signal sources of various kinds that you use to provide the driving inputs for your physical system. SOURCES are also sometimes used to provide other features of an equation that involve time functions that are *completely known* before the simulation is run. In our example, the time-varying damper requires a specific time function for its simulation. Of the various SOURCES which SIMULINK provides, we will discuss only nine of the most commonly needed. These are shown in Fig. 6-15. Six of these are accessed from the SOURCES icon on the main menu and the other three from the EXTRAS icon. When you double-click on the SOURCES icon you will get a new detailed SOURCES menu which will include six of the items shown in Fig. 6-15, plus some

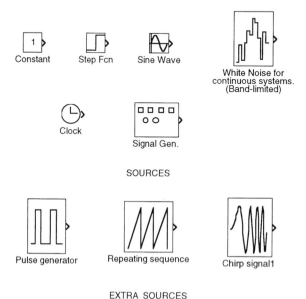

Figure 6-15 Often-used sources in SIMULINK.

Solution Methods for Differential Equations   **375**

more which we here ignore. If, for example, you needed one or more *constants* as inputs in your simulation, you would now "click and drag" one of these icons from the SOURCES menu onto your "empty" block diagram window, "dropping" it, for the time being into any convenient location. That is, at the beginning, we simply gather up all the different individual icons we need; we will *later* arrange and connect them properly. Recall that you only need to initially get *one* of each icon; you can later copy it as many times as needed.

Once icons are on your block diagram window, several manipulations are easily possible. Sometimes you want icons to be larger than their original size. This is especially desirable for those icons which display a drawing or number inside them, and which might be too small to clearly show this or to show it at all. To "zoom" any icon to a larger or smaller size, first click on it anywhere; this will create some little black squares ("handles") at the corners. Carefully put the cursor on one of these squares, click and hold, and drag the icon to the size you want. Don't make the icons *too* big, however, because you may not have enough space for your entire diagram. (It *is* possible to use *several* screens to depict a large system, but we don't want to deal with this here.)

Some icons require that you assign numbers to various features. For the STEP FUNCTION icons you need to assign how large a step you want and at what time you wish for it to occur. For any icons which require such numerical input, you first double-click anywhere on the icon and a dialog box will appear. This box *usually* is self-explanatory with regard to what you need to do, but sometimes it is not, and you have to get help from the manual or the on-line HELP system. For the STEP FUNCTION icon, it is easy to see how to enter the initial value (say, -1.0), the final value (say, $+2.0$), and the time of occurrence (say, 3.23). The time of occurrence (called "step time") is in whatever time units you decided to use in your equations.

Many practical problems involve sine waves ("frequency response"), so we often need the SINE WAVE icon. Again, you can use as many as you need, and each requires the setting of numerical values for the amplitude, frequency (radians/ time), and phase angle (radians). When we need explicitly the independent variable (time) of our equations, we use the CLOCK icon. We always need it if we want to plot MATLAB graphs versus time. In the time-varying damper of our example system, we would use the clock icon to also help generate the needed function of time. If you double-click on this icon a clock *window* appears, which when opened will give a running display of time as your simulation proceeds. You may or may not want to use this feature; too many windows can get confusing.

The CONSTANT icon is very simple; it provides a constant value of your choice, settable in the way described above. The SIGNAL GENERATOR is similar to electronic oscillators you may have used in lab work. It can be set up to provide sine, square, and sawtooth waves, with a frequency and amplitude of your choice. As with any other icon, you can use as many of these on your diagram as you need. It also has a random noise generator, but you usually will *not* want to use this. A "better" random signal generator is available separately, the "white noise for continuous systems" of Fig. 6-15. This provides a random signal with a Guassian ("normal") distribution function, and has frequency content only up to a frequency of your choice.

Three more useful signal sources are available by double-clicking on EXTRAS on the main menu and then on SOURCES. The PULSE GENERATOR creates a series of rectangular pulses at regular intervals and allows you to choose the pulse width and the time between pulses. REPEATING SEQUENCE allows you to create a periodic function with a period and waveform of your choice. The waveform is specified by giving a list of discrete points, so you don't need a mathematical function to describe it, you could use experimental data. It interpolates linearly between given points, so lots of points are needed if you want relatively smooth functions. You may be able to save some effort here by using MATLAB's SPLINE function to generate many points from a shorter list of "manually entered" points. The CHIRP signal provides a "sine wave" of fixed amplitude but time-varying frequency. Such waveforms are used routinely in lab vibration testing or other dynamic tests, so we may want to simulate them. You can learn about the CHIRP signal by clicking on its icon.

We have now explained all the sources commonly needed and now move on to the icons used to describe the *system*. To know which icons will be needed, one must of course always derive the *equations* describing the system. These equations will tell us which icons to use and how to arrange them on the block diagram. SIMULINK separates system icons into LINEAR and NONLINEAR. In Fig. 6-16 we show those linear icons that are most commonly used; there are others which we here ignore. The SUM icon allows us to "add up" the various terms on the right-hand side of our differential equations, once we have isolated the highest-derivative term on the left-hand side, as in Fig. 6-14. One such icon is needed for each equation, if we have a set of simultaneous equations. If you double-click on a sum icon that you have moved from the menu into your block diagram, you will get a dialog box allowing you to choose how many inputs you want and the algebraic sign for each input.

The GAIN icon allows us to multiply any signal by a positive or negative constant ("gain") of our choice. Double-click on it to set its value, which will then appear in the icon *if* it has been sized large enough. (Even if it *doesn't* appear, the value *will* be used as you wished.) The INTEGRATOR icon is the "heart" of the simulation since it is here that the numerical integration of the equation occurs. Double-click on it to set the initial value (initial condition in the equations) for that integrator. Remember, the initial value is the value of the *output* signal of the integrator, just before the driving inputs are applied. The *kind* of integrator used (Euler, Runge-Kutta, etc.), start and stop times, and the step size are selected later, using the SIMULATION item in the main menu bar. You should try to *avoid* using

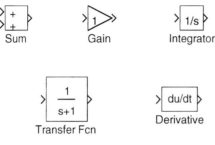

Figure 6-16 Often-used linear modules in SIMULINK.

Solution Methods for Differential Equations

the DERIVATIVE block because numerical differentiation is a *noise-accentuating* operation and is difficult to do accurately. If differentiation *seems* to be needed, there often are tricks available to avoid it, some of which we will explain as the need arises. If we are unable to avoid it, and if the signal to be differentiated is *known* to be "smooth," differentiation may be successful.

The TRANSFER FUNCTION block allows you to simulate any linear, constant-coefficient system for which you have derived a transfer function. The icon is labeled with the simplest transfer function, $1.0/(1.0s + 1)$, but by double-clicking on it you get a dialog box which allows you to insert *any* polynomial in s into the numerator and denominator. For example, if we type into the numerator box provided, the vector [1.23 -3.55 1.0] and into the denominator box [4.22 1.67 -2.34 1.0], we would be simulating a system with transfer function

$$\frac{Q_o}{Q_i}(s) \triangleq G(s) = \frac{1.23s^2 - 3.55s + 1.0}{4.22s^3 + 1.67s^2 - 2.34s + 1.0} \tag{6-77}$$

Recall that a transfer function represents a single differential equation relating a single input and a single output, with all initial conditions equal to zero. Simulation software, SIMULINK included, does not allow initial conditions *other* than zero when you are using the TRANSFER FUNCTION block. We usually prefer to avoid use of transfer functions in favor of simulating directly from the "physical" set of simultaneous equations, which allow easy and correct setting of the initial conditions. There definitely will be, however, many cases where we use transfer functions to speed our diagraming and keep the diagram compact, being always careful to use the concept correctly.

We are now ready to start discussing the NONLINEAR blocks shown in Fig. 6-17. This is again a selection of the most commonly used, from a longer list. The ABSOLUTE VALUE block is self-explanatory; the output signal is the absolute value of the input signal. It has many uses. Our example of Fig. 6-12 will use it to model the square-law damper; see Eq. (6-74). We used it in Chapter 5, on the SCR amplifier model. For any signal $y(t)$, the expression $y(t)/\text{abs}(y(t))$ produces a signal which is either $+1$ or -1 and can be used to adjust the algebraic sign of some other

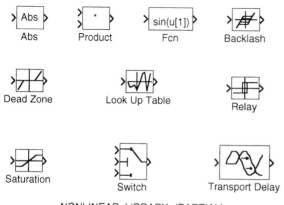

Figure 6-17 Often-used nonlinear modules in SIMULINK.

signal, such as a dry friction force. The PRODUCT block allows us to multiply together any number of input signals. Double-click on this icon to set the number of inputs desired (default is 2). To *divide* one signal, say y2, by another, say y1, in SIMULINK use the FUNCTION block (explained next) to form 1/y1 and then multiply y2 by 1/y1.

The FUNCTION (fcn) block provides a very general nonlinear capability which duplicates the math function operations used in command-line simulation software. Any function that can be written in the C language, which is what SIMULINK uses, can be invoked with this block. (Remember that C uses *many* of the same symbols for functions and operations as Fortran, which may be more familiar to you.) That is, if you wanted to form a term (needed in one of your equations) that has the form $3 \sin(x) + 2e^{-x} - 5.67x^2$, and the signal x was available somewhere in your block diagram, you could use the FUNCTION block to do this. Just double-click on the icon and type into the box provided the function $3.0^* \sin(u) + 2.0 * \exp(-u) - 5.67u^2$. Then connect the signal x to the input of this block. Notice that we *must* use the symbol u to represent the input to the block, *no matter what its "real" name might be in our equations*.

The BACKLASH block models "lost motion" or hysteresis in mechanisms. For example, the tooth of a mating gear normally does not completely fill the space between the two meshing teeth. When we rotate one gear, the other does not move at all, until we close up this airspace or backlash. When the driving gear reverses, the driven gear stays *where it last was* until the airspace is closed up; then motion of the driven gear recommences. A dialog box allows us to choose numbers for the airspace and also the location of input and output at time zero. DEAD ZONE provides a simple "dead space" in a mechanism or action. It doesn't cause the hysteresis effect present in BACKLASH. When outside the dead zone, the "slope" of the output/ input ratio is $+1.0$ or -1.0.

One of the most useful nonlinear elements is the LOOKUP TABLE. It allows us to specify almost any functional relationship between the input and output signals because it does not require any *mathematical formula*, only sets of "x, y" data points, such as we might measure in an experimental calibration in the lab. A dialog box allows you to enter as many x, y points as you wish; linear interpolation is used for values needed between given points. Use "lots" of points if you require relatively smooth curves, but you can use MATLAB's spline function, as we mentioned above with the REPEATING SEQUENCE block, to ease this task. Input points are given in a vector in one box and corresponding output points in another; there must of course be an equal number of input and output points. If you expand this icon to sufficiently large size, a small version of the actual graph will appear inside it, allowing a visual check of the correctness of your setup.

RELAY provides a simple model of "on-off" type devices, such as the controls for a house heating system. A dialog box allows us to set four numerical values: input for ON, input for OFF, output when ON, output when OFF. Selecting an "input for ON" value greater than the "input for OFF" value models hysteresis, whereas selecting equal values models a switch with a threshold at that value. RELAY is also useful as a general simulation tool for changing algebraic signs of equation terms when certain events occur, and for activating or disabling terms in an equation. SATURATION is widely useful in realistic models since many physical effects which are nearly linear will reach a limiting value (saturation) if the input gets

Solution Methods for Differential Equations 379

too large. One common example is the output voltage of any electronic amplifier. When the input voltage gets too large, the output no longer increases. In fluid valves, such as the servovalve of Fig. 5-33, when the spool has completely uncovered the port, further spool motion causes *no* increase in fluid flow. In electromagnetic actuators using iron-core coils, when the coil current increases to the point where it has completely aligned the magnetic domains in the iron, further current increases result in *no* increase in magnetic field or force. In this icon we are allowed to independently set both the lower and upper limiting values; they need not be the same. SWITCH lets us "steer" one of two signals to the output, depending on the value of a third signal. In the icon, the "middle" signal is the control and the two "outer" signals are the alternate paths. If the middle signal is equal to or greater than zero, the upper signal goes to the output; otherwise the lower signal goes there.

TRANSPORT DELAY really shouldn't be in the *nonlinear* library; it is a *linear* operation. Its inclusion in a system *does*, however, usually make the overall equations analytically *unsolvable*, just as for most nonlinear elements, so this is probably why the SIMULINK software engineers decided to put it in this library. As long as it functions properly, of course we don't really care *where* we go to find it. Transport delay is also called transport lag, discrete delay, or dead time. Its operation is exactly that of the dead time element that we discussed in Fig. 6-6. The mathematical definition is quite simple: The output is an *exact* duplicate of the input, but it "happens" only after a certain time delay, called the dead time or transport delay. Dead time has many practical uses. Suppose you are steering a robot Moon vehicle from earth, watching a TV monitor receiving pictures from a camera on the vehicle. Steering commands that you send to the Moon arrive there with a delay of 240,000 miles/186,000 miles/sec = 1.29 seconds, since radio signals travel at about 186,000 miles/sec. Pictures received from the Moon by your TV are also delayed by the same amount. These delays make the human operator's steering task quite difficult, especially if the vehicle moves rapidly. As another example, consider the control of sheet thickness in an aluminum rolling mill. The gage for measuring sheet thickness is of necessity located 15 feet from the rolls, where changes in thickness are accomplished by moving the rolls closer together or farther apart. If the sheet moves at, say, 15 ft/sec, the gage only hears about thickness changes 1.0 seconds after they really happen. This delay can cause instability in the feedback control system used to control thickness.

Having described the most-used signal sources and system blocks, we are now ready to discuss the icons used to generate graphs of the signals of interest to us in our system. SIMULINK itself provides some graphing capability, but most users will prefer to use MATLAB's graphing tools, which are easily accessible from SIMULINK. The two SIMULINK graphs that you might want to try are accessed from the main menu under EXTRAS and then DISPLAYS. The GRAPH SCOPE (Fig. 6-18) will make a running plot versus time of any signal connected to its input. When you double-click on it, you can set the max and min limits on the vertical scale and also the range of time covered. Since you often *don't know* the max and min values of signals before you make the first run, these settings usually involve some trial and error. If you connect, say three graph SCOPES to three separate signals, you will get only *one* graph, but it will have the three curves superimposed on it. Unfortunately they will all share *the same* scale, so some may look very small. You can see that GRAPH SCOPE has a number of inconvenient features, which is why

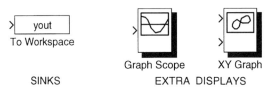

Figure 6-18 Often-used readout modules in SIMULINK.

you may prefer to do MATLAB graphs. Perhaps the only advantage of GRAPH SCOPE is that the graph appears automatically and you can watch the signal develop from time zero as the simulation proceeds. The MATLAB graphs are available only *after* the simulation run is complete. Similar in behavior to GRAPH SCOPE, XY GRAPH allows "cross-plotting" of any two signals connected to its two inputs.

If you prefer to do MATLAB graphs, then each signal you want to use must be sent to a TO WORKSPACE icon, available from the main menu under SINKS. When you double-click on this icon you can choose a name for that signal and the maximum number of points that you want stored for later use. I routinely request 5000 points, which is more than the 1000 or 2000 that are usually produced in a simulation. If you reserve too few points, your graphs will be incomplete, so I "over-reserve" to prevent this. We will give more details on MATLAB graphing shortly. Be sure to send the time signal (from the CLOCK icon) to the workspace; you always need it.

At this point we have described all of the most-used icons that you need to drag onto your new file window in preparation for creating your simulation. I usually position all of these along the bottom edge of the window and drag them up, one by one, to the locations I want. We generally start with "steps 1, 2, and 3" as in Fig. 6-14. We now need to discuss how to interconnect icons, that is, how to "wire up" the diagram. "Source" icons always have an output connection point, "sink" icons have an input connection point, and "system" icons have *both* an input and output connection point (see Figs. 6-15 to 6-18). Arrange your blocks from left to right and up and down to suit the needs of your equations. This usually involves some trial and error to fit things into a confined space. It is not hard to make adjustments in location as they are needed.

To actually make connections, move the cursor to an output connection point and click and hold on it. (You can use either the right or left mouse button, but the right button seems to be more "versatile"; try them both to find your preference.) Now drag the cursor to the desired input connection point and release the mouse button. This action should "draw" the desired line; it may take a little practice to get it right. Note that the connecting lines have *arrowheads* which show the direction of signal flow. If you need to orient blocks *vertically* or *right to left*, first click on the icon to select it (the little black "handles" appear), then click on OPTIONS in the main menu bar, and then ORIENTATION. Select the option you want and the icon will be reoriented.

While drawing connecting lines, you can pause at any point, click and hold again, and take off in another direction with your line. While diagrams with lines going off at various angles will run OK, most people prefer to make all lines exactly

Solution Methods for Differential Equations

horizontal or exactly vertical. If a line has been established and you want to "branch off" from it, position the cursor at the desired branching point and click and hold the right mouse button. Now drag away from the branch point and you should see your new line developing. To delete a line, select it by clicking on it to make the "handles" appear. Then go to EDIT and CUT to delete the line. Icons are similarly deleted. If you have some line "corners" which are not 90°, click and hold the right button on the corner (a circle appears) and then drag it until you get the "square" corner.

You should now be ready to try an example; let's return to the 3-mass sytem of Fig. 6-12 and Eqs. (6-74) to (6-76). Begin as in Fig. 6-14 and then complete the diagram as in Fig. 6-19. If you want to try out GRAPH SCOPE, you can add one or more of these to Fig. 6-19. You can send the same signal to *both* GRAPH SCOPE and TO WORKSPACE if you wish. To make any diagram easier to understand and debug, you should routinely *label* all the significant signals and blocks with names that relate to the physical system. This is easily done using SIMULINK's "note block" feature. Just click where you want the label to appear and start typing. When you enter the label it usually jumps slightly to a new position. Click and hold on the label and you can drag and drop it exactly where you want it. Figure 6-19 shows many such labels. Now that you know about the details of SIMULINK graphical "programming," it might be useful to review some of our earlier examples, which were done with minimal explanation (Figs. 2-3, 2-18, 2-32, 2-51, 2-55, 2-59, 3-9, 3-34, 3-39, 5-18, 5-21, 5-29, 5-31, 5-37).

When a simulation diagram is complete, as in Fig. 6-19, to run it you go to the main menu bar and select SIMULATION and then PARAMETERS. A dialog box appears where you select start time (generally 0), stop time (to suit your problem),

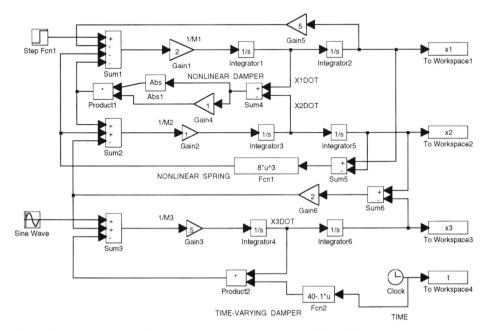

Figure 6-19 Simulation diagram for mechanical system of Fig. 6-12.

integrator type (use RK-5 for most problems), and minimum and maximum computing step size. The RK-5 integrator is a variable-step-size type, but I generally start every new simulation with a fixed step size, only going to the variable-step feature after I am sure that I am getting reasonable results. To force the RK-5 integrator to be fixed-step, just enter the max and min step sizes as the same value. Choose the step size to be about 1/1000 or 1/2000 of the total problem time if you have nothing else to go on. If oscillations of known frequency are expected, make the step size about 1/10 of the oscillation period. Once you are getting reasonable results, you should increase and decrease the step size to make sure you are in a range where halving or doubling step size seems to have no effect on the results. If you let RK-5 decide on the step size and how it changes during a run (by making max and min step sizes *different* numbers), watch out for large step sizes which may give accurate points, but "jagged" graphs because of coarse point spacing.

You could now run the diagram of Fig. 6-19, using a stop time of 50 and fixed step size of 0.02. This produces data which can be MATLAB plotted as in Fig. 6-20. We now want to say a little about MATLAB graphing, as it applies to SIMULINK. MATLAB is very popular as a general-purpose computing environment, so you may already know a lot about it. I must assume here that you know *nothing*, so "experts" will please bear with me. MATLAB has *extremely* comprehensive and complex graphic capability, but we here want to discuss only the "bread-and-butter" features useful in SIMULINK. In Fig. 6-19 we have sent time t and mass displacements x1, x2, and x3 to the workspace, so we can plot the displacements versus time. Once a simulation run is complete, you switch from your diagram window to the MATLAB

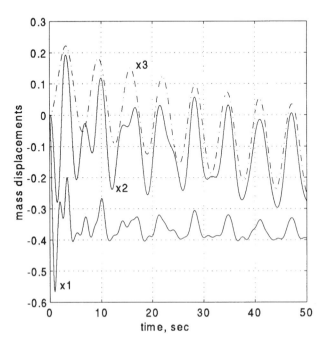

Figure 6-20 Results of simulation of Fig. 6-19.

Solution Methods for Differential Equations   **383**

command window, because that is where the graphing commands will be entered. To get Fig. 6-20 we would enter

```
plot(t,x1,t,x2,t,x3,'-.'); grid; xlabel('time,seconds')
ylabel('mass displacements')
```

The label x1 on the curve is obtained with

```
set(gca,'drawmode','fast')
gtext('x1')
```

When you enter the gtext statement, the graph reappears with a cross-hair cursor which you move with the mouse. Move the cursor to where you want the label to begin and click the mouse to drop the label there. Repeat for the x2 and x3 labels. If we had wanted a title at the top of the graph we could use

```
title('3-mass system response')
```

Let's now go over the most-used graphing commands. The line types and colors used for each curve can be selected as follows. Available colors are yellow (y), magenta (m), cyan (c), green (g), blue (b), white (w), and black (k), with yellow being the default. Line types are solid (–), dotted (:), dashdot (-.), and dashed (- -), with solid being the default. The default line width is 0.5/72 inch, denoted by .5 in the 'linewidth' command. Thus to use a red, dashed line of width 1.0/72 for the variable x1 we would enter

```
plot(t,x1,'r--','linewidth',1)
```

If you want a plot with only point symbols but no connecting lines, the symbols (o) circles, (.) dots, (x) x's, (+) crosses, (*) asterisks, are available. Thus plot (t,x1,'b+') produces a graph with unconnected blue crosses, one at every computed point, which is usually not desirable. (Default symbol size is 1/72 of an inch. Six sizes from 1/72 to 6/72 can be selected with 'markersize'.) Such plots look better if we don't plot every point. To reduce the number of plotted points, first find out how many *total* points there are with

```
length(x1)
```

whereupon the computer returns

```
ans = 2500
```

To plot, say, every 50th point, create a new variable newx1 with the statement newx1=x1(1:50:2500) and a new time variable newt with the statement newt=t(1:50:2500). You can now get your new plot with plot(newt, newx1,'b+','markersize',2) if you want blue crosses twice as large as the default.

If you want to have a curve with *both* symbols and a drawn line, you have to superimpose a symbol plot on a line plot (either can be done first). Having entered the commands for the first plot and observed that this graph is as you wanted, you then enter the command hold on, which holds the first graph on the screen. (This command is useful for all kinds of superpositions, not just our present application.) You then enter the statements for the second graph, whereupon the superposition occurs.

We mentioned that MATLAB graphs with multiple curves all share the *same* vertical scale, which usually will make some of the curves too small. As soon as you see this happening, you can "fix" it by simply plotting some convenient multiple of the "small" variable. Then all the curves can more nearly extend over the full vertical scale, making them easier to read. (Of course the shared scale can have the correct *units*, say pounds, for only *one* of the curves.) For example, if y1 is about one-third the size of y2 on a shared graph, just use plot(t,3*y1,t,y2). Another way to plot multiple curves uses completely separate graphs, like a multichannel lab recorder. Now each curve fills its scale and has correct units, but each curve is now smaller, since they are all "stacked" on the same sheet of paper. Remember that in dynamic systems we usually want to see all the variables in one view and on a common time scale, since time correlation of different variables tells us a lot about system behavior.

To get the "stacked" plots just mentioned, use the "subplot" command. For example, subplot(3,2,p) divides the paper into six separate graphing areas, two columns of three rows each, and prepares to plot something in area p, where p is counted from left to right in each row, starting at the top row. Thus subplot(3,2,5) will do its plotting in the first column, third row location. We can get such a plot for the data of Fig. 6-20 as follows.

    subplot(3,1,1),plot(t,x1); subplot(3,1,2),plot(t,x2);
    subplot(3,1,2),plot(t,x3)

The result is shown in Fig. 6-21. We could of course have used the usual title, xlabel, gtext, r--, etc. commands in each of the plot statements.

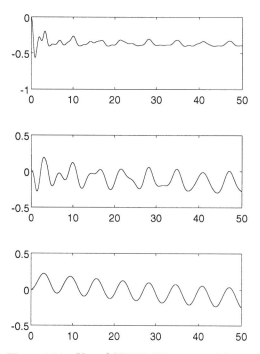

Figure 6-21 Use of SUBPLOT command for multiple graphs.

Solution Methods for Differential Equations *385*

If you want to only plot a selected subarea of the total range of computed values, the *axis* command may be used. Suppose you had earlier computed and plotted your values for a time range of 0 to 10 seconds and this produced a range of 0 to 14 for some variable of interest. Now you decide to examine the time range from 2 to 3 seconds and the variable range from 4 to 7 more closely. To use the entire graphing area to display the restricted ranges just stated, with the original graph on the screen, you would use the command `axis([2 3 4 7])`. Another way to "zoom" a graph is to use only part of your points in the plot statement. Thus `plot(t(1:100),x1(1:100))` plots only the first 100 points of your complete data set.

Printing simulation diagrams, text commands, or graphs follows normal Windows procedures and usually presents no problems. Since simulation diagrams tend to get "busy," I always expand them to full-screen size before printing them so I can carefully check that all the desired details are present. When graph printing size is left to default, SIMULINK makes graphs which use up most of the 8.5 by 11 inch paper. To control this printing size, use the command

```
set(0,'defaultfigurepaperposition',[1.5 1.5 4.5 4.5])
```

The numbers (inches) inside the brackets are: [left edge, bottom edge, width, height] and can of course be numbers other than I used above. When printing graphs of reduced size, be sure to check the final result to make sure that no features near the edges have been "lost" or that text has gotten too small to easily read.

6-6 SPECIFIC DIGITAL SIMULATION TECHNIQUES

The above discussion has outlined the most commonly used features of SIMULINK in a general way. Remember that other software packages for continuous system simulation languages (CSSLs) will have very similar capabilities and methods of usage. Learning the details of actual practical application is perhaps best done by working out many specific examples, such as those done earlier in this text (Figs. 2-3, 2-18, 2-32, 2-51, 2-55, 2-59, 3-9, 3-34, 3-39, 5-18, 5-21, 5-29, 5-31, 5-37, 6-19) and many more which will appear in later chapters. However, it may be of some use to gather together in one place, and explain, some specific techniques which are used over and over in simulation studies.

Generation of Input Signals. Every simulation problem involves some signals which are not unknowns but rather are given as known functions of time. These could be external driving signals (input forces, input voltages, etc.) or parameters that vary with time in a known way (a damper that gradually heats up, reducing its B value). Some known functions of time are obviously handled by the step, sine, and signal generator (sine, square, triangle) modules. Other functions of time given by a known formula over their entire course can be handled by sending the time signal (from the clock icon) into the nonlinear block called "Fcn", and entering the desired function into that block. For example, a parabolic input $5t^2$ would be given by sending t into a Fcn block with `5*u^2` entered in it.

A very general capability is available by sending t into a LOOKUP TABLE. Here you don't even need a formula; you could use data points from a lab measurement. When entering points into the input and output vectors, you may want to use the MATLAB function called SPLINE to get a smooth curve from a small number of manually entered points. Suppose you had 11 time points and 11 corresponding function values. If you use these directly in the lookup table it will interpolate linearly, giving a "curve" made up of straight-line segments. To get a smooth curve that goes exactly through all 11 points, proceed as follows in MATLAB.

```
x=0:.1:1.0;
y=[0 .053 .077 .097 .118 .143 .177 .227 .310 .480 1.00];
xinterp=0:.01:1.0);
yinterp=spline(x,y,xinterp);
```

Now, when you go to set up your LOOKUP TABLE, use the names xinterp for the input vector and yinterp for the output vector. You of course can use any number of interpolating points, I used 101 just for this example. You could also MATLAB plot yinterp versus xinterp before using them in the lookup table, just to check for correctness.

When an input has a "formula" description but the formula changes for different segments of time (as in Fig. 6-22a), the following technique may be useful. The first segment is easily obtained by sending t into a Fcn block set up to square its input. However, we need to "turn off" this parabola after $t = 1.0$. One way to do this is to multiply it by a step function which is 1.0 until $t = 1$ and zero thereafter. We then need to add another step function which is "on" only for $1 < t < 4$. Finally, we form $(5 - t)$ from t, multiply this by a step which is "on" for $4 < t < 5$ and add this to our sum. All these operations are shown in fig. 6-22b. This use of step functions to "turn on and off" other signals is of general utility.

We occasionally want to use random signals in our simulations. Sometimes these are main inputs, such as the random wind gust forces felt by an aircraft wing. Other times we superimpose a small random input on a deterministic input (say a sine wave) to model the "noise" present in all real-world signals. Figure 6-15 showed a random signal source but we did not explain any details there. If you double-click on this icon, you can enter three numbers which determine the characteristics of the random signal. First, all random number generators use a *seed value* to start the operation. If you always used the same seed value, the sequence of random numbers generated would always be the same. If we want to *repeat* a simulation run as a check on earlier calculations, we need to use the same seed value as we did earlier, so always keep track of your seed value—SIMULINK does not. If you need several random signals in a simulation and you want them to be unrelated (uncorrelated), be sure to use *different* seed values for each. You can use most any integer values for seed values; I have decided personally to use five-digit odd integers.

Having selected a seed value, you can next pick a number for the *noise variance*, which determines the "size" of your random signal. The larger the variance, the larger the signal. Usually you will need to set this by trial and error, until you get a graph that meets your needs. Finally, you need to set the *sample time*, which determines how rapidly the signal varies with time. The shorter the sample time, the more rapid the variation. The output signal "jumps" from one random number to another at time increments equal to the chosen sample time, giving a "random amplitude square wave." Most physical random signals are "smooth" rather than "square-cornered," so we often send this signal into a low-pass filter to give a more

Solution Methods for Differential Equations

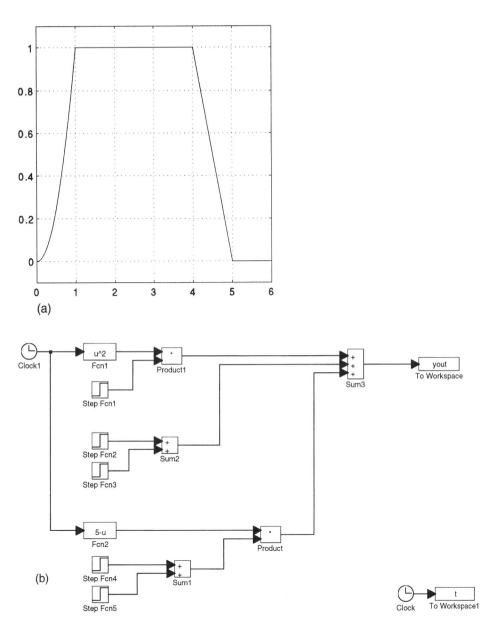

Figure 6-22 Use of step functions to create segmented inputs.

realistic signal. A second-order transfer function $1/(\tau^2 s^2 + 1.4s + 1)$ is often a reasonable filter, with τ chosen to be about half the sample time. In any case, we adjust the noise generator and the filter until we get a signal that meets our needs. Figure 6-23 shows some typical results.

Side-by-Side Comparisons. We regularly need to decide how simple or how complex a model will be needed to get useful results. Often this choice is between a

388 **Chapter 6**

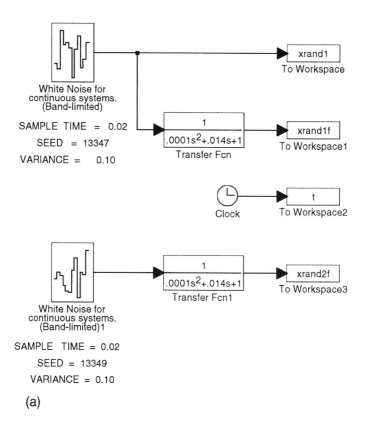

(a)

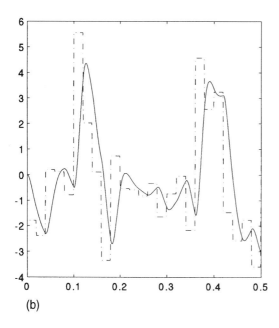

(b)

Figure 6-23 Simulation of random inputs

Solution Methods for Differential Equations 389

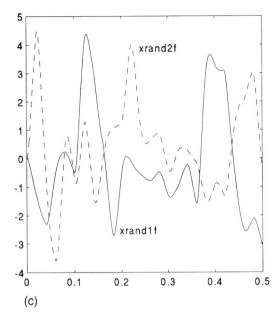

(c)

Figure 6-23 *(continued).*

linearized approximation or a more correct nonlinear model. Linearized models often are adequate for a certain range of input amplitudes but "go bad" for other ranges. How do we find these ranges? In design, we usually formulate several alternative designs and develop adequate models for each. We then need to decide which design is "the best" and will be further developed. When a given design has been chosen, we want to "optimize" it by finding the best combination of parameter values. Often we study one parameter at a time, to see what is its effect on system performance. All the above situations can be efficiently studied using a "side-by-side" comparison of simulation models.

That is, we construct the separate models to be compared on the same simulation screen and run them "in parallel" (at the same time). It is then easy to plot graphs comparing variables of interest from the different models, and use these results in making the needed decisions. When we have a single model and want to study the effect of changing a parameter, the variation of this parameter can often be "automated" in the simulation, to speed our analysis. For example, in the obsolete language CSMP, if you wanted to make five runs with five different values of the parameter B, you just used a statement PARAM B=(3.,4.,7.,8.,9.,) and five runs, with all the requested tables and graphs, would be automatically made. Current languages provide similar facilities with varying degrees of convenience. In SIMULINK, one method goes as follows. In Fig. 6-24a we show a diagram for a simple second-order linear equation where we wish to vary two of the parameters, called B and KS. In the gain blocks where numbers for these parameters ordinarily appear, we instead insert letters, B and KS. This is done exactly as when you insert numbers; double-click, get the dialog box, type in the letters you wish. In this diagram you see an icon called OUTPORT, which we have not previously used or discussed. It is obtained from the CONNECTIONS menu. You need to provide

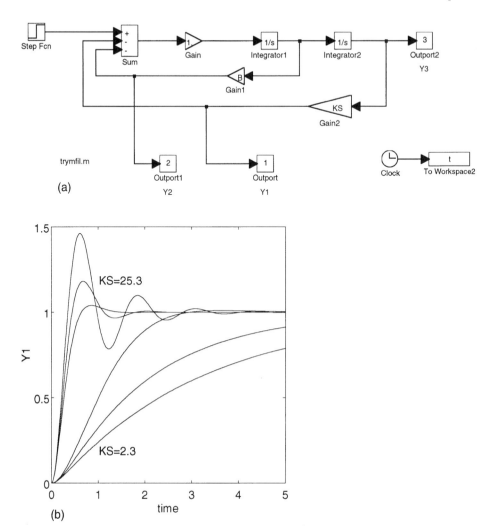

Figure 6-24 Generating repeated runs with varying parameter values.

an OUTPORT for each signal that you want to plot; in the example I have selected three such signals. When you double-click on an OUTPORT you are asked to assign it a number; I have used the numbers 1, 2, and 3. When you later plot these signals, they will be part of an array called y. Y1 will be referred to as y(:,1), which means all the rows of the first column. This list has the successive values of this signal, one for each time value. Similarly Y2 will be y(:,2) and Y3 will be y(:,3). On the simulation diagrams, the letters Y1, Y2, and Y3 did *not* appear automatically. I used the usual "NOTES" method for typing them into the diagram. You don't *have* to do this, but it helps one remember that these signals *must* be called y's when we plot them.

Once the diagram has been set up, to run it, you *don't* use the menu (SIMULATION, START) as we usually do. Rather, you command the run from

Solution Methods for Differential Equations   **391**

the MATLAB command window, so you now need to switch to it. In this window
you enter the following command.

```
hold off; for i=1:3; B=2.5*i; for j=1:11:12; KS=2.3*j;...
[t,x,y]=rk45('filename',[0 5],[0 0],[.001 .01 .01]);...
plot(t,y(:,1)); hold on; end; end; hold off
```

In the first line we are asking for six combinations of B and KS: two values (2.3 and
25.3) of KS paired with three values (2.5, 5.0, 7.5) of B. This is done using two nested
FOR LOOPS in the usual way. The "hold off" statement should always be used to
cancel any "hold on's" that you might have used earlier and forgotten about. In the
second line we always use [t,x,y] = rk45 if we are using the default Runge-Kutta 45
integrator, but what is *inside* the parentheses changes for each problem. When you
set up and saved your simulation diagram, you had to assign a name to it; this must
be used where I show 'filename'. The first [ ] has two entries which are the starting
time (usually 0) and the stopping time, in this case 5. The second [ ] has as many
entries as you have integrators in your diagram, and contains the initial values for
those integrators, in the same order as the integrators are numbered on your diagram
(this diagram numbering *is* automatic). In my example, both I.C's are zero. The last
[ ] has three entries; the first is the accuracy tolerance which appears by default when
you run from the menu, but must be manually entered when you run from the
command line. Use 0.001 until you find out you need to change it. The next two
entries are the min and max step sizes. I have recommended using *fixed* step sizes
when starting any new simulation, so I have here used 0.01 and 0.01. In the last line, I
have decided to plot, versus time, the signal called Y1 on the diagram. The "hold
on" causes the six curves to be superimposed. The two "end's" terminate the respective FOR LOOPS, and the "hold off" is again a "safety feature" to make sure no
"hold on's" are still lurking anywhere to mess up any subsequent graphing. Figure
6-24b shows the results of this procedure. You should be able to adapt this example
to most of your needs for this type of parameter variation.

Event-Controlled Switching.   We showed above how to use step functions and multipliers to turn signals on and off at known *times*. Sometimes we need such "switching" operations based, not on time, but when certain variables reach certain critical
values. One way to implement such operations uses the RELAY and PRODUCT
blocks. Figure 6-25 shows a simple example. The upper RELAY is set up so that its
output is +1 when the input is positive and −1 when it is negative. The lower
RELAY is set so that its output is +1 when the input is positive and 0 when it is
negative. We want the switching actions to occur when the input signal x crosses the
value +5, so we use a summer to create this bias. In Fig. 6-25 I have intentionally left
the curves unlabeled; you should have no difficulty in discerning which graph is
which.
      Sometimes a simulation requires the selection of alternative signal paths or
parameter values when a variable reaches some critical value. In Fig. 6-26a two
pumps are used to quickly and accurately fill a tank of 10.0 ft^2 area to a desired
level of 10.0 feet. A "coarse" pump supplies 50 ft^3/min and a "finishing" pump
supplies 10. The coarse pump is used alone until the level reaches 9.5 feet, whereupon
we switch to the finishing pump to accurately complete the filling process, turning

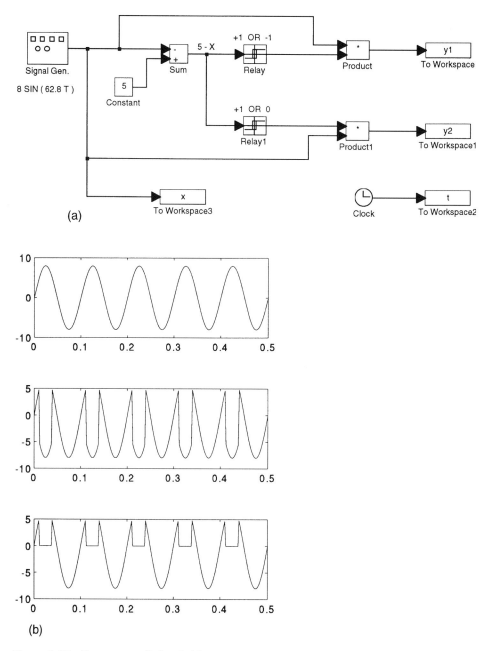

Figure 6-25 Event-controlled switching.

this pump off when the level reaches 9.99 ft. (What practical aspect of pump behavior leads us to command pump turnoff at a level less than 10.00, such as 9.99?) A tank level sensor measures the level h and tells the pumps when to switch. The SIMULINK SWITCH block connects the upper input to the output as long as the middle input is positive; otherwise it connects the bottom input to the output.

Solution Methods for Differential Equations

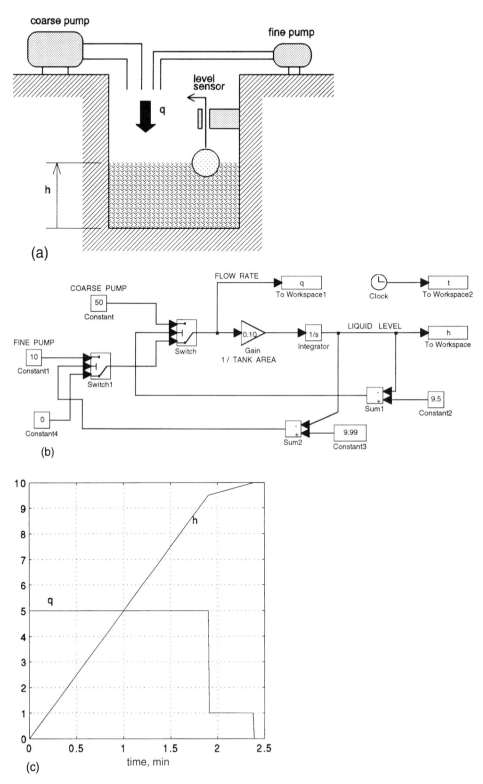

Figure 6-26 Use of switching in a tank-level control system.

We thus want to switch from the coarse pump to the fine when (9.5-h) crosses zero, and turn the fine pump off when (9.99-h) crosses zero. Note that SWITCH provides selection between only two signals, but by "cascading" several SWITCH blocks, we can select among any number of signals. Figure 6-26b shows the simulation diagram, and Figure 6-26c the response of tank level h and flow rate q. This simple problem by itself hardly needs computer power, but its simulation might be useful if the tank system were part of some larger process.

6-7 SIMULATION SOFTWARE WITH AUTOMATIC MODELING

"Classical" continuous system simulation languages like CSMP, ACSL, and SIMULINK require that the engineer derive the differential equations describing the system to be studied. This task is "automated" in software products like WORKING MODEL[3] and ADAMS.[4] Here the engineer draws "schematic" rather than block diagrams. These schematic diagrams resemble the actual system and are made up of standard physical components such as masses, springs, dampers, rotating joints, sliding joints, pulleys, etc. Such software is usually limited to a narrow range of physical hardware, such as mechanical systems, whereas languages like ACSL and SIMULINK, since they work with equations, can be used on *any* kind of system, including those that might be physically impossible. The advantage of software like WORKING MODEL and ADAMS is mainly in efficiency, since no time is spent deriving or writing equations, and in realism, since the graphical user interface displays an animation of the machine simulated. Animation features can of course be added to languages like ACSL and SIMULINK but they are not the basic operating method.

While the cost-effectiveness of automatic-modeling software makes it rightly popular in real-world engineering, in a student textbook such as this, we prefer to "force" the student to develop skill in modeling, assumption making, and equation writing by using the less automated software. We feel that these skills are really necessary background to later industrial use of the more advanced software. However, if a curriculum sufficiently develops these student skills in the earlier years, it might be defensible to use the more advanced software in "capstone" courses in the senior year. Many software products of this type are available for different application areas such as stress analysis, heat transfer, fluid flow, electronic circuits, etc. and there is no question that they are widely used in practical engineering. One product of particular interest in system dynamics focuses on hydraulic and pneumatic systems.[5]

For systems that are mainly mechanical, WORKING MODEL is quite easy to learn and use, but concentrates on two-dimensional mechanics. ADAMS can handle the more complex three-dimensional motions, but requires considerable effort to master. We here will give a very cursory discussion of WORKING MODEL. Figure 6-27a shows a screen from WORKING MODEL where I have set up a

---

[3]Knowledge Revolution, 66 Bovet Road, Suite 200, San Mateo, CA 94402, 415-574-7777.
[4]Mechanical Dynamics, 2301 Commonwealth Blvd., Ann Arbor, MI 48105, 313-994-3800.
[5]HyPneu, BarDyne Inc., 5111 N. Perkins Road, Stillwater, OK 74075, 405-743-4337.

Solution Methods for Differential Equations

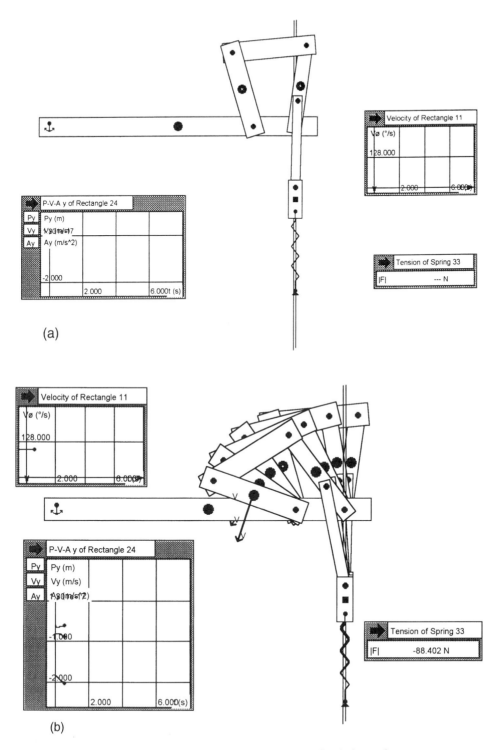

Figure 6-27 Simple example of the WORKING MODEL simulation software.

six-bar "drag link" mechanism, as used in some mechanical presses. One starts such a setup with a blank screen and a menu bar with icons for rigid bodies of various shapes, springs, dampers, pin joints, sliding joints, etc. An "anchor" icon when attached to any link, prevents any motion of that link. The large horizontal link is so anchored and serves as the frame of the machine. I then drew some rectangles of desired sizes and pin-jointed them to each other and the frame, as needed. The "output" of the mechanism is the block which slides in a vertical slot—a spring being used to represent the opposing force of the object being punched in the press. The input link's motion is defined by giving it a large moment of inertia and initial angular velocity, modeling the flywheel action of a real press. Starting from this given input condition, the software will derive all the needed equations and integrate them to give all the forces and motions associated with the various links. In addition to providing tables and graphs of all the unknowns, the software automatically displays an animation of the motions, showing a complete cycle of the press or any desired portion of a cycle, as in Fig. 6-27b.

While the display shows all links as simple rectangles, we can separately input masses and moments of inertia which correspond to the real parts, which will usually not be rectangular. It is even possible to detail complex parts in separate CAD software and import these shapes into WORKING MODEL. While linear springs and dampers are the defaults, we can easily define, using formulas, nonlinear elements. If gravity forces are important, these are easily included, even unusual cases like moon gravity. System variables can be displayed as line graphs, bar graphs, and numerical values. Vector displays of velocity and acceleration of any selected point are available. While most applications do not require any programming or equation derivation, a programming language called WORKING MODEL BASIC is provided for those users needing such a facility.

While the efficiency of such software makes it appealing for many practical engineering situations, its "black box" nature requires that the user carefully evaluate results to make sure that the assumptions and methods being used "behind the scenes" do not violate any physical features of the system being studied. This warning of course applies, in varying degree, to *all* engineering software, not just this type.

6-8 STATE-VARIABLE NOTATION

In reading about system dynamics and its various applications you may encounter terminology associated with *state-variable* concepts. Since this ties in somewhat with simulation methods I want to present here enough background that you can interpret these materials and relate them to the presentation in this text. While state-variable methods include nonlinear and time-varying–coefficient effects, they are most commonly applied to linear, time-invariant models. The general form of the state-variable description of such systems is given by

$$\dot{x}(t) = Ax(t) + Bu(t) \tag{6-78}$$

$$y = Cx(t) \tag{6-79}$$

Solution Methods for Differential Equations

These are *matrix differential equations*, a compact way of writing the set of first-order simultaneous equations:

$$\begin{bmatrix} \dot{x}_1(t) \\ \dot{x}_2(t) \\ \vdots \\ \dot{x}_n(t) \end{bmatrix} = \begin{bmatrix} a_{11} & a_{12} & \cdots & a_{1n} \\ a_{21} & a_{22} & \cdots & a_{2n} \\ \vdots & & & \vdots \\ a_{n1} & a_{n2} & \cdots & a_{nn} \end{bmatrix} \begin{bmatrix} x_1(t) \\ x_2(t) \\ \vdots \\ x_n(t) \end{bmatrix} + \begin{bmatrix} b_{11} & b_{22} & \cdots & b_{1r} \\ b_{21} & b_{22} & \cdots & b_{2r} \\ \vdots & & & \vdots \\ b_{n1} & b_{n2} & \cdots & b_{nr} \end{bmatrix} \begin{bmatrix} u_1(t) \\ u_2(t) \\ \vdots \\ u_r(t) \end{bmatrix}$$
(6-80)

$$\begin{bmatrix} y_1(t) \\ y_2(t) \\ \vdots \\ y_m(t) \end{bmatrix} = \begin{bmatrix} c_{11} & c_{22} & \cdots & c_{1n} \\ c_{21} & c_{22} & \cdots & c_{2n} \\ \vdots & & & \vdots \\ c_{m1} & c_{m2} & \cdots & c_{mn} \end{bmatrix} \begin{bmatrix} x_1(t) \\ x_2(t) \\ \vdots \\ x_n(t) \end{bmatrix}$$
(6-81)

The column vector **x** is an *n*-dimensional *state vector* in the *state variables* $x_1, x_2, \ldots, x_n$, **u** is an *r*-dimensional *control vector*, and **y** is an *m*-dimensional *output vector*. Matrix **A** is called the *system matrix* and is $n \times n$, **B** is the $n \times r$ *control matrix* and **C** is the $m \times n$ *output matrix*. Figure 6-28 shows the conventional block diagram used to represent such multiple-input–multiple-output systems; the broad arrows (as contrasted with the single lines in our usual block diagrams) denote the flow of *multiple* signals. The state variables of a system, interpreted in terms of simulation concepts, are the outputs of all the integrators used in the simulation of the system itself. This is true even for nonlinear and/or time-varying–coefficient systems.

The symbol **I** represents the *identity matrix* while $1/s$ is our usual Laplace operator for integration:

$$1/s \begin{bmatrix} 1 & 0 & \cdots & 0 & 0 \\ 0 & 1 & \cdots & 0 & 0 \\ \cdots & \cdots & \cdots & \cdots & \cdots \\ 0 & 0 & \cdots & 1 & 0 \\ 0 & 0 & \cdots & 0 & 1 \end{bmatrix} = \begin{bmatrix} 1/s & 0 & \cdots & 0 & 0 \\ 0 & 1/s & \cdots & 0 & 0 \\ \cdots & \cdots & \cdots & \cdots & \cdots \\ 0 & 0 & \cdots & 1/s & 0 \\ 0 & 0 & \cdots & 0 & 1/s \end{bmatrix}$$
(6-82)

Thus $[\mathbf{x}] = (1/s)[\mathbf{I}][\dot{\mathbf{x}}]$ merely shows how the *x*'s are obtained by integrating the $\dot{x}$'s.

The "feedback system appearance" of Fig. 6-28 is due solely to the form of Eqs. (6-80); the actual physical system may or may not be a feedback control system. Note also that the *actual* unknowns are the *x*'s; the *y*'s are merely linear combinations ($y_1 = c_{11}x_1 + c_{12}x_2 + \ldots + c_{1n}x_n$, etc.) of the *x*'s that are easily found once the *x*'s have been solved for.

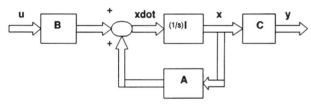

Figure 6-28 Conventional symbology for multiple-input, multiple-output systems.

EXAMPLE: THREE-MASS PROBLEM

Let's use the mechanical system of Fig. 6-12 to illustrate the state-variable matrix notation. We first have to replace any nonlinear or time-varying springs or dampers with linear elements since the matrix methods are limited to such systems. Choosing some specific numerical values (different from our earlier choices), the equations are:

```
.1*X1DOT2 = F1-5*X1-8*(X1-X2)-2*(X1DOT-X2DOT)
.2*X2DOT2 = 8*(X1-X2)+2*(X1DOT-X2DOT)-20*(X2-X3)
.25*X3DOT2 = F3+20*(X2-X3)-15*X3DOT
```

Since the physical equations are second-order while only first-order equations are permitted by the state-variable format, we simply *define*

$$X1 \triangleq X1 \quad X2 \triangleq X2 \quad X3 \triangleq X3$$
$$X4 \triangleq V1 \triangleq X1DOT \quad X5 \triangleq V2 \triangleq X2DOT \quad X6 \triangleq V3 \triangleq X3DOT$$

It is also clear that the input forces play the role of "controls" u, so we let

$$U1 \triangleq F1 \quad U2 \triangleq F3$$

We can then write

$$\dot{x}_1 = (0)x_1 + (0)x_2 + (0)x_3 + (1)x_4 + (0)x_5 + (0)x_6$$
$$\dot{x}_2 = (0)x_1 + (0)x_2 + (0)x_3 + (0)x_4 + (1)x_5 + (0)x_6$$
$$\dot{x}_3 = (0)x_1 + (0)x_2 + (0)x_3 + (0)x_4 + (0)x_5 + (1)x_6$$
$$\dot{x}_4 = (-130)x_1 + (80)x_2 + (0)x_3 + (-20)x_4 + (20)x_5 + (0)x_6 + 10F1 \qquad (6\text{-}83)$$
$$\dot{x}_5 = (40)x_1 + (-140)x_2 + (100)x_3 + (10)x_4 + (-10)x_5 + (0)x_6$$
$$\dot{x}_6 = (0)x_1 + (80)x_2 + (-80)x_3 + (0)x_4 + (0)x_5 + (-60)x_6 + 4F1$$

or, in matrix form

$$
\begin{bmatrix} \dot{x}_1 \\ \dot{x}_2 \\ \dot{x}_3 \\ \dot{x}_4 \\ \dot{x}_5 \\ \dot{x}_6 \end{bmatrix}
=
\begin{bmatrix}
0 & 0 & 0 & 1 & 0 & 0 \\
0 & 0 & 0 & 0 & 1 & 0 \\
0 & 0 & 0 & 0 & 0 & 1 \\
-130 & 80 & 0 & -20 & 20 & 0 \\
40 & -140 & 100 & 10 & -10 & 0 \\
0 & 80 & -80 & 0 & 0 & -60
\end{bmatrix}
\begin{bmatrix} x_1 \\ x_2 \\ x_3 \\ x_4 \\ x_5 \\ x_6 \end{bmatrix}
+
\begin{bmatrix}
0 & 0 \\
0 & 0 \\
0 & 0 \\
10 & 0 \\
0 & 0 \\
0 & 4
\end{bmatrix}
\begin{bmatrix} u_1 \\ u_2 \end{bmatrix}
$$

$$(6\text{-}84)$$

Finally, if we are interested in say, the spring force 8*(X1-X2) and the damper force -15*X3DOT, we might wish to define two outputs y1 and y2 as these forces. Then Eq. (6-81) becomes

$$
\begin{bmatrix} y_1 \\ y_2 \end{bmatrix}
=
\begin{bmatrix}
-8 & 8 & 0 & 0 & 0 & 0 \\
0 & 0 & 0 & 0 & 0 & -15
\end{bmatrix}
\begin{bmatrix} x_1 \\ x_2 \\ x_3 \\ x_4 \\ x_5 \\ x_6 \end{bmatrix}
\qquad (6\text{-}85)
$$

Solution Methods for Differential Equations **399**

Recall that in the analytical solution of linear, constant-coefficient differential equations, by either the D-operator or Laplace transform methods, we always at some point need to find numerical values for the roots of the system characteristic equation. We have suggested that use of a software *root finder* for this purpose. With the state-variable method we can now offer an alternative approach. The *eigenvalues* of the **A** matrix (system matrix) of Eq. (6-84) are the same as the roots of the characteristic equation. Numerical algorithms, *different* from the root-finding algorithm, for getting matrix eigenvalues are available in most mathematical software packages, such as MATLAB or MATHCAD. Since root finders and eigenvalue routines are *different* approximate calculation methods, it is a useful check to run both. The eigenvalue routines tend to be more reliable than the root finders when numerical difficulties arise.

EXAMPLE: ROOT FINDER VERSUS EIGENVALUES

Let's apply both a root finder and an eigenvalue routine to our present mechanical system example. Assuming we have the A matrix of Eq. (6-84), MATLAB has a command for generating the characteristic polynomial; just enter `c= poly(a)`. The computer returns the coefficients of the polynomial, from which we easily construct the characteristic equation. It turns out to be the same as Eq. (6-5) since that example had been generated from our present mechanical system. We can now apply the root finder with `roots(c)` and get the same roots shown just below Eq. (6-5). To get the eigenvalues of the A matrix, enter `eig(A)`. The computer returns the "same" set of values as obtained from the root finder. Actually there is a *very* slight difference, but you have to show more digits to see it. Repeat the above calculations after entering `format long`, which gives results to 14 decimal places. Even now, only two of the roots differ in the 14th decimal place; evidently our example is "well behaved" numerically. It is possible to make up examples where roots and eigenvalues give *significantly* different results, but fortunately most real physical systems *don't* present such numerical problems.

BIBLIOGRAPHY

Cellier, F., *Continuous System Modeling*, Springer, New York, 1991.

Doebelin, E. O., *Control System Principles and Design*, Wiley, New York, 1985. This text uses simulation as a major tool of design and includes about 35 detailed simulation examples.

Doebelin, E. O., *System Modeling and Response: Theoretical and Experimental Approaches*, Wiley, New York, 1980. Chapter 5 covers general simulation methods; examples are sprinkled throughout the text.

Gardner, M. F., and J. L. Barnes, *Transients in Linear Systems*, vol. 1, Wiley, New York, 1942. ("Old but good.")

Kaplan, W., *Operational Methods for Linear Systems*, Addison-Wesley, Reading, Mass., 1962.

Kaplan, W., *Elements of Differential Equations*, 2nd ed., Addison-Wesley, Reading, Mass., 1964.

Lawden, D. F., *Mathematics of Engineering Systems*, Wiley, New York, 1954.

Raven, F. H., *Mathematics of Engineering Systems*, McGraw-Hill, New York, 1966.

400   **Chapter 6**

Speckhart, F. H., and W. L. Green, *A Guide to Using CSMP*, Prentice-Hall, Englewood Cliffs, N.J., 1976. CSMP is "obsolete" but good books on *practical* simulation are rare, and since all CSSLs are "the same," the methods discussed are *not* obsolete.

Timothy, L. K., and B. E. Bona, *State Space Analysis: An Introduction*, McGraw-Hill, New York, 1968.

PROBLEMS

These problems are intended mainly to develop some familiarity with the *basic mechanics* of equation solution by analytical and simulation means. Use of these methods in the context of analysis and design of practical systems is deferred to upcoming chapters, where we gradually progress to consideration of more complex and realistic engineering applications.

6-1. Classify the following equations according to the categories
Linear with constant coefficients
Linear with time-varying coefficients
Nonlinear with constant coefficients
Nonlinear with time-varying coefficients

a. $3\dfrac{dx}{dt} = 5x + 7 \qquad x(0+) = -2$

b. $3\dfrac{dx}{dt} = 7t + 5x \qquad x(0+) = 0$

c. $e^{-5t} - 2\dfrac{d^2x}{dt^2} = 5x \qquad x(0+) = 0, \ \dot{x}(0+) = 0$

d. $3\dfrac{dx}{dt} + 5\dfrac{x}{|x|}\sqrt{|x|} - 20 = 2\sin(6.28t) \qquad x(0+) = 15$

e. $9760x^3 - 2000 = -\dfrac{d^2x}{dt^2} - 50\dfrac{dx}{dt^2} - 50\dfrac{dx}{dt} \qquad x(0+) = 0, \ \dot{x}(0+) = 0$

f. $(1.0 - 0.01t)5\dfrac{d^2x}{dt^2} + 3\left(\dfrac{dx}{dt}\right)^3 + 12\sin(2x) = 2 + 3t \qquad x(0+) = 4,$

$$\dot{x}(0+) = 0$$

g. $2\dfrac{d^2x}{dt^2} + 0.1\dfrac{dx}{dt} + 5x = 4 \qquad x(0+) = 1, \ \dot{x}(0+) = 0$

h. $2\dfrac{d^2x}{dt^2} + 20\dfrac{dx}{dt} + 5x = 4 + 3t \qquad x(0+) = 1, \ \dot{x}(0+) = -1$

Solution Methods for Differential Equations   **401**

i. $5\dfrac{d^4x}{dt^4} + 4\dfrac{d^3x}{dt^3} + 10\dfrac{d^2x}{dt^2} + 5\dfrac{dx}{dt} + 3x = 0$ all i.c's are zero except

                                                                       $x(0+) = 2$

j. $4\dfrac{dx}{dt} + 3x - 7y = 2t$ $x(0+) = 1$

     $2x + 7\dfrac{dy}{dt} - 8y = 0$ $y(0+) = 0$

k. $6\dfrac{d^2x}{dt^2} + 3\dfrac{dx}{dt} + 5y = \sin 2t$ $y(0+) = -2$

     $2\dfrac{dy}{dt} + 4y - 6x = 0$ $x(0+) = 0, \ \dot{x}(0+) = 1$

l. $8\dfrac{d^2x}{dt^2} + 2\dfrac{dx}{dt} + 10x - 2\dfrac{dy}{dt} - 10y = 2t - 2$

     $3\dfrac{d^2y}{dt^2} + 2\dfrac{dy}{dt} + 15y - 2\dfrac{dx}{dt} - 10x = 3\sin t$ all i.c's are 0

6-2.   Find complementary solutions for problem equations above:
     a.   6-1a   b.   6-1b   c.   6-1c   d.   6-1g   e.   6-1h
     f.   6-1i   g.   6-1j   h.   6-1k   i.   6-1l

6-3.   Find particular solutions only for the equations of problem 6-2.

6-4.   Get complete solutions, using the *D*-operator method, for the equations of problem 6-2, but do not find the constants of integration.

6-5.   Using the *D*-operator method, get complete specific solutions, including the constants of integration, for the equations of problem 6-2.

6-6.   Repeat problem 6-5, but use the Laplace transform method. The initial conditions just before inputs are applied can be taken the same as those given for $t = 0+$.

6-7.   Set up simulation diagrams for the equations of problem 6-1. Use SIMULINK notation or that for your locally available software. Do *not* run the simulations.

6-8.   Repeat problem 6-7 but *do* run the simulations.

6-9.   a. Modify the simulation diagram of Fig. 2-2b for a spring which has $f = 5x + 3x^3$.
      b. Run the simulation of part (a).

6-10. a. Modify the simulation diagram of Fig. 2-18 so that the applied force is the square of that given.
      b. Run the simulation of part (a).

6-11. a. Modify the simulation diagram of Fig. 2-51 so that the load torque is given by $T = 0.0000533 \ (\text{rpm})^2$. If the average engine torque is still $50\,\text{ft-lb}_f$, find the initial steady speed. Where is this number needed in the simulation?
      b. Run the simulation of part (a).

6-12. a. Modify the simulation diagram of Fig. 2-55 to get graphs of the spring force and damper force. Why would these be of interest?

402   **Chapter 6**

b. Run the simulation of part (a) to find out whether you can improve the accuracy for the given input to ±2% by adjusting the numerical values of system parameters.

6-13.   For the optimum damper simulation of Fig. 2-59, see whether you can improve the "jerky" force of Fig. 2-62 by using seven holes, rather than five. Use your own judgment on hole location and size.

6-14.   For the inductor simulation of Fig. 3-9, use the SPLINE function method to get a more realistic (smoother) curve of inductance variation with current (Fig. 3-8b). Run the simulation with this improved model and comment on the results.

6-15.   For the computer-aided system simulation of Fig. 3-34, change the sine-wave input to a triangular wave of the same frequency and amplitude and change the sample time to 0.004 seconds. Let the A/D converter have 8 bits, and the D/A 5 bits. Run the simulation and comment on the results.

6-16.   For the digital servo system simulation of Fig. 3-39, make the power amplifier model more realistic by including a saturation nonlinearity at its output. Set the current limit at ±5 amps and use all the other numbers as for Fig. 3-42. Run the simulation and comment on the results. Does the stability "fix" used for Fig. 3-43 work with the saturation present?

6-17.   For the induction motor simulation of Fig. 5-18, delete the pulse load torque but add the effect of a sudden change in motor voltage V_s at $t = 0.40$ seconds. Let V_s drop instantly from 180 to 150 volts, and assume that this voltage change causes an instantaneous change in motor torque as given by Eq. 5-43. Run the simulation and comment on the results.

6-18.   For the stepping motor simulation of Fig. 5-21, we want to study how rapidly we can ask the motor to accelerate to a constant average speed without losing any steps. Delete the disturbing torque feature and the sine wave command angle. Replace the sine wave command angle with a command that corresponds to a fixed acceleration until an average speed of 5.0 rad/sec is reached. Find out how large you can make this acceleration before steps are lost. Repeat for average speeds of 10 and 15 rad/sec.

6-19.   In problem 6-18, make the command angle correspond to a "duty cycle" which consists of:
    a.   Constant acceleration for a time interval t_a
    b.   Constant velocity for a time interval t_v
    c.   Constant deceleration [same magnitude as part (a)] for a time interval t_a, bringing velocity back to zero

We want to use this duty cycle to accomplish a move of 100 steps (180°) in minimum time without losing any steps. Use the simulation plus any useful hand calculations to solve this problem.

6-20.   Using words and carefully drawn waveform sketches, explain clearly in detail how the PWM amplifier simulation of Fig. 5-29 works.

6-21.   Repeat problem 6-19 for the SCR amplifier simulation of Fig. 5-31.

6-22.   Problem 5-19.

6-23.   Problem 5-20.

6-24.   Problem 5-21.

7

FIRST-ORDER SYSTEMS

7-1 INTRODUCTION

We now begin a detailed and systematic consideration of certain combinations of the basic elements from the mechanical, electrical, fluid, and thermal areas. Since we found analogous behavior quite common when considering elements, we should not be surprised to encounter it also in systems. This commonality allows a great efficiency in our study of system response since knowledge of the characteristics of a particular *class* of systems is immediately applicable to any member of that class, irrespective of its physical nature. For example, if we once master the step response of a "generic" system, we need never repeat that mathematical work again. Whether a system is mechanical, electrical, fluid, thermal, or mixed will make no difference; their step responses will all follow exactly the same formula if they all belong to the same class. Design techniques learned for one example of a class may be fruitfully carried over to physically different systems of the same class. Every time we encounter a new example of a familiar class, it reinforces our understanding of system behavior and design possibilities.

Two classes of systems, the so-called *first-order* and *second-order*, are found to be of fundamental importance. Many practical devices and processes will be found to fit one of these two classes; thus they are important in their own right. Furthermore, we shall find that more complex systems may profitably be considered in terms of *combinations* of simple first- and second-order types. Thus a temperature control system might include a combination of a second-order process, a first-order heater, and a first-order temperature sensor. The basic importance of these two simple systems rests partly on the fact that, for linear, constant-coefficient models, the "natural" response (complementary solution) is determined by the roots of the characteristic equation. Except for the rare case of repeated roots, only two basic forms of solution, one for real roots and another for complex root pairs, can occur, no matter how high the equation order might get in a complicated system. The basic first-order system is described with a single real root while the underdamped second-order system has a single pair of complex roots. A complicated system will thus in most cases have a natural response made up of terms, each of which is either like a first-order response or a second-order response. Thus, as we get familiar with simple

403

404   **Chapter 7**

first- and second-order systems, we also get useful background for our later study of
more general systems.

7-2 MECHANICAL FIRST-ORDER SYSTEMS

While we will shortly define the *generic* first-order system, it may be better to start
with concrete examples. Mechanical examples may be particularly appropriate, even
for nonmechanical engineers, since the variables involved (forces and motions) are
more directly sensed and understood by humans than, say, voltages and currents.
Figure 7-1 displays some combinations of masses, springs, and dampers which will
all be found to be first-order systems. We should immediately note that in addition to
showing the sketch, we must also be specific about the variable chosen to be the
input (given) and output (unknown) in our system definition. That is, until we define
the input and output, we *can't tell* what type of system we are dealing with. Thus in
Fig. 7-1, I have labeled each input with subscript i and each output with subscript o.
Having done this I can then *guarantee* that the differential equation relating input
and output will be exactly the same form for each of the examples shown. We
already begin to see the efficiency mentioned earlier; the 11 examples of Fig. 7-1
are all mathematically identical and it would be wasteful of time and money to treat
each of them as if it were a new problem.

While these examples are described with the pure and ideal mass, spring, and
damper elements, each of them can be related to one or more practical engineering
devices which could be reasonably modeled as shown. To convince you that study of
this material is worth your time and effort, we want to always relate our discussions
to "bread-and-butter" engineering concerns. The model of Fig. 7-1a is directly
applicable to the design and use of machines which have movable slides and car-
riages—machine tools such as lathes and milling machines, precision coordinate
measuring machines, assembly machines, etc. The driving force f_i (or torque T_i) is
provided by some kind of motor (electric, hydraulic, pneumatic, etc.) which is to
move the slide (whose mass is M) at some desired speed dx_o/dt or to some desired
position x_o. Slides on machines may be supported and guided by lubricated bearing
surfaces called "ways"; The damper B may represent viscous friction due to shearing
the lubricating oil film. Since both rotary and translational motions are found in
machines, Fig. 7-1a shows models for each. It also shows a combined rotary and
translational system since it is quite common to use a rotary motor to power a
translational machine motion.

The model of Fig. 7-1b is sometimes used to represent a real spring which has
significant energy dissipation. Since a pure spring element has no losses, we must add
a damper to our model to provide for this. (An even more realistic spring model
would also add a mass, but this model would be a *second-order*, not a first-order
system.) Figure 7-1c might serve as a model for a *delayed-action* (or mechanical lag)
mechanism. Analysis will show that motions applied at x_i are reproduced at x_o if
they are "slow," but will be delayed if they are rapid. We will use this mechanism
shortly to enhance the stability of a hydraulic motion control system used as a design
example in this chapter. If we want to measure translational or rotary velocity, the
systems of Fig. 7-1d produce an output displacement proportional to input velocity.
This displacement could be read directly on a calibrated scale, or applied to an

First-Order Systems

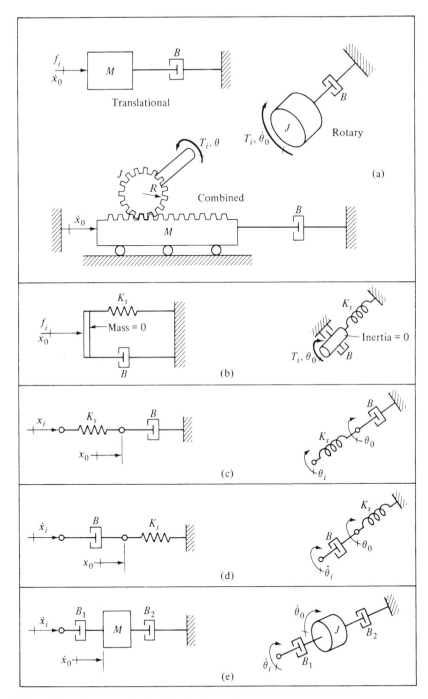

Figure 7-1 Some mechanical first-order systems.

406   **Chapter 7**

electrical displacement sensor if we wanted a voltage output. The rotary system of Fig. 7-1e could represent a mechanical drive in which a power source running at speed θ_i is used to move a load of inertia J and friction B_2 through a fluid coupling modeled as a damper B_1.

Preliminaries to Equation Setup.   Let us use the translational system of Fig. 7-1a as a vehicle for illustrating some general concepts useful in mechanical system analysis. We recall first that the mass and damper elements shown represent an idealized, but hopefully useful, model of some real, practical system and that it is necessary for analysts to make judgments, as best they can, that this model would be adequate for their purposes. The ultimate test of such judgments is experimental testing of the real system. If the predictions of the analysis are largely confirmed by such experiments, then our judgment will have been justified, and if a related problem arises in the future, we feel more confident about making similar assumptions. On the other hand, if experiments do *not* agree with predictions, we carefully study the measured data to try to discover those aspects of system behavior which we did not sufficiently understand. We may then be able to improve our analysis model so that it will agree more with reality.

    It is necessary to be clear from the outset which physical variable will be taken as system input and which as output. This decision generally requires consideration of the interfaces between the subsystem being studied and the overall system of which it is a part. In our present example the force f_i might be provided by a driving motor of some sort and used to control the speed $v_o \triangleq dx_o/dt$ of the moving mass, which speed might be measured by some instrument used in a control system for the overall machine. It would thus be natural to consider the force f_i as an input which causes the velocity v_o. We next must become more specific about the input and output signals with regard to such things as *coordinate systems* and *sign conventions*. These are important considerations which must be dealt with *before* we begin to write equations. Since both force and velocity are vector quantities they can in general take arbitrary directions in space. We here consider, as part of the definition of our model, that forces and motions are constrained to a single axis; that is, the model is one-dimensional. With this restriction it is still necessary to decide on an origin and positive direction for the single coordinate x_o. That is, if we later find out that $x_o = -5.3$ inches at some instant of time, we have to know the answer to the questions "5.3 inches from where?" and "5.3 inches in which direction?"

    These choices may be made freely, but in some cases one choice is more convenient than another. In Fig. 7-2a we note that with no force applied, there is no preferred position for the mass M. It will sit wherever it is put; thus the choice of an origin for the coordinate x_o is arbitrary and no advantage accrues to a particular choice. The translational system of Fig. 7-1b is an example where one particular choice *is* preferred over all others, even though *any* choice could be made, and would give correct results. In this system, when the input force is zero, the spring will assume one definite position, that corresponding to its "free length." If we choose this as the $x_o = 0$ position, when it comes time to write an expression for the spring force, our formula will be the simplest possible, so it makes sense to choose this origin. In Fig. 7-2b no such preference exists, so we can locate the origin anywhere and pay no price. It *is* necessary, however, that this origin from which displacement is measured have *zero* absolute motion; that is, the displacement x_o (measured

First-Order Systems

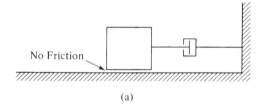

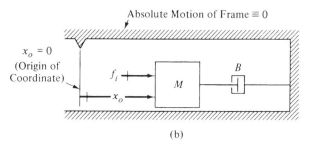

Figure 7-2 Coordinate origins and sign conventions.

relative to its origin) is an *absolute* displacement. (Since any system resting on the earth must participate in the earth's own motions, the notion of a "fixed point" to use as a reference is theoretically somewhat elusive, but for the vast majority of problems it is conventional, and adequately accurate, to consider the earth "fixed.") The reason we are so concerned about the "fixity" of our reference points is that our analyses are rooted in Newton's law of motion $F = MA$, which *requires* that the acceleration A be the *absolute* acceleration. The choice of a positive direction for x_o is completely arbitrary and there is generally little if any advantage to choosing one direction or the other; however, a choice must be made, and it must be made at the beginning of the analysis, not the end.

Once a positive direction for x_o has been chosen, the positive directions for velocity and acceleration *must* be chosen the same as for displacement, because calculus defines these quantities in terms of displacement and time. Since time does not run backward, Δt is always positive, so in the definition $v \triangleq \lim \Delta x / \Delta t$, a positive Δx must cause a positive velocity. Similar reasoning can be made for acceleration. All forces which might act on a mass must be taken positive in the same direction as was displacement of that mass, not because of calculus but because of physics. Newton's law requires that a positive force produce a positive acceleration (mass is by definition positive). If you push on an object it moves in the same, not the opposite, direction. The input force f_i in motion problems can always be carried through the equation setup as an arbitrary function of time, $f_i(t)$. Of course when it comes time to *solve* the equation, then we must have a *specific* function of time. In Fig. 7-2b we have taken x_o positive to the right (symbol ↦) so velocity, acceleration, and all forces are also positive to the right. I prefer the symbol ↦ to the more common → since it is less ambiguous.

Writing the System Equation. Analysis of any system is based upon proper application of the appropriate physical laws. For mechanical systems Newton's law is fundamental, even though it may appear in various forms such as D'Alembert's principle or energy methods. It is generally best to first express the applicable law in *word* (rather than equation) form. For our present example we might say to ourselves (or actually write out on a homework paper), "At any instant of time, the summation of forces acting on the mass must equal the product of the mass and its instantaneous acceleration $\ddot{x}_o$." The concept of the free-body diagram is widely useful in mechanics problems as an aid to properly enumerating and expressing the various forces which enter into the summation. One draws a sketch of the body to which Newton's law is to be applied and carefully indicates *all* of the forces which impinge on the body in the direction of motion being considered.

To make sure that you are including *all* the pertinent forces, recall that fundamentally only two types of forces can act on a body: forces due to actual *contact* with another body (solid or fluid), and the "mysterious" gravitational, magnetic, and electrostatic forces which can cause a mass to move with *no physical contact whatever*. Also note that, if you have been introduced to the *inertia force method* of solving dynamics problems, the inertia force used there is *not* a real force, but rather a *fictitious* force added to the real forces to make a dynamics problem into an equivalent statics problem. Thus in using Newton's law $\Sigma F = MA$, you *never* include an inertia "force" in the summation on the left-hand side; inertia force is *not* a real force that can move a mass initially at rest. It has *no* place in an analysis which directly uses $\Sigma F = MA$. You thus need to account only for direct contact forces and gravitational, magnetic, and electrostatic forces in the summation of forces. (We are here restricting ourselves to the "macroscopic" world and thus don't get involved with other types of forces which exist at atomic and nuclear levels.)

In Fig. 7-3a all the forces acting on M are shown. The gravity force (weight of M) is just balanced by the reaction of the support; thus the net force in the vertical direction is zero, and no motion occurs in this direction. We can thus employ the simpler free-body diagram of Fig. 7-3b for the x_o direction only. The only forces acting in this direction are the direct-contact forces due to the input force source and the force of the damper on the mass. We can now write Newton's law in equation form as

$$\sum \text{forces} = (\text{mass})(\text{acceleration}) \tag{7-1}$$

$$f_i - B\dot{x}_o = f_i - Bv_o = M\ddot{x}_o = M\dot{v}_o \tag{7-2}$$

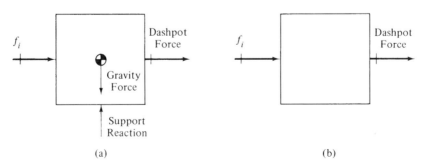

Figure 7-3 Free-body diagrams.

First-Order Systems   **409**

It is important to be careful about the algebraic signs of the various forces. Getting a sign wrong in a differential equation does *not* just cause a mass to move in the opposite direction from what it should. Rather, a wrong sign can often cause the model to predict *infinite* motion (an *unstable system*), rather than a stable motion! To avoid such situations we must always check each equation term to make sure that it accurately represents the physical facts *for all possible directions of motion.* That is, the motion is usually the unknown in the equation, so we *don't know* whether it will go to the left, right, or stand still. Thus whatever term we write into the equation for a force, it must be correct for *all* possible motion directions.

The sign of an input force, such as f_i, is always taken positive at this stage since at this point it is a general, rather than specific, function of time. When we go to *solve* the equation, then f_i must be made specific and we then can tell what its proper sign should be. As for the damper force, we know from Chapter 2 that the magnitude of a viscous damper force is Bv_o and that it *opposes* the velocity. Since our equation must agree with these known facts, let's check it for *all possible* velocities, because we don't know at this stage what the velocity is doing. If v_o is positive (mass moving to the right), the damper opposes with a force (on the mass) which is to the left, a negative force; thus $-Bv_o$ should be negative, which it is (B is understood to be positive). If v_o were negative (mass moving to the left), the force of the damper on the mass is to the right (positive), which again agrees with $-Bv_o$. Finally, if $v_o = 0$, the damper exerts no force on the mass. We see that, no matter whether the velocity is $+$, $-$, or 0, our damper force term in Eq. (7-2) is correct in magnitude and direction. If you have in the past not used such a systematic procedure to get correct algebraic signs in your equations, I recommend that you now adopt it, not just for damper forces but for *every* term in every differential equation. Having satisfied ourselves as to the correctness of signs, we now arrange the equation with terms in the unknown on the left (in decreasing order of the derivatives), and the given inputs on the right:

$$M \frac{dv_o}{dt} + Bv_o = f_i \tag{7-3}$$

You should get into the habit of *always* arranging your equations in this systematic way. Engineers and mathematicians talk about the "left-hand side" and the "right-hand" side; these terms are *meaningless* unless we all agree to arrange our equations in the standard way.

As another example let us analyze the rotational version of Fig. 7-1c. At the "junction" of the spring and damper (the θ_o location) the torque of spring and damper must be identical:

$$K_s(\theta_i - \theta_o) = B\dot{\theta}_o \tag{7-4}$$

$$B \frac{d\theta_o}{dt} + K_s\theta_o = K_s\theta_i \tag{7-5}$$

If you asked, without supplying any other information, some mathematicians to solve equation Eqs. (7-3) and (7-5) they would say that these two are "the same" equation, and that only one needs to be solved. As engineers, the physical difference between the two mechanical systems is clear to us, but we should also not waste our time "doing the same math twice." The two physical systems belong to the same class, first-order systems, and we should focus on *class* behavior, not "reinvent the wheel" for each new example.

When a model for a mechanical system assumes that neglecting mass is a reasonable approximation, we sometimes have difficulty setting up the differential equation correctly, because we are used to doing a free-body study on the mass when we write Newton's law. If you have had such difficulty, let me suggest a way to force such problems into the more familiar mold, using the previous example as an illustration. There, the spring and damper were treated as pure (massless) devices. Let's now place a *fictitious* mass with moment of inertia J at the junction of spring and damper, as in Fig. 7-4. We can now do a conventional free body of J and write

$$\sum \text{torques} = (\text{moment of inertia})(\text{angular acceleration}) \tag{7-6}$$

$$+K_s(\theta_i - \theta_o) - B\dot{\theta}_o = J\ddot{\theta}_o \tag{7-7}$$

and since $J \equiv 0$,

$$B\frac{d\theta_o}{dt} + K_s\theta_o = K_s\theta_i$$

which of course agrees with our earlier result. This use of fictitious masses at the equation setup stage often helps in getting correct algebraic signs. As usual, the signs in Eq. (7-7) must be justified by physical reasoning. Positive directions for angles θ_i and θ_o may be arbitrarily chosen; in fact we could take θ_i positive clockwise and θ_o positive counterclockwise if we wished. However, since θ_i is the input which *produces* θ_o, it may be less confusing, once the positive direction for θ_i is chosen, to pick θ_o the same way. Then a positive θ_i will in steady state produce a positive θ_o, which, though not a necessity, may be convenient. That is, certain choices may be mathematically equivalent but one or the other may be practically preferable. Once a positive direction for θ_o is chosen, then $\dot{\theta}_o$, $\ddot{\theta}_o$, and torques on J *must* conform to this choice. If $\theta_i > \theta_o$, then the spring exerts clockwise (+) torque on J; if $\theta_i < \theta_o$, the torque is counterclockwise (−); thus the torque term $+K_s(\theta_i - \theta_o)$ correctly represents the physical facts. Similarly, $-B\dot{\theta}_o$ gives the correct damper torque for any situation. The origins for angular displacements θ_i and θ_o can be chosen at any desired locations; however, the spring torque can be written as $K_s(\theta_i - \theta_o)$ only if the origin for θ_i is identical to that for θ_o so that the torque goes to zero when $\theta_i = \theta_o$.

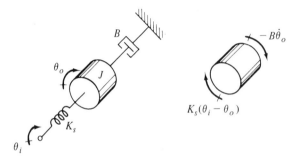

Figure 7-4 Use of fictitious inertia.

First-Order Systems 411

The Generic First-Order System and its Step Response. Having shown a large number of mechanical first-order systems and having derived the system equations of two of these, it is now appropriate to define the generic first-order system and begin a study of its response to various types of inputs. As we have pointed out before, ideally one would use as input functions for analysis and design the actual input forces or motions encountered in the operation of the real system; however, these are generally somewhat unpredictable and peculiar to each specific application. Since we are at this point trying to obtain *general* information about system response, we use simple "standard" inputs such as steps and sine waves. As we progress to second-order and more complex systems, the step and sinusoidal response will serve as useful benchmarks for comparison. When designing a specific system, if lab or field testing has obtained time histories of typical inputs, our simulation methods allow us to use these to make our design more realistic. However, early stages of design will often use simple standard inputs.

For a standard step input we assume that the system is initially in equilibrium with both input and output at the zero level, when the input suddenly jumps up to some constant value, at which it remains thereafter. We now define the generic first-order system, with input q_i and output q_o, as a system whose equation is

$$a_1 \frac{dq_o}{dt} + a_0 q_o = b_1 \frac{dq_i}{dt} + b_0 q_i \qquad (7\text{-}8)$$

where the a's and b's are parameters which are assumed constant. Any system whose equation fits this pattern is, by definition, a first-order system. Later in this chapter we will show some physical examples of systems which include *all* the terms shown above, but we will now concentrate on those systems which have $b_1 \equiv 0$. This subclass is by far the most common and important, so we spend most of our time getting you familiar with its behavior. While it appears that it takes three constants (a_1, a_0, b_0) to define such a system, we can always divide through by any one of these to reduce the number of *essential* constants to only two. This reduction in the number of essential parameters is something engineers *always* try to do. It simplifies system design because design involves the determination of an *optimum* set of numerical values for the parameters. This optimization requires that we study all possible *combinations* of parameters, and the number of such combinations will be smaller if there are fewer parameters to deal with.

While we could divide through our equation by any of the parameters, it is conventional to divide through by a_0 to get

$$\frac{a_1}{a_0} \frac{dq_o}{dt} + q_o = \frac{b_0}{a_0} q_i \qquad (7\text{-}9)$$

Whenever you encounter a new example of a first-order system, you should *always* reduce its equation to this standard form *immediately*. Then define the two standard parameters K and τ by

$$\tau \triangleq \frac{a_1}{a_0} \triangleq \text{system time constant} \qquad (7\text{-}10)$$

$$K \triangleq \frac{b_0}{a_0} \triangleq \text{system steady-state gain, or static sensitivity} \qquad (7\text{-}11)$$

412   **Chapter 7**

The equation now assumes the compact *standard form*

$$\tau \frac{dq_o}{dt} + q_o = Kq_i \tag{7-12}$$

It thus requires only two parameters, K and τ, to completely describe any first-order system of this subclass. The physical significance of these parameters, and the reasons for defining them as we did, will become apparent shortly. Whenever you encounter a new physical device or other system whose equation conforms to (7-8) with $b_1 \equiv 0$, you should *immediately* define K and τ and convert to the standard form (7-12). All the results which we are about to develop pertaining to first-order systems are then instantly available. Operational, sinusoidal, and Laplace transfer functions are easily found in the usual ways from the differential equation, the Laplace transfer function being

$$\frac{Q_o}{Q_i}(s) = \frac{K}{\tau s + 1} \tag{7-13}$$

Applying these concepts to the system of Fig. 7-2 we get

$$\frac{M}{B} \frac{dv_o}{dt} + v_o = \frac{1}{B} f_i \tag{7-14}$$

$$\tau \triangleq \frac{M}{B} = \frac{(\text{lb}_f\text{-sec}^2)/\text{inch}}{\text{lb}_f/(\text{inch/sec})} = \text{sec} \qquad K \triangleq \frac{1}{B} = \frac{\text{inch/sec}}{\text{lb}_f} \tag{7-15}$$

$$(\tau D + 1)v_o = Kf_i \tag{7-16}$$

$$\frac{V_o}{F_i}(s) = \frac{K}{\tau s + 1} \tag{7-17}$$

Note that the units of the time constant τ are seconds. The time constant will *always* have the units of time; whatever time unit you chose when you set up the differential equation. The steady-state gain K has the units of output (velocity) over input (force). It will *always* have the units of the system output quantity divided by the system input quantity. These facts should be used with each new application to check for errors.

Let's now find the step response of all first-order systems defined by Eq. (7-12). This is easily done with either the classical operator method or Laplace transform. Using Laplace transform and the fact that the output is given to be zero just before the step input is applied, we get

$$\tau[sQ_o(s) - 0] + Q_o(s) = KQ_i(s) = \frac{Kq_{is}}{s} \tag{7-18}$$

$$Q_o(s) = \frac{Kq_{is}}{s(\tau s + 1)} \tag{7-19}$$

$$q_o(t) = Kq_{is}(1 - e^{-t/\tau}) \tag{7-20}$$

The result can be applied to *any* first-order system, such as that of Fig. 7-2, where the step response would be

First-Order Systems

$$v_o(t) = K f_{is}(1 - e^{-t/\tau}) \tag{7-21}$$

Thus, whenever you want the step response of *any* first-order system, put its equation in standard form and then immediately use the result of Eq. (7-20). Figure 7-5a shows a plot of input force and output velocity for our example system. To get a *universal* step response plot which will apply to *all* first-order systems we nondimensionalize both the horizontal (time) axis and the vertical (output) axis. This is done by plotting $q_o/(Kq_{is})$ versus t/τ, as in Fig. 7-5b. This graph applies for *any* values of K, τ, and q_{is}.

Let's now point out some important characteristic features of the step response of any first-order system. As time goes by, the output (such as the mass velocity v_o in our example) asymptotically approaches a steady-state value given by Kq_{is} (Kf_{is} in our example). While this final value theoretically requires an *infinite* time to achieve, in actual practice after a time equal to four or five time constants has elapsed the output is very close to the steady-state asymptote. Practical design requires that we choose some *specific* criterion for comparing competitive systems with regard to the speed of approach to the new steady state after a step input. The *5% settling time* is a commonly used specification, and the table of Fig. 7-5b shows that, for all first-order systems, the output will have settled within 5% of the final value in three time

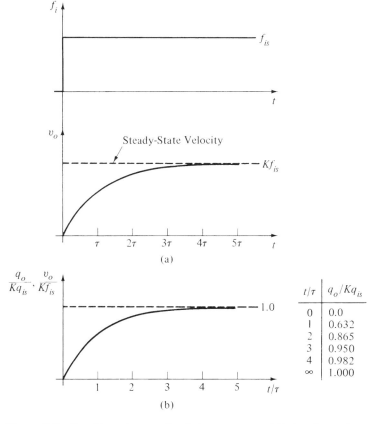

Figure 7-5 Specific and generalized step response of first-order systems.

constants. This fact is worth memorizing; all first-order systems, after a step input of any size, settle within 5% of the final value in 3τ. For example, a first-order temperature sensor with a time constant of 0.02 second, after being plunged into a hot fluid, will read the fluid temperature with 95% accuracy in 0.06 second. While we have at this point examined only the step response, it turns out that the time constant is the proper criterion of speed of response for *any* type of input, even random inputs. The smaller the time constant, the faster the response to any input.

In Eq. (7-12), if you let τ get smaller and smaller, approaching zero, the system equation becomes the *algebraic* equation $q_o = Kq_i$. Now the response of q_o to q_i is *instantaneous*, no matter what form q_i might take. Such a system is called a *zero-order* system. While no real system can have instantaneous response, the zero-order model *is* useful in practical work. When a real system has some components which are *much* faster than some of the other components, then a zero-order model for the "fast" components may be valid. Perhaps the most common example of this is an electronic amplifier used in an electromechanical, electrofluid, or electrothermal system. Here the mechanical, fluid, or thermal elements may be so much slower than the amplifier that no serious system design errors will be caused by treating the amplifier's response as instantaneous. Note, however, that if we use the same amplifier in an *all-electronic* system, then its response very likely can *not* be treated as zero-order.

Having established the time constant as the indicator of *speed* in every first-order system, we turn now to the significance of K, the steady-state gain. When the new steady state after a step input is achieved, the output will have come to a value Kq_{is}. That is, the steady-state output is K times the size of the step input. Note also that K has *no effect* on how rapidly the new steady state is achieved; this is governed entirely by τ. In linear systems, speed of response is generally defined in terms of how long the system takes to reach some given *percentage* of its steady-state output. This is possible because reponse is proportional to stimulus; doubling the input, for example, also doubles the steady-state output. Thus the time to reach the same *actual value* of output may be different for a large step input than for a small, but the time to reach the same *percentage* of final value will be the same for all sizes of steps (see Fig. 7-6). Using this viewpoint, the speed of response for a first-order system is determined entirely by the numerical value of its time constant, since the percentage of steady-state response is given by

$$\frac{q_o}{Kq_{is}} = 1 - e^{-t/\tau} \tag{7-22}$$

Clearly, if τ is, for example, cut in half, identical values of $q_o/(Kq_{is})$ will be achieved in one-half the time; thus the speed of response is inversely proportional to τ. We can thus summarize the significance of the two basic first-order system parameters by saying that K is an indication of *how much* steady-state output will be produced for each unit of input, and that τ will determine *how fast* that steady state will be reached.

While we have emphasized the fundamental importance of K and τ, we should not lose sight of the fact that any design changes in a particular first-order system can only be accomplished by adjusting the values of the *physical* parameters, such as M and B in our example system. That is, we now know how K and τ affect system response and can thus use their definitions [Eq. (7-15)] to change M and B to achieve

First-Order Systems

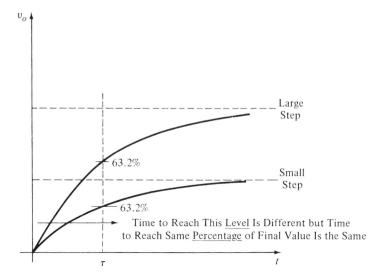

Figure 7-6 Definition of speed-of-response.

performance adjustments. We see that since $K \triangleq 1/B$, the steady-state velocity caused by a given input force can be changed only by changing the damping, mass has no effect on this at all. Speed of response, on the other hand, depends on *both* M and B since $\tau \triangleq M/B$. If we change B, both speed of response and steady-state value are changed, while changing M affects only speed of response.

While we have concentrated on finding the response of the output variable (v_o in our example), we can of course find the response curves of any other system variables that might be of interest (see Fig. 7-7). The displacement x_o of the mass is obtained from

$$x_o = x_o(0) + \int_0^t v_o\, dt = Kf_{is}t + Kf_{is}\tau(e^{-t/\tau} - 1) \tag{7-23}$$

where we have taken the initial displacement $x_o(0)$ to be zero. Note that after a transient period during which $e^{-t/\tau}$ is dying out, the displacement becomes asymptotic to the straight line $Kf_{is}(t - \tau)$. (If the force f_{is} is left on, a translational damper must sooner or later encounter mechanical stops and cause the mass to stall. However, the rotational version allows continuous unimpeded motion such that the output angular displacement can actually "approach infinity" as indicated by the equations.)

A mathematical description of system behavior as given by the graphs of Fig. 7-7 should generally be interpreted in physical terms as a means of checking the plausibility of the results and reinforcing our intuitive feelings about the system. We might put it this way: The suddenly applied force f_{is} causes a sudden acceleration of the mass M; however, as the acceleration acts over time and produces some velocity, the dashpot develops an opposing force which reduces the net accelerating force available. As the velocity builds up, the dashpot force approaches the external driving force more closely, and the acceleration approaches zero. The system thus approaches asymptotically a terminal velocity given by the ratio of applied force

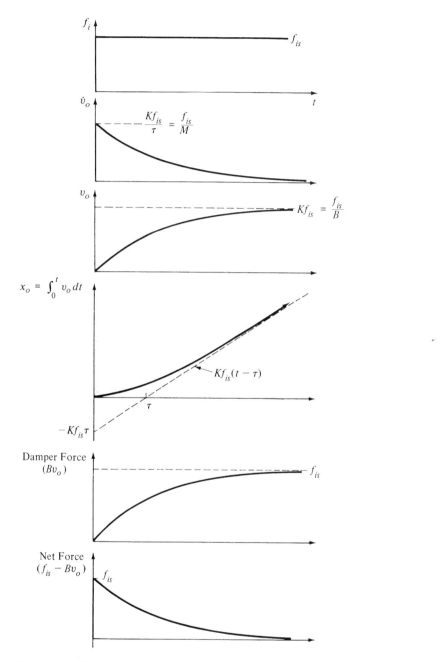

Figure 7-7 Mass/damper system step response.

f_{is} and the damper constant B. After three or four time constants have gone by, transient effects in all the variables have practically disappeared and the system is in steady-state operation. If we define speed of response as the speed with which the system gets into steady state, system speed can be increased only by reducing the time constant $\tau = M/B$. We may thus try to reduce M, increase B, or both. If we reduce

First-Order Systems 417

M, the system steady-state gain $K = 1/B$ is unaffected; an increase in B will, however, reduce the gain.

Perhaps the main reason for defining speed of response as we have done is that the inputs to our systems are often in the nature of *commands* which the output quantity is to reproduce or follow. In a differential equation, the command input appears on the right-hand side and is the term which produces the particular solution portion of the output. This portion of the response is sometimes called the *forced* part, while the complementary solution (which is present whether there is a driving input or not) is called the *natural* part of the response. If the forced portion of the response corresponds to the command input, and the natural part is of a form that eventually disappears [such as $e^{-t/\tau}$ in Eq. (7-21)], then the faster the natural part disappears the faster the system response conforms to the command input. In a first-order system this leads to the conclusion that a small time constant denotes a fast system.

Experimental Step-Input Testing. Since theoretical models always involve assumptions which are not exactly true, we often want to run lab tests to determine whether our theory adequately describes actual behavior. Perhaps the simplest tests are those using a step input. We here need to apply and measure a "sudden" change in the input quantity and measure a time history of the response (output) variable. For the system of Fig. 7-2 let's assume that f_i is applied by a translational electric motor whose moving part is attached to the mass. The mass of the motor's moving part is included in the value of M. By suddenly applying a fixed current to the motor armature, the magnetic force (which is our f_i) suddenly jumps up. Actually, *perfect* step changes in any physical variable are impossible since a sudden change in energy level implies a source of *infinite* power. Realistic step-input testing requires only that the input quantity change "much faster" than the output can respond. Then the actual test will be a good approximation of a perfect step test. In the case of our electric motor, the inductance and resistance of the armature form a first-order electrical system with time constant L/R seconds (L in henrys, R in ohms). We are hoping that this time constant will be much shorter than the M/B mechanical time constant we are trying to test. If we have no previous experience with this apparatus, we probably should also include a motor *current* measurement in our experiment to verify that the current rises much faster than does the mass velocity.

Assuming that the motor force is sufficiently "steplike" to give a valid test, we might record a velocity-time trace as in Fig. 7-8. A quick visual check for first-order behavior can be accomplished by fitting ("by eye") a straight line to the response curve at the origin ($t = 0$). Theory shows that the slope at this point is Kq_{is}/τ. If this initial slope is extended until it intersects the steady-state asymptote, the intersection will occur exactly at $t = \tau$. The table of Fig. 7-5b shows that 63.2% of the total change has been accomplished at this time. If these two methods of locating τ agree quite closely, this is preliminary evidence of first-order behavior.

A much more critical check is available with a little extra work. We are really trying to verify that our measured response curve is the exponential curve of Eq. (7-21). The human eye and brain are not very expert in comparing one curve with another but they are quite good at deciding whether points fall on a straight line. We thus use the ancient device of "curve rectification," where the data is suitably

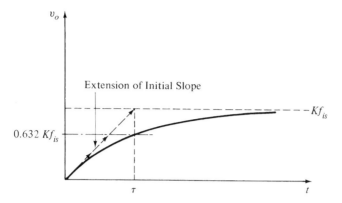

Figure 7-8 First-order system response characteristics.

transformed so that perfect exponential data now plots as a straight line on the transformed axis. To this end we define a quantity Z by

$$Z \triangleq \log_e\left(1 - \frac{q_o}{Kq_{is}}\right) = \log_e(e^{-t/\tau}) = -\frac{t}{\tau} \qquad (7\text{-}24)$$

A plot of Z versus time t is thus a straight line whose slope is $-1/\tau$. Figure 7-9 shows how this procedure is applied to a general first-order system with input q_i and output q_o. Since q_o at any instant, and its final value Kq_{is} are known from the measurements, we can compute as many values of Z as we wish from the relation

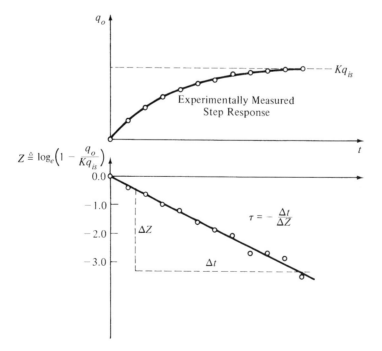

Figure 7-9 Experimental modeling by step testing.

First-Order Systems

$Z = \log_e (1 - q_o/Kq_{is})$ and plot against t as shown. If the data points can be well fitted with a straight which goes through the origin, then the real system is behaving very nearly as a first-order system, and an accurate value of τ is obtained from the slope of the line. The numerical value of K is obtained by simply dividing the final value of q_o by the known value of input q_i. Note that such testing gives values of K and τ, but does *not* give information about how these are related to physical parameters such as M and B. If the component being tested is considered acceptable "as is" (requires no adjustment or redesign) and is to be part of a larger system, then knowledge of only K and τ may be sufficient. In fact, *system designers* often purchase "ready-made" parts for their systems and may not even attempt a theoretical analysis, relying only on lab testing to get the numbers (such as K and τ) needed for system design. *Component designers*, who work for the manufacturer from whom we buy such components, must of course have a much more detailed knowledge of their devices in order to intelligently design them.

Computer Simulation. While computer simulation is hardly necessary for isolated linear first-order systems, we do need to spend a little time here since simulation *will* be useful for larger systems which include first-order components. Figure 7-10a shows SIMULINK simulations of the system of Fig. 7-2 with $M = 5.0$ kg and $B = 50.0$ N/(m/sec), making $K = 0.02$ (m/sec)/N and $\tau = 0.10$ second. A step input force of 50.0 N gives a steady-state velocity of 1.0 m/sec. Two simulations of this system are shown, one which shows internal details and another (more compact) which provides only input force and output velocity.

In the upper part of this diagram we simulate the differential equation in detail, using our usual procedures. Here we can "lay our hands on" details such as the damper force, net force, and acceleration, and physical parameters M and B are directly entered. We can also integrate velocity to get displacement, if that is of interest. Such a simulation might be used in the detail design of components such as the damper, where knowledge of the damper force would be needed to design its parts for adequate strength.

In the lower part of the simulation diagram we show a compact version, such as might be used if this system were a subsystem of a larger process or machine. Here we use the transfer function icon to simulate the system with a single block. Now M and B are not individually entered; we instead enter numbers for K and τ. Also, internal details such as damper force are *not* available to us; they would probably be of no interest when we have moved to the design of the larger process. Of course, both simulations must give identical results for the output velocity, which is called vo in the detailed simulation and votf in the transfer function simulation. Figure 7-10b verifies this and also shows some results available from the detailed simulation.

These two levels of simulation are not peculiar to first-order systems. You should be aware of these different approaches when simulating *any* system, since each has its own advantages at various stages of system design. The detailed form is most useful when designing the subsystem itself, while the more compact transfer function form would be used when designing the larger machine or process of which the subsystem is only a part.

Design Example: Electric Motor Drive for a Machine Slide. We want to use a brushless dc rotary motor to drive a translational slide, using a rack-and-pinion

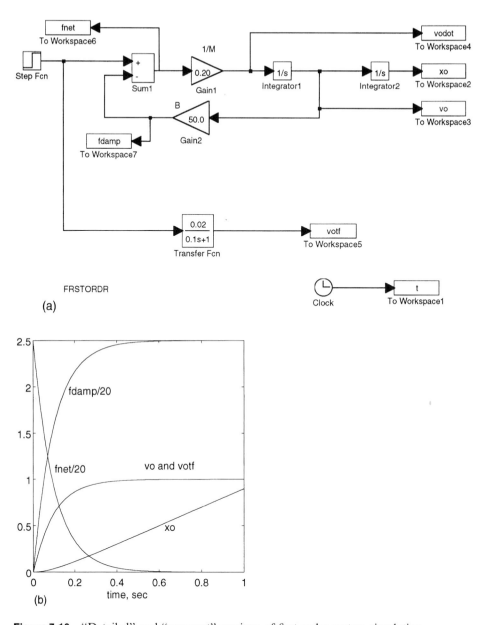

Figure 7-10 "Detailed" and "compact" versions of first-order system simulation.

arrangement as in Fig. 7-1a. If you have no experience with gearing, please accept the fact that the rotary pinion gear and the translational rack behave exactly like a friction drive ("no teeth") that does not slip. That is, the translational displacement x_o is given by $R\theta$, where angle θ is in radians and R is the radius of the cylindrical friction member, or the so-called *pitch radius* of the toothed gear. Actually this gear model assumes that the gears have no backlash ("lost motion") and that the gear teeth do not deflect under load. Neither of these assumptions is exactly true but experience has shown that they are adequate for many practical systems.

First-Order Systems

When the two motions θ and x_o are related by an algebraic (rather than differential) equation, we can use the equivalent system techniques of Fig. 2-43 to speed our analysis. Since our major interest is in the translational velocity v_o, our equivalent system will refer the rotary inertia J to the translational member using $M_{eq} = J/R^2$. The driving torque T_i is referred to the translation member as a force T_i/R. The equivalent translational system will then have the equation

$$\frac{T_i}{R} - Bv_o = \left(M + \frac{J}{R^2}\right)\frac{dv_o}{dt} \tag{7-25}$$

$$\tau \triangleq \frac{M + J/R^2}{B} \qquad K \triangleq \frac{1}{RB} \tag{7-26}$$

$$(\tau D + 1)v_o = KT_i \tag{7-27}$$

This equivalent first-order system can be analyzed or simulated directly; however, it conceals some details which might be of interest in some design studies. For example, detail design of the gear teeth requires knowledge of the force, call it f_g, which they transmit.

To deal with the gear force explicitly we use two free-body diagrams, one for the pinion and one for the rack. At any instant a tooth on the pinion and the mating tooth on the rack feel the same force magnitude f_g but with opposite direction. As in the friction-drive equivalent, this gear force acts horizontally at the pitch radius R. You should now draw the two free bodies with the force f_g pushing one way on the pinion and the opposite way on the rack. You may be wondering what direction to give these forces, but note that this force is an *unknown*, so we don't yet know its direction. Thus you can show this force acting in *either* direction on the pinion, so long as you show the force on the rack in the *opposite* direction. If you don't believe that this is OK, try it both ways. You will see that it makes no difference. If you have drawn the two free bodies you will see the truth of the equations

$$T_i - f_g R = J\frac{d^2\theta}{dt^2} \qquad f_g - Bv_o = M\frac{dv_o}{dt} \tag{7-28}$$

In these equations I have taken f_g to the right on M; if you chose the opposite direction, that would also be correct, so long as you apply f_g on the pinion in the opposite direction. If you don't believe this, try it both ways, to see that it really makes no difference.

The simulation diagram of Fig. 7-11 uses the above two equations and the kinematic relations $x_o = R\theta$, $v_o = R\omega$, and $dv_o/dt = R\alpha$, where ω is the angular velocity and α is the angular acceleration. Note that all the motions and forces are accessible from this simulation, in particular the gear force. If you run this simulation on SIMULINK you will get a message that it includes an *algebraic loop*. This means that there is a "circular" path in the diagram that allows instantaneous signal propagation ("all algebra, no integrations"). In this example the algebraic loop goes from the signal ALPHA, through gains R and M, through the summer to FG, and back to ALPHA through gain R, the torque summer and gain $1/J$. Algebraic loops are not errors but they require the simulation software to do some "extra" calculations. Most software does all this automatically and simply lets you know that it is happening. No special action is required on your part and the results can be used in

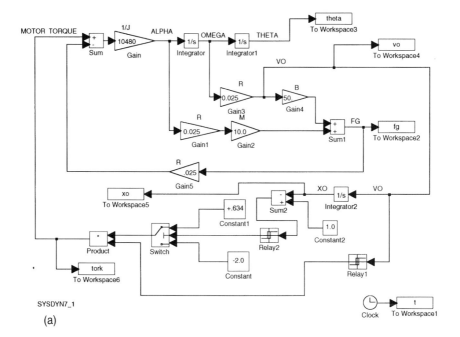

(a)

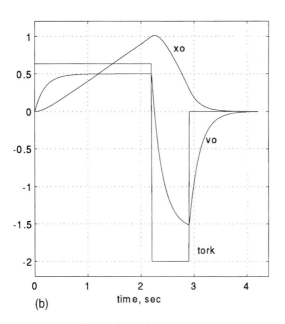

(b)

Figure 7-11 Simulation of motor-driven machine slide.

First-Order Systems   **423**

the normal way. If you want to understand more of the background for algebraic loops, such information is available in the literature.[1]

Our design problem has the following constraints. The mass M of the slide, 10 kg, has been fixed by other considerations and cannot be changed. We must be able to accelerate the mass from rest to a steady velocity of 0.5 m/sec, reaching 95% of this steady velocity in 0.6 second or less. When the displacement reaches 1.0 meter, we must return the displacement to zero, arriving there with zero velocity. This maneuver is to be accomplished by first reversing the motor and then turning it off, letting it coast to rest at zero displacement, taking no more than 2 seconds for this part of the cycle. The deceleration during this motion reversal should not exceed 1g (9.8 m/sec^2). The motor is to be chosen from a series of 19 brushless servo motors available from a vendor[2] (see Table 7-1). Space limitations require that the pinion have a radius no greater than 25 mm.

The upper part of Fig. 7-11a implements Eqs. (7-28), while the lower part uses some "logic" to give the switching actions needed in the motion reversal described above. Our design procedure uses direct calculation from established first-order system results when possible, and simulation for those aspects not readily handled analytically. The rotary inertia J is made up of motor inertia plus the inertia of the pinion gear. We have not yet chosen the motor but the pinion inertia can be quickly estimated if we assume that it is steel and essentially a cylinder of 25-mm radius and 25-mm thickness. This rough estimate will shortly be justified when we see that J/R^2 for the pinion is *much* smaller than M in Eq. (7-26). Using standard formulas for moment of inertia we find J for the pinion is about 0.000087 kg-m^2, making J/R^2 about 0.14 — a contribution of only about 1% of M.

Since we want 95% response in 0.6 second, we know that the time constant must be 0.2 second or less. This gives the following relation from Eq. (7-26):

$$\frac{10 + 1600J}{B} = 0.2 \tag{7-29}$$

We also have

$$KT_m = \frac{1}{RB} \, T_m = \frac{40T_m}{B} = 0.5 \, \text{m/sec} \tag{7-30}$$

The parameters J, B, and motor torque T_m are open to choice but must be chosen so as to satisfy these equations. Our approach will be to first choose a motor from the series available. This fixes J and the maximum torque allowed for that motor. Once J is fixed the required B comes from Eq. (7-29). Then (7-30) gives the needed T_m. If the chosen motor can supply this torque without exceeding its allowable current, we have a potential solution to our problem. Economics, however, dictates that we should strive for the smallest (cheapest) motor that meets our needs, so we need to then try the next smaller motor in the series. By proceeding in this fashion we should be able to find a solution which is technically and economically sound. Since

---

[1]Bennett, B. S., *Simulation Fundamentals*, Prentice-Hall, New York, 1995, pp. 99, 133; SIMULINK User's Guide, 1993, pp. 2–29; ACSL Reference Manual, 1987, pp. 3–5, 4–26.
[2]Brushless Servomotors, Electric Products and Controls, Vickers, Inc., 5435 Corporate Drive, Suite 350, P.O. Box 302, Troy, MI, 810-853-1000.

424 **Chapter 7**

at this point we have no "feel" for how large a motor will be needed, we might start by trying the smallest motor in Table 7-1.

Its inertia, converted to SI units, is $8.49 \times 10^{-6} \text{kg-m}^2$ and, using this in Eq. (7-29) gives $B = 50.1 \text{ N}/(\text{m/sec})$. Equation (7-30) then gives a motor torque requirement of 0.626 N-m, which converts to 5.59 in-lb$_f$. This motor is capable of only 1.85 in-lb$_f$ so our initial motor choice will not meet our requirements. Scanning down the motor table it becomes clear that motor inertia has little effect in Eq. (7-29) until we get to rather large motors. This means that B can be taken as about 50 and T_m as about 5.6, which means that motor 5 will probably be the smallest motor to meet our needs. Actually, motor 4 might also be worth considering since the maximum torque given in the table is for continuous duty, whereas our motion is of an intermittent nature. Motor sizing usually comes down to *heating* considerations and a motor can tolerate an intermittent torque which is considerably larger than its rated continuous torque so long as overheating does not occur.

Since we have based our torque calculations on the steady speed requirement and have not yet considered what is needed to decelerate the load and return it to

Table 7-1

Motor	J	K_t	T_m	R	L	W	P	R_t	C_t
1	0.075	1.95	1.85	33.0	5.4	2.2	0.15	2.60	0.500
2	0.128	2.04	3.36	12.3	3.3	2.64	0.25	2.25	0.650
3	0.168	2.21	4.43	7.8	2.2	2.97	0.31	2.16	0.800
4	0.575	2.30	4.43	8.5	9.6	5.28	0.45	1.62	0.950
5	0.797	2.30	8.85	2.9	4.8	6.38	0.70	1.44	1.200
6	1.239	2.30	17.7	1.1	2.4	8.36	1.10	1.18	1.770
7	2.478	3.54	17.7	1.60	4.6	10.34	1.20	0.97	2.540
8	3.983	3.54	35.4	0.60	2.3	14.74	1.90	0.79	3.000
9	5.487	3.54	53.1	0.35	1.5	18.70	2.80	0.61	3.600
10	6.992	3.54	70.8	0.25	1.1	22.00	3.30	0.53	4.100
11	14.60	5.13	53.1	0.90	4.8	25.3	2.20	0.60	5.600
12	24.34	5.13	106.2	0.30	2.8	36.3	3.40	0.50	7.600
13	34.08	5.13	159.3	0.20	1.9	47.3	4.70	0.42	9.600
14	43.81	5.13	212.4	0.13	1.5	58.3	5.00	0.35	11.60
15	110.6	7.08	212.4	0.16	1.2	68.2	7.20	0.32	13.70
16	147.8	7.08	318.6	0.12	0.94	85.8	8.50	0.28	16.70
17	192.1	7.08	424.8	0.065	0.56	103.4	12.0	0.24	19.70
18	272.6	7.08	637.2	0.049	0.47	138.6	14.2	0.19	25.70
19	354.0	7.08	849.6	0.027	0.28	176.0	20.0	0.16	32.00

$J \triangleq$ rotor inertia, 10^{-3} inch-lb$_f$-sec^2
$K_t \triangleq$ motor torque constant, inch-lb$_f$/amp
$T_m \triangleq$ maximum motor torque, continuous duty, locked rotor, temperature rise 65°C, inch-lb$_f$
$R \triangleq$ winding resistance at 20°C, ohms
$L \triangleq$ winding inductance, milli-henrys
$W \triangleq$ motor weight, lb$_f$
$P \triangleq$ maximum output power, continuous duty, 65°C temperature rise, kW
$R_t \triangleq$ thermal resistance, motor to air, °C/W
$C_t \triangleq$ thermal capacitance, kJ/°C

First-Order Systems

zero displacement, we now tentatively select motor 5 and turn to simulation to deal with the "return motion" part of our requirement.

Using the inertia of motor 5 we find that $B = 50.7$ and required motor torque is 0.634 N-m (5.61 in-lb_f), well within this motor's capabilities. The upper part of Fig. 7-11a uses these numbers and Eqs. (7-28). The lower part is some "logic" which uses a displacement signal XO from a sensor to detect when we reach xo = 1.0 m and, at that point, switch the motor torque from +0.634 to some negative value. This negative torque is applied until the velocity reaches a negative value where, if we make the motor torque zero, the velocity will decay to zero while the displacement returns to zero. The magnitude of the negative torque, and the velocity at which we make motor torque zero are both found by trial and error, using simulation. Our specifications require that this deceleration phase take no more than 2 seconds and impose no more than 9.81 m/sec^2 deceleration.

Figure 7-11a shows one set of numbers which meets these criteria. The negative torque shown is −2.0 N-m but the velocity for turning off the motor is "hidden" inside the icon for Relay 1. This velocity is −1.511 m/sec. Actually, both Relays 1 and 2 contain some subtleties, shown in Fig. 7-12a and b, which you need to be aware of if you try to duplicate this simulation. Both these relays use the hysteresis capability of this icon to get the switching actions to work properly. When you put hysteresis into a relay, the point at which it switches ON is made different from the point at which it switches OFF. Relay 1 is used to turn the motor off at a selected negative velocity, but should *not* disable the motor earlier. In Fig. 7-12a I made this "switch on" point a small negative value (−0.001 m/sec), so that when we started the simulation (where vo starts at 0.0), this relay would be ON, providing the signal 1.0 to the Product icon. Since the Switch icon outputs either +0.634 or −2.0 (depending on whether xo has exceeded 1.0 yet), multiplying this signal by the output of Relay 1 (either 1.0 or 0.0) makes the motor torque +0.634, −2.0, or 0.0. The "Switch off" point for Relay 1 is found by trial and error, as explained earlier.

For Relay 2 (Fig. 7-12b) the "switch on" point is chosen a little less than 1.0 (0.999), so that this relay will be ON when we start the simulation, because the icon

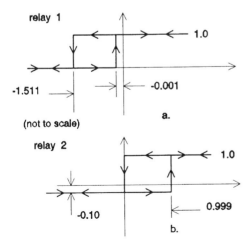

Figure 7-12 Details of relay actions.

Constant 2 is already sending a 1.0 signal to this relay. The "switch off" point is made 0.0 but the OFF signal (-0.1) is made negative (any negative value would do) to activate the switching action of the Switch icon.

As seen in Fig. 7-11b, it took a torque of -2.0 N-m (17.7 in-lb$_f$) to return the mass to the zero displacement location in the 2 seconds allowed for that maneuver. (Figure 7-11a doesn't send acceleration to the workspace for plotting, but this feature is easily added. We find that the peak acceleration is slightly more than the allowed $1g$.) Since the -2.0 N-m torque took almost the allowed 2 seconds, the decelerating torque cannot be reduced much below this value, which is almost twice the motor's continuous rating. We could of course go to the next larger motor, but motor 5 may well be adequate, since motors can usually tolerate considerable overloads (200 or 300%) if they are not steady. The limit of 200 to 300% overload must be observed to prevent excessive armature current (even if it is momentary) from *demagnetizing* the motor's permanent magnets. Since the final sizing of a motor usually depends on heating considerations, we should try to get an estimate of these effects.

Motor heating is caused mainly by the i^2R power dissipated in the windings. Figure 7-11a again doesn't show such a calculation but it is easily added, using the motor's torque constant to relate current to torque. By integrating this heating power over the motion cycle, we can get the total thermal energy input to the motor for that cycle. The thermal time constant of a motor is roughly given by the product of its thermal resistance and thermal capacitance, about 4.8 hours for motor 5. For such a slow response, we can estimate the motor temperature rise by using the average heating rate. Let's assume that the motion cycle of Fig. 7-11b is repeated as fast as possible, about 1 cycle every 4 seconds. The thermal energy calculation obtained from the simulation as described above gives the energy per cycle as 155 watt-sec. If this is averaged over 4 seconds we get an average heating rate of 38.8 watts. Since the motor's thermal resistance is given as 1.44°C/watt, we estimate the motor's steady-state temperature rise as about 56°C. This is less than the rated 65°C for continuous duty, so the motor will probably be OK for this application. Actually, machine and system designers who use electric motors will, after a rough analysis as we have presented, consult with the motor manufacturer to get the benefit of their experienced application engineers and avoid unforseen pitfalls. Some motor manufacturers also provide, usually free, sizing software for motors and amplifiers to potential customers. Some of this software[3] is quite versatile in accommodating many different types of mechanical loads and duty cycles, and this is a great aid in motor selection.

Motion Control by Feedback: An Alternative Design. The motion control system we have just designed would be called an "open-loop" system. Its accuracy in providing the desired acceleration, steady velocity, and return to zero displacement depend on the numerical values of all the system parameters staying fixed at the design values. When some or all of the parameters *cannot* be depended on to stay nearly constant, and/or the accuracy requirements become stringent, open-loop systems may not

---

[3]EMERSize Motor Sizing Software, Emerson Electronic Motion Controls, 1365 Park Road, Chanhassen, MN 55317, 612-474-1116.

First-Order Systems 427

meet specifications. Then the designer will consider closed-loop ("feedback") systems. For our present example, the motor, amplifier, and mechanical load might remain the same, but we would add a sensor to continuously measure load displacement, not just a limit switch to reverse the motor when the displacement reaches 1.0 meter. We also would add a means to enter a command voltage, and a comparator (summing amplifier) to compare the desired displacement with the measured actual displacement.

When we turn to a feedback system, we no longer depend on the damper to transduce the motor torque into our desired steady velocity. Since the damper is actually a source of inefficiency (energy supplied to it is converted into heat), we now might want B to be as small as possible, representing just the unavoidable bearing friction of the moving parts. Suppose this B value is one-tenth of the value used in the open-loop system, that is, 5.0 N/(m.sec). We still have a first-order mechanical system but the parameters are now $K = 8.0$ (m/sec)/(N-m) and $\tau = 2.0$ seconds. In the simulation diagram of Fig. 7-13a, we show this mechanical system together with the needed sensor (1.0 volt/m), comparator, and command voltage source, all connected in a feedback configuration. Also included is electronic dynamic compensation. If you duplicate this simulation but *don't* include this compensation, you will find that you are unable to meet the specifications no matter what value you try for the only adjustable parameter, the amplifier gain.

Books devoted to feedback system design show many different compensation techniques that can be used in different situations when the "bare bones" feedback system cannot be adjusted to meet specifications. In our present example, the technique called *cancellation compensation*[4] turns out to be useful. This can also be interpreted as proportional plus derivative control. Here we include in our system a (usually electronic) dynamic compensator of so-called *lead-lag* form, $(\tau_1 s + 1)/(\tau_2 s + 1)$ (see Fig. 7-13b for one way to actually build such a device). We choose the number τ_1 to match (and thus *cancel*) the time constant of the mechanical system, and make τ_2 much (as much as 10 times) smaller. I tried one such compensator in the system and still could not meet specifications. I then added a second compensator as shown, and by adjusting amplifier gain, got the acceptable response of Fig. 7-13c. Note that because of the "cancellation" effect, the two compensators in cascade with the mechanical system have an overall transfer function of $8/(0.05s + 1)$; thus the feedback system "thinks" it is controlling a mechanical system with *this* transfer function, not the actual mechanical system. This cancellation compensation trick is used to speed up the response of many practical open-loop and closed-loop systems, including sensors[5] such as hot-wire anemometers and thermocouples. Since such compensators are approximate differentiators, they may not be usable in real systems unless the level of random noise (present in every real system) is sufficiently low. Noise problems can be studied at the design stage with simulation and verified by lab testing once actual hardware is in hand.

We should point out the versatility of our feedback system, compared with the earlier open-loop design. The displacement command voltage xi was given the tri-

---

[4]E. O. Doebelin, *Control System Principles and Design*, Wiley, New York, 1985, p. 571.
[5]E. O. Doebelin, *Measurement Systems*, 4th ed., McGraw-Hill, New York, 1990, p. 951.

428   Chapter 7

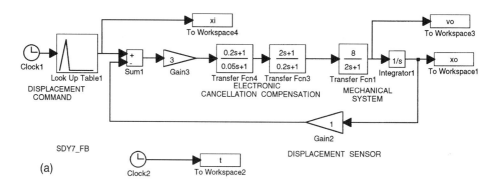

(a) SDY7_FB

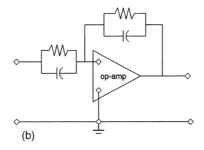

(b)

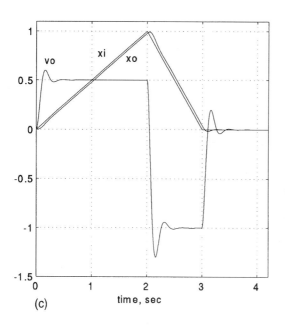

(c)

Figure 7-13 Feedback system with dynamic compensation ("cancellation compensation") to improve speed of response.

First-Order Systems

angular shape of Fig. 7-13c to give us the desired steady-state velocity, and the return to zero displacement required by our specifications. By tailoring the shape of this xi command, we can produce a wide variety of motion cycles *without adjusting any other system parameters*. If we implement this system digitally, the changes in command xi are strictly *software*, not hardware, changes and can be easily and quickly made, giving our machine the benefits of so-called *flexible automation*. That is, if our manufacturing process requires frequent changes to produce different products ordered by customers, we can accommodate these customer needs efficiently.

Optimum Step Response Using a Nonlinear Approach. If a first-order system is to be used with step inputs of various sizes, a linear system is not really optimum if our design goal is to get the fastest possible response for every size command. This is because a small command will only use a fraction of the force or torque capability of the motor. That is, a motor is sized so that the *largest* command will just use the maximum capability of the motor. For any *smaller* command, the motor provides a proportionately smaller torque, so it is underutilized. To get *optimum* (fastest) response for every size command, we can apply the maximum available torque *initially*, use a velocity sensor to tell us when we have reached the desired velocity, and then switch the motor to the smaller torque needed to sustain that velocity as a steady state. With this scheme, we are using the full capability of the motor whenever possible and, for any command less than the maximum, the response speed will be significantly faster than for a strictly linear system.

Figure 7-14a shows one possible implementation of such a nonlinear system. The mechanical load is taken as the rotary system of Fig. 7-1a, with gain of 5.0 (rad/sec)/(N-m) and time constant of 0.10 second. If we drove this system directly with step inputs of various sizes, every such input would cause a response which settled within 5% in 0.30 seconds. Using the nonlinear scheme explained above, we can show that *much* faster responses are possible for commands less than the maximum. We assume a motor with torque constant of 3.54 (N-m)/amp and maximum allowable torque of 21.24 N-m, making the maximum allowable current 6.0 amps. The output angular velocity is measured by an instrument with a sensitivity of 0.0565 volts/(rad/sec). This number is chosen so that the sensor signal will equal the command velocity vcom when the velocity just equals that commanded. That is, (vcom)(1.0(3.54)(5.0)(0.0565) = vcom. If we then compare the sensor signal with vcom, using the summer shown, we can use the summer output as a switching signal for the icon Switch 1. This icon selects one of two inputs, depending on whether the velocity has reached the commanded value or not. Until it reaches the commanded value, the motor uses its maximum allowable current and torque. When the commanded value is reached, the motor "throttles back" to provide exactly the torque needed to sustain that particular velocity. The sensing, comparing, and switching used in our simulation are all possible with real equipment, so this scheme can be practically implemented.

Figure 7-14b shows responses to commands of four different sizes. When the command is the largest designed for, we get the familiar linear system exponential, asymptotic response, which achieves 95% of the final value in 0.30 second. For smaller commands the system is nonlinear; response is *not* proportional to input. Rather, the response curve follows that for the *maximum* input until the desired velocity is reached, whereafter the velocity stays constant at the commanded

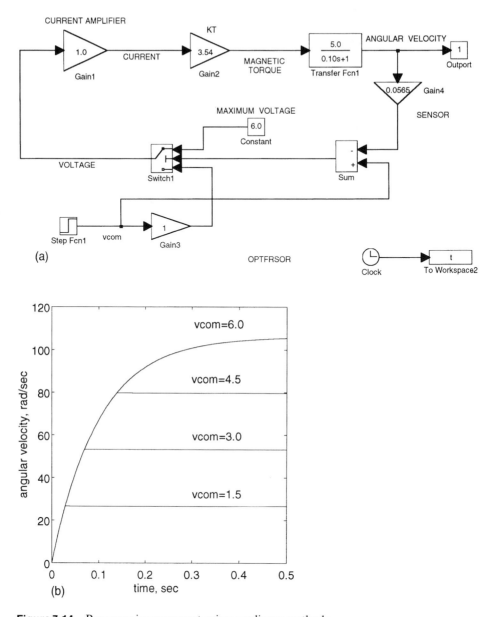

Figure 7-14 Response improvement using nonlinear methods.

value. The approach to steady state is *not* asymptotic and the 95% response time is shorter than 0.30 second. For the smallest input shown, vcom = 1.5, the 95% response time is about 0.03 second, a great improvement. Similar nonlinear techniques have been used in head-positioning servo systems used in computer hard-disk drives.[6] While linear systems are "mathematically optimum" in that they can be

---

[6]E. O. Doebelin, *Control System Principles and Design*, Wiley, New York, 1985, pp. 327–331.

First-Order Systems

431

analytically solved, they generally are *not* optimum in terms of engineering performance, so we should always consider various nonlinear techniques which might improve performance, and/or lower cost. "Before simulation" such studies were rarely carried out because nonlinear analysis methods were often cumbersome and inaccurate. With fast, inexpensive, and easy-to-use simulation software readily available, designers are today free to explore many alternative design concepts whether they are linear or not.

7-3 RAMP, SINUSOIDAL, AND IMPULSE RESPONSE OF FIRST-ORDER SYSTEMS

While study of step response characteristics has revealed much about the behavior of first-order systems, it is useful to also be aware of the response to a few other "standard" inputs.

Ramp Response. We define a ramp input as a signal which starts at a fixed level (usually taken as zero) and then grows at a fixed (usually positive) rate, giving a graph which is an upward-sloping straight line. While our main interest is in the ramp response of *generic* first-order systems, we still prefer to begin with a specific example which has practical applications. The rotary system of Fig. 7-1d has been used in automobile speedometers and is of general utility as a rotary speed measuring instrument or an analog computing element for taking the first derivative of rotary displacement. Its differential equation is easily found as

$$B(D\theta_i - D\theta_o) - K_s\theta_o = JD^2\theta_o = 0 \tag{7-31}$$

$$\frac{B}{K_s}\frac{d\theta_o}{dt} + \theta_o = \frac{B}{K_s}\dot{\theta}_i \tag{7-32}$$

$$\tau \triangleq \frac{B}{K_s} \quad \text{seconds} \qquad K \triangleq \frac{B}{K_s} \quad \frac{\text{rad}}{\text{rad/sec}}$$

$$(\tau D + 1)\theta_o = K\dot{\theta}_i = K\omega_i \tag{7-33}$$

For any steady velocity ω_{is} the output displacement θ_o is $K\omega_{is}$, so an electrical displacement sensor measuring θ_o gives a measurement of velocity; that is, $\omega_i = \theta_o/K$. After a step change in velocity, this instrument will read the new velocity with 95% accuracy in 3τ.

If velocity is not steady but rather increasing in a ramplike fashion, we might want to know how accurate our instrument is under such conditions. Taking $\omega_i = \alpha_i t$, where α_i is a known constant, we get

$$(\tau D + 1)\theta_o = K\omega_i = K\alpha_i t \tag{7-34}$$

which is easily solved using either the *D*-operator or Laplace transform methods to get

$$\theta_o = K\alpha_i\tau e^{-t/\tau} + K\alpha_i(t - \tau) \tag{7-35}$$

This solution can be plotted and/or we can use the simulation of Fig. 7-15a to get the desired graph. We see that *no matter how long we wait*, the instrument will never read the correct velocity, it will have a *steady-state error* equal to $-\alpha_i\tau$. Note that for a given α_i this error is proportional to τ, so we again strive for a small time constant to reduce this error. From the solution we can also see another interpretation: Once in steady-state, the instrument at any instant reads what the true velocity was τ seconds ago. Thus the defect in this instrument for a ramp input can be expressed as a steady-state error or a steady-state time lag. While *all* first-order instruments can, for step inputs, achieve any desired dynamic accuracy simply by waiting long enough, for ramp inputs an error persists no matter how long we wait to read the instrument. This example shows the general need to examine inputs other than steps if we want full knowledge of a system's behavior.

We mentioned above that our example device can also be used as a mechanical differentiator for rotary displacements. This can be made more obvious by writing

$$(\tau D + 1)\theta_o = K\omega_i = KD\theta_i$$

$$\frac{\theta_o}{D\theta_i}(D) = \frac{K}{\tau D + 1} \tag{7-36}$$

We see that if τ is small enough (relative to the rapidity of variation of θ_i) then the output will be an accurate representation of the derivative of the input.

We stated earlier that a small τ is needed for fast response to *any* input and have now verified this for steps and ramps. Perhaps the most general and realistic input is a random one, so let's use simulation to show that here also small τ is best. We can use our present measuring instrument example for this study, and Fig. 7-16a is set up to demonstrate this. We generate a random input as we have before and pass it through a second-order system used as a filter to "round off its square corners," making it more realistic. This signal is then sent to two first-order instruments, one of which has a time constant (0.10 second) which is too long for accurate measurement and another which has adequate speed. We see these statements verified in Fig. 7-16b, where qo1 is only a rough approximation to qi, while qo2 is a much better measurement.

The measurement error, which for qo2 is the vertical distance between qi and qo2, is actually even less than it appears visually. In most practical measuring situations, if the instrument output is an accurate reproduction of the size and shape of the measured input, the fact that it might be *delayed* by a certain time interval is usually of no consequence. The curve of qo2 in fact seems to exhibit such a delay. It can be shown[7] that for *any* order instrument and any form of input, when it is used in its dynamically accurate range, its output will exhibit a time delay. For a first-order instrument, this delay is actually the time constant, as we saw in Fig. 7-15b for the special case of a ramp input. We demonstrate this behavior using the lower part of the simulation in Fig. 7-16a. There we insert a Transport Delay equal to τ (0.01 second) and then compute the measurement error for both the undelayed and delayed cases. Figure 7-16c shows that when we remove the effect of instrument delay, the measured value qo2 and the instrument input qidel agree very closely. The

---

[7]E. O. Doebelin, *Measurement Systems*, 4th ed., McGraw-Hill, New York, 1985, p. 184.

First-Order Systems

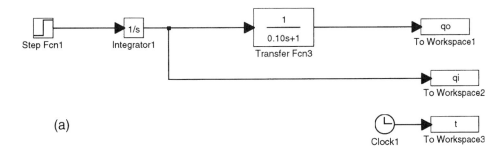

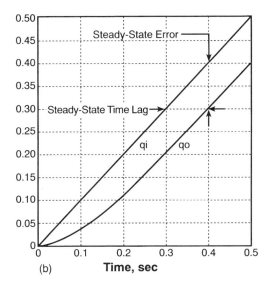

Figure 7-15 Ramp response of first-order systems.

graphs of err2*5 and err1*5 clearly show the improvement possible when we recognize and correct for the time delay effect.

Sinusoidal Response (Frequency Response). We have emphasized the importance of system frequency response from the earliest chapters in this book, so we certainly want to become familiar with it for first-order systems. If we make our system input sinusoidal with amplitude q_{i0} and frequency ω rad/sec, the system equation becomes

$$(\tau D + 1)q_o = Kq_i = Kq_{i0}\sin\omega t \tag{7-37}$$

Recall that for frequency response we are interested only in the sinusoidal steady state, which is achieved after transients die out; that is we want the *forced*, not the natural, part of the response. We could use either Laplace transform or the D-operator method to get this result. In this case the D-operator method is perhaps preferable since it separately gets the forced (particular) solution without having to bother with the transient. The method of undetermined coefficients gives us the particular solution as

$$q_{op} = A\sin\omega t + B\sin\omega t \tag{7-38}$$

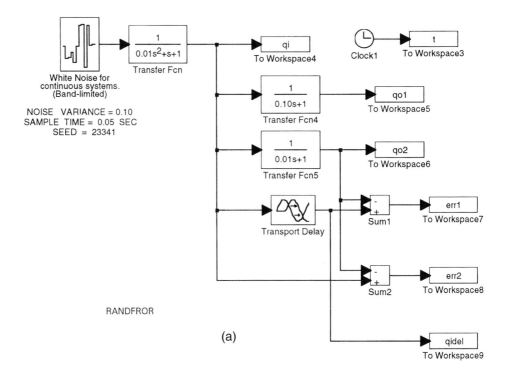

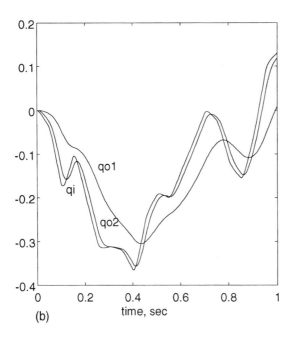

Figure 7-16 Use of random input to verify time-delay behavior.

First-Order Systems

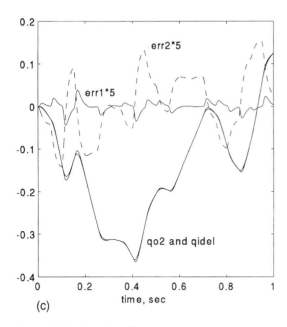

Figure 7-16 (*continued*)

$$\tau(\omega A \cos \omega t - \omega B \sin \omega t) + A \sin \omega t + B \cos \omega t = K q_{i0} \sin \omega t \tag{7-39}$$

$$-\omega B \tau + A = K q_{i0} \qquad \omega A \tau + B = 0 \tag{7-40}$$

$$A = \frac{K q_{i0}}{\omega^2 \tau^2 + 1} \qquad B = -\frac{\omega \tau K q_{i0}}{\omega^2 \tau^2 + 1} \tag{7-41}$$

and thus

$$q_{op} = \frac{K q_{i0}}{\omega^2 \tau^2 + 1} \sin \omega t - \frac{\omega \tau K q_{i0}}{\omega^2 \tau^2 + 1} \cos \omega t \tag{7-42}$$

This may be simplified by using the trig identity

$$A \sin \alpha + B \cos \alpha \equiv \sqrt{A^2 + B^2} \sin \left(\alpha + \tan^{-1} \frac{B}{A} \right)$$

to give finally

$$q_{op} = \frac{K q_{i0}}{\sqrt{\omega^2 \tau^2 + 1}} \sin [\omega t + \tan^{-1}(-\omega \tau)] \tag{7-43}$$

This same result is given much more quickly using the sinusoidal transfer function:

$$\frac{q_o}{q_i}(i\omega) = \frac{K}{i\omega \tau + 1} = \frac{K \angle 0°}{\sqrt{\omega^2 \tau^2 + 1} \angle \tan^{-1}(\omega \tau)} = \frac{K}{\sqrt{\omega^2 \tau^2 + 1}} \angle \tan^{-1}(-\omega \tau) \tag{7-44}$$

That is,

$$\text{Amplitude ratio} \triangleq \frac{K}{\sqrt{\omega^2 \tau^2 + 1}} \qquad \text{Phase angle } \phi \triangleq \tan^{-1}(-\omega\tau) \qquad (7\text{-}45)$$

For a given K and τ the amplitude ratio and phase angle curves appear as in Fig. 7-17a. As frequency ω approaches zero, the amplitude ratio approaches the steady-state gain K and the phase angle approaches zero. For increasing frequency the amplitude ratio decreases toward zero while the phase angle approaches $-90°$ asymptotically. If the system time constant is decreased, at any frequency the amplitude ratio will be larger and the phase angle less lagging. The universal curves of Fig. 7-17b are obtained by plotting the amplitude ratio of q_o/Kq_i (rather than q_o/q_i) against $\omega\tau$ (rather than ω):

$$\frac{q_o}{Kq_i} = \frac{1}{\sqrt{(\omega\tau)^2 + 1}} \; \underline{/\tan^{-1} -(\omega\tau)} \qquad (7\text{-}46)$$

When we want accurate curves for specific values of K and τ we can use math software, such as MATLAB, to ease the work. For example, if a first-order tem-

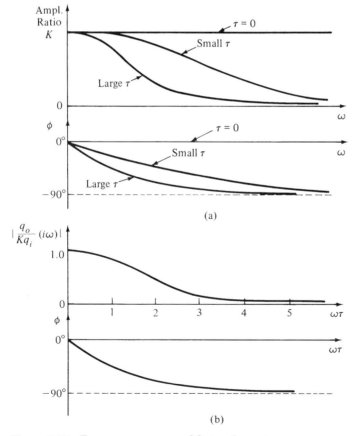

Figure 7-17 Frequency response of first-order systems.

First-Order Systems

perature sensor with electrical output has K = 3.65 volt/°F, $\tau = 0.01$ second, and we are interested in frequencies from 0 to 1000 rad/sec, a MATLAB session might go as follows.

```
K=3.65; T=0.01; w=0:1:1000; s=w*1i; tf=K./(T.*s+1);
ar=abs(tf); phi=57.3.*angle(tf);
subplot(2,1,1); plot(w,ar); grid;
xlabel('frequency,rad/sec'); ylabel('ampl ratio, volt/deg F')
subplot(2,1,2); plot(w,phi); grid; ylabel('phase angle, deg')
```

Figure 7-18 shows the results of this calculation. If we want to see the sine waves plotted against time we can use SIMULINK to send a sine wave of selected frequency and amplitude into a transfer function icon set up for a first-order system. Figure 7-19 shows both the input sine wave (amplitude 1.0) and the output sine wave for a system with K = 1.0 and $\tau = 0.1$ second, for the frequencies 1.0, 10.0, and 100.0 rad/sec. At 1.0 rad/sec, the input and output are almost identical (top graph), while at 10.0 rad/sec the output amplitude has dropped to 0.707 and the phase angle is $-45°$. At 100.0 rad/sec the output amplitude is about 0.10 and the phase angle is about $-84°$. Note that simulation of the differential equation gets *both* the transient and steady-state solution. The transient is hardly visible for frequencies of 1.0 and 10.0 rad/sec because the graph time scale obscures this event which is over in about 0.3 second. For the graph with 100.0 rad/sec frequency, the first few cycles of the "sine wave" show the transient dying out, since the trace is not yet symmetrical about the zero line.

Logarithmic Frequency-Response Plotting. Years ago, the advantages of plotting frequency-response curves on logarithmic coordinates were recognized, so the prac-

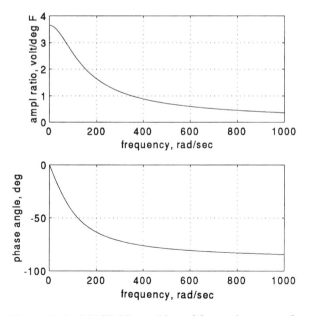

Figure 7-18 MATLAB graphing of first-order system frequency response.

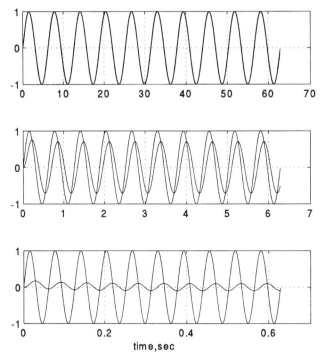

Figure 7-19 Simulation results for first-order system sinusoidal input.

tice is very common today. This technique applies to any sinusoidal transfer function, not just first-order systems. For such graphs (sometimes called Bode plots in honor of their inventor) the amplitude ratio is plotted in decibels. If the actual amplitude ratio is a number N, its decibel equivalent is given by

$$\text{Decibel value of } N \triangleq dB \triangleq 20 \log_{10} N \tag{7-47}$$

Note that an actual amplitude ratio of 1.0 is 0 dB, 10.0 is 20 dB, 0.10 is -20 dB, 100 is 40 dB, and .01 is -40 dB. These few values are worth memorizing since the method is in such wide daily use.

The basic advantages of this plotting scheme are revealed by applying it to first-order systems. If $K = 1.0$, we have

$$\text{Amplitude ratio in dB} = 20 \log_{10} \frac{1}{\sqrt{(\omega\tau)^2 + 1}} = -20 \log_{10} \sqrt{(\omega\tau)^2 + 1} \tag{7-48}$$

For very low frequencies we have $(\omega\tau)^2 \ll 1$ and thus

$$dB \approx -20 \log_{10} 1 = 0 \tag{7-49}$$

while for high frequencies $(\omega\tau)^2 \gg 1$ and

$$dB \approx -20 \log_{10} \omega\tau = -20 \log_{10} \tau - 20 \log_{10} \omega \tag{7-50}$$

When amplitude ratio in dB is plotted on a linear scale, against frequency ω on a logarithmic scale, we get a curve which is asymptotic at low frequency to a straight

First-Order Systems   **439**

horizontal line at 0 dB, Eq. (7-49), and at high frequency to a straight line of slope -20 dB/decade, Eq. (7-50), where a *decade* is any 10-to-1 frequency range. An *octave* is any 2-to-1 frequency range. These low and high frequency asymptotes meet at $\omega = 1/\tau$, which frequency is called the *breakpoint frequency*. When the steady-state gain is K rather than 1.0, the low-frequency asymptote is horizontal at the value $20 \log_{10} K$ dB, rather than 0.0 dB.

We want to now show a quick manual technique for sketching such graphs. Such a skill is still useful for several reasons. Before we produce accurate graphs on our computer, a quick "back-of-an-evenlope" manual sketch helps us choose good frequency ranges to enter into the computer, and also serves as a check on the computer results. Sometimes we are away from any computer and need to make some quick estimates. Finally, in preliminary design studies, a good understanding of how the curves are affected by design changes is very useful. For all these reasons, you should develop your ability to quickly "rough out" such graphs. A systematic procedure goes as follows.

1. Procure three-cycle semilog graph paper. (Three cycles is a 1000-to-1 frequency range, which has been found to be adequate for most practical systems.) *Which* 1000-to-1 range to use (0.1–100, 1–1000, 10–10000, etc.) depends on the particular system. Label the frequency axis with the range you choose.

2. Convert the steady-state gain K to its dB value and use this number to help you lay out a dB scale on the vertical (linearly divided) graduations. Leave room for about a 40-dB range on this scale. Also leave room at the bottom of the same graph sheet for a phase angle graph that will go from 0 to $-90°$. That is, the amplitude ratio and phase angle graphs must appear on *the same* sheet, with the same frequency axis. Most of the uses of such graphs require that the user see both graphs simultaneously.

3. Draw the low-frequency asymptote as a horizontal straight line at the dB value of K.

4. Locate the breakpoint frequency $\omega_b = 1/\tau$ on the low-frequency asymptote. Draw the high-frequency asymptote as a straight line sloping downward at -20 dB/decade. This is most easily done by finding a point 1 decade above the breakpoint and 20 dB below the low-frequency asymptote.

5. The straight-line asymptotes just drawn are sufficient for many rough calculations. If more accurate curves are needed, correct the asymptotes as follows. At the breakpoint, the correction is always -3 dB. One octave above and below the breakpoint, the correct is -1 dB. Two octaves above and below the breakpoint the correction is $-\frac{1}{4}$ dB, which is close enough to the asymptotes to ignore any further corrections.

6. Sketch the phase angle curve using the fact that, at the breakpoint, the phase angle is always $-45°$, and the low- and high-frequency asymptotes are $0°$ and $-90°$. The table of Fig. 7-20 gives four other angle values that are sufficient to complete the phase angle curve. This figure also illustrates the entire procedure.

Since frequency-response plotting using these Bode plots is so common, MATLAB provides a quick way. For the system of Fig. 7-18:

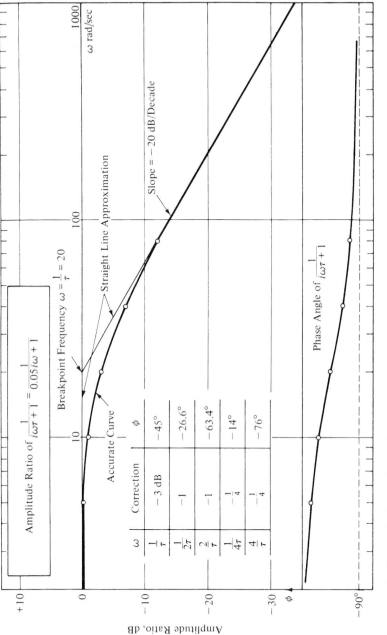

Figure 7-20 Logarithmic plotting of first-order system frequency response.

First-Order Systems

`num=[3.65]`	A vector giving the numerator coefficients of the transfer function
`den=[0.01 1]`	A vector giving the denominator coefficients of the transfer function
`bode(num,den)`	

This works with *all* transfer functions, not just first-order. Figure 7-21 shows the graphs that result from the above three statements. The frequency range is automatically chosen for you. If you don't like that, use `bode(num,den,w)` where w is a vector containing a list of desired frequencies.

Experimental Modeling Using Frequency-Response Testing. We earlier showed how lab testing using step inputs could verify that a system was essentially first-order and also get numerical values for K and τ. Even stronger evidence of system behavior is available if we are willing to invest the greater effort needed to do sinusoidal testing. This type of testing is so common and useful that specialized test equipment for running and analyzing the tests is available from several manufacturers. Even if such specialized gear is not available to you, valid tests can often be run using simple sinewave input generators and a two-channel recorder to record system input and output.

We excite the system with a sinusoidal input, wait for transients to die out and the sinusoidal steady-state to be established, and then measure amplitude ratio and phase angle. We then repeat this for many different frequencies, covering the frequency range of interest to us. We plot the data logarithmically since this allows the best checking of conformance to specific theoretical models, such as the first-order systems of this chapter. Figure 7-22 shows a typical set of such measured data. Note

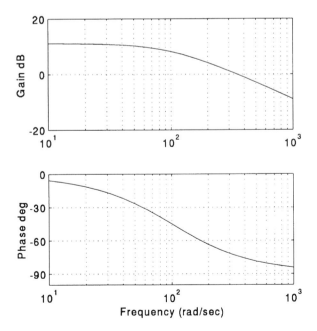

Figure 7-21 Use of MATLAB command BODE NUM,DEN).

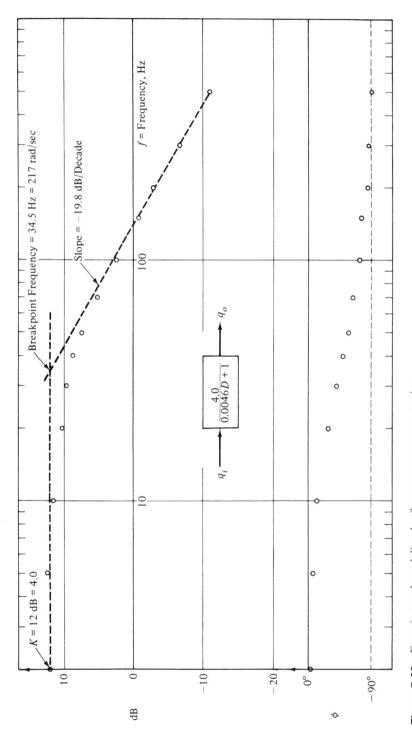

Figure 7-22 Experimental modeling by frequency-response testing.

First-Order Systems   **443**

first that we now plot frequency in Hz (cycles/sec) rather than in rad/sec. This is more natural for experimental data and also recognizes that standard lab equipment nearly always uses this frequency unit. Our Bode plots can be graphed and interpreted exactly as before, with one exception. When we locate the *breakpoint frequency* as a means of identifying τ, we *must* convert to rad/sec when we use the formula $\tau = 1/\omega_b$.

When we use such plots to verify first-order behavior we are looking for several features. First, the low-frequency asymptote of the amplitude ratio curve should be horizontal and at high frequency the curve should approach a straight line with a slope of $-20\,\text{dB/decade}$. The entire curve should not exhibit a peak anywhere. By drawing the two asymptotes we can locate their intersection, which is taken as the breakpoint frequency, from which we can estimate τ. The phase angle curve should be asymptotic to $0°$ at low frequency and to $-90°$ at high, and should decrease monotonically. At the breakpoint frequency the phase should be near $-45°$. If all these criteria are reasonably met we can conclude that our system is essentially first order, at least over the measured frequency range. The low-frequency asymptote gets us a number for K and the breakpoint frequency identifies τ. The example data of Fig. 7-22 meet the above criteria quite well and we thus conclude that the system is well modeled as first order, with $K = 4.0$ and $\tau = 0.0046$ second. As was also the case with step-function testing, we are *not* able, using only such test results, to relate K and τ to *physical* parameters of our system.

We have just explained "manual and visual" methods for processing the measured data to find K and τ. More "scientific" methods are available and these make the results less sensitive to human variability. For example, the MATLAB signal processing TOOLBOX provides a least-squares curve fitting procedure called INVFREQS which can be used for *any* transfer function that is a ratio of polynomials. The measured values of the sinusoidal transfer function, given at each frequency as a complex number with real and imaginary parts, are entered as a vector called h; the frequencies (rad/sec) are entered as a vector called w. You must choose the *form* of model which you wish fitted; the software will then find the "best" coefficient values to fit the measured data. That is, if you want to try a model with a transfer function of the form

$$G(s) = \frac{b_1 s + b_0}{a_3 s^3 + a_2 s^2 + a_1 s + a_0} \tag{7-51}$$

you would enter a first-order numerator (nb=1) and a third-order denominator (na=3) into the MATLAB statement

```
[b,a]=invfreqs(h,w,nb,na)
```

The software then computes the "best" values of the polynomial coefficients and returns them as vectors b and a. You can then plot the frequency-response curves for this fitted model and compare them with the measured values. If the fit is not good enough you can repeat the process with a different form of model, hopefully finding a satisfactory fit after a few iterations. If you are thinking that it might be clever to *start* with a model that has "lots of a's and b's" and let the software tell you which coefficients are "zero," this approach can be dangerous, since it sometimes leads to ill-conditioned numerical processes. Instead, use everything you know about your data to make the best and simplest initial guess at the form of the model and go to complicated models only when simpler ones are found wanting.

Impulse Response of First-Order Systems. We have earlier (Fig. 3-4) encountered the concept of impulse functions when considering the output (current) response of a capacitor to a step input voltage, and later observed this phenomenon with analogous elements and inputs. We now consider the response of a system to an impulse function applied as the *input*. Recall that an ideal impulse has infinite "height," infinitesimal duration, but a finite and definite area. No real physical variable can behave in precisely this fashion; however an approximation sufficiently close for many practical purposes is often possible. Furthermore, the *theoretical* aspects of system response to impulsive inputs are of considerable importance.[8]

Let's consider an impulsive force input to the translational system of Fig. 7-1a, whose simulation diagram is shown in Fig. 7-23. The system equation would be

$$M\dot{v}_o + Bv_o = f_i = A_i\delta(t) \tag{7-52}$$

where $\delta(t)$ is the symbol for a unit impulse function (an impulse of area 1.0). For $t > 0$, the right-hand side is zero, since the impulse is over in an infinitesimal time; thus the equation becomes

$$M\dot{v}_o + Bv_o = 0 \tag{7-53}$$

$$v_o = Ce^{-t/\tau} \tag{7-54}$$

To find C we need the initial condition $v_o(0^+)$, which is the velocity *just after the impulse has occurred*. The simulation diagram is helpful in finding this initial condition; in fact, such diagrams would be helpful in understanding the behavior of

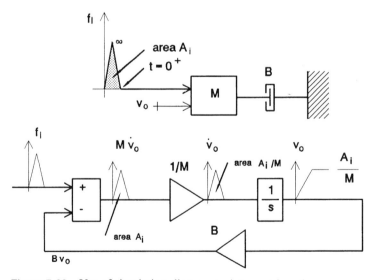

Figure 7-23 Use of simulation diagram to interpret impulse response.

[8] E. O. Doebelin, *System Modeling and Response*, Wiley, New York, 1980, chap. 3.

First-Order Systems

differential equations even if computer simulation had never been invented. We "track" the propagation of the input force through the diagram and see that, since the "feedback" signal Bv_o is of finite size and can thus, at the summer input, be neglected relative to the "infinite" f_i, the signal $M(dv_o/dt)$ is the same as f_i, for $0 < t < 0^+$. Thus $M(dv_o/dt)$ and dv_o/dt are both impulses, but with different areas. Since an integrator produces at its output the *area* of its input signal, the signal v_o must at time $= 0^+$ be given by

$$v_o(0^+) = \int_0^{0^+} \frac{f_i}{M} \, dt = \int_0^{0^+} \frac{A_i \delta(t)}{M} \, dt = \frac{A_i}{M} \tag{7-55}$$

The complete solution of Eq. (7-54) is thus

$$v_o = \frac{A_i}{M} e^{-t/\tau} = \frac{KA_i}{\tau} e^{-t/\tau} \tag{7-56}$$

which is graphed in Fig. 7-24 for both this specific system and the general first-order. *It is clear that the response of an initially motionless system to an impulsive force of area A_i is identical to the response of an unforced system with an initial velocity of magnitude A_i/M, or in the general case, KA_i/τ.* This same conclusion can be reached more quickly, but with less physical insight, using Laplace transform.

$$V_o(s) = \frac{K}{\tau s + 1} F_i(s) = \frac{A_i K}{\tau s + 1}$$

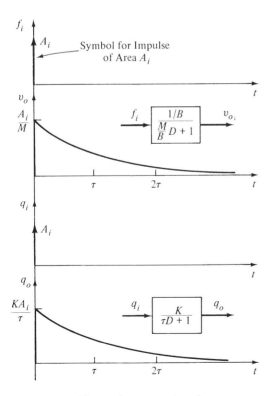

Figure 7-24 First-order system impulse response.

$$v_o(t) = \frac{KA_i}{\tau} e^{-t/\tau}$$

To see how the conditions of an impulsive input may arise (approximately) in a real-world situation, consider the force input shown in Fig. 7-25. If its area FT were kept constant as T approached zero, F would approach infinity and the force itself would approach an impulse. For $0 \leq t \leq T$ the input is a step of size F; thus the response is

$$v_o = KF(1 - e^{-t/\tau}) \tag{7-57}$$

At $t = T$ we have

$$v_o = KF(1 - e^{-T/\tau}) \tag{7-58}$$

For $t > T$ the system equation is

$$(\tau D + 1)v_o = 0 \tag{7-59}$$

with initial condition

$$v_o(T) = KF(1 - e^{-T/\tau}) \tag{7-60}$$

Thus

$$v_o = Ce^{-t/\tau}$$

$$KF(1 - e^{-T/\tau}) = Ce^{-T/\tau}$$

$$C = \frac{KF(1 - e^{-T/\tau})}{e^{-T/\tau}} = KFe^{T/\tau} - KF$$

$$v_o = KF(e^{-(t-T)/\tau} - e^{-t/\tau}) \tag{7-61}$$

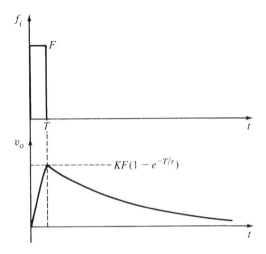

Figure 7-25 First-order system response to rectangular pulse.

First-Order Systems 447

If we now let $FT = A_i$ and keep A_i constant as $T \to 0$, Eq. (7–61) becomes

$$v_o = \frac{KA_i}{T} (e^{-(t-T)/\tau} - e^{-t/\tau}) \tag{7-62}$$

As $T \to 0$, we get $v_o \to 0/0$, an indeterminate form, so we apply L'Hospital's rule to get

$$v_o = \frac{KA_i}{\tau} e^{-t/\tau} \tag{7-63}$$

which agrees with Eq. (7-56). The actual response (7-62) approaches the perfect-impulse response (7-63) more closely as T becomes small *compared to* τ. That is, T does not have to be small in absolute terms, only *relative* to the time constant of the system under study. This can be seen by doing some numerical examples or, in general terms, by noting that the Taylor series expansion for $e^{T/\tau}$ is

$$e^{T/\tau} = 1 + \frac{T}{\tau} + \frac{1}{2}\left(\frac{T}{\tau}\right)^2 + \frac{1}{6}\left(\frac{T}{\tau}\right)^3 + \cdots \tag{7-64}$$

For $T/\tau = 0.1$, for example,

$$e^{T/\tau} = 1 + 0.1 + 0.005 + 0.00017 + \cdots$$

thus we might neglect all but the first two terms, giving in Eq. (7-62)

$$v_o = KA_i \frac{e^{-t/\tau}(e^{T/\tau} - 1)}{T} \approx KA_i \frac{e^{-t/\tau}(1 + T/\tau - 1)}{T} = \frac{KA_i}{\tau} e^{-t/\tau} \tag{7-65}$$

which we again see to be the perfect-impulse response. Thus for a rectangular pulse of duration the order of $\tau/10$ or less, the system acts nearly as if driven by a perfect impulse with the same area as the actual pulse. It can be shown[9] that this is true for *any* shape of pulse; if the pulse is short enough, only its net area (*not* its shape) is of any consequence. The reference also shows that similar rules hold for *all* linear systems, not just the first-order which we have here studied. For more complex systems, the criterion for judging whether a pulse is short enough to treat as a perfect impulse is no longer the system time constant, since the speed of such systems cannot be described in terms of just one number. However, practical criteria are available from the theoretical studies referenced. While not as general, simulation methods can also be used to reach such judgments. Since simulation *cannot* handle infinite (perfect impulse) input signals, we must implement such studies using the proper *initial conditions*, as we showed above for the first-order example. Figure 7-26 shows results from such a simulation for a first-order system with a time constant of 0.1 second and a rectangular pulse input of duration 0.01 second. The *exact* impulse response is produced, not with an impulse input, but with an initial condition of 0.50. The actual system response (initial condition zero) rises rapidly (but *not* instantly) to nearly 0.5 at $t = 0.01$. The two simulations are run "side by side" to allow easy graphical comparison of the two outputs. We see, as expected, that the actual system response is quite close to the perfect impulse response.

---

[9]Ibid., pp. 77–81.

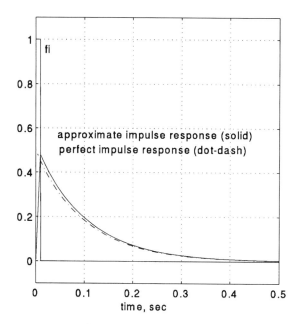

Figure 7-26 Comparison of response to exact and approximate impulse inputs.

7-4 VALIDATION OF LINEARIZED APPROXIMATIONS USING SIMULATION

We have used the Taylor series linearization scheme a number of times to enable us to obtain analytical approximate solutions for nonlinear systems. It is clear that such approximations get inaccurate if the system variables move too far from the chosen operating point, but it is not obvious how far is "too far." Some appreciation of the accuracy of linearization is useful to us, since it is so widely applied. We can develop such an appreciation by comparing analytical linearized solutions with "exact" nonlinear solutions obtained by simulation. You might be saying: "Why bother with approximations when we can easily get exact results by simulation?" You should remember that simulation gets only *numerical* results, never any *formulas* that show *relations* among physical parameters that are so useful in design. Linearized analytical solutions can provide such relations.

Let's explore this situation using a nonlinear version of the mass/damper system of Fig. 7-2. Many commercial dampers more nearly follow a "square-law" damping relation than the linear Bv which we often assume. The system differential equation is then

$$f_i - Cv_o|v_o| = M\dot{v}_o \tag{7-66}$$

$$C \triangleq \text{damping coefficient} \quad \frac{\text{newtons}}{(\text{m/sec})^2}$$

Let's assume that a constant force $f_{i,0}$ has been applied long enough that a constant velocity $v_{o,0}$ has been achieved, thus defining an operating point for our linearization.

First-Order Systems **449**

In the neighborhood of this operating point we can linearize the damping force in the usual way:

$$Cv_o|v_o| \approx C[v_{o,0}^2 + 2v_{o,0}(v_o - v_{o,0})] \qquad (7\text{-}67)$$

$$M\dot{v}_o + (2Cv_{o,0})v_o \approx f_i + Cv_{o,0}^2 \qquad (7\text{-}68)$$

This linearized model allows us to define a time constant in letter form as

$$\tau \triangleq \frac{M}{2Cv_{o,0}} \qquad (7\text{-}69)$$

This useful design relation shows how speed of response depends on M, C, and the operating point $v_{o,0}$. Simulation provides no such relations.

To investigate the accuracy of our linearization we compare its predictions with the exact results, using simulation for *both* calculations. We study two step inputs, one "small" and the other "large," since we know that linearization accuracy degrades as the larger input drives the system farther from the operating point. We will use a technique called *perturbation analysis* for this study since it is very common for linearized analyses of all kinds of systems. We define total, operating point, and perturbation values of our variables as follows.

Total value $\triangleq$ operating-point value + perturbation value

$$f_i \triangleq f_{i,0} + f_{i,p} \qquad (7\text{-}70)$$

$$v_o \triangleq v_{o,0} + v_{o,p} \qquad (7\text{-}71)$$

Equation (7-68) can then be rewritten as

$$M\dot{v}_{o,p} + Cv_{o,0}^2 + 2Cv_{o,0}v_{o,p} \approx f_{i,0} + f_{i,p} \qquad (7\text{-}72)$$

and since $f_{i,0} = Cv_{o,0}^2$,

$$M\dot{v}_{o,p} + 2Cv_{o,0}v_{o,p} \approx f_{i,p} \qquad (7\text{-}73)$$

$$\tau \triangleq \frac{M}{2Cv_{o,0}} \qquad K \triangleq \frac{1}{2Cv_{o,0}} \qquad (7\text{-}74)$$

$$(\tau D + 1)v_{o,p} \approx Kf_{i,p} \qquad (7\text{-}75)$$

To do our simulation, let's take some numerical values as $M = 10.0\,kg$, $C = 25.0\,N/(m/sec)^2$, $f_{i,0} = 100.0\,N$, and let $f_{i,p}$ be 10 N for the small step change and 50 N for the large. In the upper part of Fig. 7-27a the integrator is given an initial condition ($v_{o,0}$) of 2.0 m/sec and the step input starts at 100 N ($f_{i,0}$), giving an equilibrium condition. At $t = 0.10$ second (chosen for convenience) the step input jumps up to 110 (or 150) to start the dynamic response of the nonlinear simulation. In the linearized simulation (lower part of Fig. 7-27a), we use a transfer function to compute $v_{o,p}$ and then add $v_{o,0} = 2.0$ to it to get v_o itself. Step Fcn2 jumps up from 0 to 10 (or 50) at $t = 0.10$ second.

Figure 7-27b and c shows that the linearized model is almost perfect near the operating point but deviates more and more as we move away from it. For the small

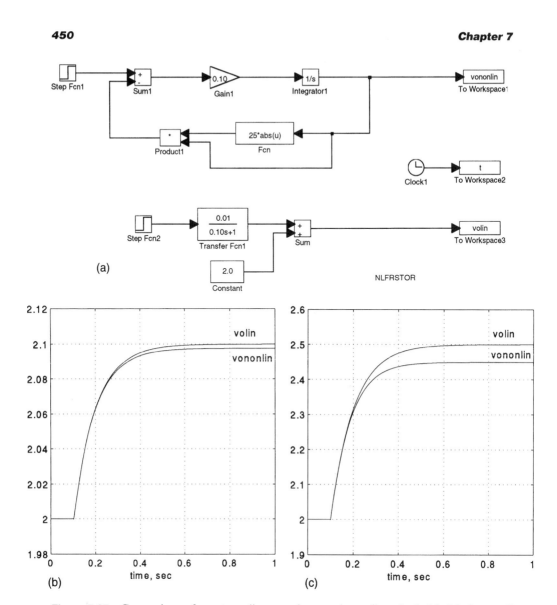

Figure 7-27 Comparison of exact nonlinear and approximate linearized. Models for small and large inputs.

step change the accuracy is quite good for the entire response, but for the large change the error may not be acceptable. This kind of comparison simulation can of course be used for any kind of linearized system and any kind of input signal and is very helpful in establishing some "feel" for the validity of linearized models.

7-5 ELECTRICAL FIRST-ORDER SYSTEMS

Since we have used mechanical first-order systems also as a vehicle for introducing the characteristics of *general* first-order systems, this section can concentrate on the

First-Order Systems   **451**

electrical aspects almost exclusively. We begin with a discussion of basic analysis techniques which apply to *all* electrical circuits, not just first-order systems. Our viewpoint is that of the *non*electrical engineer who needs some basic understanding of circuit behavior and some simple analysis tools to deal with the electrical and electronic portions of mixed systems such as data acquisition systems, motion-control systems, computer-aided automotive engines, etc. Specialized and advanced techniques which electrical engineers use to deal with complex electrical circuits are not presented.

General Circuit Laws and Sign Conventions.   Just as Newton's law is basic to the analysis of all mechanical systems, so are the *Kirchhoff's laws* (voltage loop law, and current node law) basic to electrical circuits. Knowing how to use these laws, and combining this with knowledge of the current/voltage behavior of the circuit elements (*R*, *L*, *C*, diodes, transistors, op-amps, etc.) used in a particular device, one can always formulate and analyze a circuit model which can be solved to obtain the desired information. The *voltage loop law* is merely a statement of an intuitive truth; it requires no mathematical or physical "proof." If, at any instant of time, we choose some point in a circuit and then trace out a loop along any chosen path which returns to the original point, keeping track of all voltage drops or rises encountered along that path, the net potential difference must clearly be zero, since we have returned to the very same point. For actual application to circuit analysis this law can be stated and used in at least three forms:

1. The summation of voltage drops around a closed loop must be zero at every instant.
2. The summation of voltage rises around a closed loop must be zero at every instant.
3. The summation of the voltage drops around a closed loop must equal the summation of the voltage rises at every instant.

None of the statements has any particular relative advantage; however, most people tend to choose one and then stick with it. For no reason which I can now recall, I years ago "latched onto" the first of the above three, so it is used in this book unless there is some good reason not to.

The *current node law* is based on the physical fact that at any *point* ("node") in a circuit there can be no accumulation of electric charge. That is, when we draw circuit diagrams we connect the elements (*R*, *L*, *C*, etc.) with "wires" which are considered perfect conductors (*devoid* of any *R*, *L*, or *C*). *Real* connecting wires are represented on circuit diagrams as some combination of *R*, *L*, and *C*, and these perfect conductors. Thus when we refer to "points" or "nodes" in a circuit we are really referring to locations on these (fictitious) perfect conductors, and these *cannot* accumulate any charge. Since current is defined as the "flow of charge" (time rate of change of charge), we may say that, at any instant of time:

1. The summation of currents into a node must be zero. Or, alternatively,
2. The summation of currents out of a node must be zero. Or, still another way,
3. The summation of currents into a node must equal the summation of currents out.

452   **Chapter 7**

Just as with the voltage loop law, any of the above statements of the current node law may be applied in circuit analysis. Depending on how you set up the sign conventions for the various currents in a specific circuit, one or the other of the three forms may seem most natural.

In mechanical systems we needed sign conventions for forces and motions; in electrical systems we need them for voltages and currents. The assumed positive direction of a current is indicated by the symbol $\mapsto$ and may in general be chosen arbitrarily at the beginning of an analysis. Later, when the solution for the unknown current has been obtained, for example, $i = 2.8 \sin 377t$, at any instant when i is a positive number we know that the current is actually going in the direction given by the sign-convention arrow. If i should later be a negative number, the current is *opposite* to the arrow. If the assumed positive direction of a current has not been specified at the *beginning* of a problem, an orderly analysis is quite impossible, and any results obtained can not be properly interpreted, since the meaning of positive and negative currents is undefined. For voltages, the sign conventions consist of $+$ and $-$ signs at the terminals where the voltage exists. Which terminal receives the $+$ sign is again an arbitrary choice made at the beginning of an analysis. When the solution is obtained, if the voltage is at some instant a positive number, then the actual polarity is the same as that shown by the sign-convention marks. If the voltage is negative, the actual polarity is *opposite* to that shown by the sign-convention marks.

Once sign conventions for all the voltages and currents have been chosen, combination of Kirchhoff's laws with the known voltage/current relations which describe the circuit elements leads us directly to the system differential equations. While practitioners of circuit analysis have developed many systematic and specialized techniques to speed analysis of complicated circuits, these are beyond the scope of this text and are not really necessary or desirable for the relatively simple systems which are our main concern. We should also say again that everything said so far in this section is of course quite general and *not* restricted to the first-order examples which are the subject of this chapter.

Practical Examples of Electrical First-Order Systems.   In Fig. 7-28 we show eight examples of circuits which are first-order systems of the same type we have so far emphasized, that is, $(\tau D + 1)q_o = Kq_i$. Recall that one cannot define system type from just the diagram; we must also specify which signal is the input and which the output. This has again been done with the subscripts i and o. As usual we want you to know that these example circuits are not just "made-up exercises" but represent real-world devices with important practical applications. The circuit of Fig. 7-28a finds use as a low-pass filter, which eliminates high-frequency "noise" from a low-frequency desired signal. It also is an approximate integrator, which is used, for example, to integrate the acceleration signal from an accelerometer so as to get a velocity signal. Among its other uses is as a model for the dynamic response of a length of cable used to connect two electrical devices. Cable dynamics are often neglected when the cable is short and/or the transmitted signal is low-frequency. When this is *not* the case, cable dynamics must be considered.[10]

---

[10]E. O. Doebelin, *Measurement Systems*, 4th ed., McGraw-Hill, New York, 1990, pp. 844–846.

First-Order Systems

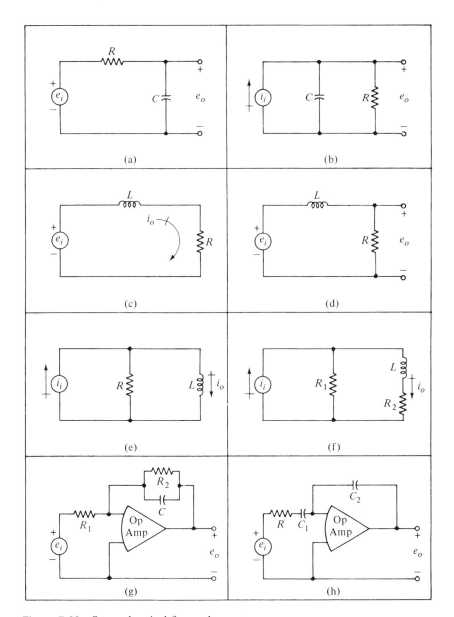

Figure 7-28 Some electrical first-order systems.

Figure 7-28b shows a circuit used to model piezoelectric sensors (pressure, force, or acceleration transducers) and photodiodes (optical sensing devices used to transduce light energy into an electrical signal). The sensors mentioned all produce a current related to the physical signal to be measured, thus a current, rather than voltage, source is appropriate. The circuits of Fig. 7-28c, d, e, and f are used as models for portions of various electromechanical devices which employ magnetic fields. Strong magnetic fields usually require the use of iron cores, making the inductance large and nonnegligible. The inductance here is considered a parasitic effect in the sense that we don't "design in an inductor," and the inductance effect is often undesirable since it reduces speed of response. The op-amp circuits of Fig. 7-28g and h are "active" versions of the "passive" circuit of Fig. 7-28a, and are useful as low-pass filters and approximate integrators. The presence of the op-amp allows attainment of some numerical values that would be difficult or impossible with the passive version.

Analysis of Passive and Active Low-Pass Filters. Let's illustrate the application of the general concepts of circuit analysis outlined above, by consideration of the circuit of Fig. 7-28a, a passive low-pass filter. We might begin by showing, in Fig. 7-29, this circuit embedded in a larger system, as it might ordinarily arise in practice. The input comes from an amplifier whose output circuit is modeled as an ideal voltage source in series with a 10-ohm resistance. A digital voltmeter (or digital data acquisition system) with a very high input resistance (10 MΩ) is connected to our circuit's output. While it is technically possible to analyze this entire assemblage of hardware as one system, it is often advantageous to attempt some initial simplification based on judgment and experience. Since beginners in circuit analysis (as in any other field) have little of either of these two attributes to draw on, they may tend to make few simplifications and instead choose to deal with a rather complex model. In these days of computer-aided analysis it might seem that a complex model should not cause too much worry; however, a model *more complex than really necessary* will always be undesirable, since it obscures basic understanding and wastes time and money.

In our present example we simplify the circuit by combining the source resistance R_s with R_1 into a total resistance R, and by treating the output terminals across C as an open circuit (infinite resistance) rather than, as is actually the case, shunted by a 10-megohm resistor. The latter simplification is also an approximation, but will be quite accurate because the "load" resistor R_L is so large *relative* to $(R_1 + R_s)$ that the current it draws from the circuit may be neglected *relative* to the current flowing in R and C. The judgment necessary to make such simplification decisions "on the

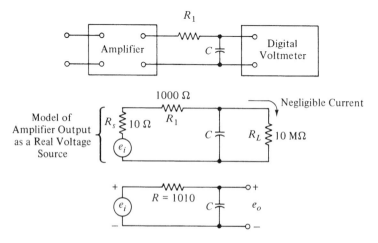

Figure 7-29 Formulation of simplified circuit model.

First-Order Systems

spot" comes from experience, both with experimental tests on real systems and also with analysis of similar systems in the past.

With the above background in mind we return to analysis of our circuit as an "isolated" entity. In Fig. 7-30a we initially choose the positive sense of e_i as shown; we could of course have chosen the reverse. For e_o we also have a free choice; however, once e_i's positive direction has been fixed there may be some incentive to choose e_o's such that a positive constant e_i causes a positive e_o. This is not *necessary* but may be desirable and is what was done in Fig. 7-30a. Since we assume no current flowing in the output terminals, there is only one current in the system and we call it i. Its positive direction is open to choice but we are again influenced by a desire to

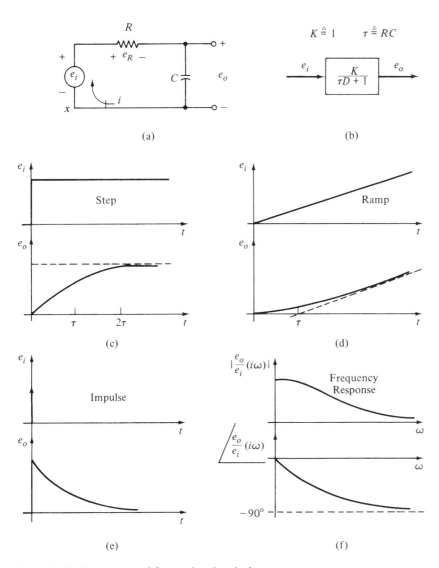

Figure 7-30 Responses of first-order electrical system.

456   **Chapter 7**

have a positive e_i cause a positive i and thus choose i positive in the direction shown. The voltage across the resistor is called e_R and is chosen positive as shown (again the reverse choice could have been made) so that a positive i causes a positive e_R. Since the capacitor voltage is also e_o there is no need to define it again.

Because this circuit has only one current, Kirchhoff's voltage loop law seems more appropriate than the current node law. In applying this law one must choose a *point* in the circuit at which to start, the *loop path* to be followed, and the *direction* of traversing the loop. More complex circuits require that *several* loops be used, to generate the several simultaneous equations needed to describe the circuit. Our present circuit has one obvious loop. Once a loop is defined, we can choose the starting point and direction of traversal as we wish. The starting point is totally arbitrary but we can choose the direction of traversal to ease our equation writing. Since I earlier expressed a preference for keeping track of voltage *drops*, I also prefer to "walk around" the loop in the direction of assumed positive current. You don't *have* to do this, but if you do, then the voltage drops for all the passive elements (R, L, C) will always be positive; $+iR$, $+L(di/dt)$, $+(1/C)\int i\,dt$. If you "standardize" on a fixed routine like this (or some other which you might prefer) then there is less chance of making mistakes in equation writing and also the work goes faster.

We have spent considerable time on "preliminaries" but these are always necessary and speed the later work. Starting at point x (which could be chosen anywhere on the loop) in Fig. 7-30a and traversing the loop clockwise (direction of positive current), Kirchhoff's voltage loop law gives

Summation of voltage drops at any instant $= 0$

$$-e_i + iR + \frac{1}{C}\int i\,dt = 0 \tag{7-76}$$

Note that since we are summing drops, the input voltage e_i enters with a minus sign, according to its chosen sign convention. That is, when e_i is, say, -5 volts, the drop, going clockwise from point x across the voltage source, is $+5$ volts. While our interest is in the relation between input e_i and output e_o, we will achieve this indirectly by first relating i to e_i and then using the fact that $e_o = e_c = (1/C)\int i\,dt$. This is not the only way to solve this problem, but it shows that often we need to solve for currents even if our ultimate interest is in voltage relations. Using D operators

$$\left[R + \frac{1}{CD}\right]i = e_i \qquad i = \left[\frac{CD}{RCD+1}\right]e_i \tag{7-77}$$

$$e_o = \left[\frac{1}{CD}\right]i = \left[\frac{1}{CD}\right]\left[\frac{CD}{RCD+1}\right]e_i \tag{7-78}$$

$$\frac{e_o}{e_i}(D) = \frac{1}{RCD+1} = \frac{K}{\tau D+1} \tag{7-79}$$

$$K \triangleq 1 \quad \frac{\text{volts}}{\text{volt}} \qquad \tau \triangleq RC \quad \text{seconds} \tag{7-80}$$

First-Order Systems

We see that this clearly fits our definition of a first-order system and thus all the standard responses of Fig. 7-30 are immediately available from our earlier work on generic first-order systems; *no new calculations are needed.*

The circuit just analyzed could be described as a *passive* low-pass filter, passive because it includes no power sources or amplifiers. An alternative design with the same generic first-order response is shown in Fig. 7-31. This is an *active* low-pass filter since it uses an op-amp (which has its own power supply). Recall that two basic op-amp assumptions are that the node n is essentially at ground potential (0 volt) and that the amplifier input current is negligible. Applying Kirchhoff's current node law at node n we get

$$i_{R1} + i_{R2} + i_C = 0 \qquad \frac{e_i - 0}{R_1} + \frac{e_o - 0}{R_2} + C\frac{d}{dt}(e_o - 0) = 0 \qquad (7\text{-}81)$$

$$(R_2 C D + 1)e_o = -\left(\frac{R_2}{R_1}\right)e_i \qquad \frac{e_o}{e_i}(D) = \frac{K}{\tau D + 1} \qquad (7\text{-}82)$$

$$K \triangleq -\frac{R_2}{R_1} \frac{\text{volts}}{\text{volt}} \qquad \tau \triangleq R_2 C \quad \text{seconds} \qquad (7\text{-}83)$$

The negative value of steady-state gain K means that an input voltage of positive polarity produces an output voltage of negative polarity. If this is unacceptable we can connect an op-amp inverter at e_o. Its transfer function of -1 gives the combined circuit a gain of $+R_2/R_1$. Remember that most op-amps are cheap and small; popular integrated circuit devices may provide 4 op-amps on one "chip" at a total price of about a dollar. Circuit designers thus don't hesitate to use "extra" op-amps to provide desired functions. Let's assume our active filter circuit includes the inverter, giving the positive gain desired.

Design Example: Low-Pass Filter. We now want to design two low-pass filters, one passive and one active, to meet the same performance specification. Suppose our input signal comes from a temperature sensor designed to measure fluctuating temperatures of frequency content up to about 1 Hz. The sensors are located in a region

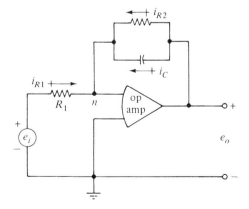

Figure 7-31 Active low-pass filter.

458   **Chapter 7**

of our factory that unavoidably has ac power lines and motors, and lab tests show that our desired temperature signals (about 0.200 volt) are "contaminated" with 60-Hz "noise" of about 0.10 volt amplitude. We want our filter, connected between the sensor and the input to our data-acquisition system, to reduce the noise to about 0.005 volts without disturbing the desired signals "too much." Based on the frequency-response graph of Fig. 7-30f, we want the amplitude ratio to be $0.005/0.100 = 0.05$ at a frequency of 60 Hz. That is, the noise signal must be reduced to 1/20 of its original amplitude.

$$0.05 = \frac{1}{\sqrt{(\omega\tau)^2 + 1}} = \frac{1}{\sqrt{(377\tau)^2 + 1}} \qquad \tau = 0.0530 \text{ seconds} \qquad (7\text{-}84)$$

A filter with this time constant squelches the noise as desired but we now need to check whether it messes up the desired temperature signal. *Perfect* measurement systems are zero-order systems, with flat amplitude ratios for all frequencies. *Real* measurement systems must always be allowed to deviate somewhat from perfect flatness; about 5% is a typical specification. We must thus check our filter's amplitude ratio at the highest temperature-signal frequency, that is, at 1 Hz.

$$\text{Amplitude ratio} = \frac{1}{(\sqrt{(6.28 \cdot 0.0530)^2 + 1}} = 0.949 \; \frac{\text{volts}}{\text{volt}} \qquad (7\text{-}85)$$

This design just barely meets a 5% dynamic error specification. If the allowable error were *less* than 5%, note that our design task becomes *impossible* since the required τ is now less than 0.0530 and such a value cannot possibly meet the noise-reduction requirement. Actually, when such conflicting conditions arise, we can then consider more sophisticated low-pass filters and these will very often meet our needs. The next more complex filter would be a *second-order* system, which gives us more design freedom in discriminating between desired and undesired frequencies.

Whether we use the passive or active circuit, there are an *infinite* number of R, C combinations which would give us the desired τ of 0.0530 second. We must of course choose *one* specific combination if we plan to actually build this filter. Several considerations help us make this choice. First, when we connect the temperature sensor to the filter, the filter should not draw too much power from the sensor (this is called the "loading" problem). This requirement can be satisfied by making the filter input resistance high, relative to the sensor output resistance. Let's at this point state that our sensor is a resistance thermometer type, with a nominal resistance of 100 ohms. A rule of thumb often used to "rough out" the design of electrical systems which consist of a "chain" of interconnected components is to make the resistance ratio at least about 10-to-1 at each interface. In our present example this suggests that R (for the passive filter) and R_1 (for the active filter) should be at least 1000 ohms. If we actually used 1000 ohms then C would be $0.0530/1000 = 53\,\mu\text{F}$. A 53-$\mu$F capacitor, depending on the specific type, can be quite large, heavy, and expensive. Since C gets smaller as R gets larger, and since larger R is actually *better* from a "loading" standpoint, we should use a larger R, perhaps about 10,000 ohms.

We now should note that for the active circuit, the loading problem is related to R_1 while the time constant depends on R_2. That is, this more complex circuit is more versatile since it has more adjustable parameters. We can, in fact, make R_1 the

First-Order Systems   **459**

10,000 ohms needed to control loading and make R_2 say, 100,000 ohms to make C even smaller. We then get the additional feature that the "filter" is now also an amplifier with a gain of 10 volts/volt. This makes our temperature measurement system 10 times as sensitive, which may (or may not) be useful.

This simple design example makes clear that design problems never have a single "answer" and that many, often conflicting, requirements must be carefully balanced by the designer.

Design Example: Approximate Integrator. The same circuits we have just used as low-pass filters are also applicable to the task of signal integration, needed in some instrumentation systems. While motion transducers are available for displacement, velocity, acceleration, and jerk, acceleration sensors (accelerometers) are the most widely used, for several reasons.[11] One reason is that if we have a voltage proportional to acceleration, we can get also velocity and displacement signals by *integrating* the acceleration signal, once to get velocity and once more to get displacement. If, instead, we had used a displacement sensor, to get velocity and acceleration we would need to use a *differentiator*. Whether done by analog or digital means, integration is always preferred over differentiation since integration *smooths* any high-frequency noise that accompanies the desired signal whereas differentiation accentuates the noise.

While "theoretically exact" integrators can be constructed and are used in some applications (see Fig. 3.29b), approximate integrators are sometimes preferred. One reason for this is that the "exact" integrators have more severe "drift" problems than do the approximate. That is, if the exact integrator's input signal (e_1 in Fig. 3.29b) is exactly zero, the output signal will gradually (or rapidly) move away from zero, whereas it should stay exactly at zero. This defect is caused by subtle imperfections that are present, to some degree, in every op-amp. When an integrator uses high sensitivity and/or must be used over long time periods, this drift behavior may not be tolerable.

Since vibration-measuring systems are generally designed to operate over a specific range of vibration frequencies, let's assume for this example that we need an integrator that works well from 5 to 500 Hz. We first need to show that a first-order system will behave as an approximate integrator. For many dynamic systems, not just first-order, when we want to show under what conditions a certain behavior is approximately achieved, frequency-response methods are the best tool. For a generic first-order system,

$$\frac{q_o}{q_i}(i\omega) = \frac{K}{i\omega\tau + 1} \tag{7-86}$$

Now if $\omega\tau \gg 1$, we can write

$$\frac{q_o}{q_i}(i\omega) \approx \frac{K}{i\omega\tau} \qquad \frac{q_o}{q_i}(D) \approx \frac{K}{\tau D} \qquad q_o \approx \left[\frac{K}{\tau D}\right]q_i \qquad q_o \approx \frac{K}{\tau}\int q_i\, dt \tag{7-87}$$

We see that as long as $\omega\tau \gg 1$, we have an approximate integrator, the quality of the approximation depending on how much greater $\omega\tau$ is than 1.0. Since for a given

---

[11] E. O. Doebelin, *Measurement Systems*, 4th ed., McGraw-Hill, New York, 1990, p. 323.

system τ is fixed, it is clear that the approximation is least accurate for the lowest frequency; thus if we design τ for the lowest frequency, the integration will be more nearly perfect at all higher frequencies. Because the "sensitivity" of the integrator is K/τ, it is also clear that if we increase τ to extend the accurate frequency range to lower frequencies, we pay the price of reduced sensitivity (less output voltage).

These generic results can of course be applied to the circuits of Fig. 7-28a and g, our approximate electrical integrators. To proceed with the design, we next need to choose how much error we can tolerate in our approximation, since this will set the value of $\omega\tau$ that we want at the lowest frequency, 5 Hz = 31.4 rad/sec. Suppose we require that, at 5 Hz, the amplitude ratio of the approximate integrator must be 99% of the amplitude ratio of an exact integrator with the same K value. This calculation shows that for such "99% accuracy," $\omega\tau = 9.9$, so in our case we want $\tau = 9.9/31.4 = 0.3153$ sec. As in the previous low-pass filter design, there are an infinite number of R, C combinations that will give the needed value of τ, so other constraints must be invoked in order to define specific R and C values. In addition to those discussed in the filter design, let's add another. Suppose that at 5 Hz the full-scale input voltage coming from our accelerometer has an amplitude of 5 volts and that we want this input signal to cause an integrator output which is also 5 volts, since we are reading the output of the integrator on a meter which is itself 5 volts full scale.

For the passive integrator the amplitude ratio at 5 Hz is very nearly $1/(31.4 \times 0.3153) = 0.101$ volts/volt; thus a 5-volt input will give only a 0.505-volt output, too small to accurately read on our 5-volt full-scale meter. We would thus need an external amplifier to "make up the difference." For the active integrator, such amplification is "built in," since K for this integrator is R_2/R_1. We thus require $R_2/R_1 = 5/0.505 = 9.09$. Combining this constraint with those explained in the low-pass filter example, we can arrive at actual values for the two resistors and the capacitor. As usual, a number of combinations would be satisfactory, one of which might be $R_1 = 10,000$ ohms, $R_2 = 90,900$ ohms, $C = 3.47\,\mu\text{F}$.

Bode plots of the exact and approximate integrators give additional insight into the nature of the approximation. Our "accuracy" design was based entirely on amplitude ratio, phase angle was not considered at all. Such designs are usually satisfactory, but one should certainly check them more comprehensively. The Bode plots of Fig. 7-32 show amplitude ratio and phase angle for both an approximate and a perfect integrator with $K = 1$ and $\tau = 0.3153$ second. Note that the dB curve for a perfect integrator is a perfect straight line (*no* corrections needed) with a slope of -20 dB/decade. We see that the phase angle of the approximate integrator approaches the perfect $-90°$ value as frequency gets higher. Perhaps a more convincing test is to simulate the time-domain behavior using typical input signals. This is done in Fig. 7-33, where we sum three sine waves of different frequencies to produce an "acceleration signal" for input to our integrators, whose output would then be a velocity signal. Both integrators, perfect and approximate, use the same gain (1.0 in our example) for a fair comparison.

Note that the perfect integrator is followed by a high-pass filter, which has a sinusoidal transfer function which is zero at 0 frequency and approaches 1.0 at high frequency. This filter, which is selectable at the input of most oscilloscopes by switching the instrument to "ac coupling", would be used in most vibration measurements to force the displayed signal to have an average value of zero. Acceleration and

First-Order Systems

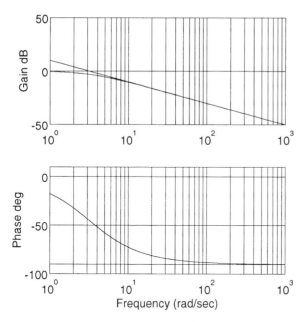

Figure 7-32 Frequency response of approximate and exact integrators.

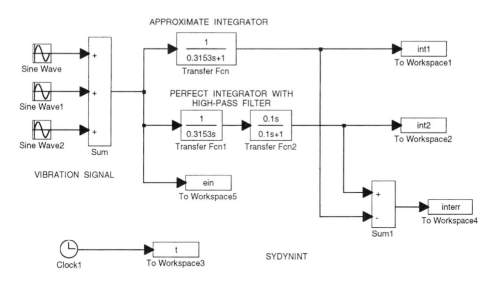

Figure 7-33 Simulation study of approximate and exact integrators.

velocity signals of vibrating machine parts *must* have a long-term average value of zero; otherwise the displacement would gradually drift away, whereas we know that such parts "stay put," vibrating around a fixed average position. For this reason, vibration "meters," such as oscilloscopes, are usually set up to force the average values to be zero. Since a high-pass filter has zero amplitude ratio at zero frequency, it suppresses the average value, which has zero frequency. For the desired measurement range of 5 to 500 Hz, we select the filter time constant (0.10 second in our

462   **Chapter 7**

example) so that the filter's sinusoidal transfer function is close to 1.0 $\underline{/0°}$. Thus the filter is effectively "not there" for this frequency range.

In Fig. 7-34a the vibration signal is the sum of three sine waves, each of amplitude 1.0, at frequencies 5, 7, and 9 Hz, giving a typical input signal in the low-frequency range. Simulation shows that the approximate integrator is initially in considerable error, but after about three time constants ($3 \times 0.3153 = 0.95$ second) it is in good agreement with the (filtered) exact integrator. The *instantaneous* error signal interr is not graphed for you. If you reproduce this simulation and plot interr this error seems excessive, but it is actually acceptable. This is because, as we have mentioned before, measurement systems need only to reproduce the correct size and shape of the measured signal. A time shift (which can cause large instantaneous errors) is usually of no consequence.

In Fig. 7-34b we explore the high-frequency behavior by inputting a vibration signal with three frequencies (200, 350, and 500 Hz), each with amplitude of 100. Now the two integrators are in "perfect" agreement just after $t = 0$, but seem to be diverging as time increases. Note also that both of them do *not* show a zero average value in this time range. Actually, because the time range is so short (it *must* be to clearly show such high frequency data), the averaging effects have not yet had time to make themselves felt. If we run the simulation to longer times we find that, at first, the error gets worse, but as we go beyond $3\tau = 0.95$ second, the two integrators agree almost perfectly (better than at low frequency). Also, both integrators will show a zero average value at these longer times. Note that "long" here is only a few seconds, so that in the real measurement system, our oscilloscope displays "good" values almost as soon as we turn it on.

Design Example: Optical Sensor. The optoelectronic component called a *photodiode detector* is widely used to convert signals in the form of light energy to related electrical signals. Information which we are about to present was obtained from a publication of one of the large suppliers of optical components and systems.[12] These devices produce a *current* proportional to the wattage (power) in the light signal input to the photodiode, thus the electrical circuitry will involve a *current source*. The proportionality factor ("sensitivity") of this optical/electrical conversion depends on the wavelength of the incoming light, a typical value at, say, 800 nanometers wavelength is about 0.5 amps/watt. Typical light signals have microwatts of power so we are talking about microamps of current. Actually, the devices are quite linear over large ranges of optical power input, so working currents may range from a few hundred nanoamps to a few hundred milliamps.

Both active and passive circuits are used to obtain output voltage signals from these sensors. The simplest passive circuit is that of Fig. 7-28b, where the current source i_i is the photocurrent (proportional to incoming light power in watts). The R and C shown are *not* separate components "wired in" by the designer, but rather "parasitic" effects inherent in the photodiode. The C is called the *junction capacitance* and the R is called the *shunt resistance*. The junction capacitance varies with the reverse bias ("battery" voltage of a few volts applied to the photodiode), getting

---

[12]Melles Griot Optics Guide 1995/1996, pp. 66-1 to 69-12, Melles Griot, 1770 Kettering Street, Irvine, CA 92714, 1-800-835-2626.

First-Order Systems

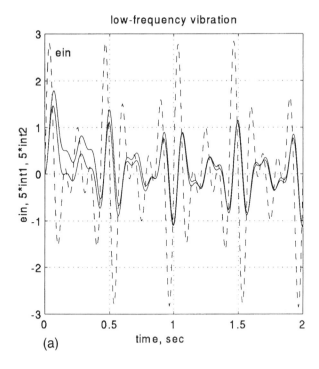

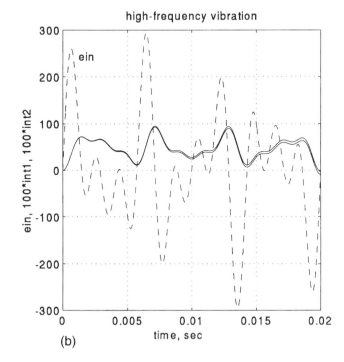

Figure 7-34 Simulation results for low- and high-frequency vibrations.

smaller as the bias is increased. Small capacitance is desirable for fast response, since there is an RC time constant. While reverse bias slightly complicates the circuit, it is often used to get faster response. Typically, a device which has $C = 180\,\text{pF}$ with no reverse bias will have 43 pf with 5 volts of reverse bias, giving more than a 4-to-1 speedup. The shunt resistance varies significantly with temperature, so practical circuits will negate this defect by wiring in an intentional resistor in parallel with the diode's shunt resistance. This intentional resistor is made about 100 times smaller than the diode's own shunt resistance; thus the parallel combination will effectively be dominated by the intentional resistor, which is *not* temperature sensitive. Since these two resistors are in parallel, the circuit of Fig. 7-28b still applies—we just use a number for R which is the parallel combination. Typical shunt resistances of these diodes are in the range 4 to 300 MΩ, so the effective R in our circuit would be about 1/100 of these values.

In Fig. 7-35 we can apply the current node law at node n to get

Sum of currents into $n = 0$

$$i_i - i_C - i_R = 0 \qquad i_i = C\frac{de_o}{dt} + \frac{e_o}{R} \qquad (7\text{-}88)$$

$$(RCD + 1)e_o = Ri_i \qquad \frac{e_o}{i_i}(D) = \frac{K}{\tau D + 1} \qquad (7\text{-}89)$$

$$K \triangleq R \qquad \tau \triangleq RC \qquad (7\text{-}90)$$

Most circuits can be analyzed in several ways. Let's use an impedance approach on this same circuit.

$$i_i = \frac{e_o}{Z_{\text{total}}} = \frac{e_o}{\dfrac{(R)(1/CD)}{R + 1/CD}} = \frac{e_o}{R/(RCD + 1)} \qquad \frac{e_o}{i_i}(D) = \frac{R}{RCD + 1} \qquad (7\text{-}91)$$

Analyzing circuits in several different ways gives a check on correctness and also develops your ability to pick the easiest method for new problems.

Let's look at a specific device to get a feel for the numbers. The Melles Griot silicon photodiode 13 DSI 005 has junction capacitance of 72 pF when the reverse bias is zero. The shunt resistance is 60 MΩ, so we "wire in" an intentional parallel resistor of 0.60 MΩ, making the parallel combination 0.594 MΩ. The complete circuit thus has a first-order response with $\tau = 42.7\,\mu\text{sec}$. If we use a shutter to apply a

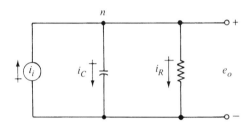

Figure 7-35 Passive circuit for photodiode light sensor.

First-Order Systems

step input of laser light from a red (wavelength = 632.8 nm) helium-neon laser of 0.10-mW power to this detector circuit, what will be the electrical output signal? Checking the Melles Griot reference cited above, we find that at this wavelength the photodiode sensitivity is 0.40 amps/watt, so the current input will be a step of size $0.40(0.0001) = 0.00004$ amps. Using our generic step response facts we can say that the output voltage will rise exponentially to $0.00004(0.594 \times 10^6) = 2.376$ volts, taking 128 μsec to get 95% of the way.

Photodiodes are also used in active circuits, using op-amps as in Fig. 7-36, to achieve better speed and linearity, and to reduce noise levels. Note that the diode shunt resistance and junction capacitance *have no effect* on circuit performance because they are "short-circuited" at the op-amp input. That is, the voltage across the op-amp input is essentially zero (our standard op-amp assumption), so no voltage exists across R_{sh} and C_j, so no current flows through them. Thus *all* the photocurrent goes into the op-amp feedback path:

$$i_i = i_R + i_C = \frac{e_o - 0}{R_{fb}} + C_{fb} \frac{d}{dt}(e_o - 0) \tag{7-92}$$

$$\frac{e_o}{i_i}(D) = \frac{R_{fb}}{R_{fb}C_{fb}D + 1} \tag{7-93}$$

We again see the generic first-order response, so all our previous results apply also to this circuit. Note that performance now depends not on the parasitic R_{sh} and C_j of the photodiode, but on R_{fb} and C_{fb}, "wired in" components which we can select from a wide range to get improved performance. (When we said that the diode "parasitics" had *no* effect on performance this was not totally correct. They don't affect the above analysis, but they *do* show up in studies of the random noise effects in the system. These noise analyses can be found in the Melles Griot reference but are beyond our scope here.)

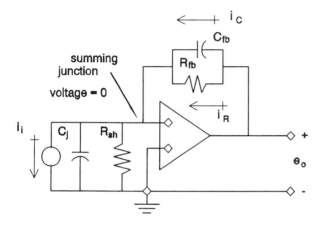

Figure 7-36 Active (op-amp) circuit for photodiode light sensor.

7-6 ELEMENTARY AC CIRCUIT ANALYSIS AND IMPEDANCE METHODS

In Eq. (7-91) we used an impedance method which may have been unfamiliar to you, though it may have seemed intuitively correct since you know that parallel resistors combine as $R_1 R_2/(R_1 + R_2)$. Using Fig. 7-37 we want to develop rules for combining arbitrary impedances (not just resistors) in series and parallel. For the series combination, both Z_1 and Z_2 must carry the same current i; thus

$$e_{total} = iZ_1(D) + iZ_2(D) = [Z_1(D) + Z_2(D)]i \tag{7-94}$$

and since impedance is always defined as the ratio voltage/current,

$$Z_{total}(D) = \frac{e_{total}}{i}(D) = Z_1(D) + Z_2(D) \tag{7-95}$$

Thus the rule for combining impedances in series is to simply add them, just as pure resistances add up in dc circuits. Note that this may be done with either operational, $Z(D)$, sinusoidal, $Z(i\omega)$, or Laplace, $Z(s)$, impedances. Applying this method to the circuit of Fig. 7-30a, the impedance "seen" by the voltage source e_i is the series combination of R and C:

$$Z(D) = R + \frac{1}{CD} = \frac{RCD + 1}{CD} = \frac{e_i}{i}(D) \tag{7-96}$$

This agrees with Eq. (7-77), which was obtained directly from Kirchhoff's voltage-loop law, a different method. Note that our result, derived for the combination of *two* series impedances, can immediately be extended to *any* number of series impedances. How?

For the parallel combination of Fig. 7-37, both Z_1 and Z_2 have the same voltage drop; thus

$$i_{total} = i_1 + i_2 = \frac{e}{Z_1(D)} + \frac{e}{Z_2(D)} = \left[\frac{1}{Z_1(D)} + \frac{1}{Z_2(D)}\right]e \tag{7-97}$$

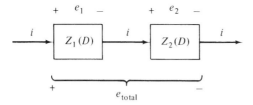

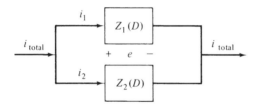

Figure 7-37 Series and parallel combinations of impedances.

First-Order Systems   **467**

$$Z_{\text{total}}(D) = \frac{e}{i_{\text{total}}}(D) = \frac{Z_1(D)Z_2(D)}{Z_1(D) + Z_2(D)} \tag{7-98}$$

The parallel combination is thus the product over the sum of the individual impedances, again just as for resistances in dc circuits. When extending this result to more than two impedances, don't fall into the trap

$$Z_{\text{total}}(D) = \frac{Z_1(D)Z_2(D)Z_3(D)}{Z_1(D) + Z_2(D) + Z_3(D)} \tag{7-99}$$

which "follows the pattern" of Eq. (7-98) but is *wrong*, as can immediately be seen from its dimensions ohms^2, where we know that all impedances, whether operational, sinusoidal, or Laplace, *must* have dimensions of ohms. The *correct* relation for more than two parallel impedances can be obtained by successive application of Eq. (7-98) to pairs of impedances, replacing each successive pair by its single equivalent before combining it with the next impedance. Perhaps easier to remember and implement is the extension of Eq. (7-97), that is,

$$\frac{1}{Z_{\text{total}}} = \frac{1}{Z_1} + \frac{1}{Z_2} + \cdots + \frac{1}{Z_n} \tag{7-100}$$

This results is correct; the reciprocal of impedance is called the *admittance* of the circuit.

The impedance combination methods just developed are useful for quickly finding transfer functions and differential equations for both simple and complicated circuits. The sinusoidal form is widely used in ac circuit analysis, and we now want to develop some methods and terminology from that area. The frequency response of electric circuits is of particular interest since much of our electrical power is generated, transmitted, and utilized in the form of sinusoidal waves, that is, alternating current and voltage (ac). Commercial power in the United States is at the frequency of 60 Hz, whereas some other countries use 50 Hz. In "mobile" applications such as aircraft and ships, where the ac power line is not available and ac power must be "manufactured" on the spot, higher frequencies such as 400 Hz may be used. It can be shown that higher ac frequency allows smaller size and weight (important in any vehicle) while simultaneously giving improved performance, such as speed of response in dynamic systems. For ac measurement and control systems, dynamic response is roughly limited to about one-tenth of the power frequency. Thus a 60-Hz ac motion-control system might have good motion response to commands of about 6 Hz or less, while a 400-Hz system would perform well to about 40 Hz.

In Eq. (7-96) let's consider voltage e_i as an input which produces current i as output. The quantity $(i/e_i)(D)$ is the admittance and its sinusoidal version is given by

$$\frac{i}{e_i}(i\omega) = \frac{i\omega C}{i\omega RC + 1} = \frac{\omega C}{\sqrt{(\omega RC)^2 + 1}} \quad \underline{/(90° - \tan^{-1} \omega RC)} \tag{7-101}$$

If $e_i = E \sin \omega t$,

$$i = \frac{E\omega C}{\sqrt{(\omega RC)^2 + 1}} \sin(\omega t + 90° - \tan^{-1} \omega RC) \overset{\Delta}{=} I \sin(\omega t + \phi) \tag{7-102}$$

468 Chapter 7

We now want to calculate the instantaneous power p supplied by the source to the circuit. Recall the definition of electrical power,

$$p \triangleq ei = e_i i = (E \sin \omega t)(I \sin(\omega t + \phi)) \tag{7-103}$$

Using appropriate trig identities this leads to

$$p = \frac{EI}{2}(\cos \phi - \cos(2\omega t + \phi)) \tag{7-104}$$

which is plotted in Fig. 7-38. Since the average value of $\cos(2\omega t + \phi)$ is zero, the average power into the circuit is

$$p_{\text{avg}} = \frac{EI \cos \phi}{2} \tag{7-105}$$

The instantaneous power varies cosinusoidally around its average value at a frequency 2ω, just *twice* the frequency of the impressed voltage (and the resulting current). During any one cycle, power flows into the circuit from the source for a portion of the time and is returned to the source from the circuit the rest of the time. Average power is that which is actually "used up" by the circuit and is what the electric company charges for. The angle ϕ by which the current leads the voltage is called the *power factor angle*, and $\cos \phi$ is called the *power factor*. In general, the angle ϕ may be "leading" (between $0°$ and $+90°$) or "lagging" (between $0°$ and $-90°$); thus the power factor is between 0 and 1.

Many measurements and calculations in ac systems employ the so-called *effective values* of current and voltage. The effective value of a current or voltage is defined as that *constant* value which would produce the same average power in a resistor as would the actual *time-varying* voltage or current. An effective value exists for *any* waveform whatever, including random voltages and currents. A general formula for calculating the effective value of a time-varying voltage is given by

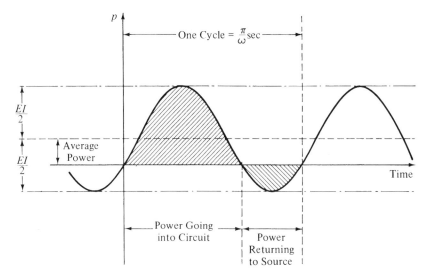

Figure 7-38 Power in ac circuits.

First-Order Systems

$$\text{Average power for a constant voltage } E = \frac{E^2}{R}$$

$$\text{Average power for a time-varying voltage } e = \frac{1}{T} \int_0^T \frac{e^2}{R} \, dt$$

$$\text{Effective value } E_{\text{eff}} \overset{\Delta}{=} \sqrt{\frac{1}{T} \int_0^T e^2 \, dt} \tag{7-106}$$

This formula can be easily implemented in simulation for any time-varying voltage. For random voltages, the "sample time" T must be long enough that the computed effective value has "converged" to a reasonably steady value. For any periodic signal, one complete period is sufficient to get the exact value. For a single sine wave the integral is easily carried out analytically to give

$$E_{\text{eff}} = \frac{E}{\sqrt{2}} = 0.707E \tag{7-107}$$

Thus the effective value [also called the *root-mean-square* (RMS) value] of a sinusoidal voltage is 0.707 times its peak value. It is easy to show that the same relation, $I_{\text{eff}} = 0.707I$, holds for currents. Since most ac voltmeters and ammeters are calibrated to read effective values, we must be careful to multiply their readings by 1.414 if we want the peak values. Also, most ac meters do *not* actually carry out the squaring and averaging operations required in the definition. To save cost, they instead *rectify* the waveform (take its absolute value, *not* its square) and use a low-pass filter (*not* an integrator) to get the average. They then apply a correction factor to produce the reading displayed. For pure, single sine waves, this method gives exactly the effective value. For *any other* waveform the result is approximate. You should thus always be careful to find out whether the meter you are using is a "true RMS" meter (more expensive) or an "ordinary" ac meter.

ac Circuit Analysis Example. Let's apply some of the above ideas to the circuit of Fig. 7-28a with $R = 10,000 \, \Omega$, $C = 0.10 \, \mu F$ when e_i is taken as the 115 V, 60 Hz available at a standard "wall plug." We first convert the 115 V (which is RMS) to the peak value $E = (1.414)(115) = 163$ V to get $e_i = 163 \sin 377t$. (It is possible to compute entirely with RMS values, and in fact most electrical power engineers do that; however, most of the readers of this text will be nonelectrical engineers. For this audience, I feel it is more important to understand the physical details than to perform routine rapid calculations. In my opinion this goal is better reached using both peak and RMS values. When you see a waveform on an oscilloscope you always "see" *peak* values, *not* RMS.)

If we are interested in the output voltage,

$$\frac{e_o}{e_i} (i377) = \frac{1}{\sqrt{(0.377)^2 + 1}} \; \angle \; \tan^{-1} -0.377 = 0.935 \; \angle -20.6° \quad \frac{\text{volts}}{\text{volt}} \tag{7-108}$$

$$e_o = (163)(0.935) \sin(377t - 20.6°) = 152 \sin(377t - 20.6°) \quad \text{volts} \tag{7-109}$$

470   **Chapter 7**

The effective value of e_o would be given as $(0.707)(152) = 108$ V. This same result could be obtained from the effective values directly, i.e., $(0.935)(115) = 108$. To find the current,

$$\frac{i}{e_i}(i377) = \frac{(377)(10^{-7})}{\sqrt{(0.377)^2 + 1}} \quad \underline{/(90° - \tan^{-1} 0.377)}$$

$$= 0.0000352 \quad \underline{/69.4°} \quad \frac{\text{amps}}{\text{volt}} \tag{7-110}$$

and thus

$$i = (163)(0.0000352)\sin{(377t + 69.4°)}$$
$$= 0.00575 \sin{(377t + 69.4°)} \quad \text{amps} \tag{7-111}$$

The effective value of i would be 4.06 mA. The power factor angle is $+69.4°$, giving a power factor of 0.352 (leading), and an average power of $(163)(0.00575)(0.352)/2 = 0.165$ watt. To find the impedance $Z = (e_i/i)(i\omega)$ at $\omega = 377$ rad/sec, we just take the reciprocal from Eq. (7-110),

$$Z(i\omega) = \frac{e_i}{i}(i\omega) = \frac{1}{0.0000352 \quad \underline{/69.4°}} = 28400 \quad \underline{/-69.4°} \quad \text{ohms} \tag{7-112}$$

Note again that impedances (being the ratio of voltage to current) *always* have the units of ohms, irrespective of the combination of R, C, and L they may represent, or whether they are operational, sinusoidal, or Laplace versions. To get impedances in ohms, always take R in ohms, C in farads, L in henries, and frequency in rad/sec. If you use these standard units, then any time constants (RC or L/R) that might appear will always be in seconds.

7-7 FLUID FIRST-ORDER SYSTEMS

Figure 7-39 shows various fluid systems which arise in practical applications and which might reasonably be modeled as first-order systems. The first four are tanks (three open and one pressurized) in which the liquid level is determined by a pressure or flow rate input. Such arrangements arise in many process plants (refineries, chemical plants, power plants, food processing, etc.) where tanks are used for storage, mixing of fluids, heating, chemical reaction vessels, etc. Most such operations employ automatic feedback control systems to manage tank storage, inflow, and outflow. Design[13] of these control systems requires dynamic models of all the components, including of course the tanks themselves. Figure 7-39e and f shows useful models for pressure-measuring systems[14] in which a length of tubing connects the pressure to be measured (p_i) to the chamber of a pressure transducer. It turns out that these "plumbing" effects are usually more significant than the dynamics of the transducer itself. In Fig. 7-39e the fluid is incompressible and the bellows represents

---

[13] E. O. Doebelin, *Control System Principles and Design*, Wiley, New York, 1985, pp. 37, 208, 260, 347, 441.

[14] E. O. Doebelin, *Measurement Systems*, 4th ed., McGraw-Hill, New York, 1990, pp. 473–489.

First-Order Systems

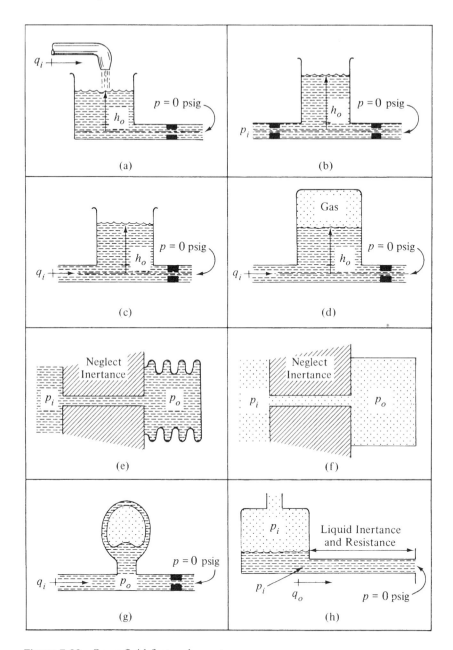

Figure 7-39 Some fluid first-order systems.

the compliance of the transducer elastic element, while in Fig. 7-39f the compressible fluid is itself the dominant compliance. Without the accumulator in Fig. 7-39g, sudden changes in flow rate q_i would cause sudden (possibly damaging) rises in pressure p_o. The accumulator helps to reduce this effect. The system of Fig. 7-39h is intended mainly for the study of the flow response of a liquid-filled pipe to driving pressure.

472   **Chapter 7**

Basic Laws Useful for Equation Setup. For systems involving fluid compliances (tanks, accumulators) the conservation of mass is generally useful in deriving system equations. Over a time interval dt, one can always write

Mass in − mass out = additional mass stored   (7-113)

When the fluid is considered incompressible, one can substitute volume for mass in this relation. If a flow path branches, the instantaneous summation of flow rates in and out must be equal, entirely analogous to the current node law in electric circuits. This holds fundamentally for *mass* flow rates, but again the assumption of incompressibility extends it to volume flow also. For those fluid systems which are obviously in the form of "circuits" (loops) so that one can start at a chosen point, follow a loop back to the starting point, and keep track of all pressure drops, the summation of pressure drops at any instant must be zero, just as in Kirchhoff's voltage loop law. Many fluid systems, however, do *not* "recirculate" the fluid, and attempts to manipulate them into circuit form, while possible, do not really simplify the analysis or give a better physical understanding of behavior. A direct approach which simply equates the driving pressure difference between any two points to the summation of the pressure drops due to resistance and inertance is generally preferable.

Linearized and Nonlinear Analysis of a Tank/Orifice System. In Fig. 7-39, the fluid resistances shown could be either linear or nonlinear. By linearizing any nonlinear resistances for small changes about a chosen operating point, we are able to define time constants and steady-state gains in letter form, with the usual design advantages. Knowing such studies to be approximations, we often check them against simulations of the "exact" nonlinear system, to build our judgment of when such approximations can legitimately be used. Let's carry out such a study for the system of Fig. 7-39a, where we assume the resistance to be a sharp-edge orifice (nonlinear). Since pressure in this system is determined entirely by the height of liquid in the tank, we treat the fluid as incompressible. For example, each foot of water in a tank represents 0.433 psi of pressure; $(62.4 \, lb_f/ft^3)/(144 \, in^2/ft^2) = 0.433 \, psi/ft$. Thus a tank would have to be *very* high before the pressure at the bottom would be enough to cause significant compression of the water. We further assume the discharge pipe is short enough to neglect its resistance and inertance, both of which are proportional to length. Inertance and resistance effects in the tank (think of it as a very large vertical pipe) will amost certainly be negligible since its large area means the velocity dh_o/dt and acceleration d^2h_o/dt^2 will be very small. (Recall that resistive pressure drops are related to velocity and are small when velocity is small. Similarly for inertial pressure drops and acceleration.)

Our analysis model is thus made up of the compliance of the tank, the orifice resistance, and a volume-flow-rate source $q_i \, ft^3/sec$. Conservation of mass (also volume here, because of incompressibility) gives

Volume in − volume out = additional volume stored

$$q_i dt - K_{or}\sqrt{h_o}\,dt = A_T\,dh_o \qquad (7-114)$$

First-Order Systems

$$\text{Instantaneous orifice flow rate} = A_{or}C_d\sqrt{\frac{2g\,\Delta p}{\gamma}} = \sqrt{2g}A_{or}C_d\sqrt{\frac{\gamma h_o - 0}{\gamma}}$$

$$= K_{or}\sqrt{h_o}$$

$$(7\text{-}115)$$

where

$$g \triangleq \text{local acceleration of gravity, ft/sec}^2$$
$$C_d \triangleq \text{orifice discharge coefficient}$$
$$A_T \triangleq \text{tank cross-section area, ft}^2$$
$$\gamma \triangleq \text{fluid specific weight, lb}_f/\text{ft}^3$$
$$A_{or} \triangleq \text{orifice area, ft}^2$$
$$h_o \triangleq \text{tank level, ft}$$

(Much as we might prefer to use only "metric" SI units, most current catalogs for fluid-handling equipment in the United States quote specifications in British units, so I feel obligated to continue to prepare the reader for both kinds of units by doing some problems in each.) The system differential equation is then

$$A_T\frac{dh_o}{dt} + K_{or}\sqrt{h_o} = q_i \qquad (7\text{-}116)$$

We will use on this problem the perturbation method I earlier recommended for most linearization studies, beginning by defining the total, operating point, and perturbation quantities as follows:

$$h_o \triangleq h_{o,0} + h_{o,p}$$

$$q_i \triangleq q_{i,0} + q_{i,p}$$

$$\sqrt{h_o} \approx \sqrt{h_{o,0}} + \frac{1}{2\sqrt{h_{o,0}}}(h_o - h_{o,0}) = \sqrt{h_{o,0}} + \frac{1}{2\sqrt{h_{o,0}}}h_{o,p} \qquad (7\text{-}117)$$

We assume that initially the inflow q_i was constant for a long time at the value $q_{i,0}$, so that tank level had become steady at $h_{o,0}$, where $h_{o,0} = q_{i,0}^2/K_{or}^2$. Equation (7-116) may now be rewritten as

$$A_T\frac{dh_{o,p}}{dt} + K_{or}\left(\sqrt{h_{o,0}} + \frac{1}{2\sqrt{h_{o,0}}}h_{o,p}\right) = q_{i,0} + q_{i,p} \qquad (7\text{-}118)$$

Simplification finally gives Eq. (7-119), which we see is our familiar first-order model. Faster response is obtained by reducing tank area, increasing K_{or}, and/or operating the tank at a lower level. Higher gain K is obtained with smaller K_{or} and/or operating the tank at a higher level (note that tank area has *no* effect on K). It is interesting that attempts to make the process "more sensitive" (increase K) result in *slower* response. While not a universal rule, we will see over and over in our system dynamics studies this tradeoff between sensitivity and speed of response. Nature just seems to be intent on requiring engineers to compromise between conflicting desirable goals.

$$\left(\frac{2A_T\sqrt{h_{o,0}}}{K_{or}}\right)\frac{dh_{o,p}}{dt} + h_{o,p} = \left(\frac{2\sqrt{h_{o,0}}}{K_{or}}\right)q_{i,p} \qquad (7\text{-}119)$$

$$\tau\frac{dh_{o,p}}{dt} + h_{o,p} = Kq_{i,p} \qquad (7\text{-}120)$$

$$\tau \triangleq \frac{2\sqrt{h_{o,0}}A_T}{K_{or}} \quad \text{sec} \qquad K \triangleq \frac{2\sqrt{h_{o,0}}}{K_{or}} \quad \frac{\text{ft}}{\text{ft}^3/\text{sec}} \qquad (7\text{-}121)$$

At this point we can define the various kinds of transfer functions, the Laplace form being

$$\frac{h_{o,p}}{q_{i,p}}(s) = \frac{K}{\tau s + 1} \qquad (7\text{-}122)$$

We should emphasize that this technique of defining perturbation variables and rewriting the differential equation in terms of them is of very general applicability in linearized analyses of all kinds and should *always* be employed if transfer functions are to be used. Otherwise, when you get the linearized system equation you will find on the right-hand side not only the input variable, but also a constant term, which will confuse the definition of the transfer function. A major application area for system dynamics is in the development of models for the components of automatic control systems. Here transfer functions and their associated block diagrams are widely used tools. In the class of feedback controls called *regulators*, the controlled variable must, in the face of uncontrollable disturbances, be kept close to a chosen desired value. If such systems are well designed, all the variables stay fairly close to certain fixed values, validating the main assumption of linearization.

Numerical Example: Nonlinear and Linearized Response of Tank/Orifice System to Step and Sine Inputs. To get some feeling for the accurate range of the above linearized analysis we compare the exact nonlinear response (obtained by simulation) with the linearized response, for perturbations of two different types and two different sizes. The simulation requires use of numerical values, so let's choose:

$$A_T = 1.20\,\text{ft}^2 \qquad\qquad C_d = 0.60$$
$$A_{or} = 0.001056\,\text{ft}^2 \qquad h_{o,0} = 2.0\,\text{ft}$$

These numbers make $q_{i,0} = 0.00722\,\text{ft}^3/\text{sec}$. Let's take the perturbation $q_{i,p}$ to be a step change of $-0.00037\,\text{ft}^3/\text{sec}$ for the small change and -0.00212 for the large. Figure 7-40 shows that the linearization gives almost perfect results for the small step change but deviates considerably for the larger one. (I leave the formulation of the simulation diagram, which is quite similar to others we have shown, to the end-of-chapter problems.)

If we now take the same size of perturbation, but make it sinusoidal at a frequency of 0.002 rad/sec (about 52 min/cycle), i.e., $q_{i,p} = 0.00037\sin(0.002t)$ or for the larger change, $0.00212\sin(0.002t)$, we get the results of Fig. 7-41a and b. Again, the smaller perturbation gives excellent results; however, now even the large one is quite good. This is because, while the sine *input* is as large as the step was, the *output* is much smaller. This is mainly due to the frequency being high enough that the amplitude of h_o has attenuated a lot, making the perturbation in h_o smaller than

First-Order Systems

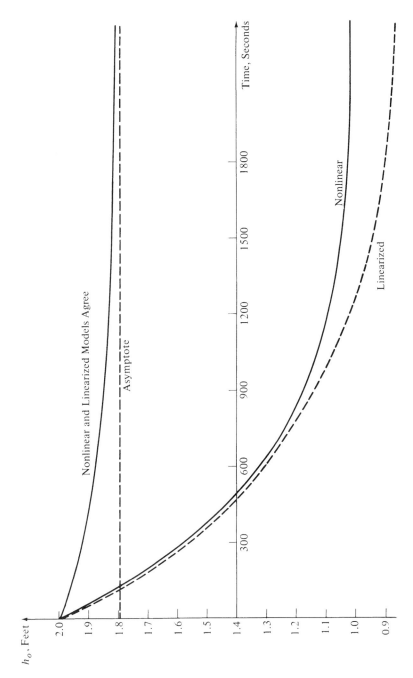

Figure 7-40 Comparison of exact nonlinear and approximate linearized. Step responses for tank/orifice system.

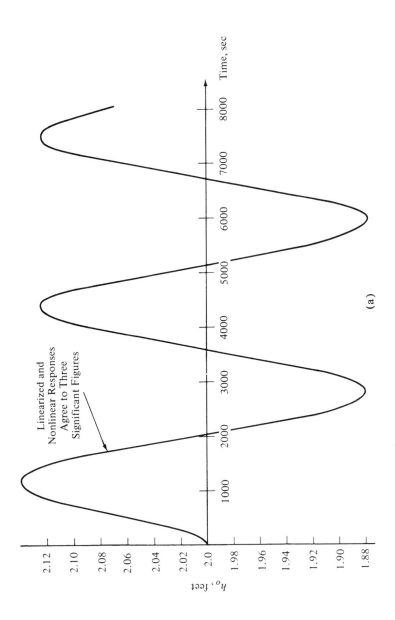

(a)

First-Order Systems

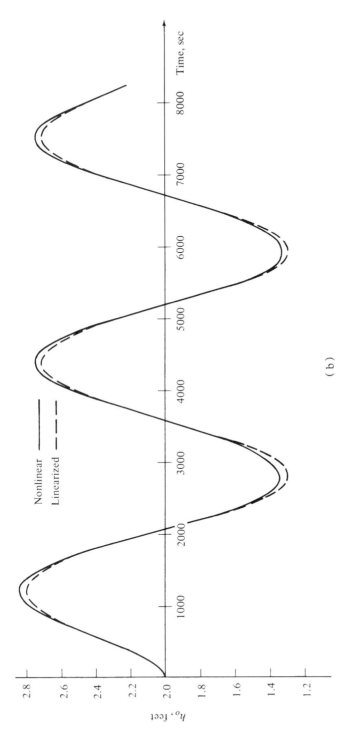

Figure 7-41 Comparison of exact nonlinear and approximate linearized sinusoidal responses for tank/orifice system.

it would be for a *constant* $q_{i,p}$ of the same magnitude. With a time constant of 667 seconds, the breakpoint frequency is $1/667 = 0.0015$ rad/sec, so it is clear that at 0.002 rad/sec the normalized amplitude ratio will be less than the 0.707 (-3 dB) that we know occurs at the breakpoint of every first-order system. At a frequency of 0.002 the normalized amplitude ratio is 0.60, making the actual amplitude ratio $0.60K = 334$ and the output amplitude for the large input 0.708 ft, which agrees with the graph of Fig. 7-34b.

Design Example: An Accumulator Surge-Damping System. In Fig. 7-39g we assume the pipe short enough to neglect inertance and resistance there. Since the accumulator is a large intentional compliance we can neglect the (small) compressibility of the fluid itself. We consider resistance only at the flow restriction, which we will assume is linear. Such hydraulic systems work with pressures up to several thousand psi, so the pressure due to the height of oil in the accumulator will be negligible (3 feet of oil give about 1 psi). While not necessary, we can draw a fluid circuit diagram as in Fig. 7-42 to show clearly that flow q_i branches into the accumulator and the resistance discharging to atmosphere, giving

$$q_i = q_C + q_R = C_f \frac{dp_o}{dt} + \frac{p_o - 0}{R_f} \qquad (7\text{-}123)$$

where

$C_f \triangleq$ accumulator compliance, ft^3/psi
$R_f \triangleq$ fluid resistance, psi/(ft^3/sec)

$$\tau \frac{dp_o}{dt} + p_o = Kq_i \qquad (7\text{-}124)$$

$$K \triangleq R_f \; \frac{\text{psi}}{\text{ft}^3/\text{sec}} \qquad \tau \triangleq R_f C_f \; \text{sec} \qquad (7\text{-}125)$$

To show how such a system can reduce pressure surges due to flow transients, let us assume an R_f of 10,000 psi/(ft^3/sec) with a steady q_i of 0.02 ft^3/sec, giving a steady p_o of 200 psi. If now a flow transient of rectangular pulse form as in Fig. 7-43 occurs, and if no accumulator were present, the peak pressure would be 800 psi. Such pressure "spikes" can cause mechanical damage and undesirable noise. We wish to choose an accumulator which will reduce this pressure peak to 300 psi. The response of a first-order system to a rectangular pulse input was solved for in Fig. 7-25, so we

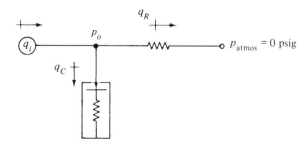

Figure 7-42 Accumulator surge-damping system.

First-Order Systems

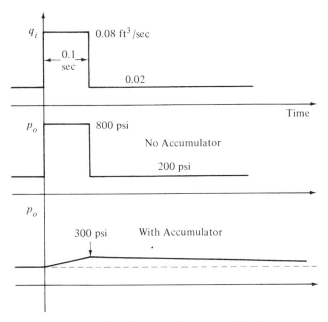

Figure 7-43 Benefits of accumulator surge-damping system.

can use that result without rederiving it. The peak pressure, above the steady 200 psi is given by

$$(R_f)(0.06)(1 - e^{-0.1/\tau}) = 100 \tag{7-126}$$

$$1 - e^{-0.1/\tau} = 0.167 \qquad \tau = 0.546 \text{ sec}$$

$$C_f = \frac{\tau}{R_f} = 5.46 \times 10^{-5} \frac{\text{ft}^3}{\text{psi}} \tag{7-127}$$

We thus need an accumulator which will displace 9.43 in^3 of fluid for each 100 psi of pressure. Of course an accumulator "softer" than this will reduce the peak pressure even more; however, there may be reasons, such as excessive size, weight, or cost, which would discourage use of a larger accumulator.

7-8 THERMAL FIRST-ORDER SYSTEMS

While, in practice, thermal devices and processes are very common and commercially important, the variety of *simple basic* first-order thermal systems which one can display is more limited than was the case for mechanical, electrical, and fluid systems. This is partly because there are only *two* thermal elements instead of the three found in these other system types; there are just fewer combinations possible. So far as basic laws available for equation setup are concerned, conservation of energy is clearly the most generally useful. For any system, over a time interval dt, we can always write

$$\text{Energy in} - \text{energy out} = \text{additional energy stored} \tag{7-128}$$

In Fig. 7-44 we show a solid body at temperature T_o immersed in a fluid at temperature T_i. This configuration is an adequate model for many practical situations, including the response of temperature-sensing instruments (thermometers, thermocouples, resistance temperature detectors, etc.), heat-treating of metal parts, heating and cooling food, etc. We assume that the Biot number is such that the temperature of the solid body may at any instant be taken as uniform throughout, and that the surface coefficient of convective heat transfer is uniform over the surface and constant with time. The heat transferred between the fluid and the solid is all stored in (or removed from) the solid, so we may write over a time interval dt:

Energy in (or out) = additional energy stored (or removed) (7-129)

$$hA_s(T_i - T_o)\,dt = MC\,dT_o \tag{7-130}$$

where

$h \triangleq$ surface heat transfer coefficient, watts/(m²-°C)

$A_s \triangleq$ surface area for heat transfer, m²

$M \triangleq$ mass of solid body, kg

$C \triangleq$ specific heat of solid body, j/(kg-°C)

Manipulation of Eq. (7-130) leads to

$$\frac{T_o}{T_i}(s) = \frac{K}{\tau s + 1} \tag{7-131}$$

$$K \triangleq 1.0 \ \frac{°C}{°C} \qquad \tau \triangleq \frac{1}{hA_s} MC = R_t C_t \quad \text{sec} \tag{7-132}$$

where R_t, C_t are thermal resistance and capacitance. We see that fast response of T_o to T_i requires large values of h and A_s, and small values of M and C. Interpreted in terms of thermal resistance R_t and thermal capacitance C_t, fast response requires these both to be small, just as R and C in electrical first-order systems must be small.

As an application of this general model, consider the mercury-in-glass thermometer of Fig. 7-45. The temperature which it *indicates* is the mercury temperature T_o, which will equal the fluid temperature T_i only for steady-state conditions. Its dynamic response can be found from the above type of model if we make some additional assumptions:

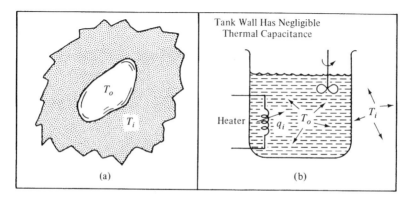

Figure 7-44 Some thermal first-order systems.

First-Order Systems

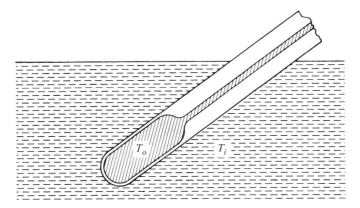

Figure 7-45 Mercury-in-glass thermometer.

1. Heat loss from the mercury bulb up the thermometer stem and then to the surrounding air is negligible.
2. Mass of mercury in the bulb does not change much as mercury rises in the capillary tube.
3. All material properties are constant.
4. Glass wall of bulb is thin enough that its energy storage is negligible.

The heat transfer between T_i and T_o is governed by a thermal resistance made up of a fluid film on the outside, the glass wall, and a fluid film on the inside; thus we should replace h in Eq. (7-130) by U, the *overall* heat transfer coefficient. In most cases U is dominated by the outside film coefficient, since the glass wall is thin and mercury is a very good conductor. A typical laboratory thermometer has a cylindrical bulb about $\frac{1}{8}$ by $\frac{1}{2}$ inch, giving a surface area of about $0.2\,\text{in}^2$ and a volume of about $0.006\,\text{in}^3$. For mercury $\rho = 0.491\,\text{lb}_m/\text{in}^3$ and $C = 0.033\,\text{Btu}/(\text{lb}_m\text{-}°F)$. The film coefficient on the outside surface varies greatly with the fluid and its flow velocity; 2 Btu/(hr-ft²-°F) for still air and 500 for rapidly moving water being indicative of the general range. These numbers give time constants ranging from 0.5 to 125 seconds, which agree quite well with step-function test measurements.

When a thermometer is used in a stationary fluid, the heat transfer is by free, rather than forced, convection and the film coefficient itself depends significantly on the temperature difference, giving a nonlinear system. A typical formula for h might be

$$h = K_h |T_i - T_o|^{1/4} \qquad (7\text{-}133)$$

which makes Eq. (7-130) nonlinear as follows:

$$A_s K_h |T_i - T_o|^{0.25}(T_i - T_o) = MC\frac{dT_o}{dt} \qquad (7\text{-}134)$$

We could linearize this in our usual (Taylor series) way, but another approach is more common in heat transfer problems, and can also be used in other applications. It may give better results when the excursions from the operating point are too large for the Taylor series method. It simply replaces the variable heat transfer coefficient by some kind of *constant* average value. If we can estimate the range that $(T_i - T_o)$

will cover, we could use its average value to compute an average h. Another way would be to compute h at each extreme of the range and average those two h values. Perhaps most accurate, but more work, would be to use calculus to compute the *true* average value over the given range.

Consider the example of Fig. 7-46, where we withdraw a thermometer from a 170°F bath to cool in still 70°F air. As an example of the above type of linearization, let's take the equation to be

$$\frac{dT}{dt} = 0.00376(T - 70)^{0.25}(70 - T) \tag{7-135}$$

If we compute h at the midpoint (120°F) of the temperature range and use this as our "average" value for linearization, the nonlinear term $0.00376(T - 70)^{0.25}$ becomes 0.01. If instead we compute the nonlinear term at the two extremes of the range and take the average we get 0.006. Using calculus to get the true average we get 0.0095. Figure 7-46 shows that good results are obtained using the "midpoint" method, but it is clear that the "calculus" method would also work well, since 0.0095 and 0.01 are so close.

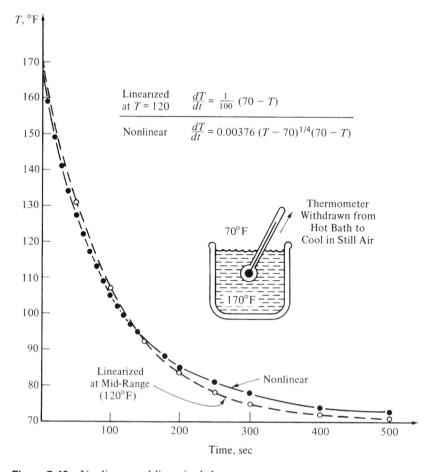

Figure 7-46 Nonlinear and linearized thermometer responses.

First-Order Systems   **483**

Systems with Several Inputs. The heated tank of Fig. 7-44b will give us some more experience specifically with thermal systems and also will introduce some useful general ideas with regard to systems whose output is influenced by more than one input. A dynamic system can in general have any number of inputs and outputs. So far we have emphasized those with a single input and output because our focus was on the "standard" first-order system.

We show a stirrer in the tank to allow us to assume a uniform temperature throughout the fluid at any instant. The heat added by the "churning" action of the stirrer is assumed negligible. The heater (which could be electrical, a steam coil, or some other type) supplies a heat input at a rate q_i watts. Note that watts are the proper SI unit, even for a *steam* heater. Temperature T_i of the tank's surroundings (so-called "ambient" temperature) may vary with time, as could the heating rate q_i. In a temperature control system for such a tank, $T_i(t)$ might be a random disturbing effect such as outdoor temperature, while $q_i(t)$ would be manipulated by a controller to fight against such disturbances and/or to cause the tank temperature to follow a commanded desired value. Conservation of energy gives us

$$q_i dt - UA(T_o - T_i)dt = MC\, dT_o \tag{7-136}$$

$$\tau \frac{dT_o}{dt} + T_o = (K_q)q_i + (K_T)T_i \tag{7-137}$$

where

$$\tau \triangleq MC/UA, \text{ sec}$$

$$M \triangleq \text{mass of liquid in tank, kg}$$

$$C \triangleq \text{specific heat of liquid in tank, j/(kg-°C)}$$

$$U \triangleq \text{overall heat transfer coefficient, watts/(m}^2\text{-°C)}$$

$$A \triangleq \text{heat transfer area, m}^2$$

$$K_q \triangleq 1/UA, \text{ °C/watt} \triangleq \text{temperature/heat-flow gain}$$

$$K_T \triangleq 1.0, \text{ °C/°C} \triangleq \text{temperature/temperature gain}$$

In Eq. (7-137) if we are given q_i and T_i as known functions of time, we can find the resulting tank temperature T_o; that is, T_o is determined by the simultaneous action of the two inputs. To get transfer functions and block diagrams, we employ the principle of superposition (valid for linear or linearized systems) to separate the effects of q_i and T_i. To get the transfer function $(T_o/q_i)(D)$ we momentarily consider T_i to be zero in Eq. (7-137) and get

$$\frac{T_o}{q_i}(D) = \frac{K_q}{\tau D + 1} \tag{7-138}$$

and then consider q_i to be zero to get

$$\frac{T_o}{T_i}(D) = \frac{K_T}{\tau D + 1} \tag{7-139}$$

In drawing the block diagram we must now superimpose the two effects to properly represent Eq. (7-137). This is done in Fig. 7-47 in two possible ways and can clearly be applied to *any* system with two inputs, and obviously extended to *any* number of inputs. The circle with a cross in it is the standard symbol for a *summing junction* and

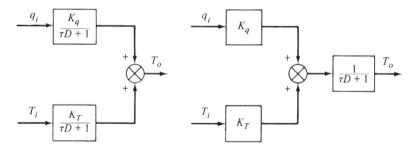

Figure 7-47 Two-input systems: alternative forms of block diagram.

can be extended to any number of inputs, with whatever + and − signs the equation demands.

7-9 MIXED FIRST-ORDER SYSTEMS

Having shown many examples which utilize only elements from *one* of our basic areas (mechanical, electrical, fluid, thermal), it is now appropriate to consider those first-order systems which use elements from *several* of these fields. Of the many possible examples we choose one each from the "mixed-media" fields of electromechanics, hydromechanics, and thermomechanics.

Electromechanical Open-Loop Speed Control. There are many industrial applications where it is necessary to control the speed (rpm) of a rotating machine part, such as a shaft. We may want to hold the speed steady in the face of disturbing torques, or vary the speed accurately according to some command. Electric motors are widely used for such tasks and the simplest control systems are the *open-loop* type which we now analyze.

A common arrangement is that of Fig. 7-48, where a dc motor with a fixed field is controlled by varying the voltage applied to its armature circuit. The mechanical load is modeled as inertia and viscous friction. We are interested in the response of the load speed ω_o to a control voltage input e_i and a disturbing load torque T_i. Since the field is fixed (either permanent magnet or wound), only the armature circuit need

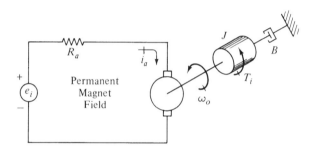

Figure 7-48 dc motor and load.

First-Order Systems   **485**

be analyzed. Armature inductance is usually negligible (just to be sure, check your motor's specifications to verify this in each case) so we model the armature circuit with resistance R_a, which includes both the motor's resistance and that of the voltage source (often an amplifier). This circuit also includes the motor's back emf. Kirchhoff's voltage loop law gives at every instant

$$-e_i + i_a R_a + K_E \omega_o = 0 \tag{7-140}$$

where

$$K_E \omega_o \triangleq \text{voltage drop due to motor back emf}$$

$$K_E \triangleq \text{motor back emf constant, volts/(rad/sec)}$$

Turning now to Newton's law at the motor shaft we get

$$K_T i_a + T_i - B\omega_o = J\dot{\omega}_o \tag{7-141}$$

where

$$K_T \triangleq \text{motor torque constant, (n-m)/amp}$$

$$B \triangleq \text{combined viscous damping of motor and load, (n-m)/(rad/sec)}$$

$$J \triangleq \text{combined inertia of motor and load, kg-m}^2$$

Recall that K_E and K_T have identical numerical values when expressed (as above) in standard SI units, but have *different* values if you use other unit systems.

The algebraic signs of the terms $K_E\omega_o$ and $K_T i_a$ may not be self-evident. We freely choose the positive sense of e_i, but once this is done it is convenient (though not necessary) to choose the positive direction of ω_o such that a positive e_i will cause a positive ω_o. To see how this was done, mentally "clamp" the motor shaft in Eq. (7-140), making $\omega_o = 0$. We see then that a positive e_i causes a positive i_a; a result of our choice for the i_a sign convention. Now in Eq. (7-141) also take $T_i = 0$. It is then clear that a positive i_a will cause a positive acceleration $\dot{\omega}_o$ and thus a positive speed ω_o if the shaft is released; thus our sign conventions do, in fact, give a positive ω_o for a positive e_i, as desired, and the choice of sign on $K_T i_a$ is justified. To justify the sign of $K_E\omega_o$ we must invoke our knowledge, from basic physics, of the *physical* nature of a motor back emf. This is that the back emf must *oppose* the voltage which caused the motion which is producing the back emf. Since a positive e_i tends to produce a positive ω_o, the signs of e_i and $K_E\omega_o$ *must* be opposite in Eq. (7-140), which they are.

Equations (7-140) and (7-141) form a simultaneous set which describes our physical system. If e_i and T_i are considered known inputs, we see that there are two unknowns, ω_o and i_a. We could solve for either or both of these; since our primary interest is in the load *motion* we choose to eliminate i_a in favor of ω_o. From Eq. (7-140),

$$i_a = \frac{e_i - K_E\omega_o}{R_a} \tag{7-142}$$

making (7-141)

$$\frac{K_T}{R_a}(e_i - K_E\omega_o) + T_i - B\omega_o = J\dot{\omega}_o \tag{7-143}$$

and giving finally

$$(\tau D + 1)\omega_o = (K_{TG})T_i + (K_{EG})e_i \tag{7-144}$$

where

$$\tau \triangleq \frac{J}{(BR_a + K_T K_E)/R_a} \quad \text{sec}$$

$$K_{TG} \triangleq \text{system speed/torque gain} \triangleq \frac{R_a}{BR_a + K_T K_E} \quad \frac{\text{rad/sec}}{\text{n-m}}$$

$$K_{EG} \triangleq \text{system speed/voltage gain} \triangleq \frac{K_T}{BR_a + K_T K_E} \quad \frac{\text{rad/sec}}{\text{volt}}$$

Note that the system's basic parameters τ, K_{TG}, and K_{EG} depend on *both* electrical and mechanical quantities. From

$$\tau = \frac{J}{B + K_T K_E / R_a}$$

we see that $K_T K_E / R_a$ provides a "viscous" damping effect entirely analogous to B; even if $B = 0$, *damping is still present.* The two system gains tell us how much the speed will change when there are changes in input voltage or disturbing torque. If we should need to get i_a (say for motor heating calculations) it is easily obtained from Eq. (7-142) once we have ω_o from (7-144). Any of these calculations are of course most quickly obtained by simulation; however, the analytical results are a great aid in system preliminary design.

Electromechanical Closed-Loop (Feedback) Speed Control. In the open-loop speed control system just discussed, speed errors caused by disturbing torques could in principle be corrected by manual adjustment of input voltage. If disturbances are not too severe and accuracy requirements not too great, such systems may be satisfactory. Otherwise, a more sophisticated approach using feedback principles may be required. Because of the importance of automatic control in all branches of engineering, most undergraduate students will have an entire course in this area, usually somewhere after a system dynamics course. We want to give a brief "preview" of some basic ideas from this area since they give strong motivation to our system dynamics studies. That is, a preliminary to all control system design is the dynamic modeling of every component in the control system.

In the open-loop system discussed above, in the absence of disturbing torque T_i the load speed ω_o is apparently completely under the control of voltage input e_i, and it would seem that setting e_i at a particular value would guarantee a certain speed. This would be true if all system parameters were absolutely constant; however in real systems, quantities such as B, R_a, and K_T are all subject to drift with time, temperature, etc.; thus highly accurate control should not be expected from such an arrangement. That is, the load torque which we analyzed is *not* the only disturbance acting to cause inaccurate speed control. To overcome these difficulties the feedback concept suggests that if we are interested in controlling speed we should *measure* what the speed actually is, *compare* this actual value to a desired value, and if they differ, *adjust* the voltage e_i in such a fashion as to reduce the error. Figure 7-49 shows one possible scheme for implementing this plan.

The load speed is measured by a *tachometer generator* directly coupled to the load shaft. (In a digital version of our system, a *tachometer encoder* might instead

First-Order Systems

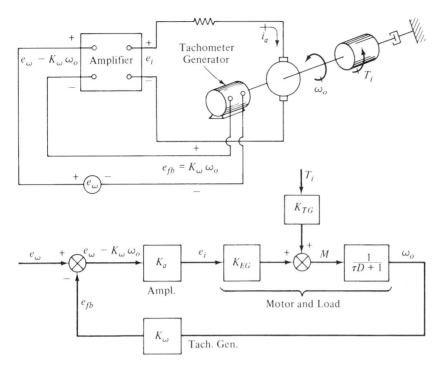

Figure 7-49 Speed control by feedback.

be used.) The tachometer generator produces a "dc" voltage accurately proportional to shaft speed, while a tachometer encoder produces a pulse train whose frequency is proportional to shaft speed. Our tachometer generator voltage is compared with a speed reference voltage e_ω obtained from an accurately regulated voltage source. This comparison is accomplished by simply connecting the two voltages in series, with polarity such that they oppose each other. This connection is called "series opposition" or "bucking." An alternative method of comparison would be to use an op-amp-type summing amplifier. The "error voltage" $(e_\omega - K_\omega \omega_o)$ is applied to an amplifier of gain K_a whose output supplies the armature voltage for the motor.

Usually we get our system equation *first* and *then* draw a block diagram. In control systems it is quite common to select components with known transfer functions, arrange them in a block diagram which will implement our design concept, and then write out the equation for the complete system, simply by "looking at" the diagram. Let's practice this useful skill on our present example. The *principle* behind all such manipulations is that we use the transfer functions to write out relations among the variables, but we only "keep" in our equations those variables which represent the desired value (e_ω in our case), the disturbances (T_i in our case), and the controlled variable (ω_o in our case). Starting at the left in Fig. 7-49 we write the error voltage, *not* as $(e_\omega - e_{fb})$ but rather $(e_\omega - K_\omega \omega_o)$, since our interest is in ω_o, not e_{fb}. The signal e_i is thus $K_a(e_\omega - K_\omega \omega_o)$ and the signal M is given by $K_{EG}K_a(e_\omega - K_\omega \omega_o) + K_{TG}T_i$, and since $\omega_o = (1/(\tau D + 1))M$ we may write

$$[K_{EG}K_a(e_\omega - K_\omega\omega_o) + K_{TG}T_i]\frac{1}{\tau D + 1} = \omega_o \tag{7-145}$$

"Multiplying" through by $(\tau D + 1)$ to "clear of fractions" gives

$$(\tau_{cl}D + 1)\omega_o = (K_{\omega_o})\omega_{o,\text{des}} + (K_{T_i})T_i \tag{7-146}$$

where

$$\tau_{cl} \triangleq \text{closed-loop time constant} \triangleq \frac{\tau}{1 + K_{EG}K_aK_\omega}$$

$$K_{\omega_o} \triangleq \text{closed-loop speed gain} \triangleq \frac{1}{1 + 1/K_{EG}K_aK_\omega}$$

$$K_{T_i} \triangleq \text{closed-loop torque gain} \triangleq \frac{K_{TG}}{1 + K_{EG}K_aK_\omega}$$

$$\omega_{o,\text{des}} \triangleq \text{desired speed} \triangleq \frac{e_\omega}{K_\omega} \quad \frac{\text{rad}}{\text{sec}} \tag{7-147}$$

While the system with feedback (also called "closed-loop" system) is still a first-order system, some remarkable improvements have occurred. These changes all depend on making the quantity $K_{EG}K_aK_\omega$, called the *loop gain*, a large number relative to 1.0. Fortunately this can generally be done, although great care must be used since excessive loop gain will cause instability, as we saw in Fig. 3-42. Much of feedback control design theory[15] has to do with various "tricks" which allow one to use large loop gain without causing instability. (Our present model does *not* predict this instability since we have neglected several dynamic effects which, while unimportant at low and medium loop gains, become critical at very high loop gains.)

To see what the above-mentioned improvements consist of, let's assume that we are able to make the loop gain as large as 20. The time constant τ_{cl}, which determines the speed of response of the complete feedback system, is now $\tau/21$; the system is 21 times as fast as the open-loop system, although the motor and load have not been changed in any way! Such great improvements can often be realized in practical systems, though they will be limited to commands that don't cause saturation (nonlinear behavior) or overheating in the amplifier and/or motor. In addition to speed, we also get great improvements in steady-state accuracy. We can study these effects by noting that the steady-state response to a constant desired speed can be expressed by the ratio

$$\left[\frac{\omega_o}{\omega_{o,\text{des}}}\right]_{\text{steady state}} = K_{\omega_o} \tag{7-148}$$

This result follows directly from the particular solution to Eq. (7-146) for a step input of $\omega_{o,\text{des}}$.

Ideally, we want the actual speed to be exactly equal to the desired speed. This requires a closed-loop speed gain of exactly 1.0, which requires an impossible loop gain of infinity; however, *perfect* accuracy is never required in practical systems, so

[15] E. O. Doebelin, *Control System Principles and Design*, Wiley, New York, 1985.

First-Order Systems   489

large, but finite loop gain is always acceptable. Actually, as long as the closed-loop speed gain is *constant*, we can get the actual speed equal to what we want without having infinite loop gain, by using the following simple trick. If the closed-loop speed gain were, say, 0.90, and we wanted an actual speed of 100 rad/sec, we would just set $\omega_{o,\text{des}}$ to $100/0.90 = 111.1$. That is, we ask for a larger speed than we actually want, because we know we get only 90% of what we ask for.

The practical problem with such a scheme is that the gain is neither precisely known nor constant, due to uncertainties and drifts in the physical parameters. By comparing open-loop and closed-loop systems when parameters change, we want to now show that feedback makes our systems much more tolerant of such real-world complications. For simplicity, let's take K_T, K_E, R_a, and B all equal to 1.0, while $K_a K_\omega = 40$. This makes $K_{EG} = 0.5$, the loop gain $= 20.$, and the closed-loop speed gain $= 0.952$. Suppose now that a temperature change causes the damping B to change to 0.5, a 50% drop. The new K_{EG} is now 0.667 and the new closed-loop speed gain is 0.964. The open-loop system experiences an error of $(0.667 - 0.500)/0.500 = 33\%$ while the closed-loop has only $(0.964 - 0.952)/0.952 = 1.2\%$. This type of insensitivity to parameter changes (but not the same numbers) would be seen for all the parameters *except* the sensor's K_ω, and is one of the great advantages of all feedback systems, so long as loop gain is high. Uncertainties, drifts, and nonlinearities of most of the hardware have much less effect in closed-loop systems than in open-loop. Sensor errors are *not* reduced by feedback; if your measured value is off by 20% the controlled variable will be in error by about the same percentage. This fact underlines the vital importance of accurate sensing in any control system. If you can't measure it accurately, you can't control it accurately. Finally, we see from the expression for K_T that the effects of disturbing torques are also 21 times smaller for the feedback system.

This simple example hopefully will impress the reader with the power of the feedback concept and give some idea as to why it is used in literally thousands of different types of applications in all fields of engineering.

Hydromechanical Systems: A Hydraulic Dynamometer.   In lab testing of machines which produce shaft power, we often need an adjustable load for the power source to drive. Such *absorbtion dynamometers* come in several forms; we here study one which uses a positive-displacement hydraulic pump. In Fig. 7-50 the pump's shaft is directly coupled to the output shaft of the power source, which provides torque T_i at speed ω. The pump outflow is sent to an adjustable valve, modeled as flow resistance R_{f2}. When this valve is "wide open" the pump's flow is only slightly impeded, the pump output pressure p_o is small, and the pump exerts little load on the tested power source. If we want to test the power source at higher levels, we simply close the valve more, causing higher pressure and pump torque. Thus with a simple pump and valve we have a conveniently adjustable load for testing internal combustion engines, electric motors, turbines, etc. Since the pump may be used at high pressure, we include its leakage effect as flow resistance R_{f1}.

We will take system input as the torque T_i. Outputs include pump/load speed, pump pressure, and pump flow rate. Let's choose pressure p_o, leaving the others to the end-of-chapter problems. While most valves would behave as nonlinear orifices, we choose the simpler linear model and again leave the nonlinear study for the problems. Newton's law at the pump shaft gives

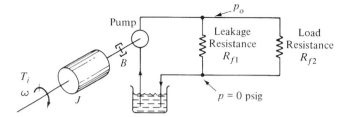

Figure 7-50 Hydraulic dynamometer.

$$T_i - (p_o - 0)D_p - B\omega = J\dot\omega \tag{7-149}$$

If the connecting pipes are short we can probably neglect fluid compliance and inertance. Then conservation of mass (volume) gives

Pump volume flow rate = flow through the two flow resistances

$$D_p\omega = \frac{p_o}{R_{f1}} + \frac{p_o}{R_{f2}} = p_o\frac{R_{f1}+R_{f2}}{R_{f1}R_{f2}} = \frac{p_o}{R_{ft}} \tag{7-150}$$

Total flow resistance $\triangleq R_{ft} \triangleq \dfrac{R_{f1}R_{f2}}{R_{f1}+R_{f2}}$

Eliminating ω in favor of p_o gives

$$\frac{p_o}{T_i}(s) = \frac{K}{\tau s + 1} \tag{7-151}$$

$$\tau \triangleq \frac{J}{B + D_p^2 R_{ft}} \quad \text{sec} \tag{7-152}$$

$$K \triangleq \frac{D_p R_{ft}}{B + D_p^2 R_{ft}} \quad \frac{\text{psi}}{\text{in-lb}_f} \tag{7-153}$$

We again clearly have a basic first-order system and all our generic results are immediately applicable. Note in Eq. (7-153) that $D_p^2 R_{ft}$ plays the same role as mechanical viscous damping and thus the system is damped even if $B = 0$. Since all the power absorbed from the tested power source is degraded into heat by the fluid resistances (mainly the valve), such dynamometers will generally require the oil reservoir to include some kind of cooler, to keep the oil at allowable temperatures (often below 170°F).

Hydromechanical Systems: Open-Loop Hydraulic Speed Control. When we need to control rotary speed, electric drives are widely used, but hydraulics may be more appropriate for certain applications. Figure 7-51 shows a typical arrangement, using a proportional valve to direct oil flow from a constant-pressure power supply to a rotary hydraulic motor which drives the mechanical load. By opening the valve wider, we can increase the motor/load speed. As in electric drives, open-loop systems are adequate for simpler applications, but closed-loop hydraulic systems may be needed when high performance is required. We here consider the open-loop system

First-Order Systems

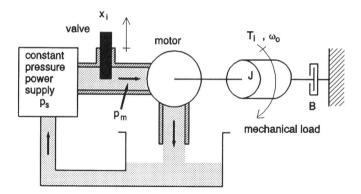

Figure 7-51 Open-loop hydraulic speed control.

only; the extension to closed-loop operation should be clear from our earlier electrical example.

We want to treat the valve realistically, as a nonlinear orifice of variable area, but since we want transfer functions, we will linearize about an operating point. Figure 7-51 shows the valve rather schematically, the actual construction is unimportant for analysis purposes so long as our flow formula is essentially correct. For many practical valves the valve volume flow rate is given by the orifice flow equation (4-37).

$$\text{Valve volume flow rate} = K_f f(x_i) \sqrt{\Delta p} \qquad (7\text{-}154)$$

where

$$K_f \triangleq C_d \sqrt{\frac{2}{\rho}}$$

$$\text{Flow area} = \text{function of valve position} = f(x_i) \qquad (7\text{-}155)$$

For the small valve position changes used in our linearization, the flow area can be taken proportional to valve position, irrespective of valve-port geometry, so our flow rate formula can be written as

$$\text{Valve volume flow rate} = K_v x_i \sqrt{\Delta p} = K_v x_i \sqrt{p_s - p_m} \qquad (7\text{-}156)$$

If the connecting pipes are short, fluid inertance and compliance will usually be negligible, and we can set the valve flow rate equal to the motor displacement flow rate plus the motor leakage flow rate, to get our first system equation. A second equation is obtained from Newton's law, where the algebraic sum of motor torque, disturbing torque T_i, and viscous damping torque determine motor/load acceleration at every instant.

To establish the needed operating point for linearization, let's assume that the valve has been at a fixed position $x_{i,o}$ and the disturbing torque at a fixed value $T_{i,o}$ for a "long time," so that motor pressure p_m has stabilized at $p_{m,o}$ and motor/load speed has stabilized at $\omega_{o,o}$. We now make small perturbations $x_{i,p}$ in valve position and $T_{i,p}$ in disturbing torque, causing perturbations $p_{m,p}$ and $\omega_{o,p}$. We linearize the valve flow rate using the multivariable formula from Eq. (2-18).

$$k_v x_i \sqrt{p_s - p_m} \approx K_v \left[x_{i,o} \sqrt{p_s - p_{m,o}} + \sqrt{p_s - p_{m,o}} (x_{i,p}) - \frac{x_{i,o}}{2\sqrt{p_s - p_{m,o}}} (p_{m,p}) \right]$$

(7-157)

Using the flow and Newton's law equations with all variables expressed as the sum of operating point and perturbation values, noting that all the operating-point terms cancel out (leaving only the perturbation variables), and eliminating $p_{m,p}$ in favor of $\omega_{m,p}$ gives after some manipulation

$$(\tau D + 1)\omega_{o,p} = K_x(x_{i,p}) + K_T(T_{i,p})$$

(7-158)

where

$$\tau \triangleq \frac{J}{B + D_m^2 R_t} \quad \text{sec} \qquad K_x \triangleq \frac{K_v\sqrt{p_s - p_{m,o}}}{\frac{B}{D_m R_t} + D_m} \quad \frac{\text{rad/sec}}{\text{in}}$$

$$K_T \triangleq \frac{1}{B + D_m^2 R_t} \quad \frac{\text{rad/sec}}{\text{in-lb}_f}$$

(7-159)

and

$$R_v \triangleq \frac{2\sqrt{p_s - p_{m,o}}}{K_v x_{i,o}} \quad \text{valve linearized flow resistance,} \quad \frac{\text{psi}}{\text{in}^3/\text{sec}}$$

(7-160)

$$R_t \triangleq \frac{R_l R_v}{R_l + R_v} \qquad R_l \triangleq \text{motor leakage resistance}$$

(7-161)

The derivation of these results is left for the end-of-chapter problems.

Recall the meaning and utility of the two gains. If, for example, $K_x = 540$ (rad/sec)/inch, then a step change of $x_{i,p}$ of 0.05 inch from the operating point will cause a steady-state speed change of $(0.05)(540) = 27$ rad/sec. If K_T is, say, 2.3 (rad/sec)/(in-lb$_f$), then a disturbing torque step change of 5 in-lb$_f$ away from its operating point, will cause a steady-state speed change $(5)(2.3) = 11.5$ rad/sec. If the time constant were, say, 0.58 second, then these steady-state values would be "95% complete" in $(3)(0.58) = 1.74$ seconds. If we wanted to use this same hardware in a closed-loop speed control system, the above model could be used directly. We would need to add a speed sensor, and if this had an electrical output, a summing amplifier and a reference voltage source to enter our desired speed in the form of a proportional voltage. The valve would now need to be actuated electrically, from the amplifier output, and such servovalves are readily available. If the purpose of our speed control system was to maintain a constant speed in the face of disturbing torques, this constant speed would be chosen as the operating point for our linearization.

Thermomechanical Systems: Thermal Expansion Actuators. We will give two examples of thermomechanical systems, one for each "direction" of energy conversion. In thermal expansion actuators, we input thermal energy and produce output motion and force. In friction brakes we input mechanical energy and output heat flow and temperature.

While electrical, hydraulic, and pneumatic actuation methods are much more common, actuation by thermal expansion does serve a number of practical

First-Order Systems

applications.[16] Figure 7-52 shows the operating principle of most thermal actuators. A waxlike material of high thermal expansion fills the cylinder chamber to the left of the actuator piston. Buried inside is a small electrical resistance heater, usually powered directly from the ac power line through a suitable on-off switch. When the heater is turned on, the expansion material's temperature T_o rises and the material expands, driving the piston to the right with displacement x_o. If the heater is turned off, heat will flow from the actuator to its surroundings (at temperature T_i) and the material will contract. This contraction will *not* by itself cause the piston to retract, so a "return spring" is used to keep the piston in contact with the contracting expansion material and thus move x_o in the reverse direction. Some thermal actuators use an expansion material which changes phase from solid to liquid (and the reverse) to get large expansion coefficients. These actuators require a different analysis since the expansion material does *not* change temperature while it is changing phase, but expansion is still occurring. This analysis is not more difficult, but does not lead to a simple linear first-order model, so we restrict our study to expansion materials which do not change phase.

Some thermal actuators are used in an "on-off" mode, where the intention is to move the piston to its full-stroke position against a fixed stop when the heater is switched on. Here the final temperature (reached if the heater is left on) is high enough that the piston reaches its full-stroke stop before this temperature is reached. The expansion material still *tries* to expand, but is restrained by the chamber walls. Forces *internal* to the actuator develop due to this restrained expansion. These forces are not of interest to us here, but could be calculated if we knew the elastic properties of the chamber and the expansion material. If the actuator piston rod feels an external load force f_L, the displacement x_o would be reduced by these same elastic effects. Our analysis below assumes no such load force, so the elastic effects need not be considered.

We assume that the "lump" of expansion material (mass M and specific heat C) is at a uniform temperature T_o at every instant, and that heat transfer from the T_o region to the surroundings at T_i is characterized by an overall heat transfer coefficient U and a heat transfer area A. Also, expansion is assumed to occur instantly with temperature rise, according to the relation $x_o = (K_{Tx})(\Delta T_o)$. The return-spring

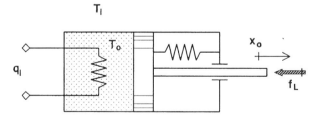

Figure 7-52 Thermal actuator.

---

[16]TCAM Technologies, Inc., 33800 Curtis Boulevard, Suite 114, Eastlake, OH 44095, 216-942-2727. T. M. Kenny, A Guide to Self-Actuating Control Valves, *Plant Engineering and Maintenance*, June 1993. Eltek Thermoactuator, Stajac Industries, Inc., 155 Fisher Ave., P.O. Box 187, Eastchester, NY 10709, 800-441-4014.

force causes elastic deflections in the same manner as the load force discussed above, but this spring is usually not very stiff, so the effect is small. Actually, if we run a steady-state experiment to measure K_{Tx} by plotting x_o versus T_o, with the return spring installed, the effect of the spring will be taken into account. If you have to estimate parameter values from theory, before any hardware is available for test, be sure to note that the expansion coefficient is the *relative* expansion of the expansion material with respect to the chamber walls.

Having stated the major assumptions we can proceed with an analysis based on conservation of energy.

$$q_i dt - UA(T_o - T_i) dt = MC\, dT_o \tag{7-162}$$

$$x_o = K_{Tx}(T_o - T_i)$$

$$(\tau D + 1)x_o = Kq_i \tag{7-163}$$

$$\tau \triangleq \frac{MC}{UA} \text{ sec} \qquad K \triangleq K_{Tx} UA \; \frac{\text{inches}}{\text{watt}} \tag{7-164}$$

We have here assumed that the ambient temperature T_i is constant at "room temperature" and that the displacement x_o is considered to be zero when the actuator is at room temperature. Figure 7-53 shows a SIMULINK simulation where I have included a saturation nonlinearity to represent a mechanical stop at 0.20 inch. For low heating rate inputs, the maximum temperature is low enough that the maximum displacement is less than 0.20 inch and the stop is not encountered. Figure 7-54a shows results for such a case, using typical numbers for system parameters. We see a symmetrical response for actuator extension and retraction, with the time constant serving its usual purpose as an indicator of response speed.

When such actuators are used for "on-off" applications, the heating rate is chosen so as to cause the piston to "bottom out" against the stop, as in Fig. 7-54b. Note here that "speed of response" involves *more* than just the first-order time constant. For complete actuator extension, the response is faster than one would

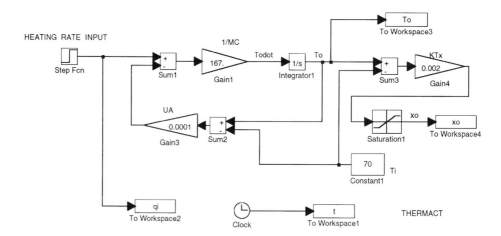

Figure 7-53 Simulation diagram for thermal actuator.

First-Order Systems

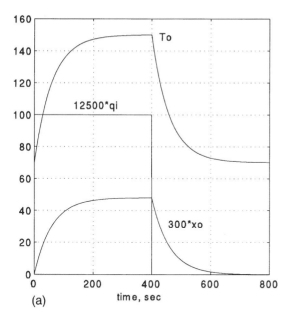

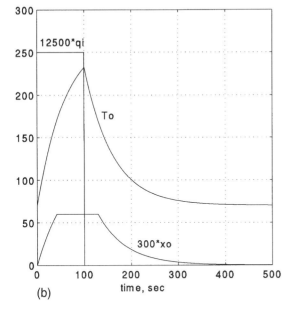

Figure 7-54 Simulation results for thermal actuator.

expect from the time constant, since we can use a high heating rate to make the early part of the exponential curve very steep. For complete retraction, a similarly high *cooling* rate is not available, since it is limited by the fixed T_i, so retraction speed is slower, and *is* properly described by the time constant.

Some of the parameters in thermal actuators can be estimated fairly accurately from basic material properties and part dimensions, whereas others might need to be measured in lab tests. We have already mentioned the possibility of a static calibration experiment to determine K_{Tx} by measuring and plotting x_o versus T_o. A dynamic test with a step input of heating rate and plotting of x_o versus time allows checking of first-order behavior (see Fig. 7-9) and determination of an accurate τ value. Good M and C values are available from material properties tables and dimensions, but heat transfer coefficient U is always quite uncertain. Having τ and MC, we can "back out" a numerical value for UA. At this point, we would have enough numbers to implement a simulation as in Fig. 7-53. Heat transfer area A may seem to be accurately available from dimensions, but actuator geometry may not be as simple as our Fig. 7-52 suggests, so a computed A value may be quite uncertain. If we *do* estimate A, then, having already found UA, we can get a U value. Of course, if a simulation as detailed as Fig. 7-53 is not necessary, Eq. (7-163) shows that we can "get by" with only τ and K. The gain K can be accurately measured from a "static calibration" experiment where we use several q_i values (easily measured electrically with a wattmeter), measure the resulting steady displacements, and fit this data with a straight line.

Thermomechanical Systems: A Simple Friction Brake. Friction brakes and clutches are widely used in machinery and often exhibit thermal problems which must be analyzed to allow needed design modifications. Excessively high temperatures can cause "brake fade," where the available friction force or torque drops off as repeated or lengthy brake applications cause the friction coefficient to drop. Temperature variations can also cause or exacerbate annoying and destructive brake "squeal." Thermal analysis of brakes can be quite complex so we limit ourselves here to only some simplified studies.

In Fig. 7-55 we show a simple drum-type brake, where a force f_i pushes the brake shoe against the rotating drum, creating a friction force and torque. In some brake applications the intent is to slow or stop the rotating drum; in others the drum rotation is at constant velocity and the brake's purpose is to create a constant drag torque on a machine member, perhaps to maintain a desired tension when winding a roll of paper or other material. In either case, the interface between the drum and the shoe is considered as a heat-flow source q equal to the frictional power (friction force

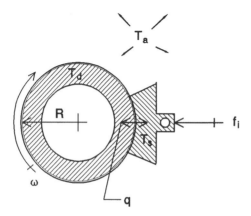

Figure 7-55 Thermomechanical system: A simple friction brake.

First-Order Systems 497

times velocity) dissipated at the interface. This heat flow divides into two parts, one going into the drum and one into the shoe. A common assumption[17] partitions the heat flow as

$$\text{Shoe heat flow} = q \; \frac{1}{1 + \sqrt{\dfrac{(ck\rho)_d}{(ck\rho)_s} \dfrac{A_d}{A_s}}} \triangleq qf_s \qquad \text{Drum heat flow} = q(1 - f_s)$$

(7-165)

where $(ck\rho) =$ (specific heat)(conductivity)(density) for drum d and shoe s. The shoe contact area is A_s, while A_d is the drum "swept" area, that is, drum circumference times shoe width.

The simplest thermal model assumes "one lump each" for the shoe and the drum; that is, we assume the temperature of each is uniform throughout at every instant. More correct models divide the bodies into a number of layers, each layer having its own temperature. Thermal finite-element software could divide each body into hundreds of lumps and solve for each temperature. Only the simplest model will result in a first-order system, so that is what we analyze here. It would work best for a "thin" drum and shoe of high conductivity, and for "long" times after a transient input. It will underestimate temperatures near the friction interface and for "early" times. We can write a conservation of energy equation for each body:

$$(1 - f_s)(f_i\mu)R\omega - U_d A_d(T_d - T_a) = M_d C_d \frac{dT_d}{dt} \tag{7-166a}$$

$$f_s(f_i\mu)R\omega - U_s A_s(T_s - T_a) = M_s C_s \frac{dT_s}{dt} \tag{7-166b}$$

There is no coupling between these equations; each can be dealt with separately. We are using SI units here so that the frictional heat generation rate $(f_i\mu)R\omega$ which is in N-m/sec is consistent with the heat transfer and storage terms in watts. These two equations are easily manipulated into our standard first-order form.

$$(\tau_d D + 1)T_d = (K_d)f_i + (K_a)T_a \tag{7-167}$$

$$(\tau_s D + 1)T_s + (K_s)f_i + (K_a)T_a \tag{7-168}$$

$$K_a \triangleq 1.0 \; \frac{°C}{°C} \quad K_d \triangleq (1 - f_s)\frac{\mu R\omega}{U_d A_d} \; \frac{°C}{\text{newton}} \quad K_s \triangleq f_s \frac{\mu R\omega}{U_s A_s} \tag{7-169}$$

$$\tau_d \triangleq \frac{M_d C_d}{U_d A_d} \; \text{sec} \quad \tau_s \triangleq \frac{M_s C_s}{U_s A_s} \tag{7-170}$$

The ambient temperature T_a would usually be treated as constant, but need not be. For a step input of force f_i we can use our generic first-order results to predict temperatures if the application is one where drum speed ω is constant. If the brake causes the drum to slow down, we need a separate Newton's law equation to calcu-

---

[17]D. C. Sheridan, J. A. Kutchey, F. Samie, Approaches to the Thermal Modeling of Disk Brakes, SAE Paper 880256, 1988.

498   Chapter 7

late speed as a time-varying quantity, and use this for ω in the two gains. This is easily done with simulation, which could also nicely handle a friction coefficient which varies with temperature (many do), so long as we know *how* it varies. Our single-lump model is rather crude but our analysis shows the principles involved, which are easily extended to models with more lumps.

7-10 FIRST-ORDER SYSTEMS WITH "NUMERATOR DYNAMICS"

At the beginning of this chapter we pointed out that there are several types of linear first-order systems and that we would emphasize one of these because it is most common and important. We now need to briefly familiarize you with two other types which are less common, but still have practical significance. The generic forms of these two types of first-order systems are:

$$\frac{q_o}{q_i}(s) = \frac{K\tau s}{\tau s + 1} \tag{7-171}$$

$$\frac{q_o}{q_i}(s) = \frac{K(\tau_1 s + 1)}{\tau s + 1} \tag{7-172}$$

The presence of the s (or D, or $i\omega$) in the numerator of the transfer functions gives rise to the "numerator dynamics" terminology. Various physical systems with different practical applications have these forms of transfer functions, and we now show one example of each type.

The simple circuit of Fig. 7-56 is easily analyzed using ac impedance methods.

$$e_o = iR = \frac{e_i}{R + 1/i\omega C} R = \frac{i\omega RC}{i\omega RC + 1} \tag{7-173}$$

$$\frac{e_o}{e_i}(i\omega) = \frac{Ki\omega\tau}{i\omega\tau + 1} \qquad K \triangleq 1.0 \; \frac{\text{volt}}{\text{volt}} \qquad \tau \triangleq RC \; \text{sec} \tag{7-174}$$

and thus

$$\frac{e_o}{e_i}(D) = \frac{K\tau D}{\tau D + 1} \tag{7-175}$$

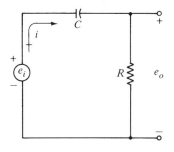

Figure 7-56 High-pass filter: a system with "numerator dynamics."

First-Order Systems

$$\tau \frac{de_o}{dt} + e_o = K\tau \frac{de_i}{dt} \qquad (7\text{-}176)$$

Note how, in this problem, we *started* with the frequency response (ac circuit theory) and "worked backward" to the differential equation. You may (or may not) prefer this approach and find it quicker and/or easier than starting with the differential equation. As usual, we can use the step response and the frequency response to get acquainted with the dynamic behavior of such systems. For the step response we set e_i in Eq. (7-176) equal to a constant e_{is}, and since the derivative of a constant is zero, we get for all $t > 0$,

$$\tau \frac{de_o}{dt} + e_o = 0 \qquad (7\text{-}177)$$

which has the complete solution

$$e_o = Ce^{-t/\tau} \qquad (7\text{-}178)$$

To find C we need to know $e_o(0^+)$. Whereas for the first-order systems emphasized earlier in this chapter the output at $t = 0^-$ and $t = 0^+$ is the same, for this type of system there is a *sudden* change in e_o when the step input is applied. Since capacitor C is assumed initially uncharged, when e_i jumps up to e_{is}, there being no voltage drop across C, Kirchhoff's voltage loop law tells us that e_o instantly jumps up to e_{is}. The complete specific solution is thus

$$e_o = e_{is}e^{-t/\tau} \qquad (7\text{-}179)$$

which is shown in Fig. 7-57a. This same result can of course be found using Laplace transforms, so you might do this right now, to establish (or reinforce) the good habit of checking calculations with alternative methods whenever possible. While this circuit (and the generic class, Eq. (7-171), to which it belongs) responds instantly to step inputs, its output decays to a zero steady-state, achieving 95% of the change in 3τ. The step response is thus quite different from the first-order systems we have emphasized in this chapter.

That such systems have practical utility as *high-pass filters* becomes apparent from the frequency response (see Fig. 7-57)

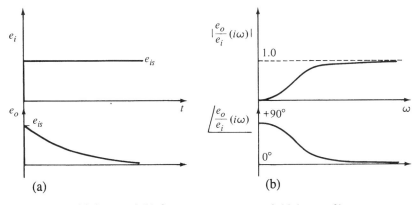

Figure 7-57 (a) Step and (b) frequency responses of high-pass filter.

$$\frac{e_o}{e_i}(i\omega) = K \frac{i\omega\tau}{i\omega\tau + 1} = \frac{K\omega\tau \;\angle 90°}{\sqrt{(\omega\tau)^2 + 1} \;\angle \tan^{-1}(\omega\tau)}$$

$$= \frac{K\omega\tau}{\sqrt{(\omega\tau)^2 + 1}} \;\angle 90° - \tan^{-1}(\omega\tau))$$

(7-180)

which for low frequencies approaches 0 $\angle 90°$ and for high frequencies 1 $\angle 0°$. That is, a high-pass filter rejects constant and low-frequency inputs but passes high-frequency inputs essentially unchanged. A common example of such a circuit is found in most oscilloscope input networks. There is generally a switch (labeled "ac coupling") which allows connection of a high-pass filter between the scope input terminals and the vertical deflection amplifiers. The time constant τ is chosen so that the range of frequencies where $(e_o/e_i)(i\omega) \approx 1 \;\angle 0°$ starts at about 2 Hz. Thus any signal with frequency content above 2 Hz is accurately measured, signal components between 0 and 2 Hz are distorted, and steady (dc) values are completely wiped out.

Such action is useful, for example, when a pressure transducer and oscilloscope are used to study pressure fluctuations in the output of a reciprocating air compressor. This pressure signal has a large mean value (say 100 psi) while the fluctuations are small (say 5 psi). If you turn up the scope sensitivity to get a good look at the small oscillations, the large mean value will deflect the picture completely off the screen, since the scope zero-suppression control has a limited range. Now, if we switch in the high-pass filter, the large mean value is completely blocked and we can easily turn up the sensivitity to fill the whole screen with the oscillations. This technique is of course not limited to oscilloscopes but applies to all kinds of data-acquisition systems. We might finally mention that this generic type of system is also useful as approximate *differentiating* devices, but we leave the details of this application for the end-of-chapter problems.

Turning now to the generic type of Eq. (7-172), we use the *pneumatic phase-lag compensator* of Fig. 7-58 as our example. If at some time you study feedback control theory you will find that a standard tool for improving performance uses dynamic compensators with transfer functions like Eq. (7-172). Depending on the selection of numerical values, such compensators can improve speed of response, steady-state accuracy, or both. When the control system uses pneumatic hardware, pneumatic

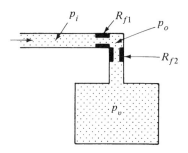

Figure 7-58 Pneumatic phase-lag compensator.

First-Order Systems   **501**

compensators may be appropriate. This form of transfer function is also achievable in other physical forms, such as electrical, mechanical, and digital software.

Industrial pneumatic process control systems often use a standard pressure range, such as 3 to 15 psig. That is, all pressure signals will be found somewhere in this range. Also, the systems are relatively slow and the fluid medium (air) is contained in metal tubing and chambers which tend to stay at "room" temperature. These operating conditions allow us to invoke useful simplifying assumptions that treat the gas as incompressible and isothermal. That is, we will assume constant density and temperature, and small pressure changes around a chosen operating point, usually taken as the midpoint (9 psig) of the standard operating range, unless there is some reason to choose otherwise. Note that in the perfect gas law density formula $\rho = p/RT$, the pressure p and temperature T are both *absolute* values, thus a 1-psi pressure change is *not* a $\frac{1}{9} = 11\%$ change but rather a $1/(9 + 14.7) = 4\%$ change. Similarly, a 20F° temperature change is not a $\frac{20}{70} = 29\%$ change but rather a $20/(70 + 460) = 4\%$ change. Our approximate analysis should thus be fairly accurate.

An additional assumption is that the space containing p_o is so small that negligible air is stored there and thus the flow rate through resistance R_{f1} is at all times equal to that through R_{f2}, giving

$$\frac{p_i - p_o}{R_{f1}} = \frac{p_o - p_v}{R_{f2}} \tag{7-181}$$

(The flow resistances are defined for volume flow rate.) The assumption of negligible storage can be checked by assuming it is *not* negligible, getting a new model (it won't be first-order), and comparing the frequency response with that of our present model. You would find that our simplified model matches the more correct one so long as the system never "sees" input signals with frequency content beyond a certain value. If this upper limiting frequency is higher than any present in the actual application, the simpler model is adequate and, in fact, *preferred*. This method of validating assumptions by *not* making them and then comparing the two competing models is a powerful tool that is made practical by the availability of easy-to-use simulation software, as we have demonstrated several times already in this text.

Conservation of mass for the tank volume V gives

Mass inflow during time interval dt = mass storage during time interval dt

$$\rho \frac{p_i - p_v}{R_{f2}} \, dt = dM = \frac{V}{RT} \, dp_v \tag{7-182}$$

We here have used the perfect gas law to relate mass storage to tank pressure, taking V/RT as a constant. The air density ρ is also taken constant at the operating point values of temperature and pressure. Our interest is in relating small changes in input pressure signal p_i to the resulting small changes in output pressure p_o, so we choose to eliminate p_v from our equations, using

$$\left(\frac{R_{f2}V}{\rho RT} D + 1\right)p_v = (\tau_1 D + 1)p_v = p_o \tag{7-183}$$

$$p_i - p_o = \frac{R_{f1}}{R_{f2}}\left(p_o - \frac{1}{\tau_1 D + 1}p_o\right) = \frac{R_{f1}}{R_{f2}}\left(\frac{\tau_1 D}{\tau_1 D + 1}\right)p_o \tag{7-184}$$

$$(\tau D + 1)p_o = K(\tau_1 D + 1)p_i \tag{7-185}$$

$$\frac{p_o}{p_i}(D) = K\frac{\tau_1 D + 1}{\tau D + 1} \tag{7-186}$$

$$K \overset{\Delta}{=} 1.0 \qquad \tau_1 \overset{\Delta}{=} \frac{R_{f2}V}{\rho RT} \qquad \tau \overset{\Delta}{=} \tau_1\left(1 + \frac{R_{f1}}{R_{f2}}\right) \tag{7-187}$$

In the generic form (7-172), the two time constants can take on any values at all, but in our example, Eq. (7-187) clearly shows that τ must be larger than τ_1. When this is the case, the device is called a *phase-lag compensator*, and in control systems is used as an approximate proportional plus integral controller.

We again want to find step and frequency responses. For a step input, $p_i = p_{is} = $ a constant, and Eq. (7-185) becomes for $t > 0$

$$(\tau D + 1)p_o = p_{is} \tag{7-188}$$

whose complete solution is

$$p_o = Ce^{-t/\tau} + p_{is} \tag{7-189}$$

We again need $p_o(0^+)$ to find C, and we can find it from Eq. (7-181), which holds at every instant, including $t = 0^+$. Note that p_v is still zero at 0^+ since it takes a finite time for a finite flow rate to build up a tank pressure (the tank is an "integrator"). Thus

$$\frac{p_{is} - p_o(0^+)}{R_{f1}} = \frac{p_o(0^+)}{R_{f2}} \qquad p_o(0^+) = \frac{p_{is}}{1 + R_{f1}/R_{f2}} \tag{7-190}$$

which makes Eq. (7-189)

$$p_o = p_{is}\left[1 + \left(\frac{1}{1 + R_{f1}/R_{f2}} - 1\right)e^{-t/\tau}\right] \tag{7-191}$$

We see again (Fig. 7-59a) that p_o experiences a sudden jump at $t = 0$, but now it only jumps "part way," proceeding the remaining amount on an exponential curve. The frequency response

$$\frac{p_o}{p_i}(i\omega) = \frac{i\omega\tau_1 + 1}{i\omega\tau + 1} \tag{7-192}$$

is shown in Fig. 7-59b. Since τ_1 is always greater than τ in this example [see Eq. (7-187)], the phase angle is always lagging, giving rise to the name "phase-lag compensator."

Other systems, called *phase-lead compensators*, can have $\tau_1 > \tau$, and in fact we have already used two of these (the electrical version) in the system of Fig. 7-13. Many items of hardware that we need to use in dynamic systems have *lagging* behavior which may make them too slow for a certain application. We can often speed up the overall system by cascading the unavoidable lagging component with a suitably chosen lead compensator. Analysis (left for the problems) of these devices

First-Order Systems

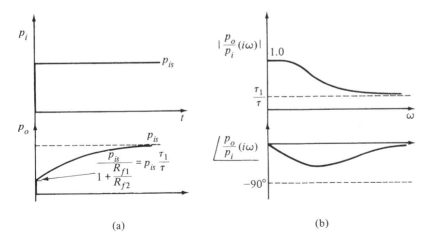

Figure 7-59 (a) Step and (b) frequency responses of phase-lag compensator.

gives the step and frequency reponses of Fig. 7-60. The step response again jumps up instantly, but now it overshoots and returns to steady-state on an exponential curve. The frequency response shows an amplitude ratio that increases with frequency before leveling off. This increase can counteract the *decrease* which is typical of "slow" hardware which we may want to speed up.

Design Example Showing Where System Dynamics Fits in the Overall Design Sequence. While people with engineering degrees end up doing a wide variety of tasks when they leave school, most engineering educators consider *design* as the major focus of the curriculum. Every course should contribute, as appropriate, to this central goal; creating new and useful products and services for society. While a system dynamics course of necessity spends a lot of time on modeling and analysis,

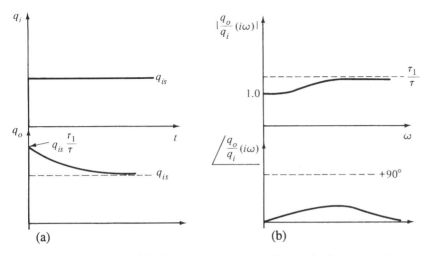

Figure 7-60 (a) Step and (b) frequency responses of phase-lead compensator.

we try wherever feasible to show links with design, and we have already done this in numerous instances. I want to use this present example to discuss the design process in somewhat more depth than we have up till now.

Entire books[18] have been written on this subject. If you don't have time to study an entire book, condensed versions[19] can give you the essentials rather quickly. Our present example must of course be even more brief. The design of *all kinds* of engineered systems can usefully be thought of as proceeding through several stages. These stages can be defined and organized in various ways, one of which is:

1. Conceptual design
2. Substantive design
3. Detail design

In *conceptual design* we start with sometimes vague specifications describing what functions we want the new product or service to perform. We then try to conceive of "all" the possible ways to achieve these functions, allowing our minds wide latitude and not insisting, at this point, that ideas be totally practical. Usually a design *team* will work together to benefit from the cross-fertilization of the various experiences and viewpoints. At some point this "brainstorming" must cease and the list of possibilities narrowed by "shooting" down those which can't survive the arguments of the design team about various practical issues. The output of this design stage is a shorter list of concepts which, at least at this stage, can't be shot down as being obviously unworthy of further study.

To further narrow the list, we now need to engage in *substantive design*. Here each alternative concept is refined and analyzed so that we understand it better and can begin to compare its performance with our specifications. This of course requires physical analysis, but we still try to avoid dealing with extreme detail. That is, if our system uses a pneumatic cylinder, we may need to decide on its diameter, stroke, and air pressure, but we *don't* consider details such as wall thickness to prevent bursting, seals to prevent leakage, lubrication, etc. During substantive design our list of feasible concepts is further narrowed as we discover problems and possibilities that were not apparent earlier. The output of this stage of design will usually be a *decision* to further develop only one of the concepts. If our system has significant dynamic aspects, the methods of system dynamics will most often be useful in this *substantive* design stage. In our upcoming example, substantive design shows that we require, as part of our overall system, a mechanical spring/damper system with a certain time constant.

The scope of *detail design* is subject to some interpretation, but usually involves the choice of materials and dimensions which guarantee that our system will function properly and not fail prematurely. This is the level of design that is considered by most textbooks on "machine design," perhaps better called "machine component design." In our example, we will need to design a specific spring and damper that will realize the needed time constant and also meet other practical requirements.

---

[18]G. Pahl and W. Beitz, *Engineering Design: A Systematic Approach*, Springer-Verlag, New York, 1988.

[19]E. O. Doebelin, Apparatus Design and Construction, chap. 6 in *Engineering Experimentation*, McGraw-Hill, New York, 1995.

First-Order Systems

The above brief overview of design cannot do justice to the subtelties of this complex process, but will at least allow us to proceed with the example and appreciate its more general significance. Our example is a system for controlling the position of a rotating mechanical load in response to electrical voltage commands. The conceptual design stage would consider "all" possible ways of accomplishing this, such as electric, pneumatic, and hydraulic motors of various kinds, different types of motion sensors, etc. To keep this example within the size and scope of this text, we decide not to explore the conceptual stage in detail but rather just present its *results*, a decision to further investigate a specific concept.

The design concept which we now wish to carry into the substantive stage of design is shown in Fig. 7-61. This scheme was actually used to position antiaircraft guns aboard naval vessels, and is discussed in greater detail in the literature.[20] Its principle could of course be considered for *any* rotary motion control system. While we can't here get into detailed performance specifications, one requirement was that there be no theoretical position error when the guns were tracking moving aircraft. While "moving aircraft" can move in many different ways, it has been found that tracking a constant angular velocity with no position error seems to result in adequate performance in real combat. This would be the situation if the enemy aircraft flew at constant speed in circles concentric with the gun location. While enemy pilots

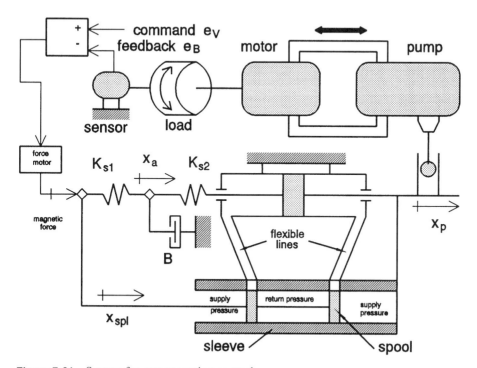

Figure 7-61 System for rotary motion control.

---

[20] E. O. Doebelin, *Control System Principles and Design*, Wiley, New York, 1985, pp. 464–468.

506   **Chapter 7**

can not be relied upon to be so accommodating, their *actual* maneuvers turn out to be surprisingly well modeled by our simplified criterion.

In Fig. 7-61 the use of a variable-displacement pump to drive a fixed-displacement rotary motor connected to the load is not a novel concept; it has been used in many applications for many years. Similarly, the use of a valve-controlled piston actuator to stroke the pump is routine practice. What *was* original in this application was a mechanical delay mechanism using two springs and a damper to implement a combination of positive and negative feedback, resulting in a stable system with zero position error for constant-velocity commands. This scheme was probably an "invention" rather than being mathematically founded, but, given the invention, we can of course analyze it mathematically. An electrical position sensor, summing amplifier to compare commanded and actual positions in terms of voltages, and electromagnetic force motor to stroke the servovalve are also conventional techniques.

Before starting a mathematical study, let's try to understand system operation in purely physical terms. Consider initially a simplified system in which the spring/damper system is *not* present, but the force motor has an internal spring. A command voltage e_v produces a proportional current at the output of the summing amplifier, which is the transconductance type. This current produces a magnetic force, which acts against the force motor's spring to produce valve-spool motion x_{spl}. This opens the valve to apply supply pressure to the actuator piston, causing it to move (x_p) to the right and thus stroke the pump. The pump, driven at constant speed by an electric motor, now starts pumping oil to the motor, which rotates the load toward the commanded position. Note that the valve sleeve is directly connected to the piston, so when x_p goes to the right, the sleeve motion tends to *close* the valve ports. This is a mechanical feedback which implements the relation $x_v = x_{spl} - x_p$, where x_v is the actual opening of the valve port. If the spool were disconnected from the force motor and given a step input $x_{spl,s}$ to the right, the piston would also move to the right, but *only* until x_p became equal to $x_{spl,s}$, which *closes* the valve and stops any further motion. We can thus think of spool motion as a command which produces an equal x_p motion, which is pump stroke. This relation between spool motion and pump stroke is desirable since we don't want a fixed error signal to result in an increasing pump stroke, which would soon reach its limit. The system just described, without the spring/damper system, is called a type I or single-integral control and would work, but would *not* give a zero error for a constant-velocity command.

By adding the spring/damper system and getting proper numerical values, the system of Fig. 7-61 becomes a type II or double-integral control, which we will see *does* gives the desired zero position error for a constant-velocity command. The SIMULINK simulation of Fig. 7-62 will be used to study this system, and it also serves as a block diagram representing the schematic diagram of Fig. 7-61. The valve/piston dynamics relating valve opening x_v and piston motion x_p neglect some dynamic effects which must be included in very fast systems but which we here ignore. The simple integrator dynamics *do* properly model the main effect; a step input of valve opening *does* result in a ramp output of piston displacement. Similarly for the pump/motor model; a step input of pump stroke *does* result in a ramp output of motor rotation angle, which is the behavior of an integrator.

First-Order Systems

While we don't want to get into too much practical detail, we might indicate where the "gains" of 40 and 50 for the two integrators come from. The valve/piston gain would have the units of (inches per second of piston velocity) per (inch of valve opening). This number could be estimated from theory, using formulas presented in Chapters 4 and 5 of this text. If the hardware is available, we could easily run a lab test to get a more accurate value. Since the pump, motor, and valve/piston are commercially available as a unit, let's assume that we purchase such a unit and measure these two gains in lab tests. To select such a unit from those available, we *do* need to do some rough "sizing" calculations. That is, how "big" a pump and motor do we require for our application?

The sizing of "power" components in dynamic systems is often done, not from dynamic performance specifications, but rather from basic steady-state requirements. These provide an initial choice of hardware, which is then analyzed for dynamic performance. If this analysis shows the first choice to be inadequate, we then make a new choice of hardware and re-analyze. For our motion-control system some useful steady-state requirements involve estimated maximum loads and speeds. Suppose that an analysis of system requirements showed that the maximum motor torque needed was about 50 ft-lb$_f$ and the maximum speed about 500 rpm. From Chapter 5 we know that motor torque = (motor displacement) (motor pressure drop). Available pump/motor units are always rated for maximum allowable pressure; let's assume this is 1000 psi. The needed motor displacement is then $(50)(12)/1000 = 0.60$ in^3/rad. To drive such a motor at 500 rpm, recall from Chapter 5 that motor flow rate = (motor displacement) (motor speed). Thus the needed flow rate is given by flow rate = $(0.60)(500)(6.28)/60 = 31.4$ in^3/sec. The pump must be able to supply such a flow rate when at maximum stroke.

Let's assume the pump will be driven by an ac induction motor running at about 1750 rpm when driving the pump. We know that such motors do not change speed much for quite a range of loads. The flow rate formula for a pump is the same as that for a motor, so the required pump displacement is given by $(31.4)/(1750 \times 0.105) = 0.17$ in^3/rad. These pump/motor calculations allow us to, at least tentatively, select a unit from those available and then get from the vendor estimates of the two gains (40 and 50) needed in our analysis. We can also quickly estimate the electric motor size. The motor torque and speed will be the same as those of the pump, so the motor's nominal horsepower rating will be about $(1750 \times 0.105)(1000 \times 0.17/12)/550 = 4.7$ hp.

Having given some explanation of where some of the numbers in Fig. 7-62 come from, we next need to analyze the spring/damper system to get the transfer functions used in this same diagram. From Fig. 7-61 we can write the following equations:

$$f_m - K_{s1}(x_{spl} - x_a) = M_{spl}\ddot{x}_{spl} = 0 \tag{7-193}$$

$$K_{s1}(x_{spl} - x_a) + K_{s2}(x_p - x_a) - BDx_a = M_a\ddot{x}_a = 0 \tag{7-194}$$

where f_m is the magnetic force and we have assumed all mass negligible. Since x_a is of no interest to us we eliminate it and finally get

$$(\tau D + 1)x_{spl} = K_x x_p + K_f(\tau_1 D + 1) \tag{7-195}$$

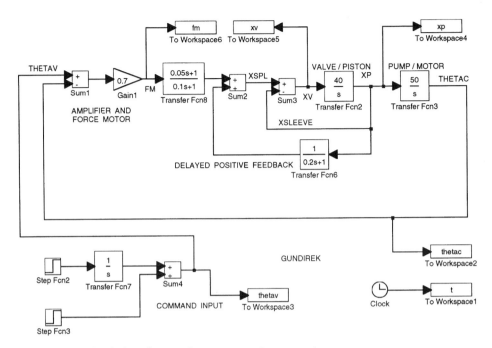

Figure 7-62 Simulation diagram for rotary motion-control system.

$$K_f \triangleq \frac{K_{s1} + K_{s2}}{K_{s1} K_{s2}} \quad \frac{\text{inch}}{\text{lb}_f} \qquad K_x \triangleq 1.0 \frac{\text{inch}}{\text{inch}} \qquad \tau \triangleq \frac{B}{K_{s2}} \quad \text{sec}$$
$$\tau_1 \triangleq \frac{B}{K_{s1} + K_{s2}} \quad \text{sec} \tag{7-196}$$

At this point we are able to draw Fig. 7-62, but some of the numbers used there would not yet be known. These are the delay time constant τ, the spring constant ratio K_{s1}/K_{s2}, and the amplifier/force-motor gain K_{afm}. We will use simulation, together with some dynamic system performance specifications to find acceptable values of these parameters. Since the model is entirely linear, we could also use the well-developed tools of analytical control system design to estimate these values, but even then we would still use simulation to refine the analytical design and get accurate response numbers. Since this text presumes no background in control theory, we proceed entirely by simulation.

If τ_1 is made much less than τ, the lead-lag term after FM in the simulation diagram becomes quite lagging, which is undesirable, so K_{s1} should not be made large relative to K_{s2}. Let's tentatively make the two spring constants equal and see whether that choice causes any trouble. Because gain K_f is "in series" with the amplifier/force-motor gain K_{afm}, we don't show it separately and lump it into that gain; adjusting this one gain has the same effect as adjusting either or both. As there are only two parameters to "play with" (τ and K_{afm}), and simulation results appear so quickly, even "blind guessing" quickly gets us to values that are at least stable. The performance specifications mentioned earlier require that, after a step input of 1 radian, the load position must "settle" to the correct value within 1 second. Also, for a constant-velocity (ramp) input, the position error must settle to zero within 1

First-Order Systems

second. For simulation we combine these two inputs, applying a step at time $= 0$ and an additional ramp at $t = 1.5$ seconds.

Figure 7-63 shows this command and the resulting very oscillatory response for $\tau = 0.05$ second; smaller τ's are even worse. Trying larger τ's we finally get, with $\tau = 0.20$ sec, the response of Fig. 7-64, which meets the specifications (all numbers are as in Fig. 7-62). From this graph the "error" appears visually to go to zero for both the step and the ramp, but one can't really see numerical values. We can easily plot the error itself, which will then appear graphically to a scale large enough to get numbers. Further investigation shows that τ's larger than 0.20 sec will also work, as will a range of amplifier gains. This is not unusual; most design problems don't have a single unique solution. The error plot of Fig. 7-65 shows that very large τ values do *not* meet specifications.

The analysis just completed is typical of the substantive design stage. We were able to get numerical values for some important system parameters even though we used rather gross descriptions of some of the components. We can now proceed to the detail design stage, say for the spring and damper used to realize the time constant τ, for which we now have a numerical value, 0.20 second. There are of course an *infinite* number of combinations of B and K_{s2} which will give a τ of 0.20 second, so some more information is needed to narrow this choice. The force motor can only produce a certain maximum force, and this force should be able to open wide the servo valve. Suppose the valve that "comes with" the hydraulics package we have selected has a full stroke of about 0.10 inch and the force motor can produce at most 1 pound of force. In Fig. 7-61 a step input of force causes a sudden spool

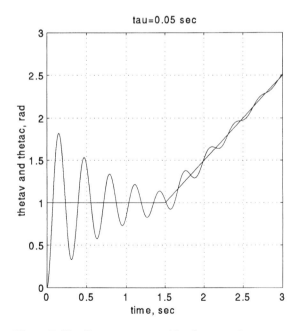

Figure 7-63 Response to combined step and ramp commands: too small time constant gives excessive oscillation.

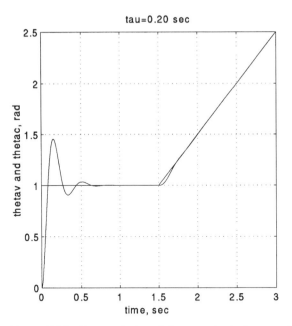

Figure 7-64 Proper choice of time constant gives good response.

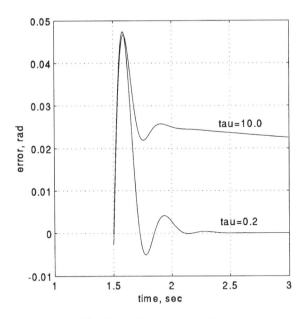

Figure 7-65 Too-large time constant slows return to zero error.

motion given by force/K_{s1}, so K_{s1} (and thus K_{s2}) should be about 10 lb$_f$/in. There is then only one B value, 2.0 lb$_f$/(in/sec), that will give the desired τ.

In the referenced system, which was actually built, the spring K_{s2} was implemented as a cantilever beam, and the damper was made according to the design of

First-Order Systems

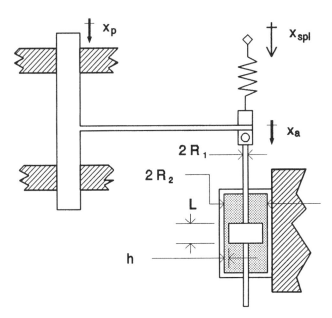

Figure 7-66 Construction details of spring and damper.

Fig. 2-22, with an arrangement as in Fig. 7-66. Available space dictated that the beam be no longer than 2 inches, and that the damper occupy no more space than a 1-inch cube. These constraints allow us to design a specific spring and damper as follows. We will try to use a spring grade of stainless steel for our beam, to protect against corrosion. A materials handbook provides a modulus of elasticity E of about 30×10^6 psi and a yield point stress of about 100,000 psi, for cold-worked material in thin sections. Figure 2-12 gives us the needed spring constant formula:

$$K_{s2} = \frac{Ebh^3}{4L^3} \qquad (7\text{-}197)$$

Let's use all of the available space, making $L = 2.0$ inch. There are then an infinite number of combinations of beam width b and thickness h that will give us the desired spring constant. To narrow this choice we next consider the *stress* tending to cause mechanical failure. If you have not had a course in strength of materials, you will have to accept on faith the fact that the largest stress in a beam of this type is that at the "built-in" end of the beam, caused by bending. For the maximum load of 1 pound, this stress is given by

$$\text{Stress} = \frac{12}{bh^2} \qquad (7\text{-}198)$$

If we now decide to use a stress safety factor of 2.5, making the working stress 40,000 psi, we can solve for b and h, giving the values $b = 0.237$ in. and $h = 0.0356$ in.

In the damper design formula, Eq. (2-34), we again start by deciding to use all of the alloted space. Most dampers use silicone damping fluids specially formulated for that purpose. These are available in a very wide range of viscosities, as can be seen from the charts in Appendix A. Since B is directly proportional to viscosity, our design approach will be to estimate the needed dimensions, solve for the required viscosity, and then check the charts to see whether such a fluid is commercially available. To fit within the 1-inch cube of space available, let's choose $R_2 = 0.30$ in., $R_1 = 00.5$ in., $L = 0.30$ in., and $h = 0.01$ in. These choices leave some room for cylinder wall thickness and make the piston rod large enough to prevent buckling.

Very small values of clearance h present several practical problems. First, the manufacturing tolerances on the parts get difficult; if h were intended to be 0.001 and the shop "misses" it by 0.001, there is a 100% error in h. Temperature changes could also cause thermal expansion, again causing large percentage changes in h. Tiny dirt particles could cause jamming. We don't however want to use excessively large h values since this leads to large viscosity. Some of the very high viscosity fluids are more like grease, making it hard to properly fill the cylinder. The chosen h of 0.01 is a tentative compromise of these various considerations. It might be necessary to build and test some dampers to resolve some of these questions.

Using the dimensions chosen above, we find that the viscosity needed for $B = 2.0 \, lb_f/(in/sec)$ is $1.46 \times 10^{-5} \, lb_f$-sec/in^2. This value is not near either extreme of the viscosity charts, so such a fluid is definitely available. It is also clear that we could use somewhat larger h values, if that became necessary, without requiring excessively viscous fluid.

This completes our little design exercise. Though it has not come to grips with all the complexities of industrial practice I hope it is helpful in putting into some perspective the role of system dynamics in the overall scheme of system design. That is, system dynamics is able to develop numerical values which meet system performance requirements for "composite" parameters such as gains and time constants. These composite parameters always depend on several (or many) specific physical parameters, such as spring constants, resistances, damping coefficients, etc. Detail design attempts to realize the needed values of the composite parameters, using standard techniques developed in component design courses. Sometimes this will not be possible. Just because system dynamics says that we must have a time constant of 0.0001 second in a certain component does not mean that such a component can actually be constructed.

One important feature which we are not able here to emphasize properly is the integration of product design with the "design" of the manufacturing process which actually produces the parts. Manufacturing engineers must be part of the design team *right from the start*. If manufacturing questions are "put off" till a later time, product design engineers will not make the best decisions about which alternative designs to further develop. The parallel development of product design and process design is called *concurrent engineering* and has led to improved quality and profitability for companies which adopt this practice over the earlier method of doing product design *first*, and only then considering manufacturing.

BIBLIOGRAPHY

All the bibliography entries for Chapter 1 have material pertinent to Chapter 7.

Andersen, B. W., *The Analysis and Design of Pneumatic Systems*, Wiley, New York, 1967.
de Silva, C. W., *Control Sensors and Actuators*, Prentice Hall, Englewood Cliffs, N.J., 1989.

First-Order Systems   **513**

Doebelin, E. O., *System Modeling and Response: Theoretical and Experimental Approaches*,
    Wiley, New York, 1980.
Doebelin, E. O., *Control System Principles and Design*, Wiley, New York, 1985.
Doebelin, E. O., *Measurement Systems*, 4th ed., McGraw-Hill, New York, 1990.
Gibson, J. E., and F. B. Tuteur, *Control System Components*, McGraw-Hill, New York, 1958.
Merritt, H. E., *Hydraulic Control Systems*, Wiley, New York, 1967.
Mills, A. F., *Basic Heat and Mass Transfer*, Irwin, Chicago, 1996.
Nachtigal, C. L., ed., *Instrumentation and Control*, Wiley, New York, 1990.
Rizzoni, G., *Principles and Applications of Electrical Engineering*, 2nd ed., Irwin, Chicago,
    1996.
Watton, J., *Fluid Power Systems*, Prentice Hall, New York, 1989.

PROBLEMS

7-1.   Define mathematically the generic linear, constant-coefficient first-order system and then show its three special cases.

7-2.   Why is just a schematic sketch of a physical system insufficient to decide its order? Use examples.

7-3.   For the translational systems of Fig. 7-1b:
   a. Derive the system differential equation relating the indicated output quantity to the indicated input quantity.
   b. Put the equation in standard form and define the standard parameters. Display operational, sinusoidal, and Laplace transfer functions and show an overall block diagram.
   c. Using SIMULINK or other available simulation, draw a simulation diagram which uses the *physical* parameters and does *not* use the transfer function icon. Also show a simulation diagram which uses the standard parameters and the transfer function icon.

7-4.   Repeat problem 7-3 for the rotational system of Fig. 7-1b.

7-5.   Repeat problem 7-3 for the translational system of Fig. 7-1c.

7-6.   Repeat problem 7-3 for the rotational system of Fig. 7-1c.

7-7.   Repeat problem 7-3 for the translational system of Fig. 7-1d.

7-8.   Repeat problem 7-3 for the rotational system of Fig. 7-1d.

7-9.   Repeat problem 7-3 for the translational system of Fig. 7-1e.

7-10.   Repeat problem 7-3 for the rotational system of Fig. 7-1e.

7-11.   Sketch freehand, but carefully on graph paper, the step response of first-order systems with time constants of 0.5, 2.0, and 5.0 seconds (all on one set of axes). What is the initial rate of change of the output in each case?

7-12.   Sketch freehand, but neatly, the logarithmic frequency-response curves for first-order systems with:
   a. $K = 1.0$, $\tau = 2.5\,\text{min}$ c. $K = 0.2$, $\tau = 3.5\,\text{hr}$
   b. $K = 3.4$, $\tau = 0.05\,\text{sec}$ d. $K = 0.01$, $\tau = 0.001\,\text{sec}$

7-13.   In the combined system of Fig. 7-1a take:
       $M = 1.0\,\text{slug}$ $B = 20.0\,\text{lb}_f/(\text{ft-sec})$

$$J = 0.1 \text{ lb}_f\text{-ft-sec}^2 \quad T_i = 10.0 \text{ ft-lb}_f \text{ step}$$
$$R = 0.10 \text{ ft}$$

a. In steady state, what is the velocity of M and the angular velocity of J?
b. How long does it take to reach 95% of steady-state speed?
c. It is suggested that M be lightened to increase response speed (decrease the time to reach steady state). What is the maximum improvement to be achieved in this way? If this is not enough, suggest other changes that would help (the steady-state speed must remain the same). Give specific numerical values which would result in a doubling of response speed.
d. Using the numbers originally given, but with input torque as in Fig. P7-1, draw a SIMULINK simulation diagram to compute system response.
e. Using whatever simulation software is available, run the simulation of part (d) and find the time to reach 95% of steady-state speed.

7-14. In the translational system of Fig. 7-1b the spring is nonlinear with force $= 25x_o + x_o^2 \text{ lb}_f$, x_o in inches, while the linear damper has $B = 25 \text{ lb}_f/(\text{in/sec})$.
  a. Find the linearized system time constant for small motions near $x_o = 0$ and also near $x_o = 5.0$ inches.
  b. Draw a SIMULINK diagram to study the response of both the linearized and nonlinear systems in a single run. The force has been steady at -10 lb before $t = 0$ and jumps to $+10$ lb at $t = 0$.
  c. Repeat part (b) for a force going from 200 to 300 lb.
  d. Using available simulation software, run the study of part (b) and compare exact and linearized results.
  e. Repeat part (d) for part (c).

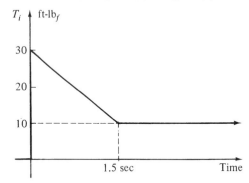

Figure P7-1

7-15. Explain clearly *in words* how the systems of Fig. 7-1e respond to step inputs.

7-16. List the three types of noncontact forces that might appear in Newton's law. Discuss the choice of sign conventions for forces and motion quantities. What assumption is usually made about the coordinate origin in "earth-bound" mechanics problems?

7-17. Obtain Eq. (7-20) using the D operator, rather than Laplace transform, method.

7-18. We often use 3τ as the 95% settling time. What if we instead wanted a 99% settling time? How many time constants must we wait to satisfy this criterion?

First-Order Systems   **515**

7-19.   Under what conditions may we treat a piece of hardware as a zero-order system?

7-20.   Show that the op-amp circuit of Fig. 7-13b has the transfer function needed in Fig. 7-13a. Choose R and C values to actually get the numbers used in Fig. 7-13a.

7-21.   The cancellation compensation used in Fig. 7-13a can never be perfect, because the numerator time constant of the compensator can never *exactly* match the denominator time constant of the device being compensated. They are both physical devices whose parameters have uncertainty and drift due to environmental effects, such as temperature. Duplicate the simulation of Fig. 7-13a, but then study the effect of imperfect cancellation by letting the compensator numerator be $1.8s + 1$ (10% low) and then $2.2s + 1$ (10% high), instead of $2s + 1$. See what effect this has on system response and discuss your results.

7-22.   The system of Fig. 7-13a meets specifications but assumes totally linear operation without any of the nonlinear effects associated with real hardware. One such is the current limiting used in real amplifiers, which limits the maximum torque available from the motor. Show how you can use this simulation diagram to find the maximum torque existing for the motion cycle of Fig. 7-13b. Use all numbers as in Fig. 7-13a. Assume that in a real system the torque is limited to 50% of the peak torque you found. Add a torque limiter at this value to your simulation and re-run it. Compare system response with that of Fig. 7-13b.

7-23.   Derive Eq. (7-35).

7-24.   Obtain the result of Eq. (7-43) using Laplace transform methods.

7-25.   Draw a SIMULINK diagram which could be used to get the results of Fig. 7-19. Using available simulation software, get these results.

7-26.   Use the MATLAB `bode(num,den)` command to get logarithmic frequency-response graphs for the systems of problem 7-12.

7-27.   Curve-fitting software, such as the MATLAB INVFREQS explained for Eq. (7-51), can be tested by supplying it with "manufactured" data, for which we know the correct answer. Make up a perfect first-order system with a known K and τ of your choice, compute sinusoidal transfer function values at 15 selected points, and then "contaminate" this data by adding small random values to each point. Supply this data to INVFREQS and see what K and τ values it suggests. Repeat this experiment with larger random values. Discuss your results.

7-28.   Do a simulation for the situation of Fig. 7-26. Then let fi have several different waveforms of your choice, but all must have an area of 0.01 and must be zero after $t = 0.01$. Show that the shape of the waveform is immaterial, only its area matters.

7-29.   Design a passive first-order low-pass filter to attenuate noise at 400 Hz by 20-to-1. The desired signals have frequencies in the range 0 to 10 Hz. Will these signals be badly distorted by the filter? Do the needed calculations to justify your answer. How would you decide on a specific R and C for your filter?

7-30.   For the circuit of Fig. 7-23c:
a.   Derive the system differential equation relating the indicated output quantity to the indicated input quantity.

b. Put the system equation in standard form and define the standard parameters. Display operational, sinusoidal, and Laplace transfer functions and show an overall block diagram.

c. Using SIMULINK or other available simulation, draw a simulation diagram which uses the *physical* parameters and does *not* use the transfer function icon. Also show a simulation diagram which uses the standard parameters and the transfer function icon.

7-31. Repeat problem 7-30 for the circuit of Fig. 7-23d.

7-32. Repeat problem 7-30 for the circuit of Fig. 7-23e.

7-33. Repeat problem 7-30 for the circuit of Fig. 7-23f.

7-34. Repeat problem 7-30 for the circuit of Fig. 7-23h.

7-35. Design a passive approximate integrator which will be 95% accurate for frequencies higher than 10 Hz. How would you decide on actual R and C values?

7-36. A photodiode as in Fig. 7-36 is used to transduce optical signals in a fiber optic system into electrical signals. The highest frequency optical signal is at 100 kHz and we must reproduce it with 95% amplitude accuracy. The device receiving e_o has characteristics which require that $R_{fb} = 1000$ ohms. Find the value of C_{fb} needed to meet these requirements.

7-37. Get expressions for the operational impedance $Z(D)$ and sinusoidal impedance $Z(i\omega)$ for the circuits of:

    a. Fig. P7-2   c. Fig. P7-4
    b. Fig. P7-3   d. Fig. P7-5

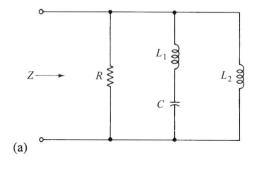
(a)

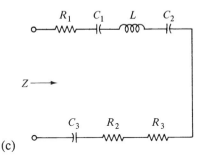
(c)

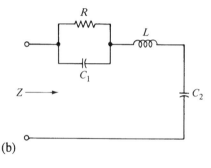
(b)

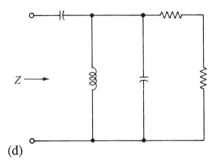
(d)

Figure P7-2 to 5

First-Order Systems   **517**

Use the impedance to get the differential equation relating the terminal current to the terminal voltage.

7-38.   Using the known behavior of L and C at very high and very low frequencies, show simplified versions of circuits:

    a.   Figure P7-2 at very low frequencies
    b.   Figure P7-2 at very high frequencies
    c.   Figure P7-3 at very low frequencies
    d.   Figure P7-3 at very high frequencies
    e.   Figure P7-4 at very low frequencies
    f.   Figure P7-4 at very high frequencies
    g.   Figure P7-5 at very low frequencies
    h.   Figure P7-5 at very high frequencies

7-39.   Take all $R = 100$ ohms, all $C = 1\,\mu F$, all $L = 0.01\,H$ and apply a 60-Hz sinusoidal voltage of 110 volts *effective* value to the circuit terminals. Find the current, power factor angle, power factor, and average power for the circuits of:

    a.   Fig. P7-2   c.   Fig. P7-4
    b.   Fig. P7-3   d.   Fig. P7-5

7-40.   The circuit of Fig. 7-28c uses the varistor of Fig. 3-2c as the resistance element and has $L = 1.0\,H$. Using SIMULINK or other available simulation find the current if:

    a.   Voltage e_i is a step of $+10$ volts from zero.
    b.   Voltage e_i is steady at $+10$ volts and then drops in step fashion to zero.
    c.   Voltage e_i is steady at $+8$ volts and then jumps to $+10$ volts. Also do a linearized analytical solution and compare with the exact result.
    d.   Voltage $e_i = 9 + 1 \sin 1000\,t$ volts, t in seconds. Also do a linearized analytical solution and compare with the exact result.

7-41.   Show a simulation diagram for computing the RMS value of *any* waveform. Most ac voltmeters do *not* use this exact method, but rather the approximate method discussed just after Eq. (7-107). For a sine wave, the average of the absolute value can be shown to be 0.637 times the peak value. Using this fact, show a simulation diagram for an ac meter of this approximate type. Use a first-order low-pass filter to do the needed averaging, since this is how it is done in commercial meters. Run both simulations for perfect sine wave input and comment on the results. Then make up several periodic waveforms of different shapes, and compare the results from the two meters.

7-42.   Explain what is meant by a perturbation analysis and why we use this method.

7-43.   In the circuit of Fig. 7-28f take $L = 0.01$ Henry, $R_1 = 1000$ ohms, and $R_2 = 500$ ohms. If i_i is a rectangular pulse of duration T, what is the largest T for which the input may be approximated as an impulse?

7-44.   For the fluid system of Fig. 7-39b:

    a.   Derive the system differential equation relating the indicated output quantity to the indicated input quantity, treating flow resistances as linear. Display the transfer function.

518   **Chapter 7**

    b. Taking the flow resistances as nonlinear (orifices), do a perturbation analysis for small changes about an equilibrium operating point to get a linearized system equation. Give the transfer function for the input and output perturbations.

    c. Show a SIMULINK simulation diagram for the case where the flow resistances are nonlinear (orifices).

7-45. Repeat problem 7-44 for the system of Fig. 7-39c.

7-46. For the system of Fig. 7-39d:

    a. Assume the gas is at constant temperature. Note that the pressure at the bottom of the tank is caused by *both* h_o and gas pressure. Compute the *total* compliance of this tank and then linearize it.

    b. Assume a linear flow resistance and, using the linearized compliance from part (a), derive the system differential equation relating h_o to q_i. Put it in standard form and define the standard parameters.

    c. Show a SIMULINK simulation diagram for this system, using the nonlinear relations for both the compliance and the resistance (orifice).

7-47. For the system of Fig. 7-39e:

    a. Assume linear resistance and compliance and derive the system differential equation.

    b. If the tube is 0.08 inch in diameter and 5 inches long and the liquid has viscosity $0.005\,lb_f\text{-sec}/ft^2$, what is the largest tolerable compliance if the time constant cannot exceed 0.05 second?

    c. Estimate how large $(p_i - p_o)$ could get before turbulent flow occurred. Fluid weighs $60\,lb_f/ft^3$.

7-48. For the system of Fig. 7-39f:

    a. Using the linearized compliance for an isothermal process and assuming a linear resistance, derive the differential equation relating p_o to p_i. How do the results change if an adiabatic process is assumed?

    b. Show a SIMULINK simulation diagram for this system, using a nonlinear isothermal compliance.

    c. Repeat part (b) using a nonlinear adiabatic compliance.

7-49. For the system of Fig. 7-39h:

    a. Assuming a linear flow resistance, derive the differential equation relating q_o to p_i.

    b. After a step input pressure is applied, how long does it take to achieve steady flow? For water with viscosity $2 \times 10^{-5}\,lb_f\text{-sec}/ft^2$, specific weight $62.4\,lb_f/ft^3$, flowing in a 0.1-ft diameter pipe 10 feet long, how long does it take? What is the largest step pressure for which laminar flow would exist? How would the analysis change for pressures larger than this?

7-50. Working directly from the block diagram of Fig. 7-47, set up a block diagram which has the heat loss $q_L = UA(T_o - T_i)$ as an output signal. From this diagram get the transfer function $(q_L/q_i)(s)$ and $(q_L/T_i)(s)$. What would the response of q_L to a step input of T_i look like?

7-51. Using simulation, verify the graphs of Fig. 7-46.

First-Order Systems

7-52. Carry out the calculus average value calculation mentioned just after Eq. (7-135).

7-53. For the system of Fig. 7-44b, suppose the thermal effect of the stirring device is *not* negligible and we have to find a transfer function relating T_o to stirrer torque. How would you model the stirrer if you wanted to make this new transfer function first order? Make the needed assumption and derive the requested transfer function. If q_i were known, how would you decide whether the stirrer effect could be neglected?

7-54. In the combined system of Fig. 7-1a let a force f_i be applied directly to M (T_i is still present). Get differential equations, block diagrams, and transfer functions showing how x_o is produced by f_i and T_i.

7-55. In the system of Fig. 7-1c, if x_i is a step input, find the force which must be provided by the motion source.

7-56. In the circuit of Fig. 7-28f, if i_i is a step input, find the voltage across the current source. Get an expression for the power taken from the source.

7-57. For the circuit of Fig. 7-28b get the transfer function relating capacitor current to i_i.

7-58. In the system of Fig. 7-1e, let a torque T_i act directly on J (θ_i is still present). Get differential equations, transfer functions, and block diagram showing how θ_o is produced by θ_i and T_i.

7-59. In the system of Fig. 7-1c, if the input is sinusoidal, what is the average power drawn from the motion source in steady state? What is the torque?

7-60. In the system of Fig. 7-39b, add an inflow q_i to the top of the tank. Assume linear flow resistances and find differential equations, transfer functions, and block diagram showing how h_o is produced by p_i and q_i.

7-61. In the system of Fig. 7-48, get differential equations, transfer functions, and block diagram showing how i_a is produced by e_i and T_i.

7-62. For the circuit of Fig. P7-6 get differential equation, transfer function, step response, and frequency response.

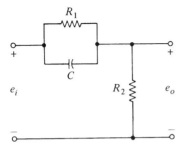

Figure P7-6

7-63. For the system of Fig. P7-7 get differential equation, transfer function, step response, and frequency response.

7-64. Verify Eqs. (7-158) to (7-161).

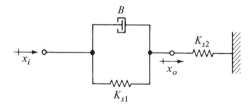

Figure P7-7

7-65. Modify the system of Fig. 7-51 to implement *closed-loop* speed control. Model the added components so that the closed-loop system is first-order, using Fig. 7-49 for guidance. Get the closed-loop system differential equation and put it in standard form.

7-66. Get transfer functions for the systems of the following figures, put them in standard form and define the standard parameters. The x's are displacements.

    a. P7-8a   b. P7-8b   c. P7-8c   d. P7-8d   e. P7-8e
    f. P7-8f   g. P7-8g   h. P7-8h   i. P7-8i   j. P7-8j
    k. P7-8k

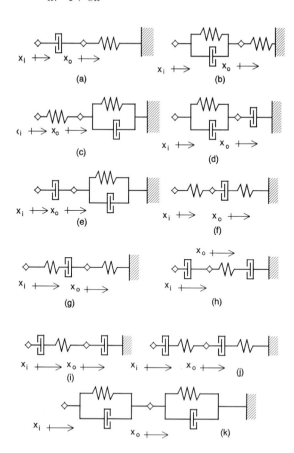

Figure P7-8

8

SECOND-ORDER SYSTEMS AND MECHANICAL VIBRATION FUNDAMENTALS

8-1 INTRODUCTION

As we pointed out at the beginning of Chapter 7, because only two types of roots (real and complex) can arise in the characteristic equations of linear systems of arbitrarily high order, and since real roots exhibit behavior characteristic of first-order systems while complex root pairs behave like second-order systems, it is useful and efficient to become intimately familiar with the responses of these two classes of systems. That is, there are many practically important devices and processes that are well modeled by one of these two classes, and the more complicated systems can be considered as combinations of several first and/or second-order forms.

At this point, having completed Chapter 7, you should be very familiar with the behavior of the three types of first-order systems discussed there. None of these types is capable of *free oscillation* or *forced resonance*, features of extreme importance regularly observed in physical systems, both as useful phenomena to be exploited in design or as potentially dangerous problems. The underdamped second-order system is the simplest form which exhibits such behavior. As we did with first-order systems, we will study the generic form, the four basic physical types, and various mixed-media examples. Now, however, the mechanical form takes on a special practical significance. If you were to take a beginning course (or pick up an introductory text) on the subject of *mechanical vibration*, the first few chapters of such a book would be devoted entirely to a study of a simple second-order system. In fact, all the basic phenomena of even quite complicated vibrating systems are exhibited in this simple system.

We have earlier stated that we consider system dynamics, in addition to its intrinsic value, as a common basic foundation for later specialized courses (required or elective) on topics such as measurement, control, vibration, acoustics, electromechanics, etc. In the case of vibration, we go even further. Our treatment in this chapter will provide sufficient breadth and depth to serve as the *only* vibration coverage needed by nonspecialists. Clearly, many mechanical engineers will want

522   **Chapter 8**

to pursue this topic further, with additional reading and/or courses. Our coverage in this chapter will then allow such study to build on an established foundation and thus pursue greater breadth and depth than would otherwise be possible.

8-2   SECOND-ORDER SYSTEMS FORMED FROM CASCADED FIRST-ORDER SYSTEMS

Because many practical machines and processes are formed by joining available components or subsystems, we want to first look at how second-order systems can arise in this way. This will also allow us to develop the important system concept of *loading effect*.

The most general form of second-order system encountered in practice has an equation of the form

$$a_2 \frac{d^2 q_o}{dt^2} + a_1 \frac{dq_o}{dt} + a_0 q_o = b_2 \frac{d^2 q_i}{dt^2} + b_1 \frac{dq_i}{dt} + b_0 q_i \tag{8-1}$$

however the most common and important special case, which we will emphasize, is given by

$$a_2 \frac{d^2 q_o}{dt^2} + a_1 \frac{dq_o}{dt} + a_0 q_o = b_0 q_i \tag{8-2}$$

Just as in first-order systems, a widely accepted *standard form* of (8-2) has been defined and should generally be employed. As in first-order systems we choose to divide by a_0 to get

$$\frac{a_2}{a_0} \frac{d^2 q_o}{dt^2} + \frac{a_1}{a_0} \frac{dq_o}{dt} + q_o = \frac{b_0}{a_0} q_i \tag{8-3}$$

$$\left(\frac{D^2}{\omega_n^2} + \frac{2\zeta D}{\omega_n} + 1 \right) q_o = K q_i \tag{8-4}$$

where

$$\omega_n \triangleq \sqrt{\frac{a_0}{a_2}} \triangleq \text{undamped natural frequency} \quad \frac{\text{rad}}{\text{time}} \tag{8-5}$$

$$\zeta \triangleq \frac{a_1}{2\sqrt{a_2 a_0}} \triangleq \text{damping ratio, dimensionless} \tag{8-6}$$

$$K \triangleq \frac{b_0}{a_0} \triangleq \text{system steady-state gain (sensitivity)} \tag{8-7}$$

Remember that K always has the dimensions of the output quantity divided by the input quantity. The various transfer functions are now immediately available, for example,

$$\frac{q_o}{q_i}(s) = \frac{K}{\dfrac{s^2}{\omega_n^2} + \dfrac{2\zeta s}{\omega_n} + 1} \tag{8-8}$$

Second-Order Systems and Mechanical Vibration

Whenever a new physical form of second-order system is first encountered, you should *immediately* define K, ζ, and ω_n, and change over to the standard form to gain all the benefits of standardization, as we have seen with first-order systems. The significance of the standard parameters will be developed shortly.

Cascaded Subsystems: The Loading Effect. Second-order system models may arise naturally when a complete system is first physically analyzed as an entity. They also arise when two *components* of a system, each individually modeled as first-order, are connected in *cascade* to form a larger overall system. By cascade connection we mean that the output of the first component is connected as the input of the second, as in Fig. 8-1. Proceeding by formal mathematics, we are tempted to write

$$(\tau_2 D + 1)q_{o2} = K_2 q_{i2} = K_2 \frac{K_1}{(\tau_1 D + 1)} q_{i1} \qquad (8\text{-}9)$$

and thus

$$(\tau_2 D + 1)(\tau_1 D + 1)q_{o2} = [\tau_1 \tau_2 D^2 + (\tau_1 + \tau_2)D + 1]q_{o2} = K_1 K_2 q_{i1} \qquad (8\text{-}10)$$

which we see fits the second-order form. Unfortunately this result, and the general method used, may be quite wrong.

The defect in this analysis lies in the fact that, when the two first-order models are analyzed as *isolated devices*, certain physical assumptions are normally made which are, at least partially, *violated* when the two systems are connected. This problem exists for *all* situations where separately analyzed subsystems are connected, not just our first-order example. The violation consists of the second system withdrawing from the first some *power* which was not accounted for in the model of the first system. Thus the transfer function $(q_{o1}/q_{i1})(D)$ is *different* when the second system is connected, and the overall transfer function $(q_{o2}/q_{i1})(D)$ is *not* just a simple product of the two isolated transfer functions. This phenomenon is called a *loading effect*;[1] in some cases it is slight enough to neglect, while in others it is so large that we must reanalyze the *complete* system "from scratch" rather than trying to employ models derived for the isolated subsystems. The electrical example of Fig. 8-2 should help you understand the general situation. There, circuit 1 was analyzed as an iso-

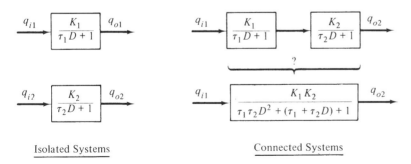

Figure 8-1 Second-order systems formed by cascading first-order systems.

---

[1] E. O. Doebelin, *System Modeling and Response*, Wiley, New York, 1980, chap. 7, Subsystem Coupling Methods; *Measurement Systems*, 4th ed., McGraw-Hill, New York, 1990, pp. 74–88.

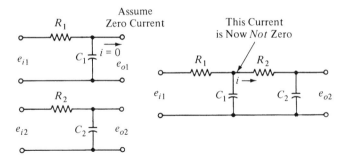

Figure 8-2 Electrical example of the loading effect.

lated device. That is, with nothing connected at the output, it is reasonable to model the output condition as an open circuit, with no current flowing there. This assumption leads to a simple first-order model for this circuit. When we connect circuit 2's input as the output of circuit 1, we violate the no-current assumption, which invalidates the relation we had derived relating e_{o1} to e_{i1}.

The references also show some methods for describing subsystems more completely than just giving a transfer function. It turns out that a *complete* description, capable of accounting for any loading effects, requires the specification of *three* relationships, not just the one transfer function. In one such method, you must obtain for each subsystem:

1. The unloaded transfer function
2. The subsystem input impedance
3. The subsystem output impedance

If these three quantities can be theoretically calculated or measured in the lab, we can then compute overall transfer functions for coupled subsystems without getting any errors due to the loading effect. These subsystem coupling methods are actually quite useful in engineering practice since we often want to predict overall system behavior from analysis or measurements on the isolated subsystems.

For example, jet engines are manufactured in, say, Cincinnati, Ohio, while the aircraft is made in Seattle, Washington. When the engine and aircraft are joined at final assembly, we don't want any unforeseen vibration problems to crop up at that late date. Thus, analysis or measurements such as the three listed above would be made by the engine maker on the engine and by the aircraft manufacturer on the airframe. The overall system would then be "assembled" *analytically* by combining the two sets of data, giving a prediction of how the overall system woudl behave, but *without* the need to physically join the subsystems. If this study revealed an overall vibration problem, the analysis methods also show how one or both of these subsystems needs to be changed so that the overall behavior is acceptable. This allows correction of design defects much earlier in the product cycle, when such changes are more easily and cheaply made.

These advanced coupling methods are beyond the scope of this text, but we wanted to at least make you aware of them. In this text we will just analyze the complete system from scratch and compare these correct results with the approximations obtained by simple multiplication of the isolated transfer functions.

EXAMPLE: LOADING EFFECT IN TWO MECHANICAL FIRST-ORDER SYSTEMS

To give the details of a concrete example, consider the system of Fig. 8-3, which is composed of two subsystems from Fig. 7-1. Considered individually, these two subsystems would each be simple first-order systems. Recall that the first system might represent a machine tool slide being positioned by a motor providing the force f_i; the second system could represent a velocity-measuring device which we wish to attach to the slide to measure its speed. If we simply multiply the two isolated first-order transfer functions we get for the total system

$$\frac{x_{o2}}{f_{i1}}(D) = \frac{B_2/(B_1 K_{s2})}{\frac{M_1 B_2}{K_{s2} B_1} D^2 + \frac{M_1 K_{s2} + B_1 B_2}{K_{s2} B_1} D + 1} \tag{8-11}$$

which will be a good approximation only if the loading effect is small. To discover the nature of the loading effect we analyze the connected system "from scratch" without using the isolated first-order models at all. We can write two Newton's law equations:

$$f_{i1} - B_1 \dot{x}_{o1} - B_2(\dot{x}_{o1} - \dot{x}_{o2}) = M_1 \ddot{x}_{o1} \tag{8-12}$$

$$B_2(\dot{x}_{o1} - \dot{x}_{o2}) = K_{s2} x_{o2} \tag{8-13}$$

Note in Eq. (8-12) the presence of the force in damper 2; this force was *not* present when system 1 was analyzed as an isolated device.

Since we are interested in $(x_{o2}/f_{i1})(D)$ we eliminate x_{o1} by substituting

$$\dot{x}_{o1} = \left(\frac{B_2 D + K_{s2}}{B_2}\right) x_{o2} \tag{8-14}$$

from (8-13) into (8-12) to get

$$f_{i1} - \frac{B_1}{B_2}(B_2 D + K_{s2}) x_{o2} - B_2 \left(\frac{B_2 D + K_{s2}}{B_2 D} - 1\right) \dot{x}_{o2} = M_1 \left(\frac{B_2 D + K_{s2}}{B_2}\right) \dot{x}_{o2} \tag{8-15}$$

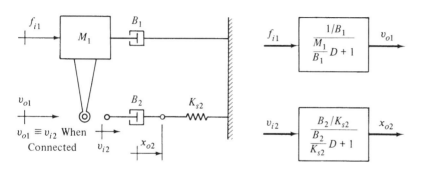

Figure 8-3 Mechanical example of the loading effect.

This leads to

$$\frac{x_{o2}}{f_{i1}}(D) = \frac{B_2/(K_{s2}(B_1+B_2))}{\dfrac{M_1 B_2}{K_{s2}(B_1+B_2)} D^2 + \dfrac{M_1 K_{s2}+B_1 B_2}{K_{s2}(B_1+B_2)} D + 1} \tag{8-16}$$

which may be compared directly with its approximation, Eq. (8-11). We see that if B_2 is small *compared to* B_1, then (8-11) becomes a good approximation to (8-16). For example, if all parameters are 1.0 except $B_2 = 0.05$ we get

$$\text{Assuming no loading:}\quad \frac{x_{o2}}{f_{i1}}(D) = \frac{0.0500}{0.0500 D^2 + 1.05 D + 1}$$

$$\text{Exact analysis:}\quad \frac{x_{o2}}{f_{i1}}(D) = \frac{0.0476}{0.0476 D^2 + 1.00 D + 1}$$

Figure 8-4 compares responses of these two models for a random input. We see that here ignoring the loading effect does not cause significant error. This would *not* be true if B_2 were large relative to B_1. We should also point out that when the loading effect is negligible, the "internal" variable (v_{o1} in our example) may be calculated with good accuracy from the "isolated" transfer function $(v_{o1}/f_{i1})(D) = (1/B_1)/((M_1/B_1)D + 1)$. That is, the output of the first system is negligibly affected by the presence of the second.

Because system design often consists of assembling a new arrangement of existing components or subsystems, the loading question frequently arises. To save time and effort, we prefer whenever feasible to use already-derived transfer functions

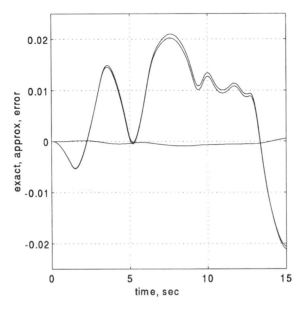

Figure 8-4 Simulation results for numerical values where loading effect is small.

Second-Order Systems and Mechanical Vibration **527**

that might be available from vendors who sell us components. Sometimes we can insert so-called *buffer amplifiers* between two components and thereby allow multiplication of the isolated transfer functions. This is quite common with electrical devices. The buffer amplifier has high input impedance, low output impedance, and its own power supply. It can thus drive the second component without stealing much power from the first. Buffer amplifiers are available in technologies other than electrical, the pneumatic version being in wide use.[2] You should always be on the lookout for potential loading situations and satisfy yourself that the effect can be neglected, controlled with buffer amplifiers, or treated with one of the subsystem coupling techniques referenced earlier. Otherwise it may be necessary to analyze the complete system as an entity, rather than trying to use existing component descriptions.

8-3 MECHANICAL SECOND-ORDER SYSTEMS

We have just seen how second-order systems can arise from coupling of first-order systems. They also commonly arise in their own right when a complete physical system is not obviously "manufactured" from first-order "components." We can generate a number of practically useful systems from the first-order examples of Fig. 7-1 by deciding to develop "more accurate" models. For example, in Fig. 7-1b, the assumption of negligible inertia would be accurate only for certain operating conditions; suppose we wish to find criteria for judging rationally when we are allowed this simplification. By comparing responses with and without inertia we can formulate such criteria. We will of course at the same time be solving the response problem for those systems in which inertia is *not* negligible, as in Fig. 8-5a. In Fig. 8-5b we consider a more accurate model of Fig. 7-1c, in which the moving part of the damper has inertia. Similarly, Fig. 8-5c depicts Fig. 7-1d when the mass of the damper cylinder and/or the spring is included. By considering the springiness of the rod connecting the damper B_1 to the mass M in Fig. 7-1e, we get the second-order system of Fig. 8-5d. Finally, Fig. 8-5e is a version of Fig. 7-1d in which the springiness in the damper and damping in the spring are included in the model. All the systems of Fig. 8-5 will lead to relations of the form of Eq. (8-2) between the indicated input and output quantities.

Step Response and Free Vibration of Second-Order Systems. We will be using the system of Fig. 8-5a as our example for developing general response characteristics of second-order systems and the fundamentals of mechanical vibration. Our preference for this example rests on the fact that this configuration is widely accepted as the simplest system for introducing the basic concepts of the important field of mechanical shock and vibration. In Fig. 8-6 we reorient the system so that motion is now in the vertical direction, so as to illustrate the method of treating gravitational forces. These forces are most conveniently dealt with by choosing the origin of the displacement coordinate x_o to coincide with the location of the mass when it is at rest, with only its weight W and the spring force acting on it. These two forces clearly must

---

[2] E. O. Doebelin, *Control System Principles and Design*, Wiley, New York, 1985, p. 335.

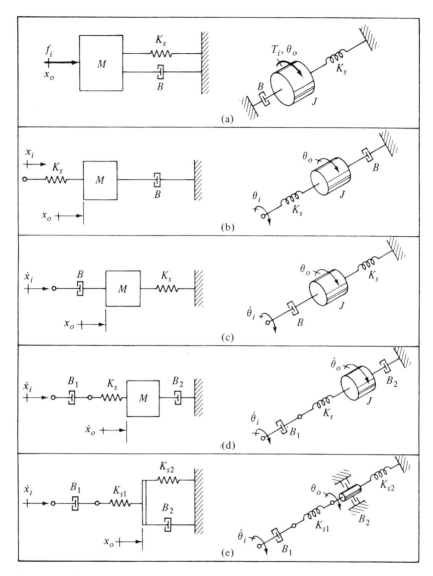

Figure 8-5 Some mechanical second-order systems.

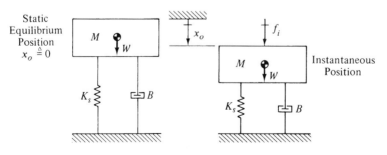

Figure 8-6 Basic vibrating system: vertical motion.

Second-Order Systems and Mechanical Vibration **529**

balance each other at this point, so we know that the spring force must be $-W$ when $x_o = 0$, and thus it can be written in general as $-W - K_s x_o$. Newton's law then gives

$$f_i + W - (W + K_s x_o) - B\dot{x}_o = M\ddot{x}_o \tag{8-17}$$

We now see the reason for choosing the origin as we did; it makes the gravity force "disappear" from the equation, giving

$$M\ddot{x}_o + B\dot{x}_o + K_s x_o = f_i \tag{8-18}$$

It is not *necessary* to choose the origin as we did, but this choice does simplify the solution. For *any* "vertical" vibration problem, no matter how many masses, springs, and dampers might be in the model, when the system is at rest with no driving forces applied, each mass will naturally settle into a static equilibrium position. It is generally best to choose these static equilibrium positions as the coordinate origins for displacement of each mass.

Equation (8-18) clearly fits the standard form (8-2) so we immediately define

$$\omega_n \triangleq \sqrt{\frac{K_s}{M}} \quad \frac{\text{rad}}{\text{time}} \qquad \zeta \triangleq \frac{B}{2\sqrt{K_s M}} \qquad K \triangleq \frac{1}{K_s} \quad \frac{\text{meter}}{\text{newton}} \tag{8-19}$$

to get

$$\left(\frac{D^2}{\omega_n^2} + \frac{2\zeta D}{\omega_n} + 1 \right) x_o = K f_i \tag{8-20}$$

Whether we use Laplace transform or D-operator methods of equation solution, we must find the two roots of the characteristic equation, which can of course be done in letter form from the quadratic equation. These roots are

$$-\zeta\omega_n + \omega_n\sqrt{\zeta^2 - 1} \quad \text{and} \quad -\zeta\omega_n - \omega_n\sqrt{\zeta^2 - 1} \tag{8-21}$$

Several important cases can be defined:

Undamped: $B = 0$, $\zeta = 0$, roots $= \pm i\omega_n$

$$x_{oc} = C \sin(\omega_n t + \phi) \tag{8-22}$$

Underdamped: $0 < B < 2\sqrt{K_s M}$, $0 < \zeta < 1.0$, roots $= -\zeta\omega_n \pm i\omega_n\sqrt{1 - \zeta^2}$

$$x_{oc} = C e^{-\zeta\omega_n t} \sin(\omega_n\sqrt{1 - \zeta^2}\, t + \phi) \tag{8-23}$$

Critically damped: $B = 2\sqrt{K_s M}$, $\zeta = 1.0$, roots $= -\zeta\omega_n, -\zeta\omega_n$

$$x_{oc} = C_1 e^{-\zeta\omega_n t} + C_2 t e^{-\zeta\omega_n t} \tag{8-24}$$

Overdamped: $B > 2\sqrt{K_s M}$, $\zeta > 1,0$, roots $= -\zeta\omega_n \pm \omega_n\sqrt{\zeta^2 - 1} \triangleq -\frac{1}{\tau_1}, -\frac{1}{\tau_2}$

$$x_{oc} = C_1 e^{-t/\tau_1} + C_2 e^{-t/\tau_2} \tag{8-25}$$

Note that the *undamped* and *critically damped* cases have only a single member while the *underdamped* and *overdamped* are *families*, with an infinite number of members.

The undamped case gives us an interpretation of one of our standard parameters, ω_n, the undamped natural frequency. We see from the complementary solution that, if we could remove all friction, then the mass would vibrate at a frequency equal to the undamped natural frequency. Since, in the real world, frictionless sys-

530
                                                                    Chapter 8

tems are impossible (they violate the second law of thermodynamics), we can never observe and measure ω_n directly in lab tests of viscous-damped systems. For real-world underdamped systems the frequency which we *can* observe in a transient lab test is $\omega_n(1 - \zeta^2)^{0.5}$. Since there are ways to measure ζ, we can calculate back to ω_n even though we can't observe it directly. We will also show shortly that ω_n *is* the frequency at which the frequency-response exhibits a $-90°$ phase shift, and we do have instruments for measuring phase angle. Finally, if the system has perfect *coulomb* (dry) friction, then it has been shown theoretically for this nonlinear system that transient oscillations *do* occur at the undamped natural frequency, and we can thus measure it directly.

We also can now see reasons for defining the damping ratio ζ as we did. If a second-order system is to exhibit free (unforced) oscillations, the damping ratio must be less than 1.0. This important "dividing line" at $\zeta = 1.0$ is called *critical damping*. Here the damping coefficient is equal to the value, $2(K_sM)^{0.5}$, that makes $\zeta = 1.0$. We shall see shortly that a critically damped system is the *fastest* second-order system that does *not* overshoot and oscillate. This feature is sometimes wanted in certain applications.

Since we want to develop as much familiarity with basic vibration theory as is feasible in a system dynamics text, let's next look at *free vibration*. In Chapter 1 we organized system inputs into various categories, such as "external driving" and "initial energy storage." By free vibration we mean a situation where there are *no* external forces f_i, but the system can still vibrate if we give it some initial energy. This can be done in two ways since our system has two energy storage elements, the mass and the spring. Our discussion will be clearer if we now consider the "horizontal" version (Fig. 8-5a) of our system, where the spring is at its free length when in the static equilibrium condition with no external force applied. If the spring is now given an initial deflection, this gives it some elastic (potential) energy, which will cause the system to vibrate when released. Alternatively we could leave the spring at its free length, but give the mass some initial velocity (kinetic energy), and it would again vibrate when released. Of course, we could also do *both*, by stipulating an initial displacement *and* an initial velocity.

We now want to analytically solve for the motion when either or both initial energies are present, since this is by definition the *free vibration* motion. While we are at it, we can also include a driving force in the form of a step input. This is *not* a free vibration, but it is an important input and the solutions are similar. Since superposition holds in this linear system, if we get a result for all three inputs simultaneously, we can easily get the response to individual terms or various combinations by "crossing out" the terms we don't want. We will work out the underdamped case since it is of most interest. Our desired results can be found by either the *D*-operator or Laplace transform methods. Let's here use Laplace transform. Laplace transforming Eq. (8-20) with both initial conditions nonzero and a step input force f_{is} gives

$$\left[\frac{s^2}{\omega_n^2} + \frac{2\zeta s}{\omega_n} + 1\right] X_o = \frac{K}{s} f_{is} + \left[\frac{s}{\omega_n^2} + \frac{2\zeta}{\omega_n}\right] x_o(0) + \left[\frac{1}{\omega_n^2}\right] \dot{x}_o(0) \qquad (8\text{-}26)$$

We need to manipulate this into the forms given in our Laplace transform table, as follows:

Second-Order Systems and Mechanical Vibration

$$a \overset{\Delta}{=} \zeta\omega_n \qquad b \overset{\Delta}{=} \omega_n\sqrt{1-\zeta^2} \qquad b_0 \overset{\Delta}{=} \sqrt{a^2+b^2} = \omega_n$$

$$X_o = \left[\frac{K\omega_n^2}{s[(s+a)^2+b^2]}\right]f_{is} + \left[\frac{s+2\zeta\omega_n}{[(s+a)^2+b^2]}\right]x_o(0) + \left[\frac{1}{[(s+a)^2+b^2]}\right]\dot{x}_o(0)$$

(8-27)

The table can now be used directly to get the time response.

$$x_o(t) = \left[K + \frac{K}{\sqrt{1-\zeta^2}} e^{-\zeta\omega_n t} \sin(\omega_n\sqrt{1-\zeta^2}t - \phi_1)\right]f_{is}$$

$$+ \left[\frac{1}{\sqrt{1-\zeta^2}} e^{-\zeta\omega_n t} \sin(\omega_n\sqrt{1-\zeta^2}t + \phi_2)\right]x_o(0)$$

$$+ \left[\frac{1}{\omega_n\sqrt{1-\zeta^2}} e^{-\zeta\omega_n t} \sin(\omega_n\sqrt{1-\zeta^2}t)\right]\dot{x}_o(0) \qquad (8\text{-}28)$$

$$\phi_1 \overset{\Delta}{=} \tan^{-1}\frac{\omega_n\sqrt{1-\zeta^2}}{-\zeta\omega_n} \qquad \phi_2 \overset{\Delta}{=} \tan^{-1}\frac{\omega_n\sqrt{1-\zeta^2}}{\zeta\omega_n}$$

This result allows us to get the response to any of the three individual inputs, or any combination of them. Note that, irrespective of which input or combination is applied, only one frequency of vibration, $\omega_n(1-\zeta^2)^{0.5}$, appears. This frequency, which *can* be directly observed and measured in transient lab tests, is called the *damped natural frequency*, $\omega_{n,d}$.

EXAMPLE: INITIAL ENERGY STORAGE

Graphs of these time responses can be obtained from the above equations, or perhaps more conveniently from simulation. Let's take an example system with $M = 1\,\text{kg}$, $K_s = 1\,\text{N/m}$, and $B = 0.2\,\text{N/(m/sec)}$. This makes the undamped natural frequency 1 rad/sec, the damping ratio 0.20, and the damped natural frequency 0.98 rad/sec. Let's choose our initial displacement and velocity such that each corresponds to the *same* initial energy storage. That is, we want

$$\text{Initial spring energy storage} = \frac{K_s x(0)^2}{2} = \text{initial mass energy storage}$$

$$= \frac{M\dot{x}(0)^2}{2}$$

For our chosen numbers, the two initial energies will be equal if we choose the initial displacement numerically equal to the initial velocity. Let's take $x_o(0) = 1.0\,\text{m}$ and $(dx_o/dt)(0) = 1.0\,\text{m/sec}$; each initial energy will then be 0.50 joule. For our step input force, if we choose it to be 1.0 newton, then the *final* energy storage for this case will also be 0.50 joule.

A simulation for this situation was run, giving the results of Fig. 8-7, where the upper graph shows the motion for an initial displacement, the middle graph for an initial velocity, and the lower graph for a step input force (zero initial velocity and displacement). Just as in the analytical solution, the simulation would allow us to try

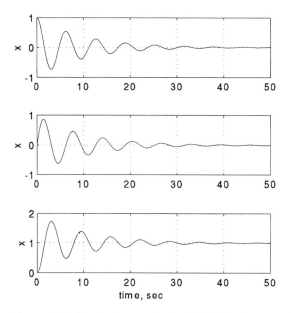

Figure 8-7 Simulation results for initial displacement, initial velocity, and step force input.

any combination of these three inputs. If we want to see how the initial energy is dissipated in the first two cases, simulation allows an easy calculation of the instantaneous stored energy, which starts out at 0.50 joule and goes toward zero as the damper gradually degrades the mechanical energy into heat. Figure 8-8a shows these results; we see that the energy dissipation curves are different for initial displacement and initial velocity, though they start and end at the same values.

On Fig. 8-8b we show the input energy taken from the force source, and the instantaneous stored energy in the system, for the case of the step input force. The difference between these two energy values is the energy dissipated by the damper, shown in the third curve. We see that the force source puts in a peak energy about 3.4 times that finally stored in the spring when the mass finally comes to rest. Whether the system is activated by initial energy storage or an external driving force, since both the spring and mass can store and give up energy, the energy "flows" back and forth between these two elements as the vibration occurs, giving the oscillatory curves shown.

Having introduced some concepts about free vibration, we now return to a more detailed treatment of the step response of second-order systems, which is of general interest. The *undamped* case solution could be obtained separately but is also available from Eq. (8-28) by setting ζ equal to zero.

$$x_o = Kf_{is}(1 - \cos \omega_n t) \tag{8-29}$$

Figure 8-9 shows that this case exhibits a 100% overshoot. That is, if the input force were applied very slowly, the output displacement would gradually approach Kf_{is} from below. The very same force, when applied suddenly, causes a peak displace-

Second-Order Systems and Mechanical Vibration

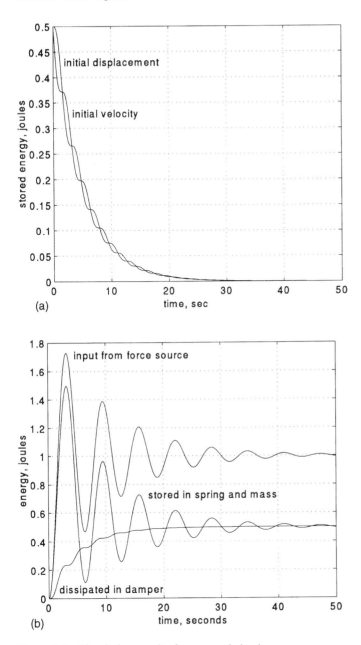

Figure 8-8 Simulation results for energy behavior.

ment *twice* this large. This means that the peak spring force is also twice the "static" value, causing high stresses and possible failure. While systems with no friction are not possible in the real world, some real systems have very low friction, the order of $\zeta = 0.005$. Such systems have overshoots very nearly as large as for $\zeta = 0$.

The solution for the *critically damped* system is easily found using our methods for repeated roots.

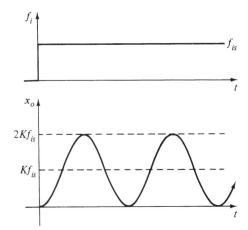

Figure 8-9 Step response of undamped second-order system.

$$x_o = Kf_{is}[1 - (1 + \omega_n t)e^{-\omega_n t}] \tag{8-30}$$

The term $\omega_n t$ goes to infinity, but is squelched by the stronger negative exponential, so the total response stays finite, as the graph of Fig. 8-10 shows. *Overdamped* systems have response curves of similar shape, but are all slower than the critically damped. The overdamped solution [see Eq. (8-25)] is

$$x_o = Kf_{is}\left[1 - \frac{\zeta + \sqrt{\zeta^2 - 1}}{2\sqrt{\zeta^2 - 1}} e^{-t/\tau_1} + \frac{\zeta - \sqrt{\zeta^2 - 1}}{2\sqrt{\zeta^2 - 1}} e^{-t/\tau_2}\right] \tag{8-31}$$

When ζ is very large [B very large relative to $2(K_s M)^{0.5}$], the second-order system will respond very nearly like a first-order system because the damping term in the equation greatly overshadows the inertia effect. For example, if $\zeta = 10.0$,

$$x_o = Kf_{is}(1 - 1.002e^{-0.05\omega_n t} + 0.002e^{-19.95\omega_n t}) \approx Kf_{is}(1 - e^{-0.05\omega_n t}) \tag{8-32}$$

If inertia were *completely* neglected in the model, it would be first-order with $\tau = B/K_s$. For $\zeta = 10$,

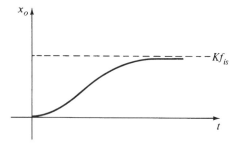

Figure 8-10 Step response of critically damped second-order system.

Second-Order Systems and Mechanical Vibration   **535**

$$\zeta = \frac{B}{2\sqrt{K_s M}} = \left(\frac{1}{2}\sqrt{\frac{K_s}{M}}\right)\left(\frac{B}{K_s}\right) = 10$$

$$\tau = \frac{B}{K_s} = 20\sqrt{\frac{M}{K_s}} = \frac{20}{\omega_n} \tag{8-33}$$

This agrees with Eq. (8-32) to the number of significant figures carried there. Since no real system is free of inertia, analyses such as this are quite useful in deciding when inertia has a negligible effect and may be eliminated from the system model. In our example, a first-order model was adequate for the step response. It might not be for inputs with stronger high-frequency content. Frequency response studies would show that significant differences in amplitude ratio and phase angle exist at high frequency between the first-order approximation and the second-order exact models. If an input has strong frequency content in this range, the first-order model would be inadequate.

EXAMPLE: DESIGN OF PACKAGE CUSHIONING FOR DROPPED PACKAGES

As an example drawing on our results for free vibration with initial energy storage in the form of mass kinetic energy, we will consider the design of "cushions" for products shipped in boxes. One can expect that such boxes may be dropped one or more times during their travels. There is actually quite a bit of advanced technology used to design packaging for such situations, and several companies which specialize in this work.[3] As part of their testing programs, shippers will actually include three-axis accelerometers and data recorders in typical packages and send them by various modes of transportation to document the vibration environment characteristic of that mode (air, truck, train, ship, etc.). As the package is handled (and perhaps thrown or dropped), all the shock data is being recorded for later analysis. This provides vital information to packaging designers.

Let's assume that the object being shipped can be treated as a single mass and that the foam plastic cushioning which surrounds it acts as a combined spring and damper. When the surrounding box is dropped from a certain height, the whole package will build up a velocity before it strikes the floor below. We will assume that the surrounding box does not itself bounce when it hits the floor but rather goes instantly to zero velocity. This situation can be considered to be modeled by the ideal case where the surrounding box is at all times stationary and the cushioned object has a certain downward velocity at time zero, as shown in Fig. 8-11.

While various damage criteria might be used to design package cushioning, let's take a simple viewpoint and require that the maximum acceleration of the cushioned object cannot exceed a certain value when the box is dropped from a specified height. You may be rightly concerned that when a box is dropped, it needn't always land "flat" on the floor, allowing an infinite variety of shock patterns, depending on the orientation of the box when it strikes the floor. Test data such as that described above is helpful in dealing with such complications. Let's assume that such data shows that the "flat" landing of the box is a worst case. We would then design for this and assume that we would be safe for all other possibilities.

---

[3]Lansmont Corp., 5 Harris Court, Bldg. N, Monterey, CA 93940-5739, 408-655-6600.

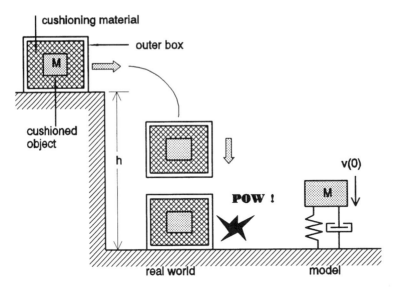

Figure 8-11 Package cushioning design problem.

The design principle of package cushioning is fairly simple. If a falling object strikes a hard surface, its velocity is brought to zero very quickly, giving a large deceleration and possible damage. If instead we provide a soft cushion (a spring), the decelerating force builds up more gradually and can bring the velocity to zero with much smaller peak force and less damaging peak deceleration. In Fig. 8-11 the outer box will experience a large deceleration but the cushioned object is protected. The design problem is mainly one of selecting a spring soft enough to give an acceptable peak deceleration, but not so soft that its static and dynamic deflections are excessive.

Let's take a cushioned object of weight $W = 10\,\text{lb}_f$ falling through a height h of 3 feet. We want the peak deceleration to be no more than $50\,g$'s ($1610\,\text{ft/sec}^2$). Air resistance during the drop will have little effect, so we assume a "free fall," which gives a velocity of $(2gh)^{0.5}$ and a kinetic energy of Wh at the time of impact. This kinetic energy will be totally converted to spring elastic energy as the spring deflects to its peak displacement. At this peak displacement, the upward spring force on the mass creates the peak acceleration. Combining all these facts we can write

$$\frac{K_s x_{\text{peak}}^2}{2} = W(h + x_{\text{peak}}) \approx Wh \tag{8-34}$$

$$M\ddot{x}_{\text{peak}} = \text{max spring force} = K_s x_{\text{peak}} = \sqrt{2WhK_s} \tag{8-35}$$

$$K_s = \frac{M^2 \ddot{x}_{\text{peak}}^2}{2Wh} = \frac{W[\ddot{x}_{\text{peak}}/g]^2}{2h} \tag{8-36}$$

$$x_{\text{peak}} = \frac{2h}{\ddot{x}_{\text{peak}}/g} \qquad x_{\text{static}} = \frac{2h}{(\ddot{x}_{\text{peak}}/g)^2} \tag{8-37}$$

Second-Order Systems and Mechanical Vibration **537**

For the numbers given earlier, the required spring constant works out to be $4667 \, lb_f/$ ft, with a static, with a static deflection of 0.029 inch and a peak deflection of 1.44 inches, which do not seem excessive.

While the design procedure concentrates on the spring, cushioning materials (such as plastic foam) also exhibit damping properties, which is why we included a damper in our model. While the spring effect of a piece of foam plastic can be easily estimated from a static force/deflection lab test, the damping properties are more subtle and models more complex than that of Fig. 8-11 are sometimes needed. We can't pursue such details here so we stick with the simpler model and do a simulation, using the numbers already given, and trying various B values, starting at zero. For $B = 0$ we get exactly the performance designed for; the peak acceleration is exactly $50 \, g$'s, but the vibration "goes on forever." As we add damping we see that even small B values are beneficial; the peak acceleration is less than $50 \, g$'s and the vibration dies out. For a B value of $20 \, lb_f/(ft/sec)$ the vibration dies out in about 1 cycle and the peak acceleration is about 1300, better than required. (See Problem 8-30 for more details.)

Significance of K, ζ, and ω_n. When designing second-order systems for various practical applications we need to know what effect parameter changes have on system behavior. What happens when we change K, ζ, or ω_n? The effect of steady-state gain, when defined as we have always done, is always the same, no matter what order the system. Because K always appears as a multiplying factor on the input quantity of a transfer function, it effect is always a simple scaling up or down of the response. Thus K has no effect on speed of response when we define it in the usual way; how long it takes to accomplish a given percentage of the total change in going from one steady state to another.

The other two parameters both effect speed, but ω_n is by far the most influential. Whether we have the undamped, underdamped, critically damped, or overdamped cases, the equation solutions [Eqs. (8-22) to (8-25)] show that wherever ω_n appears, it appears as the product $\omega_n t$. This means that if we, say, double ω_n, the same stage of the response will occur in exactly one half the time. Thus the undamped natural frequency is a direct and proportional indicator of speed. If you want to speed up any second-order system by a multiplying factor n, you need to increase ω_n by that same factor, keeping ζ constant. This holds for *any* kind of input whatever. Once we discover this feature, we can plot nondimensional step response curves as in Fig. 8-12. There we plot q_o/Kq_i versus $\omega_n t$. These can be used for *any* values of K, ζ, and ω_n (graphical interpolation is used for ζ values not plotted).

We have earlier seen some examples of the effect of damping ratio on step response. Figure 8-12 summarizes all these features. If an application can tolerate no overshooting, we need $\zeta \geq 1.0$. For example, in a machine tool control system if we overshoot the "target" dimension, metal is cut away which cannot be replaced. For underdamped systems the first overshoot (peak value) is often of practical interest and may be found by standard calculus maximizing methods to depend only on ζ, according to the formula

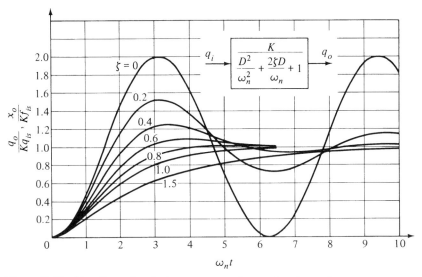

Figure 8-12 Nondimensional step response of second-order systems.

$$\text{Percent overshoot} = 100\,\frac{q_{o,\text{peak}} - Kq_{is}}{Kq_{is}} = 100 e^{-\pi\zeta/\sqrt{1-\zeta^2}} \quad (8\text{-}38)$$

which is plotted in Fig. 8-13. For a fixed value of ω_n, the speed of response, as measured by the time required for the output to settle within a chosen plus and minus tolerance around the final value, is determined by ζ. Using this idea of a *settling time* as a speed of response indicator, we see in Fig. 8-12 that ζ's which are either too large or too small take longer to settle, and thus an *optimum value* of ζ should exist. This depends on the tolerance band chosen, but for bands in the neighborhood of $\pm 5\%$ the optimum range of ζ is about 0.55 to 0.75. Many measuring instruments, which are required to respond quickly, are second-order systems and have ζ designed to be about 0.65.

DESIGN EXAMPLE: HIGH-SPEED SCALE FOR PACKAGING CONVEYOR

Boxes of food products on a conveyor system must be weighed rapidly to check whether they are within a certain tolerance range required by regulatory agencies. Such scales are essentially second-order systems as in Fig. 8-6. Displacement x_o is measured with an electrical transducer whose dynamics are known to be much faster than the mechanical system of the scale. Transducer output voltage is proportional to displacement and therefore to weight, at least for steady-state. Boxes appear at the weighing station at the rate of 5/sec and we must get a weight measurement accurate to $\pm 5\%$.

Using a damping ratio of 0.6 (within the suggested optimum range and already plotted in Fig. 8-12), we see that the needed accuracy is attained at an $\omega_n t$ value of about 6. Let's assume we can use for the weighing operation about half of the 0.2 second available for each package. This tells us that we need an undamped natural frequency of at least 60 rad/sec. Suppose the packages weigh 1 pound and that the displacement sensor has a full-scale range of 0.05 inch. This makes the needed spring

Second-Order Systems and Mechanical Vibration

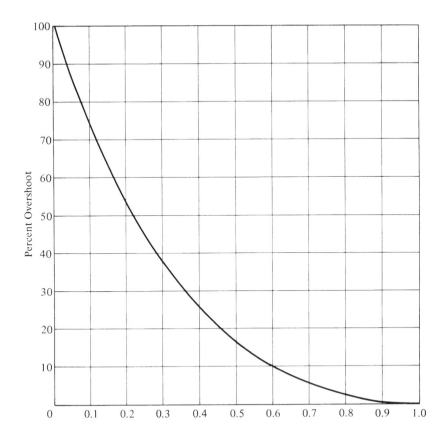

Figure 8-13 Effect of damping ratio on overshoot.

constant 20 lb$_f$/inch, and for the required natural frequency, the total mass must be less than 0.067 slug, making the total weight 2.1 pounds. Since the package weighs 1 pound, the moving parts of our scale should weigh less than 1.1 pounds. Detail design of the scale would reveal whether this mass requirement can be met. It doesn't seem outlandish. If the actual mass turned out to weigh, say, 0.83 pound, then our scale will be a little faster than really needed. The B value would then be 0.25 lb$_f$/(in/sec).

8-4 LAB TESTING SECOND-ORDER SYSTEMS USING STEP INPUTS

When numerical values of model standard parameters are to be found by experiment on an actual system, step-function testing is often useful, particularly if the system is underdamped. Perhaps the simplest test consists of giving the system an initial deflection and then releasing it to perform free vibrations. This could be considered an initial energy input, or a negative step change in force. We could also strike the mass with a hammer of some sort; this would approximate an initial velocity input. We have seen from our equation solutions that all these test methods would produce, for underdamped systems, an oscillatory motion at the damped natural frequency, which we could measure by counting and timing cycles. To get the best accuracy, use as many cycles as can be distinctly observed, not just one cycle.

Having found a number for $\omega_{n,d}$, if we could measure ζ, then we could get ω_n. Various methods for finding ζ exist. It is often possible to use lab tests which are good approximations to step inputs or initial displacements. That is, the measured response curves "look like" the theoretical curves. If this is so, we can measure the percent overshoot of the first peak (percent of step input or percent of initial displacement) and use Fig. 8-13 to estimate the damping ratio. When a system is excited by a momentary pulse (such as a blow from a hammer) whose detailed waveform is unknown, if the system is essentially second-order and the damping is light, a response similar to Fig. 8-14 will be observed. Initially the response curve will probably *not* look like a "textbook" second-order curve, because all real-world vibrating systems actually have an *infinite number* of natural frequencies, as will be demonstrated in Chapter 10. A sharp blow tends to excite *several* of these natural frequencies, but those of higher frequency will quickly disappear ("damp out"), leaving the single frequency associated with a simple second-order model. These high-frequency "wiggles" appear at the beginning of the response curve and we ignore their presence, using only the later portions of the curve for our measurements. (One can sometimes suppress these high-frequency effects by using a "soft" hammer, such as a rubber mallet, rather than a hard metal hammer.)

Note in Fig. 8-14 that, once the response has become "smooth," at any peak the velocity is zero and the displacement is some value which we could call $x_{o,0}$, and

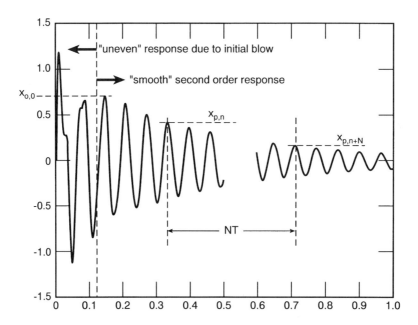

Figure 8-14 Lab tests to find parameter values for second-order model.

Second-Order Systems and Mechanical Vibration **541**

treat the response thereafter as a release from an initial deflection, allowing use of the initial displacement solution of Eq. (8-28). Applying standard calculus maximization methods to this solution, we find that peaks occur at the times

$$0, \quad \frac{2\pi}{\omega_{n,d}}, \quad \frac{4\pi}{\omega_{n,d}}, \quad \frac{6\pi}{\omega_{n,d}}, \quad \text{etc.}$$

and are spaced at a time interval of $2\pi/\omega_{n,d}$, which is called the *period* of the damped oscillation. These peaks of the damped sine wave do *not* coincide with the peaks of $\sin(\omega_{n,d}t + \phi)$, which occur at times

$$\frac{\pi/2 - \phi}{\omega_{n,d}}, \quad \frac{\pi/2 - \phi + 2\pi}{\omega_{n,d}}, \quad \frac{\pi/2 - \phi + 4\pi}{\omega_{n,d}}, \quad \text{etc.}$$

The *spacing* of these peaks is the same as that of the damped wave and also the same as the spacing of the zero crossings, so we can measure the period T (and thus the damped natural frequency) either from peaks or from zero crossings. If we now substitute the peak times of the damped wave into the displacement solution we can compute the *amplitude* of any peak. Taking the ratio of any two successive amplitudes we find that this ratio is *constant*; each successive peak reduces from its predecessor by the same multiplying factor. If the second peak is 0.90 times the first, then the eighth peak is 0.90 times the seventh. The formula for this ratio turns out to be

$$\frac{x_{p,n+1}}{x_{p,n}} = e^{\frac{-2\pi\zeta}{\sqrt{1-\zeta^2}}}$$

$$\text{Logarithmic decrement} \triangleq \delta \triangleq \log_e\left[\frac{x_{p,n}}{x_{p,n+1}}\right] = \frac{2\pi\zeta}{\sqrt{1-\zeta^2}} \tag{8-39}$$

$$\text{For } N \text{ cycles: } R \triangleq \frac{\log_e[x_{p,n}/x_{p,n+N}]}{2N\pi} = \frac{\zeta}{\sqrt{1-\zeta^2}} \tag{8-40}$$

In a lab test, by measuring the ratio of two peaks that are N cycles apart we can compute R and then use Fig. 8-15 to find the damping ratio.

When $0.6 < \zeta < 1.0$, few oscillations are available and both ζ and $\omega_{n,d}$ are hard to measure accurately. One approach might be to estimate them as best you can and then run a simulation with those values and compare to the measured time-response curve. If the system really is close to second-order, a few trials and adjustments should provide sufficiently accurate estimates. For overdamped systems no oscillations at all occur, frustrating the methods explained above. For *severe* overdamping, we can of course model the system as first-order and find a single time constant as shown in Chap. 7. For less severe overdamping, methods to estimate the two time constants needed are available.[4] We will also show shortly that *frequency-response* testing, although more work, can get accurate models for *all* levels of damping.

[4]E. O. Doebelin, *Measurement Systems*, 4th ed., p. 193.

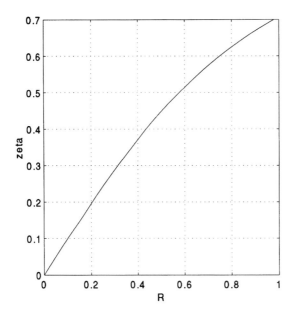

Figure 8-15 Graphical aid for getting numerical value for damping ratio.

Detecting Nonviscous Damping in Transient Testing. In real mechanical systems, viscous damping rarely occurs uncontaminated by more complex and nonlinear frictional effects. When friction is caused by bearings and/or a real damper it is often a combination of viscous and dry (Coulomb) types. If a structure, like the frame of a machine, has no intentional dampers, the damping is small and tends to be proportional to displacement rather than velocity. We briefly discussed this so-called "structural damping" in Fig. 2-32. "Air resistance" tends to be proportional to the square of velocity. All these, and other, frictional effects may be present alone or in combination in real systems, making accurate modeling difficult. Transient testing such as with step inputs can sometimes give some clues as to what forms of damping are present, and we now want to use simulation to briefly study this question.

A simulation such as Fig. 8-16 lets us try individual damping forms and various combinations. The simplest form of Coulomb or dry friction assumes a constant friction force and produces a distinctive pattern in the response curve which allows us to diagnose the presence of this type of friction in real machines. When this is the only type of friction present, it has been proven theoretically[5] that the decay envelope of a free vibration is a *straight line*, rather than the exponential curve of viscous damping. This is verified in the simulation result of Fig. 8-17. When running lab tests of this kind, be on the lookout for decay envelopes that are nearly straight lines; they indicate the presence of simple Coulomb friction.

When Coulomb and viscous friction are present in comparable amounts, the decay envelope is neither exponential nor straight line, but visually appears similar to exponential. This makes the separation of the two effects and assignment of numer-

---

[5]W. T. Thomson, *Mechanical Vibrations*, Prentice-Hall, New York, 1948, p. 55.

Second-Order Systems and Mechanical Vibration

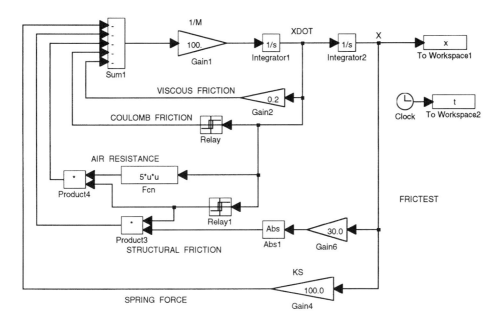

Figure 8-16 Simulation for studying frictional effects.

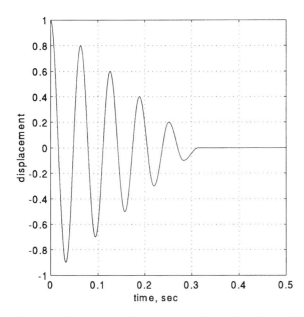

Figure 8-17 Straight-line decay envelope of coulomb friction.

ical values difficult. If the damping is light and many cycles of vibration are available, one can measure the logarithmic decrement δ at several points along the response curve. For pure viscous damping, δ should be the same no matter where you measure it. If δ varies, this is evidence of the presence of damping forms other than viscous. One could then simulate various combinations of, say, viscous and

Coulomb friction and try to match the measured response curve to find numerical values for the two friction components.

When structural damping is the only form present, it may not be immediately obvious since the decay curve (see Fig. 8-18) visually resembles the exponential of viscous damping. In fact, for light damping, it has been shown[6] that the logarithmic decrement for structural damping is constant, giving an exponential decay curve, as is the case for viscous friction. While a discrete physical friction force cannot be identified for structural damping, it can be modeled by a fictitious force which produces the proper effect. Conventionally this is done by defining the damping force as follows.

$$\text{Damping force} \triangleq \left[\gamma K_s \frac{-\dot{x}}{|\dot{x}|} \right] x \tag{8-41}$$

$\gamma \triangleq$ dimensionless structural damping factor

The reference shows that a numerical value for γ can be found by measuring the logarithmic decrement δ and using the formula $\gamma = \delta/\pi$.

8-5 RAMP INPUT RESPONSE OF SECOND-ORDER SYSTEMS

If we have defined the input of our system to be a translational or rotational displacement, then a ramp input means that we apply a constant velocity at the

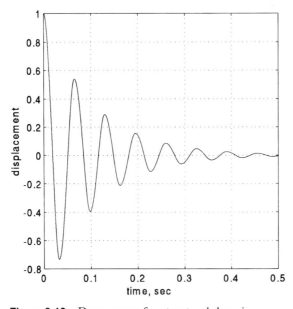

Figure 8-18 Decay curve for structural damping.

[6]Ibid., p. 57.

Second-Order Systems and Mechanical Vibration
545

input location. We will use the rotational system of Fig. 8-5b as an example and make $\theta_i = \omega_{is}t$, where ω_{is} is a step input of angular velocity. The system equation is then

$$J\ddot{\theta}_o + b\dot{\theta}_o + K_s\theta_o = K_s\theta_i = K_s\omega_{is}t \tag{8-42}$$

with initial conditions $\theta_o = \dot{\theta}_o = 0$ at $t = 0^+$. The standard parameters are

$$K \triangleq 1.0 \;\; \frac{\text{rad}}{\text{rad}} \qquad \omega_n \sqrt{\frac{K_s}{J}} \;\; \frac{\text{rad}}{\text{time}} \qquad \zeta \triangleq \frac{B}{2\sqrt{JK_s}} \quad \text{dimensionless} \tag{8-43}$$

giving

$$\left[\frac{D^2}{\omega_n^2} + \frac{2\zeta D}{\omega_n} + 1 \right] \theta_o = K\omega_{is}t \tag{8-44}$$

Using either D-operator or Laplace transform methods, the underdamped solution is

$$\theta_o = K\omega_{is}t - \frac{2\zeta\omega_{is}K}{\omega_n}\left[1 - \frac{e^{-\zeta\omega_n t}}{2\zeta\sqrt{1-\zeta^2}} \sin(\omega_n\sqrt{1-\zeta^2}t + \phi) \right]$$

$$\tan\phi = \frac{2\zeta\sqrt{1-\zeta^2}}{2\zeta^2 - 1} \tag{8-45}$$

Whether overdamped, critically damped, or underdamped, the steady-state (forced) response is given by

$$\theta_{o,\text{ss}} = K\omega_{is}t - \frac{2\zeta\omega_{is}K}{\omega_n} \tag{8-46}$$

$$\text{Steady-state output velocity} = \dot{\theta}_{o,\text{ss}} = K\omega_{is} \tag{8-47}$$

For the present example, $K = 1.0$, so the output shaft at steady state turns at the same speed as the input. However, it lags behind in angular position by an amount $2\zeta\omega_{is}K/\omega_n = B\omega_{is}/K_s$. To check the correctness of this result we note that if θ_o is rotating at speed ω_{is}, the damper B requires a torque $B\omega_{is}$ and this torque can come only through the spring, which must deflect $B\omega_{is}/K_s$ radians to provide this torque. The inertia J has no effect at all in steady state, since it takes no torque to drive a pure inertia at a *constant* velocity. You should cultivate this habit of checking mathematical results by physical reasoning. It not only helps in catching mistakes but also gives a physical feel for the system that enhances design skills.

Transient behavior is affected by ζ and ω_n in a manner similar to that for a step input of displacement; for a given ζ the speed with which the steady state is achieved is directly proportional to ω_n, while for a given ω_n the value of ζ controls the overshoot and degree of oscillation. Figure 8-19 shows the ramp-input response for a general, underdamped second-order system with input q_i and output q_o.

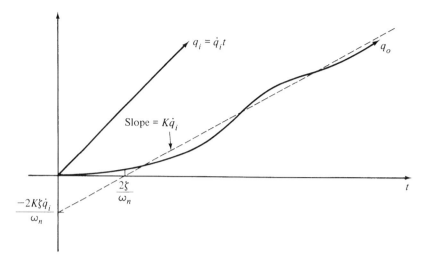

Figure 8-19 Ramp response of second-order systems.

8-6 FREQUENCY RESPONSE OF SECOND-ORDER SYSTEMS

While the initial energy and step force responses studied so far are useful in practical problems involving suddenly applied forces or "shocks," the system frequency response is directly applicable to situations where oscillatory forces produce continuous vibrations which might cause mechanical damage or failure, or lead to acoustic noise objectionable to humans. A prime cause of mechanical failure is *fatigue failure* due to accumulation of too many cycles of too-high stress. Perhaps the most common sources of oscillatory forces which cause vibration problems are the centrifugal forces associated with rotating machine parts that are not perfectly balanced. An unbalanced part rotating at a specific speed creates vertical and horizontal force components which are sine waves with a frequency equal to the rotary speed. An imperfectly balanced motor running at 1800 rpm creates 30-Hz sinusoidal forces. We will now study the frequency response of the system of Fig. 8-6. This analysis will apply to *any* sinusoidal force, not just those due to rotating unbalanced parts.

As always, the frequency response is most efficiently found from the sinusoidal transfer function.

$$\frac{x_o}{f_i}(i\omega) = \frac{q_o}{q_i}(i\omega) = \frac{K}{\frac{(i\omega)^2}{\omega_n^2} + \frac{2\zeta(i\omega)}{\omega_n} + 1} = \frac{K}{\sqrt{\left[1 - \left(\frac{\omega}{\omega_n}\right)^2\right]^2 + \frac{4\zeta^2\omega^2}{\omega_n^2}}} \underline{/\phi} \quad (8\text{-}48)$$

$$\phi \triangleq \tan^{-1}\frac{2\zeta}{\frac{\omega}{\omega_n} - \frac{\omega_n}{\omega}}$$

Note that, in contrast to our results for step, ramp, and initial energy inputs, a *single* formula suffices for all values of ζ. For the undamped case, the amplitude ratio in

Second-Order Systems and Mechanical Vibration 547

Eq. (8-48) goes to infinity when the driving frequency ω is equal to the undamped natural frequency ω_n. This is called the *resonance* phenomenon and tells us that very small driving inputs, if applied at a certain frequency, can cause very large outputs. Physically, the driving input is being applied in exact synchronism with the natural motion and thus builds up the amplitude in the same manner as someone pushing a swing is able to add a little energy each cycle and achieve a large final motion with relatively small forces. For real systems, which always have at least a little damping, the output can't go to infinity but it can get very large. Also, the exciting frequency needn't be *exactly* at the peak location; large outputs are produced over a *range* of frequencies nearby to the peak.

We are of course interested in the frequency of peak response (called the resonant frequency) and also the amplitude ratio at that frequency, so we apply standard calculus maximization methods to find

$$\text{Frequency of peak forced response} \triangleq \omega_p = \omega_n \sqrt{1 - 2\zeta^2} \qquad (8\text{-}49)$$

$$\text{Peak amplitude ratio} = \frac{x_o}{f_i}(i\omega_p) = \frac{q_o}{q_i}(i\omega_p) = \frac{K}{2\zeta\sqrt{1 - \zeta^2}} \qquad (8\text{-}50)$$

Since an input q_i applied statically ($\omega = 0$) gives an output Kq_i, the *resonant magnification factor* is $1/(2\zeta(1 - \zeta^2)^{0.5})$. For $\zeta = 0.01$, for example, a driving input can produce 50 times the output at resonance that it would statically! While resonance often appears as a dangerous problem which must be solved, it has also been used intentionally as a design principle. For example, radio receivers use lightly damped resonant circuits to pick up weak electrical signals, and vibrating conveyors use resonant mechanical systems to reduce the power needed to drive them. When we later look at more complex systems we will find that they have *several* resonant frequencies, but the behavior near a particular resonance is much like that near the single resonance of our simple second-order system. This shows again how familiarity with basic first- and second-order systems prepares us for more complicated situations.

From Eq. (8-49) we see that a resonant peak exists only if the damping ratio is less than 0.707. For damping greater than this the amplitude ratio decreases monotonically to zero as ω goes to infinity, thus there is no magnification. Note also that the frequency of peak response shifts farther below ω_n as ζ increases, and that this peak frequency is *not* the same as $\omega_{n,d}$, the frequency of damped natural oscillations. Also, for a step input, overshoots occur if $\zeta < 1.0$, whereas for sinusoidal input a resonant peak occurs only if $\zeta < 0.707$. These facts are important but easily confused, so be sure that you commit them to memory. That is, be certain that you are clear on the three important frequencies (ω_n, $\omega_{n,d}$, and ω_p) and the two critical values of ζ (1.0 and 0.707).

Figure 8-20 shows the frequency-response curves for the undamped system, while Figs. 8-21 and 8-22 show families of curves for various values of damping. Figures 8-21 and 8-22 plot against a "normalized" frequency ω/ω_n, which is possible since, in Eq. (8-48), ω appears everywhere divided by ω_n. The logarithmic curves ("Bode plots") of Fig. 8-22 have some important features worth memorizing as aids to freehand sketching useful in preliminary design and as quick checks on computer results. (Note the MATLAB graph's use of the term "gain" for what I prefer to call

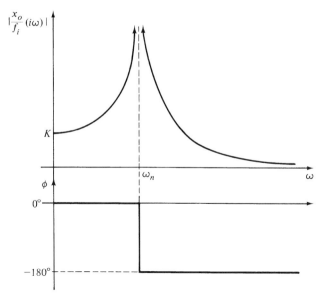

Figure 8-20 Frequency response of undamped second-order system.

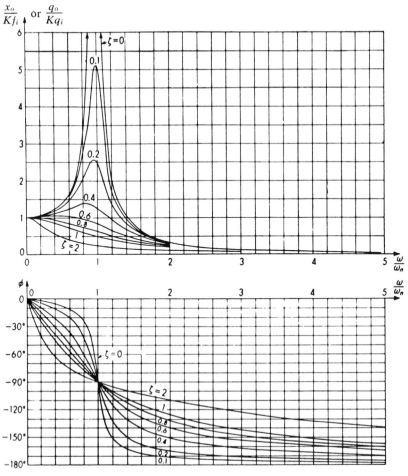

Figure 8-21 Nondimensional frequency-response curves for second-order systems.

Second-Order Systems and Mechanical Vibration

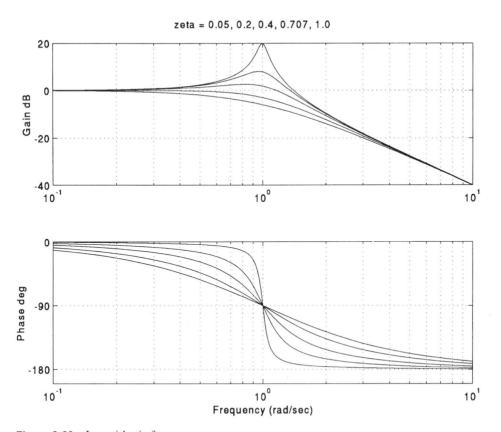

Figure 8-22 Logarithmic frequency-response curves.

amplitude ratio. This "gain" nomenclature is quite common, but amplitude ratio is much more descriptive of the physical situation. In this text the term gain is used as *steady-state gain*, a *number* that is a multiplying factor on a transfer function and has a simple and useful physical meaning.)

The amplitude-ratio curves have two straight-line asymptotes, just as we found for first-order systems, but now the high-frequency asymptote has a $-40\,db/decade$ slope, rather than the -20 of the first order. The breakpoint frequency where the two asymptotes meet is exactly at ω_n, for all values of damping. Rough Bode plots are thus easy to sketch. Just locate ω_n and draw the two asymptotes, using a -40-db point one decade above ω_n to align the high-frequency asymptote. Just as in first-order systems, the steady-state gain K, converted to db, simply raises the entire curve by that amount. Having drawn the two asymptotes, we need some "corrections," which depend on ζ, near the breakpoint. Usually, just locating the peak properly is sufficient for rough work, so get that data from Eqs. (8-49) and (8-50) and then "eyeball" the corrected curve. For the phase angle curve, only a few features are worth memorizing. For all values of damping, the low-frequency asymptote is $0°$ and the high is $-180°$. All curves go exactly through $-90°$ at ω_n, and this is the only physical measurement directly correlated with the *undamped* natural frequency. That is, when running a frequency-response lab test, when the phase-angle meter reads

550   *Chapter 8*

$-90°$, the testing frequency is the natural frequency. This of course holds perfectly only for perfect second-order systems and perfect instruments.

We learned from the step response that ω_n is a direct and proportional indicator of system speed, and this is also seen in the frequency response. From the Bode plot straight-line asymptotes we can see that the output amplitude is about K times the input amplitude for frequencies up to ω_n, but drops off toward zero for higher frequencies. Thus if we want our system to respond well to higher frequencies, we need to raise ω_n accordingly. Since ω in Eq. (8-48) appears everywhere divided by ω_n, this speed effect is directly proportional to ω_n. If we, say, double ω_n, then the same amplitude ratio will now be found at exactly twice the driving frequency. The range of frequencies for which the amplitude ratio is "nearly flat" is called the *bandwidth* of the system. A common specification defines "nearly flat" as flat within -3 db, which is about 30% below the low-frequency asymptote. To significantly increase the bandwidth of our system, we need to raise ω_n.

8-7   VIBRATION ISOLATION AND TRANSMISSIBILITY

The frequency-response curves, with small damping values, clearly show the dangers of resonance. They also show how the effects of vibration can often be *reduced*, and we want to explore this design possibility in this section. No matter what the value of ζ, when the exciting frequency is well above the natural frequency, the motion is much *less* than the same force would produce if applied statically. For example, when the driving frequency is one decade above the natural frequency (10 times ω_n), the output amplitude is 100 times smaller (40 db below its low-frequency value; see Fig. 8-22). The design principle of vibration isolation is thus to make the natural frequency *much lower* than the exciting frequency. If a motor runs at 1800 rpm (30 Hz) and vibrates excessively, mount it on springs such that the natural frequency is, say, 6 Hz. Let's now explore this idea in more detail.

When vibration of a floor-mounted machine is excessive, several undesirable phenomena occur. The machine itself vibrates, perhaps interfering with its operation or reducing its useful life. Also, a vibratory *force* is applied to the floor, causing it to vibrate. This vibration may be transmitted to nearby floor-mounted machines, computers, instruments, etc., perhaps interfering with their operation. Finally, all these vibrating objects produce acoustic noise, which may cause annoyance or hearing damage to nearby humans. Let's consider first the vibratory motion of the machine itself. Equation (8-48) is directly useful here since the output is the displacement of mass M, which represents the machine's mass.

DESIGN EXAMPLE: VIBRATION ISOLATION OF ELECTRIC MOTOR

Consider the system of Fig. 8-23, where an electric motor weighing $80\,\text{lb}_f$ is mounted on a steel beam. Using the formula of Fig. 2-12c, the spring constant at the center of the beam, where the motor is mounted, is calculated to be $K_s = 8400\,\text{lb}_f/\text{inch}$. This gives

Second-Order Systems and Mechanical Vibration

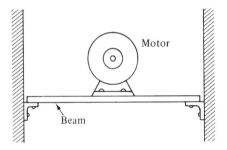

Figure 8-23 Electric motor vibration problem.

$$\omega_n = \sqrt{\frac{8400}{\frac{80}{386}}} = 201 \text{ rad/sec} \qquad f_n = \frac{\omega_n}{2\pi} = 32.1 \text{ Hz} \qquad (8\text{-}51)$$

Since no intentional damper exists in this (and most such) systems, the damping is due to effects such as friction in bolted joints, metal hysteresis, air damping, etc. and is practically impossible to accurately calculate from theory. To estimate the damping and also check the theoretical frequency calculation, a vibration pickup is mounted on the beam and its output recorded as the beam is struck with a large rubber mallet. Data analysis shows the damping to be very light (as expected) but not exactly viscous, since the measured logarithmic decrement varies somewhat as the vibration decays. Taking an average of ζ values measured at large and small amplitudes we get $\zeta \approx 0.01$. For such small damping the damped natural frequency is easily measured and turns out to be 29.2 Hz, which number can also be used for ω_n. The lower actual frequency is due to the inertia of the beam itself (neglected in our model) and the reduced stiffness of the beam spring because its ends are not perfectly fixed as the theory assumes. The beam inertia effect can be well approximated by adding a fraction of the beam's mass to that of the motor, using theoretical results available in the vibration literature;[7] for a beam with fixed ends the fraction is $\frac{3}{8}$. The flexibility of the beam end attachments is usually difficult to assess without running a complicated finite-element analysis, which is hard to cost-justify unless the device will be mass produced, or the application involves potential loss of life or unusual financial loss.

Suppose the motor rotor weighs 40 lb, has the mass center 0.01 inch away from the center of rotation, and turns at 1750 rpm. Due to the unbalance the exciting force is given by

$$f_i = MR\omega^2 \sin 183t = \tfrac{40}{386}(0.01)(183^2)\sin 183t = 34.8 \sin 183t \text{ lb}_f \qquad (8\text{-}52)$$

The motor speed of 1750 rpm coincides exactly with the peak frequency of 29.2 Hz, giving the worst possible situation. For the ζ value given, the resonant magnification factor is 50 and the vibration amplitude is given by

$$\text{Amplitude of } x_o = (50)\left(\frac{34.8}{8400}\right) = 0.207 \text{ inches} \qquad (8\text{-}53)$$

---

[7]C. M. Harris and C. E. Crede, eds., *Shock and Vibration Handbook*, 1st ed., McGraw-Hill, New York, 1961, pp. 1–13.

This vibration is quite violent and would undoubtedly be unacceptable. A number of design changes that would improve this situation are possible, based on our knowledge of vibration phenomena.

1. Run the motor at a different speed, to avoid the resonance. If the motor controls allow a speed change and if the driven machine operates properly at the new speed, this is a simple solution.
2. Change the system natural frequency by
   a. Stiffening the spring, or
   b. Softening the spring, or
   c. Increasing the mass, or
   d. Decreasing the mass.
3. Add a damper to the system.
4. Balance the motor rotor to a higher specification.

Method 2a puts the operating frequency much below the natural frequency and makes the vibration amplitude about the same as the static deflection caused by the unbalance force. From Fig. 8-22 we can see that an operating frequency about 20% of the natural frequency gives this condition. To increase the natural frequency by a factor of 5 requires a 25-to-1 increase in the spring stiffness. If this is feasible, the vibration amplitude would then be reduced to about $34.8/((25)(8400)) = 0.00017$ inch, a *very* small amplitude. If the above degree of spring stiffening is *not* feasible, it looks like we can still get great improvements with lesser stiffening. Increasing the natural frequency by only a 2-to-1 ratio requires only a 4 times stiffening of the spring, but now we are in a frequency range on Fig. 8-22 where there is some amplification (about 40%). Now the estimated amplitude would be about $(34.8)(1.4)/((4)(8400) = 0.00145$ inch, still quite small. Feasibility of any such spring redesigns requires beam details such as form, dimensions, and materials. These are beyond our scope here.

Method 2b puts the operating frequency above the natural frequency, and Fig. 8-22 now shows that the amplitude can be made much smaller than the static deflection caused by the unbalance force. You should note, however, that softening the spring *increases* the static deflection, so the *net* change in amplitude may or may not be the desired reduction. If we want the new amplitude ratio to be, say, 10 times smaller than the static case, we want the -20-db point in Fig. 8-22, which is where the operating frequency is about 3 times the natural frequency. This requires a softening of the spring by about 9-to-1. The predicted vibration amplitude would then be $(34.8)(0.1)/(8400/9) = 0.0038$ inch, a great improvement over the original resonant condition. Whenever we significantly soften springs, we need to check the static deflection due to the dead weight of the mass. In this case it is 0.086 inch, which is not excessive.

Method 2c increases the mass, which is often feasible if space is available. Just as with softening the spring, increasing the mass lowers the natural frequency. Let's lower it by the same amount we used for the spring softening: 3-to-1. This requires a 9-to-1 increase in mass, making the new weight 730 pounds, which does not really seem practical. We can use less than this mass if we accept less amplitude reduction. On Fig. 8-22, if we only want to reduce the amplitude to its "static" value, it appears that the natural frequency need be reduced only to about 70% of its original value.

Second-Order Systems and Mechanical Vibration **553**

This requires the new mass to be about twice the original, and makes the vibration amplitude about $34.8/8400 = 0.0041$ inch.

Method 2d requires a reduction in mass, and this is often *not* possible. In our case the motor's size and weight have been determined to meet certain performance requirements, so we would probably not be allowed to simply substitute a lighter motor. In applications where the mass *can* be reduced, the calculations are similar to those we did for the stiffer spring, except that the "static" amplitude ratio will now *not* be affected by the mass change.

Method 3 adds a damper, which is often feasible if space is available. From Fig. 8-22 it appears that we want a ζ value of about 0.7 or greater if we want to hold the vibration amplitude about equal to the "static" value. We don't have "perfect" values for M and K_s, but the values we do have are good enough to allow an estimate of the damper's required B value, which comes out to be about 58 lb$_f$/(in/sec). If such a damper is technically and economically feasible, it would give a vibration amplitude of 0.0041 inch.

Method 4 is often technically feasible, at least up to a point. Rotating objects can be balanced to various degrees of perfection, with corresponding costs. Extremely fine balancing may not be justified if temperature, wear and other influences will degrade it as time goes by.

This example has used some basic results from vibration theory to explore the design possibilities for vibration reduction. As in most design situations, a number of potential solutions are available and these must be technically and economically evaluated before a decision is reached.

Force Transmissibility. Our example above was focused on reducing the vibration amplitude of the mass. Another important problem deals with the *force* transmitted into the floor on which the machine is mounted. Suppose we wish to reduce this force. The floor feels the sum of the spring force and the damper force, so we want to get a transfer function in which the input is the exciting force and the output is the sum of the spring and damper forces.

$$f_i - K_s x_o - B\dot{x}_o = M\ddot{x}_o$$

$$f_i - M\ddot{x}_o = K_s x_o + B\dot{x}_o \triangleq \text{floor force} \triangleq f_f = f_i - MD^2 x_o$$

$$= f_i - MD^2 \left[\frac{x_o}{f_i}(D) \right] f_i \tag{8-54}$$

$$\text{Force transmissiblity} \triangleq \frac{f_f}{f_i}(D) = \frac{K(\tau D + 1)}{\left[\dfrac{D^2}{\omega_n^2} + \dfrac{2\zeta D}{\omega_n} + 1 \right]} \tag{8-55}$$

$$K \triangleq 1.0 \; \frac{\text{newton}}{\text{newton}} \qquad \tau \triangleq \frac{2\zeta}{\omega_n} \; \text{sec} \tag{8-56}$$

This transfer function is an example of a second-order system with "numerator dynamics." Such systems will be treated generically later in this chapter, but we here give a "preview" since we want to develop the vibration aspect at this point.

Figure 8-24 shows the frequency-response graphs for Eq. (8-55) for a few selected damping values. The amplitude ratio in this graph is called the *force transmissibility*. All the amplitude-ratio curves stay above 1.0 (0 db) until the frequency reaches 1.414 times the natural frequency; thus the transmitted force *exceeds* the exciting force until that point. Thus to reduce the force felt by the floor to less than the exciting force, we must arrange to have the natural frequency less than 70.7% of the exciting frequency. With respect to damping, we note two conflicting trends. At high frequency, where force isolation is possible, this isolation is improved with lighter damping. This light damping unfortunately gives more force *magnification* at lower frequencies, especially at the peak. This dilemma is of practical concern because the exciting force in real systems usually is not a single sine wave but rather has a range of frequencies. If the exciting force were at only a single frequency, the high peak associated with small damping would be of no concern. The tradeoff between high and low damping can not be realistically resolved until we have a practical problem where the actual input frequency spectrum can be estimated. For *any* value of damping, it is clear from Fig. 8-24 that force isolation improves as the exciting frequency exceeds the system natural frequency more and more.

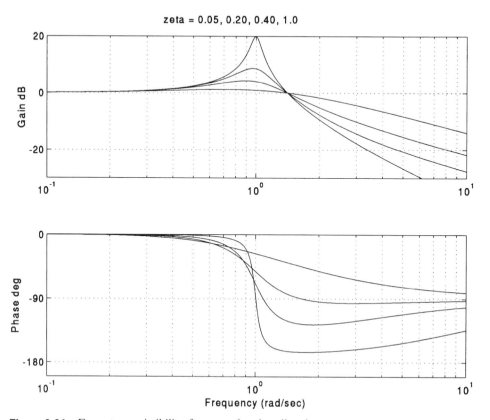

Figure 8-24 Force transmissibility for second-order vibrating system.

Second-Order Systems and Mechanical Vibration

Motion Transmissibility. Sometimes our vibration problem involves a given input *motion* rather than an input force. A common and important example is found in automotive suspension systems, where the tires follow the "bumps in the road," but we want to give the passengers a smooth ride. We want to *isolate* them from the road roughness, which means we want a suspension system with low *motion transmissibility*. Other examples are found in factories or labs where we want to isolate sensitive equipment from existing floor vibrations. Figure 8-25, drawn for the suspension situation, can serve for all such motion-input problems. Newton's law gives

$$B(\dot{x}_i - \dot{x}_o) + K_s(x_i - x_o) = M\ddot{x}_o \tag{8-57}$$

$$\text{motion transmissibility} \triangleq \frac{x_o}{x_i}(D) = \frac{K(\tau D + 1)}{\dfrac{D^2}{\omega_n^2} + \dfrac{2\zeta D}{\omega_n} + 1} \tag{8-58}$$

$$K \triangleq 1.0 \frac{\text{meter}}{\text{meter}} \qquad \tau \triangleq \frac{2\zeta}{\omega_n} \text{ sec} \tag{8-59}$$

We see that this motion transmissibility is exactly the same as the force transmissibility of Eq. (8-55), so Fig. 8-24 will serve for this case also. To isolate the mass ("car body") from the motion input at the wheel, we want the natural frequency to be well below the frequency of the input motion. Note that for a given road profile, the input frequency increases as we drive the car faster, so the isolation gets better at higher speeds.

If passengers or cargo are more sensitive to velocity or acceleration, we might want to define motion transmissibilities for these quantities, in addition to the displacement transmissibility used in our example. These are easily obtained from Eq. (8-58), but we leave the details for the end-of-chapter problems.

Rotating Unbalance. We earlier discussed the vibration caused by rotating unbalanced machine parts, using Eq. (8-48), which applies to any kind of sinusoidal input force. Most vibration texts also provide another analysis which is designed specifically for rotating unbalance, and we now want to develop this result. Figure 8-26 will be used as an analysis diagram. The main difference from our earlier rotating unbalance study is that we provide for the fact that the magnitude of the unbalance

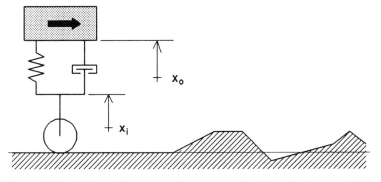

Figure 8-25 Automotive suspension as example of motion transmissibility.

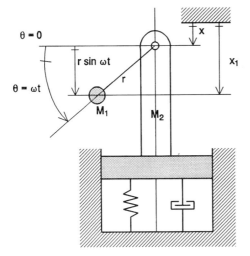

Figure 8-26 Analysis model for vibration due to rotating unbalance.

force *increases* as the frequency (rotary speed) increases, whereas before, our equations and graphs did not include this effect. Also, the rotating unbalance participates in the translational motion of the main mass. To analyze this system we consider two moving masses, M_1 and M_2. The motion of M_1 is considered a superposition of the translational motion of the main mass M_2 and the rotation of M_2; that is,

$$x_1 = x + r\sin\omega t$$

and thus

$$\ddot{x}_1 = \ddot{x} - r\omega^2 \sin\omega t \tag{8-60}$$

In addition to the spring and damper forces, M_2 feels a force from contact with M_1 at the rotary joint. We want only the x component of this force, which is found as follows.

$$x \text{ force on } M_1 = M_1\ddot{x}_1 = M_1(\ddot{x} - r\omega^2 \sin\omega t)$$

$$x \text{ force of } M_1 \text{ on } M_2 = -M_1(\ddot{x} - r\omega^2 \sin\omega t) \tag{8-61}$$

To "cancel out" the gravitational forces in this vertical vibration problem we choose the origin for x as the static equilibrium position, with $\theta = 90°$. Newton's law for M_2 then gives

$$-K_s x - B\dot{x} - M_1(\ddot{x} - r\omega^2 \sin\omega t) = M_2\ddot{x} \tag{8-62}$$

$$(M_1 + M_2)\ddot{x} + B\dot{x} + K_s x = M_1 r\omega^2 \sin\omega t \triangleq f_i \tag{8-63}$$

This last equation is essentially the same as Eq. (8-18) except that now the input force is explicitly restricted to the unbalance force of the rotating mass. When we look at frequency response, the frequency ω appears as part of the input force, so we can't really form a proper transfer function since that would require that we be able to choose the input force amplitude independently. We still would like to display graphs

Second-Order Systems and Mechanical Vibration

that show how output amplitude varies with frequency, but these graphs will *not* be transfer functions, they will be graphs of output amplitude. That is, the dimensions will be meters of displacement, not meters/newton.

While the earlier transfer function graphs (Fig. 8-21) show the amplitude ratio going to zero at high frequency, our new amplitude graph will *not* got to zero, because the unbalance force input goes to infinity for high frequency. That is, when the output/input ratio goes to zero, but the input goes to infinity, the product can be nonzero. The graphs found in most vibration texts plot a dimensionless or normalized amplitude rather than the amplitude itself. When the frequency approaches infinity, we can show that the displacement amplitude approaches the value $M_1 r/(M_1 + M_2)$, which we will call x_∞. It is then easy to show that

$$\frac{|x|}{x_\infty} = \frac{(\omega/\omega_n)^2}{\sqrt{[1 - (\omega/\omega_n)^2]^2 + 4\zeta^2 \omega^2/\omega_n^2}} \tag{8-64}$$

This relation is graphed in Fig. 8-27. Calculus maximization methods show that a peak exists if $\zeta < 0.707$, the peak frequency is $\omega_n/(1 - 2\zeta^2)^{0.5}$, and the peak value is $1/(2\zeta(1 - \zeta^2)^{0.5})$.

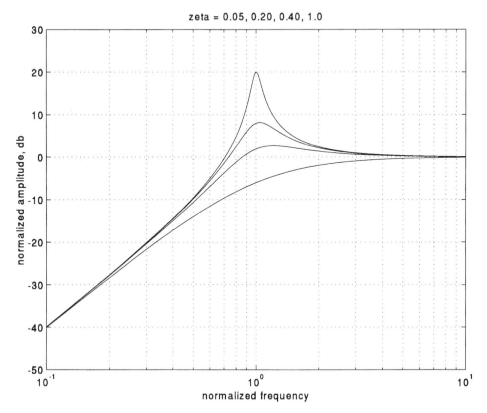

Figure 8-27 Normalized vibration amplitude for rotating unbalance system.

Acceleration to Operating Speed: "Transient Resonance." For those situations where we have designed the natural frequency to be lower than the operating frequency, so as to get vibration isolation, we need to consider what happens when we start or stop such a machine. That is, as the speed rises toward the operating value, we must pass through the resonant peak, which might be quite high if we have small damping. We intuitively guess that we may be OK if we pass through the resonance region "quickly enough." Let's do a simulation to get some information on this problem.

The most realistic simulation would model the rotary dynamics of the motor and load (such as we have done before with, say, an induction motor) to simulate the acceleration to final speed. We will take a simpler "generic" approach which still brings out the essence of the problem. The rotary first-order system of Fig. 7-1a, when subjected to a step input torque T_{is}, accelerates from rest to a final steady speed T_{is}/B, and we can adjust the time taken to reach steady speed with the time constant J/B. By simulating this simple system such that angular acceleration α and angular velocity ω are accessible, we get the signals needed to form the unbalance force in our vibration simulation. By using different time constants we can go through the resonance region quickly or slowly.

We can base our simulation on Eq. (8-63) except that the exciting force must be changed to take into account the angular acceleration, as shown in Fig. 8-28. From that figure we find that f_i in Eq. (8-63) is now

$$f_i = M_1 r \omega^2 \sin\theta - M_1 r \alpha \cos\theta \tag{8-65}$$

Let's use numbers related to our earlier electric motor example:

Final operating speed $\omega = 1750$ rpm $= 183$ rad/sec
System natural frequency $\omega_n = \omega/3 = 61$ rad/sec
$\zeta = 0.01$
$M_1 = M_2 = \frac{40}{386}$
$r = 0.01$ inch
$K_s = 8400/9 = 933$ lb$_f$/inch

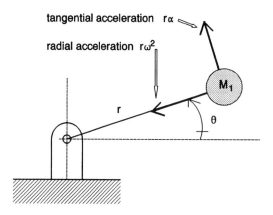

Figure 8-28 Analysis diagram for acceleration of rotating unbalance.

Second-Order Systems and Mechanical Vibration

In the SIMULINK simulation diagram of Fig. 8-29, the vibrating system is modeled with a second-order transfer function block using the above numbers to establish K, ζ, and ω_n. To generate the input force f_i we get the needed ALPHA and OMEGA from the "J, B" rotary system simulation. In Fig. 8-29 I have set this system time constant at $J/B = \frac{5}{2} = 2.5$ seconds. This means that 95% of steady speed will be reached in 7.5 seconds. Having α and ω available, Eq. (8-65) is implemented as shown, using function blocks and multiplier blocks.

The "two parts" of f_i are put through gain blocks set at 1.0 and then summed to form the input to the vibrating system. These two gains are set at 0.0 when I want to get just the steady-state situation. I also provide separately the steady-state input force $34.7\sin(183t)$, also with a gain. This gain is set at 0.0 when running the acceleration study and 1.0 when I want to see just the steady state. These features are desirable because the steady state is approached asymptotically, and one thus has to run for rather long times to see it on the "acceleration" simulation. This creates some graphing problems; parts of the graph are "squeezed together" and become unintelligible. By running the separate "steady-state" simulation, we can get an accurate steady-state result quickly.

Figure 8-30 shows the vibratory displacement of the mass as the motor accelerates, using all the numbers shown in Fig. 8-29. If we ran the motor *steadily* at the natural frequency, the exciting force would be $3.86\,\text{lb}_f$ and the amplitude would be 0.207 inch. In Fig. 8-30 we pass through this resonant condition so quickly that the peak amplitude is only about 0.02 inches. To the casual eye, Fig. 8-30 seems to have reached steady state at $t = 3$ seconds, but this is not the case—the motor is still accelerating because our time constant was chosen as 2.5 seconds. The apparent steady amplitude is the result of the simultaneous increase in unbalance force and

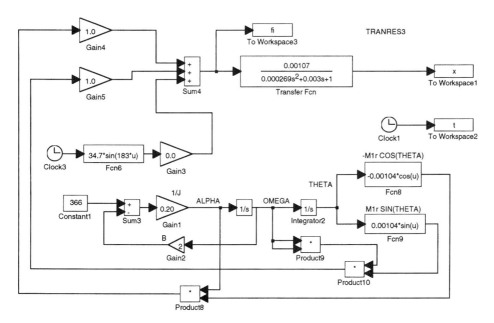

Figure 8-29 Simulation diagram for acceleration of rotating unbalance.

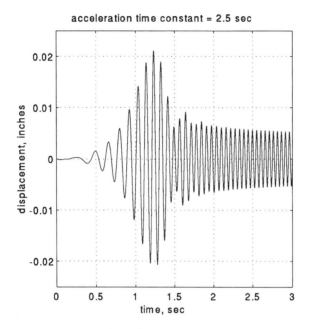

Figure 8-30 Rapid acceleration through resonance prevents large resonant vibration.

decrease in amplitude ratio as speed increases. To see the *real* steady-state situation we would have to run the simulation (and thus Fig. 8-30) out to about 10 seconds, but then the graph gets unreadable, as mentioned above. To see the steady state at a proper graph scale, we would set Gain4 and Gain5 to zero and Gain3 to 1.0.

To see the effect of slower acceleration to the final operating speed I re-ran the simulation with an acceleration time constant of 10 seconds, giving the graph of Fig. 8-31. We see that this slower acceleration does not cause a significantly larger peak vibration (it's still about 0.02 inch) but now there are many more *cycles* of this vibration as we pass through resonance. This means that any acoustic noise associated with the vibration would be more noticeable and we would be accumulating more cycles of possibly damaging fatigue stress.

8-8 IMPULSE RESPONSE OF SECOND-ORDER SYSTEMS

Recall from our treatment of impulse response of first-order systems that perfect impulses cannot occur in the real world, but that short-duration transient inputs *can* be treated approximately as perfect impulses with the same net area. For this approach to yield a good approximation for first-order systems, we found that the duration of the transient input must be less than one-tenth the system time constant. Recall also that when the input duration satisfies this requirement, *the shape of the input is of no consequence, only the net area matters*. Finally, the response to a perfect impulse is exactly the same as the system's response to a properly chosen initial condition.

Second-Order Systems and Mechanical Vibration

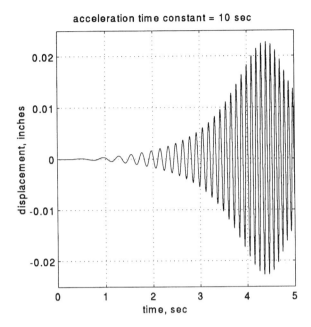

Figure 8-31 Slow acceleration causes many cycles of fatigue stress.

For second-order systems, we can use Laplace transform to obtain the impulse response for the underdamped case as follows.

$$\left[\frac{s^2}{\omega_n^2} + \frac{2\zeta s}{\omega_n} + 1\right] Q_o = KQ_i = KA_i \qquad (8\text{-}66)$$

where $A_i \triangleq$ area of impulse.

$$q_o = \frac{KA_i \omega_n}{\sqrt{1-\zeta^2}} e^{-\zeta\omega_n t} \sin(\omega_n \sqrt{1-\zeta^2}\, t) \qquad (8\text{-}67)$$

Figure 8-32 shows a plot of a non-dimensional version of this response, including also the critically damped and overdamped cases. (It is not hard to show that this impulse response is identical to the system's response to an initial value $(dq_o/dt)(0)$ equal to $KA_i\omega_n^2$.) For the mechanical vibrating system example which we have been carrying along, Eq. (8-67) becomes

$$x_o = \frac{A_i}{\sqrt{K_s M(1-\zeta^2)}} e^{-\zeta\omega_n t} \sin(\omega_n \sqrt{1-\zeta^2}\, t) \qquad (8\text{-}68)$$

For second-order systems, the requirement for a short-duration pulse of any shape to give nearly the same response as a perfect impulse of the same net area is that the duration be less than about one-tenth of the oscillation period $2\pi/\omega_n$. You can check this out using simulation, as requested in one of the end-of-chapter problems.

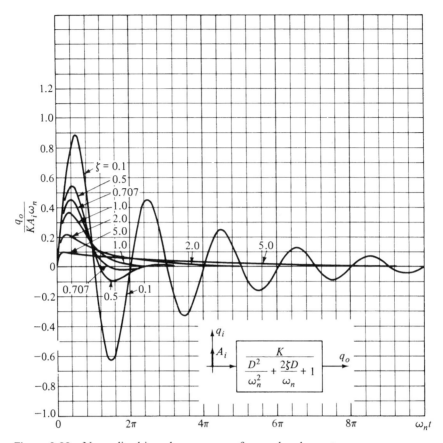

Figure 8-32 Normalized impulse response of second-order systems.

8-9 ELECTRICAL SECOND-ORDER SYSTEMS

Having now covered many of the characteristics of generic second-order systems and showed some useful results for mechanical examples, we can in this section concentrate on the strictly electrical aspects of electrical second-order systems.

A Passive Low-Pass Filter. Figure 8-33 shows eight examples of electrical circuits which all have exactly the same generic second-order relation, Eq. (8-4), between the indicated output and input variables. The circuit of Fig. 8-33a is useful as a low-pass filter with a sharper cutoff than its first-order relative of Fig. 7-28a. Low-pass filters are intended to allow signals with frequencies below a certain range to pass through largely unaffected, while rejecting (attenuating) signals with frequency above this range. When the pass-band and the reject-band are widely spread, say, pass 0 to 10 Hz and reject 1000 to ∞, then even a simple first-order filter can do a good job. In many practical applications, however, it is advantageous to be able to discriminate between quite closely spaced frequencies, and filters approaching the perfect cutoff characteristics are needed.

Second-Order Systems and Mechanical Vibration

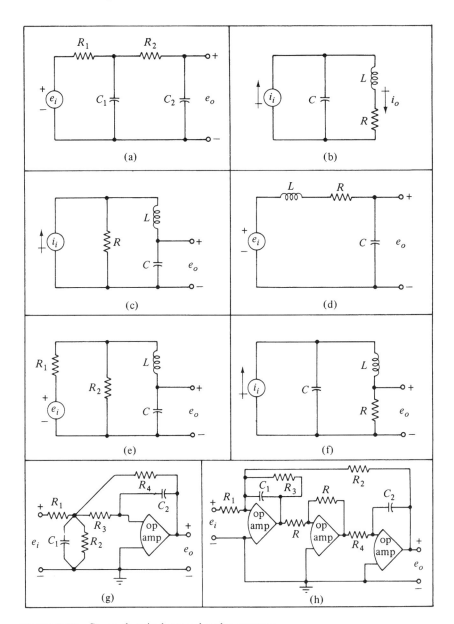

Figure 8-33 Some electrical second-order systems.

The ideal low-pass filter (which is *impossible* to realize) would have an amplitude ratio of 1.0 for frequencies below the designed cutoff frequency and 0.0 for frequencies above, as shown in Fig. 8-34. Also shown there are first-order and second-order filters designed to attenuate frequencies above 1 rad/sec by at least 20-to-1 (amplitude ratio = 0.05). The second-order filter is critically damped; that is, it is two first-order systems with equal time constants, cascaded. If we define the range of frequencies which we wish to pass as that for which the amplitude ratio is between 1.0 and 0.95, Fig. 8-34 shows that the second-order filter's range extends to

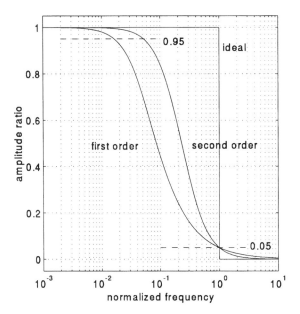

Figure 8-34 Frequency response of ideal and real low-pass filters.

about 0.05 times the cutoff frequency, while the first-order extends to only about 0.015. Once one has chosen the order of a low-pass filter, two alternative design approaches may be used. One can design for a specified attenuation at a specified frequency and then check to see how far the pass band extends, as was done in Fig. 8-34. Alternatively, one can design for a desired amplitude ratio (say, 0.95) at the highest pass-band frequency and then check to see at what frequency the attenuation reaches a desired value (say, 0.05 amplitude ratio). For the most stringent requirements, filters as high as eighth order are not uncommon.[8]

Let's analyze the circuit of Fig. 8-33a, shown in more detail in Fig. 8-35. Our goal is to get a differential equation relating output voltage e_o to input voltage e_i. Even though no one is asking us about the *currents* in this circuit, we usually have to deal with them in order to get the voltage relation we really want. This is often the case, so we generally identify and label the various currents in any circuit we are working with, even though information about these currents may not be the final goal. We are assuming that our circuit is *not* connected to some other circuit at the output terminals, so those terminals are legitimately considered to carry no current. If there *is* a circuit connected there, we need to worry about a *loading problem*, as discussed earlier in this chapter.

When labeling currents (this is done *before* writing any equations), try to minimize the number of currents used, by applying Kirchhoff's current node law at any branch points. That is, in Fig. 8-35, we *could* label three currents: i_1 from *a* to *b*, i_2 from *b* to *e*, and i_3 from *b* to *c*. This would not be wrong, but it makes it look like we have three unknowns (the currents), and therefore we search for three equa-

---

[8] E. O. Doebelin, *Measurement Systems*, 4th ed., pp. 777–790.

Second-Order Systems and Mechanical Vibration

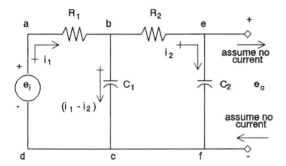

Figure 8-35 Second-order low-pass filter.

tions. There are actually only *two* unknown currents, $i_3 = i_1 - i_2$, so i_3 is not really a third unknown. Thus we label the current from *b* to *c* as $(i_1 - i_2)$, *not* i_3. One can of course solve such problems either way, but I think you will find it advantageous to apply the current-node law at the diagram-drawing stage rather than later at the equation-writing stage.

The reason that we have defined the currents at all is that our plan for solving the problem is in two stages: Find i_2 in terms of e_i and *then* get what we really wanted (e_o) from $e_o = (1/C_2 D)i_2$. There are, as usual, several other ways to solve this problem, but this method does not require remembering any special "tricks" and is about as quick as any other method. Another advantage of finding the currents is that once you have them, you can get almost "everything else" that might be of interest in a circuit. That is, the voltages across any *R*, *L*, or *C* can be found immediately when we know the currents through these elements.

To get two unknown currents we need two independent equations and these can be obtained from Kirchhoff's voltage loop law, applied to the proper loops. Before writing these or any other equations we must first choose sign conventions for all the currents and voltages of interest. Recall that these choices are arbitrary, but some choices are more convenient than others. In Fig. 8-35 the assumed positive polarity for e_i is as shown; it *could* have been chosen just the opposite. Once we *have* made this choice, however, then the choices for the positive senses of the currents and the output voltage are influenced (though not fixed) by that initial choice. That is, having chosen the positive direction for e_i, it is then *convenient* (not necessary) to choose the positive directions for the currents to be the same as would be caused by a positive e_i, and similarly for e_o. Such choices were made in Fig. 8-35.

When thinking about applying Kirchhoff's voltage loop law in Fig. 8-35, most people tend to first "see" two obvious loops: *abcda* and *befcb*. We will in fact use those two loops to derive the two equations needed to solve for our two unknown currents, but several other loops are possible and correct. We could, for example, take the path *abefcda* or "figure-eights" like *abcfebcda*; these would all give us correct equations, because *no matter what path you take*, if you keep careful track of all the voltage drops along your chosen path and return to your starting point, the sum of the voltage drops must be zero. Since each different path generates a different equation, we can usually generate *more* equations than are really needed to solve for a fixed number of unknowns. You only need as many equations as you have unknowns, but the equations must be *independent*. If you travel *all* the possible

566   **Chapter 8**

loops and generate *all* the possible equations, some of them will not be independent
of others.

Thus, when we select loops to use in a problem we hope that we will generate
an independent equation for each loop. Circuit analysis specialists (*not* readers of
this book) have developed various clever ways to generate *only* the set of equations
necessary to solve a particular problem. I want you to be aware of the existence of
such tools, but they are not really necessary or desirable for the relatively simple
problems encountered by non-electrical engineers when they deal with systems that
are only partly electrical. We thus will choose our loops intuitively and generally
encounter no difficulty. You should be reassured that if you choose a "wrong" loop,
you *dont't* get a wrong answer. If you have two unknowns and get two equations, but
one of them is *not* independent because you chose a "wrong" loop, you don't get a
wrong answer, you get *no* answer; that is the set of equations is unsolvable. When
you discover this, you simply then try some other loop until you are successful. If
you travel three loops when only two are needed, again, you don't get a wrong
answer. You simply find that one of the three equations wasn't really needed. We
are thus relatively safe in adopting a circuit analysis scheme that is not as "scientific"
as it might be.

Once we decide which loops to use, we then have to choose a starting point and
a direction of travel as we go around each loop. The starting point can be chosen
arbitrarily; it makes no difference at all. The direction is also arbitrary, but one
direction may be more convenient than the other. Earlier in this book I revealed
my personal preference for the version of Kirchhoff's voltage loop law that keeps
track of voltage *drops*. This choice then leads me to always select the direction of
loop traversal the same as the assumed positive direction for the current of that loop.
This is *not* necessary, but it does make the drop across any R, L, or C a positive term
in the equation, which establishes a simple pattern that reduces mistakes. In Fig. 8-35
I thus chose to go "clockwise" (the assumed positive direction of i_1) in loop *abcda*
and similarly in loop *befcb*.

Having reviewed these important preliminaries, we can now write the two
Kirchhoff voltage loop equations:

$$i_1 R_1 + \frac{1}{C_1 D}(i_1 - i_2) - e_i = 0 \tag{8-69}$$

$$i_2 R_2 + \frac{i_2}{C_2 D} + \frac{1}{C_1 D}(i_2 - i_1) = 0 \tag{8-70}$$

If these two equations were not independent, we would find them unsolvable as we
proceeded. They are, however, independent and we encounter no difficulty in solving
them. If you had chosen only one of these two loops and then done for your second
loop, say, the loop *abefcda*, those two loops would also work. We could at this point
solve for either of the currents, but our real interest is in e_o, which requires only that
we get i_2. For two equations in two unknowns, solution by substitution and elimination is quick and easy; however, we instead use determinants to get more practice in
this technique since it is almost a necessity for even slightly more complicated problems. Note that if we wanted *only* to simulate this system, we would work directly
from Eqs. (8-69) and (8-70); *none* of the upcoming manipulations need be carried
out.

Second-Order Systems and Mechanical Vibration   **567**

Using the D-operator notation we can treat the two equations as if they were algebraic and write

$$(R_1 C_1 D + 1)i_1 + (-1)i_2 = (C_1 D)e_i \tag{8-71}$$

$$(-C_2)i_1 + (C_1 C_2 R_2 D + C_1 + C_2)i_2 = 0 \tag{8-72}$$

Here and in general, when we find D-operators in the denominator of expressions we multiply through the whole equation by some term which clears the D's from the denominator. In (8-69) we multiplied through by $C_1 D$ and in (8-70) by $C_1 C_2 D$ to get (8-71) and (8-72). Now, by determinants

$$i_2 = \frac{\begin{vmatrix} R_1 C_1 D + 1 & C_1 D e_i \\ -C_2 & 0 \end{vmatrix}}{\begin{vmatrix} R_1 C_1 D + 1 & -1 \\ -C_2 & C_1 C_2 R_2 D + C_1 + C_2 \end{vmatrix}} \tag{8-73}$$

$$C_2 D e_o = \frac{C_1 D e_i}{R_1 R_2 C_1 C_2 D^2 + (R_1 C_1 + R_1 C_2 + R_2 C_2)D + 1}$$

$$\frac{e_o}{e_i}(D) = \frac{1}{R_1 R_2 C_1 C_2 D^2 + (R_1 C_1 + R_1 C_2 + R_2 C_2)D + 1} \tag{8-74}$$

Since this is clearly our standard second-order form, we define

$$K \triangleq 1 \; \frac{\text{volt}}{\text{volt}} \qquad \omega_n \triangleq \sqrt{\frac{1}{R_1 R_2 C_1 C_2}} \; \frac{\text{rad}}{\text{time}}$$

$$\zeta \triangleq \frac{R_1 C_1 + R_1 C_2 + R_2 C_2}{2} \sqrt{\frac{1}{R_1 R_2 C_1 C_2}} \tag{8-75}$$

It turns out that ζ must for this circuit always exceed 1.0, which we show as follows:

$$\zeta^2 = \frac{Z}{4} + \frac{1}{4Z} + \frac{Y}{2} + \frac{W}{2} + YW + \frac{1}{2} \tag{8-76}$$

where $\quad Z \triangleq \dfrac{R_1 C_1}{R_2 C_2} \qquad Y \triangleq \dfrac{R_1}{R_2} \qquad W \triangleq \dfrac{C_2}{C_1} \tag{8-77}$

The minimum value of $(Z/4 + 1/4Z)$ occurs at $Z = 1$ and is $\frac{1}{2}$. Since Y and W must both be positive, (8-76) shows that ζ^2 approaches 1.0 from above if $Z = 1.0$ and Y and W both approach zero. Note that these conditions are precisely those which would give the cascade combination of two first-order filters with identical time constants and negligible loading effect. Equation (8-76) also shows that ζ may be made as large as we wish.

While it is perfectly correct to define ω_n as in (8-75), with ζ always > 1.0 it is clear that there can never be any natural oscillations in this circuit. It is in fact *generally* true that *any* passive circuit (one without amplifiers) must have *both* inductors and capacitors to be capable of free oscillations. That is, we require *two* forms of energy storage in order for the energy to shuttle back and forth between them in an oscillatory fashion. Similarly in mechanical systems; both springs and masses must be present to have free vibrations. Springs and dampers or masses and dampers lead only to real roots, not the complex ones needed for oscillation. Since inductors have many practical limitations, the capability of active circuits, such as those using op-amps, for producing oscillatory behavior with only R and C is of considerable importance, because oscillatory behavior has many practical applications.

Since this system will *always* be overdamped, we may prefer to use for the step response the form of Eq. (8-31), where we define two time constants rather than ζ and ω_n. The differential equation can then be written as

$$(\tau_1 D + 1)(\tau_2 D + 1)e_o = (\tau_1\tau_2 D^2 + (\tau_1\tau_2)D + 1) = Ke_i \tag{8-78}$$

which shows that the frequency response can be calculated or graphed as a superposition of two first-order terms.

$$\frac{e_o}{e_i}(i\omega) = \frac{1}{(i\omega\tau_1 + 1)(i\omega\tau_2 + 1)} \tag{8-79}$$

Figure 8-34 showed a second-order filter with two *identical* time constants; with the present circuit this can be closely approached but not quite reached since it requires $\zeta = 1.0$. It *can* be done with two first-order systems and a buffer amplifier or with op-amp circuits. When the two time constants are not equal the frequency response is as in Fig. 8-36. If we use software graphing techniques such as the Bode plotting of MATLAB that we have shown before, there is no need to factor the second-order term into two first-orders.

Series Resonant Circuit. Since the circuit above is capable of only overdamped response, we now wish to show one which can display the whole range of damping behaviors, the circuit of Fig. 8-33d. It can serve as a model for a number of practical applications. When R is very small and L and C are chosen to give a desired natural frequency, the circuit is said to be "tuned" or series resonant. This means that it will magnify signals of a narrow range of frequencies, relative to those with higher or

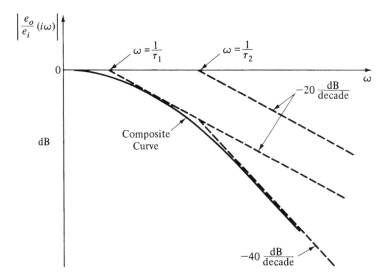

Figure 8-36 Second-order low-pass filter with unequal time constants.

Second-Order Systems and Mechanical Vibration   **569**

lower frequencies, and can thus "pick out" a certain frequency. This is exactly what is needed in tuning a radio receiver to a station operating at a known frequency.

Most circuits can be analyzed in several different ways, so let's use an impedance approach for this one, rather than the Kirchhoff law method used on the low-pass filter. The impedance "seen" by the voltage source is

$$Z(D) = LD + R + \frac{1}{CD} = \frac{LCD^2 + RCD + 1}{CD} = \frac{e_i}{i}(D) \tag{8-80}$$

and since $e_o = (1/CD)i$, we get

$$\frac{e_o}{e_i}(D) = \frac{1}{LCD^2 + RCD + 1} = \frac{K}{\dfrac{D^2}{\omega_n^2} + \dfrac{2\zeta D}{\omega_n} + 1} \tag{8-81}$$

$$K \triangleq 1 \ \frac{\text{volt}}{\text{volt}} \qquad \omega_n \triangleq \sqrt{\frac{1}{LC}} \ \frac{\text{rad}}{\text{time}} \qquad \zeta \triangleq \frac{RC}{2\sqrt{LC}} \tag{8-82}$$

From the expression for ζ we see that in this circuit the full range of behavior from undamped to overdamped is theoretically possible by adjusting R. Of course, just as friction B cannot really be made zero in mechanical systems, resistance R cannot be zero in a real circuit, and thus $\zeta = 0$ cannot be realized even if we do not intentionally "wire in" a resistor, because all inductors have parasitic resistance, as do the "wires" themselves. When we look at op-amp circuits we will find that $\zeta = 0$ can be realized with no inductors at all and with nonzero resistors present.

For some applications we might be interested in the current i rather than voltage e_o as the output. For ac operation we would have $e_i = E_i \sin \omega t$ and the sinusoidal impedance would be of interest.

$$Z(i\omega) = i\omega L + \frac{1}{i\omega C} + R = R + i\left(\omega L - \frac{1}{\omega C}\right) \tag{8-83}$$

Considering an input voltage of fixed amplitude but adjustable frequency the maximum current will occur when the impedance is a minimum. Since the resistive component of impedance is constant at R, we must minimize the reactive component $(\omega L - 1/\omega C)$. This will clearly be zero when

$$\omega L = \frac{1}{\omega C} \qquad \omega^2 = \frac{1}{LC} \qquad \omega = \sqrt{\frac{1}{LC}} \tag{8-84}$$

The peak current will thus occur when $\omega = (1/LC)^{0.5}$ irrespective of the value of R. Note that this "current resonance" differs from the "voltage resonance" of Eq. (8-81) in several ways. First, $(e_o/e_i)(i\omega)$ exhibits no peak at all unless $\zeta < 0.707$, while current *always* has a peak. Second, the current peak is always at the undamped natural frequency ω_n while the voltage peak (if there is one) is at ω_p. Also, Eq. (8-83) shows that, at current resonance, e_i and i are precisely in phase with each other since the impedance is purely resistive at this one frequency. That is, the inductive reactance ωL and the capacitive reactance $1/\omega C$ just cancel each other. Figure 8-37 shows how impedance varies with frequency.

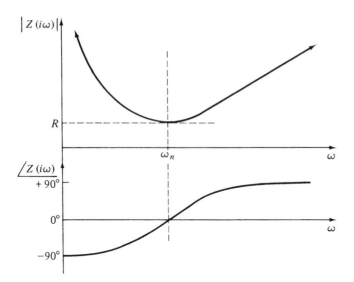

Figure 8-37 Impedance variation of series-resonant circuit.

AC POWER NUMERICAL EXAMPLE

In this text I tend to emphasize electrical applications from the instrumentation and control fields since these are major areas of system dynamics interest. We do not, however, want to completely neglect the classical area of "ac power," so let's do an example which brings out some useful concepts from this field, namely *power factor and power factor correction*. Consider the circuit of Fig. 8-33d but with no capacitor. Let $R = 22\,\Omega$, $L = 0.1$ H, and e_i taken to be the 110 volt RMS, 60 Hz from a "wall plug." We have

$$Z(i\omega)|_{\omega=377} = 22 + i37.7 = 43.6 \;\underline{/+59.8°} \text{ ohms} \tag{8-85}$$

The RMS current is thus $110/43.6 = 2.52$ amp and the power factor is $\cos 59.8° = 0.503$. The product of RMS voltage and current is called the *volt-amperes* and is 277, while the actual power consumed is the product of volt-amperes and power factor, in this case 139.5 watts. We can check this by computing the power I^2R in the resistor, which should also be 139.5 watts since the inductance does not actually consume any power. In ac power circuits such as those including motors, transformers, induction furnaces, etc., where loads are inductive and the power factor lagging, one often adds capacitance to bring the power factor closer to 1.0. This is called *power factor correction* and is encouraged by the power companies by offering reduced rates. To get a power factor of 1.0 in our present circuit we need to add a capacitor such that

$$\frac{1}{377C} = 37.7 \qquad C = 70.4\,\mu\text{F} \tag{8-86}$$

Then $Z = 22\;\underline{/0°}$ ohms, the current is 5 amp RMS, and both the volt-amperes and the consumed power are 550 watts. We see that with the power factor adjusted to 1.0, the same line voltage can now supply more power to the load.

Second-Order Systems and Mechanical Vibration

Actually, in most practical applications the inductive "load" (series combination of L and R) must remain connected directly across the power line even after the power-factor-correcting capacitor is added; thus it must be connected in parallel as in Fig. 8-38. For this circuit

$$Z(D) = \frac{e}{i}(D) = \frac{(1/CD)(R+LD)}{1/CD + R + LD} = \frac{LD + R}{LCD^2 + RCD + 1} \tag{8-87}$$

and the frequency response

$$Z(i\omega) = \frac{i\omega L + R}{(1 - LC\omega^2) + iRC\omega} \tag{8-88}$$

will have a zero phase angle (power factor $= 1.0$) if the numerator phase angle equals that of the denominator.

$$\tan^{-1}\frac{\omega L}{R} = \tan^{-1}\frac{RC\omega}{1 - LC\omega^2} \qquad \frac{\omega L}{R} = \frac{RC\omega}{1 - LC\omega^2} \tag{8-89}$$

$$C = \frac{L}{R^2 + L^2\omega^2} = \frac{0.1}{484 + 1420} = 52.5\,\mu\text{F} \tag{8-90}$$

If this capacitor is connected in parallel, the total impedance at $\omega = 377\,\text{rad/sec}$ is $86.5\,\underline{/0°}$, making the current drawn from the power line $110/86.5 = 1.27\,\text{amp RMS}$. Since the load is connected across the power line exactly as if C were not there, the *load* current and power are identical with our original case; however, the current drawn from the power line has been reduced from 2.52 amp to 1.27.

This reduction in line current is really the main reason for correcting the power factor to near 1.0 and explains why the power company encourages customers to do this by granting better rates if they do. Basically, generating and transmission equipment is sized to handle a required *current*; thus if currents can be kept down while still providing the required consumed power, the equipment is more efficiently utilized. When the power factor is far from 1.0, large amounts of power are pumped into the load in one part of the cycle but most of it is returned to the line in the next part. Thus the current is large without really providing much useful power to the load.

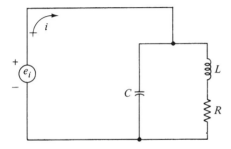

Figure 8-38 Capacitor used to correct power factor for inductive load.

Band-Pass Filters. We have earlier discussed both high-pass and low-pass filters. The tuned circuit of Fig. 8-33d which we have analyzed can be used as a *band-pass filter* if we take the output voltage across the resistor rather than the capacitor. The transfer function is then

$$\frac{e_o}{e_i}(D) = R\frac{i}{e_i}(D) = \frac{RCD}{LCD^2 + RCD + 1} = \frac{\frac{2\zeta}{\omega_n}D}{\frac{D^2}{\omega_n^2} + \frac{2\zeta D}{\omega_n} + 1} \qquad (8\text{-}91)$$

This transfer function will approach zero for both low and high frequencies, but will pass signals of intermediate frequency. If we make ζ small (by making R small), we get a sharply tuned peak at frequency ω_n. Such a circuit "passes" signals in a band of frequencies close to the peak but rejects signals with either lower or higher frequency content. This behavior is useful in a number of practical applications. Using the MATLAB bode(num,den,w) command, we can get the frequency-response curves of Fig. 8-39. The amplitude ratio at the peak appears to be 1.0 for any value of ζ and this can be seen theoretically by noting that the impedance at the peak is purely R.

To illustrate the practical use of band-pass filters we employ the simulation of Fig. 8-40. There a desired signal $1.0 \sin 1.0t$ is submerged in random noise, making its presence visually undetectable in the unfiltered signal called *xi*. By passing the noisy signal though a band-pass filter tuned to 1 rad/sec, the desired signal is clearly extracted (see Fig. 8-41). Note from the trace of the filtered signal that we must *wait* a few cycles before the desired signal fully appears. This "waiting time" depends on the filter's ζ value. Smaller ζ's give sharper tuning but require longer "waiting" before the signal becomes accurately available.

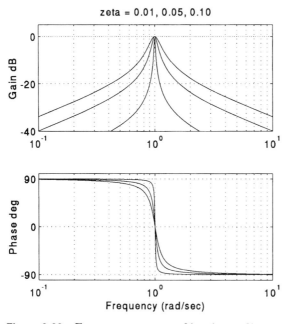

Figure 8-39 Frequency response of band-pass filter.

Second-Order Systems and Mechanical Vibration

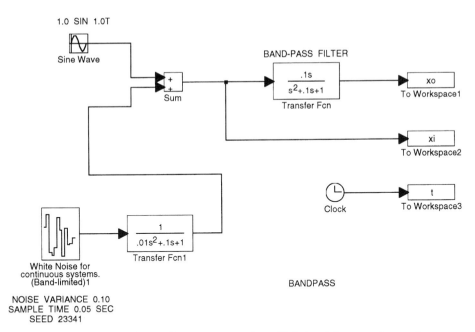

Figure 8-40 Use of band-pass filter to extract sinusoidal signal from random noise.

Notch Filters. Whereas band-pass filters select a narrow band of frequencies, *notch filters* reject one frequency completely and attenuate frequencies in a band around the "notch." Figure 8-42 shows the configuration of a symmetrical optimized[9] notch filter. Analysis (left for the end-of-chapter problems) yields

$$\frac{e_o}{e_i}(D) = \frac{R^2C^2D^2 + 1}{R^2C^2D^2 + 4RCD + 1} \tag{8-92}$$

The frequency response shows that when $\omega = 1/RC$, the numerator goes exactly to zero, which means that an input signal at this frequency produces *no* output at all. Figure 8-43 shows the frequency-response graphs. Notch filters have a number of practical uses. Most sensitive voltage recorders and *x-y* plotters use notch filters in their input circuits to reject the 60-Hz noise which often contaminates signals from sensors used for lab experiments. Such recorders often are used for signals with frequency content well below 60 Hz, so the fact that the notch filter distorts signals somewhat below 60 Hz is not a real problem.

Another common application is in feedback-type motion control systems. Sometimes the mechanical load being positioned has a lightly damped resonance (load inertia and a "springy" shaft) which cannot be redesigned. To avoid the mechanical resonance problem, the control electronics are designed with a notch filter tuned to the known mechanical frequency. Thus the resonant load never "feels" any torque signals from the motor that contain the "bad" frequency.

---

[9] J. E. Gibson and F. B. Tuteur, *Control System Components*, McGraw-Hill, New York, 1958, pp. 46–52.

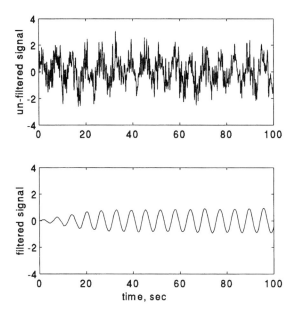

Figure 8-41 Simulation results for band-pass filtering.

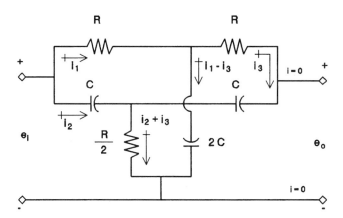

Figure 8-42 Circuit for notch filter.

Figure 8-44 shows a simulation to demonstrate this design principle. A lightly damped mechanical system is driven by a torque from an electric motor. In one case the amplifier signal to the motor is applied directly to the load, and in the other case the signal is first passed through a notch filter tuned to the mechanical resonant frequency. By running these two cases "side by side" we can see how the filter improves the response.

Two kinds of inputs will be tested: a step input, and a random input which has strong frequency content near the mechanical resonance. Figure 8-45 displays the step input test and clearly shows that the filter prevents any violent vibrations of the load. We also graph the output of the notch filter to show what the filter does to

Second-Order Systems and Mechanical Vibration 575

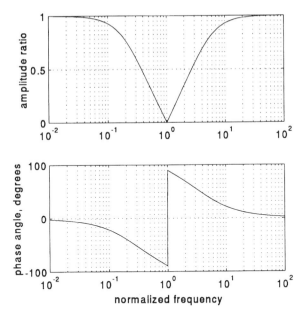

Figure 8-43 Frequency response of notch filter.

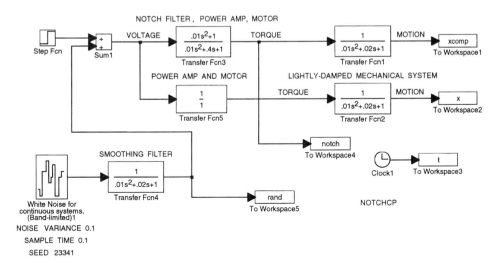

Figure 8-44 Use of notch filter to solve mechanical resonance problem.

produce the desired effect. In Fig. 8-46 we again see the beneficial effect when a random input is applied. We should also note that in many control systems the various types of filters we have discussed (low-pass, high-pass, band-pass, and notch) are implemented in digital *software* (algorithms) rather than the hardware we have shown. The inner workings of these digital filters usually *don't* have to be designed by the motion-control engineer. They are provided as "black boxes" by the software engineers who design commercial motion-control software. This software

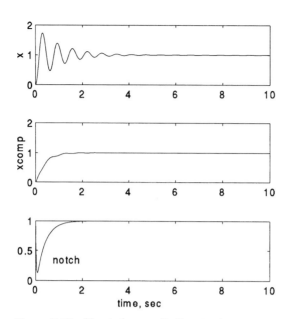

Figure 8-45 Simulation results for step input.

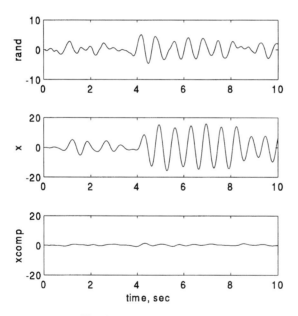

Figure 8-46 Simulation results for random input.

usually requires the user to merely enter numerical values for the filter parameters needed, such as the frequency of the notch in a notch filter.

Op-Amp Circuits. Many filters, controllers, and compensators used in instrumentation and control systems use op-amps to gain various design and performance

Second-Order Systems and Mechanical Vibration

advantages not available from passive circuits. Figure 8-33g and h shows two op-amp circuits that exhibit the standard second-order behavior. As mentioned earlier, such active circuits can give oscillatory response even though no inductors whatever are involved. The circuit of Fig. 8-33h uses three op-amps connected as summing integrator, inverter, and integrator, with proper feedback to the summer, much as would be done in an analog computer. While this circuit is flexible and easily adjusted to a wide range of parameters, it is rather expensive and wasteful of components, and thus might not be preferred for some special-purpose applications where cost was a factor and flexibility less significant. Since Fig. 8-25g also uses a configuration different from that (Fig. 3-30a) which we have assumed up to now, let us analyze it to show the techniques involved.

Figure 8-47 shows this circuit with parameter values as recommended by an op-amp manufacturer.[10] While the circuit appears complex at first glance, it yields readily to application of the current node law at the location N.

$$i_b + i_i + i_a = i_C + i_R \tag{8-93}$$

The voltage e_a at node N is found from

$$e_a = -i_b R = \left(-\frac{C}{\gamma} De_o\right) R \tag{8-94}$$

since the capacitor of value C/γ has one end at e_o and the other at the op-amp input terminal, which we recall to be a virtual ground, that is, nearly at zero potential. Since the op-amp input current is also treated as zero, the current in C/γ must also go through R to reach N. Since the right end of R is at zero potential, N must be at potential $-i_b/R$. We can now express all the currents in Eq. (8-93) in terms of e_i and e_o and thereby get the desired relation between e_i and e_o:

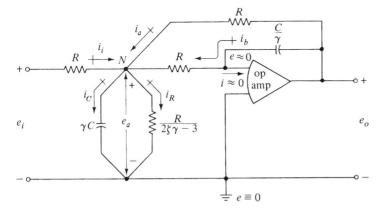

Figure 8-47 Second-order active circuit using a single op-amp.

---

[10] The Lightning Empiricist, G. A. Philbrick Researches, Vol. 13, Nos. 1–4, Dedham, Mass., 1965.

$$i_a = \frac{e_o - e_a}{R} = \frac{\left(1 + \dfrac{RC}{\gamma} D\right) e_o}{R} \qquad i_i = \frac{e_i - e_a}{R} = \frac{e_i + \dfrac{RC}{\gamma} De_o}{R} \tag{8-95}$$

$$i_C = \gamma C De_a = -RC^2 D^2 e_o \qquad i_R = \frac{e_a}{R/(2\gamma\zeta - 3)} = -\frac{C(2\gamma\zeta - 3)}{\gamma} De_o \tag{8-96}$$

Substitution in (8-93) leads to

$$(R^2 C^2 D^2 + 2\zeta RCD + 1)e_o = -e_i \tag{8-97}$$

$$\frac{e_o}{e_i}(D) = \frac{K}{\dfrac{D^2}{\omega_n^2} + \dfrac{2\zeta D}{\omega_n} + 1} \tag{8-98}$$

$$K \triangleq -1 \frac{\text{volt}}{\text{volt}} \qquad \omega_n \triangleq \frac{1}{RC} \frac{\text{rad}}{\text{sec}} \qquad \zeta \triangleq \zeta \tag{8-99}$$

DESIGN EXAMPLE: OP-AMP CIRCUIT

In designing a circuit of this type, recommendations given in the reference are helpful. These include

$$\gamma \geq \frac{3}{2\zeta} \qquad \zeta \geq 0.87 \sqrt{\frac{f_n}{P_{gb}}} \tag{8-100}$$

where $f_n \triangleq \omega_n/2\pi$ and $P_{gb} \triangleq$ *gain-bandwidth product* of the op-amp, in Hz, a performance specification commonly quoted for op-amps and typically about 10^6 Hz for good-quality amplifiers. To judge whether the circuit will load the device supplying the input voltage e_i we must know the input impedance at these terminals; it is approximately R. Using these guidelines let us design a circuit with $f_n = 100$ Hz, $\zeta = 0.1$, and an input impedance of 50,000 ohms. This input impedance would generally cause acceptably small loading of the circuit providing input to the op-amp circuit if that circuit's output impedance was less than about 5000 ohms (our usual "10-to-1" rule). Since R must be 50,000 to get the desired input impedance, Eq. (8-99) gives

$$C = \frac{1}{(50,000)(628)} = 0.0319 \times 10^{-6} \text{ farad} = 0.0319 \, \mu\text{F} \tag{8-101}$$

while (8-100) gives $\gamma \geq 15$, let's try $\gamma = 30$. Then the two required capacitors are $\gamma C = 0.955 \, \mu\text{F}$ and $C/\gamma = 0.0016 \, \mu\text{F}$ while the resistor $R/(2\zeta\gamma - 3) = 16,600$ ohms. From (8-100) the gain-bandwidth product for our op-amp must be at least 7500, which is easily met.

The circuit type just discussed is useful for natural frequencies in the range of about 0.01 to 100,000 Hz and a wide range of ζ values from underdamped to overdamped. For very small or zero (undamped) values of ζ the conventional "analog computer" type circuit (Fig. 8-33h) may, however, be more appropriate.

Second-Order Systems and Mechanical Vibration **579**

In these circuits $\zeta = 0$ is achieved by severing a feedback path completely, rather than setting a component to a particular numerical value. Thinking of the circuit as an analog computer, it is set up to solve the second-order differential equation with zero damping. This approach is quite practical and is used to build sinusoidal oscillators of very pure waveform. In Fig. 6-9, for example, we would disconnect the feedback path through the coefficient B, making B exactly zero. The signal generator would also be disconnected. To generate an undamped sine or cosine wave we would just apply a nonzero initial condition to either the first integrator or the second integrator. These initial conditions would be just like giving an undamped spring-mass system some initial energy and then "letting it go." It oscillates sinusoidally "forever." (Practical oscillators that use this principle and must operate for long periods of time add a nonlinear feature which holds the amplitude constant indefinitely.)

8-10 FLUID SECOND-ORDER SYSTEMS

Figure 8-48 displays some examples of fluid systems whose behavior, with respect to the labeled input and output quantities, will be found to follow the basic second-order equation (8-2). In Fig. 8-48a we see an example of cascaded first-order systems in which no loading effect whatsoever occurs. That is, the addition of the second tank has no influence at all on the response of the first. When connected as in 8-48b, however, the usual loading effect *does* occur. Figure 8-48c and d again might represent pressure-measuring systems as did their first-order counterparts in Fig. 7-39; however, we now no longer neglect fluid inertance in the tube. The dynamic response of pressure sensing systems is of great practical interest and is covered in depth in specialist texts.[11]

Let us first analyze the two-tank system of Fig. 8-48b. In most such tank systems the inertance of the liquid can be neglected since flow accelerations are quite small. Since the pressure at the bottom of the tanks is determined by gravity effects, it is usually small enough to neglect liquid compressibility. Using these assumptions, the only significant fluid elements are the compliances of the tanks and the resistance in the "pipes." The "pipe" resistance can also include resistance effects of constrictions or valves, which commonly appear in tank systems. Interconnected tanks appear regularly in power plants, refineries, chemical processes, food processing, etc. and often are part of automatic control systems. The design of the overall control system requires knowledge of tank system dynamics.

Having recognized the presence of a loading effect, we avoid the possible loading errors by analyzing the entire system as an entity, rather than trying to use available results for single tanks. During a time interval dt, a volume inventory (really conservation of mass) for the left tank gives

$$q_i \, dt - \left[\frac{\gamma(h_1 - h_o)}{R_{f1}} \right] dt = A_{T1} \, dh_1 \qquad (8\text{-}102)$$

---

[11] E. O. Doebelin, *Measurement Systems*, 4th ed., pp. 473–489.

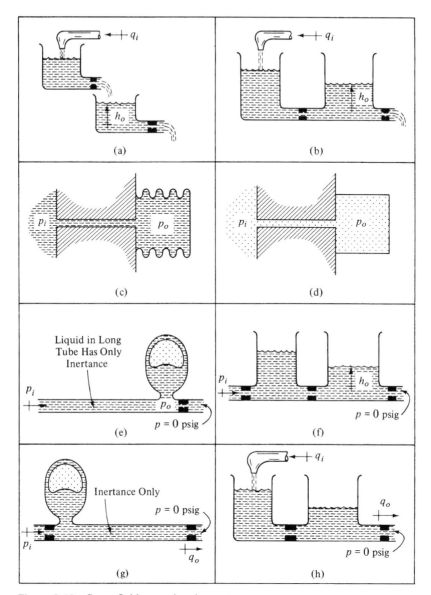

Figure 8-48 Some fluid second-order systems.

where

$\gamma \triangleq$ specific weight of liquid
$h_1 \triangleq$ level in left tank
$A_{T1} \triangleq$ cross-section area of left tank
$R_{f1} \triangleq$ fluid resistance between tanks

Similarly, for the right tank,

$$\left[\frac{\gamma(h_1 - h_o)}{R_{f1}} - \frac{\gamma h_o}{R_{f2}}\right] dt = A_{T2}\, dh_o \tag{8-103}$$

Second-Order Systems and Mechanical Vibration   **581**

where

$$R_{f2} \triangleq \text{outlet fluid resistance}$$
$$A_{T2} \triangleq \text{cross-section area of right tank}$$

The fluid resistances are being treated as linear; nonlinear resistances could be analyzed as such or linearized in our usual fashion. The tank compliances are also considered linear, as would be the case for any tank of prismatical shape. Nonlinear compliances (such as for spherical tanks) could be analyzed as such or linearized. When tanks are under automatic control which tries to maintain a desired level, linear or linearized models are appropriate. If nonprismatical tanks experience large level changes, nonlinear compliances should be used and response determined by simulation of the nonlinear differential equations.

Proceeding with the linear model assumed in our equations we next put them in operator form as preparation for use of determinants.

$$\left(\frac{R_{f1} A_{T1}}{\gamma} D + 1\right) h_1 + (-1) h_o = \left(\frac{R_{f1}}{\gamma}\right) q_i$$

$$\left(-\frac{R_{f2}}{R_{f1} + R_{f2}}\right) h_1 + \left(\frac{A_{T2} R_{f1} R_{f2}}{\gamma (R_{f1} + R_{f2})} D + 1\right) h_o = 0 \tag{8-104}$$

We could now easily solve for either of the two tank levels. Our interest here is in h_o so we use determinants to finally get

$$\left(\frac{R_{f1} R_{f2} A_{T1} A_{T2}}{\gamma^2} D^2 + \frac{A_{T1} R_{f1} + A_{T1} R_{f2} + A_{T2} R_{f2}}{\gamma} D + 1\right) h_o = \left(\frac{R_{f2}}{\gamma}\right) q_i$$

$$\tag{8-105}$$

EXAMPLE: USING VARIOUS CHECKING METHODS TO FIND ERRORS

We have several times earlier shown methods to check derived results for possible errors, but I want here to give more details on a variety of methods which can be applied to all kinds of results from system dynamics studies. Three useful categories are:

1. Dimensional checks
2. Special case checks
3. Limiting case checks

We will now apply each of these types to our present tank problem, but be sure that you realize that these methods are *general* and can usually be applied to many other kinds of systems. While, as professional engineers, we strive to do our work accurately, we must admit that mistakes can occur. Important results should always be checked by ourselves and then by a colleague, since it is easy to miss our own errors even if we go over the work many times. Knowing that our work will be checked by others and that we will check theirs, it is only common courtesy to make sure that we present our results in a neat and well-organized fashion.

Dimensional checks. Checking for dimensional errors is perhaps the most common type of checking, but it can take several useful forms. In Eq. (8-105) the left and right sides must have the same dimensions. If you write out the derivatives rather than using the D-operator form, then the dimensions of, say, dh_o/dt are clearly meters/seconds. If you instead stay with the D-operator form, be sure that you recall that the D-operator is *not* dimensionless; it has the dimensions of $1/\text{time}$ (D^2 would be $1/(\text{time})^2$, etc.). If you prefer the Laplace transform methods, again s is *not* dimensionless; it has dimensions $1/\text{time}$. (Sometimes we need to use for D or s the dimensions radians/time, which is correct since radians are dimensionless.) Checking each term on the left- and right-hand sides of Eq. (8-105) verifies that they all have the same net dimensions, namely meters, the same as the term h_o. When a coefficient, such as that on the Dh_o term, has a sum of terms, these must all have the same dimensions. Thus if the first term had come out $A_{T1}R_{f1}^2$, we would suspect a mistake right then.

Special case checks. Another type of checking involves evaluation of special cases for which the answer is obvious or easily found. Often the steady-state response to a step input provides such a case. For q_i equal to a constant q_{is}, Eq. (8-105) predicts (particular solution) that h_o will become steady at the value $q_{is}R_{f2}/\gamma$. Does this make sense? If h_o is steady at this value, the outflow from the right tank would be q_{is}, exactly the same as the inflow q_i to the left tank. Since inflow = outflow corresponds to an equilibrium condition, h_o could indeed remain steady at $q_{is}R_{f2}/\gamma$, verifying at least the steady-state aspect of Eq. (8-105).

Limiting case checks. Another type of checking lets selected parameters become zero or infinity and then determines whether the general equation collapses into a simpler form corresponding to a known result. This limiting case check may be applied to the flow resistance R_{f1} by letting it become zero in Eq. (8-105), giving

$$\left(\frac{R_{f2}(A_{T1} + A_{T2})}{\gamma}D + 1\right)h_o = \left(\frac{R_{f2}}{\gamma}\right)q_i \tag{8-106}$$

A little reflection shows this to be the equation of a *single* tank of area $(A_{T1} + A_{T2})$ discharging through a resistance R_{f2}. When R_{f1} is zero, the levels in the two tanks are always identical; thus the two tanks are really equivalent to one with an area equal to the sum of the individual areas, and we see that Eq. (8-105) appears to handle this limiting case correctly. Finally, let's set $R_{f1} = 0$ (same as one big tank) and $R_{f2} = \infty$ (the outflow from the right tank is completely shut off). The relation between q_i and h_o should now be that of a pure integrator, since all the inflow is captured by the tank. To check Eq. (8-105), first divide through by R_{f2} and then set $R_{f1} = 0$, $R_{f2} = \infty$ to get the result

$$\frac{A_{T1} + A_{T2}}{\gamma}Dh_o = \frac{1}{\gamma}q_i \tag{8-107}$$

$$h_o = \left(\frac{1}{A_{T1} + A_{T2}}\right)\frac{q_i}{D} = \frac{1}{A_{T1} + A_{T2}}\int q_i\, dt \tag{8-108}$$

While these various checking methods do not *guarantee* the validity of a result, they are most helpful in discovering mistakes and establishing confidence in computed values. All engineers make mistakes; good engineers discover and correct theirs before they become disasters. Remember, of course, that such checks only verify a model within the *assumptions* made at the outset. If the assumptions are far from the truth, a "correct" model based on the assumptions is of little value. There is always a *hierarchy* of models for any physical system, ranging from the simple and

Second-Order Systems and Mechanical Vibration   **583**

crude to the complex and accurate. At any stage of system design it is important to select from this hierarchy the *simplest* model which is adequate for the needs of that stage.

Let's now complete our treatment of Eq. (8-105) by putting it into standard form.

$$\frac{h_o}{q_i}(D) = \frac{K}{\dfrac{D^2}{\omega_n^2} + \dfrac{2\zeta D}{\omega_n} + 1}$$ (8-109)

$$K \triangleq \frac{R_{f2}}{\gamma} \quad \frac{\text{m}}{\text{m}^3/\text{sec}} \qquad \omega_n \triangleq \frac{\gamma}{\sqrt{R_{f1}R_{f2}A_{T1}A_{T2}}} \quad \text{rad/sec}$$

$$\zeta \triangleq \frac{A_{T1}R_{f1} + A_{T1}R_{f2} + A_{T2}R_{f2}}{2\sqrt{R_{f1}R_{f2}A_{T1}A_{T2}}}$$ (8-110)

By analogy with Eq. (8-75) we can show that here also ζ cannot be less than 1.0; oscillatory behavior is impossible. This is basically due to our neglecting all inertial effects in the fluid. Inertia of the fluid in the pipe connecting the two tanks could conceivably lead to oscillatory behavior, but only if fluid resistance were very small, such as might be associated with pipe wall friction for a low-viscosity liquid. An example is found in U-tube manometers filled with mercury, where the "tanks" are the two tubes that form the U. Such manometers are invariably underdamped and require inclusion of fluid inertia for accurate modeling.[12] For concentrated flow resistances such as orifices or partially open valves, the pressure drops due to inertia will normally be negligible compared to the resistive drops and our overdamped model will be valid. If desired, Eq. (8-109) can be cast into the two-time-constant form as was done in (8-78).

For our last example of fluid second-order systems, let's consider the tank/ tubing system of Fig. 8-48d, in which the fluid medium is a gas. The tank and tube walls are considered rigid, and the tube volume is small compared to the tank, so that fluid compliance effects are predominantly in the tank. For sufficiently low frequencies we can use a single-lump model of inertance and resistance to characterize the fluid in the tube, thus the overall model will have compliance, inertance, and resistance (one of each). In addition to its usefulness as a model for the dynamic response of gas-pressure measuring systems, this configuration also corresponds to the *Helmholtz resonator* of classical acoustics, and is thus applicable to the study of certain problems in noise reduction. Due to the low viscosity of most gases, such systems will generally be oscillatory. If the oscillation frequency predicted by our single-lump model becomes too high, we approach the regime where distributed-parameter models (or else multilump models) become necessary for accuracy.

A simple criterion for judging the validity of our model consists of comparing the characteristic lengths of the system (the length of the tube and the greatest transverse dimension of the tank) with the wavelength associated with the propaga-

---

[12]Ibid., pp. 446–455.

584   **Chapter 8**

tion of pressure waves in the fluid medium. If system dimensions are small compared with this wavelength, then our simple model should be reasonably accurate. That is, our model *neglects* the effects of wave propagation and we need to check whether this is a good assumption. From physics you may recall a general formula relating frequency and wavelength for all kinds of wave propagation phenomena (electromagnetic waves, elastic waves in solids, pressure waves in fluids, etc.).

$$\text{Wavelength } \lambda = \frac{\text{velocity of propagation}}{\text{frequency of the wave}} \triangleq \frac{c}{f} \tag{8-111}$$

For gases, the velocity of propagation ("speed of sound") c is given by

$$c = \sqrt{kgRT} \quad \frac{\text{ft}}{\text{sec}} \tag{8-112}$$

where

$k \triangleq$ ratio of specific heats for the gas

$g \triangleq$ 32.2 ft/sec

$R \triangleq$ gas constant

$T \triangleq$ gas absolute temperature, $°R$

For air at 70°F, for example,

$$c = \sqrt{(1.4)(32.2)(53.3)(530)} = 1120 \; \frac{\text{ft}}{\text{sec}} \tag{8-113}$$

Thus if the maximum dimension of a system were, say, 2 feet, and if we accept a commonly suggested "10-to-1" rule that says the wavelength should be at least 10 times the characteristic length, our model should be good for frequencies less than 56 Hz. Such "10-to-1" rules are based on the assumption that a spatial sine wave of a certain wavelength can be reasonably approximated by a 10-segment waveform which changes in 10 small steps, rather than smoothly. That is, the approximation assumes that the fluid properties and variables can be treated as *constant* over a distance of $\frac{1}{10}$ wavelength or less. Thus our piece of tubing can be treated as a single lump of fluid if its length is less than $\frac{1}{10}$ wavelength at the highest frequency for which we want to use the model.

In addition to wave propagation considerations, we also have to deal with some nonlinearities in this system. We want to treat fluid properties such as density as constants, whereas they actually vary. This problem is dealt with by restricting our model to small percentage changes away from a selected operating point. We assume that p_i and p_o are both initially constant at some value p_m, when p_i changes in an arbitrary fashion, but with small magnitude compared to p_m. For instance, if $p_m = 100$ psia we might restrict p_i to the range 90 to 110 psia, that is, a $\pm 10\%$ range about the mean value. Such restrictions are necessary since for gases the fluid compliance, inertance, and resistance are all nonlinear and a linearized model can be accurate only for small changes about some operating point. [These assumptions are good for acoustic systems, since the pressure oscillations (sound waves) are a *very* small fraction of the mean (atmospheric) pressure.]

Since small tube pressure drop $p_i - p_o$ tends to encourage laminar flow, we take the inertance and resistance for this condition. The compliance of the tank depends

Second-Order Systems and Mechanical Vibration **585**

on the type of compression process assumed; for rapid oscillation there is not enough time for much heat transfer to occur and the adiabatic (no heat flow) process is a good model. For such a process the compliance dV/dp is given by V/kp, which nonlinear term we approximate by assuming p fixed at p_m. Our model thus has elements

$$I_f = \frac{16\rho L}{3\pi D_i^2} \qquad R_f = \frac{128\mu L}{\pi D_i^4} \qquad C_f = \frac{V}{kp_m} \tag{8-114}$$

where

$$L \stackrel{\Delta}{=} \text{tube length}$$
$$D_i \stackrel{\Delta}{=} \text{tube inside diameter}$$
$$\mu \stackrel{\Delta}{=} \text{fluid viscosity}$$

Note that:

1. C_f really varies with p_o but is assumed constant at a value corresponding to p_m.
2. The density ρ actually varies from one end of the tube to the other at any instant of time, and also with time, due to changing p_i and p_o. We assume it constant at a value corresponding to p_m and initial temperature T.
3. To treat viscosity as constant, we assume the fluid in the tube remains at the initial temperature T. For the assumed adiabatic process in the tank, we could compute a varying temperature from our predicted values of tank pressure, but this calculation is not needed in our model for predicting tank pressure. The tank temperature change would also be quite small for the small pressure changes required in our model.

If we now consider p_{ip} and p_{op} to be perturbations $(p_i - p_m)$ and $(p_o - p_m)$ we may write the instantaneous volume flow rate q as

$$q = \frac{(p_{ip} - p_{op})}{R_f + DI_f} \quad \frac{\text{ft}^3}{\text{sec}} \tag{8-115}$$

For the tank compliance,

$$\int q \, dt = \frac{q}{D} = C_f p_{op} \quad \text{ft}^3 \tag{8-116}$$

thus

$$\frac{p_{ip} - p_{op}}{R_f + DI_f} = C_f D p_{op} \tag{8-117}$$

and finally

$$(C_f I_f D^2 + C_f R_f D + 1)p_{op} = p_{ip} \tag{8-118}$$

This is clearly second-order with

$$K \stackrel{\Delta}{=} 1.0 \quad \frac{\text{psi}}{\text{psi}} \qquad \omega_n \sqrt{\frac{1}{C_f I_f}} = \frac{D_i}{4}\sqrt{\frac{3\pi k p_m}{L\rho V}} \qquad \zeta \stackrel{\Delta}{=} \frac{16\mu}{D_i^3}\sqrt{\frac{3VL}{\pi\rho k p_m}} \tag{8-119}$$

Chapter 8

EXAMPLE: PRESSURE-MEASURING SYSTEM DYNAMICS

As a numerical example, consider an air system with a tank of volume $2\,in^3$, a tube with $D_i = 0.2$ inch and $L = 3$ inch, with $p_m = 14.7$ psia and temperature 75°F. The viscosity of air at 14.7 psia and 75°F is found from a table of air properties to be $3.8 \times 10^{-7}\,lb_f\text{-sec}/ft^2$ and the density is calculated from the perfect gas law as

$$\rho = \frac{p_m}{gRT} = \frac{14.7 \times 144}{32.2 \times 53.3 \times 535} = 0.00231 \; \frac{lb_f\text{-sec}^2}{ft^4}$$

$$\omega_n = \frac{0.2}{48} \sqrt{\frac{3 \times 3.14 \times 1.4 \times 14.7 \times 144}{(\frac{3}{12}) \times 0.00231 \times 1.4 \times (\frac{2}{1728})}} = 852 \; \frac{rad}{sec} \qquad f_n = 136\,Hz \qquad (8\text{-}120)$$

$$\zeta = \frac{16 \times 3.8 \times 10^{-7}}{(0.2/12)^3} \sqrt{\frac{3 \times \frac{2}{1728} \times \frac{3}{12}}{3.14 \times 0.00231 \times 1.4 \times 1.4 \times 14.7}} = 0.00836 \qquad (8\text{-}121)$$

If this system were being used to model a pressure-sensing system, the tank would represent the internal volume of a pressure transducer, whose sensitive diaphragm would be in contact with the tank pressure p_o. The tube would be a pressure-transfer tube transmitting the pressure to be measured, from its location to the transducer location. Such tubes are actually undesirable from the viewpoint of dynamic accuracy; we would much prefer to mount the pressure diaphragm in direct contact with p_i. However some transducers and/or installation setups do not allow this preferred "flush-diaphragm" arrangement and the tube is unavoidable. When you purchase a pressure transducer, the manufacturer often quotes a natural frequency, which may be rather high, let's say 10,000 Hz. When used in a situation such as we have just analyzed, this number is quite *meaningless*. The diaphragm itself may be capable of responding to a few thousand hertz, but the pneumatic dynamics of the tube and volume prevent such high frequencies from ever reaching the diaphragm. Thus the *installed* dynamics of pressure sensing systems are always worse than those of the transducer diaphragm itself, and often *much* worse. Thus analyses of the fluid system dynamics such as we present here are vital to accurate design of pressure-measuring systems.[13]

For the conditions of the above calculation, the speed of sound is 1334 ft/sec, so the wavelength at 136 Hz would be 9.8 feet, much longer than our 3-inch tube, so the single-lump model we use should be OK. Experiments have shown that this model, when used within its restrictions, is quite accurate in predicting the natural frequency, usually being with ±10% of measured values. The damping estimate is much worse, being invariably too low. In fact, the calculation of damping for any oscillatory fluid system is subject to considerable uncertainty, and experimental testing is often necessary if accurate values are important. Fortunately, the useful frequency range of pressure-measuring systems of this type depends much more on the natural frequency than on the damping ratio.

---

[13] Ibid., pp. 455–489.

Second-Order Systems and Mechanical Vibration

The (linearizing) assumption of constant ω_n actually suffers *less* from nonlinearity than we might expect. In Eq. (8-119), note that p_m appears explicitly in the numerator and implicitly in ρ of the denominator. Thus p_m effectively *cancels out*, which, at least partially, makes ω_n less sensitive to the mean pressure and pressure amplitude. Physically, this is due to the pressure affecting *both* the inertance (mass) and stiffness (spring) in the same way [see Eq. (8-114)].

8-11 THERMAL SECOND-ORDER SYSTEMS

In contrast with mechanical, electrical, and fluid systems, which can have two types of energy storage and thus may exhibit natural oscillations, thermal systems have only one type of storage element and oscillation is theoretically impossible. By taking enough "lumps" of resistance and capacitance we can get a system equation of arbitrarily high order, however the roots of the characteristic equation will always be found to be real, preventing occurrence of natural oscillations.

Improved Tank Heating Model. In Fig. 8-49 we reconsider the system of Fig. 7-44b for those applications in which the energy storage of the heater can not be neglected, as it was in that earlier model. This heater energy storage can become significant (relative to that of the fluid) when the product of temperature rise and thermal capacitance of the heater becomes sufficiently large. Such a condition may arise from large heater capacitance, resistance, or both. To take these effects into account, we must write conservation of energy equations for both the heater mass and the fluid mass.

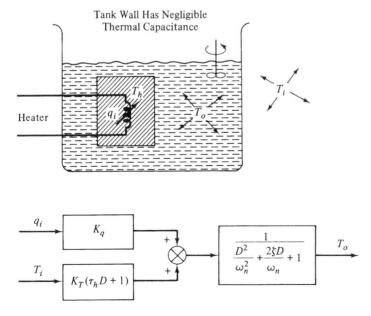

Figure 8-49 Thermal second-order system with two inputs.

Chapter 8

$$q_i dt - U_h A_h (T_h - T_o)\, dt = M_h c_h\, dT_h \tag{8-122}$$

$$U_h A_h (T_h - T_o)\, dt - U_w A_w (T_o - T_i)\, dt = M_f c_f\, dT_o \tag{8-123}$$

where

$U_h \overset{\Delta}{=}$ overall heat transfer coefficient between heater and fluid

$A_h \overset{\Delta}{=}$ surface area of heater

$T_h \overset{\Delta}{=}$ heater temperature

$M_h \overset{\Delta}{=}$ heater mass

$c_h \overset{\Delta}{=}$ specific heat of heater material

$U_w \overset{\Delta}{=}$ overall heat transfer coefficient between fluid and environment

$A_w \overset{\Delta}{=}$ surface area of vessel

$T_o \overset{\Delta}{=}$ fluid temperature

$M_f \overset{\Delta}{=}$ fluid mass

$c_f \overset{\Delta}{=}$ fluid specific heat

Our two simultaneous equations may be written as

$$(\tau_h D + 1)T_h + (-1)T_o = \left(\frac{1}{U_h A_h}\right) q_i \tag{8-124}$$

$$\left(-\frac{U_h A_h}{U_w A_w}\right)T_h + \left(\tau_f D + 1 + \frac{U_h A_h}{U_w A_w}\right)T_o = T_i \tag{8-125}$$

where

$$\tau_h \overset{\Delta}{=} \frac{M_h c_h}{U_h A_h} \qquad \tau_f \overset{\Delta}{=} \frac{M_f c_f}{U_w A_w}$$

These two τ's are legitimately defined as time constants since they do have the dimensions of time; however, they will *not* be the time constants of the complete system, as found from the system characteristic equation. We do not *need* to define them at all, but they do make the upcoming expressions more compact.

As usual we can now reduce our set of simultaneous equations to one equation in the desired unknown T_o, using substitution and elimination or determinants. If we plan only to *simulate* this system we should work directly from Eqs. (8-124) and (8-125); none of the upcoming manipulations are necessary or desirable. Proceeding by determinants we get

$$T_o = \frac{\begin{vmatrix} \tau_h D + 1 & \dfrac{q_i}{U_h A_h} \\[2ex] -\dfrac{U_h A_h}{U_w A_w} & T_i \end{vmatrix}}{\begin{vmatrix} \tau_h D + 1 & -1 \\[2ex] -\dfrac{U_h A_h}{U_w A_w} & \tau_f D + 1 + \dfrac{U_h A_h}{U_w A_w} \end{vmatrix}} \tag{8-126}$$

Expanding the determinants and then "cross-multiplying" gives

Second-Order Systems and Mechanical Vibration

$$\left[\tau_f\tau_hD^2 + \left(\tau_f + \tau_h + \frac{M_hc_h}{U_wA_w}\right)D + 1\right]T_o = (\tau_hD + 1)T_i + \left(\frac{1}{U_wA_w}\right)q_i \qquad (8\text{-}127)$$

which is of the form

$$\left[\frac{D^2}{\omega_n^2} + \frac{2\zeta D}{\omega_n} + 1\right]T_o = K_T(\tau_hD + 1)T_i + K_qq_i \qquad (8\text{-}128)$$

$$K_T \triangleq 1.0 \; \frac{°F}{°F} \qquad K_q \triangleq \frac{1}{U_wA_w} \; \frac{°F}{Btu/sec} \qquad (8\text{-}129)$$

$$\omega_n \triangleq \sqrt{\frac{1}{\tau_f\tau_h}} \; \frac{rad}{sec} \qquad \zeta \triangleq \frac{\tau_f + \tau_h + M_hc_h/U_wA_w}{2\sqrt{\tau_f\tau_h}} \qquad (8\text{-}130)$$

Since we know that the system will be overdamped we may wish to rewrite Eq. (8-128) in the two-time-constant form of (8-78), factoring the quadratic to find the two system time constants, which will *not* be τ_f and τ_h [see Eq. (8-127) to deduce under what circumstances τ_f and τ_h *will* be close to the system time constants]. Note from Eq. (8-128) that the response to heating rate is our standard second-order form while the response to "outside" temperature has numerator dynamics, and thus requires a separate solution.

Accelerated Coffee Cooling. There are several ways to speed up the cooling of a cup of overly hot beverage such as coffee. One method provides a nice example of the utility of simulation for certain classes of problems; it involves repeated transfers of a metal spoon between the hot coffee and the cool room air. While the spoon is in the coffee, it is absorbing some of its heat, and when we remove the spoon to the cool air, this heat is removed from the spoon, whereupon we repeat this cycle over and over. To "optimize" this process we might be interested in discovering how long the spoon should remain in each medium to accomplish the greatest cooling in the shortest time. While our coffee application may seem a little frivolous, it is related to serious industrial heat transfer processes in devices called *regenerators*, thus our study satisfies both our everyday curiosity and also results in a professional benefit. Figure 8-50 shows an idealized version of the coffee-cooling situation. The coffee cup is modeled as a single thermal capacitance C_c for the mass of coffee and a single overall thermal resistance R_{ca} between coffee temperature T_c and air temperature T_a, with air temperature assumed constant. The metal spoon is modeled as a single thermal capacitance C_s at temperature T_s. We need *two* different thermal resistances associated with the spoon; R_{sa} when the spoon is giving up heat to the air, and R_{sc} when the spoon is taking up heat from the coffee. Our simulation must provide "logic" for properly switching between these two resistances when the spoon is moved from one location to the other. We can now write two conservation of energy equations as follows.

$$\frac{((T_a \text{ or } T_c) - T_s)}{R_{sa} \text{ or } R_{sc}} = C_s\frac{dT_s}{dt} \qquad (8\text{-}131)$$

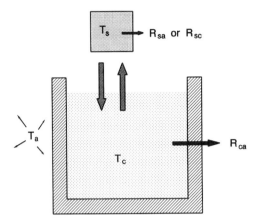

Figure 8-50 Coffee-cooling example.

$$\frac{(T_s - T_c) \text{ or } 0}{R_{sc}} + \frac{(T_a - T_c)}{R_{ca}} = C_c \frac{dT_c}{dt} \tag{8-132}$$

We see from these equations that some logic will also be needed to select the proper temperatures to use in the equations as the spoon moves from one medium to the other.

Figure 8-51 shows a SIMULINK simulation diagram for this system; it uses the SWITCH module to implement the logic needed to select the proper temperatures and thermal resistances. (If we were using a command-line type of simulation language this logic would be handled with some kind of IF statement.) The

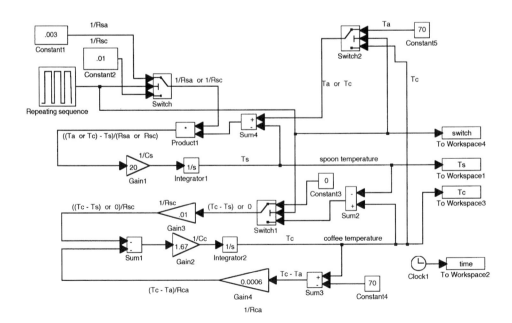

Figure 8-51 Simulation diagram for coffee-cooling problem.

Second-Order Systems and Mechanical Vibration

SWITCH modules are activated by the output signal of the REPEATING SEQUENCE module. This is where we select the timing cycle for moving the spoon between coffee and air. Since SWITCH modules change state when the middle input changes sign, we could use for our repeating sequence any waveform which changed sign at the desired times. I have here chosen a square wave going between +1 and −1 for this signal. The "duty cycle" is 5 seconds in the coffee and 10 seconds in the air, but this is easily adjusted to search for the optimum cycle. Air temperature is taken as 70°F and the various R's and C's are given some arbitrary numerical values. Figure 8-52 shows some results for the numbers given in Fig. 8-51.

Our system equations, being linear with constant coefficients, could be solved analytically; however, it would be extremely tedious and error-prone. Whenever the spoon moved from one medium to the other we would have to change equations to suit the new conditions, using the final condition of the previous equation as the initial conditions of the new equation. All this is possible, but not very pleasant, particularly if we try to carry through the parameters as letters rather than numbers. With simulation, the solution is quick and easy, though we can't of course work with letters. We can, however, evaluate different sets of numerical values quite quickly and thus find an optimum design if one exists.

8-12 MIXED SECOND-ORDER SYSTEMS

Of the many possible examples that might be shown we select one hydromechanical system and one electromechanical.

Hydraulic Material-Testing Machine: Resonance Put to Good Use. In our earlier coverage of vibration we concentrated our study of the resonance phenomenon on situations where it presented a danger. Our present example will instead show how

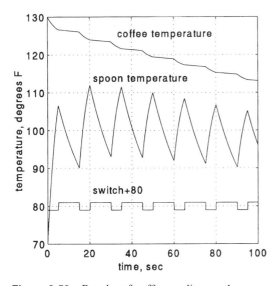

Figure 8-52 Results of coffee-cooling study.

designers used resonance to improve the performance of the world's largest dynamic materials testing machine.[14] This machine is installed at the Federal Institute of Materials Research in Berlin, Germany, and has a static force capability of ± 20 MN (4.4 million pounds) with a dynamic force capability of ± 13 MN (2.86 million pounds). The machine is basically an electronically controlled servohydraulic system in which a servovalve supplies high-pressure oil to a large-diameter piston/ cylinder, which applies the force to the test specimen. The reference describes a typical specimen as a 600-mm-diameter welded tube 4 m long, used as part of an off-shore drilling rig. To fatigue test the welds, a typical sinusoidal force of ± 4 MN at 15 Hz was superimposed on a static load of 7 MN and run for up to 15 million cycles. By running at resonance, much higher frequencies (15 Hz instead of 0.5) could be used and the machine energy consumption reduced by factors as large as 50-to-1. Without resonant operation such testing can consume about 5000 kW of power, which is prohibitive.

The idea behind resonant operation is that the metal test specimen acts like a very stiff spring connected to the piston of the hydraulic cylinder, which has mass, thus creating a mechanical system with a natural frequency and not much damping. By driving the servovalve electrically at this resonant frequency, large specimen forces can be created with small exciting forces. A major problem is that specimens with widely varying spring constants must be tested, while the machine's piston has a fixed mass, giving a wide range of natural frequencies which are not really selectable by the machine operator. An obvious solution is to simply attach additional masses to the piston until you get a natural frequency compatible with the machine's capabilities, which allow frequencies in the range 7 to 30 Hz. This turns out to usually not be practical since the needed masses can be the order of 200,000 kg!

The solution to this problem was to get the added mass effect using the fluid inertance of oil in "resonance tubes" which are attached to the hydraulic cylinder rather easily. It turns out that the area ratio of the hydraulic cylinder to the resonance tube provides a multiplying effect on the mass. That is, if the tube has an area, say $\frac{1}{50}$ of the cylinder, the mass of the oil in the tube is magnified 2500 times in its effect on the piston dynamics. Thus 100 kg of oil can have the same effect on the natural frequency as 250,000 kg of metal weights attached to the piston. By varying the number, length, and diameter of the tubes, we have a flexible and convenient method for adjusting the mass to suit a wide range of specimen stiffness.

Figure 8-53 shows the essential features of this system. The upper servovalve/ actuator controls the static force in the main cylinder, which has a diameter of about 37 inches and applies the force to the test specimen. This large actuator system is "tuned" to the desired frequency by connecting a proper set of resonance tubes and accumulators, as shown. The dynamic force is controlled with the lower (small) servovalve/actuator, which applies the dynamic force f_i, causing piston and specimen displacement x. This actuator need supply only relatively small forces because it is driving a lightly damped second-order system at resonance. Our analysis will develop the differential equation relating x to f_i and the needed formulas to properly design

---

[14]F. W. Neikes and D. Schone, Largest Dynamic Test System Performance Enhanced with Modern Digital Controls and Software, 1996, MTS Systems Corp., 14000 Technology Drive, Eden Prairie, MN, 55344-2290, 1-800-944-1687.

Second-Order Systems and Mechanical Vibration 593

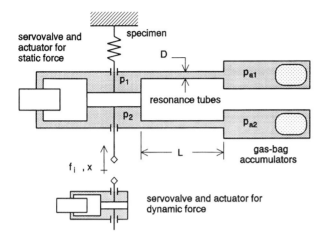

Figure 8-53 Hydraulic materials-testing machine: resonance put to practical use.

the resonance tubes. The characteristics of the servovalves are not involved in this study since the upper unit is simply holding a commanded *constant* force and the lower unit is considered just as a source for the force f_i.

Writing a Newton's law for the main piston mass M we get

$$-K_s x - B\dot{x} + A_p(p_2 - p_1) = M\ddot{x} \qquad (8\text{-}133)$$

where

$A_p \triangleq$ piston net area
$K_s \triangleq$ specimen spring stiffness
$B \triangleq$ damping associated with piston motion
$M \triangleq$ mass of all solid parts that move with the piston

To get the additional needed equations we consider the motion of the liquid in each resonance tube. A useful simplifying assumption takes the liquid to be incompressible, and thus the liquid motion is determined "kinematically" by the piston motion. That is, when the piston moves, the liquid "has no place to go" except to flow into or out of the resonance tubes. The gas-bag accumulators are precharged to a pressure high enough so that the liquid is always in compression. For example, when the piston is moving up in Fig. 8-53 the chamber containing p_2 is getting larger and the accumulator pressure p_{a2} urges the liquid in its resonance tube to "fill up" this growing space. Thus, during an oscillation, no "air spaces" ever open up in the liquid.

Since this system has been designed and operated based on the above assumption, and predictions have been found to be accurate, we can proceed with some confidence. Conservation of volume (really mass) gives us for the volume flow rates q_1 and q_2 in the two tubes

$$q_1 = A_p \dot{x} \qquad q_2 = -A_p \dot{x} \qquad (8\text{-}134)$$

The liquid in the tubes is assumed to have fluid resistance and inertance, but negligible compliance (incompressible). The inertance (mass) effect of the liquid is of course the operating principle of this design, so we certainly must include it, and in fact *design* it to get the desired effect. The fluid resistance (friction) is actually undesirable since it, together with other frictional effects, will limit the height of the resonant peak and thus require more exciting force. All the frictional effects in such a system are difficult to predict theoretically, so we provide for them in our simulation, but wait for experimental testing to provide working numerical values. Applying the "fluid version" of Newton's law to each tube's liquid we can write

$$p_1 - p_{a1} - R_f q_1 = I_f \dot{q}_1 \qquad p_2 - p_{a2} - R_f q_2 = I_f \dot{q}_2 = -I_f \dot{q}_1 \qquad (8\text{-}135)$$

Because of the system's symmetry, the inertance and resistance are taken equal in the two tubes. For the two accumulators, each with fluid compliance C_a, we get

$$C_a \dot{p}_{a2} = q_2 = -A_p \dot{x} \qquad C_a \dot{p}_{a1} = q_1 = A_p \dot{x} \qquad (8\text{-}136)$$

Combining the fluid equations we can write

$$p_1 = I_f A_p \ddot{x} + R_f A_p \dot{x} + \frac{A_p}{C_a} x \qquad (8\text{-}137)$$

$$p_2 = -\left(I_f A_p \ddot{x} + R_f A_p \dot{x} + \frac{A_p}{C_a} x\right) = -p_1 \qquad (8\text{-}138)$$

We now have a complete set of equations and can draw the simulation diagram of Fig. 8-54. The numbers there are for a specimen with $K_s = 2.77 \times 10^7 \text{ lb}_f/\text{inch}$. The frictional values B (100 $\text{lb}_f/(\text{in}/\text{sec})$) and R_f (0.002 psi/(in^3/sec)) were estimated from experimental tests on the actual system. Fluid inertance was calculated for

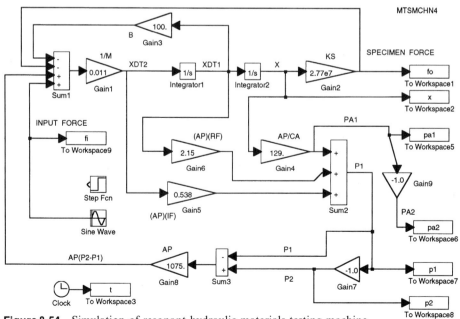

Figure 8-54 Simulation of resonant hydraulic materials-testing machine.

Second-Order Systems and Mechanical Vibration 595

resonance tubes 138 inches in length and 5.3 inches in diameter, giving $I_f = 0.0005\,(\text{lb}_f\text{-sec}^2)/\text{in}^5$. The mass M is found by weighing the parts to be 91.2 $(\text{lb}_f\text{-}\text{sec}^2)/\text{in}$ while the piston area is 1075 in². The reference did not directly give a value for accumulator compliance C_a, but I was able to indirectly estimate it at about 8.34 in³/psi.

Using the step input force of Fig. 8-54 rather than the sine wave, we can find the damped natural frequency, which should be very close to the resonant peak frequency. This turned out to be about 150 rad/sec (23.9 Hz) and we can then use this frequency for our sinusoidal input. For very lightly damped systems, the peak is quite sharp, so we need to search carefully to locate it, by trying a few frequencies in the neighborhood of 150. In the real system, the natural frequency will drift around somewhat, so the testing machine has a control system which checks for "perfect" resonance and adjusts the driving frequency to always be at the peak. For our simulation, Fig. 8-55a shows the initial buildup of the resonant oscillation when the input force amplitude is 1.0 pounds. The steady-state oscillation is pretty well established after 2 or 3 seconds, as we see in Fig. 8-55b. The output force there appears to be about 38 times the input force, so we have achieved a magnification of 38. In the simulation, one can reduce the friction without limit and thereby get huge magnifications, but the real system must of course live with whatever frictional effects are actually present; they can't be arbitrarily reduced.

If you duplicate this simulation in SIMULINK you will get a message that an algebraic loop exists, but the software handles it with no problems. This loop can be avoided by manipulating the equations to eliminate the pressures and get a single equation in motion x only.

$$\left[(M + 2A_p^2 I_f)D^2 + (B + 2A_p^2 R_f)D + \left(K_s + \frac{2A_p^2}{C_a}\right)\right]x = f_i \qquad (8\text{-}139)$$

This equation can be simulated directly, but of course provides no direct information on any of the fluid variables. It does have the advantage of clearly showing spring, mass, and damping effects of the mechanical and fluid parts of the system. In particular, we see that the fluid equivalent mass is $2A_p^2 I_f$, which allows us to easily design the inertance to get a desired total mass and natural frequency. We also see that the fluid equivalent damping is $2A_p^2 R_f$, which can be directly compared with B. Finally, the accumulators have an equivalent spring effect of $2A_p^2/C_a$. The reference says that this term is negligible relative to the specimen spring stiffness K_s, being "2 orders of magnitude" smaller. I used this statement and the known spring stiffness to estimate C_a, since it was not directly given in the reference.

dc Motor Control by Field and Armature. While dc motors used in motion-control systems most commonly employ permanent magnet fields and armature control (see Fig. 7-48), some applications control the motor by manipulating the strength of a wound field, or use combined field/armature control.[15] Figure 8-56 shows a dc motor with a separately excited field and a constant-current armature supply. Manipulation of the field voltage allows control of the motor's motion. We will analyze this system and find that it is essentially linear with constant coefficients. If we then also manip-

---

[15]F. J. Bartos, DC drives still stand and deliver, *Control Engineering*, March 1996, pp. 56–60.

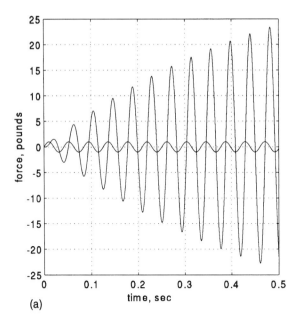
(a)

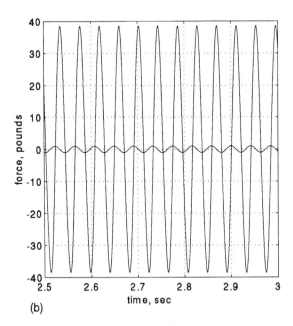
(b)

Figure 8-55 (a) Transient buildup in resonant testing machine; (b) steady-state resonant amplification.

Second-Order Systems and Mechanical Vibration

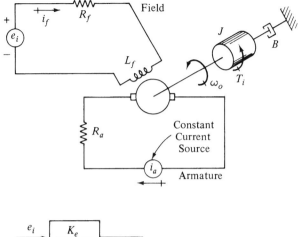

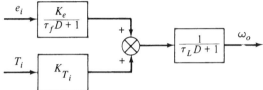

Figure 8-56 DC motor control by field and armature.

ulate the armature current as a control input, more versatile control is achieved, but the system now has a time-varying coefficient and analytical solution is not possible. We can use the same Taylor series technique used on nonlinear systems to approximate this equation, or, of course, we can simulate the exact equations. Finally, the armature supply can be changed from a current supply to a voltage supply, either constant or variable, giving two more versions of this system.

For the simplest version (constant armature current) we can write

$$i_f R_f + L_f \frac{di}{dt} - e_i = 0 \tag{8-140}$$

$$(\tau_f D + 1)i_f = \frac{1}{R_f} e_i \qquad \tau_f \triangleq \frac{L_f}{R_f} \tag{8-141}$$

We see that the field current (and thus the field strength) is determined entirely by the field voltage e_i, independent of what might be going on in the armature circuit. Thus we can solve directly for i_f as soon as e_i is specified. (Actually, a phenomenon called "armature reaction" effects the field when armature current flows, but this effect is usually small enough to neglect.) Since we assume a *current* source for the armature, no armature circuit equation is needed, and armature resistance, inductance, and back emf play no role. (This will *not* be true when we later consider a *voltage* source for the armature.) Turning to Newton's law for the rotating load we get, using Eq. (5-1),

$$T_i + K_{Twf} i_f i_a - B\omega_o = J\dot{\omega}_o \tag{8-142}$$

$$(\tau_L D + 1)\omega_o = K_{T_i} T_i + \frac{K_{Twf} i_a}{B} i_f \tag{8-143}$$

$$\tau_L \triangleq \frac{J}{B} \quad \text{sec} \qquad K_{T_i} \triangleq \frac{1}{B} \quad \frac{\text{rad/sec}}{\text{n-m}} \tag{8-144}$$

Since i_f is already known from (8-141),

$$(\tau_L D + 1)\omega_o = K_{T_i} T_i + \frac{K_{Twf} i_a}{B} \left[\frac{1/R_f}{(\tau_f D + 1)} \right] e_i \tag{8-145}$$

$$(\tau_L D + 1)(\tau_f D + 1)\omega_o = K_{T_i} T_i + K_{e_i} e_i \qquad K_{e_i} \triangleq \frac{K_{Twf} i_a}{BR_f} \quad \frac{\text{rad/sec}}{\text{volt}} \tag{8-146}$$

To draw a block diagram we need the transfer functions

$$\frac{\omega_o}{e_i}(D) = \frac{K_{e_i}}{\tau_f \tau_L D^2 + (\tau_f + \tau_L)D + 1} \tag{8-147}$$

$$\frac{\omega_o}{T_i}(D) = \frac{K_{T_i}(\tau_f D + 1)}{(\tau_L D + 1)(\tau_f D + 1)} = \frac{K_{T_i}}{\tau_L D + 1} \tag{8-148}$$

Note that the "cancellation" of $(\tau_f D + 1)$ makes the response of speed to the external torque T_i a first-order type, while the response to field voltage is second-order. This somewhat unusual result may be attributed to the "one-way" nature of the coupling between the field circuit and the load motion. Application of field voltage directly produces torque, which obviously influences load motion; however, application of torque T_i, while influencing load motion, has *no* effect on what is happening in the field circuit, as Eq. (8-140) shows. Thus we should not expect the speed/torque transfer function to involve τ_f. Note also that the second-order response of Eq. (8-147) arises as the product of two first-order terms, and that no loading effect is present, because of the "one-way coupling" present.

If we now modify the system of Fig. 8-56 to allow armature current to be an adjustable control input rather than a fixed parameter, we get more versatile control, but a linear differential equation with a time-varying coefficient, a class of equations usually analytically unsolvable. We can of course treat this exactly with simulation, but are then denied any useful design relations that come from an analytical solution. The offending term in the equation, while not nonlinear, can be approximated with the same Taylor series technique we regularly use for nonlinear terms. The field circuit equation is of course unaffected, but the Newton's law is changed as follows:

$$T_i + K_{Twf} i_a i_f - B\omega_o = J\dot{\omega}_o \tag{8-149}$$

Since $i_f(t)$ can be solved for before trying to solve for ω_o, the term $i_f i_a$ has a time-varying coefficient. (Actually, the product $i_f i_a$ could be considered *one* time-varying input, since both currents would be known as functions of time, however we want to use e_i and i_a as *separately adjustable* control signals. Also, if this system is made part of a feedback-type motion-control system, then i_a becomes a system *unknown* rather than a given input.) Let's thus proceed with our approximate analytical treatment.

$$i_a i_f \approx i_{ao} i_{fo} + i_{ao} i_{fp} + i_{fo} i_{ap} \tag{8-150}$$

Second-Order Systems and Mechanical Vibration

Here we have defined our usual operating-point and perturbation values of the currents, and can now rewrite the Newton's law as

$$(T_{io} + T_{ip}) + K_{Twf}(i_{ao}i_{fo} + i_{ao}i_{fp} + i_{fo}i_{ap}) - B(\omega_{oo} + \omega_{op}) \approx J\dot{\omega}_{op} \qquad (8\text{-}151)$$

If you now write the steady-state version of Newton's law and subtract it from (8-151), you get

$$T_{ip} + (K_{Twf}i_{ao})i_{fp} + (K_{Twf}i_{fo})i_{ap} - B\omega_{op} = J\dot{\omega}_{op} \qquad (8\text{-}152)$$

$$(\tau_L D + 1)\omega_{op} = K_{T_i}T_{ip} + \frac{K_{e_{i2}}e_{ip}}{\tau_f D + 1} + K_{i_a}i_{ap} \qquad (8\text{-}153)$$

$$K_{e_{i2}} \triangleq \frac{K_{Twf}i_{ao}}{BR_f} \qquad K_{i_a} \triangleq \frac{K_{Twf}i_{fo}}{B} \qquad (8\text{-}154)$$

which allows us to draw the block diagram of Fig. 8-57.

The two other versions of this system which we mentioned earlier (let the armature supply be a *voltage* source, either fixed or variable), are left as end-of-chapter problems. They can be treated exactly with simulation or "linearized" using the same techniques we just applied to the current-source problem.

8-13 SYSTEMS WITH NUMERATOR DYNAMICS

Our emphasis in this chapter has been on the simplest and most common form of second-order system, that defined by Eq. (8-3). The most general form, Eq. (8-1), admits of several special cases. If b_1 and/or b_2 are present, we say the system has numerator dynamics. We have already in this chapter encountered some such examples—Eqs. (8-55), (8-58), (8-64), (8-80), (8-87), (8-91), (8-92), (8-128). In this section we present two examples specifically chosen to illustrate this type of dynamic behavior.

Automobile Handling Dynamics. Vehicles of all kinds (cars, airplanes, ships, trains, satellites, etc.) are technological products of great economic importance to modern

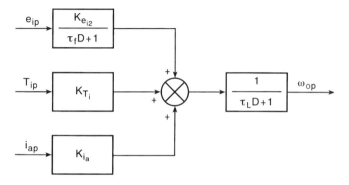

Figure 8-57 Motor speed control using both field and armature inputs.

600   **Chapter 8**

society. Their dynamic behavior is often of critical importance and is an interesting application of system dynamics methods. We choose here to introduce some simple models of automobile "handling" behavior, since this vehicle is the most familiar to most of us. There are many interesting dynamic problems "internal" to the automobile (engine balancing, engine control, transmission shifting, etc.) but we here focus on motions of the "entire" vehicle. Engineering groups in automobile companies separate these problems into three major groups:

1. Performance (acceleration and braking)
2. Ride (vertical vibration, pitching and rolling)
3. Handling (vehicle maneuvering)

Let's now develop the simplest handling model that still predicts behavior close to what is measured in road tests.

While there is some interaction between the three types of motion listed above, experience has shown that useful results can be achieved by considering them one at a time, so we will consider the handling motions as if the others were not present. The car will be treated as a single mass which can translate forward and sideways and rotate ("yaw") about a vertical axis, always staying level in a horizontal plane. (The next more complicated model treats the car as a two-mass system, with the body free to roll relative to the frame.[16] This apparently simple change complicates the model considerably.) The handling (also called "lateral-directional") motions of the car are caused by three main inputs: driver steering, road unevenness, and wind forces. Again, we choose to consider these separately, and study only the driver's steering input.

We consider an automobile proceeding down a straight and flat road at a constant speed V m/sec when the driver initiates a steering maneuver by turning the front wheels through a small angle δ. This will cause the car to deviate from its original straight path and we are interested in this motion. Experiments and more comprehensive analysis have shown that for maneuvers which are not too violent (sidewise acceleration less than about $0.4\,g$'s) a linearized study gives results which compare well with actual behavior. Under our assumptions, the only forces available to cause the motion are the horizontal forces of the road on the tires, and thus on the car. A key to the rational analysis of automobile steering dynamics is an understanding of how a pneumatic tire develops a sidewise force (called the "cornering force").

The slip angle concept provides this understanding and its discovery, years ago, was a significant breakthrough. In Fig. 8-58, when the center plane of the tire and the velocity of its axle (viewed from above) are aligned, no sidewise force is developed at the tire/road interface. To develop a cornering force, a slip angle β_t must exist and experiments show that if β_t is less than about $6°$ the cornering force is proportional to β_t. The proportionality factor K_t is called the cornering stiffness and is the order of 220 to 880 n/degree. Its numerical value is found by experiment on special tire-testing machines.

---

[16]E. O. Doebelin, *System Modeling and Response*, chap. 12, Vehicle Dynamics.

Second-Order Systems and Mechanical Vibration

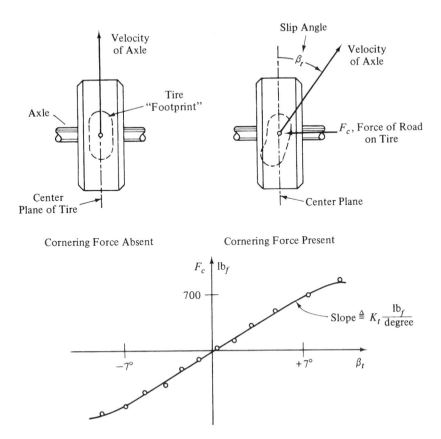

Figure 8-58 Automotive tire cornering force and slip angle.

Since the car's motion is a combination of rotation about a vertical axis and sidewise translation, superimposed on the constant-velocity forward motion, the pertinent physical laws are

$$\sum F_Y = MA_Y \tag{8-155}$$

$$\sum T_Z = J_Z \ddot{\psi} \tag{8-156}$$

Figure 8-59 shows that A_Y is the sidewise acceleration of the center of mass of the entire car, and ψ (called the yaw angle) denotes the angular displacement of the car centerline about the vertical Z axis. The moment of inertia of the entire car about the Z axis is called J_Z and is found by experiment, as described in Fig. 2-39. Because the direction of the velocity V need not coincide with that of the car centerline, the angle β (called the sideslip angle) is defined as the angle between them. Assumption of small angles allows us to approximate the sidewise acceleration A_Y in terms of $d\psi/dt$ and $d\beta/dt$ as shown in Fig. 8-59.

It is next necessary to express the forces and torques in Eqs. (8-155) and (8-156) using the cornering stiffness of the tires and the slip angles. Note that the four tire/road interfaces are the *only* places where forces act on the car, since we have

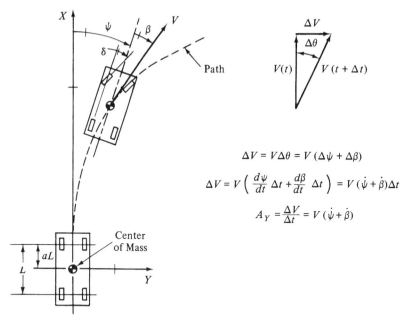

Figure 8-59 Kinematics of simple vehicle handling model.

neglected aerodynamic forces. (The largest air force is the drag opposing the forward motion. This has *no* effect since we assume a constant forward velocity; propulsion forces just balance all frictional forces. The dynamic air drag which opposes sidewise velocity and yaw rotation is small enough to neglect because these velocities are also small.) At the front wheels the sidewise velocity is $V \sin \beta + aL(d\psi/dt) \approx V\beta + aL(d\psi/dt)$, since $\sin \beta \approx \beta$ for small angles. If the steer angle δ were zero, the front wheel slip angle β_{tf} would be given by

$$\beta_{tf} \approx \tan \beta_{tf} = \frac{V\beta + \dot{\psi}aL}{V} = \beta + \frac{\dot{\psi}aL}{V} \tag{8-157}$$

Since δ is not zero, the actual slip angle will be

$$\beta_{tf} \approx \beta + \frac{\dot{\psi}aL}{V} - \delta \tag{8-158}$$

Note that the steering input is taken as the *front-wheel angle*, not the driver's steering wheel angle; thus we ignore any dynamics between these two locations. More sophisticated models include these effects. In road testing, measuring devices are connected to the front wheel angle, so our simpler model *can* be compared with test results.

While real steering systems intentionally make the two front wheel angles slightly different ("Ackerman steering"), the difference is small and largely cancelled out by using the measured *average* of the two front wheel angles as our δ when comparing theory to experiment. Treating the two front wheels as identical, we can write the total front wheel force as

$$F_{cf} = 2K_t \left(\beta + \frac{\dot{\psi}aL}{V} - \delta \right) \tag{8-159}$$

Second-Order Systems and Mechanical Vibration   **603**

$K_t \triangleq$ cornering stiffness of each front tire

While the cornering stiffness of the rear tires might be slightly different from that of the front, and it is not difficult to provide this feature in our model, we opt for simplicity and take the front and rear values equal at K_t. Thus for the rear wheels

$$F_{cr} = 2K_t\left(\beta - \frac{(1-a)L\dot\psi}{V}\right) \tag{8-160}$$

where we note that the effect of rotation $d\psi/dt$ now *subtracts* from the side velocity $V\beta$.

We are now ready to substitute into (8-155) and (8-156) to get

$$\sum F_Y = -2K_t\left(\beta + \frac{\dot\psi aL}{V} - \delta\right) - 2K_t\left(\beta - \frac{\dot\psi(1-a)L}{V}\right) = MV(\dot\psi + \dot\beta) \tag{8-161}$$

$$\sum T_Z = -2K_t\left(\beta + \frac{\dot\psi aL}{V} - \delta\right)aL + 2K_t\left(\beta - \frac{\dot\psi(1-a)L}{V}\right)(1-a)L = J_Z\ddot\psi \tag{8-162}$$

It is conventional to treat yaw rate $d\psi/dt$ and sideslip angle β as the outputs in this set of equations, with steer angle δ of course being the input. Note from Fig. 8-59 that, once we have $d\psi/dt$ and $d\beta/dt$, A_Y is easily obtained, whereupon one integration gets us sidewise velocity and one more gets us sidewise displacement Y. Several transfer functions can then be defined. If we, say, eliminate β from (8-161) and (8-162), we can get

$$\frac{\dot\psi}{\delta}(D) = \frac{K_{\dot\psi}(\tau_{\dot\psi}D + 1)}{\dfrac{D^2}{\omega_n^2} + \dfrac{2\zeta D}{\omega_n} + 1} \tag{8-163}$$

$$K_{\dot\psi} \triangleq \frac{2K_t}{MV(1-2a) + \dfrac{2K_tL}{V}} \quad \frac{\text{rad/sec}}{\text{rad}} \qquad \tau_{\dot\psi} \triangleq \frac{aVM}{2K_t} \quad \text{sec} \tag{8-164}$$

$$\omega_n \triangleq \sqrt{\frac{2K_tL\left[MV(1-2a) + \dfrac{2K_tL}{V}\right]}{MVJ_Z}} \quad \frac{\text{rad}}{\text{sec}} \tag{8-165}$$

$$\zeta \triangleq \frac{2K_tJ_Z + MK_tL^2(2a^2 - 2a + 1)}{\sqrt{2VJ_ZMK_tL\left[MV(1-2a) + \dfrac{2K_tL}{V}\right]}} \tag{8-166}$$

The location of the car's mass center, relative to the center of the wheelbase L, turns out to be a critical parameter. If distance aL is greater than 0.5L (mass center *behind* center of wheelbase), then there will be a forward speed V_{crit} above which the car becomes *unstable*. That is, if you obtain the system's second-order characteristic equation and find its two roots, they will both be real, but one will be positive. A positive root gives an increasing exponential in the complementary solution, which forces all variables to tend toward infinity. For $a > 0.5$, the critical speed is found to be

$$V_{\text{crit}} = \sqrt{\frac{2K_t L^2}{ML(2a-1)}} \tag{8-167}$$

Since this result is well known, vehicle designers generally proportion vehicles so that $a < 0.5$. Then there is *no* speed V that gives unstable roots. Note, however, that if you put some heavy passengers in the back seat and then load up the trunk with bags of cement, you can *move* the center of mass toward the rear, perhaps defeating the designers plans.

When a vehicle *does* exhibit such instability the behavior is roughly as follows. If you were going down the road at a speed greater than V_{crit}, if you momentarily steered to the left (to avoid an obstacle in the road) and then returned the steering wheel to "neutral" *and held it there*, the car would "steer itself" into a tighter and tighter turn until it ran off the road or overturned. Of course a non-suicidal driver would never let this occur! As soon as the car started to deviate, the driver would correct for this and keep the the car under control. That is, even though the *machine* is unstable, the human body and brain are capable of stabilizing the human/machine system. Of course, we do not want to force drivers to keep busy with stabilizing unstable vehicles and thus neglecting other important driving tasks, so we choose to keep $a > 0.5$ in our designs.

As usual, if we want to simulate this system, we should work from the original simultaneous equations (8-161) and (8-162). Figure 8-60 shows a SIMULINK dia-

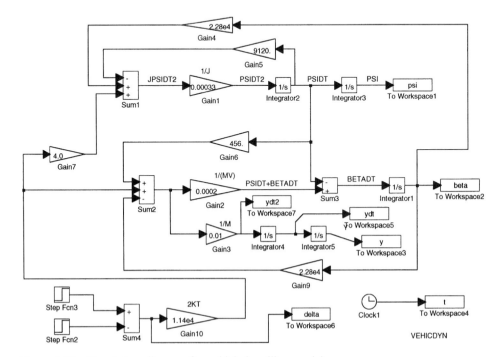

Figure 8-60 Simulation diagram for vehicle handling model.

Second-Order Systems and Mechanical Vibration **605**

gram for a particular set of numerical values. From this one diagram we can get yaw angle ψ and its first and second derivatives, sideslip angle β and its first derivative, and sidewise displacement, velocity, and acceleration. Any steering input could be used; I used a positive step, followed by a negative step, and then a return to zero. This is what one would do in changing lanes or moving to a new lateral position on the road. The numerical values used in Fig. 8-60 come from the following basic parameters.

$M = 100.\text{slugs}$
$J_z = 3000.\ \text{slug-ft}^2$
$V = 50.\ \text{ft/sec}\ (34.1\ \text{mph})$
$K_t = 5700.\ \text{lb}_f/\text{rad} \qquad L = 10.0\ \text{ft} \qquad a = 0.40$

Figure 8-61a shows the steering input and the response of yaw and sideslip angles. All these angles are small enough to satisfy our small-angle approximations. Figure 8-61b shows the sidewise displacement, velocity, and acceleration. Note that a steering input of only 0.02 radian (1.15°) causes a final sidewise displacement of about 4.2 feet. The peak acceleration is about $\pm 4\ \text{ft/sec}^2$, well below our linear model's limit of $0.4\,g$'s.

To demonstrate the instability, Fig. 8-62 shows the response when we change a to 0.6, giving a critical speed of 75.5 ft/sec. By running the car at $V = 100.\ \text{ft/sec}$ and using the same steering input as in Fig. 8-61a, we see that with the steering wheel held fast at 0.0 degrees, the car's displacement "takes off" exponentially, reaching about 50 feet in 4 seconds.

Several other transfer functions can be obtained from the basic equations.

$$\frac{\beta}{\delta}(D) = \frac{K_\beta(\tau_\beta D + 1)}{\dfrac{D^2}{\omega_n^2} + \dfrac{2\zeta D}{\omega_n} + 1} \tag{8-168}$$

$$\frac{\ddot{Y}}{\delta}(D) = \frac{K_{\ddot{Y}}\left(\dfrac{D^2}{\omega_{n,n}^2} + \dfrac{2\zeta_n D}{\omega_{n,n}} + 1\right)}{\dfrac{D^2}{\omega_n^2} + \dfrac{2\zeta D}{\omega_n} + 1} \tag{8-169}$$

It is perhaps surprising that a machine as complex as the automobile can be usefully modeled with the simplicity of Eqs. (8-161) and (8-162), but most of the results predicted are quite close to measured behavior. When maneuvers are severe enough to cause distinctly nonlinear behavior, and/or when other details ignored in our model become of interest, more complex models and simulations are necessary and have been developed. One such[17] model has 17 degrees of freedom (our model has 2), includes details such as the dynamics of antilock braking systems, and operates in "real-time." The "real-time" feature means that the simulation computer and software run fast enough (integration step size $= 855\,\mu\text{sec}$) that the simulation can

---

[17] Real-Time, Seventeen-Degree-of-Freedom Motor Vehicle Simulation, Report AB 11006, 1990, Applied Dynamics International, Inc., 3800 Stone School Road, Ann Arbor, MI 48108-2499, 313-973-1300.

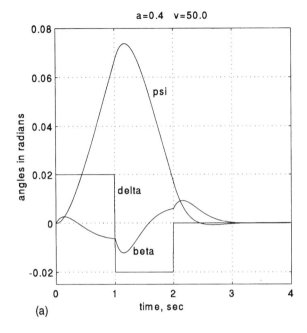
(a)

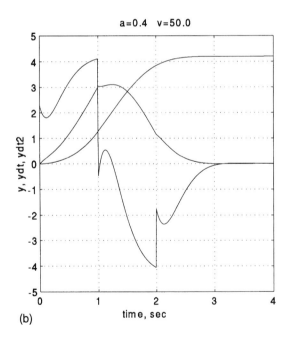
(b)

Figure 8-61 Results of vehicle handling study.

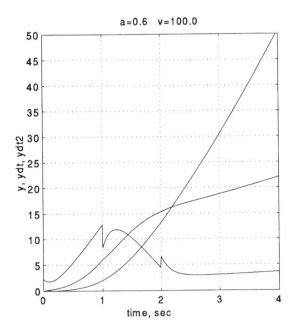

Figure 8-62 Car model predicts instability for high speeds.

be connected to real hardware, giving a so-called "hardware-in-the-loop" simulation. That is, before any hardware has been built, we must simulate "everything." As, say, an antilock braking system or active suspension system is developed and built, this real hardware can replace its computer simulation, giving a simulation which is part computer and part real hardware. The computer and hardware are interconnected using sensors, actuators, D/A and A/D converters. The advantage of such an arrangement is that the hardware part of the simulation requires no assumption making or equation writing; the *real* behavior of that part of the system is obtained.

Leadlag Dynamic Compensator (Approximate Proportional Plus Derivative Plus Integral Control). The most widely used control law for all kinds of feedback control systems is the so-called "PID" controller: a combination of proportional, integral, and derivative control. Since its three modes of control are each separately adjustable, it can be "tuned" to meet the needs of many applications. It can be implemented mechanically, pneumatically, hydraulically, electronically, or as is most common today, in digital software. An approximate version called a *leadlag compensator* is adequate for some applications, and can be realized as a rather simple passive electrical circuit. Such analog models are often useful for preliminary design purposes even when the final implementation will be in digital software. That is, we use analog design tools to estimate numerical values and then rely on software "black boxes" to implement our concept digitally.

Figure 8-63 shows the simplest passive circuit for a leadlag compensator. An impedance approach is probably the quickest route to the transfer function we want.

$$e_o(D) = i(D)Z_1(D) = i(D)\left[R_2 + \frac{1}{C_2 D}\right] = i(D)\left[\frac{R_2 C_2 D + 1}{C_2 D}\right] \quad (8\text{-}170)$$

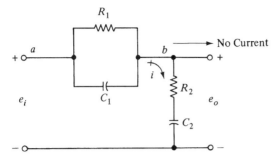

Figure 8-63 Passive circuit leadlag compensator.

$$i(D) = \frac{e_i}{Z_{\text{total}}} = \frac{e_i}{\dfrac{(R_1)\left(\dfrac{1}{C_1 D}\right)}{R_1 + \dfrac{1}{C_1 D}} + R_2 + \dfrac{1}{C_2 D}} \quad (8\text{-}171)$$

$$i(D) = \frac{C_2 D(R_1 C_1 D + 1) e_i}{R_1 C_1 R_2 C_2 D^2 + (R_1 C_1 + R_2 C_2 + R_1 C_2) D + 1} \quad (8\text{-}172)$$

$$\frac{e_o}{e_i}(D) = \frac{(R_1 C_1 D + 1)(R_2 C_2 D + 1)}{R_1 C_1 R_2 C_2 D^2 + (R_1 C_1 + R_2 C_2 + R_1 C_2) D + 1} \quad (8\text{-}173)$$

$$\frac{e_o}{e_i}(D) = \frac{(\tau_1 D + 1)(\tau_2 D + 1)}{(\tau_3 D + 1)(\tau_4 D + 1)} \quad (8\text{-}174)$$

$$\tau_1 \triangleq R_1 C_1 \qquad \tau_2 \triangleq R_2 C_2 \qquad \tau_3 \tau_4 \triangleq R_1 C_1 R_2 C_2$$

$$\tau_3 + \tau_4 \triangleq R_1 C_1 + R_2 C_2 + R_1 C_2$$

In (8-173) the numerator is clearly overdamped second-order and since the denominator has the same ω_n and a larger D term ($\tau_3 + \tau_4 > \tau_1 + \tau_2$) the denominator is also overdamped, as we would expect from a passive R-C circuit without inductors.

In practical applications τ_3 is often chosen to be some fraction of τ_1, say $\tau_3 = k\tau_1$, which then makes $\tau_4 = \tau_2/k$. The compensator then has two sections, the section involving τ_1 and τ_3 being a lead compensator and the section with τ_2 and τ_4 being a lag compensator. (Sometimes the restriction of this circuit that the time constant ratios of the two sections *must* be identical is not acceptable. We can then go to an op-amp version of the circuit which allows independent choice of these ratios.) In an application which we are about to show, we want $\tau_1 = \tau_2 = 0.02\,\text{sec}$ and $k = 0.125$, making $\tau_3 = 0.0025\,\text{sec}$ and $\tau_4 = 0.16\,\text{sec}$. To choose actual R and C values we find that we have three equations in four unknowns, allowing an infinite number of designs. Since the impedance presented to the source e_i has a minimum value of R_2 at high frequency we might wish to choose R_2 sufficiently large so as to not load the source excessively. Let's assume $R_2 = 10{,}000$ ohms is adequate. Then there is only one solution for the remaining parameters: $R_1 = 61250\,\Omega$, $C_1 = 0.327\,\mu\text{F}$, $C_2 = 2.0\,\mu\text{F}$.

The step response of such a system is found from its differential equation

Second-Order Systems and Mechanical Vibration

$$[\tau_3\tau_4 D^2 + (\tau_3 + \tau_4)D + 1]e_o = [\tau_1\tau_2 D^2 + (\tau_1 + \tau_2)D + 1]e_i \quad (8\text{-}175)$$

using either Laplace transform or D-operator methods. Let's pursue the D-operator solution since it develops some physical understanding of the behavior. For e_i a step input, the right-hand side becomes the number 1.0 for any $t > 0$ and thus the particular solution is clearly $e_{op} = 1.0$. The complementary solution will have the standard form for overdamped second order; however, the initial conditions will *not* conform to those of a system without numerator dynamics, so we cannot use any previous solutions. Using physical reasoning from the circuit diagram, at $t = 0^+$ the capacitor C_1 is still uncharged since it takes a finite current a finite time to charge a capacitor. Since there is thus no voltage *drop* across C_1, the potential at point b (which is the output voltage e_o) must be the same as at a; thus e_o instantly jumps up to e_i, which is one of our needed initial conditions at $t = 0^+$.

The current i must be finite since it must go through the resistor R_2; in fact $i(0^+)$ is e_i/R_2. Thus, initially, all the current i goes through C_1 (*none* of it through R_1) and through R_2 and C_2. To find $De_o(0^+)$, our other needed initial condition, we note that $e_o = e_i + e_{ab}$ and thus $De_o = De_i + De_{ab}$. At $t = 0^+$, e_i has become constant; thus $De_i = 0$ and $De_o = De_{ab}$. The voltage rise e_{ab} is the voltage across C_1, and since C_1 carries the current $i = e_i/R_2$, we have $De_{ab} = -i/C_1 = e_i/(R_2C_1)$ and thus $De_o(0^+) = -e_i/(R_2C_1)$. Applying these two initial conditions we finally get

$$e_o = 1 - \frac{\tau_1 + \tau_2 - \tau_3 - \tau_4}{\tau_4 - \tau_3}e^{-t/\tau_3} + \frac{\tau_1 + \tau_2 - \tau_3 - \tau_4}{\tau_4 - \tau_3}e^{-t/\tau_4} \quad (8\text{-}176)$$

Figure 8-64 shows the general shape of this step response. Figure 8-65 shows the logarithmic frequency-response curves ("Bode plots") for a different set of tau values. The amplitude ratio is approximated using just the straight-line asymptotes, and is just a combination of the individual first-order terms.

To see a typical application of a leadlag compensator, consider Fig. 8-66. A basic feedback control system with a step input desired value of 1.0 is shown in the upper part of this figure. We want the controlled variable to follow this command as accurately as possible. The three first-order terms represent three pieces of hardware whose design can not be changed. Gain1 represents the loop gain K of the feedback

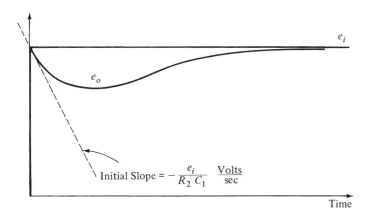

Figure 8-64 Step response of leadlag compensator.

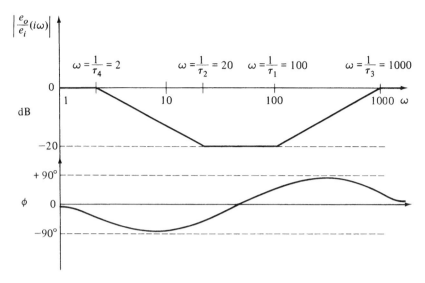

Figure 8-65 Frequency response of leadlag compensator.

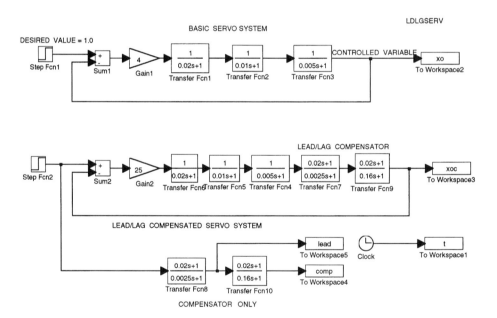

Figure 8-66 Basic servo control system and compensated version.

system, a number we try to make as large as possible, without making the response too oscillatory or unstable. By trial and error we find that $K = 4$ gives the response shown as "basic system" in Fig. 8-67. Any larger value of K makes the overshoot and oscillation worse than that shown, which would be unacceptable. This response is thus the best that can be done with this basic system. Its steady-state response is 0.80, a 20% error relative to the desired value of 1.0. It also takes about 0.2 second to "settle" after the step input.

Second-Order Systems and Mechanical Vibration

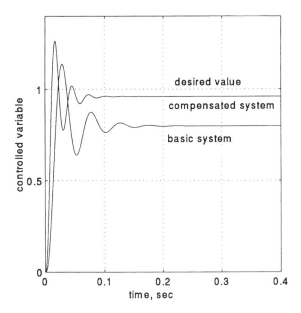

Figure 8-67 Compensation improves both speed and accuracy.

If this steady-state error and speed of response were inadequate, no further improvement is possible in the basic system, since we have only one adjustable parameter, loop gain K. By adding a leadlag compensator (combination of a lead compensator, which can speed things up, and a lag compensator, which can increase steady-state accuracy) in the lower part of the figure, we find that we can raise K to about 25 before oscillation gets too strong. Figure 8-67 shows that the compensated system is about twice as fast, and about 6 times as accurate in steady state as the basic system. These improvements are possible because the compensator has four adjustable parameters and we now have more "design freedom" when we try to meet difficult specifications. If you take a course in feedback control systems you will learn *analytical* methods for estimating the needed values of compensator parameters to accomplish desired response improvements. These theoretical values are used as the starting points in simulations which "fine-tune" the parameter values. Without such background theory, simulation can become aimless groping, especially if many parameters must be chosen. Since a system dynamics course usually comes *before* a controls course, I cannot presume that readers would have this theoretical background, thus I have emphasized the simulation, which *does* clearly show that properly designed compensators can dramatically improve performance.

In a simulation of a strictly linear system such as this, the *location* of the compensator in the "chain" of components is immaterial, however in the actual system, the compensator is usually placed "at the left" (rather than at the right, as in Fig. 8-66). This puts it in the "low power" portion of the system, where it can usually be implemented electronically, as part of the amplifier which also provides the loop gain adjustment. If the compensator is implemented in digital software, then it must be located in the computer which serves as system controller.

At the very bottom of Fig. 8-66 I provide for showing the step response of the compensator itself. This part of the simulation has nothing to do with the compensated system, it is provided to verify the theoretical results of Fig. 8-64 for some typical numerical values.

Simulation software usually provides a "ready-made" transfer function module to directly simulate functions with numerator dynamics, thus avoiding the need to differentiate the input signal. It may be useful to see how such modules are internally configured, should the need arise for you to provide such capability. Let's use for an example the form

$$\frac{q_o}{q_i}(s) = \frac{n_2 s^2 + n_1 s + n_0}{d_2 s^2 + d_1 s + d_0} \tag{8-177}$$

It appears we must twice differentiate the input signal q_i. We have several times earlier warned of the general undesirability of differentiation, due to its noise accentuation problems. Fortunately this can always be avoided using the following approach. Rearrange Eq. (8-177) as

$$\frac{q_o}{q_i}(s) = \left[\frac{q_a}{q_i}(s)\right]\left[\frac{q_o}{q_a}(s)\right] \tag{8-178}$$

$$\frac{q_a}{q_i}(s) = \frac{1}{d_2 s^2 + d_1 s + d_0} \qquad \frac{q_o}{q_a}(s) = n_2 s^2 + n_1 s + n_0 \tag{8-179}$$

A simulation using *only* integrators (*no* differentiators) may then be set up as in Fig. 8-68. This same technique is easily applied to other cases with numerator dynamics, should you ever need to deal directly with this situation.

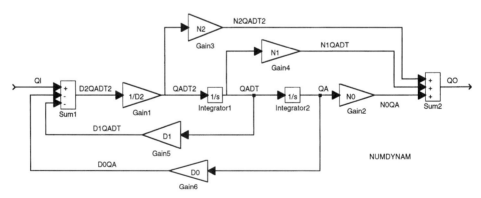

Figure 8-68 Simulation technique for numerator dynamics.

BIBLIOGRAPHY

Andersen, B. W., *The Analysis and Design of Pneumatic Systems*, Wiley, New York, 1967.
Doebelin, E. O., *Control System Principles and Design*, Wiley, New York, 1985.
Doebelin, E. O., *Measurement Systems*, 4th ed., McGraw-Hill, New York, 1990.

Second-Order Systems and Mechanical Vibration **613**

Doebelin, E. O., *System Modeling and Response: Theoretical and Experimental Approaches*, Wiley, New York, 1980.

de Silva, C. W., *Control Sensors and Actuators*, Prentice Hall, Englewood Cliffs, N.J., 1989.

Ellis, J. R., *Vehicle Dynamics*, London Business Books Ltd., London, 1969.

Gibson, J. E., and F. B. Tuteur, *Control System Components*, McGraw-Hill, New York, 1958.

Gillespie, T. D., Fundamentals of Vehicle Dynamics, SAE, Warrendale, PA 15096-0001, 1992.

Harris, C. M., and C. E. Crede, eds., *Shock and Vibration Handbook*, McGraw-Hill, New York, 1961.

Inman, D. J., *Engineering Vibrations*, Prentice Hall, Englewood Cliffs, N.J., 1996.

Krause, P. C., and O. Wasynczuk, *Electromechanical Motion Devices*, McGraw-Hill, New York, 1989.

Leonhard, W., *Control of Electrical Drives*, Springer, New York, 1985.

Merritt, H. E., *Hydraulic Control Systems*, Wiley, New York, 1967.

Millikan, W. F., and D. L. Milliken, Race Car Vehicle Dynamics, SAE, Warrendale, PA 15096-0001, 1995.

Mills, A. F., *Basic Heat and Mass Transfer*, Irwin, Chicago, 1996.

Nachtigal, C. L., ed., *Instrumentation and Control*, Wiley, New York, 1990.

Rizzoni, G., *Principles and Applications of Electrical Engineering*, 2nd ed., Irwin, Chicago, 1996.

Rothbart, H. A., ed., *Mechanical Design and Systems Handbook*, McGraw-Hill, New York, 1964.

Steeds, W., *Mechanics of Road Vehicles*, Iliffe and Sons, Ltd, London, 1960.

Watton, J., *Fluid Power Systems*, Prentice Hall, New York, 1989.

PROBLEMS

8-1. Do a detailed "loading study" of the circuit of Fig. 8-2, similar to what was done for the system of Fig. 8-3.

8-2. Fill in all the steps between Eqs. (8-26) and (8-28).

8-3. Get Eq. (8-28) using the *D*-operator solution method.

8-4. Using available simulation software, get the results of Fig. 8-7.

8-5. For the rotary system of Fig. 8-5a:

    a. Write the differential equation, put it in standard form and define the standard parameters. State the operational, sinusoidal, and Laplace transfer functions and show a block diagram, using standard parameters.

    b. Draw a simulation diagram using basic physical parameters (use integrators, *not* a transfer-function block). Draw a simulation diagram using standard parameters (use integrators, *not* a transfer-function block). Draw a simulation diagram using standard parameters and a transfer-function block.

8-6. Repeat Problem 8-5 for the translational system of Fig. 8-5b.

8-7. Repeat Problem 8-5 for the rotary system of Fig. 8-5b.

8-8. Repeat Problem 8-5 for the translational system of Fig. 8-5c.

8-9. Repeat Problem 8-5 for the rotary system of Fig. 8-5c.

8-10. Repeat Problem 8-5 for the translational system of Fig. 8-5d.

8-11. Repeat Problem 8-5 for the rotary system of Fig. 8-5d.

614   **Chapter 8**

8-12.   Repeat Problem 8-5 for the translational system of Fig. 8-5e.

8-13.   Repeat Problem 8-5 for the rotary system of Fig. 8-5e.

8-14.   Modify the combined system of Fig. 7-1a by adding both a rotary and a translational spring in such a way that the system is now second-order. Then repeat Problem 8-5.

8-15.   The rotary system of Fig. 8-5b may be used as a model for the shaft (K_s) and cutting blade (J) of a rotary lawnmower. The shaft is 0.5 inch in diameter and 3 inches long, the blade is a rectangular flat 0.1 by 2 by 18 inches; both are steel. If the engine providing the driving motion runs at 1000 rpm, would torsional vibration problems be expected? The motion θ_i may contain fluctuations at frequencies up to 3 times the engine speed.

8-16.   A more accurate model of the lawnmower of Problem 8-15 places another inertia J_1 at the other end of the spring, to represent the motor inertia. The input is now a driving torque T_i, applied to J_1. Derive the differential equation for this new model and find the natural frequency. If J_1 is $0.05\,\text{in-lb}_f\text{-sec}^2$, and other conditions are as in Problem 8-15, are vibration problems predicted?

8-17.   For the accelerometer of Fig. 2-35a, get the differential equation relating input acceleration x_i to output displacement x_o. If the proof mass weighs 0.05 pound, what spring constant is needed to give 0.1 inch x_o for a steady acceleration of $10\,g$'s? Recommend a numerical value for damping coefficient B, and explain your choice. With this damping value, estimate the range of frequencies for which the instrument would measure properly if the input acceleration were sinusoidal.

8-18.   Design package cushioning (see Fig. 8-11) for the following specifications.

    Cushioned object weight $= 10\,\text{n}$
    Drop height $= 0.5\,\text{m}$
    Maximum allowable acceleration $= 20\,g$'s

Check your design with a simulation that includes damping, but set the damping to zero. Then find what B value it takes to make vibration die out in about 1 cycle.

8-19.   Equation (8-34) uses an approximation. Get results similar to (8-35) to (8-37) *without* making this approximation. Discuss conditions under which the approximation seems to be acceptable.

8-20.   Sketch logarithmic frequency-response curves for second-order systems with:

    a. $K = 10$, $\zeta = 0.5$, $\omega_n = 25\,\text{rad/sec}$
    b. $K = 0.04$, $\zeta = 0.1$, $\omega_n = 15,000\,\text{rad/sec}$

8-21.   Repeat Problem 8-5 for the circuit of Fig. 8-33b.

8-22.   Repeat Problem 8-5 for the circuit of Fig. 8-33c.

8-23.   Repeat Problem 8-5 for the circuit of Fig. 8-33e.

8-24.   Repeat Problem 8-5 for the circuit of Fig. 8-33f.

8-25.   Repeat Problem 8-5 for the circuit of Fig. 8-33h.

8-26.   In Fig. 8-33e, what is the impedance seen by the voltage source e_i? Find the steady-state sinusoidal current drawn from the source if $R_1 = 1000\,\text{ohms}$, $R_2 = 800\,\text{ohms}$, $L = 0.2\,\text{H}$, $C = 0.5\,\mu\text{F}$, and $e_i = 100 \sin 377t$ volts (t in seconds).

8-27.   The system shown in Fig. P8-1 is proposed as a means of differentiating e_i signals with frequency content up to about 1.0 Hz. The noise filter is necessary since

Second-Order Systems and Mechanical Vibration

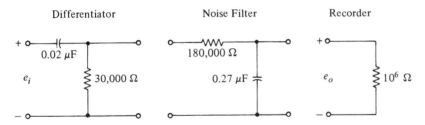

Figure P8-1

e_i also contains small amplitude but high-frequency (50 Hz and above) noise components.

    a. Get transfer functions for the differentiator and noise filter separately.
    b. Assuming the e_o terminals open circuit (recorder not connected), plot logarithmic frequency-response curves for $(e_o/De_i)(i\omega)$, assuming no loading effect between differentiator and noise filter.
    c. To check the validity of the no-loading assumption, get $(e_o/De_i)(i\omega)$ *exactly*, by analyzing the whole circuit (including recorder) as an entity.
    d. Does the circuit perform its intended functions?

8-28. Repeat Problem 8-5 for the system of:
    a. Fig. 8-48a   b. Fig. 8-48c   c. Fig. 8-48f
    d. Fig. 8-48g   e. Fig. 8-48h

8-29. In Fig. 8-48b, treat the level in the left tank as the output quantity, get the system differential equation, put it in standard form and define the standard parameters.

8-30. The energy-type analysis of the system of Fig. 8-11 is the standard method of treating such problems, but more detailed analysis is sometimes needed. In the text's analysis the box is assumed to strike the floor and instantly go to, and remain at, zero velocity. A more correct model, shown in Fig. P8-2, considers the mass of the box and the springiness and damping of the floor. Write the system differential equations which model the behavior during the free fall and floor impact. Draw and explain a simulation diagram for this system. Then run this simulation, using the same numerical values as for the text example, plus:

$$K_{s,f} = 50{,}000 \text{ lb}_f/\text{ft}$$
$$B_f = 900 \text{ lb}_f/(\text{ft/sec})$$
$$M_b \text{ weighs } 10 \text{ lb}_f$$

Now try some other values for $K_{s,f}$ and B_f to see the effect, and discuss your results.

8-31. In Fig. 8-48b, add an inflow $q_{i1}(t)$ to the right tank, giving a system with two inputs. Get differential equations, transfer functions, and block diagrams showing the response of *both* tank levels to *both* inflows. Put transfer functions in standard form and define standard parameters.

8-32. Review the text's Design Example: High-Speed Scale for Packaging Conveyor. Discuss in detail how the requirements on the scale's construction and the sensor's characteristics change if we need to deal with faster and faster conveyor speeds.

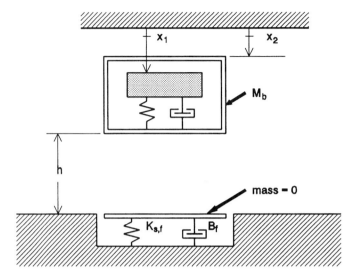

Figure P8-2

8-33. Derive Eqs. (8-39) and (8-40).

8-34. Modify the simulation of Fig. 8-16 to model the motion of a 3000-pound car coasting to a stop from an initial speed of 60 mph, with viscous, coulomb, and air resistance friction acting. At 60 mph the three forms of friction each provide an equal amount of force, the total friction force being 150 pounds. Run this simulation and get graphs of the individual friction forces, car deceleration, velocity, and position.

8-35. Get. Eq. (8-45) by both the Laplace transform and D-operator methods.

8-36. Derive Eqs. (8-49) and (8-50).

8-37. Modify the motion-transmissibility results of Eq. (8-58) to get transfer functions in which the output is:

    a. Velocity
    b. Acceleration

Obtain Bode plots (similar to Fig. 8-24) for these two transfer functions.

8-38. The "car" of Fig. 8-25 is run over the bump of Fig. P8-3 at various speeds. Set up a simulation of this situation, using displacement as the output. Take $M = 100$ slugs, $K_s = 6000\,\text{lb}_f/\text{ft}$, and B the value required for critical damping. Run speeds of 10, 40 and 70 mph. Then try values of B larger and smaller than

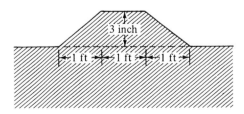

Figure P8-3

Second-Order Systems and Mechanical Vibration 617

critical damping, trying to find a single value that is "best" for all the speeds. Discuss your results.

8-39. Just before Eq. (8-64), derive the value of x_∞ given there.

8-40. Modify the simulation of Fig. 8-29 to model the acceleration of an inertia-plus-viscous friction load by:

    a. An ac induction motor.

    b. A dc motor with fixed field and a step input of armature current.

    c. A dc motor with fixed field and a step input of armature voltage.

    d. A dc motor with fixed armature current and a step input of field voltage.

    e. A dc motor with fixed armature voltage and a step input of field voltage.

    f. A variable-displacement hydraulic pump driving a fixed-displacement motor, as in Fig. 7-62, but without the delayed positive feedback. Input is a step of XSPL.

8-41. Show that the impulse response of Eq. (8-67) is identical with the response to a properly chosen initial velocity.

8-42. Check the statement immediately following Eq. (8-68) by using simulation and input pulses of several different shapes.

8-43. A voltage in a measurement system is the sum of a desired component $5.0 \sin 10t$ and a noise $1.0 \sin 377t$ volts, t in seconds. We want to compare the filtering performance of first-order (Fig. 7-28a) and second-order (Fig. 8-33a) low-pass filters. After filtering, we want the noise component to have an amplitude which is 1% of the desired component. From loading considerations, R in the first-order filter and R_1 in the second-order are to be $100,000 \, \Omega$. In the second-order filter, $R_2 = 10R_1$ and $C_2 = 0.10C_1$. Find the capacitor values needed to meet the specification. Then check each filter to see how much it distorts the desired signal. Discuss the acceptability of each filter. Check all these results with a simulation.

8-44. Draw a simulation diagram for the system of Eqs. (8-69) and (8-70).

8-45. Use Eqs. (8-71) and (8-72) and determinants to get $(i_1/e_i)(D)$.

8-46. For the band-pass filter of Fig. 8-40, duplicate the simulation and use it to study the "waiting time" effect mentioned in the text, by trying various ζ values.

8-47. Derive Eq. (8-92) for the notch filter.

8-48. Using the hints following Eq. (8-101), design a second-order op-amp circuit with $\zeta = 0.0$ and a natural frequency of 100. Hz. How would you decide on specific R and C values?

8-49. Obtain results analogous to Eq. (8-127), but with the heater temperature as the output quantity.

8-50. Solve Eq. (8-128) for a step input of T_i, using:

    a. The D-operator method

    b. The Laplace transform method

8-51. Duplicate the simulation of Fig. 8-51 and try different duty cycles to see if an optimum cycle exists.

8-52. For the system of Fig. 8-53, if the specimen has a spring constant of $1.87 \times 10^6 \, lb_f/inch$, design resonance tubes to give an operating frequency of 15 Hz. All other numbers are as in the text example. Note that a given fluid inertance may be achieved with an infinite number of combinations of tube length and diameter. Speculate on how one would decide on a *specific* length and diameter.

8-53. For the system of Fig. 8-56, derive the system equations when:
  a. The armature supply is a fixed voltage source.
  b. The armature supply is a variable voltage input.

If the equations are not linear with constant coefficients, do an approximate linearized perturbation analysis, and also a simulation diagram for the exact equations.

8-54. Derive Eqs. (8-163) through (8-166).

8-55. Derive Eq. (8-167).

8-56. Derive Eq. (8-168).

8-57. Derive Eq. (8-169).

8-58. Show an op-amp circuit that allows completely independent choice of all four time constants in Eq. (8-174).

8-59. Obtain Eq. (8-176) using Laplace transform methods.

8-60. Duplicate the simulation of the compensated system in Fig. 8-66. Use *only* the lead part of the leadlag compensator and get the step response for various values of loop gain. Then use *only* the lag part of the leadlag compensator. Discuss these results.

8-61. In Fig. P8-4 a thermometer is inserted into a thin-wall metal well containing fluid with mass M_w and specific heat c_w. The well wall is considered a pure resistance and the thermometer bulb contains fluid with mass M_t and specific heat c_t. Get the differential equation relating T_o to T_i.

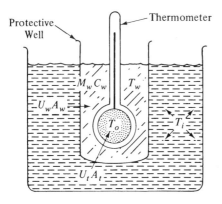

Figure P8-4

8-62. The system of Fig. 5-8 is made specifric in Fig. P8-5. Treating the system as strictly linear, get differential equations, transfer functions, and block diagrams showing how outputs Δp, Q_L, and ω are produced by input torque T.

8-63. While armature-controlled dc motors are most common in industrial motion-control systems, some applications, such as electric vehicles,[18] may use the tradi-

---

[18] R. J. Valentine, Electric vehicle motor power control, *Motion*, May/June 1996, pp. 2–20.

Second-Order Systems and Mechanical Vibration

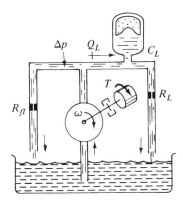

Figure P8-5

tional series-connected dc motor configuration. Figure P8-6a shows this arrangement, where the field circuit and armature circuit carry the same current.

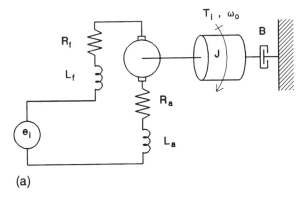

(a)

Figure P8-6a

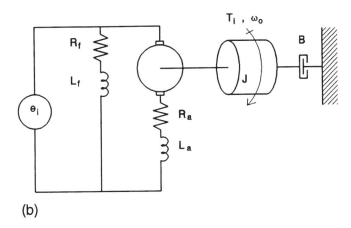

(b)

Figure P8-6b

620   **Chapter 8**

    a. Write the equations for this system, assuming the driven load is inertia plus viscous friction, with an external load torque input.

    b. For *steady-speed* operation, combine the equations of part (a) to get a relation between the torque available from the motor, and motor speed. Sketch the general shape of this torque versus speed curve.

    c. For the dynamic equations, linearize these to get transfer functions and a block diagram showing how load speed is produced by the two inputs, input voltage e_i and load torque T_i.

    d. Draw and explain a simulation diagram for the exact nonlinear dynamic equations.

8-64. Repeat Problem 8-63 for the shunt-connected dc motor of Fig. P8-6b.

9

GENERAL LINEAR SYSTEM DYNAMICS

9-1 INTRODUCTION

It is probably clear to the reader at this stage that the modeling of systems comprised of large numbers of elements, whether mechanical, electrical, fluid, or thermal, will lead to high-order differential equations. As long as linear (or linearized) constant-coefficient models are adequate, the analytical methods (D-operator or Laplace transform) of Chapter 6 will yield information on the nature of system behavior, even though computer simulation may be desirable for evaluating specific numerical cases. If "exact" (not linearized) variable-coefficient and/or nonlinear models are deemed necessary, then simulation is usually the only practical route for system analysis and design.

If we are to be true to the word "general" in the chapter title, we must restrict ourselves to constant-coefficient linear models; a general mathematical theory that is practically useful exists *only* for this class of system descriptions. Most of the chapter is devoted to this limited but important area. However, as has been demonstrated in all the earlier chapters, we don't want to limit our design perspective so severely; it rules out many nonlinear designs which are superior to the linear. Simulation software will usually be the main tool in these situations, but familiarity with linear theory often provides approximate insights which can be helpful in the initial stages of design, even if the final model is strongly nonlinear. Chapter 9 will thus include some nonlinear system examples, even though a *general* theory of such systems is largely lacking.

With regard to deriving or setting up the system equations, no new physical laws are needed; we simply apply the same laws "more times." That is, a mechanical system with 10 masses still yields to Newton's law. We just apply it, one mass at a time, generating a set of 10 simultaneous differential equations. Similarly for electrical, fluid, thermal, or mixed systems. When systems become *very* intricate, "bookkeeping" matters such as knowing when a complete set of equations has been obtained, and just how many unknowns there really are, may complicate matters. Geometric considerations, as in three-dimensional motion, can lead to complications. Systematic methods such as *Lagrange's energy approach* for mechanical systems and *network topology* for electrical circuits are available to treat some of

621

622   **Chapter 9**

these situations but are beyond the scope of this introductory text. Finite-element software for dealing with complex mechanical, electrical, fluid, thermal, and mixed systems is in wide use. Most of this software requires long learning times for the user to become proficient, and presents the potential danger of misuse by the inexperienced. It has, however, become invaluable for dealing with ever more accurate models of practical devices and processes. Such software generally does *not* require the user to actually derive and write out the equation set. Rather the user merely *describes* the system in terms of geometry, material properties, and input forces, voltages, etc. The software then automatically assembles and solves the appropriate set of equations. Training in the proper use of such software is beyond the scope of this text.

With regard to solving the sets of equations for the restricted class of problems addressed by this text, as has been mentioned often before, our familiarity with basic first- and second-order systems is good preparation for the general linear case. We know from Chapter 6 that high-order equations lead to an equally high number of roots of the characteristic equation, but these roots are either real ("like a first-order system") or complex pairs ("like a second-order system"). The equation solution thus holds no basic surprises; we simply see *several* exponential or damped-oscillatory terms (rather than just one) in the complementary solution. The roots of the characteristic equation also provide information on the *stability* of our system, vital information for all feedback control systems. For external driving inputs, the method of undetermined coefficients usually gets us the particular ("forced") solution. We will shortly illustrate these concepts from Chapter 6 with a few concrete examples.

Throughout the text we have used frequency-response methods as major tools in system analysis and design. We have used the sinusoidal transfer function for many useful calculations, but have proven its validity only in some special cases such as first–order systems. In this chapter we establish its correctness for the general case, and also develop a matrix-frequency-response calculation method which efficiently computes (from the original simultaneous equation set) all the sinusoidal transfer functions of a system with several inputs and several outputs. We also want to present some methods which extend frequency response ideas to signals more complicated than pure sine waves. Fourier series will be used to deal with all kinds of periodic, nonsinusoidal inputs. A brief discussion of Fourier transform will illustrate how transients of arbitrary waveform can also be included in our frequency-domain thinking. System and signal modeling by lab testing, using spectrum analyzers, will also be explained.

Finally, a number of examples, both linear and nonlinear, will be presented, to illustrate the application of these tools to systems more complex than those treated earlier in this text. As usual, the analytical methods will be augmented with simulation techniques to further develop your skill in combining these two types of problem-solving tools in the most effective ways.

9-2 SYSTEM MODELING AND EQUATION SETUP

As we just said above, no new physical laws or techniques are needed to set up the equations for complex systems; however, we do need to provide some concrete

General Linear System Dynamics   **623**

examples for practice. Figure 9-1 shows four different examples of systems which will be found to be higher than second order, and thus good candidates for study in this chapter. We will use the mechanical system of Fig. 9-1a to illustrate the general situation, leaving the other three for your practice in end-of-chapter problems.

Applying Newton's law to each of the three masses gives us three simultaneous equations which can quickly be put into the form we recommend as preparation for applying determinants.

$$(M_1 D^2 + B_1 D + K_{s1})x_1 + (-B_1 D - K_{s1})x_2 + (0)x_3 = f_{i1} \tag{9-1}$$

$$(-B_1 D - K_{s1})x_1 + (M_2 D^2 + B_1 D + K_{s1} + K_{s2})x_2 + (-K_{s2})x_3 = f_{i2} \tag{9-2}$$

$$(0)x_1 + (-K_s)x_2 + (M_3 D^2 + B_2 D + K_{s2} + K_{s3})x_3 = f_{i3} \tag{9-3}$$

If we wanted to include initial conditions as inputs (in addition to the external forces shown) we would use the Laplace transform approach. Then the right-hand sides of the above equations would include terms involving the initial displacements and initial velocities of the three masses, and of course we would be using s's instead of D's. Let's assume that we are interested only in the response to external driving forces, and that all initial conditions are zero. Then Eqs. (9-1) through (9-3) would, using Laplace transforms, merely substitute s's for D's. It should be clear that deriving the equation set for a system with *any* number of masses leads to a larger set of equations, but does *not* present any new difficulties or require any new methods.

If the external driving forces were given as known functions of time, using simulation, we could solve for the unknown motions directly from the above equation set. If we instead choose to work analytically, then the simultaneous equation set must first be reduced to single equations in single unknowns, usually by use of determinants. While this reduction *could* be carried out in letter form, the resulting expressions become very cumbersome. Also, at the *next* step of solution, use of letter form becomes not cumbersome but *impossible*, because the system characteristic equation will be sixth degree, and we stated in Chapter 6 that polynomial equations of degree higher than 4 can be solved for their roots only when the coefficients are known numbers. For these reasons we now decide to assign some numerical values before proceeding further:

$$M_1 = M_2 = M_3 = 1.0\,\text{kg}$$
$$B_1 = B_2 = 1.0\,\text{N/(m/sec)}$$
$$K_{s1} = K_{s2} = K_{s3} = 1000.\ \text{N/m}$$

We can now get single equations in a single unknown for each of the unknown displacements x_1, x_2, and x_3 by working out the appropriate determinants. Each such single equation will have all three of the driving forces as inputs on the right-hand side. That is, the motion of any one mass is caused by a superposition of the effects of each of the three inputs. We can define and use three transfer functions from each of the three single equations, giving a total of nine transfer functions, relating the three inputs to the three outputs. Of the nine transfer functions, three will be identical with three others, so there will be only six *different* transfer functions. This is a consequence of the *reciprocity* phenomenon associated with symmetric system matrices.

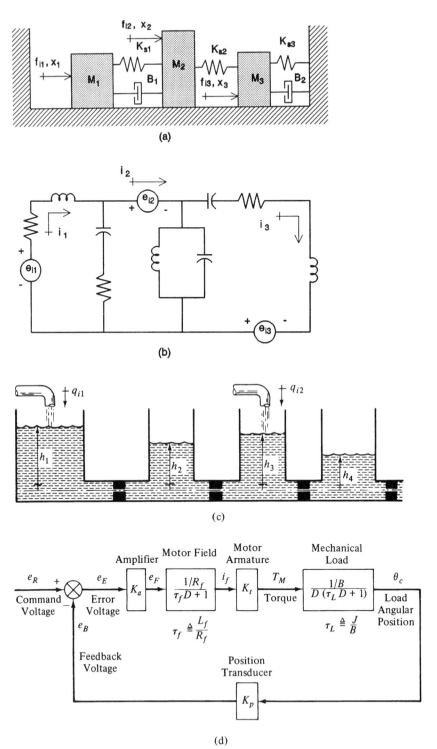

Figure 9-1 Examples of systems higher than second-order.

General Linear System Dynamics 625

$$\frac{x_1}{f_{i2}}(D) \equiv \frac{x_2}{f_{i1}}(D) \qquad \frac{x_1}{f_{i3}}(D) \equiv \frac{x_3}{f_{i1}}(D) \qquad \frac{x_2}{f_{i3}}(D) \equiv \frac{x_3}{f_{i2}}(D) \tag{9-4}$$

This reciprocity effect can be used to good advantage in at least two ways. When grinding out analytical results, if they don't agree with Eqs. (9-4), we start looking for mistakes. When running lab tests, such as sinusoidal transfer functions, it is wise to *both* apply force at location 1 and measure motion at location 2, and also apply force at location 2 and measure motion at location 1. If these two measured transfer functions don't agree reasonably well, our system may not be well modeled with a symmetric matrix, and/or our measurement system may not be reliable.

Let's carry the solution process further by getting the single equation for x_1, leaving the other two equations for the end-of-chapter problems.

$$x_1 = \frac{\begin{vmatrix} f_{i1} & -D - 1000 & 0 \\ f_{i2} & D^2 + D + 2000 & -1000 \\ f_{i3} & -1000 & D^2 + D + 2000 \end{vmatrix}}{\begin{vmatrix} D^2 + D + 1000 & -D - 1000 & 0 \\ -D - 1000 & D^2 + D + 2000 & -1000 \\ 0 & -1000 & D^2 + D + 2000 \end{vmatrix}} \tag{9-5}$$

Expanding the two determinants (I did it "by hand" and then checked the result with the symbolic processor MAPLE which is part of my MATHCAD software) we get

$$(D^6 + 3D^5 + 5002D^4 + 8000D^3 + 6.001e6D^2 + 2e6D + 1e9)x_1$$
$$= (D^4 + 2D^3 + 4001D^2 + 4000D + 3e6)f_{i1}$$
$$+ (D^3 + 1001D^2 + 3000D + 2e6)f_{i2} + (1000D + 1e6)f_{i3} \tag{9-6}$$

If the three input forces were now given as known functions of time that satisfied the requirements of the method of undetermined coefficients, we could solve for the displacement x_1. We would need to also get the six roots of the characteristic equation. Using the MATLAB "roots" command, these turn out to be:

$$-0.7416 \pm i56.9713 \qquad -0.6939 \pm i39.4301 \qquad -0.0644 \pm i14.0737$$

Each of these root pairs can be thought of as a second-order system with a certain ω_n and ζ, where the real part of the root is $-\zeta\omega_n$ and the imaginary part is $\omega_n(1 - \zeta^2)^{0.5}$. For example, the first root pair would have

$$\zeta\omega_n = 0.7416 \qquad \omega_n\sqrt{1 - \zeta^2} = 56.9713$$

$$\zeta = 0.0130 \qquad \omega_n = 56.98 \,\frac{\text{rad}}{\text{sec}} \tag{9-7}$$

If we were to actually carry out the analytical solution, the Laplace transform method would be preferred because the D-operator method requires the separate calculation of the initial ($t = 0^+$) values of x_1 and its first five derivatives, which are not all necessarily zero. The Laplace transform solution could proceed directly from the "s version" of Eq. (9-6), once the transforms of the input forces had been inserted, because the initial ($t = 0^-$) conditions were all earlier assumed to be zero. Here the input force functions would be restricted to those which have Laplace transforms. While it is useful to understand this analytical background, it would be rare today that such solutions (by either method) would actually be carried out.

626   **Chapter 9**

Instead, we use simulation from the original set of simultaneous equations, as we will shortly show.

The above analysis *is* useful in showing us the form of the various transfer functions relating the input forces and output motions. These are easily obtained from Eq. (9-6):

$$\frac{x_1}{f_{i1}}(D) = \frac{D^4 + 2D^3 + 4001D^2 + 4000D + 3e6}{D^6 + 3D^5 + 5002D^4 + 8000D^3 + 6.001e6D^2 + 2e6D + 1e9} \tag{9-8}$$

$$\frac{x_1}{f_{i2}}(D) = \frac{D^3 + 1001D^2 + 3000D + 2e6}{D^6 + 3D^5 + 5002D^4 + 8000D^3 + 6.001e6D^2 + 2e6D + 1e9} \tag{9-9}$$

$$\frac{x_1}{f_{i3}}(D) = \frac{1000D + 1e6}{D^6 + 3D^5 + 5002D^4 + 8000D^3 + 6.001e6D^2 + 2e6D + 1e9} \tag{9-10}$$

Similar results can be obtained relating x_2 and x_3 to the three input forces. The sinusoidal versions of all these transfer functions can of course be used to compute and graph the frequency response relating any output to any input. We will shortly pursue this and other frequency response topics for this example and also in general.

9-3 STABILITY

We have earlier in this text encountered examples of unstable systems, but have not presented a general discussion. Stability considerations are an important part of control system theory and are covered in detail in books[1] devoted to that subject. While stability is a major question in feedback control systems, it also arises in noncontrol applications, as we saw in the vehicle handling dynamics study at Eq. (8-167). Some other noncontrol problems which require stability considerations include machine-tool chatter associated with cutting and grinding operations, hydrostatic bearing oscillations, flutter of aircraft components, safety-valve pulsations, and all self-excited vibrations.

A complete and comprehensive stability theory is available only for system models in the form of ordinary linear differential equations with constant coefficients. The stability criterion here is quite simple and direct. For a system to have what is defined as *absolute stability*, all the roots of the system characteristic equation must lie in the left half of the complex plane. If there is even a single real root or real part of a complex root pair that is positive (lies in the right-half plane), then the complementary solution must tend toward infinity, which is the definition of instability. Note that the *particular* solution of the equation has nothing to do with stability, according to this definition. Thus, this class of systems is either stable or unstable, *irrespective* of how we might be driving the system with external forces, voltages, etc. The *slightest* disturbance of any form is sufficient to start the unstable behavior, which then tends toward infinity because a positive exponential (like, say $e^{+0.367t}$) grows without bound.

---

[1]E. O. Doebelin, *Control System Principles and Design*, Wiley, New York, 1985, chaps. 6, 8.

General Linear System Dynamics

The prediction of our linear model for the unstable situation is a response which goes to infinity, which is of course impossible in the real world, so we might question the validity of the entire analysis. Actually, these methods have been found in practice to be quite reliable in predicting the *onset* of unstable behavior, which is really all that is needed for practical design. Once the unstable behavior starts, the response does grow, but not to infinity. Rather, the larger values of the response variable (motion, voltage, temperature, etc.) carry the system away from the region near the operating point of the linearized model into regions of *nonlinear* behavior, where the linear model no longer holds. For an unstable system which exhibits oscillations that initially grow, the nonlinear effects will either cause a system failure (springs break, motors burn out) or else the finite power supply causes the oscillations to "level off" at a fixed amplitude, called a *limit cycle*. Either of these two behaviors is of course unacceptable, so the predictions of the linear stability theory are definitely of practical use in system design.

As mentioned above, no comprehensive and practical stability theory is available for all systems that cannot be accurately linearized near a chosen operating point. Certain practically important classes of nonlinear systems yield to a stability criterion called the *describing function*.[2] While stability is strictly a *system* consideration for linear systems, stability of nonlinear systems can also depend on the system *inputs*. A nonlinear system can be stable for small inputs but unstable for large, something that can never happen in a linear system.

When we have available all the numerical values for the parameters of a linear system model, finding numerical values for the roots of the characteristic equation is not difficult, as we saw in Eq. (9-6). The three pairs of complex roots found there all had negative real parts, so instability is not possible. It would be useful to have available a stability criterion that would warn us of instability at the *design* stage, where all the parameters have not yet been given numerical values. Actually, such a criterion was developed in the 1870s by E. J. Routh. It predicts how many roots will not be in the left half plane, but does not give their numerical values. Even specialist control texts do not *prove* the Routh criterion, since that proof is quite complicated, so we here merely show how to use the criterion.

To apply the Routh criterion you must first get the system characteristic equation, with either literal or numerical values for all its coefficients, as in Eq. (6-3). The coefficients must then be arranged in a "triangular" array, according to some specific rules. The first two rows of the array are always formed as follows,

$$a_n \qquad a_{n-2} \qquad a_{n-4} \qquad a_{n-6} \qquad a_{n-8}$$
$$a_{n-1} \qquad a_{n-3} \qquad a_{n-5} \qquad \cdots \qquad \cdots$$

carrying these first two rows until you "run out" of coefficients. Then form a third row.

$$b_1 \qquad b_2 \qquad b_3 \qquad b_4 \qquad \cdots$$

from the first two, using the rule

---

[2]Ibid., chap. 8.

$$b_1 \triangleq \frac{a_{n-1}a_{n-2} - a_n a_{n-3}}{a_{n-1}} \qquad b_2 \triangleq \frac{a_{n-1}a_{n-4} - a_n a_{n-5}}{a_{n-1}}$$

$$b_3 \triangleq \frac{a_{n-1}a_{n-6} - a_n a_{n-7}}{a_{n-1}} \tag{9-11}$$

and continuing in this established pattern until the b's become zero. A *fourth* row is then constructed from the second and third rows in exactly the same way as the third was constructed from the first two. By continuing this row-forming process until all zeros are obtained, a triangular array is formed.

The procedure is actually much quicker and simpler than the above general instructions suggest. While the criterion is most useful for literal coefficients, let's first do an example with numerical coefficients. Suppose we form the array for a system with characteristic equation

$$D^5 + 3D^4 + D^3 + 5D^2 + D + 1 = 0 \tag{9-12}$$

$$
\begin{array}{cccc}
1 & 1 & 1 & 0 \\
3 & 5 & 1 & 0 \\
-\frac{2}{3} & \frac{2}{3} & 0 & \\
+8 & 1 & 0 & \\
+\frac{3}{4} & 0 & & \\
1 & 0 & & \\
0 & & &
\end{array}
$$

While it is necessary to form the entire array, *stability is determined by the first column only. The number of changes of algebraic sign in this column is equal to the number of roots that are not in the left half-plane.* Going down the first column, our example exhibits two changes of sign ($+$ to $-$ and then $-$ to $+$), so the Routh criterion labels this system as having two unstable roots. We can in this example easily check this prediction by finding the actual roots, which turn out to be

$$-3.1622 \qquad 0.1927 \pm i1.1535 \qquad -0.1116 \pm i0.4677$$

Part of the complementary solution for this system would thus be $Ce^{+.1927t} \sin(1.1535t + \phi)$, which clearly goes to infinity with oscillations of ever-increasing amplitude.

To show how the Routh criterion is helpful in control system design, consider the position-control system ("servomechanism") of Fig. 9-1d. The system equation relating the controlled variable (angle θ_C) to the command input (voltage e_R) is written directly from the block diagram as

$$(e_R - K_p \theta_C) \frac{K_a K_t/(R_f B)}{D(\tau_f D + 1)(\tau_L D + 1)} = \theta_C \tag{9-13}$$

$$(\tau_f \tau_L D^3 + (\tau_f + \tau_L)D^2 + D + K)\theta_C = \frac{K}{K_p} e_R \tag{9-14}$$

$$K \triangleq \text{loop gain} \triangleq \frac{K_a K_t K_p}{R_f B} \quad \frac{\text{volts/sec}}{\text{volt}}$$

General Linear System Dynamics **629**

Since stability depends entirely on the characteristic equation, we see that the only parameters pertinent to stability are the loop gain and the two time constants. We know from earlier text examples of feedback control systems that we generally try to make loop gain as high as possible, since it improves both speed and accuracy. The Routh array is

$$
\begin{array}{ccc}
\tau_f \tau_L & 1 & 0 \\
\tau_f + \tau_L & K & 0 \\
\dfrac{\tau_f + \tau_L - K\tau_f \tau_L}{\tau_f + \tau_L} & 0 & 0 \\
K &
\end{array}
$$

Because τ_f, τ_L, and K are normally positive numbers, it is clear that a sign change (and thus instability) can result only if

$$
K > \frac{\tau_f + \tau_L}{\tau_f \tau_L} = \frac{1}{\tau_f} + \frac{1}{\tau_L} \tag{9-15}
$$

This inequality is of great practical importance since it puts an upper limit on K and also shows what design changes are needed if a higher K is necessary. We see that our desire for large K must be limited by stability considerations. If, for example, $\tau_f = 0.01$ and $\tau_L = 0.10$ second, K must be less than 110. If a K of 200 were needed for accuracy requirements, we see that we must somehow change the motor and/or load to reduce τ_f, τ_L, or both.

The above inequality does not tell us the "best" value for K, only an upper limit. Texts on control systems show that a good choice for a "starting value" of K is about 40% of the value which just causes instability. This gives a "safety factor" (called *gain margin*) of 2.5. Final choice of K will be done with simulation, but this rule of thumb gives the simulation a good starting point.

Let's close our discussion of stability with an example that is *not* a control system. Figure 9-2 shows an analysis diagram for a fluid/mechanical system which might represent either a hydrostatic air bearing, or an air-cushion vehicle. These two examples vary greatly in size and physical appearance, but turn out to operate on essentially the same principle, and thus yield to the same analysis. Air, at a constant supply pressure p_s flows through a fluid resistance into a region of volume V, which is at a pressure p. Before the supply pressure is "turned on," the housing of mass M, under the effect of its weight, has moved down to contact the base plate, thus closing off the outflow passage around the periphery of the housing. When we turn on the air supply, pressure p starts to rise, since there is an inflow from p_s and no outflow. Pressure p, acting over a housing area A, produces an upward force which tries to raise the housing. When the pressure force just exceeds the housing weight (plus any external force f_i that might be present) the housing starts to rise, opening up some outflow area at the housing periphery. The volume V now has *both* an outflow and an inflow, and when these two flows become equal, the system will come to an equilibrium state, with the housing stationary at a fixed value of height h. The two flows *will* become equal because, as the pressure rises, the outflow area increases as h increases. Thus the pressure p "adjusts itself" to exactly the value that balances inflow and outflow.

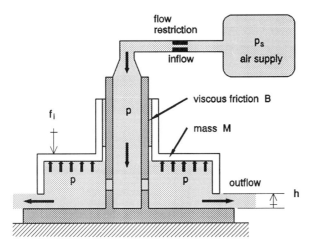

Figure 9-2 Hydrostatic air bearing or air-cushion vehicle.

The equilibrium state just described qualitatively will be taken as the operating point for a linearized perturbation model and analysis. We will not present every physical detail of this modeling since we have earlier done similar studies. All the variables in the upcoming equations are to be considered as the *perturbation* values, *not* the actual values. The two basic laws applied are the conservation of mass for the chamber of volume V, and Newton's law for the vertical motion of the housing. Let's first express the perturbations in the inflow and outflow quantities. The mass inflow rate perturbation depends only on the perturbation of pressure p, since the flow area and supply pressure are fixed.

$$\dot{m}_i = -K_{pi}p \tag{9-16}$$

Note that a positive pressure perturbation causes a negative perturbation in inflow rate. The positive coefficient K_{pi} can be estimated by linearizing a theoretical relation from steady-flow fluid mechanics or can be more accurately determined from steady-flow pressure/flow-rate experiments on the actual flow resistance if this piece of hardware is available.

The outflow at the housing periphery depends on the pressure drop across this flow restriction and the flow area. A linearized model (either theoretical or experimental) would take the form

$$\dot{m}_o = K_{po}p + K_{ho}h \tag{9-17}$$

An increase in pressure causes a greater outflow rate, as does an increase in gap h, since this flow area is proportional to h. For the chamber of volume V, assuming constant values of temperature T, volume V, and gas constant R, the perfect gas law gives

$$\frac{dp}{dt} = \frac{RT}{V}\frac{dm}{dt} = \frac{RT}{V}(\dot{m}_i - \dot{m}_o) = \frac{RT}{V}(-K_{pi}p - K_{po}p - K_{ho}h) \tag{9-18}$$

Turning now to Newton's law and assuming any frictional effects associated with the mass's vertical motion to be viscous, we get

$$-f_i - B\dot{h} + Ap = M\ddot{h} \tag{9-19}$$

General Linear System Dynamics   **631**

where A is the area over which the pressure acts to produce a vertical force on the housing. We now have two equations in the two unknowns h and p, so we could solve for either or both, but one at a time. Since this is a stability study, and stability depends only on the *system* characteristic equation, which is always the same no matter *which* unknown we are dealing with, an equation in either h or p would be acceptable.

If we choose to eliminate p in favor of h, the resulting single equation is

$$\frac{MV}{ARTK_{ho}}\,\dddot{h} + \frac{MRT(K_{pi}+K_{po})+BV}{ARTK_{ho}}\,\ddot{h} + \frac{B(K_{pi}+K_{po})}{AK_{ho}}\,\dot{h} + h$$
$$= -\frac{K_{pi}+K_{po}}{AK_{ho}}\left[\frac{V}{RT(K_{pi}+K_{po})}\frac{df_i}{dt}+f_i\right] \tag{9-20}$$

For cubic characteristic polynomials $(a_3 D^3 + a_2 D^2 + a_1 D + a_0)$, Routh's criterion produces only a single inequality which must be satisfied to guarantee stability: $a_0 a_3 < a_1 a_2$. We can apply this to *any* cubic characteristic polynomial, including our present example:

$$\frac{MV}{ARTK_{ho}} < \left[\frac{MRT(K_{pi}+K_{po})+BV}{ARTK_{ho}}\right]\left[\frac{B(K_{pi}+K_{po})}{AK_{ho}}\right] \tag{9-21}$$

$$AK_{ho} < \left[\frac{RT}{V}(K_{pi}+K_{po})+\frac{B}{M}\right](K_{pi}+K_{po})B \tag{9-22}$$

This inequality can be used in various ways. For example, if all the parameters have been given numerical values except damping B, the inequality shows that, for stability, B must be larger than some definite value. The effect on stability of all the other parameters can be similarly deduced, as is explored in an end-of-chapter problem.

A portable classroom demonstration apparatus as in Fig. 9-2 can be easily and cheaply constructed, using a bottle of compressed air (say about 90 psig) as the supply. In the one I made, I used porous bronze bearings for the sliding joint, and an aluminum housing of about 3-inch diameter. If I oiled the bearings just before the demonstration, the B value was large enough to give stability, If I instead carefully wiped all the oil off the bearings, B was small enough to give instability in the form of an impressively noisy limit cycle vibration. The bottle-gas supply will of course gradually lose pressure, violating our assumption of constant supply pressure. This is actually an advantage for the demonstration since we thus discover the effect of different supply pressures on system behavior.

9-4 GENERALIZED FREQUENCY RESPONSE

Throughout this text we have been using the sinusoidal transfer function method to calculate system frequency response, but have only *proven* its validity in a few special cases, elsewhere simply accepting it on faith. We now want to establish its validity in general terms and then also extend the logarithmic plotting methods to linear systems of arbitrary complexity.

For a general linear system with constant coefficients and no dead times, the differential equation relating a chosen output/input pair has the form

$$(a_n D^n + a_{n-1} D^{n-1} + \cdots + a_1 D + a_0) q_0 = (b_m D^m + b_{m-1} D^{m-1} + \cdots b_1 D + b_0) q_i \quad (9\text{-}23)$$

If $q_i = A_i \sin \omega t$, the method of undetermined coefficients tells us that the forced (steady-state) solution must have the form $q_0 = A \sin \omega t + B \cos \omega t$, which can always be written as $q_0 = A_0 \sin(\omega t + \phi)$.

Our analysis uses the rotating-vector (also called "phasor") method of representing sinusoidal quantities. From a basic trigonometric identity,

$$A e^{i\theta} \equiv A(\cos \theta + i \sin \theta) = A \cos \theta + iA \sin \theta \quad (9\text{-}24)$$

The complex number given by the right-hand side is shown in Fig. 9-3. We wish to represent both q_i and q_o in this way, so we let $A = A_i$ and $\theta = \omega t$ for q_i and $A = A_o$ with $\theta = \omega t + \phi$ for q_o. For a given frequency ω the two phasors rotate at a constant angular velocity, always maintaining the fixed angle ϕ between them. We will need to be able to differentiate phasors with respect to time, so we note that

$$\frac{d}{dt}(A_i e^{i\omega t}) = (i\omega) A_i e^{i\omega t} \qquad \frac{d}{dt}(A_o e^{i(\omega t + \phi)}) = (i\omega) A_o e^{i(\omega t + \phi)} \quad (9\text{-}25)$$

and furthermore,

$$\frac{d^n}{dt^n}(A_i e^{i\omega t}) = (i\omega)^n A_i e^{i\omega t} \qquad \frac{d^n}{dt^n}(A_o e^{i(\omega t + \phi)}) = (i\omega)^n A_o e^{i(\omega t + \phi)} \quad (9\text{-}26)$$

so that differentiating n times is the same as multiplying by $(i\omega)^n$.

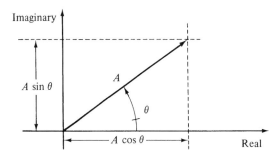

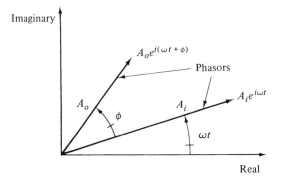

Figure 9-3 Phasor representation of sine waves.

General Linear System Dynamics 633

We are now ready to examine Eq. (9-23) when both q_i and q_o are in sinusoidal steady state. Then *every term* in that equation will be sinusoidal and representable by an appropriate phasor. In fact we may write

$$a_n(i\omega)^n A_o e^{i(\omega t + \phi)} + \cdots + a_1(i\omega)A_o e^{i(\omega t + \phi)} + a_0 A_o e^{i(\omega t + \phi)}$$
$$= b_m(i\omega)^m A_i e^{i\omega t} + \cdots + b_1(i\omega)A_i e^{i\omega t} + b_0 A_i e^{i\omega t} \tag{9-27}$$

Equation (9-27) will lead directly to the sinusoidal transfer function but we must first show that when (9-27) holds, the basic differential equation (9-23) also holds, since these are *not* the same equation. Equation (9-27) is a complex algebraic (not differential) equation and may be split up into real and imaginary parts. Since a complex equation is satisfied only if *both* the real and imaginary parts are separately satisfied, we will show that satisfaction of the imaginary parts guarantees that the basic differential equation is also satisfied. This is done by examining each term; however, two on each side will be sufficient to establish the pattern for the whole equation.

$$\text{Im}\,[a_1(i\omega)A_o e^{i(\omega t + \phi)}] = a_1\omega A_o \cos(\omega t + \phi) = a_1 D q_o \tag{9-28}$$

$$\text{Im}\,[a_0 A_o e^{i(\omega t + \phi)}] = a_0 A_o \sin(\omega t + \phi) = a_0 q_o \tag{9-29}$$

$$\text{Im}\,[b_1(i\omega)A_i e^{i\omega t}] = b_1\omega A_i \cos\omega t = b_1 D q_i \tag{9-30}$$

$$\text{Im}\,[b_0 A_i e^{i\omega t}] = b_0 A_i \sin\omega t = b_0 q_i \tag{9-31}$$

Clearly, all the other terms will reduce in the same way, thus we can use Eq. (9-27) and be assured that the system differential equation will be satisfied. Thus

$$[a_n(i\omega)^n + \cdots + a_1(i\omega) + a_0]A_o e^{i(\omega t + \phi)} = [b_m(i\omega)^m + \cdots + b_1(i\omega) + b_0]A_i e^{i\omega t} \tag{9-32}$$

$$\frac{A_o e^{i(\omega t + \phi)}}{A_i e^{i\omega t}} = \frac{A_o}{A_i}\,e^{i\phi} = \frac{A_o}{A_i}\,(\cos\phi + i\sin\phi) = \frac{A_o}{A_i}\,\underline{/\phi}$$

$$= \frac{b_m(i\omega)^m + \cdots + b_1(i\omega) + b_0}{a_n(i\omega)^n + \cdots + a_1(i\omega) + a_0} \triangleq \frac{q_o}{q_i}\,(i\omega) \tag{9-33}$$

This last result is of course the sinusoidal transfer function, whose general validity we have now established.

While software plotting methods such as MATLAB's BODE(NUM,DEN), which we have used before, will work directly from transfer functions in the form of a numerator polynomial over a denominator polynomial, there are some advantages to also developing a method which works with these polynomials expressed as the product of simple first- and second-order terms and a steady-state gain. In some applications, such as feedback control systems, where the system is designed as a set of connected hardware components, this "factored" form appears naturally. When it does not, one can always use root finders to factor the numerator and denominator polynomials. The transfer function then takes the form

634   **Chapter 9**

$$\frac{q_o}{q_i}(s) = \frac{K(\tau_1 s + 1) \cdots \left[\dfrac{s^2}{\omega_{n1}^2} + \dfrac{2\zeta_1 s}{\omega_{n1}} + 1\right] \cdots}{(\tau_a s + 1) \cdots \left[\dfrac{s^2}{\omega_{na}^2} + \dfrac{2\zeta_a s}{\omega_{na}} + 1\right] \cdots} \tag{9-34}$$

where we used as many first-order and second-order terms as needed to express the two polynomials.

While Eq. (9-34) has an arbitrary number of factors, if we show the validity of our method for a single pair of factors, it obviously holds for any number. We want to show that if we use our logarithmic (db) version of amplitude ratio, the multiplication of terms in Eq. (9-34) becomes simple *addition*. For a single pair of factors we can write

$$\frac{q_o}{q_i}(s) = G_1(s)G_2(s) \tag{9-35}$$

and thus

$$\frac{q_o}{q_i}(i\omega) = G_1(i\omega)G_2(i\omega) = (M_1 \underline{/\phi_1})(M_2 \underline{/\phi_2}) = M_1 M_2 \underline{/\phi_1 + \phi_2} \tag{9-36}$$

Now, applying the definition of the decibel to the amplitude ratio, we get

db value of the overall amplitude ratio $= 20 \log_{10}(M_1 M_2)$

$$= 20 \log_{10} M_1 + 20 \log_{10} M_2 = \text{sum of the individual db values} \tag{9-37}$$

Thus if we graph db amplitude ratio curves for any number of individual factors, the overall amplitude ratio is obtained by simple graphical addition. Equation (9-36) shows that the phase angle curves *also* simply add up.

Before personal computers and dynamic system software became so common, this graphical curve construction method was a vital skill of designers of dynamic systems. Today we carry this skill only to the point of rough freehand curve sketching, not accurate curve plotting, since we have plotting software to do this better and faster. The skill of rough sketching is *still*, however, useful in several ways, thus we urge you to cultivate it. First, it gives us a quick check on the general correctness of any computer graphs we might produce. If you have *no idea* what a particular curve should look like, you will never detect the human graphing errors that unfortunately do occur with computer methods. Perhaps even more important, when we *design* new systems (rather than just analyze existing ones), it is important to "have a feel" for how a particular piece of hardware or parameter value will change the frequency response. These rough graphs often serve this need admirably.

The technique is best shown with a specific example, which is kept simple, but illustrates all the steps in the general method. Suppose our transfer function is

$$\frac{q_o}{q_i}(s) = \frac{1.0(0.05s + 1)}{\left[\dfrac{s^2}{1.0^2} + \dfrac{(2)(0.05)s}{1} + 1\right]\left[\dfrac{s^2}{5^2} + \dfrac{(2)(0.05)s}{5} + 1\right]} \tag{9-38}$$

whose frequency response is graphed in Fig. 9-4. The curves shown there were obtained "by hand" in just a few minutes by the following stepwise procedure.

General Linear System Dynamics

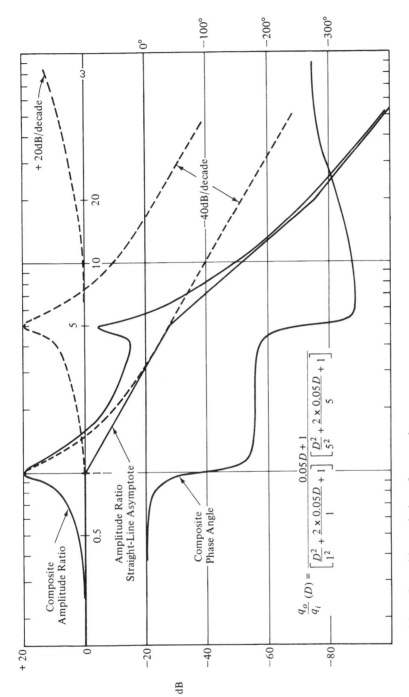

Figure 9-4 Manual graphing technique for complex systems.

1. Note that (9-38) contains two second-order terms with ω_n's of 1 and 5 rad/sec, and one first-order term with $\tau = 0.05$ second. We thus have three breakpoint frequencies: 1, 5 and 20 rad/sec. These frequencies indicate the range of frequency that our graph should encompass—somewhat below the lowest breakpoint and somewhat above the highest. Such breakpoints will also be apparent in every example.

2. Procure 3-cycle semilog graph paper and lay out a frequency scale which will include the breakpoints with "a little to spare" on both the low- and high-frequency sides. Experience has shown that 3-cycle paper usually will be adequate to show the interesting features of any system's frequency response, since it covers a 1000-to-1 range.

3. The vertical scale for amplitude ratio (db) must also be selected; it depends on the K (steady-state gain) value and the expected range of amplitude ratios. In our example, $K = 1.0$, and the "baseline" for the upcoming curves is thus the 0-db line. For the more general case, convert the K value to db and use *this* value as the horizontal baseline for sketching the curves. This baseline locates the "middle" of the db scale, but we still have to choose the actual scale; how many db will each inch or centimeter of graph paper represent? This can be estimated by checking the peak values of any second-order terms that might be present, using the rough approximation that the peak value is about $1/2\zeta$ [Eq. (8-50) gives the exact relation]. For our example, both peaks would be about $1/0.10 = 10$, which is 20 db, so we need to scale our graph to accommodate this. (When second-order terms appear in the *numerator*, their db curves are mirror images of those for denominator terms, and their phase angles are positive rather than negative. Thus our example values, if for a numerator term, would have *valleys* of -20 db rather than peaks of $+20$.) On the lower end of the db scale, we know that second-order terms drop off at a rate of -40 db/decade, and our lowest breakpoint is at 1 rad/sec, so we need to provide a scale that reaches to about -80 db, since our frequency scale goes to 100 rad/sec. As in any graphing, there is no unique choice of scales that we *must* use. The above hints will, however, help in choosing an *adequate* scale. Remember that we are mainly interested in quick and rough sketches, so use some preliminary layouts to help you in getting acceptable final scales.

4. Now locate the breakpoints on your baseline and draw in the straight-line asymptotes for each first- or second-order term, using the known slopes of -20 db/decade and -40 db/decade ($+20$ and $+40$ if the terms are in the numerator). Get the composite asymptote by simply adding the individual curves. Roughly correct this asymptote at the breakpoints by using a -3 db correction for first-order terms and a $+1/2\zeta$ (db value) peak correction for second-order ($+3$ db and $-1/2\zeta$ for numerator terms).

5. For the phase-angle curve, recall that first-order terms go from 0 to $-90°$, with exactly $-45°$ at the breakpoint (0 to $+90°$, $+45°$ for numerator terms) and second-order go from 0 to $-180°$, with exactly $-90°$ at the breakpoint (0 to $+180°$, $+90°$ for numerator terms). For our example we see that the first-order numerator has its breakpoint at a frequency well beyond the second-order denominator terms, so the composite phase angle will never go positive. Also the high-frequency asymptote must be

General Linear System Dynamics 637

$90 - 180 - 180 = -270°$. All these facts allow you to choose a good phase angle scale and roughly sketch the actual curve. Remember that amplitude ratio and phase angle curves should appear *together* on the same sheet of paper, with the phase angle at the bottom.

I hope you will invest a little practice time in personally developing the above rough sketching capability. Once you have sketched a few examples on semilog graph paper you will, if necessary, be able to draw useful graphs on a piece of plain white paper ("back of an envelope"), so that you can use this method anytime and anyplace that you might need it.

The graphing of the factored-form transfer function, as compared with the polynomial form, helps us understand certain features of system response. The two second-order terms in Eq. (9-38) both have equally light damping, so our earlier experience with individual second-order systems would lead us to believe that their resonance behavior would be equally "bad" in this more complex system. The graphing procedure, however, clearly shows that, once we pass the lowest natural frequency, any higher-frequency resonances will be *depressed* because their amplitude ratio is plotted "on top of" the -40 db/decade asymptote of the first natural frequency. Thus the first peak *is* at about $+20$ db (magnification of 10) but the second is only about -5 db (magnification of 0.56). A curve plotted by the computer, directly from the polynomial transfer function will of course show the correct values of all the peaks, but will not *explain* why the higher-frequency peaks are small even though they have equally small ζ's.

Let's do one more example which will reveal some further insights into general system behavior and also use some helpful software tools. The vibrating system of Fig. 9-1a has Eq. (9-8) as the relation between force and displacement at mass number 1. We could use Eq. (9-8) directly to compute this frequency response, but will instead convert this transfer function into the factored form, to gain the understanding just illustrated above. To carry out this conversion, we will employ some computer aids that are part of MATLAB (other commercial software provides similar tools). The first step is to use a root-finding procedure to decompose the numerator and denominator polynomials into first- and second-order factors. In MATLAB we would write

```
cn=[1 2 4001 4000 3e6]; numerator coefficients
cd=[1 3 5002 8000 6.001e6 2e6 1e9]; denominator
rn=roots(cn) request numerator roots
rn= -0.5000+54.7700i computer returns roots
    -0.5000-54.7700i
    -0.5000+31.6188i
    -0.5000-31.6188i
rd=roots(cd) request denominator roots
rd= -0.7416+56.9713i computer returns roots
    -0.7416-56.9713i
    -0.6939+39.4301i
    -0.6939-39.4301i
    -0.0644+14.0737i
    -0.0644-14.0737i
```

638   **Chapter 9**

We see that the numerator factors into two lightly damped second-order terms while the denominator has three such terms. The natural frequencies of all these terms range from about 14 rad/sec to about 57 rad/sec.

Next we want to convert the second-order roots into second-order factors in our standard form, Eq. (8-4). This could be done "by hand" but MATLAB provides the needed computer aids. The command POLY multiplies the two root factors to give a single second-order polynomial, with the leading coefficient equal to 1.0, which is *not* our standard form. To get *our* standard form this polynomial must be divided by the trailing term. We also need to request MATLAB's long format to get enough significant digits displayed in the results.

```
format long     requests 14 decimal places in results
rn12=rn(1:2,1) selects the first two roots from rn
rn12=   -0.5000+54.7701i computer displays the
        -0.5000-54.7701i requested roots
pn12=poly(rn12) request second-order polynomial
pn12=    1.0000 computer displays the
         1.0000 three coefficients
         3000.0
pn12=pn12./pn12(3,1) converts coefficients to our
                     standard form
pn12=    0.0003333 computer displays coefficients
         0.0003333 in our standard form
         1.0000000
```

The above procedure can be repeated for each pair of roots in the numerator and denominator. (I have not shown above all of the 14 decimal places which were actually displayed.)

At this point we would have Eq. (9-8) expressed as the product of two standard second-order terms in the numerator, divided by the product of three standard second-order terms in the denominator. The steady-state gain K of the system is easily found as the ratio of the trailing terms in the numerator and denominator [set $D = 0$ in Eq. (9-8)], which gives $K = 3e6/1e9 = 0.003\,\text{m/N}$. Let's now use MATLAB to plot the individual terms of this transfer function and also the composite. For clarity in this example, I will ignore the K term since it just shifts the whole amplitude-ratio curve vertically by -50.46 db. In an actual application, the K value would easily be included in the graph. In the MATLAB statements below, I have rounded off the numerical values of the coefficients for easy manual data entry. One could of course modify the MATLAB "program" so that the coefficients would be taken from our earlier calculations (with 14 decimal places) rather than entering them manually in truncated form.

```
w=[logspace(0,2.3,500)]'; defines a set of 500 frequencies
                          from 1 to 200 rad/sec
s=w.*1i; forms s=iw for use in transfer functions
n1=.001.*s^2+.001.*s+1; computes first numerator term
n2=.0003333.*s.^2+.0003333.*s+1; second numerator term
d1=1./(.0003080.*s.^2+.0004569.*s+1); first denominator term
d2=1./(.0006429.*s.^2+.0008924.*s+1); second denominator term
```

General Linear System Dynamics

```
d3=1./(.005049.*s.^2+.0006507.*s+1); third denominator term
mn1=20.*log10(abs(n1));  db amplitude ratio, first num term
mn2=20.*log10(abs(n2));  db amplitude ratio, second num term
md1=20.*log10(abs(d1));  db amplitude ratio, first den term
md2=20.*log10(abs(d2));  db amplitude ratio, second den term
md3=20.*log10(abs(d3));  db amplitude ratio, third den term
semilogx(w,mn1,'--')     plot first num term as a dashed line
hold on;semilogx(w,mn2,'--')    plot second num term
semilogx(w,md1,'--';semilogx(w,md2,'--');semilogx(w,md3,'--')
mtf=mn1+mn2+md1+md2+md3; add up terms to get composite curve
semilogx(w,mtf,'linewidth',1)   plot composite curve as
                                a heavy, solid line
```

The statement HOLD ON stipulates that all succeeding curves be plotted on the same axes as were used for the preceding graph; thus we superimpose all the individual (dashed line) component curves and also the final composite (solid heavy line) curve. Figure 9-5 shows the results of the above calculations. (Recall that we have "left out" the steady-state gain of -50.46 db; thus our curves are all referenced to the 0-db line.) As in our earlier example, we again see that the higher-frequency resonances are suppressed, even though they "individually" are lightly damped with

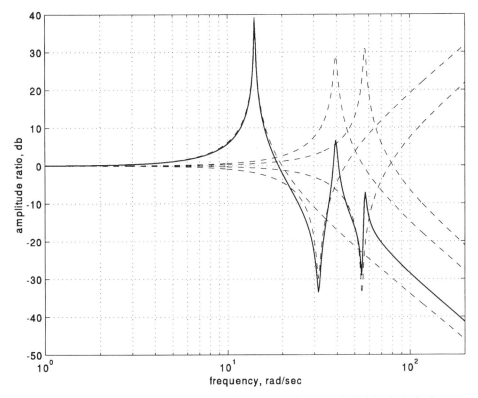

Figure 9-5 Building up the composite (solid) response from the individual (dashed) terms.

Chapter 9

high peaks. We now also see that each numerator term contributes a deep "valley" near its natural frequency. These are sometimes called *antiresonances*. Finally, since the composite curve closely follows the individual curve for the first denominator term out to about 20 rad/sec, we could *simplify* this model to just that single term, *if the input force had little frequency content beyond 20 rad/sec*. This is just another example of how frequency response concepts are useful in judging how complicated a model is necessary in a particular application.

We did not bother to compute and plot phase angles in the above MATLAB program, but this feature would be easily added. Also, be sure you understand that all these frequency response techniques are applicable to *all* kinds of linear system models, electrical, fluid, thermal, and mixed. They are not limited to the mechanical vibration systems we have used to explain them.

9-5 MATRIX FREQUENCY RESPONSE

If you want to actually see transfer functions such as Eqs. (9-8) through (9-10), you must go through the determinant expansion procedure we used to generate them. If, on the other hand, you only want to compute sinusoidal transfer functions for specific parameter values and frequency ranges, it is possible to work *directly* from the original differential equation set, such as (9-1) through (9-3). A matrix calculation from linear algebra allows us to get all possible output/input frequency-response curves from this original set of equations quite efficiently.

To see how this method is used, consider a general linear system with constant coefficients, with inputs q_{ia}, q_{ib}, etc. and outputs $q_{o1}, q_{o2}, \ldots, q_{on}$. Such systems can always be represented by a set of differential equations, such as those [Eqs. (9-1) through (9-3)] for our example. If the general set of equations is Laplace transformed with zero initial conditions, and we let $s = i\omega$, we will get a set of n algebraic equations in the n unknowns $Q_{o1}(i\omega)$, $Q_{o2}(i\omega)$, etc., where the Q_o's are complex numbers, with magnitudes and phase angles. If we set a particular input Q_i to be $1.0 \, \underline{/0^\circ}$ and all the other inputs to zero, and if we set frequency ω to a specific number, then if we solve the equation set for all the Q_o's, these Q_o's will actually be the values of the sinusoidal transfer functions relating each Q_o to the chosen Q_i. That is, when $Q_i = 1.0 \, \underline{/0^\circ}$, $Q_o/Q_i = Q_o$. We can repeat this calculation for as many frequency values as we wish.

The equation set to be solved can be written as

$$A_{11}Q_{o1}(i\omega) + A_{12}Q_{o2}(i\omega) + \cdots + A_{1n}Q_{on}(i\omega) = C_{1a}Q_{ia}(i\omega) + C_{1b}Q_{ib}(i\omega) + \cdots$$
$$A_{21}Q_{o1}(i\omega) + A_{22}Q_{o2}(i\omega) + \cdots + A_{2n}Q_{on}(i\omega) = C_{2a}Q_{ia}(i\omega) + C_{2b}Q_{ib}(i\omega) + \cdots$$
$$\vdots$$
$$A_{n1}Q_{o1}(i\omega) + A_{n2}Q_{o2}(i\omega) + \cdots + A_{nn}Q_{on}(i\omega) = C_{na}Q_{ia}(i\omega) + C_{nb}Q_{ib}(i\omega) + \cdots$$

From linear algebra we know that the unknowns can be found by first obtaining the inverse of the matrix of A coefficients and then matrix multiplying this by the column vector of C coefficients that remains on the right-hand side when all Q_i's other than the selected Q_{ik} are set to zero.

General Linear System Dynamics   **641**

$$
\begin{vmatrix} A_{11} & A_{12} & \cdots & A_{1n} \\ A_{21} & A_{22} & \cdots & A_{2n} \\ \cdot & & & \\ \cdot & & & \\ \cdot & & & \\ A_{n1} & A_{n2} & \cdots & A_{nn} \end{vmatrix}^{-1} \begin{vmatrix} C_{1k} \\ C_{2k} \\ \cdot \\ \cdot \\ \cdot \\ C_{nk} \end{vmatrix} = \begin{vmatrix} Q_{o1}(i\omega) \\ Q_{o2}(i\omega) \\ \cdot \\ \cdot \\ \cdot \\ Q_{on}(i\omega) \end{vmatrix}
\tag{9-39}
$$

These matrix operations are available in a number of software packages. Let's do the system of Fig. 9-1a using MATLAB.

```
w=[logspace(0,2.3,200)]'; define frequencies from 1 to 200
s=w*i1; set s=iw for use in upcoming equations
for i=1:200 use a FOR LOOP to do the 200 frequencies
A=[s(i)^2+s(i)+1000 -s(i)-1000 0 define matrix Eq. (9-1)
  -s(i)-1000 s(i)^2+s(i)+2000 -1000            Eq. (9-2)
  0 -1000 s(i)^2+s(i)+2000]                    Eq. (9-3)
AI=inv(A); invert matrix
C1=[1 0 0]; define column vector for input f_{i1}
QO1=AI*C1; matrix multiplication
QO11(i)=QO1(1); first row of QO1 is response of x_1 to f_{i1}
QO12(i)=QO1(2); second row of QO1,  response of x_2 to f_{i1}
QO13(i)=QO1(3); third row of QO1,   response of x_3 to f_{i1}
db11=20*log10(abs(QO11); db amplitude ratio, x_1/f_{i1}
db12=20*log10(abs(QO12); db amplitude ratio, x_2/f_{i1}
db13=20*log10(abs(QO13); db amplitude ratio, x_3/f_{i1}
phi11=angle(QO11); phase angle
phi12=angle(QO12); phase angle
phi13=angle(QO13); phase angle
C2=[0 1 0]; define column vector for input f_{i2}
QO2=AI*C2; matrix multiplication
QO21(i)=QO2(1);
QO22(i)=QO2(2);
QO23(i)=QO2(3);
db21=20*log10(abs(QO21);
db22=20*log10(abs(QO22);
db23=20*log10(abs(QO23);
phi21=angle(QO21);
phi22=angle(QO22);
phi23=angle(QO23);
end  terminate the FOR LOOP
semilogx(w,db11,w,db22,'--'; plot two of the amplitude
                             ratios
grid;xlabel('frequency, rad/sec');
ylabel('amplitude ratio, db');
```

Figure 9-6 shows the graph that results from the above calculations. All the other amplitude ratio and phase angle curves could of course also be plotted. Extension of this program to computing the responses to the third input force f_{i3} should be obvious and is left for the end-of-chapter problems. When MATLAB is used in

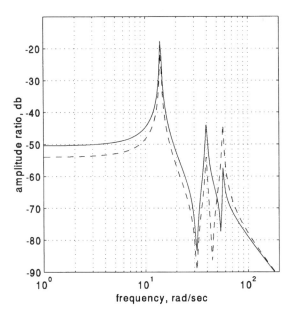

Figure 9-6 Graphical results of matrix frequency-response program—db11 (solid) and db22 (dashed).

the above "program" mode, rather than the "interactive" mode, it is usually best to put the above statements into an "*m* file." This allows easy editing and rerunning. The above example, which has a 3 × 3 system matrix, sets a pattern for the general matrix frequency-response calculation, where any number of simultaneous equations and any number of inputs can be accommodated.

9-6 TIME-RESPONSE SIMULATION

We have just shown how any desired sinusoidal transfer function calculation can be performed directly from the original set of simultaneous equations. Time-history simulations of course should usually also be based on this same set of equations, as we have seen in many previous examples. Here, of course, we are *not* limited to linear, constant-coefficient models, as is the case for the matrix frequency-response method. We have shown throughout the book many examples of such simulations, gradually developing simulation skills that can handle many different situations in all kinds of systems and problems. There is thus now no real need to discuss simulation methods for "general" systems; all the methods previously shown are applicable and may be applied to any complex system.

When a graphical user interface, such as the SIMULINK which we have largely used, is employed, the only new features that might be needed have to do with the problem of the screen being "too small" to hold all the icons for a large set of equations. In SIMULINK this is handled with a command called GROUP. With GROUP, one can select an entire screen, or any selected portion, and replace it by a single small block, which will not display all the internal details, but which will

General Linear System Dynamics

provide input and output "connections" where the user wants them. One can thus set up large simulations by interconnecting subsystems defined by use of the GROUP facility. This feature allows us to work with simulations of almost unlimited complexity, even though our monitor has a finite size screen. A command UNGROUP is available to expand portions of such a diagram when the details of that subsystem are temporarily needed. I wanted to make you aware of these capabilities, but our examples are not so complex, so I won't be presenting any actual simulations using GROUP or UNGROUP. Whatever software you might be using, its manual will provide information on the use of such features.

While no new simulation techniques are involved, I show a simulation diagram for the system of Fig. 9-1a in Fig. 9-7. To tie in with some of the analytical results we obtained earlier, I subject this system to some step input forces. The diagram provides for all three input forces, and we can of course also use any combination of initial displacements and velocities by setting up the six integrators with proper initial conditions. There are many different combinations of step input forces and initial conditions that one could examine. I show in Fig. 9-8 only one of these possibilities; all initial conditions zero, input force 3 zero, input force 1 a step of size 1, and input force 2 a step of size -2. Our earlier root finding revealed damped natural frequencies of about 14, 39 and 57 rad/sec (2.2, 6.2, 9.1 Hz), so these damped sinewave responses must appear in both the analytical and the simulation solutions. The three component solutions are, however, combined (in different ways) in each of the displacement waveforms of Fig. 9-8, and the individual damped natural frequencies are usually not obvious or accurately measurable graphically.

Looking first at motion x_1, and visually ignoring the higher frequency "wiggles," it is possible to see a large-amplitude, low-frequency component which goes through a little over 2 cycles in 1 second. The first 0.16 second of x_3 reveals

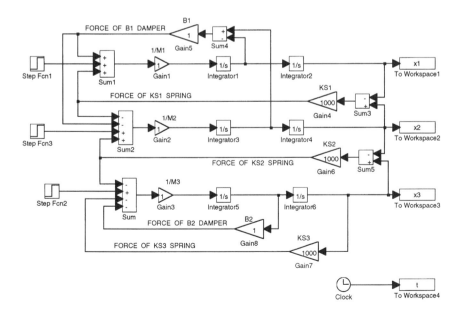

Figure 9-7 Simulation diagram for 3-mass mechanical system.

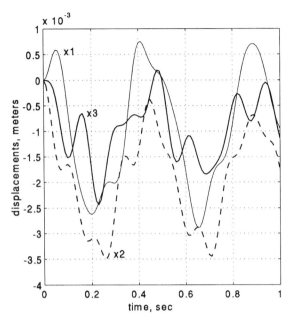

Figure 9-8 Motions of the three masses.

about one cycle of another frequency, which would be about 6 Hz. Looking at the fastest wiggles in either x_2 or x_3, one could guess at frequencies from about 10 to 15 Hz, so this component is not very clear in these response curves. We thus conclude that visual examination of transient responses of complex systems may or may not give much useful information on the frequencies actually involved. Fortunately, *frequency spectrum analysis*, which will be discussed in the next two sections, allows a more scientific identification of the frequency content of all kinds of signals.

9-7 FREQUENCY SPECTRUM ANALYSIS OF PERIODIC SIGNALS: FOURIER SERIES

While computer simulation (such as SIMULINK) allows us to find the response of very complex systems to very complex forms of driving input, it is still desirable to have available general analytical methods for such problems. These methods are necessary, not so much to grind out numerical solutions to specific problems (the computer is usually unbeatable at this) as to provide *insight* into qualitative aspects of system behavior. This kind of insight is vital for effective system design, where we must understand the effects of changes in system parameters and driving inputs. It also gives us guidance in selecting the computer studies which we should run and then interpreting these computer results.

For linear systems with constant coefficients, such generalized analytical tools are available. They consist of a system description in terms of its frequency response, and an input signal description in terms of its frequency spectrum. When both these descriptions are available, one can always calculate the system response. Furthermore, the nature of the methods is such that they give the insights mentioned above. In the immediately previous sections we have given the analytical methods for getting the frequency response of general linear systems. When the actual system is available for lab testing, there are also experimental methods which get us this same information, without the simplifying assumptions always present in any theoretical methods. We will touch briefly on these experimental techniques in section 9-9.

The problem of finding the frequency content ("frequency spectrum") of a given input signal is best treated by separating signals into certain classes, for which a certain approach is applicable. Most (but not all) signals of practical importance can be classified into three broad classes: periodic, transient, and random (see Fig. 9-9). A periodic signal is one which exhibits a definite cycle and repeats itself over and over unendingly. A transient signal has a beginning and an end. It is zero before its beginning and after its end, but may have any shape in between. Random signals continue unendingly but exhibit no predictable pattern or cycle. Frequency spectrum methods are available[3] for treating all three types of signals analytically. While theoretical analysis can assume periodic and random signals that go on "forever," when we analyze any real world signal given by experimental data, it must of necessity be a "transient"; we can only deal with a finite-length record. Thus the spectrum analyzers used in lab testing employ, for the most part, a single method of analysis for all signals.

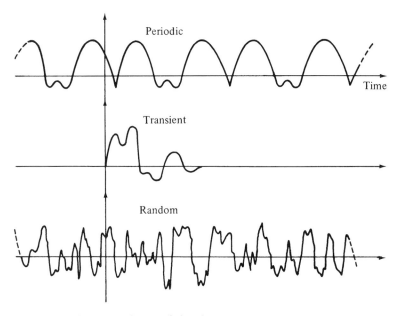

Figure 9-9 Common classes of signals.

[3] E. O. Doebelin, *System Modeling and Response: Theoretical and Experimental Approaches*, Wiley, New York, 1980, chaps. 3, 6; *Measurement Systems*, 4th ed., McGraw-Hill, New York, 1990, pp. 141–182.

In this text's treatment of spectrum analysis, we will concentrate mainly on periodic signals, with a brief discussion of transients. Periodic inputs are of considerable practical importance because they are good approximations to real-world physical variables in machinery and processes that are running in a "steady-state" cyclic condition. When a machine is started from rest, it goes through a transient time interval until it reaches steady operation. Similarly, the shutdown of a machine involves transient behavior. Between these startup and shutdown transients, the machine often operates according to a more-or-less repetitive cycle. Consider a "warmed-up" automobile entering the approach ramp of a freeway. The acceleration to freeway speed involves transient changes in car speed, engine rpm, transmission shifting, cylinder gas pressure, etc. Once on the freeway, if we, say, engage a cruise control to maintain 65 mph, the terrain is not hilly, and wind speed and direction do not vary much; after a while, all the pressures, temperatures, forces, etc. associated with rotating or reciprocating parts will be exhibiting essentially cyclic behavior. While no real-world "cycle" will be *precisely* repetitive, a mathematically periodic model may be a useful tool for dynamic analysis.

For periodic signals the mathematical tool is the *Fourier series*, which we now introduce without any proofs. It can be shown that any periodic function $q_i(t)$ which is single-valued, finite, and has a finite number of discontinuities, maxima, and minima in one cycle (conditions easily met by any real physical signal) may be represented by the Fourier series:

$$q_i(t) = q_{i,\text{avg}} + \frac{1}{L}\left[\sum_{n=1}^{\infty} a_n \cos\frac{n\pi t}{L} + \sum_{n=1}^{\infty} b_n \sin\frac{n\pi t}{L}\right] \tag{9-40}$$

$$q_{i,\text{avg}} \triangleq \text{average value of } q_i = \frac{1}{2L}\int_{-L}^{L} q_i(t)\,dt \tag{9-41}$$

$$a_n \triangleq \int_{-L}^{L} q_i(t)\cos\frac{n\pi t}{L}\,dt \tag{9-42}$$

$$b_n \triangleq \int_{-L}^{L} q_i(t)\sin\frac{n\pi t}{L}\,dt \tag{9-43}$$

In Fig. 9-10 the time for one complete cycle is called the period T, and is equal to $2L$. You may choose the location of the time origin wherever convenient or necessary for the physical problem, and the integrals in (9-41), (9-42), and (9-43) are taken over one complete cycle. You may think of the Fourier series as a curve-fitting problem in which the fitting functions must be a constant, sine waves, and cosine waves. The curve fit gets better and better as we increase n, the number of sine and cosine waves used in the fit.

EXAMPLE: SQUARE WAVE

Before going any further, let's do a simple example, the square wave of Fig. 9-11.

$$q_{i,\text{avg}} = 0 \qquad \text{by inspection}$$

General Linear System Dynamics

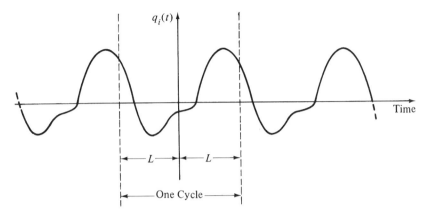

Figure 9-10 Periodic signal.

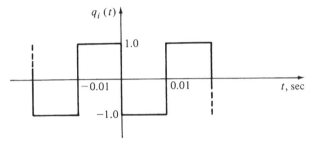

Figure 9-11 Square wave example for Fourier series calculation.

$$a_n = \int_{-0.01}^{0} \cos \frac{n\pi t}{0.01} \, dt + \int_{0}^{0.01} -\cos \frac{n\pi t}{0.01} \, dt = \frac{\sin n\pi - \sin n\pi}{100 n\pi} = 0 \quad (9\text{-}44)$$

$$b_n = \int_{-0.01}^{0} \sin \frac{n\pi t}{0.01} \, dt + \int_{0}^{0.01} -\sin \frac{n\pi t}{0.01} \, dt = \frac{\cos(n\pi) - 1}{50 n\pi} \quad (9\text{-}45)$$

Equation (9-40) then gives

$$q_i(t) = 100 \sum_{n=1}^{\infty} \frac{\cos n\pi - 1}{50 n\pi} \sin \frac{n\pi t}{0.01}$$

$$= -\frac{4}{\pi} \left[\frac{\sin 100\pi t}{1} + \frac{\sin 300\pi t}{3} + \frac{\sin 500\pi t}{5} + \cdots \right] \quad (9\text{-}46)$$

You should now freehand-sketch these first three terms to see how the series begins to approximate the square wave. Taking more terms will improve the curve fit. Waveforms with vertical jumps or discontinuities (multiple-valued sections), such as this example, are particularly difficult to fit. The series will "do the best it can"; converge to the *midpoint* of the vertical section, 0 in our case. Also, no matter how many terms we take in such an example, an "overshoot" (called Gibbs' phenomenon) will persist at "corners" of the square wave. Fortunately, most real-world periodic functions that we deal with are relatively "smooth" and their Fourier series gives a close fit, often with only a few terms.

While Eq. (9-40) indicates that, in general, both sine and cosine terms are obtained, it is desirable in most cases to combine sines and cosines of the same frequency by using the identity

$$A \cos \omega t + B \sin \omega t \equiv C \sin (\omega t + \alpha)$$

$$C \triangleq \sqrt{A^2 + B^2} \qquad \alpha \triangleq \tan^{-1} \frac{A}{B} \qquad (9\text{-}47)$$

We choose to always do this conversion because we want to use the sinusoidal transfer function to compute our results, and this transfer function is defined for inputs which are *sine*, not cosine, waves. Once this conversion has been done, we can define the *frequency spectrum* of any periodic function as a graph of the amplitudes C and phase angles α of these sine waves, both plotted against frequency ω. Figure 9-12 shows such a spectrum for our square-wave example. Note that amplitudes C are by definition always positive; if a term originally had a minus sign, this is taken care of with a $-180°$ phase angle, as in our example.

The average value is plotted at zero frequency, and again negative values are handled with a $-180°$ phase angle. The lowest-frequency sine wave is called the *fundamental or first-harmonic*; either name is correct. The frequency of this term will always be the same as the repetition rate of the original periodic function, 50 Hz or 314 rad/sec in our example. The higher-frequency terms are called respectively, the second, third, fourth, etc. harmonics, and their frequencies are always 2, 3, 4, etc. times the fundamental frequency. Depending on the symmetry of the original periodic function, certain harmonics may not be present; our example gives only the *odd* harmonics. For periodic functions of arbitrary shape, however, all the harmonics will be present. The spectrum of any periodic function is called a *discrete* spectrum. This means that it has content *only* at the discrete frequencies of the harmonics; there is *nothing* "in between." Thus you should *never* connect the amplitude values with a curve; show *only* the "spikes" at the harmonic frequencies. This is an important

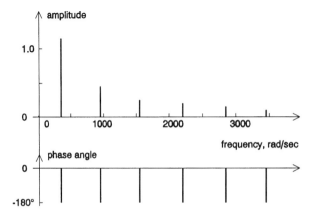

Figure 9-12 Frequency spectrum of periodic square wave: a discrete spectrum.

General Linear System Dynamics   **649**

point because the spectra of transients and random signals are *not* discrete but *continuous*; there is signal content at *every* frequency and the spectrum is properly plotted as a smooth curve, not spikes.

Once one has the Fourier series for a periodic input to a linear constant-coefficient system, the response of the system is quickly calculated if we know the system sinusoidal transfer function. When we say "response," we mean the *periodic steady-state response*. Our calculation method does *not* get us the transient which starts at the time we first apply the periodic input and ends when the negative exponentials in the complementary solution decay to zero. If you need to see this starting transient, simulation will of course get this since it gets the complete solution, starting at time zero, and gradually converging on the periodic steady state which we *are* able to calculate from the frequency-response method we are about to explain. If you are saying to yourself "Why bother with this method when we can 'get it all' with simulation?"—recall that we are here after *insight*, not number crunching.

The response of our system to a periodic input is obtained by use of the super-position theorem and the system's frequency response, together with the Fourier series for the input. The Fourier series gives the driving function as a sum of sine waves plus the average value, the superposition theorem allows us to get the total response as the sum of responses to the individual sine waves, and the frequency-response curves let us get the response to any one sine wave quickly and easily. If the Fourier series for the input $q_i(t)$ is given by

$$q_i(t) = C_0 + C_1 \sin(\omega_1 t + \alpha_1) + C_2 \sin(2\omega_1 t + \alpha_2) + \cdots \qquad (9\text{-}48)$$

we may then compute the response to each term separately if we know the amplitude ratio and phase shift of the system at $\omega = 0$, ω_1, $2\omega_1$, etc. These individual response terms are then simply added up to get the total system response, which will also be a periodic function. We emphasize again that the response found in this way is the *periodic steady-state response*, that is, the output which exists when the input has been applied for a long enough time that the system transients (natural response) have died out. Figure 9-13 shows graphically how this steady-state response is calculated for an arbitrary periodic input signal and an arbitrary system frequency response. The first few terms of this response can be written out directly from the information shown on this diagram.

Periodic steady-state output $\triangleq q_{o,ss}$

$$= C_0 A_0 + C_1 A_1 \sin(\omega_1 t + \alpha_1 + \phi_1) + C_2 A_2 \sin(2\omega_1 t + \alpha_2 + \phi_2) + \cdots \qquad (9\text{-}49)$$

Note that two factors are operating which allow us to neglect higher-frequency components of any Fourier series in this kind of calculation, and thereby deal with a *finite*, rather than infinite, number of terms. First, the amplitudes of the higher harmonics in a Fourier series always ultimately tend toward zero, since the frequency spectrum of real physical signals cannot extend to infinity. Second, the frequency response of any real system also cannot extend to infinity but must also gradually drop off; no real system can respond to infinitely fast inputs. Since the output (response) is the *product* of the input spectrum and the system amplitude ratio, when both these factors approach zero for high values of ω the response must clearly also drop off. Thus, beyond some range of frequencies, the response amplitude is a

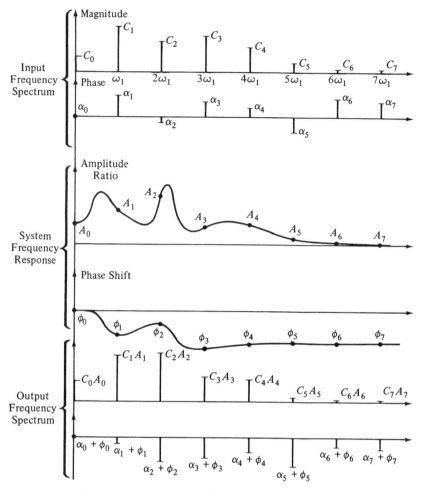

Figure 9-13 General calculation method for response of any linear system to any periodic input.

small fraction of what it is at low frequencies and may legitimately be neglected relative to the large low-frequency values.

While simple periodic functions describable by mathematical formulas (such as our square-wave example) may be Fourier-analyzed quickly and easily using Eqs. (9-40) to (9-43), in many practical problems the periodic function is complicated, or perhaps given by experimental data for which no formula is known. Since all the operations (function generation, multiplication, integration, etc.) used in the Fourier series calculation can be done *numerically* (rather than analytically), one can write a computer program to compute Fourier series to any accuracy desired. Such numerical schemes were in fact in "manual" use long before digital computers were invented. Since frequency spectrum analysis is so widely used in science and engineering, there is usually no need to write such programs yourself; they can be purchased "ready-made" and are often part of larger mathematical software packages. We next want to show how to use two different numerical approaches

General Linear System Dynamics

to calculating Fourier series. One, using SIMULINK, is a direct and obvious implementation of the defining formulas, but is rather wasteful of computer time. The other, which uses MATLAB's *Fast Fourier Transform* command, is blindingly fast but rather complex internally and thus we will ask you to simply take it on faith, rather than trying to prove to you how it works.

Let's first explain the SIMULINK method, since you will have no trouble seeing that it is valid. The Fourier series defining formulas show that the following operations need to be implemented:

1. We must define the detailed shape of our periodic waveform. This is conveniently done using the LOOKUP TABLE module. We simply enter two lists ("vectors") giving pairs of x, t points on our waveform. These points might come from a formula for the waveform (if such is available) or from an experimental graph. If we choose to use rather sparse points, we may want to produce a smoother curve using the MATLAB spline function.

2. We must integrate the waveform itself to get its average value. In the simulation diagram of Fig. 9-14 I use T for the period, which is equal to $2L$ in the defining formulas.

3. We must generate sine and cosine functions of time, at the various harmonic frequencies, multiply these by the input waveform, and then integrate the product over one complete cycle of the input waveform. The SIMULINK modules CLOCK, FUNCTION, PRODUCT, and TRANSFER FUNCTION (set up as an integrator with a gain) will provide what we need.

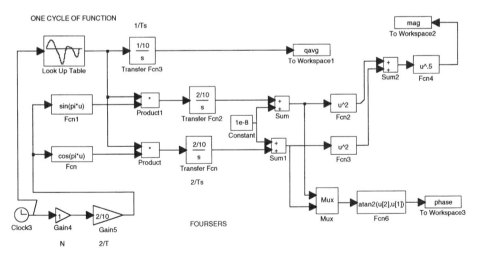

Figure 9-14 Simulation diagram for computing terms in the Fourier series.

652   **Chapter 9**

4. Since we usually want a series with only sine terms with phase angles (not both sines and cosines), our final operation is to use Eqs. (9-47) to get the series in this form.

Our simulation will give all the quantities as running functions of time, but we only need the *final* values to get the Fourier series coefficients. If, for example, we set up our integration parameters to get 501 points, and we have named the coefficients C in Eq. (9-48) "mag," the phase angle α "phase," and the average value "qavg," when the integration for any selected harmonic is finished (it takes a few seconds), we would simply go to the MATLAB command window and type:

```
[qavg(501) mag(501) phase(501)]
```

and these three results would immediately be printed. We can do this for as many harmonics as we wish.

You should now look at Fig. 9-14 to see the simulation diagram. The CLOCK module provides time t for the input waveform and the sine and cosine waves. In any specific problem, you need to insert a proper value for the waveform period T. In our upcoming example this will be 10.0 seconds. GAIN4 (n in the formulas) is set successively at 1, 2, 3, etc. as we compute the various harmonics. I used the TRANSFER FUNCTION module for the needed integrators because it allows an included gain, which saves some space on the diagram. These three integrators also must have the proper value of T inserted to suit the particular problem. Since the phase angle calculation involves division of one integral signal by another, and these signals start at 0.0, I add a tiny constant 1e-8 to prevent a division by zero. This phase angle computation also uses some techniques which may be unfamiliar. We usually use the nonlinear FUNCTION module with a single input, which is always called "u" when setting up the function. The inverse tangent function atan2(x,y) requires *two* inputs to compute the angle from its tangent, but the FUNCTION module only provides a single connection point at its input. The MUX (multiplexer) module resolves this dilemma since it accepts any number of input signals [called u(1), u(2), u(3), etc. starting at the top of the block] and provides these as "separate inputs" to the atan2(x,y) function.

When setting up your simulation parameters, I suggest you always use the RK-5 integrator, make the max and min step sizes *equal* (this makes it a fixed-step integrator) and use several hundred computing steps to get good accuracy. The calculation is always quite fast, so 501 steps (such as 0–10 seconds with 0.02 step size) might be a universally usable number. Also make the final time equal to the time for exactly one cycle of your input waveform. Let's now do an example.

EXAMPLE: EXPERIMENTAL DATA

Our example will use a waveform with period $T = 10$ seconds, giving the values shown on Fig. 9-14. Since we always start with the first harmonic, Gain4 should be set at 1, which it is in the figure. Suppose we have a lab-measured graph of a waveform for which we desire a Fourier series. We need to pick enough points from our graph to document all its "wiggles." If the graph is rather smooth, a large number of points is not really needed, but if we use this small number of points the LOOKUP TABLE module will interpolate linearly, giving a poor reproduction

General Linear System Dynamics

for the actual shape. To avoid the tedium of hand-picking a large enough number of points so that linear interpolation gives a good representation, we can use the MATLAB spline function.

If you are not familiar with splines, they are quite useful for several purposes. A common one is the computer graphing of lab data taken at discrete points that are not closely spaced. "Before computers" we would use French curves to manually draw a nice smooth curve through the measured points. This is usually desirable since we often know from theory that the physical phenomenon being measured *does* vary smoothly. If you have used computer graphing software to plot such graphs for you, you may have noticed that the graphs are not as "nice" as those drawn by hand with French curves; they are *not* smooth. By applying the spline function to sparsely spaced points, we get the equivalent of manual French curves, nice smooth graphs.

Suppose we have picked 11 points equally spaced between $t = 0$ and $t = 10$ from our periodic graph that has a period of 10 seconds. To produce a smooth curve that goes through all these points and that we can enter into our lookup table to compute its Fourier series, enter the following commands into the MATLAB command window.

```
t=0:1:10;                        11 pairs of t,f points picked
f=[0 2 2 0 -2 -4 0 .5 0 -2 0]';  off our experimental graph
tint=0:.02:10;                   define 501 new t points
fint=spline(t,f,tint);           define 501 new f points, using
                                 spline interpolation
plot(t,f,tint,fint) compare linear interpolation with spline
```

Figure 9-15 shows these two curves. We would compare the splined curve with our original measured data curve and decide if the fit was acceptable. If not, we would start over with more than 11 points, choosing them to "fill in" critical areas where the fit was not good enough (the points *don't* have to be equally spaced), and then do the spline again. If our lab data is being gathered by a computer-based digital data acquisition system, we could of course easily enter *all* the digitized measured points as they were taken. This number of points might be large enough that the spline operation would not be needed to get smooth curves.

Let's assume that the splined curve just produced is the one for which we wish to compute a Fourier series. Double-click on the LOOKUP TABLE icon and enter for the input and output vectors: [tint] and [fint]. The icon graph will then change to look like the spline curve plotted above. Whenever you enter *names*, like tint and fint, rather than actual numbers, into a lookup table, the table values will be *lost* when you leave the simulation. If you want to save both the simulation diagram *and* the table lookup values, you need to *separately* save the tint and fint numerical values. To do this, write in the MATLAB command window:

```
save filename tint fint (use any filename you want)
```

This will save in a file called filename.mat the numerical values of tint and fint. When you later want to rerun this simulation, first enter into the MATLAB command window:

```
load filename
```

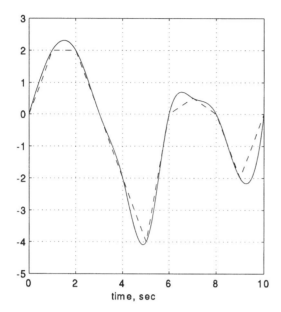

Figure 9-15 Use of lookup table and spline function to generate a periodic input (11-point curve is dashed; splined curve is solid).

If you then go to the simulation, the values needed in the lookup table will be available in the workspace, and SIMULINK will automatically get them from there.

You can now set $n = 1, 2, 3$, etc. to compute the average value and the harmonics. If you duplicate this simulation, you should get the following results.

$$q_i(t) = -0.382 + 1.2671 \sin(0.2\pi t + 1.1396) + 2.007 \sin(0.4\pi t - 0.7638)$$
$$+ 0.4234 \sin(0.6\pi t + 0.9782) + 0.4096 \sin(0.8\pi t - 0.8960)$$
$$+ 0.3382 \sin(1.0\pi t + 1.3252) + 0.0438 \sin(1.2\pi t - 0.8554) + \cdots$$

(9-50)

At this point one wants to truncate this infinite series at a finite number of terms, but still get an acceptable fit. From the amplitudes of the harmonics displayed, it seems that five harmonics might be enough. To check this, just compute this truncated series and graph it "on top of" the exact curve. This is done in Fig. 9-16 and we see that this fit is really quite close. If the small discrepancies are *not* acceptable, we can easily add a few more harmonics.

Fourier Series Calculations Using Fast Fourier Transform (FFT) Software. While the method just used above to compute the average value and harmonic terms in a Fourier series is accurate and easy to understand (if one accepts without proof the defining formulas for the Fourier series), it is not computationally efficient. Algorithms called *fast Fourier transform* routines are generally accepted as the fastest ways to compute the frequency content of all kinds of signals; periodic, transient, and random. Most computer libraries and mathematical software packages will include one or more such algorithms for your use. We will not try to prove the

General Linear System Dynamics

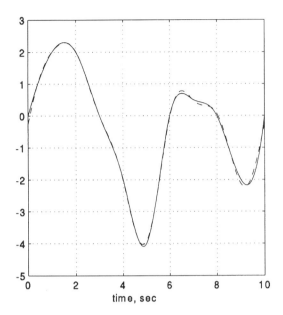

Figure 9-16 Approximation of periodic function by a truncated Fourier series (exact curve is solid; 5-term series is dashed).

validity of this method, but rather only show you how to use the version that is in MATLAB.

We will show how to use this method in general by applying it to the same function we used for our first method. This will give us a good comparison of the two techniques. While we will work this specific example, it will show a simple pattern which you can follow for *any* periodic function you might need. While MATLAB has an FFT method that allows the use of *any number* of points in describing the periodic function, all FFT routines work *much* more efficiently when the number of points n is an integer power of 2. We will here use 512 points, which you will find is adequate for most problems. More points (1024, 2048, etc.) will give better accuracy but use more computer time and memory. As before, the number of points must be adequate to document all the "wiggles" in your function.

The method will compute the average value and harmonics up to $(n-2)/2$; thus our recommended 512 points will get us 255 harmonics. As always, the lowest harmonic has a frequency equal to the repetition rate of the original periodic function, and the higher harmonics are integer multiples of this. Our earlier example had a period of 10 seconds, making its repetition rate 0.1 cycle/sec; thus the harmonics will have frequencies of 0.1, 0.2, 0.3, ... , 25.5 Hz. The MATLAB manual explanation of its FFT assumes that you are rather familiar with the subject and is not really adequate for "beginners," so we will describe a cookbook routine that you can easily follow for any problem. Let's now carry through the example, working in the MATLAB command window.

```
t=0:1:10;                        defines the same
f=[0 2 2 0 -2 -4 0 .5 0 -2 0]';  function as used before
tint=0:10/511:10;                but now with exactly
```

```
fint=spline(t,f,tint);          512 points
plot(tint,fint)          plot graph, just to make sure it's OK
fseries=fft(fint,512); computes the basic data to get the Fourier
                         series terms. You can use any name you
                         want where I used fseries. the vector
                         fseries will have 512 values, each a
                         complex number. the first value always
                         relates to the average value term.
avg=fseries(1)/512 the series average value is always given by
                fseries(1)/n. use any name you want where I
                have used avg
avg=-0.3813          computer prints the answer
mag=abs(fseries)/256 computes amplitude of each harmonic
                         (always divide by n/2)
[mag(2:10)]' request printing of first 9 harmonics, just to see
          how things are going
ans= 1.2656 computer prints out the answer.
     2.0062 compare these harmonics and the
     0.4242 average value with the results
     0.4033 from our earlier method. they
     0.3405 are not exactly the same, but
     0.0410 are quite close.
     0.0553
     0.0288
     0.0277
fr=0:.1:2.9; define harmonic frequencies for average value and
          the first 29 harmonics, to prepare for a graph
mag=[avg [mag(2:512)]]'; redefine mag to include the average
                         value, for plotting
bar(fr(1:30),mag(1:30)) request bar graph plot of amplitudes
grid; xlabel('frequency, cycles per second')
ylabel('avg value and harmonics')
fseries=[fseries]'; prepare for phase angle calculation
phase=atan2(real(fseries),imag(fseries)); compute phase
phase(2:10) print phase of first 9 harmonics, just to see some
          results
ans= 1.1417  computer prints out the answer.
    -0.7516 these results again agree quite
     1.0133  closely with those from our
    -0.8751 earlier method
     1.3745
    -0.8999
     1.0159
     1.0313
     1.1574
bar(fr(2:30),phase(2:30) request graph of first 29 harmonics
grid; xlabel('frequency, cycles per second')
ylabel('phase angle, radians')
```

General Linear System Dynamics

Figures 9-17 and 9-18 show the graphs produced by the above program. It is instructive to rerun this example using, say, 256 and 1024 points, rather than our 512, to get a feel for the effect of changing n. This is left for the end-of-chapter problems, but you would find that the results are only slightly different, since even 256 points give a good "sample" of our rather smooth periodic function, not missing any of its "wiggles."

Using Simulation to Compute Complete (Transient and Periodic Steady-State) Response of Linear or Nonlinear Systems to Periodic Inputs. Using the method of Fig. 9-13 we can compute analytically the periodic-steady-state response of any linear system to any periodic input. As we mentioned earlier, we developed this method, not so much to do actual response calculations as to give insight into the general behavior, mainly for design guidance. For example, in a vibrating system, graphs like Fig. 9-13 show which exciting frequencies are present in the input signal, and whether any of these "align" with peaks in the system's frequency response. If such alignments exist, we have a resonance phenomenon which may be dangerous and require design changes in our system or how we operate it. In Fig. 9-13 such an alignment occurs at the second harmonic, giving the output signal a large component at that frequency. The frequency spectrum of the input signal is also useful in deciding on how complex a system model is really justified. If the signal of Fig. 9-17, for example, were applied to a system, our system model need be accurate only to about 1.0 Hz, since the frequency content beyond that point appears negligible. If we did nothing but computer simulations, such insight would be largely lacking. This sort of qualitative reasoning is of course useful in all kinds of systems, not just vibration problems.

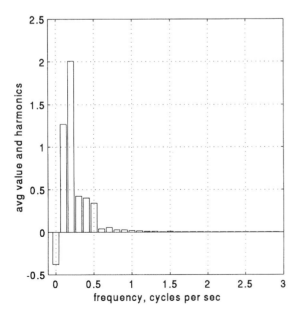

Figure 9-17 Fourier series results obtained from FFT software (magnitude).

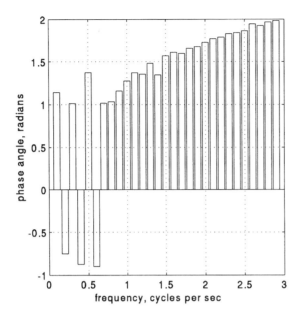

Figure 9-18 Fourier series results obtained from FFT software (phase).

When we get to that stage of design where we want to carefully examine the total time response of a system with specific numerical values, to a specific periodic input, then of course computer simulation is the tool of choice. Such a study gets us *both* the starting transient (which may have dangerous or useful features not revealed by a periodic-steady-state analytical study) and also the final steady state. Also, remember that if our system is nonlinear and/or has time-varying coefficients, then the method of Fig. 9-13 does not apply and we *must* use simulation. Let's use simulation to study the response of various systems to the same periodic function used in our previous examples. Figure 9-19 shows the "splined" input of Fig. 9-15 applied to five different linear systems and one nonlinear system.

We can use the same methods as used earlier to form one cycle of our periodic function, but now we need to repeat the function over and over. SIMULINK provides the REPEATING SEQUENCE module to accomplish this task. If you double-click on this icon you can then enter input and output vectors to define one cycle of the curve, just as we did before. In fact, if you are duplicating the presentations of this section on your own computer, you may still have the two vectors ([tint] and [fint]) in your workspace and can now use them to set up the REPEATING SEQUENCE module. Note that LOOKUP TABLE requires an input *signal*, but REPEATING SEQUENCE does not. Once you have entered one cycle of the periodic function into the module, it automatically repeats this function over and over, as long as you let the simulation run. Always enter this module *last* in building your diagram, because as soon as it is present on a diagram, all diagraming operations *slow down*, wasting your time. Once the module is in your diagram, if you want to make any changes and avoid the "slowdown," just CUT the icon, make your changes, and then PASTE it back in. Also, if you SAVE a

General Linear System Dynamics

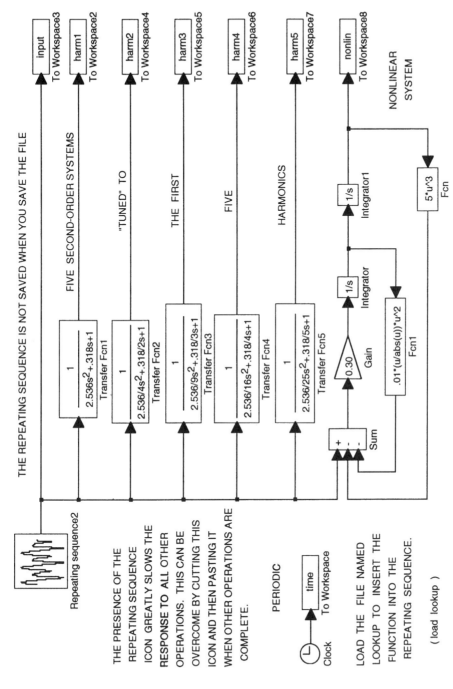

Figure 9-19 Simulation to study system response to periodic inputs.

diagram with this module in it, the actual function is *not* saved, just as we saw earlier in using the LOOKUP TABLE with vector names, rather than manually entered numbers. The solution to this difficulty is also the same as before; enter SAVE FILENAME TINT FINT to put these two vectors into a file which you can load separately whenever you want to use this simulation.

We could apply our input signal to any kind of system but choose to first look at five lightly damped second-order systems, each of which has a natural frequency "tuned" to one of the first five harmonics of our input signal. In other words, each of our systems will "resonate" with a different harmonic. Figure 9-20 shows the response for the systems tuned, respectively, to the first and second harmonics, for a time period equal to six cycles of the periodic input. In each figure, the input signal is shown dashed and the output solid. We see that the system tuned to the first harmonic takes about three cycles to build up a rather large response which appears to be close to a sine wave at the first-harmonic frequency. The system tuned to the second harmonic also gets to steady state in about three cycles but now the waveform is a bit more complex, although the second-harmonic frequency is quite clearly present. This response is even larger than harm1, since the second harmonic was the largest present in the input.

Figure 9-21 shows the responses (harm3 and harm5) of the systems tuned, respectively, to the third and fifth harmonics. These waveforms have quite complicated shapes, but the output signal is still larger, on a peak-to-peak basis, than the

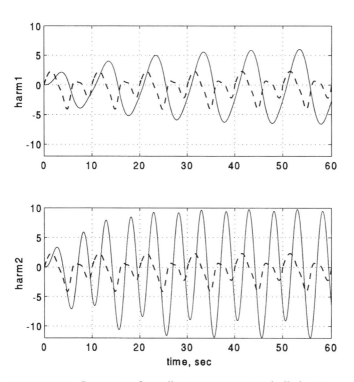

Figure 9-20 Response of two linear systems to periodic input.

General Linear System Dynamics

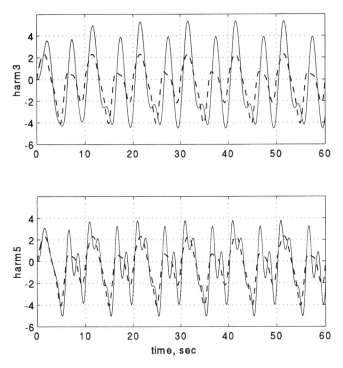

Figure 9-21 Response of two more linear systems to periodic input.

input signal, showing some kind of resonance effect. In both Figs. 9-20 and 9-21 the transient phase of the response shows no peaks higher than those present in the periodic steady state, so analytical calculations of this steady state would not overlook any dangerous features present in the transient. We can *not*, however, conclude that this will be true in *every* case. Sometimes the transient *will* have dangerous features not revealed by a periodic-steady-state response calculation, so we should always run the simulation to check for any such problems.

Our final system is a nonlinear one with a cubic spring and a square-law damper, whose response is shown in Fig. 9-22. Careful examination of the 60-second record shows that the output has *not* gone into a periodic (repetitive) motion, even though the input *is* strictly periodic. It is reasonable to ask whether perhaps such periodic behavior *would* occur if we just waited long enough. Unfortunately, unlike the situation for *linear, time-invariant* systems, there is *no* theory available to justify this conjecture. We can of course run our simulation longer, but there is no guarantee that periodic behavior will *ever* occur. If periodic behavior is not apparent for long records, we can never be 100% sure that some "disastrous" behavior is not lurking out at some long time. This uncertainty, due to lack of the needed differential equation theory, is one of the difficulties associated with the use of nonlinear models. Fortunately, sufficiently comprehensive simulation almost always gives us an adequate preview of system behavior, allowing design to proceed with some confidence.

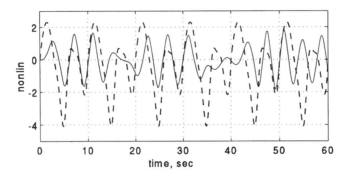

Figure 9-22 Response of nonlinear system to periodic input.

9-8 FREQUENCY CONTENT OF TRANSIENT SIGNALS: FOURIER TRANSFORM

To make the most comprehensive use of frequency-response methods and ways of thinking, we need to be able to express *all* kinds of signals in terms of their frequency content. This is possible and treated in many texts,[4] but the scope of this introductory book limits the extent of what we choose to present. Our treatment of periodic signals in the previous section is essentially complete and gives you all the tools needed for practical work. This section on transient signals will not be so complete, but it will give you a few useful tools.

Our main goal is to simply show how the frequency spectrum of any transient signal can be calculated. We will not try to explain how this spectrum can be combined with a system's frequency response to compute the output signal. That is, a method analogous to that of Fig. 9-13 *does* exist for transients, but we choose not to pursue it. Whereas the Fourier series was needed to deal with periodic signals, the *Fourier transform* is necessary to compute the frequency spectrum of transients. As with the Fourier series, we only present working methods, not proofs. While, historically, the Fourier transform was developed mathematically as a separate topic, it turns out that it is essentially identical to the Laplace transform when s is replaced by $i\omega$; that is,

$$\text{Fourier transform} \triangleq Q_i(i\omega) \triangleq \int_0^\infty q_i(t) e^{-i\omega t} \, dt$$
$$= \int_0^\infty q_i(t) \cos \omega t \, dt - i \int_0^\infty q_i(t) \sin \omega t \, dt \tag{9-51}$$

The function $Q_i(i\omega)$ is the frequency function corresponding to the transient time function $q_i(t)$. It is a complex number with a magnitude and a phase angle, and in this regard is similar to the Fourier series. Now, however, the integrals of Eq. (9-51) can be and are carried out for *any* value of frequency ω, not just the discrete values of harmonics, so the frequency spectrum is now a *continuous spectrum*, and graphs of $Q_i(i\omega)$ are smooth curves, not the spikes used for periodic-signal spectra.

---

[4]E. O. Doebelin, *System Modeling and Response*, chaps. 3, 6.

General Linear System Dynamics

Furthermore, the magnitude of $Q_i(i\omega)$ at any frequency does *not* represent the amplitude of a sine wave at that frequency, you *cannot* build up the transient by combining various sine waves, and you *cannot* use the method of Fig. 9-13 to compute the response of a system to the transient.

What we *can* say, however, is that, just like the Fourier series, when the magnitude of the Fourier transform of any transient becomes very small at high frequency, we can neglect the frequency content beyond this point—we need not continue to infinite frequency. This means that we can use Fourier transform to judge how complex a model is needed to adequately deal with any given transient. That is, the model's frequency response need be accurate *only* out to the highest frequency significantly present in the transient's Fourier transform. We have not proven this statement, but it is true and we ask you to take it on faith, or else consult the listed reference for more details. This narrow, but useful application is our only use of Fourier transform in this book.

Just as with Fourier series, when the transient has a simple form, the defining integrals can be worked out from tables of integrals, except now you may need *definite* integrals with the limits shown in Eq. (9-51). Let's use the rectangular pulse of Fig. 9-23 for our first example.

EXAMPLE: RECTANGULAR PULSE

Using the definition of Fourier transform we can write,

$$Q_i(i\omega) = \int_0^T A \cos \omega t \, dt - i \int_0^T A \sin \omega t \, dt \qquad (9\text{-}52)$$

$$= \frac{A \sin \omega T}{\omega} + i \frac{A}{\omega}(-1 + \cos \omega T) \qquad (9\text{-}53)$$

$$= \frac{\sqrt{2}A}{\omega} \sqrt{1 - \cos \omega T} \; \underline{/\alpha} \qquad (9\text{-}54)$$

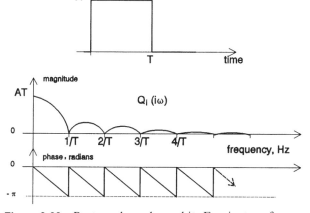

Figure 9-23 Rectangular pulse and its Fourier transform: a continuous frequency spectrum.

$$\alpha \triangleq -\frac{\omega T}{2} \tag{9-55}$$

The bottom part of Fig. 9-23 graphs the frequency spectrum of $q_i(t)$, magnitude and phase angle, as given by Eqs. (9-54) and (9-55). Several features of these graphs have important general significance. The magnitude at zero frequency is AT, the *area* of $q_i(t)$. This is *always* true for any transient, as can be seen from Eq. (9-51) with $\omega = 0$. If the time function goes both positive and negative, "area" means the *net* area. Note that (9-51) requires frequency in rad/sec, while I have converted to Hz in plotting Fig. 9-23. One can of course use either rad/sec or Hz in plotting; I here use Hz to emphasize that the frequency content of this transient goes *exactly* to zero at frequencies of $1/T, 2/T, 3/T, \ldots$ cycles/sec. One can see this by sketching sine and cosine waves of these frequencies "on top of" the graph of $q_i(t)$ and seeing that the integrals in Eq. (9-51) are both zero.

The phase angle curve does not seem to agree with Eq. (9-55), but actually is equivalent. According to (9-55) the phase angle decreases linearly with frequency, without any bound, approaching minus infinity. This is correct, but because the tangent function (from which α is defined) obeys the identity $\tan(\alpha + \pi) \equiv \tan(\alpha)$, the graph of Fig. 9-23 is also correct. Most computer software also handles this situation as we did in this graph.

A most important feature of the magnitude graph (relative to our earlier periodic-function spectra) is that our transient signal has frequency content at *every* frequency, not just at discrete frequencies. This is the difference between discrete and continuous spectra. Our example also shows that the frequency content gets smaller as we go to higher frequencies. This trend will be observed in *all* spectra of transients; thus we can use such spectra for estimating the highest frequency for which a system model must be accurate, just as we did with Fourier series for periodic inputs. Note also that much of the frequency content is concentrated below the frequency $1/T$, so $1/T$ is useful in roughly gauging where the spectrum begins to drop off. Thus if $T = 1.0$ second, we have strong content to about 1 Hz, whereas with $T = 0.001$ second, we have equally strong content to 1000. Hz. That is, *the shorter the duration of a transient, the more its spectrum extends to higher frequencies.* This feature is applicable to *general* transients, not just our simple example.

Most applications of spectrum analysis to transients involve time functions which are defined by experimental data, or are given by formulas for which the integrals of Eq. (9-51) are difficult or impossible to carry out analytically. Fortunately, all the operations of Eq. (9-51) can be done *numerically*; thus we can easily develop computer-aided methods for this problem. A "brute force" approach directly from Eq. (9-51), and similar to our Fourier series method of Fig. 9-14, is not hard to implement, getting $Q_i(i\omega)$ one frequency at a time, and then joining these discrete points with a smooth curve. This method is not computationally efficient and is left, "just for practice," to the end-of-chapter problems.

It turns out that the FFT method we used for Fourier series is also applicable, with a few detail changes, to Fourier transform calculations (it *is* called fast Fourier transform!). When used for computing the frequency spectrum of a transient we first have to "redefine" our input time signal in the following way. The transient actually

General Linear System Dynamics

665

processed by the FFT algorithm is the actual transient, which has a certain duration T, "padded" with zeros out to a time of at least $10T$. That is, the real transient goes to zero at time T, but we continue this zero level out to a much longer time, typically about $10T$. We enter this modified transient into our FFT routine, which then calculates the magnitude and phase of the frequency spectrum of the original transient. Let's now go through an example which will show you how to do it for *any* transient.

EXAMPLE: FOURIER TRANSFORM

Our example will be a rectangular pulse like that of Fig. 9-23, so that we will have an exact result to compare with. The amplitude A will be taken as 1.0 and the duration T as 0.1023 second. The modified transient will extend to 1.023 seconds, following our "$10T$" rule. As in our earlier use of FFT, we want to use an integer power of 2 as the number of points used to define the modified transient. I will use 1024 points, which will usually be OK for any problem you might do. The modified transient is formed and processed in MATLAB as follows.

```
t=[0:.001:1.023]';  define 1024 time points, 0.001 sec apart
x1=[ones(103,1)]';  define a vector with 103 ones in it
x2=[zeros(921,1)]'; define a vector with 921 zeros in it
x=[x1,x2]; define the modified transient. it is 1.0 from
           t=0 to t=0.102 and zero thereafter, and has
           1024 points
ftran=fft(x,1024); perform a 1024-point FFT, use any name you
           want where I use ftran
ftran=ftran(1:512); redefine ftran to use only the first 512
           values (always use 1/2 the points)
mag=0.001.*abs(ftran); compute magnitude of fourier
           transform. always multiply by the Δt
           between points (0.001 sec in this
           case)
phi=angle(ftran); compute phase angle of transform
freqs=[0:1/1.023:511/1.023]'; define the 512 frequencies
           that go with the transform
           values. always use 1/tfinal
           and 511/tfinal for 1024-pt
           transforms
subplot(2,1,1);plot(freqs,mag)
xlabel('frequency, cycles/sec');ylabel('magnitude')
subplot(2,1,2);plot(freqs,phi);ylabel('phase,radians')
```

When we defined the transient above, note that we *could not* duplicate the "vertical" behavior at $t = T$, and this will be one source of error in our calculations. That is, point 103 is equal to 1.0 and occurs at $t = 0.102$, while point 104 is equal to 0.0 and occurs at $t = 0.103$. Also, we only get transform values at frequency intervals of $1/1.023 = 0.9775\,\mathrm{Hz}$, so we "miss" anything that lies in between.

Figure 9-24 shows the results of the FFT calculations. The program graphing statements take the frequency to about $500\,\mathrm{Hz}$, but I have manipulated these graphs

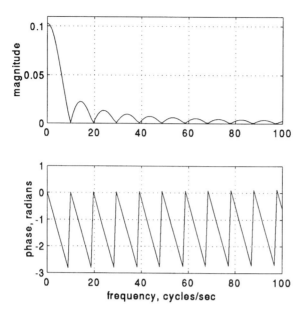

Figure 9-24 Frequency spectrum of rectangular pulse, computed by FFT software.

to show only 100 Hz maximum, for clarity. Figure 9-25 shows an exact calculation using Eqs. (9-54) and (9-55), for comparison. The agreement is quite good, with the most obvious graphical defect being in the phase angle, which doesn't clearly show the correct minimum value of $-\pi$ but rather only goes to about -2.7. If you dupli-

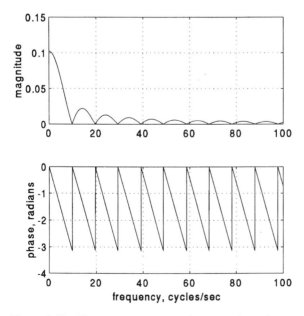

Figure 9-25 Frequency spectrum of rectangular pulse, computed by exact analytical method.

General Linear System Dynamics

cate this calculation and look at frequencies beyond 100 Hz, you will find that the phase angle exhibits larger and larger errors as you go to higher frequencies. This is mostly due to the fact that the real and imaginary parts of the transform are both getting very small and thus any calculations based on them become less and less reliable. Fortunately, this "bad" numerical behavior occurs when the magnitude has gotten so small that we probably would be ignoring this high-frequency range anyhow.

Most real-world transients that we spectrum-analyze are smoother than this example and will generally give less trouble in the FFT calculations, although the high-frequency errors just mentioned above *will* occur in every transform calculation. You should be able to follow the above "recipe" for calculating FFT frequency spectra for any transients that you might encounter. If you use software other than the MATLAB of our example, you do need to be careful to adapt to the FFT definitions of *that* software. That is, when a software product calls something an "FFT routine," you *cannot* be sure that the format and meaning of the results will be the same as for some other FFT software. Each software package will have a "correct" FFT routine, but the software developer has some freedom in how inputs and outputs are defined. It is always best to run some "test cases" for which you know the correct answer, to make sure that you are using the software as intended.

9-9 EXPERIMENTAL TESTING USING SPECTRUM ANALYZERS

We have mentioned several times before that lab testing based on frequency-response ideas is extremely common and practically useful. This is actually quite an extensive topic and is discussed at length in other texts.[5] In this introductory text we want to only introduce the basic concepts.

Spectrum analyzers are lab instruments which include digital computers specially designed to perform FFT calculations with the utmost speed and efficiency. The analyzer will usually also include an analog "front end" which accepts analog signals from external sensors and processes them before they are subjected to the digital FFT calculations. The processing consists partly of adjustable amplification to boost the small sensor voltages to, say, the ± 5-volt range required by analog-to-digital converters. There often will also be low-pass filtering, called *anti-aliasing* filtering, to remove frequency content beyond a selected range, before the signal is FFT processed. For example, if you want your FFT calculations to extend to 1000 Hz, you would prefilter the sensor signals to remove any frequency content beyond 1000 Hz. Also, the sampling rate of the A/D converter must be at least 2 times the highest analysis frequency.

Spectrum analyzers can be either single- or multiple-channel instruments. If only one analog input channel is provided, the instrument is useful only for *signal*

[5]Ibid., chap. 6.

analysis. That is, it will get us the frequency spectrum of periodic, transient, or random signals. More useful are *system analyzers*, which have at least two analog input channels. One channel accepts a sensor signal from the input of a physical system which we are testing, while a second channel accepts a sensor signal from the output of that same system.

Because a system must be "exercised" in order to study its dynamic behavior, many analyzers also provide several kinds of electrical stimulation signals. Depending on the application, we might want to use sinusoidal testing, transient testing, random signal testing, or some other specialized kind of system driving input. Since the analyzer provides such driving signals only as time-varying voltages, the experimenter must provide a suitable transducer ("actuator") to convert this voltage to the physical variable that is the system input. A similar situation exists at the system output, which will usually *not* be a voltage, so we need to provide a suitable sensor which measures the system output and converts it into a proportional voltage for entry into the analyzer's second channel. For vibrating systems, for example, the voltage-to-force transducer driving the system input could be an electrodynamic vibration shaker, the force-to-voltage sensor could be a strain gage load cell, and the system output sensor could be a piezoelectric accelerometer. Figure 9-26 shows the general configuration of such experimental testing.

The theoretical basis for measuring linear system transfer functions in frequency terms rests on the definition of the Laplace transfer function, when the general variable s is replaced by the special case $s = i\omega$:

$$G(s) \triangleq \frac{Q_o(s)}{q_i(s)}$$

$$G(i\omega) = \frac{Q_o(i\omega)}{Q_i(i\omega)} \tag{9-56}$$

That is, if we can measure the Fourier transforms $Q_i(i\omega)$ and $Q_o(i\omega)$, the ratio of these complex numbers will, at every frequency, be the sinusoidal transfer function $G(i\omega)$ relating q_o to q_i. Such calculations will of course give the sinusoidal transfer function as a table of values or the corresponding graphs, not as a mathematical formula. It is possible, however, to curve-fit these experimental curves and thereby

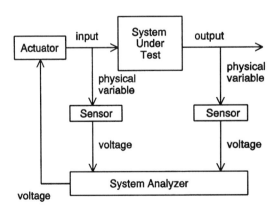

Figure 9-26 Experimental modeling using frequency spectrum analyzer.

General Linear System Dynamics   **669**

obtain transfer functions as actual formulas (ratios of polynomials in s). We have in fact earlier [see Eq. (7-51)] explained a MATLAB module called INVFREQS which performs such curve fits. Some system analyzers provide a similar feature.[6]

A critical question in practical testing is the selection of the driving signal for the system under test. One must first choose the *class* of signal (sine wave, transient, random, etc.) and then select numerical parameters for the particular test. The overriding requirement is that the driving signal have strong frequency content out to the highest frequency for which you want an accurate system model. That is, if you only "exercise" the system to 20 Hz, there is no hope of learning about its behavior at 30 Hz. Here again our FFT methods of computing frequency spectra of signals are just the tools needed. For example, if we choose a rectangular pulse as in Fig. 9-23, we see that poor results can be expected in the neighborhood of frequencies $1/T$, $2/T$, $3/T$, etc. since the input signal has *no* content at these points. This means that both input and output transforms will be near zero and their ratios will be "garbage." For this reason, we would use such an input signal only for frequencies somewhat below $1/T$. Of course we can choose T to extend the range of useable frequencies.

The field of experimental modeling of dynamic systems is a huge and fascinating one which we barely touch on here. The quoted references will give the interested reader a much more complete story. The Doebelin reference also discusses some special methods for systems which are *not* linear with constant coefficients.

9-10   DEAD-TIME ELEMENTS

We introduced the dead-time element in Sec. 3-9, where it was needed to model the computational delay in computer-aided systems. In this section we want to treat it in more general terms. Recall that this dynamic element has a rather simple behavior but does *not* fit into our much-used linear-differential-equation-with-constant-coefficients model. It thus does not usually yield to analytical treatment but is easily and accurately dealt with in simulation. Its definition is best stated in words: The output of a dead-time element is exactly the same as the input except it is delayed by a definite time interval called the dead time τ_{dt} (see Fig. 9-27). Other common names for dead time are transport lag, transport delay, and distance-velocity lag. A more mathematical definition would be as follows:

$$q_i = f(t) = \text{any time function}$$

$$q_o = f(t - \tau_{dt}) \qquad t \geq \tau_{dt} \qquad \tau_{dt} \overset{\Delta}{=} \text{dead time, sec} \qquad (9\text{-}57)$$

A long pneumatic transmission line, as used in some remote-control systems, gives a good example of dead-time behavior. Since the propagation velocity ("speed of sound") in air is about 1120 ft/sec, a step pressure p_i in Fig. 9-28 produces *no response whatever* at the receiver location p_o until 1 second has gone by. Then the

---

[6] HP 3563A, HP Dynamic Signal Analyzers, Publ. 5091-5887E. The Fundamentals of Signal Analysis AN 243. Effective Machinery Measurements Using Dynamic Signal Analyzers AN 243-1. Control System Development Using Dynamic Signal Analyzers AN 243-2. Hewlett Packard Co., 4 Choke Cherry Road, Rockville MD 20850, 301-670-4300, 1992.

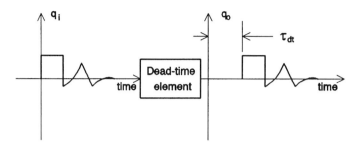

Figure 9-27 The dead-time element.

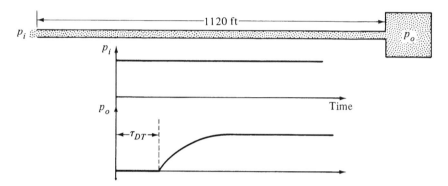

Figure 9-28 Pneumatic transmission line modeled with dead-time element.

response follows approximately a first-order curve. A model for such a system might thus be a dead-time element connected in cascade with a first-order system. Other examples of dead-time behavior include fluid-heating processes where a pipeline heater injects heat at one location and a sensor located 20 feet downstream measures the temperature. If the fluid flows at 5 ft/sec, there is a 4-second dead time between the actual change in temperature and its measured value. Rolling mills for steel and aluminum, and paper-making machines will usually include dead times since changes in process variables are measured by sensors located some distance downstream from the place where the changes actually occur. Transfer functions used to model the dynamic behavior[7] of human pilots or car drivers always include a dead time to account for delays in human sensors, nerve signal speed (varies from 0.3 to 100 m/sec) and muscle reaction time. The total dead time is about 0.12 to 0.18 second.

When a dead time appears embedded in a cascade of "ordinary" linear elements, as in Fig. 9-29a, it presents no analytical difficulties since it does nothing except shift the origin of the t axis by τ_{dt} seconds for every response variable "downstream" from it. If the overall response (q_i to q_o) is of interest, the dead time may be, for analysis purposes, shifted as in 9-29b so that all the ordinary

---

[7] E. O. Doebelin, *System Modeling and Response*, chap. 13, Human Factors in Man/Machine Systems.

General Linear System Dynamics

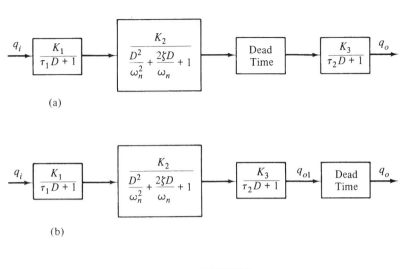

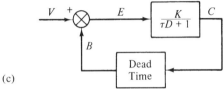

Figure 9-29 Dead times embedded in conventional systems.

dynamics precede it. Since dead time is a *linear* element, we can use Laplace transform to get its transfer function

$$\frac{q_o}{q_i}(s) = e^{-\tau_{dt} s} \tag{9-58}$$

using the delay theorem as in Eq. (6-63). As usual, we could form a *D*-operator transfer function by simply substituting *D* for *s*, but this would be useful only for drawing block diagrams, it has no analytical utility. The sinusoidal transfer function *does* have analytical uses and is easily obtained by the usual $s \to i\omega$ substitution.

$$\frac{q_o}{q_i}(i\omega) = e^{-i\omega\tau_{dt}} = \cos\omega\tau_{dt} - i\sin\omega\tau_{dt} = 1.0 \; \underline{/-\omega\tau_{dt}} \tag{9-59}$$

We see that the amplitude ratio is 1.0 for all frequencies and the phase angle decreases linearly with frequency, without any lower bound (see Fig. 9-30).

When a dead time appears in a feedback system (see Fig. 9-29c for a simple example), then real analytical difficulties arise. Using our usual method of getting differential equations directly from block diagrams, and looking for the equation relating controlled variable *C* to desired value *V*, we get

$$[V - C(t - \tau_{dt})]\frac{K}{\tau D + 1} = C$$

$$\tau\frac{dC}{dt} + C + KC(t - \tau_{dt}) = KV \tag{9-60}$$

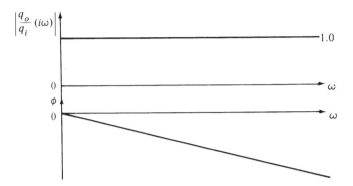

Figure 9-30 Frequency response of dead-time element.

The term $KC(t - \tau_{dt})$ makes this a *differential/difference* equation; a class *not* readily solvable by analytical methods. Approximate analytical solutions have in the past been obtained by approximating the exponential function with ratios of polynomials, which converts the equation back to our familiar linear differential equation with constant coefficients. The simplest such approximation uses the first two terms of a Taylor-series expansion to get

$$e^{-\tau_{dt}D} \approx 1 - \tau_{dt}D \tag{9-61}$$

More accurate and complicated approximations are found among the so-called *Pade approximants*, a family of polynomial ratios. By choosing a higher-order member of the family, one gets a better approximation. The first two members are

$$\frac{Q_o}{Q_i}(s) = \frac{2 - \tau_{dt}s}{2 + \tau_{dt}s} \tag{9-62}$$

$$\frac{Q_o}{Q_i}(s) = \frac{2 - \tau_{dt}s + (\tau_{dt}s)^2/6}{2 + \tau_{dt}s + (\tau_{dt}s)^2/6} \tag{9-63}$$

The quality of such approximations is best compared in terms of their frequency response. We see that both the Pade forms have exactly the correct amplitude ratio (1.0 at all frequencies), but the phase angle of the more complicated one will be closer to the ideal over a larger range of frequencies.

I should point out that in the field of feedback system design, there are analytical methods[8] that use *only* the frequency response, which we can use in its exact form, so no approximation is needed. Also, simulation handles dead times with no approximation, so we can get "exact" time responses for any system, for any set of numerical parameter values. The above approximations are thus not as much used as in earlier times, but you should at least be aware of their existence.

Finally we want to note that in the field of "process control" (feedback control of temperature, pressure, flow, tank level, etc.) a very common empirical process model, used when the process is too complex for analytical modeling, is a cascade of a first-order system and a dead time. The two parameters in the model

---

[8] E. O. Doebelin, *Control System Principles and Design*, secs. 9.6, 12.1, 12.3.

General Linear System Dynamics

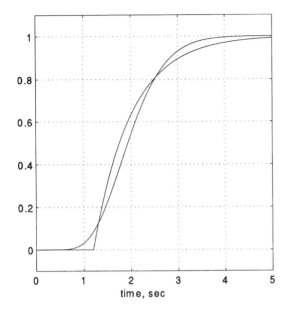

Figure 9-31 Response of 10 cascaded first-order lags approximated with one dead-time and one first-order lag.

are found from step-function tests of the actual process, and then these numbers are used to estimate settings for the controller.[9] Dead times are also sometimes used as simplified models for processes with many cascaded first-order lags, such as the hydraulic lags in distillation columns.[10] Figure 9-31 shows the step response of a cascade of 10 first-order lags ($\tau = 0.40$ sec for each) and a simplified model made up of a 1.2-sec dead time cascaded with a single first-order system with $\tau = 0.80$ second. The simple model is a quite satisfactory and practical approximation. Processes with significant dead times are among the most difficult to control, and special controllers (such as the Smith Predictor[11]) have been invented and applied to practical problems.

9-11 ANOTHER SOLUTION TO SOME VIBRATION PROBLEMS: THE TUNED VIBRATION ABSORBER

As promised earlier, we will finish this chapter with a few examples of practical systems that are more complex than those used earlier. In this section we continue our interest in the field of mechanical vibration, as an extention of the basic work in Chap. 8. When a vibration is excited by a source whose frequency is relatively fixed

---

[9] Ibid., p. 443.
[10] P. S. Buckley, *Techniques of Process Control*, Wiley, New York, 1964, p. 88.
[11] E. O. Doebelin, *Control System Principles and Design*, pp. 484–493.

and known, a device called a *tuned vibration absorber*[12] may be a viable solution. Such devices have been in use for many years and have been successfully applied to systems as small as phonograph pickups and as large as entire buildings.

Suppose we have a simple single-mass, damped vibrating system excited by a sinusoidal force whose frequency changes little and is known reasonably well. We shall show that by attaching another mass and spring (called a tuned absorber) to the original system, we can (ideally) reduce the vibration of the main mass to *zero* at the exciting frequency. Figure 9-32 shows the configuration of the total system. Our usual analysis methods lead quickly to the transfer function

$$\frac{x_m}{f}(s) = \frac{\frac{1}{K_{sm}}\left[\frac{M_a}{K_{sa}}s^2 + 1\right]}{\frac{M_a M_m}{K_{sa}K_{sm}}s^4 + \frac{M_a B}{K_{sa}K_{sm}}s^3 + \left[M_a\left(\frac{1}{K_{sa}} + \frac{1}{K_{sm}}\right) + \frac{M_m}{K_{sm}}\right]s^2 + \frac{B}{K_{sm}}s + 1}$$

(9-64)

Looking at the frequency response of (9-64), it is clear that when ω is equal to $(K_{sa}/M_a)^{0.5}$ (the natural frequency of the numerator term), the numerator is exactly *zero*, which means that no matter how much force is applied at this one frequency, the motion of the main mass will be exactly zero. Thus, we design the absorber spring and mass so that its natural frequency is the same as the exciting frequency of the force. There are an infinite number of combinations of spring and mass that will do this, but we have to pick a specific one of these. A little further analysis will help us in making this decision.

While the mathematics of Eq. (9-64) makes it perfectly clear that the main mass motion *will* be zero at the exciting frequency, it may not be obvious what makes this happen *physically*. One way to explore this is with a simulation, which we now

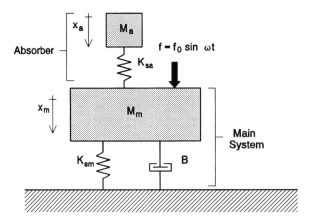

Figure 9-32 The tuned vibration absorber.

---

[12] J. C. Snowdon, Dynamic vibration absorbers that have increased effectiveness, *Trans. ASME Jour. Eng. Industry*, August 1974, pp. 940–945.

General Linear System Dynamics

discuss. Let's use numbers which make the exciting frequency, main system undamped natural frequency, and absorber natural frequency all equal to 20.0 Hz. The exciting force will be taken as $f = 5.0 \sin 126t$, lb_f.

$W_m = 12.3\, lb_f$
$W_a = 1.23\, lb_f$
$B = 1.0\, lb_f/(in/sec)$
$K_{sm} = 500.0\, lb_f/in$
$K_{sa} = 50.0\, lb_f/in$

These numbers give the main system a damping ratio of 0.125, which would result in a peak amplification of about 4.0. We leave the actual simulation diagram for the end-of-chapter problems, presenting here only results. Figure 9-33 shows that the force of the absorber's spring on the main mass quickly builds up to be *exactly equal and opposite* to the 5.0-pound exciting force, making the *net* force on the main mass zero, which results in it standing still. The lower graph makes clear that this requires that the absorber mass *does* move, with an amplitude (0.1 inch) just large enough to create the proper spring force. These facts are useful in designing practical absorbers.

Recall that we have an infinite number of choices for the absorber spring and mass, so long as their ratio gives the desired natural frequency, in our case 20.0 Hz. Now we see that if we choose a soft spring, then the absorber mass will need to have a large displacement, since we need to have a 5.0-pound spring force. In most practical machines, the space allowed for the absorber and its necessary motion are limited, so this constraint helps us choose the spring constant. Sometimes other factors are helpful. Most helicopters have at least one tuned absorber, since these vehicles are plagued with many vibration problems. In helicopters, the absor-

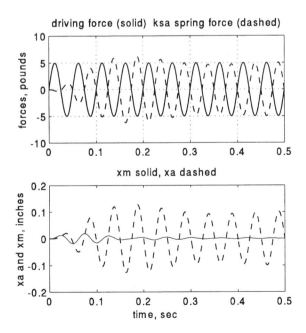

Figure 9-33 Time response of tuned vibration absorber.

ber mass is sometimes taken as the electrical *battery*, a device that has to be there anyway and can often be located to suit the vibration problem. Here, the mass would be fixed and we would choose the spring to get the needed frequency.

Using the numbers for our example, the frequency response of Eq. (9-64) is shown in Fig. 9-34. We now discover another feature of the tuned absorber! While the original main system peak at 20 Hz has been reduced to zero, we now have *two* resonant peaks on either side, with magnifications about as large as the original single peak. These two peaks should *not* be surprising; a lightly damped fourth-degree polynomial *will* have two pairs of complex roots. Note that the presence of these two peaks does *not* necessarily defeat the concept of the absorber. We stipulated at the outset that the exciting frequency should be reasonably known and fixed; thus we do not expect that there will be any exciting frequencies at the two peaks. In starting and stopping our machine, we *will* have to pass through the lower-frequency peak, but we have shown earlier (see Fig. 8-30) that this can be safely done by accelerating to the operating speed quickly enough.

The absorber spring is modeled with no damping, which is not quite correct. Some practical absorbers not only include the light damping of real springs but may also include *intentional* damping. This damping has the bad effect of preventing a perfect null at the exciting frequency, and the good effect of controlling the magnification at the two peaks. Study of these effects is left for the end-of-chapter problems.

One of the most impressive applications of the tuned absorber is in the reduction of building sway due to winds acting on very tall buildings. Figure 9-35[13] shows the system installed on the 63th floor of the Citicorp Center building in New York City. A similar system is on the 58th floor of the John Hancock Tower in Boston. The absorber mass is in the form of a huge concrete block, while the springs are nitrogen-filled gas springs. A servohydraulic control system is combined with the passive mass and spring to achieve the desired performance.

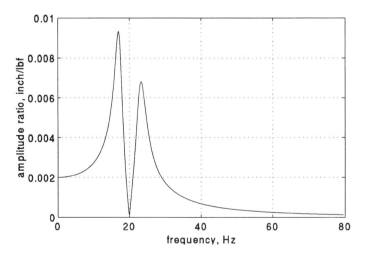

Figure 9-34 Frequency response of tuned vibration absorber

---

[13] MTS Tuned Mass Damper Systems, 1978, MTS Systems Corp., 14000 Technology Drive, Eden Prairie, MN 55344, 612-937-4000.

General Linear System Dynamics

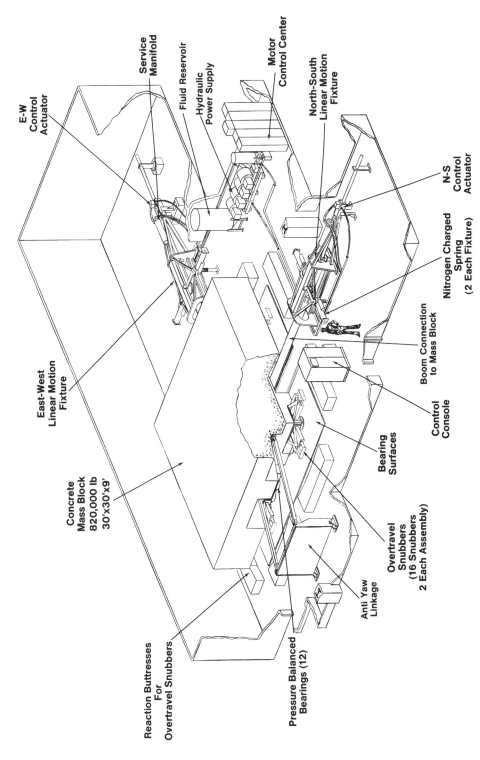

Figure 9-35 Giant tuned vibration absorber for skyscraper: MTS tuned mass damper system (63rd floor), Citicorp Center, New York.

9-12 IMPROVED VIBRATION ISOLATION: SELF-LEVELING AIR-SPRING SYSTEMS

In Sec. 8-7 we introduced some basic concepts of vibration isolation. These are adequate for many applications, but more advanced techniques are available for situations where the basic approach is not sufficient to meet requirements. A typical application of this type is the vibration-isolated benches used to set up precise optical equipment for experiments. Here the equipment mounted on the table top will be changed quite frequently, changing the payload mass. In the basic vibration isolator, this mass change causes a change in natural frequency and thus in the degree of isolation, which may not be acceptable. These mass changes may also cause the table to tilt slightly, which is also usually undesirable. Similar problems arise in other applications, such as precision machinery used in integrated-circuit wafer production. This machinery may also need to be carefully vibration isolated from floor vibrations and exciting forces from the machinery itself. Self-leveling air spring vibration isolators are a preferred solution to many problems of this type, and we now want to discuss their design and operation.

Figure 9-36 shows the essential features of these devices. We use a piston-cylinder arrangement to represent, for analysis purposes, the actual support system, which employs elastomer ("rubberlike") rolling diaphragms. These diaphragms are preferable to actual pistons and cylinders since they can be made to have much less friction. Low friction is important if we want to isolate *small* floor displacements.

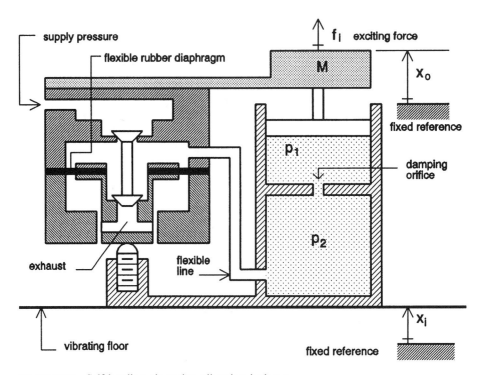

Figure 9-36 Self-leveling air spring vibration isolator.

General Linear System Dynamics   **679**

Optical tables, for instance, must isolate floor motions much less than 0.001 inch. Coulomb ("dry") friction requires a threshold force to break it loose, allowing small displacements to be transmitted straight through to the payload, defeating the isolation. Our equivalent frictionless piston/cylinder will behave essentially the same way as the diaphragm system, so we can use it for our analysis model.

The self-leveling feature is intended entirely for steady-state or static leveling, it does not attempt to fight against the high-frequency forces and motions that might be present. We can thus consider the self-leveling feature separately since it is much too slow to affect the dynamic behavior of the air spring isolators. While some manufacturers use electromechanical leveling systems, simple all-mechanical systems are adequate for many applications. These devices are simple feedback systems which sense mechanically when the table top moves vertically away from a desired location and then change the spring air pressure to return the table to the desired location. This action has two important effects. First, the table, which is usually supported by four air springs, one at each corner, is kept at the same height and level, even when equipment is added or taken away from the table top. Second, because the air pressure force of the "piston" must exactly equal the payload weight when the table is at the desired height, the spring constant of the air spring automatically *changes* when the payload mass is changed. We will shortly show that this spring constant change is *exactly* what is needed to keep the system natural frequency, and thus its isolation performance, at the chosen design value. Ordinary metal springs do *not* have this feature; the system natural frequency would change whenever the mass changed.

We will now briefly explain how the self-leveling servosystem operates, leaving a detailed analysis for the end-of-chapter problems. In Fig. 9-36 the supply/exhaust valve is shown in its neutral position, with both supply and exhaust ports closed, and pressure just sufficient to support the payload weight at the desired height. This height is manually set with the adjusting screw. Suppose now that we add some equipment to the table, causing it to initially move downward. This downward motion causes the valve supply port to open, allowing air to flow from the supply to the spring chamber, building up its pressure and upward force. This will cause the payload to stop falling and then actually rise back up to its former position. The system will not be "satisfied" until this equilibrium position is again reached, whereupon both valve seats will again be closed, but now a new, higher pressure, just sufficient to support the new payload at the "old" height, will be sealed into the spring chamber. If we had *removed* some weight from the table, the exhaust valve would initially open, dumping air from the spring to atmosphere, and a similar sequence of events would again return the table to the original height. While the operating principle of this system is not difficult to understand, as with all feedback systems, careful analysis and design are needed to prevent *instability* and to meet all other performance specifications. This analysis is left for the end-of-chapter problems. We should finally note that Fig. 9-36, for clarity, is not drawn to scale. The level-control valve is actually much smaller (about a 3-inch cube), than the air-spring system (about a 12-inch diameter cylinder, 24 inches long). The supply pressure in typical commercial systems[14] is about 100 psig.

---

[14] Pneumatic Isolation Systems, p. 6, Barry Corp., 40 Guest Street, PO Box 9105, Brighton, MA 02135, 800-227-7962.

Before starting our discussion of the air-spring dynamics we want to first consider a system which has the same form of dynamics but uses more familiar components. Figure 9-37 shows an isolation system using conventional metal springs and a viscous damper. Its configuration is, however, different from the basic isolation system we studied in Figs. 8-6 and 8-24. In that basic system we found that damping made isolation *worse* at high frequency, but was needed to control the height of the resonant peak, requiring a design compromise. This dilemma is partially removed in the improved system of Fig. 9-37. There the damper still contributes to resonance control, but at high frequency the damper becomes very "stiff," essentially connecting the bottom end of spring K_{s1} to the "floor." This creates an *undamped* vibrating system composed of spring K_{s1} and mass M, which has better high-frequency isolation than the basic system of Fig. 8-6.

Analysis (left for the end-of-chapter problems) of the system of Fig. 9-37 gives

$$\frac{x_o}{x_i}(s) = \frac{\dfrac{B}{K_{s2}}s + 1}{\dfrac{MB}{K_{s1}K_{s2}}s^3 + \left(\dfrac{1}{K_{s1}} + \dfrac{1}{K_{s2}}\right)Ms^2 + \dfrac{B}{K_{s2}}s + 1} \qquad (9\text{-}65)$$

From this result we can verify the earlier intuitive statement about this system's improved isolation at high frequency. We see that the numerator is first-order while the denominator is third. This means that at high frequency the *net* effect is a second-order term in the denominator, giving an attenuation of 40 db/decade. The basic system of Eq. (8-58) would have only a net *first*-order term in the denominator, giving only 20 db/decade of high-frequency attenuation, and thus poorer isolation. We will pursue study of this system no further here, but simply state that the air-spring system of Fig. 9-36 will have this same improved isolation behavior plus the additional advantages of self-leveling and constant natural frequency with payload mass changes. These features have made air-spring isolators popular for over 50 years.

In analyzing the air-spring system, recall that we *ignore* the presence of the self-leveling hardware since it is too slow to affect dynamic isolation behavior. In Fig.

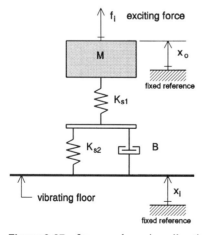

Figure 9-37 Improved passive vibration isolation.

General Linear System Dynamics   **681**

9-36, we thus consider both valve seats fixed *shut*, and the volume associated with pressure p_2 includes the small volume of the flexible line and the valve chamber. We will model the spring effect of the air in the two chambers using the adiabatic bulk modulus $B_a = kp$, since the dynamic pressure changes are too fast for much heat transfer to take place. We will do a linearized analysis, so B_a will be taken as *constant* at a value corresponding to the initial steady pressure. Our system has three unknowns, the motion x_o and the two pressures p_1 and p_2, so we will need to develop three differential equations to describe the system and solve for the unknowns. As usual for linearized analysis of nonlinear systems, we assume small changes about a specific operating point. We choose the operating point to coincide with the initial steady state of the system at time zero when the payload weight is supported motionless by the air pressure. Thus the operating point values for the two pressures are both given by W/A_p and x_o, x_i, and f_i are all defined as zero. At time equal to zero we now allow the inputs x_i and f_i to vary in any way we wish, but always with *small* changes.

Our first equation is a simple Newton's law for the payload mass M:

$$f_i + A_p p_1 = M \frac{d^2 x_o}{dt^2} \tag{9-66}$$

where A_p is the "piston" area. All pressure symbols in our equations represent not the actual pressures but the small perturbations away from the initial operating-point value. Actual air springs may require an experimental test to get a good number for the effective area A_p of the diaphragm. We would need to apply various pressures and measure the force produced, giving a nearly straight-line graph of force versus pressure. The slope of the best-fit line would give us the number for A_p.

The volume V_1, where p_1 exists, will be written as $L_1 A_p$, where L_1 is the length of this chamber. In an actual air spring the shape may not be a simple cylinder, so we would then just measure the actual volume and not try to relate it to any length. Also, we will treat the volumes as fixed constants, even though they actually vary slightly. If we don't make these assumptions the differential equations will be nonlinear and analytically unsolvable. Once we have used the approximate linearized model to help us estimate numerical values in design studies, we can always come back and investigate the nonlinear effects with simulation. During a short interval of time dt, the pressures will each vary by a small amount, and we need to relate these pressure changes to motions and flow rates. In volume 1, the pressure changes for two reasons. First, the relative motion $d(x_i - x_o)$ causes a volume change, $A_p d(x_i - x_o)$. This volume change causes a pressure change which we can relate to the bulk modulus.

Over the same time interval, there will be a flow rate through the damping orifice, adding or removing some volume from chamber 1 and again causing a pressure change. We now must model the relation between this flow rate and the pressure difference that causes it. We again assume linear behavior and define a fixed flow resistance R_f, so that the instantaneous mass flow rate is given by $(p_1 - p_2)/R_f$. The flow resistance can be estimated from fluid mechanics as soon as the detailed shape of the "orifice" is known. Some isolator manufacturers use a porous plate as the flow resistance, giving a multitude of small-diameter flow paths which encourage laminar flow. I could not find any published results which establish whether such a flow resistance is really superior to a simple single orifice. In any case, (pressure-

682   **Chapter 9**

drop)/(flow rate) experiments are necessary to get accurate values of flow resistance for whatever flow restriction is used.

The change in pressure p_1 during a time interval dt is attributed to two effects; that due to volume change caused by piston motion, and that due to volume change caused by volume inflow or outflow at the orifice:

$$dp_1 = \frac{dV_1 B_a}{V_1} = \frac{d(x_i - x_o) A_p B_a}{V_1} - \frac{(p_1 - p_2)\,dt}{R_f \rho}\frac{B_a}{V_1} \tag{9-67}$$

$$\frac{dp_1}{dt} = \left(\frac{dx_i}{dt} - \frac{dx_o}{dt}\right)\frac{A_p B_a}{V_1} - \frac{p_1}{\tau_1} + \frac{p_2}{\tau_1} \tag{9-68}$$

where

$$\tau_1 \triangleq \frac{\rho R_f V_1}{B_a} \qquad \tau_2 \triangleq \frac{\rho R_f V_2}{B_a} \tag{9-69}$$

and ρ is the density (assumed constant, corresponding to the initial steady pressure). A similar analysis for volume 2 yields our third and last equation:

$$\frac{dp_2}{dt} = -\frac{p_2}{\tau_2} + \frac{p_1}{\tau_2} \tag{9-70}$$

Having obtained three equations in three unknowns, the physical analysis is now complete. If we were going to simulate this system, it would be best to do it directly from these three simultaneous equations. If, however, we want to show that this system is analogous to the system of Eq. (9-65), we need to eliminate the two pressures and get a *single* equation just for x_o. This manipulation gives

$$\left[\frac{M\tau_2 L_1}{B_a A_p} D^3 + \frac{M L_1}{B_a A_p \tau_1}(\tau_1 + \tau_2)D^2 + \tau_2 D + 1\right] x_o = (\tau_2 D + 1) x_i \tag{9-71}$$

We see that this has *exactly* the same form as Eq. (9-65), so each coefficient in (9-71) can be equated with the corresponding coefficient in (9-65). If we had designed the system of Eq. (9-65) to be satisfactory, we could now use the numerical values of the coefficients there to establish numerical values of all the physical parameters in our air-spring system. Note that because each coefficient depends on *several* physical parameters, we may be able to get the desired coefficient values with *several* different sets of physical parameter values.

For a static force f applied downward, the air spring will deflect an amount x_o, and this force/deflection relation will define the spring constant. Using formulas already shown, we find that the spring constant is given by $A_p^2 B_a / V$. Volume V here is the *sum* of V_1 and V_2 because under static conditions, the two pressures must be exactly the same, thus the two volumes act as one. If the self-leveling system is working, pressure $p = W/A_p$, and thus the spring constant is $(kA_p/V)W$. That is, the spring constant is directly proportional to payload weight, which of course makes the natural frequency constant as payload weight changes, an advantage we claimed earlier without proof. Note that we here use the *nonlinear* form for the bulk modulus, rather than assuming it constant as we did in the dynamic analysis.

Further study of this air-spring system is pursued in the end-of-chapter problems.

9-13 ELECTROMECHANICAL ACTIVE VIBRATION ISOLATION

The air spring systems just discussed are certainly active systems since they are feedback control systems and use an external power source (the compressed air supply) to achieve their desired performance. Pneumatic servosystems of this type are, however, too slow to be effective in counteracting the floor vibrations and/or exciting force inputs which cause payload vibration. These servosystems are only useful in maintaining the paylaod level and keeping the system natural frequency constant for different payload masses. If we want to use feedback principles to dynamically fight against vibration, we must use hardware that is capable of much faster response. Various forms of electromechanical vibration isolation systems have been developed for this purpose. These can be used alone or are sometimes "piggybacked" on top of pneumatic isolators.

Figure 9-38 shows the configuration of a typical electromechanical vibration isolating system. Feedback systems of every sort always require sensors to measure the variable to be controlled, and actuators to provide the corrective effort when the variable deviates from its desired value. In Fig. 9-38, motion is the controlled variable and we use an accelerometer to measure the absolute acceleration of the payload mass M. This is not the only possibility. Various other (absolute or relative) accelerations, velocities, and displacements could conceivably be used, singly or in combinations. Our figure corresponds to a commercial system[15] which does happen to use an accelerometer as shown. To provide dynamically controllable force to counteract the undesirable vibrations, we use a magnetic force motor ("voice-coil actua-

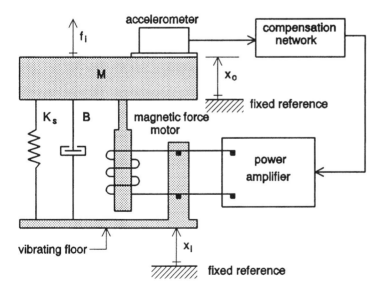

Figure 9-38 Active electromechanical vibration isolation.

[15] EVIS (Electronic Vibration Isolation System), Newport/Klinger, 1791 Deere Avenue, Irvine, CA 92714, 800-222-6440.

684   **Chapter 9**

tor") similar to those we have used earlier in motion-control systems. In many applications, because the vibrations to be suppressed are small (0.001 inch or less), relatively little force (about a pound or so) is required of the actuators.

In addition to the sensor and actuator, which *must* be present in some form, our diagram also shows a "compensation network." This is an electrical circuit (often implemented using op amps) that is required in most feedback control systems to allow achievement of the desired accuracy and stability for the complete system. The specific form and numerical values used for the compensation network vary from one system to another and are the subject of a well-developed design theory taught in courses on feedback control. The several electromechanical vibration isolation systems currently on the market do not all use the same design philosophy. The EVIS system which we study here can be best understood by first looking at an idealized version which most clearly reveals the basic principle. As with all our previous examples, we analyze a single isolator such as would be installed at one corner of a table top which was to be vibration isolated. Most practical systems would use four such isolators, one at each corner of the table. While there may be some interaction among the isolators, most of the design and analysis can be adequately presented without considering this complication.

In the basic version of Fig. 9-38, the compensation network needs to produce a signal which is the sum of an acceleration signal and a velocity signal. This could be accomplished with separate sensors measuring absolute acceleration and absolute velocity of the mass M, but in practice we can often use only an accelerometer and get the velocity signal with a simple integrating circuit. We will assume the availability of perfect acceleration and velocity signals. Using a current-output (transconductance) amplifier to drive the force motor allows the simple assumption that magnetic force is instantaneously proportional to amplifier input voltage. In a Newton's law equation for the motion of mass M this means that we include a magnetic force with a component proportional to acceleration and another proportional to velocity. Since we want these forces to *oppose* the motion, we include them with minus signs. Because the electrical circuitry makes it easy to adjust the individual magnitudes of these force components, we show them with adjustable coefficients K_m and K_b.

$$\sum \text{forces on } M = M\,\frac{d^2 x_o}{dt^2}$$

$$f_i + K_s(x_i - x_o) + B(\dot{x}_i - \dot{x}_o) - K_m\ddot{x}_o - K_b\dot{x}_o = M\ddot{x}_o \tag{9-72}$$

$$(M + K_m)\ddot{x}_o + (B + K_b)\dot{x}_o + K_s x_o = f_i + B\dot{x}_i + K_s x_i \tag{9-73}$$

The magnetic force is really applied *between* the payload and the vibrating floor, so it acts (in reverse direction) on the floor, not just on M. This does *not* affect our equations because we consider the motion x_i to be truly a *motion* input, unaffected by *any* forces. In actuality, the moving "floor" *does* feel this force, but in most cases the force is so small that it causes only imperceptible changes in x_i.

Examination of Eq. (9-73) reveals the operating principle of this type of isolator. We note that the magnetic force constant K_m appears in the equation in exactly the same way as mass M, thus it must have the same units and also *the same physical effect*. That is, we may consider K_m as "pseudo-inertia"; it acts just like real inertia

General Linear System Dynamics **685**

but we can adjust its value electrically! We can thus increase the "effective" inertia in our system *without* adding any real mass. [Actually, this feedback system design trick is usually used to *reduce* the effective inertia in motion control systems (by giving K_m a negative value, easily done electrically) to allow them to respond more quickly.[16]] In vibration isolation, as we have seen earlier, we want to *increase* mass to lower the natural frequency and thus reduce transmissibility.

A further improvement contributed by this feedback scheme is found in the damping term. We saw in some earlier examples that damping was useful in limiting the resonant peak, but made transmissibility worse at higher frequencies. In our feedback system we can now make B as small as possible (removing the "viscous coupling" between x_i and x_o that degrades isolation) but maintain a well-controlled resonant peak by providing a sufficiently large value of K_b. Another way of interpreting this is that K_b can be thought of as a damper connected between M and "the ground," so it provides damping but does *not* transmit any undesired forces from x_i to M. In the vibration isolation business this type of damping is called *inertial damping*.

We see here why feedback is used so often in designing high-performance dynamic systems of all kinds. Without feedback we have only two design parameters to adjust, the spring constant and the physical damping. This limits the performance by requiring design compromises. With feedback we have *two more* design parameters, K_m and K_b, allowing much more design freedom and thus improved performance. Equation (9-73) shows that with a given spring constant and physical mass, we can obtain as low a natural frequency as we wish by increasing the pseudo-inertia K_m, and no bad "side effects" are predicted. While significant improvements *can* be obtained, there *are* limits, but our equation does not make us aware of these. As in any feedback system, when we press for better performance we are always limited by *instability*. Our equation does not predict any instability because it is only of the second-order. It is easy to show with Routh criterion that a system equation must be of at least *third*-order if we are to even have a chance of getting valid results about stability. The reason our equation is not of higher order is that our hardware modeling neglected various dynamic effects, so that we could most clearly see the basic design concepts. For example, we neglected any dynamics in the sensor and also in the amplifier and force motor. Inclusion of some of these would raise the equation to order 3 or higher and make analysis and design more complicated, but would now allow valid stability analysis. Another "real-world" feature which limits the actual performance is the saturation nonlinearity present in any real amplifier/motor combination. That is, achieving very high performance will require very large magnetic forces, and these will not be available in a real system. This aspect of design can be dealt with using simulation by placing a limiter on the magnetic force.

Even though the model of Eq. (9-73) has these defects, it *does* give useful results so long as we don't go "too far" in trying to lower the natural frequency. If an application requires the ultimate in performance, then the model will need to be augmented with complicating features such as those just discussed above. Our usual design philosophy of using simple linear analytical methods to understand concepts and "rough out" designs, followed by simulation to check neglected dynamics and

---

[16]E. O. Doebelin, *Control System Principles and Design*, pp. 363–370.

nonlinearities, is again suggested for these applications. The system of Fig. 9-38 is given further study in some end-of-chapter problems.

9-14 AN ELECTROPNEUMATIC TRANSDUCER USING A PIEZOELECTRIC FLAPPER ACTUATOR

In Chap. 1 (Fig. 1-1) we used an electropneumatic transducer as a vehicle for explaining some basic ideas about system dynamics as a field of engineering study. That device used a permanent magnet and coil to position the flapper of a nozzle flapper, and this technology is the "classical" way to build such transducers. More recently, certain technological developments have allowed alternative designs to become competitive with this approach. The two major developments have been low-cost electrical pressure sensors based on integrated-circuit manufacturing methods, and inexpensive low-voltage piezoelectric actuators. Figure 9-39 shows the new design, as offered by one of the major manufacturers.[17]

The overall function of all electropneumatic transducers is to accept as input a command voltage (say in the range 0–10 volts) and produce a closely proportional

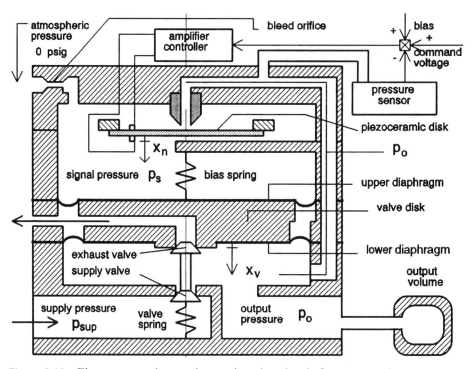

Figure 9-39 Electropneumatic transducer using piezoelectric flapper actuation.

[17] Model T7800 Series E/P Transducers, Fairchild Industrial Products Co., 3920 West Point Blvd., Winston-Salem, NC 27103-6708, 910-659-3400.

General Linear System Dynamics 687

output air pressure, usually in the range 3–15 psig. That is, a 0-volt input causes a 3-psig output and a 10-volt input causes a 15-psig output, with intermediate voltages producing proportional pressures in a closely linear fashion. Typically, linearity is within ±0.25% of full scale. These units are used where a device with an electrical output must "talk to" one with a pneumatic input. A common application is an electronic process controller sending a command to a pneumatic valve positioner, a powerful pneumatic device which can accurately stroke large valves used to control flow in pipes used in power plants, chemical plants, refineries, etc. Another application is in pneumatically actuated friction brakes or clutches used in motion-control systems.

The E/P transducer requires a constant supply pressure p_{sup} (about 20 psig) from a pressure regulator. The output pressure p_o is controlled by manipulation of a valve which can be moved to open either a path from supply to output, or output to atmospheric exhaust. When the new commanded pressure is reached, the valve goes to a neutral position where *both* the supply and exhaust ports are closed, and the desired pressure is trapped in the output volume. The valve is moved (displacement x_v) by the "valve disk" attached to two flexible rubber diaphragms. The upper diaphragm feels a force due to signal pressure p_s, while the lower feels a force due to output pressure p_o. A bias spring also exerts a force on this member. The area of the upper diaphragm is slightly larger than that of the lower diaphragm. This area ratio and the force of the bias spring cause the output pressure to always be slightly higher than the signal pressure. This relation is needed because the output pressure is also going to be used as the "supply" pressure for a nozzle-flapper device whose "output" pressure is p_s.

Output pressure exists in the output volume, below the lower diaphragm, and is also ported to the nozzle at the top of the unit and to a pressure sensor which is the feedback device for controlling the output pressure. The voltage signal from this sensor is compared with the command voltage and a bias voltage, and the output of this comparison goes to an electronic amplifier and controller which drive a piezoelectric actuator that moves the flapper. The bias voltage corresponds to a pressure of 3 psig, so that when the command voltage is 0, the output pressure will already be 3 psig. The piezoelectric actuator is a thin ceramic disk whose center deflection x_n provides the flapper motion of the nozzle-flapper device. Full-scale flapper motion requires about 100 volts from the amplifier/controller. When the flapper moves *farther* from the nozzle, more flow can pass through the nozzle, which builds up signal pressure p_s. If the flapper moves *closer* to the nozzle, reducing the inflow, the "bleed" orifice (at the top left) drains flow out of the p_s chamber, reducing this pressure. Since the bleed orifice is always passing some flow to atmosphere, the device has a continuous air consumption, no matter what output pressure might be existing. This air consumption does not represent a major cost since it is quite small, about 4–10 ft^3/hr.

Having described the major actions occuring in the E/P transducer, let's now "put it all together." Assume that the command voltage has been zero for some time and that the output pressure has become constant at 3 psig, with the exhaust valve now closed and everything in equilibrium. If we now, say, raise the command voltage to 1 volt, this causes the piezo disk to move farther away from the nozzle, and p_s starts to rise. This increases the force on the upper diaphragm, which pushes the valve stem down, opening wider the path from supply pressure to the output cham-

688   **Chapter 9**

ber. (The supply valve was *already* open, very slightly, to balance the continuous "leak" of the bleed orifice.) Air now flows into the output chamber, raising its pressure. As this pressures rises, it is sensed by the pressure sensor, whose voltage rises and subtracts from the command voltage, making the net voltage to the amplifier smaller, causing the piezo disk to move back toward the nozzle. This begins to reduce pressure p_s, which in turn allows the valve stem to move back toward the null position. The entire system will not be "satisfied" until the output pressure (and signal pressure) find a new equilibrium condition and the valve settles into the null position. This will happen at a new output pressure corresponding to the new command voltage. It should be clear from the above explanation that the E/P transducer *does* try to make the output pressure follow the command voltage, but it will require some modeling and analysis to show *how well* it performs this task.

Modeling of piezoelectric actuators was covered in Chap. 5 and showed that the relation between applied voltage and actuator displacement was basically that of a second-order system (mass, springiness, and parasitic damping). We also saw that the driving amplifier might have its frequency response reduced by the capacitance load presented by the piezoelectric device. In this application, these dynamics are largely irrelevant since the pneumatic dynamics of the volume-charging processes are much slower than the actuator dynamics. We thus model the actuator displacement x_n as instantaneous with the applied voltage. The motion x_v of the valve disk, attached diaphragms, and the valve itself also involves mass, spring, and parasitic damping effects. These dynamics are also quite fast, but probably not fast enough to neglect. Since the overall E/P transducer is marketed for applications requiring frequency response only to less than 5 Hz, we neglect inertia, making the model from applied force to valve displacement first order. This assumption, as all others, would be checked by lab testing as soon as hardware was available in the design cycle of the device.

The slowest dynamics are most likely the two volume-charging processes that determine the two pressures, p_o and p_s, with the p_o process being the slower. It involves an *external* volume associated with whatever device it is driving (often the valve positioner mentioned earlier). This external volume will usually be much larger than the p_o volume internal to the E/P transducer, and its numerical value is not known to the E/P designer, since customers will use the E/P transducer to drive an unknown variety of devices. We will see later that excessive output volume can degrade stability so we need to provide some adjustments in the E/P transducer to deal with this. We will model the volume-charging processes in a simplified manner which retains the essential features but does not relate all the parameters used to actual dimensions and fluid properties. This approach is consistent with an analysis which assumes that some lab testing will be used to get numbers for certain coefficients. We could, of course, take the time to base our analysis strictly on basic principles, but even this approach would be somewhat inaccurate due to the many assumptions that would be needed.

The chamber which contains the pressure p_s has an inflow from the nozzle flapper and an outflow through the atmospheric bleed orifice. Using the perfect gas law with assumed constant temperature T and volume V_s we can write

$$\frac{dp_s}{dt} = \frac{RT}{V_s}(\dot{m}_{si} - \dot{m}_{so}) = \frac{RT}{V_s}[(K_x x_n + K_{po}p_o - K_{psi}p_s) - (K_{pso}p_s)] \qquad (9\text{-}74)$$

Our reasoning here is that, for small perturbations, an increase in flapper opening and an increase in output pressure will cause proportional increases in inflow, while an increase in signal pressure will cause a decrease in inflow and an increase in outflow. The various "K's" could be estimated from theory or more accurately found from lab experiments. For the volume V_o containing the output pressure we have

$$\frac{dp_o}{dt} = \frac{RT}{V_o}(\dot{m}_{oi} - \dot{m}_{oo}) = \frac{RT}{V_o}(\dot{m}_{oi} - \dot{m}_{si})$$

$$\frac{dp_o}{dt} = \frac{RT}{V_o}[x_v K_{sf}(p_{sup} - p_o) - \dot{m}_{si}] \qquad (9\text{-}75)$$

We here are assuming that the valve flow rate into the p_o chamber is proportional to valve opening x_v and valve-port pressure drop $(p_{sup} - p_o)$. This model is nonlinear since it involves a product of variables x_v and p_o, so it would prevent analytical solution, but simulation is no problem. We could, of course, linearize this expression if we so chose. The outflow from the p_o chamber is exactly the inflow m_{si} to the p_s chamber, which we already expressed in Eq. (9-74).

Figure 9-40 shows a SIMULINK simulation diagram for this system. It is, of course, based on the equations and assumptions presented above, and I have included enough labeling that you should be able to relate the diagram to the equations. What is *not* clear from the diagram are the *numbers* used for the constants in the various equations. The manufacturer cannot usually be expected to provide *either* equations or numbers since these may be proprietary information, or perhaps

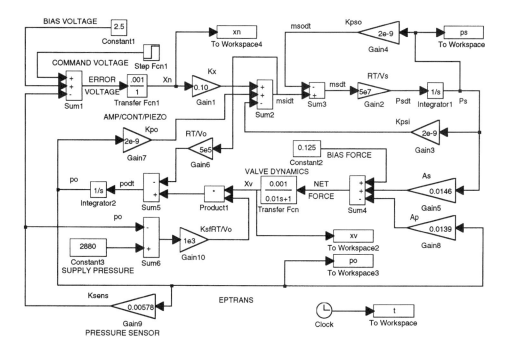

Figure 9-40 Simulation diagram for electropneumatic transducer.

690   **Chapter 9**

the manufacturer did not perform the type of analysis that we are interested in. Most manufacturers feel that their responsibility to customers is to provide transfer functions and numbers only for the *overall* device, that is, from command voltage to output pressure. Most customers are *not* interested in the "internal workings" of a product if its overall performance meets their needs. My approach to estimating numbers to use in this simulation was thus to "work backward" from the overall performance specified by the manufacturer. Even this was somewhat uncertain, because the output volume, one of the most important numbers, is determined *not* by the transducer itself, but by what device is *attached* to the transducer output port.

The specifications that I used to estimate numbers were that the overall system had approximately a second-order system response with a natural frequency of about 4 Hz and a damping ratio of about 0.4, for "typical" output volumes of "a few" cubic inches. Since the gas constant R for air is known and these systems tend to operate near "room" temperature ($530°R$), we can get a good estimate of the terms of form RT/V. In Fig. 9-40 the pressures are in pounds per square foot, the displacements in feet, forces in pounds, and time in seconds. Mass flow rates are in slugs/sec. For a volume of 1.0 in^3:

$$\frac{RT}{V} = \frac{(53.3)(530)}{(1/1728)} = 4.88 \times 10^7 \tag{9-76}$$

The transducer's midrange pressure is 9 psig, which is about 23.7 psia and 3413 lb$_f$/ft^2. The density at this pressure would be

$$\rho = \frac{M}{V} = \frac{p}{RT} = \frac{3413}{(53.3)(530)(32.2)} = 0.00375 \ \frac{\text{slugs}}{\text{ft}^3} \tag{9-77}$$

To roughly estimate K_x in Eq. (9-74) we might guess that the nozzle should be able to raise the pressure in a (typical) volume of 1 in^3 at a rate of about 5 psi/sec, if we want to have step response times of about 0.2 second for a 1-psi command.

$$\dot{p}_o = (5)(144) = \frac{RT}{V} \dot{m} = 4.88 \times 10^7 \dot{m}$$

$$\dot{m} = 1.48 \times 10^{-5} \ \frac{\text{slugs}}{\text{sec}} \tag{9-78}$$

The nozzle opening required to produce this flow rate is not obvious without some prior experience with devices of this type. The openings are generally quite small; let's use 0.010 inch. We can then estimate K_x as

$$K_x = \frac{1.48 \times 10^{-5}}{0.01/12} = 0.0176 \ \frac{\text{slugs/sec}}{\text{ft}} \tag{9-79}$$

The "pressure coefficients," such as K_{po} in Eq. (9-74) can be roughly estimated by noting that all the mass flow rate terms in this equation should be of the same order of magnitude. If they were vastly different, then the "small" ones should not really be in the model. Thus we can say that the mass flow rate term involving p_o should be the same order of magnitude as that for x_n, when we use a "typical" value for p_o.

$$p_o K_{po} = (23.7)(144)K_{po} \approx 1.48 \times 10^{-5}$$

General Linear System Dynamics

$$K_{po} \approx 4.34 \times 10^{-9} \, \frac{\text{slugs/sec}}{\text{psf}} \tag{9-80}$$

All the pressure coefficients should be roughly this size. The two diaphragm areas are easy to estimate from the known overall size of the transducer, which is about 3 inches in diameter. The upper diaphragm area A_s was taken as $0.0146 \, \text{ft}^2$ and the lower as 0.0139. The first-order valve dynamics were estimated with a gain of 0.001 feet of valve travel per pound of diaphragm differential force and a time constant of 0.01 second.

Using the above crude estimates as a starting point I ran the simulation and adjusted various values until the overall response matched roughly the behavior of the real device. Figure 9-40 uses these final numbers. In Fig. 9-40 I took the command voltage step input as zero, since the bias voltage used to get 3.0 psig for 0.0 volts command will act just like a 3-psi step input when we "turn on" the transducer. Figure 9-41a shows the response of output and signal pressures to this turn-on event. We see that the output pressure settles nicely to 3.0 psig in about 0.5 second, with a moderate and acceptable overshoot. Note that p_s rises first since the voltage input acts first on the piezo actuator, but when steady state is reached, p_s is less than p_o, as enforced by the diaphragm area ratio. In Fig. 9-41b we see the motions of the flapper and valve. The valve *appears* to go completely shut ($x_v = 0.0$) in steady state; however, it *must* be slightly open to support the continuous bleed flow. If you request *tabular*, rather than graphical output, you will see that the valve *does* remain slightly open in steady state. The graphs in psi and inches were obtained from the simulation (which uses psf and feet) by simply plotting p/144 and x*12.

We had mentioned earlier that large output volumes degrade transducer stability. In Fig. 9-42 we explore an increase of this volume by a factor of 10. Figure 9-42b shows this decrease in stability, which would usually be unacceptable. One way to regain stability is to reduce the gain of the system controller. A reduction from 0.001 to 0.0001 is seen to give the desired effect. Compared with Fig. 9-42a, the response is also slower, which should not be surprising; a large volume will take longer to "fill" than will a small one. Also note that the low-gain system loses steady-state accuracy. The final value of pressure is slightly below the desired 3.0 psig. In the commercial instrument this steady-state error is removed by adding integral control to the proportional control used in our example. Use of proportional plus integral control is studied in the end-of-chapter problems. Finally, Fig. 9-43 shows that absolutely unstable behavior will occur if we try to use excessive controller gain with very large output volumes.

Many further details of system operation could be easily studied with simulation. Large step inputs, for example, can cause "saturation" of either or both of the flow-controlling motions in the transducer. That is, a large voltage command might ask the piezoelectric actuator to move 0.050 inch, while its designed maximum stroke might be less than this. Similarly, large diaphragm forces might "command" the valve to move beyond its "wide-open" stroke. Either of these situations makes the flow behavior assumed in Fig. 9-40 incorrect, and thus the simulation does not predict accurately. Such saturation phenomena can be simulated by placing a suitable limiter on each of the motions. Then, when excessive motions are commanded, the motion, and thus the flow rate are limited to the values corresponding to the "wide-open" conditions, giving a more realistic simulation.

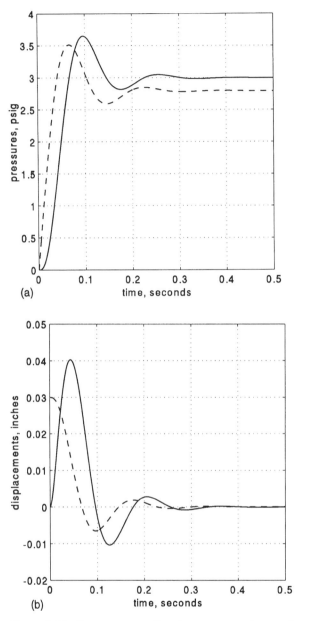

Figure 9-41 Step responses for electropneumatic transducer. (a) p_o (solid), p_s (dashed); (b) valve motion (solid), flapper motion (dashed).

While the *designer* of the E/P transducer can use the detailed simulation of Fig. 9-40 to good advantage in choosing numerical values for the various design parameters, the *user* of the transducer prefers a simplified model since the E/P transducer will be only one component in a larger complex system. The step response of Fig. 9-41a suggests that the overall device might be adequately modeled as a simple second-order system, if we would just choose the natural frequency and damping

General Linear System Dynamics

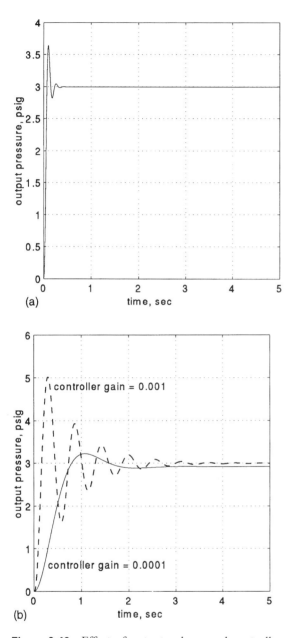

Figure 9-42 Effect of output volume and controller gain on response. (a) Controller gain equals 0.001, output volume equals V_o, (b) output volume equals $10\,V_o$.

ratio properly. This choice is easily made by adding an isolated second-order system to the simulation of Fig. 9-40 and running the two simulations simultaneously. We then plot p_o from each model and adjust ω_n and ζ till we get a reasonable "match." Figure 9-44 shows the good agreement reached with the values $\omega_n = 5.7\,\text{Hz}$ and $\zeta = 0.43$. If we lab-tested an actual transducer and tried to fit the simplest model of acceptable accuracy to the step response, we would probably find similar results.

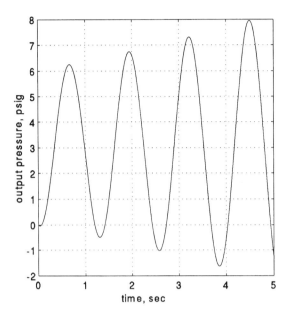

Figure 9-43 Transducer goes unstable for large volume and high gain. Controller gain equals 0.002, output volume equals 100 V_o.

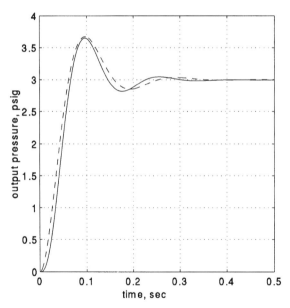

Figure 9-44 Complex transducer dynamics are well approximated with a simple second-order system. Complex simulation (solid), second-order (dashed).

9-15 WEB-TENSION CONTROL SYSTEMS

Materials (paper, plastic films, magnetic tape, aluminum foil, sheet metal, etc.) that are manufactured or processed as a continuous web often require that the tension of the moving web be controlled within certain limits. This may be required to prevent breakage, provide uniformly wound rolls of material, allow more rapid production, or increase product quality. Engineers have invented a variety of schemes for tension control of moving webs.[18] Figure 9-45 shows one possibility. Here a part of the processing machinery not shown pulls the web to the right at a speed $v_i(t)$. This speed may vary somewhat in an unpredictable fashion, causing the tension T to also vary. A weighted "dancer roll" creates the desired tension and also is used to detect when tension deviates from this value. Ideally, the input velocity v_i and the velocity v_o provided by the supply roll being unwound would be exactly equal, the dancer roll would have no vertical motion y, and the tension would be constant at $W/2$, where W is the effective weight of the dancer-roll mechanism at the y location. If, say, v_i momentarily increased, the dancer roll would rise, which motion is measured by the displacement sensor. The sensor tells the motor driving the supply roll to increase its velocity so as to match the new v_i and thus return the dancer roll to its null position. This action will tend to keep the tension constant, but one must analyze the system to see how well it can be made to work.

We will assume that the rotary inertia and friction of the dancer roll, the two support rolls, and the web of material are all negligible, making the tension T the same in all sections of the web. The motor armature and supply roll together have inertia J and viscous friction B. The supply roll moment of inertia and radius R will actually change as the material unwinds from the supply roll, but we take them as constant over short periods of time. (End-of-chapter problems explore these effects in more detail.) With these assumptions, we can start to develop the system equations. For the vertical motion of the dancer roll:

$$2T - W = \frac{W}{g} \ddot{y} \tag{9-81}$$

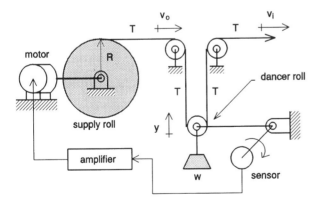

Figure 9-45 Web-tension control system using dancer roll.

---

[18] E. H. Dinger, Controlling web tension, *Machine Design*, Oct. 29, 1959, pp. 122–133; W. Gallahue, Pneumatic system controls web tension, *Automation*, May 1964, pp. 71–74.

where we have assumed negligible friction. The difference between the web velocities is related to the dancer roll motion by

$$v_i - v_o = v_i - R\omega = \frac{\dot{y}}{2} \tag{9-82}$$

The sensor produces a voltage $K_y y$, which is compared to the desired value of y (taken as 0) to produce an error voltage. This error voltage is amplified in a voltage stage and then applied to a current-output ("transconductance") power stage which supplies the PM-field dc motor with armature current. The rotary Newton's law for the supply roll is thus

$$K_T i - B\omega + RT = J\dot{\omega} \qquad i = K_a(K_y y - 0) \tag{9-83}$$

These equations can be combined to get a single equation for the overall system:

$$\left[\frac{D^2}{\omega_n^2} + \frac{2\zeta D}{\omega_n} + 1 \right] T = (K_W)W + K_{vi}(\tau D + 1)D^2 v_i \tag{9-84}$$

$$K_W \triangleq \frac{1}{2} \frac{\text{lb}_f}{\text{lb}_f}$$

$$K_{vi} \triangleq \frac{MB}{2K_a K_T K_y R} \quad \frac{\text{lb}_f}{(\text{ft/sec}^3)}$$

$$\omega_n \triangleq \sqrt{\frac{2K_a K_T K_y R}{J + MR^2}} \quad \frac{\text{rad}}{\text{sec}}$$

$$\zeta \triangleq \frac{B}{\sqrt{8K_a K_T K_y R(J + MR^2)}}$$

The particular solution for tension T reveals the steady-state performance of this tension-control system. For v_i equal to any constant V_i, or for $v_i = A_i t$ (a ramp input of velocity), $D^2 v_i = 0$ in Eq. (9-84), and the particular solution is $T = W/2$. That is, the steady-state tension is exactly what we desire. For an input velocity which increases parabolically ($v_i = G_i t^2$), there will be a steady-state error between the desired and actual tension since the particular solution is then $T = W/2 + 2G_i K_{vi}$. Of course in a practical case, ramp and parabolic increases in input velocity can not be sustained indefinitely; the machine "pulling" the web will have an upper limit on its speed and thus v_i must eventually "level off."

While the steady-state performance seems quite acceptable, we must also investigate the transient behavior. We could solve the equation analytically for the step, ramp, or parabolic input velocities discussed above, but simulation will make this task easier and also allow more freedom in pursuing more detailed design studies. Figure 9-46 shows a SIMULINK diagram for this system, including some representative numerical values. A step input of web velocity v_i is unrealistic since the machine pulling the web would be carefully brought up to speed from rest, so as to not break or overload the web material. A good simulation of this acceleration to final speed is obtained from the step response of a first-order system. By choosing

General Linear System Dynamics

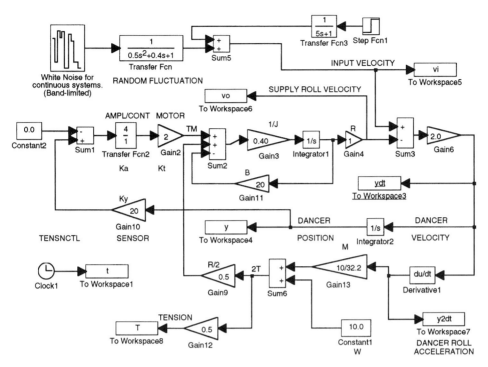

Figure 9-46 Simulation diagram for Web-tension control system.

this time constant properly, we can simulate slow or fast accelerations. In this first-order system we use a time constant of 5 seconds, a gain of 1, and a step input of 5, giving an exponential acceleration of v_i from 0 to 5 ft/sec in about 15 seconds.

Our simulation also provides for small random fluctuations in v_i, as would be typical of the real situation. The white-noise generator in Fig. 9-46 uses variance $= 0.10$, sample time of 4 seconds, and seed $= 23341$. To "smooth" the steplike random output of the noise generator we pass it through a second-order low-pass filter. The random fluctuations are "turned off" for our initial simulation runs, to more clearly evaluate the basic behavior of the system. Note also the use of the SIMULINK derivative block to obtain dancer roll acceleration from its velocity. We have routinely recommended *against* the use of the derivative operation, but in this case the velocity is quite smooth; thus the derivative operation is successful.

Figure 9-47 shows the input velocity rising gradually toward its steady state of 5 ft/sec. We see also that the output velocity initially has some oscillations, but before long becomes equal to the input. In Fig. 9-47, 50 times the difference between the two velocities is plotted, since a graph of output velocity would lie nearly "on top of" the v_i curve, and the difference between them would be hard to see. We see in Fig. 9-48 that the tension, after a few oscillations, becomes equal to the desired 5 pounds in about 2 seconds. Figure 9-49 reveals a possible problem with this design in that the dancer roll vertical position does *not* return to its null position but rather moves to about 0.6 feet and then stays there. Transient motions of the dancer roll are of course expected and allowed, but we would really prefer that it return to the null position when transients are over.

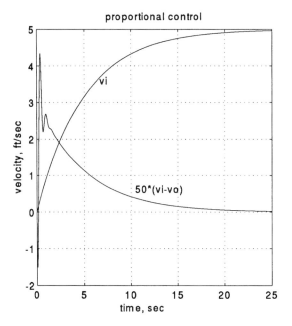

Figure 9-47 Simulation results for Web-tension control system: web velocities.

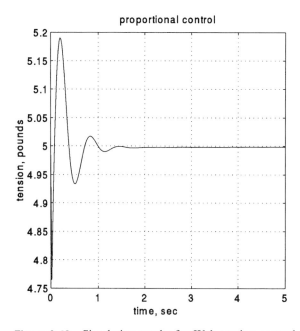

Figure 9-48 Simulation results for Web-tension control system: web tension.

General Linear System Dynamics 699

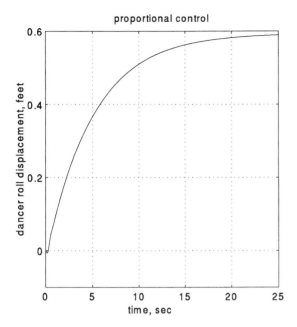

Figure 9-49 Simulation results for Web-tension control system: dancer roll position.

In Fig. 9-50a–c, we have added the random fluctuations to the input velocity. We see that, except for a starting transient, the tension never deviates more than about 1% from our desired value. The difference between input and output velocities fluctuates but is always small and has an average value near zero. Dancer roll displacement also fluctuates randomly (as would be expected for a random input), but again exhibits an undesirable average position (in "steady state") of about 0.6 feet.

Further study of tension-control systems is pursued in the end-of-chapter problems. Some questions considered there include:

1. What is the effect of using a voltage-output amplifier in place of the current-ouput (transconductance) type? Now the motor armature inductance, resistance, and back emf must be included.
2. Will adding viscous damping to the dancer roll mechanism improve the behavior?
3. It is suggested that using proportional plus integral (*PI*) control in the controller, rather than the simple proportional control of the text study, will bring the dancer roll position always to the null location in steady state. The controller transfer function would then be $(sK_a + K_I)/s$ instead of just K_a.
4. The radius R and roll inertia J will gradually change as the roll is unwound. Modify the simulation to include these effects.
5. An alternative design concept uses a pneumatically actuated friction brake in place of the motor. An E/P transducer, driven electrically from amplified sensor signals, adjusts the brake friction torque to keep tension constant. Investigate the feasibility of this concept.

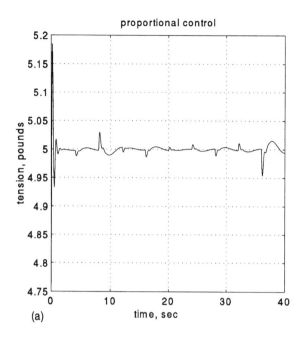

(a)

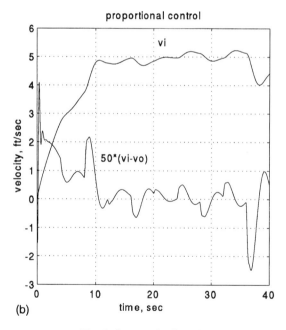

(b)

Figure 9-50 Simulation results for added random input: web tension, web velocities, dancer roll position.

6. Eliminate the dancer roll completely, and replace it with a sensor which measures tension directly. Study the feasibility of this concept, using the electric motor drive of our original example.
7. Repeat part 6 above, but use the friction brake of part 5.

General Linear System Dynamics

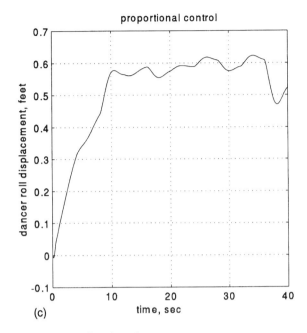

Figure 9-50 Continued.

BIBLIOGRAPHY

Andersen, B. W., *The Analysis and Design of Pneumatic Systems*, Wiley, New York, 1967.
Doebelin, E. O., *Measurement Systems*, 4th ed., McGraw-Hill, New York, 1990.
Doebelin, E. O., *System Modeling and Response: Theoretical and Experimental Approaches*, Wiley, New York, 1980.
Leonhard, W., *Control of Electrical Drives*, Springer, New York, 1985.
Merritt, H. E., *Hydraulic Control Systems*, Wiley, New York, 1967.
Mills, A. F., *Basic Heat and Mass Transfer*, Irwin, Chicago, 1996.
Rizzoni, G., *Principles and Applications of Electrical Engineering*, 2nd ed., Irwin, Chicago, 1996.
Watton, J., *Fluid Power Systems*, Prentice Hall, New York, 1989.

PROBLEMS

9-1. For the system of Fig. 9-1b:
    a. Set up the system differential equations.
    b. Using determinants, get a single differential equation for the current i_1.
    c. Repeat part (b) for current i_2.
    d. Repeat part (b) for current i_3.
    e. Draw and explain a simulation block diagram for the system.

9-2. For the system of Fig. 9-1c:
    a. Set up the system differential equations.
    b. Using determinants, get a single differential equation for the liquid level h_1.

702   **Chapter 9**

   c. Repeat part (b) for h_2.

   d. Repeat part (b) for h_3.

   e. Repeat part (b) for h_4.

   f. Draw and explain a simulation block diagram for the system.

9-3. For the system of Fig. 9-1d:

   a. Get the system differential equation, taking θ_c as the unknown.

   b. Repeat part (a) but take e_E as the unknown.

   c. Repeat part (a) but take T_M as the unknown.

   d. Draw and explain a simulation block diagram for the system.

9-4. For the system of Fig. 9-1a:

   a. Derive the equation analogous to Eq. (9-6) when x_2 is the unknown. Then get transfer functions analogous to (9-8), (9-9), and (9-10).

   b. Repeat part (a) for the unknown x_3.

9-5. a. Make up three different fifth-degree polynomials by assuming some numerical roots and multiplying out the factors. Choose the five roots so that some are unstable.

   b. Test the polynomials for stability using Routh criterion, and comment on the results.

   c. Test the root finder that is available to you by applying it to the polynomials of part (a).

9-6. Derive the equation analogous to Eq. (9-20) if p, rather than h, is taken as the unknown. Can we use this equation to check system stability? Explain.

9-7. In inequality (9-22), discuss the effect on system stability of the following parameters:

   a. A b. V c. M

   d. K_{ho} e. K_{pi} f. K_{po}

9-8. Sketch "free-hand" but neatly the logarithmic frequency-response curves for the following transfer functions:

   a. $\dfrac{q_o}{q_i}(s) = \dfrac{5(0.01s + 1)(0.2s + 1)}{(0.04s^2 + 0.2s + 1)(0.05s + 1)}$

   b. $\dfrac{q_o}{q_i}(s) = \dfrac{2.3s^2 + 0.5s + 3}{0.04s^4 + 0.22s^3 + .6s^2 + 1.4s + 2.6}$

9-9. Check your sketches of Problem 9-8 by graphing the given transfer functions using available computer software.

9-10. Extend the MATLAB program used to get Fig. 9-6 to include the input force f_{i3}.

9-11. Assume some numerical values and set up matrix frequency response programs, using MATLAB or other available software, for the systems of:

   a. Figure 9-1b.

   b. Figure 9-1c.

9-12. Find the Fourier series for the periodic waveforms shown in Figure P9-1.

General Linear System Dynamics 703

9-13. Using SIMULINK (or other available) simulation software, get the first five terms in the Fourier series for the waveforms of Fig. P9-1.

9-14. Using the method of Fig. 9-13, find the periodic steady-state response of the following systems to the periodic input given by the first seven terms of the Fourier series of Eq. (9-50).

a. $\dfrac{q_o}{q_i}(s) = \dfrac{3.45}{0.634s^2 + 0.159s + 1}$

b. $\dfrac{q_o}{q_i}(s) = \dfrac{3.45}{0.0396s^2 + 0.040s + 1}$

9-15. Using MATLAB or other available FFT software, find the Fourier series for the waveforms of Fig. P9-1.

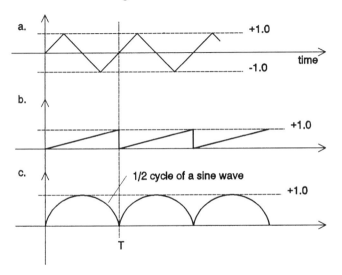

Figure P9-1

9-16. Repeat the Fourier series calculations which obtained Figs. 9-17 and 9-18, but use:
  a. 256 points
  b. 1024 points

9-17. Draw and explain a simulation block diagram which implements the Fourier transform "recipe" of Eq. (9-51). Then actually run this simulation for the transient input of Fig. 9-23, with $A = 1.0$ and $T = 1.0$. Compare your results with the exact values given by Eq. (9-53).

9-18. Treat the "splined" function of Fig. 9-15 as a transient of duration 0.01 second, rather than the 10 seconds shown there. Using MATLAB or other available FFT software, compute the Fourier transform of this transient.

9-19. Derive Eq. (9-64).

9-20. Draw and explain a simulation diagram for the system of Fig. 9-32, using the numbers given in the text. Duplicate the results of Fig. 9-33.

704   **Chapter 9**

9-21.   In the system of Fig. 9-32, add a damper between the two masses. Derive $(x_m/f)(s)$ and compare with Eq. (9-64). Using the text's numbers for this system, use simulation to explore the effect of this new damper on system behavior.

9-22.   In the system of Fig. 9-32, suppose the vibration is excited by the motion of the "floor," rather than by the external force f. Perform analysis to study whether the tuned absorber is helpful for such a situation.

9-23.   For the system of Fig. 9-32, derive $(x_a/f)(s)$.

9-24.   Consider the self-leveling servosystem used in the air-spring vibration isolator of Fig. 9-36. Assume that floor motion x_i is identically zero and that the force input f_i represents the weight of an additional "payload" mass which is placed on the table at time zero. Derive the system differential equation relating table displacement x_o to force f_i for small perturbations around an initial equilibrium condition. Assume that the valve mass flow rate in or out of the p_2 chamber is directly proportional to the displacement x_o away from the null position.

9-25.   Derive Eq. (9-65).

9-26.   Derive $(x_o/f_i)(s)$ for the system of Fig. 9-37.

9-27.   Derive Eq. (9-71).

9-28.   Derive a result analogous to Eq. (9-71) for the case where f_i is the input.

9-29.   The characteristic equation of the system of Eq. (9-65) is third-degree, so no convenient formulas are available for computing the three roots in letter form. By letting $B = 0$, we can get an approximate formula for the natural frequency of lightly damped systems. Suppose we want to design this vibration isolator to have a natural frequency of 10.0 rad/sec when the payload mass weighs $100\,\mathrm{lb}_f$. It is not clear at this point whether the two springs should have equal stiffness or have some other relation, so let's start by assuming them equal and see what we can learn from that before exploring unequal springs. To estimate a B value, neglect the s^3 term in Eq. (9-65) and then use the usual definition of second-order-system damping ratio to set ζ at 0.3. These approximations should give you a spring constant of $51.8\,\mathrm{lb}_f/\mathrm{inch}$ for each spring, and $B = 2.198\,\mathrm{lb}_f/(\mathrm{in/sec})$. Verify these values.

     a.   Using the numerical values just determined, and available computer software, plot the amplitude ratio for Eq. (9-65). Then explore the effect of damping B to see whether an optimum value exists, taking optimum to mean the system with the lowest resonant peak. If you find an optimum B, at what frequency is the amplitude ratio for this design equal to 0.01 (99% isolation)?

     b.   Now do a study to see whether *unequal* spring constants offer any benefits. To do this, try several combinations of K_{s1} and K_{s2}, but choose them so that the "series connected" spring constant $(K_{s1}K_{s2})/(K_{s1} + K_{s2})$ is always equal to the value we used with equal spring constants, that is, 25.9. You should also try adjusting the B value to see if there is any benefit. Produce the necessary graphs and discuss your results.

9-30.   By using the analogy between Eqs. (9-65) and (9-71), we can design the air-spring system to have the same behavior as a previously designed system with metal springs. Assume that the metal spring system has satisfactory behavior for a payload mass which weighs $100\,\mathrm{lb}_f$, when the two springs have equal stiffnesses of $51.8\,\mathrm{lb}_f/\mathrm{in}$

General Linear System Dynamics   **705**

and $B = 6.0\,\mathrm{lb}_f/(\mathrm{in/sec})$. Assume that space limitations dictate a chamber diameter of 3 inches or less, and that we want the two chambers to have equal lengths.

 a. Find numerical values for all system parameters not already specified. Do all these values seem reasonable?
 b. Although accurate values of mass-flow resistance R_f usually require experimental testing, we might want to do some theoretical estimates for preliminary design purposes. For the R_f found in part (a), find the diameter D needed if we use
  A. A laminar-flow sharp-edge orifice of diameter D
  B. A laminar-flow capillary tube of length 1.0 inch and diameter D

9-31. For the system of Eq. (9-73), take the weight of M to be $200\,\mathrm{lb}_f$, $K_s = 10,000\,\mathrm{lb}_f/\mathrm{in}$, and $B = 5.0\,\mathrm{lb}_f/(\mathrm{in/sec})$.

 a. Develop a SIMULINK (or other available) simulation diagram for this system.
 b. Study the response of the basic system (*no* active control) to a 1.0-lb_f step input of force.
 c. Let x_i be a sine wve of 5.0-Hz frequency and 0.0001-inch amplitude. Again find x_o.
 d. Calculate the value of K_m needed to give the active system a natural frequency of 1.0 Hz, and the value of K_b needed to get a damping ratio of 0.5. Rerun the step force and sine motion input tests you did on the basic system and discuss the improvement in isolation performance.

9-32. Obtain an equation analogous to (9-73) if the system uses a voltage (rather than current-output) amplifier. You will now have to include the force motor's inductance, resistance, and back emf in your model. Get a simulation diagram for this system and include a limiter on the amplifier output current.

9-33. Repeat Problem 9-32 if the accelerometer is (more realistically) modeled as a second-order system, and the velocity signal is obtained by integrating the accelerometer output.

9-34. Modify the simulation of Fig. 9-40 to use proportional plus integral control. That is, use for the AMP/CONT/PIEZO block the transfer function $(K_p s + K_I)/s$, instead of K_p. Adjust K_p and K_I to get a satisfactory response. Show that with integral control, the steady-state error in Fig. 9-42b is removed. The *speed* with which this error is removed increases with K_I. Show that if you try to get rid of this error *too fast*, the system goes unstable.

9-35. Modify the simulation of Fig. 9-40 to include limiters on the flapper and valve motions, to model the saturation effects present in real flow-control devices. Set the flapper limiter at ± 0.010 inch and the valve limiter at ± 0.030 inch, and rerun the simulation of Fig. 9-41.

9-36. For the system of Fig. 9-45:

 a. Using simulation, study the effect of adding a viscous damper to the dancer-roll mechanism.
 b. Study the effects of using a voltage-output, rather than current-output, amplifier to drive the motor armature.
 c. Proportional-plus-integral control should return the dancer roll to its null position after any transient. Change the simulation to include this feature

and study its effects. What happens if too much integral control is used, in an attempt to return to null too rapidly?

d. The radius R and roll inertia J will gradually change as the roll is unwound. Modify the simulation to include these effects.

e. Study, using simulation, a design concept which uses a pneumatically actuated friction brake in place of the electric motor. An E/P transducer, driven electrically from amplified sensor signals, adjusts the brake friction torque to keep tension constant.

f. Eliminate the dancer roll completely, and replace it with a sensor which measures tension directly. Study the feasibility of this concept, using the electric motor drive of the text system.

g. Repeat part (f) above, but use the friction brake of part (e) in place of the motor.

10

DISTRIBUTED-PARAMETER MODELS

10-1 LONGITUDINAL VIBRATIONS OF A ROD

The progression in model complexity (and hopefully accuracy) from first-order to second-order and then to higher (but still finite) order types has its ultimate end in the distributed-parameter model, which may be thought of as having an infinite number of infinitesimally small lumps. We have earlier, even in the first chapter, attempted to develop at least some qualitative feeling for the relations and distinctions between lumped- and distributed-parameter models. We are now in a position to do this somewhat more quantitatively and completely. Since the mathematical topic involved, partial differential equations, is a vast and complex one we cannot hope to treat it in any generality, but we will rather use two simple examples to develop some physical feeling for the concepts involved.

Our first example concerns the determination of the vibration characteristics (natural frequencies, mode shapes, etc.) of a slender rod which is initially deformed in the axial direction and then released to perform free longitudinal vibrations. In Fig. 10-1 the position coordinate x, measured from the left end of the rod, allows us to describe the location of any part of the rod. The Y and Z coordinates will be found to be unnecessary because our "slender rod" assumption will make variations in these two transverse directions negligible. To obtain the system equation we assume the rod at time $= 0$ has been deformed longitudinally in some way and is then released to perform free longitudinal vibrations with no external driving force

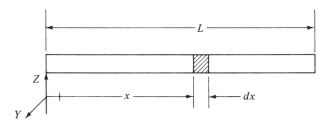

Figure 10-1 Slender rod in longitudinal vibration.

applied. We also assume no friction or damping effects either inside the rod material or at the interface between the rod surface and the medium (such as atmospheric air) in which it might be immersed.

When the rod is unstrained and at rest we define the displacement u of any transverse plane in the rod to be zero. This displacement u is actually the unknown in our system since if we know u for any station x in the bar and for any time t we have completely documented the rod's longitudinal motion. The displacement u of any plane away from its equilibrium position is thus rightly called $u(x, t)$ since it is a function of both location in the bar and time t. (This basic fact will lead us inexorably to partial differential equations since when we write derivatives of quantities which are functions of more than one variable we must write them as *partial* derivatives.) We now choose a rod element of infinitesimal length dx and at an arbitrary location x in the rod. Since this problem involves motion of bodies under the action of forces, it is natural to apply Newton's law to the element dx in hopes of getting a system equation. Since no external forces are allowed by our assumptions, the forces at the two ends of the element due to internal stresses are the only ones we need to find.

These internal forces are the product of the stresses (force per unit area) and the cross-sectional area. If the rod were not "slender" the stresses might vary over the cross section (in the y and z directions) at a given x and we would need to integrate over the area to get the total force. Our assumption of slenderness presumes a *uniform* stress at any x and we thus need merely multiply stress by area to get total force. We next recall that stress and strain are related by the material modulus of elasticity E so long as we remain within the elastic limit of the material. An analysis of Fig. 10-2 gives the relation between strain and displacement u. The definition of unit strain ε is

$$\varepsilon \triangleq \frac{\text{change in length of element}}{\text{length of element}} \tag{10-1}$$

and thus the unit strain at any location x would be $\Delta u/\Delta x$. Going to the limit this would be du/dx, but since u is a function of two variables we write it as $\partial u/\partial x$. Note

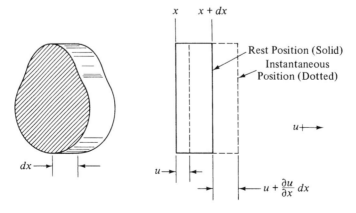

Figure 10-2 Infinitesimal element of rod.

Distributed-Parameter Models 709

that if $\partial u/\partial x$ is positive the strain is tension, while if it is negative the strain is compression.

Now the unit strain at any location is $\partial u/\partial x$ evaluated at that location; however, if we wish, we can write ε at $(x + dx)$ in terms of ε at x as follows:

$$\text{Unit strain at } (x + dx) = \text{unit strain at } x + \begin{bmatrix} \text{rate of change} \\ \text{of strain with} \\ \text{respect to } x \end{bmatrix} dx \qquad (10\text{-}2)$$

$$= \frac{\partial u}{\partial x} + \frac{\partial}{\partial x}\left[\frac{\partial u}{\partial x}\right] dx = \frac{\partial u}{\partial x} + \frac{\partial^2 u}{\partial x^2}\, dx \qquad (10\text{-}3)$$

We can now compute the stresses at x and $(x + dx)$ by using the definition of modulus of elasticity E.

$$E \triangleq \frac{\text{stress}}{\text{strain}} \qquad (10\text{-}4)$$

$$\text{Stress at } x = E\,\frac{\partial u}{\partial x} \qquad (10\text{-}5)$$

$$\text{Stress at } (x + dx) = E\left(\frac{\partial u}{\partial x} + \frac{\partial^2 u}{\partial x^2}\, dx\right) \qquad (10\text{-}6)$$

Assuming these stresses uniform over the cross-sectional area A, the forces are

$$\text{Force at } x = -AE\,\frac{\partial u}{\partial x} \qquad \left(\text{if } \frac{\partial u}{\partial x} \text{ is } +, \text{ force is } -\right) \qquad (10\text{-}7)$$

$$\text{Force at } (x + dx) = AE\left(\frac{\partial u}{\partial x} + \frac{\partial^2 u}{\partial x^2}\, dx\right) \qquad (10\text{-}8)$$

Since the element dx has mass $A\,dx\,\rho$ and its acceleration is $\partial^2 u/\partial t^2$, Newton's law gives

$$-AE\,\frac{\partial u}{\partial x} + AE\left(\frac{\partial u}{\partial x} + \frac{\partial^2 u}{\partial x^2}\, dx\right) = A\rho\,dx\,\frac{\partial^2 u}{\partial t^2} \qquad (10\text{-}9)$$

and finally

$$\frac{\partial^2 u}{\partial x^2} = \frac{\rho}{E}\,\frac{\partial^2 u}{\partial t^2} \qquad (10\text{-}10)$$

where

$$\rho \triangleq \text{material mass density}$$

Equation (10-10) is the equation of motion of this system. The unknown is $u(x, t)$ which, if solved for, will tell us how every portion of the rod moves longitudinally. That is, if you tell me which x location you are interested in, I will tell you the time history of its motion. This equation is one of the basic equations of classical physics, the *one-dimensional wave equation* and is found to apply to a number of practical problems. Actually, Eq. (10-10) alone is not sufficient to completely define, and therefore solve, the problem of rod vibration. While our force analysis for the element dx is unaffected by constraints applied at the ends ($x = 0$, $x = L$) of the rod, one would intuitively guess that such constraints *would* influence the rod motion. Thus to completely define the problem we must specify the *boundary conditions* at $x = 0$ and $x = L$. Figure 10-3 shows three possible sets of boundary conditions which would correspond to various practical configurations. We shall work out the solution for a rod free at both ends.

For a rod free at both ends, the force on the ends must at all times be zero (neglecting any force due to "bumping into" air particles). If the force is zero the stress must be zero and if the stress is zero the strain $\partial u/\partial x$ must also be zero. We may thus state the boundary conditions mathematically as

$$\text{For any value of } t: \left.\begin{array}{l} x = 0 \\ x = L \end{array}\right\} \frac{\partial u}{\partial x} = 0 \qquad (10\text{-}11)$$

We earlier stated that the vibrations are induced by an initial deformation of the rod. Mathematically speaking, at $t = 0$, $u(x, 0) = f(x)$, where $f(x)$ is a given function of x telling how the rod is initially deformed. Since we release the rod from its deformed condition with zero velocity, we also have $\partial u/\partial t = 0$ at $t = 0$ and for any x. We have now sufficiently described the situation that a solution for $u(x, t)$ should exist.

Just as in ordinary differential equations, no universal method of solution exists for partial differential equations so that one often tries some reasonable form of solution and sees if it works (satisfies equation, boundary conditions, and initial conditions). The nature of the solution assumed depends on the form of the equation, past experience with similar problems, physical intuition about the behavior of the system, etc. A product form of solution $u = f(x) g(t)$ where f and g are unknown functions is often applicable to equations of the type (10-10). An analyst knowledgeable about vibration could be even more specific and guess that u will very likely have the form

$$u = f(x) \cos \omega t \qquad (10\text{-}12)$$

Let us pursue this suggestion and see whether (10-12) actually can be the solution. Note that for $t = 0$, $u = f(x)$ and $\partial u/\partial t = \text{velocity} = -\omega f(x) \sin \omega t = 0$, which fits the initial conditions stated earlier. [Would $u = f(x) \sin \omega t$ have worked?] We must now see if (10-12) can be made to satisfy Eq. (10-10) by substituting the assumed solution into the system equation.

$$\frac{d^2 f(x)}{dx^2} \cos \omega t = \frac{\rho}{E} (-\omega^2 \cos \omega t) f(x) \qquad (10\text{-}13)$$

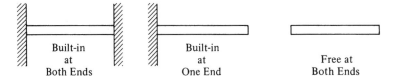

Figure 10-3   Various boundary conditions for rod.

Distributed-Parameter Models   711

$$\frac{d^2f}{dx^2} + \frac{\rho\omega^2}{E}f = 0 \tag{10-14}$$

Note that our problem has now been reduced to solving an *ordinary* (rather than *partial*) differential equation. (This is the usual pattern in solving partial differential equations no matter what specific technique is used.) In Eq. (10-14) ρ and E are known constants whereas ω (the frequency of vibration) is a constant as yet unknown. Applying our usual methods to (10-14) we get

$$\left[D^2 + \frac{\rho\omega^2}{E} \right] f = 0 \qquad D \triangleq \frac{d}{dx} \tag{10-15}$$

$$D^2 + \frac{\rho\omega^2}{E} = 0 \qquad \text{Roots} = \pm i\omega\sqrt{\frac{\rho}{E}} \tag{10-16}$$

$$f = C \sin\left(\sqrt{\frac{\rho}{E}}\,\omega x + \phi \right) \tag{10-17}$$

The constants of integration C and ϕ must be found using the boundary conditions and in this process the value of frequency ω will also come out. We may write for u

$$u = f(x) \cos \omega t = C \sin\left(\sqrt{\frac{\rho}{E}}\,\omega x + \phi \right) \cos \omega t \tag{10-18}$$

Now when $x = 0$, $\partial u/\partial x = 0$; thus

$$\frac{\partial u}{\partial x} = (C \cos \omega t) \cos\left(\sqrt{\frac{\rho}{E}}\,\omega x + \phi \right) \sqrt{\frac{\rho}{E}}\,\omega \tag{10-19}$$

$$0 = (C \cos \omega t) \sqrt{\frac{\rho}{E}}\,\omega \cos \phi \tag{10-20}$$

If we choose $C = 0$ to satisfy (10-20) we get the trivial solution $u(x, t) = 0$; thus it must be that $\cos \phi = 0$, which occurs for $\phi = \pm\pi/2, \pm 3\pi/2, \pm 5\pi/2$, etc.

Now $\partial u/\partial x$ is also zero for $x = L$ and thus

$$0 = (C \cos \omega t) \sqrt{\frac{\rho}{E}}\,\omega \cos\left(\sqrt{\frac{\rho}{E}}\,\omega L + \phi \right) \tag{10-21}$$

Again, $C = 0$ is trivial, so

$$\cos\left(\sqrt{\frac{\rho}{E}}\,\omega L + \phi \right) = 0, \quad \sqrt{\frac{\rho}{E}}\,\omega L + \phi = \pm\frac{\pi}{2}, \pm\frac{3\pi}{2}, \pm\frac{5\pi}{2}, \text{ etc.} \tag{10-22}$$

If we check the possible combinations which arise by substituting $\phi = \pm\pi/2, \pm 3\pi/2$, etc. into (10-22) (and excluding the possibility of $\omega \leq 0$ since negative frequencies have no physical interpretation) we get

$$\sqrt{\frac{\rho}{E}}\,\omega L = \pi, 2\pi, 3\pi, 4\pi, \text{ etc.} \tag{10-23}$$

and thus we have found the allowable values of vibration frequency ω (the "natural frequencies") to be

$$\omega = \frac{n\pi}{L}\sqrt{\frac{E}{\rho}} \qquad n = 1, 2, 3, \ldots \tag{10-24}$$

Our solution for u is thus

$$u = C\sin\left(\sqrt{\frac{\rho}{E}}\omega x \pm \frac{\pi}{2}, \pm\frac{3\pi}{2}, \ldots\right)\cos\omega t = C\cos\left(\sqrt{\frac{\rho}{E}}\omega x\right)\cos\omega t \tag{10-25}$$

One can use any one of the ω values of (10-24) in (10-25) to get a solution. In fact, since the differential equation is linear, the superposition principle says that a *sum* of such solutions is also a solution.

To get some feeling for the meaning of these various solutions let us consider some specific cases. Suppose we take $n = 1$ in (10-24) and use this ω value in (10-25). We get

$$u(x, t) = C\cos\left(\frac{\pi x}{L}\right)\cos\left(\frac{\pi}{L}\sqrt{\frac{E}{\rho}}t\right) \tag{10-26}$$

and for $t = 0$,

$$u = C\cos\frac{\pi x}{L} \tag{10-27}$$

This formula tells us the "shape" into which the rod must initially be deformed to give the motion of Eq. (10-26) when released. That is, up to this point we have been rather vague about the initial deformation of the rod, merely calling it $f(x)$. We see now that to get a certain natural frequency to exist *alone*, not just "any old" shape $f(x)$ may be used. Figure 10-4 is a plot of Eq. (10-27). Note that C simply determines the scale or magnitude of the oscillation of any plane in the rod; if we double C, for example, all amplitudes are doubled. At any chosen location, such as x_1, the rod vibrates longitudinally according to

$$u(x_1, t) = C\cos\frac{\pi x_1}{L}\cos\frac{\pi}{L}\sqrt{\frac{E}{\rho}}t \triangleq A_1\cos\omega t \tag{10-28}$$

a simple harmonic oscillation.

The frequencies ω given by (10-24) are called the natural frequencies of the rod; note that there are an *infinite* number of them. This is characteristic of distributed-parameter vibration models and also of the real systems which they represent; they really *do* have an infinite number of natural frequencies. Just as in lumped-parameter models, if an external driving force acts at a frequency equal to a natural frequency the undamped system builds up an infinite motion. The "dynamic deflection curve" of Fig. 10-4 is called the *mode shape*. Each natural frequency has its own characteristic mode shape; for $n = 2$ we have $\omega = (2\pi/L)\sqrt{E/\rho}$ and the mode shape of Fig. 10-5. In order to produce vibrations at a single natural frequency only, it is necessary to initially deform the rod into precisely the cosinusoidal shapes such as Fig. 10-4 or 10-5. What if the initial deformation were some *arbitrary* shape? We would then find in general that *all* the natural frequencies would be excited in varying degrees depending on the shape of the particular deformation imposed. The motion is then a superposition of the various mode shapes, each at a different amplitude. The same *Fourier series* used in Chapter 9 is also used in these sorts of problems

Distributed-Parameter Models

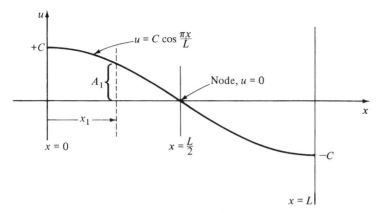

Figure 10-4 Mode shape for first natural frequency.

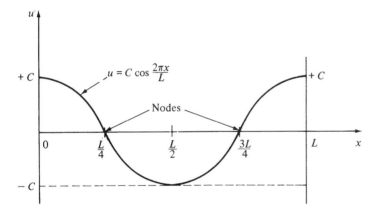

Figure 10-5 Mode shape for second natural frequency.

to "build up" an arbitrary $f(x)$ from the various cosine waves representing the natural modes.

10-2 LUMPED-PARAMETER APPROXIMATIONS FOR ROD VIBRATION

The distributed-parameter model developed and solved above is a very accurate representation of the behavior of a slender rod and predicts values which agree very closely with experimental measurements. However, the application of this type of model to practical problems, where the shapes are rarely as simple as that of our rod, encounters serious mathematical difficulties and a lumped model may be necessary. To develop some feeling for the general nature and relative merits of the two viewpoints let us now analyze the rod vibration problem from the lumped point of view. Rather than dealing with continuous distributions of mass and elasticity we will now work with discrete "massless springs" and "springless masses."

The first problem in a lumped analysis concerns how the system will be "dissected" into lumps. There are two general approaches. If the describing partial differential equation for a distributed-parameter model has been written (but *not* solved), one can apply standard finite-difference or finite element methods to create a lumped model mathematically without further explicit consideration of the physics of the problem. Either space (x, y, z) coordinates or time, or both, can be lumped. In the rod vibration problem, for example, if x is lumped but t is not, the partial differential equation becomes a simultaneous set of ordinary differential equations. If *both* x and t are lumped, the equations become strictly algebraic. In the second approach one cuts up the physical system into lumps according to some rational scheme and then applies the pertinent physical laws to each lump. This procedure will generate a set of simultaneous ordinary differential equations. If these are linear with constant coefficients, routine analytical techniques will supply exact solutions; otherwise, computer methods will most likely be used. Analog computers can handle the ordinary differential equations without lumping time; that is, t remains a continuous variable. Digital computer solution requires use of stepwise numerical integration, so t also becomes "lumped."

We will here employ the second of the two above approaches without necessarily indicating that this is always best. In cutting the system into lumps, no universal scheme can be given which will always be the "best." Any reasonable scheme will, however, give useable results which improve in accuracy as more lumps are included in the model. For our rod vibration problem we will use the following lumping scheme.

1. Divide the rod length into equal segments.
2. Lump the mass of each segment at its center of mass.
3. Connect these masses by massless springs whose spring constants are equal to those of the rod segments between the mass points.

Figure 10-6 shows how application of this scheme leads to models with one to four lumps; extension to any desired number of lumps is obvious. For a 1-lump model, if the mass is displaced from its equilibrium position no restoring forces are developed by the springs if the rod is free at both ends; thus this model is not useful for studying

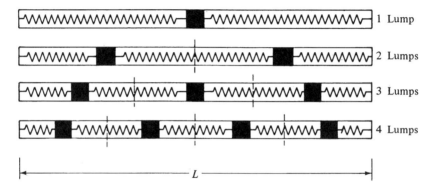

Figure 10-6 Lumping scheme for vibrating rod.

Distributed-Parameter Models

rod vibration. (Would this be true for the other boundary conditions of Fig. 10-3?) The simplest model which will give any useful information is thus the 2-lump.

We will need expressions for spring constants and masses of rod-segments of length L_s. The mass is clearly $\rho A L_s$ while Fig. 2-12i gives the spring constant as AE/L_s. For the two-mass system we set up coordinates as in Fig. 10-7. We now assume the masses are arbitrarily displaced from their neutral positions and released. Newton's law for each mass gives

$$K_{s2}(u_2 - u_1) = M_1 \ddot{u}_1 \tag{10-29}$$

$$K_{s2}(u_1 - u_2) = M_2 \ddot{u}_2 \tag{10-30}$$

which may be reduced to one equation in one unknown using our usual methods.

$$D^2(M_1 M_2 D^2 + K_{s2}(M_1 + M_2))u_1 = 0 \tag{10-31}$$

$$D^2(M_1 M_2 D^2 + K_{s2}(M_1 + M_2))u_2 = 0 \tag{10-32}$$

The roots of the characteristic equation are

$$s_1 = 0 \quad s_2 = 0 \quad s_{3,4} = \pm i \sqrt{\frac{(M_1 + M_2)K_{s2}}{M_1 M_2}} \tag{10-33}$$

making the solutions

$$u_1 = C_0 + C_1 t + C_2 \sin\left(\sqrt{\frac{M_1 + M_2}{M_1 M_2} K_{s2}} \, t + \phi_1\right) \tag{10-34}$$

$$u_2 = C_3 + C_4 t + C_5 \sin\left(\sqrt{\frac{M_1 + M_2}{M_1 M_2} K_{s2}} \, t + \phi_2\right) \tag{10-35}$$

The terms $C_0 + C_1 t$ and $C_3 + C_4 t$ refer to gross motions of the overall system corresponding to initial displacements and velocities and can be made to disappear by proper choice of initial conditions. Our main interest is in the oscillatory terms which we see exhibit only a *single* natural frequency ω given by

$$\omega = \sqrt{\frac{M_1 + M_2}{M_1 M_2} K_{s2}} = \sqrt{\frac{\rho A L}{(\rho A L/2)^2} \times \frac{2AE}{L}} = \frac{2.82}{L}\sqrt{\frac{E}{\rho}} \tag{10-36}$$

This frequency may be compared to the *first* natural frequency predicted by the "exact" distributed-parameter model, which was $(3.14/L)\sqrt{E/\rho}$. Our simplest

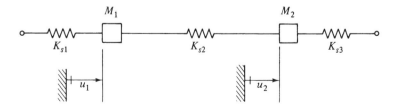

Figure 10-7 Two-lump model of rod.

lumped model thus predicts only one natural frequency (whereas an infinite number actually exist) and this one frequency is numerically somewhat inaccurate. Another aspect of the model's deficiencies lies in that we have direct results for the motion of only *2 points* in the rod, the $L/4$ and $3L/4$ points corresponding to the lumped masses. The distributed model gives results for *every* point. The "mode shape" predicted by the lumped model for equal and opposite initial displacements of M_1 and M_2 consists of three straight line segments as in Fig. 10-8 (compare with Fig. 10-4), rather than half a cosine wave.

If we now go to a 3-lump model the equations are (see Fig. 10-9)

$$(MD^2 + K_s)u_1 - K_s u_2 = 0 \tag{10-37}$$

$$K_s u_1 - (MD^2 + 2K_s)u_2 + K_s u_3 = 0 \tag{10-38}$$

$$K_s u_2 - (MD^2 + K_s)u_3 = 0 \tag{10-39}$$

which lead to

$$D^2 \left[\left(\frac{M}{K_s}\right)^3 D^4 + 4\left(\frac{M}{K_s}\right)^2 D^2 + 3\left(\frac{M}{K_s}\right) \right] u_1 = 0 \tag{10-40}$$

and identical equations for u_2 and u_3. The quartic term can be factored using the quadratic formula

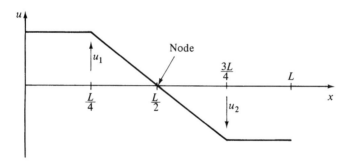

Figure 10-8 Lumped-parameter mode shape for 2-lump model.

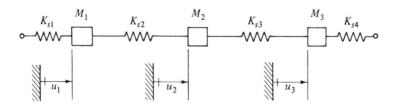

$$M_1 = M_2 = M_3 = \rho A L/3 \triangleq M \qquad K_{s2} = K_{s3} = 3AE/L \triangleq K_s$$

Figure 10-9 Three-lump model for rod.

Distributed-Parameter Models 717

$$D^4 + 4\frac{K_s}{M}D^2 + 3\left(\frac{K_s}{M}\right)^2 = 0 \tag{10-41}$$

$$D^2 = \frac{-4\frac{K_s}{M} \pm \sqrt{4\left(\frac{K_s}{M}\right)^2}}{2} = -3\frac{K_s}{M}, -\frac{K_s}{M} \tag{10-42}$$

We thus have a total of six roots

$$\left.\begin{array}{c} s_1 = 0 \quad s_2 = 0 \quad s_{3,4} = \pm i\sqrt{\frac{3K_s}{M}} = \pm i\frac{3\sqrt{3}}{L}\sqrt{\frac{E}{\rho}} \\[3mm] s_{5,6} = \pm i\sqrt{\frac{K_s}{M}} = \pm i\frac{3}{L}\sqrt{\frac{E}{\rho}} \end{array}\right\} \tag{10-43}$$

The solution for u_1 (u_2 and u_3 have the same form) is thus

$$u_1 = C_0 + C_1 t + C_2 \sin\left(\frac{3\sqrt{3}}{L}\sqrt{\frac{E}{\rho}}t + \phi_1\right) + C_3 \sin\left(\frac{3}{L}\sqrt{\frac{E}{\rho}}t + \phi_2\right) \tag{10-44}$$

Note that now *two* natural frequencies are predicted. Figure 10-10a compares these with the distributed-model and two-lump model results.

We see that the 3-lump model has not only predicted another natural frequency but has also improved the accuracy of prediction for the first frequency. The pattern should now be becoming clear. More lumps in the model produce more natural frequencies and improve the accuracy, the limiting case being an infinite number of lumps, each infinitesimally small, which is of course precisely the distributed-parameter model. Also, the mode shapes become better defined since we get direct information on more x locations (3 for a 3-lump model, 10 for a 10-lump model, etc.) and thus the linear interpolation between masses becomes more accurate.

As a final comparison of the lumped and distributed models, consider the 3-lump model of Fig. 10-10b. Analysis (left for the end-of-chapter problems) shows that

$$\frac{u_4}{f_i}(s) = \frac{\left(\frac{M}{K_s}\right)^3 s^6 + 6\left(\frac{M}{K_s}\right)^2 s^4 + 9\left(\frac{M}{K_s}\right)s^2 + 2}{2K_s s^3\left[\left(\frac{M}{K_s}\right)^3 s^4 + 4\left(\frac{M}{K_s}\right)^2 + 3\left(\frac{M}{K_s}\right)\right]} \tag{10-44a}$$

A distributed-parameter model of this same problem is available in the literature[1] and gives the result:

$$\frac{u_4}{f_i}(s) = \frac{1}{(A\sqrt{\rho E})s \tanh\left(\frac{L}{\sqrt{E/\rho}}s\right)} \tag{10-44b}$$

[1] E. O. Doebelin, *System Modeling and Response*, Wiley, New York, 1980, p. 421.

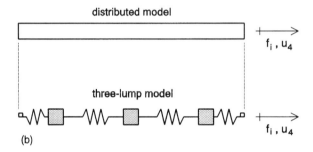

(a)

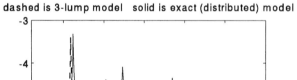

(b)

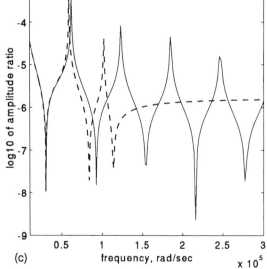

(c)

Figure 10-10 Comparison of lumped and distributed model results.

Distributed-Parameter Models

For an aluminum rod of length $L = 10$ inches and cross-sectional area $A = 0.1 \text{ in}^2$ we get the amplitude ratio curves of Fig. 10-10c. At zero frequency both models predict an infinite amplitude ratio since the free-end boundary condition makes the rod behave like a single lumped mass for this limiting case. Since the models completely neglect damping, the resonances give infinite amplitude ratios and the antiresonances go exactly to zero. The infinite amplitude ratios at resonance will of course not show up in our graphs since we compute at discrete frequencies and will thus "miss" the exact locations. The zeros at antiresonance are also missed for this reason and also because we chose to plot the logarithms of the amplitude ratios.

The graphs show that the two models agree very closely below the first antiresonance and also quite well through the first resonance and second antiresonance. The lumped model can predict only two resonances and three antiresonances, whereas the more correct distributed model predicts an infinite number of each. As we have seen many times before, a model for a real system need be accurate only below the highest frequency present in its real-world input signals. For Fig. 10-10c the 3-lump model would be usable from 0 to about 70,000 rad/sec.

10-3 CONDUCTION HEAT TRANSFER IN AN INSULATED BAR

In Fig. 10-11 a slender metal rod, initially all at temperature T_0, is buried in perfect insulation. At time $= 0$, its left end is suddenly raised to temperature T_i and held there thereafter. We wish to find the temperature-time history of any point in the rod. The slenderness of the rod again makes the problem one-dimensional; variations in the y and z directions are assumed negligible. The basic physical law here is Fourier's law of heat conduction which says that the heat flux through any cross section is proportional to the temperature gradient dT/dx at that cross section. Mathematically,

$$q_x = -kA \frac{\partial T}{\partial x} \quad \frac{\text{Btu}}{\text{sec}} \tag{10-45}$$

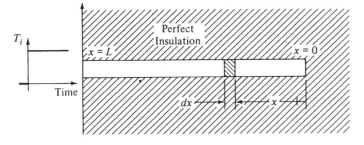

Figure 10-11 Heat conduction problem.

720   **Chapter 10**

where

$k \triangleq$ material thermal conductivity

$A \triangleq$ cross-sectional area

The temperature gradient is written as $\partial T/\partial x$ since in our problem T will be a function of *both* x and t. In the element of length dx, heat enters and leaves only at the ends since the surface is perfectly insulated; any difference between entering and leaving heat flows must show up as energy storage within the element.

We may write the heat flux at $(x + dx)$ in terms of the heat flux at x as follows.

$$\text{Heat flux at } (x + dx) = \text{heat flux at } x + \begin{bmatrix} \text{rate of change of} \\ \text{heat flux with} \\ \text{respect to } x \end{bmatrix} dx \qquad (10\text{-}46)$$

$$q_{x+dx} = -kA\,\frac{\partial T}{\partial x} + \frac{\partial}{\partial x}\left(-kA\,\frac{\partial T}{\partial x}\right)dx = -kA\,\frac{\partial T}{\partial x} - kA\,\frac{\partial^2 T}{\partial x^2}\,dx \qquad (10\text{-}47)$$

Conservation of energy now gives

Energy input rate − energy output rate = energy storage rate   (10-48)

$$-kA\,\frac{\partial T}{\partial x} - \left[-kA\,\frac{\partial T}{\partial x} - kA\,\frac{\partial^2 T}{\partial x^2}\,dx\right] = (\rho A\,dx)C\,\frac{\partial T}{\partial t} \qquad (10\text{-}49)$$

and thus

$$\frac{\partial^2 T}{\partial x^2} = \frac{\rho C}{k}\,\frac{\partial T}{\partial t} \qquad (10\text{-}50)$$

This is another common equation of classical physics, the one-dimensional diffusion equation. In solving our particular problem it will be helpful to define

$$\theta(x, t) \triangleq T(x, t) - T_i \qquad (10\text{-}51)$$

Since T_i is a constant, (10-50) becomes

$$\frac{\partial \theta}{\partial t} = a\,\frac{\partial^2 \theta}{\partial x^2} \qquad (10\text{-}52)$$

where

$$a \triangleq \text{thermal diffusivity} = \frac{k}{\rho C} \qquad (10\text{-}53)$$

Since the end $x = 0$ is perfectly insulated there can be no heat flow there and thus $\partial T/\partial x = 0$. Thus

$$\frac{\partial \theta}{\partial x}\,(0, t) = 0 \qquad (10\text{-}54)$$

and also, $\theta(L, t) = T(L, t) - T_i = T_i - T_i = 0$. Also, at $t = 0$, $T(x, 0) = T_0$ and thus

$$\theta(x, 0) = T_0 - T_i \qquad (10\text{-}55)$$

We are now again at a point where the form of the solution must be assumed. Again a product type of solution works; we assume

Distributed-Parameter Models   **721**

$$\theta(x, t) = X(x)G(t) \tag{10-56}$$

where X is a function *only* of x and G is a function *only* of t. Substituting into (10-52) gives

$$\frac{\partial \theta}{\partial t} = X(x)\frac{\partial G}{\partial t} = a\frac{\partial^2 \theta}{\partial x^2} = aG(t)\frac{\partial^2 X}{\partial x^2} \tag{10-57}$$

$$\frac{a}{X}\frac{\partial^2 X}{\partial x^2} = \frac{1}{G}\frac{\partial G}{\partial t} \tag{10-58}$$

Since the left side of (10-58) depends *only* on x and the right side *only* on t, they can be equal *only* if they both equal the same constant, call it $-a\lambda^2$. We have then the two *ordinary* differential equations

$$\frac{d^2 X}{dx^2} + \lambda^2 X = 0 \tag{10-59}$$

$$\frac{dG}{dt} + a\lambda^2 G = 0 \tag{10-60}$$

Let us solve (10-59) first. The roots of the characteristic equation are $\pm i\lambda$; thus

$$X = C_1 \sin(\lambda x) + C_2 \cos(\lambda x) \tag{10-61}$$

Now since $\theta(L, t) = 0$, Eq. (10-56) says $X(L) = 0$ and we have

$$0 = C_1 \sin(\lambda L) + C_2 \cos(\lambda L) \tag{10-62}$$

Also, if $(\partial \theta / \partial x)(0, t) = 0$, then $dX/dx = 0$ at $x = 0$, giving

$$0 = \lambda C_1 \cos(\lambda x) - \lambda C_2 \sin(\lambda x) = \lambda C_1 \tag{10-63}$$

Since $\lambda = 0$ gives a trivial solution, it must be that $C_1 = 0$ and then (10-62) requires $C_2 \cos(\lambda L) = 0$. Again, C_2 may not be zero or else $X \equiv 0$, so

$$\cos(\lambda L) = 0$$

$$\lambda = \frac{\pi}{2L}, \frac{3\pi}{2L}, \frac{5\pi}{2L}, \text{ etc. } = \frac{(2n+1)\pi}{2L} \qquad n = 0, 1, 2, \cdots \tag{10-64}$$

We finally have then $X = C_2 \cos(\lambda x)$ which we write as

$$X_n = C_n \cos(\lambda_n x) \tag{10-65}$$

to indicate that there exist an infinite number of possible solutions corresponding to all the λ's, each with a different C to go with it.

Turning to the equation in G, the solution there is clearly

$$G = Be^{-a\lambda^2 t} \tag{10-66}$$

and since we have n λ's, we again get a multiplicity of solutions which we write as

$$G_n = B_n e^{-a\lambda_n^2 t} \tag{10-67}$$

An individual solution for θ would thus be

$$\theta_n = X_n G_n = C_n \cos(\lambda_n x) B_n e^{-a\lambda_n^2 t} \tag{10-68}$$

Any *one* of these solutions cannot hope to fit the remaining initial condition $\theta(x, 0) = T_0 - T_i$, but an *infinite series* of such terms can, if each term is properly chosen. The linearity of the equation tells us that a sum of solutions will also be a solution and the Fourier series allows us to determine the proper "size" of each term. We thus write

$$\theta(x, t) = \sum_{n=0}^{\infty} A_n e^{-a\lambda_n^2 t} \cos(\lambda_n x) \tag{10-69}$$

and for $t = 0$,

$$\theta(x, 0) = T_0 - T_i = \sum_{n=0}^{\infty} A_n \cos(\lambda_n x) \tag{10-70}$$

We must now find the A_n's such that Eq. (10-70) is satisfied. Our Fourier series clearly must have only cosine terms and must "add up" to the constant $T_0 - T_i$. Note that we are fitting a Fourier series to a function which is not periodic. We can however make an interpretation that meets the needs of our present problem and also fits the usual Fourier series requirements. This is shown in Fig. 10-12. There a square wave of amplitude $(T_0 - T_i)$ and period $4L$ is shown. Note that:

1. If we can get the Fourier series it *will* add up to $(T_0 - T_i)$ over the range $0 < x < L$, just as needed in our problem.
2. By choosing the x origin as shown, the series will have *only* cosine terms, again as needed by our problem.
3. The period $4L$ is chosen since it is the period of the lowest frequency cosine wave in Eq. (10-70). That is, the period of $\cos(\pi x/2L)$ is $4L$.

Using the Fourier series formulas from Chap. 9 we have

$$A_n = \frac{1}{2L} \int_{-2L}^{2L} f(x) \cos \frac{n\pi x}{2L} dx = 4 \int_0^L (T_0 - T_i) \cos \frac{n\pi x}{2L} dx \tag{10-71}$$

$$= \frac{2(T_0 - T_i)}{L} \left[\frac{2L}{n\pi} \sin \frac{n\pi x}{2L} \right]_0^L = \frac{4(T_0 - T_i)}{n\pi} \left[\sin \frac{n\pi}{2} \right] \tag{10-72}$$

$$= \frac{4(T_0 - T_i)}{\pi}, -\frac{4(T_0 - T_i)}{3\pi}, \frac{4(T_0 - T_i)}{5\pi}, \frac{-4(T_0 - T_i)}{7\pi}, \text{etc.} \tag{10-73}$$

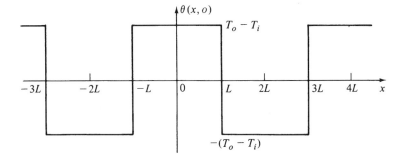

Figure 10-12 Function to be represented by a Fourier series.

Distributed-Parameter Models

The solution (10-69) may thus now be explicitly written out as

$$\theta(x, t) = (T_0 - T_i)\left[\frac{4}{\pi} e^{-\frac{a\pi^2}{4L^2}t} \cos\frac{\pi x}{2L} - \frac{4}{3\pi} e^{-\frac{9a\pi^2}{4L^2}t} \cos\frac{3\pi x}{2L} \right.$$
$$\left. + \frac{4}{5\pi} e^{-\frac{25a\pi^2}{4L^2}t} \cos\frac{5\pi x}{2L} + \cdots\right] \quad (10\text{-}74)$$

Once the location x in the rod which is of interest is chosen, (10-74) becomes a known function of time and may be computed and plotted; however, to get "perfect" accuracy an infinite number of terms is required. Unfortunately, unlike our earlier (Chap. 9) application of Fourier series where one could easily tell by examining the curve-fit whether enough terms had been taken, our present application provides no "exact" results with which to compare. The usual procedure, since the conventional series-convergence tests of calculus are not of much help, is to simply add more and more terms until the result appears to be unaffected within the number of significant figures desired.

As a numerical example, consider an aluminum rod ($a = 0.1466\,\text{in}^2/\text{sec}$) of length 10 inches, initially at $T = 0°F$ when the end is raised to $100°F$ and held there. It was found that a truncated series of as few as five terms gave good accuracy except near $t = 0$. For $x = 8$ inches, for example, a five-term series gave $T = -18°F$ at $t = 0$, whereas the correct result is of course $T = 0°F$. Going to 10 terms gave $T = 9.9°F$ at $t = 0$. Both the 5- and 10-term series almost exactly matched the "standard" curves for this problem (which are found in most heat transfer books) for $t > 15$ seconds. Unfortunately these texts do not report how many terms were

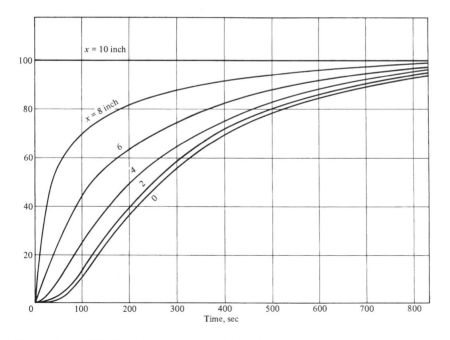

Figure 10-13 Distributed-parameter model results.

used to compute the "standard" curves. Using MATHCAD's series summation capability I found that, for $x = 8$ and $t = 0$, 21 terms gave $T = 4.1$ and 500 terms gave $T = 0.2$. Figure 10-13 shows a set of the standard curves plotted to fit our numerical values and for selected values of x. Note that Eq. (10-74) may be evaluated for any chosen x whatever.

10-4 LUMPED-PARAMETER APPROXIMATION FOR HEAT TRANSFER IN INSULATED BAR

We now model the problem of Fig. 10-11 using lumped techniques. Here the rod is divided into segments, each segment exhibiting *only* energy storage (no thermal resistance) and having a uniform temperature throughout the segment at any instant of time. Between these segments we place localized thermal resistances. As in any lumped model, there is always the question of how to define the lumped elements and also how many lumps to choose. No hard-and-fast rules are available and past experience is often the best guide. By doing lumped analyses of problems for which the "exact" (distributed-parameter) models can be solved, we have a standard against which to compare and can thus get some feeling for the nature and degree of approximation caused by the lumping. Let us do this for the aluminum rod with response as in Fig. 10-13.

Figure 10-14 shows one possible lumping scheme using six lumps. We wish to compare our results with the curves of Fig. 10-13 for $x = 2, 4, 6,$ and 8 inches; thus these points should be the *centers* of our lumps for best accuracy. For six *equal* lumps, the centers fall at $x = 5/6, 15/6$, etc., so we make two half-size lumps at the ends as shown. In any specific problem there may or may not be good arguments for "uneven" lumping. If greater temperature gradients are intuitively expected in one region of the body it may be wise to use smaller lumps there than elsewhere, for example. Once the body is sectioned, the value of capacitance assigned to that section is simply its actual capacitance, that is, Mc for the volume of material in the section. The assignment of thermal resistances is not as obvious since the resis-

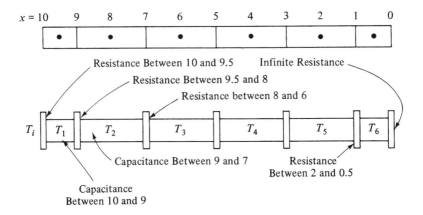

Figure 10-14 Lumping scheme for heat transfer in rod.

Distributed-Parameter Models **725**

tances are "concentrated" between the capacitances and there are several, often equally reasonable, ways of doing this. In Fig. 10-14 the rationale was, coming from T_i to the right, to select the thermal resistance at the lump interfaces such that the resistance to the left was exactly correct at the *center* of each lump. For example, the resistance between T_i and the center of the T_1 lump is precisely what it is in the actual rod; similarly for the centers of all the other lumps.

For a 0.5-inch length of rod the thermal resistance is $L/kA = 0.5/kA$ and the thermal capacitance is $Mc = \rho ALc = 0.5\ \rho Ac$. Writing conservation of energy for the lump at temperature T_1 we get

$$\left(\frac{T_i - T_1}{0.5/kA} - \frac{T_1 - T_2}{1.5/kA}\right) dt = (\rho Ac)\, dT_1 \tag{10-75}$$

$$\frac{\rho c}{k}\frac{dT_1}{dt} + \frac{8}{3}T_1 - \frac{2}{3}T_2 = 2T_i = 200 \tag{10-76}$$

and since $k/\rho c = 0.1466\ \text{in}^2/\text{sec}$ for aluminum,

$$\frac{dT_1}{dt} = 29.32 - 0.391 T_1 + 0.0978 T_2 \tag{10-77}$$

Repeating this procedure for each lump in turns gives

$$\frac{dT_2}{dt} = 0.0489 T_1 - 0.0855 T_2 + 0.0366 T_3 \tag{10-78}$$

$$\frac{dT_3}{dt} = 0.0366 T_2 - 0.0732 T_3 + 0.0366 T_4 \tag{10-79}$$

$$\frac{dT_4}{dt} = 0.0366 T_3 - 0.0732 T_4 + 0.0366 T_5 \tag{10-80}$$

$$\frac{dT_5}{dt} = 0.0366 T_4 - 0.0855 T_5 + 0.0489 T_6 \tag{10-81}$$

$$\frac{dT_6}{dt} = 0.0978 T_5 - 0.0978 T_6 \tag{10-82}$$

This set of six simultaneous equations can be solved analytically (except for solving the sixth degree characteristic equation for its roots), but simulation is much quicker and easier. A simulation was run and the results almost perfectly match the "exact" curves of Fig. 10-13, showing that our six-lump model is quite satisfactory for this problem.

While in this simple example we have our choice of distributed or lumped models, in more realistic problems the shapes of the bodies involved prevent analytical solution of the partial differential equations and lumped models become a necessity. These lumped models also easily incorporate such nonlinearities as varying, rather than constant, material properties. Further realistic features such as heat loss at the rod surface (rather than perfect insulation) are also easily incorporated and may even vary from lump to lump. The resulting sets of simultaneous nonlinear ordinary differential equations are often quite easily solved numerically with simulation software.

We do not wish to leave the reader with the impression that distributed-parameter models have negligible practical utility. While analytical solutions are limited to fairly simple shapes of bodies (prismatical rods, infinite plates, cylinders, spheres,

726   **Chapter 10**

etc.), when these solutions *can* be obtained they give great *general* insight into the nature of system behavior, which may often, at least qualititatively, be extrapolated to bodies of arbitrary shape. Thus the distributed-parameter analytical solutions give the overall theoretical framework for describing the *kinds* of behavior to be expected while the lumped numerical computer solutions give the accurate specific results needed in actual design problems. For the more complex problems, commercial finite element software is available for many different physical applications such as stress, deflection, vibration, electromagnetics, heat transfer and fluid flow. These software packages are essentially lumped modeling, however, the number of lumps is very large, sometimes many thousand. Fortunately, the definition of the lumps, and the setup and solution of the equations has been "automated" to ease application. Accurate results are, unfortunately, *not* automatic. Considerable skill and experience are required to obtain valid answers. Also, such software does not, of course, produce formulas showing relations among parameters and variables, as do analytical solutions. Rather, we get specific numerical results for specific conditions, just as in a laboratory experiment.

BIBLIOGRAPHY

Churchill, R. V., *Fourier Series and Boundary Value Problems*, McGraw-Hill, New York, 1941.
Doebelin, E. O., *System Modeling and Response*, Wiley, New York, 1980, sec. 4.1; chaps. 7, 8; secs. 9.1 to 9.4; sec. 10.1.1; sec. 11.1.
Miller, K. S., *Partial Differential Equations in Engineering Problems*, Prentice Hall, Englewood Cliffs, N.J., 1953.
Myers, G. E., *Analytical Methods in Conduction Heat Transfer*, McGraw-Hill, New York, 1971.
Rao, S. S., *Mechanical Vibrations*, 3rd ed., Addison Wesley, Reading, MA, 1995.

PROBLEMS

10-1. Using a distributed-parameter model, find natural frequencies and mode shapes for the rod of Fig. 10-3 with one end built in.

10-2. Using a distributed-parameter model, find natural frequencies and mode shapes for the rod of Fig. 10-3 with both ends built in.

10-3. Repeat Problem 10-1 but now use lumped models with:
   a. 1 lump   b. 2 lumps
   c. 3 lumps

10-4. Repeat Problem 10-2 but now use lumped models with:
   a. 1 lump   b. 2 lumps
   c. 3 lumps

10-5. Assume the rod of Fig. 10-1 is immersed in oil so that there is a viscous damping force acting on the rod surface. Analyze to show how Eq. (10-10) is changed.

10-6. Show and discuss lumped models for the situation of Problem 10-5.

Distributed-Parameter Models 727

10-7. For the rod of Fig. 10-3 with one end built in, let a force $f_0 \sin \omega t$ act on the free end. We wish to find the steady-state displacement $u(L, t)$ of the free end, using a distributed-parameter model. (*Hint*: Assume $u(x, t) = g(x) \sin \omega t$. Having found $u(L, t)$, now form the sinusoidal transfer function $(u_L/f)(i\omega)$ relating the displacement at $x = L$ to the force applied there. Plot the frequency-response curves.)

10-8. Repeat Problem 10-7 using a lumped model with two lumps. Compare the results with those of Problem 10-7.

10-9. Change the rod of Fig. 10-1 so that its diameter changes linearly with x, that is, the rod has a uniform taper from a diameter d_0 at $x = 0$ to d_L at $x = L$, $d_L < d_0$. Using a distributed-parameter model, find the system differential equation. Speculate on the "solvability" of this equation.

10-10. Repeat Problem 10-9 using a 5-lump model. Does solution of this set of equations pose any unusual problems?

10-11. Write equations for lumped models of the system of Fig. 10-11, using:

a. 1 lump   b. 2 lumps
c. 3 lumps   d. 4 lumps

10-12. Using the numerical values used in the text example (10-inch aluminum rod), solve the equations of Problem 10-11 for the unknown temperatures using SIMULINK or other simulation program. Compare results with Fig. 10-13.

10-13. Using a distributed-parameter model, reanalyze the system of Fig. 10-11 if there is a convective heat loss $hA(T - T_0)$ Btu/sec from the rod surface. The film coefficient h and ambient temperature T_0 are constant; A is the surface area of the element dx. The end at $x = 0$ is still perfectly insulated. Set up the system differential equation; do not attempt to solve it.

10-14. Repeat Problem 10-13 using the 6-lump model of Fig. 10-14. If h varied with x in a known fashion would the problem be much more difficult? Would a convective loss at the end $x = 0$ (rather than perfect insulation) cause any difficulty?

10-15. In Problem 10-14, take $h = 5$ Btu/(hr-ft^2-°F) and let the rod be 0.5 inch diameter aluminum, 10 inches long. Using SIMULINK or other simulation program, actually solve for the temperatures if the rod is all initially at 0°F when T_i jumps up to 100°F at $t = 0$.

10-16. Using lumped models, how would you handle a problem such as that of Fig. 10-11 if the material properties k, p, and c varied with x and/or T in a known fashion?

10-17. Derive Eq. (10-44a).

Appendix A

VISCOSITY OF SILICONE DAMPING FLUIDS

Mechanical dampers often use silicone fluids since they are available in a wide range of viscosities and do not change viscosity with temperature as much as petroleum-based oils do. Figures A-1, A-2, and A-3[1] give data on the "kinematic" viscosity in centistoke units and the mass density in grams/cm^3. To compute the ordinary (sometimes called dynamic) viscosity which we have used in our discussion of fluids in this text and which is used to design dampers, we have the formula

$$\text{Viscosity} \triangleq \mu = \frac{\text{lb}_f\text{-sec}}{\text{in}^2} = (\text{centistokes})(\rho)(1.45 \times 10^{-7}) \qquad \text{(A-1)}$$

$$\rho \triangleq \text{mass density, grams/cm}^3$$

In Figs. A-1 and A-2 each line represents a specific fluid compounded to give a certain viscosity. The fluids are "named" in terms of their "room temperature" (77°F) viscosity. For example, 3000 cs fluid as 3000 centistokes at 77°F. Let us compute the viscosity μ of 3000 cs fluid at 200°F as an example.

$$\mu = (1000 \text{ centistokes})(0.909)(1.45 \times 10^{-7}) = 1.32 \times 10^{-4} \frac{\text{lb}_f\text{-sec}}{\text{in}^2} \qquad \text{(A-2)}$$

---

[1] Dow-Corning Corp., Midland, Michigan.

729

Appendix A

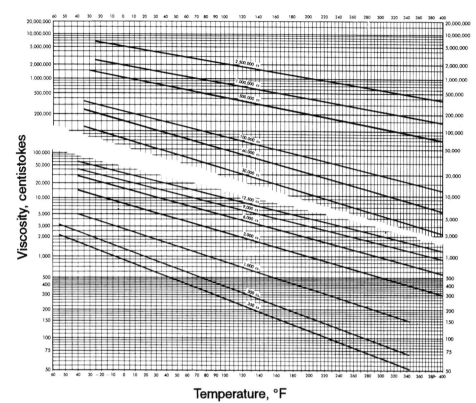

Figure A-1

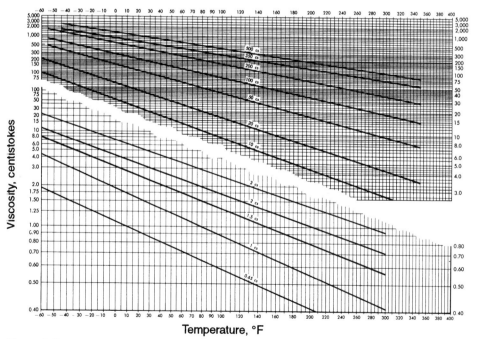

Figure A-2

Viscosity of Silicone Damping Fluids

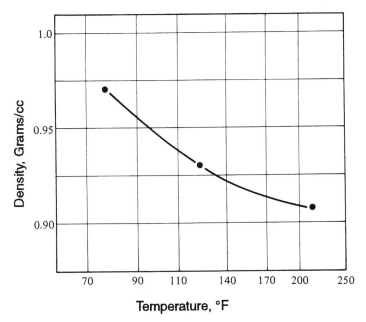

Figure A-3

Appendix B

UNITS AND CONVERSION FACTORS

At present in the United States many engineering calculations are made in units of the so-called British Engineering System (BES) whereas "scientific" (physics, chemistry, etc.) work is often carried out and reported in some form of the metric system. Since even in the metric system there is not complete uniformity, a "modernized" metric system, called The International System of Units (abbreviated SI), has been proposed and very likely will ultimately be accepted by all nations. This may take some time in the United States and one must, as an engineer, be prepared to work in both systems. In SI there are six basic units and all others are combinations of these:

Length—meter   Electric current—ampere
Mass—kilogram   Thermodynamic temperature—degree Kelvin
Time—second   Light intensity—candela

Note that force is not considered a basic unit; it is defined in terms of mass, length, and time. We will now give a brief listing of the basic units and conversions and a few of the more common derived quantities.

Length.

Multiply feet by 0.30480 to get meters   (B-1)

Mass. In most practical problems we know the *weight* of a body; that is, the force of gravity acting on it. If we know the acceleration of gravity at the location where the body was weighed, we calculate the mass as:

$$M = \text{mass} = \frac{\text{weight, pounds of force (lb}_f)}{\text{local acceleration of gravity, ft/sec}^2} = \frac{W}{g} = \text{slugs} \qquad \text{(B-2)}$$

In most cases the local value of g is close enough to the *standard* sea level value (32.174 ft/sec^2) that we do not insist on knowing the local g and just use 32.174 instead. When the mass is given in slugs, Newton's law $F = MA$ uses the force in pounds and the acceleration in feet/sec^2. When acceleration in inches/sec^2 is used, (force still in lb$_f$) the mass is $W/g = W/386.09$. This unit of mass has not been given a name; its units are of course lb$_f$-sec^2/inch.

733

Appendix B

Actually, the mass of a body (rather than its weight) is a more fundamental property since it is the same everywhere, whereas weight will vary if we take the body to a place where g is different. Mass can be measured directly with an equal arm balance and a set of standard masses by adding sufficient standard masses to one side of the balance until they balance the unknown mass in the opposite pan. Since both the standard and unknown masses are feeling the same g, its actual value has no effect. In the SI system the unit of mass is the kilogram and the conversion is

Multiply slugs by 14.594 to get kilograms (B-3)

In fluid mechanics and thermodynamics a mass unit called the pound mass (lb_m) is sometimes used.

Multiply slugs by 32.174 to get pounds mass (B-4)

Time. The standard unit of time in all systems is the second.

Force. In SI the unit of force is the newton and the conversion is

Multiply lb_f by 4.4482 to get newtons (B-5)

Energy. To interconvert mechanical and thermal energy in the BES system use:

Multiply Btu by 778.16 get foot-lb_f (B-6)

In SI,

1 newton-meter = 1 watt-second

$$= 1 \text{ joule} = 2.390 \times 10^{-4} \text{ kilogram-calorie} \tag{B-7}$$

and for conversion

Multiply ft-lb_f by 1.3557 to get newton-meters (B-8)

Power. In the BES system 1 horsepower is 550 ft-lb_f/sec.

Multiply horsepower by 42.44 to get Btu/minute. (B-9)

In SI, power is in watts, newton-meter/second or joule/sec. For conversion,

Multiply horsepower by 745.7 to get watts (B-10)

Temperature.

$^{\circ}$Kelvin $= ^{\circ}$Centrigrade (Celsius) $+ 273.16$ (B-11)

$^{\circ}$Rankine $= ^{\circ}$Fahrenheit $+ 459.69$ (B-12)

$$^{\circ}\text{Kelvin} = \left(\frac{5}{9}\right)(^{\circ}\text{Rankine}) \tag{B-13}$$

Appendix C

THERMAL SYSTEM PROPERTIES

THERMAL RESISTANCE

	Material	k Btu/(hr-ft^2-°F/ft)	Temp. at Which k Measured
Conduction	Most gases	0.005 to 0.015	
	Most liquids	0.05 to 0.4	
	Hair felt	0.021	86°F
	Rock wool	0.039	300°F
		0.050	500°F
	Brick	0.40	68°F
	Bismuth	4.7	64°F
	Mild steel	36.	32°F
		33.	212°F
	Brass	61.	32°F
	Aluminum	117.	32°F
	Silver	244.	32°F

	Situation	h Btu/(hr-ft^2-°F)	Comments
Convection	Stagnant air, 6″ dia. pipe, outside	1.	Linearized at $T_1 + T_2 =$ 320, $\Delta T = 180°F$
	15 ft/sec, air inside 4″ dia. pipe	3.	Air at atmos. pressure, 100°F
	3 ft/sec water inside 2″ pipe	700.	Water at 100°F
	3 ft/sec steam inside 8″ pipe	100.	Steam at 250 psi and 600°F
	Steam condensing on pipes	2000.	Continuous-film condensation
	Steam condensing on pipes	14000.	Dropwise condensation

	Situation	R_t °F/(Btu/hr) [Eq. (4-88)]	Comments
Radiation	1-ft length of 3″ steel pipe	0.4	oxidized pipe in large brick room $T_{\text{pipe}} = 500°F$ $T_{\text{room}} = 80°F$

THERMAL CAPACITANCE

Material		c Btu/(lb$_m$-°F)	ρ lb$_m$/ft^3
Air	0°F, 14.7 psia	0.239 (c_p)	0.0862
	1000°F, 14.7 psia	0.263	0.0272
	2000°F, 14.7 psia	0.286	0.0161
Water	32°F	1.009	62.42
	60°F	1.000	62.34
	200°F	1.004	60.13
Aluminum	32°F	0.208	169.
Copper	32°F	0.091	558.
Steel	32°F	0.11	490.
Brass	32°F	0.092	532.
Rubber	32°F	0.48	75.
Gasoline	32°F	0.50	46.
Machine oil	32°F	0.40	56.
Alcohol	32°F	0.58	49.

INDEX

A/D converter, 188
Absolute instability, 197
AC circuit analysis, 466
AC measurement and control systems, response speed, 467
AC power, 570
AC power calculations, 467
Accelerometer, 75, 104, 459
Accumulator, 237–239, 471, 593
  fluid compliance device, 237–239
Acoustical systems, 124
  linearity, 584
ACSL (Advanced Continuous Simulation Language), 33, 370
  friction simulation, 69n
  spring energy example, 37
  three-mass system example, 371
Active device, electrical, 125
Actuators, 186
  types, 193
  thermal expansion, 492
Adams integrator, 37
ADAMS software, 394
Adiabatic compliance, gas, 585
Admittance, 467
Air in hydraulic fluid, effect on motion system speed, 236
Air-cushion vehicle, 629
Aircraft hydraulic power system, 209
Algebraic loop, simulation, 421, 595
Aliasing, 187, 192, 667
Almost periodic inputs, 10
Alternator, 168, 273

Amplifiers
  class AB, 317
  electronic, inefficiency, 313
  noise effects, 158
  operational, 174
  power, linear, 317
  power, PWM (pulse-width-modulation), 291, 317–321
  power, SCR (silicon-controlled rectifier), 317, 321–323
  special for piezoelectric devices, 309
  transconductance, 195, 290, 684, 696
  with capacitance load, 307–308
Amplitude of sine wave, definition, 39
Amplitude ratio, 39
Amplitude-modulated input, 12
Analog, 149
Analog coefficient multiplier, 177
Analog computer, 150, 213, 215
Analog integrator, 177
Analog signal processing, 177
Analog simulation, 367
Analog summer, 177
Analog-to-digital converter, 188
Analogies, electromechanical, 148
Analogous systems, 5
Analogy
  fluid/electrical, 220
  fluid inertance to mass and inductance, 242
  heat flow to current flow, 257
AND gate, 126
Animation, 394
Anti-alias filter, 192, 667

737

Anti-resonance, 42, 640
Arbitrary function generator, 367
Arbitrary waveform generator, 174
Armature, of generator, 273
Armature reaction, 597
Assumptions
    effect on model validity, 582
    validation of, 501
Atomic-force microscope, 309
Audio range, 124
Automobile handling dynamics, 599
Automobile speedometer, 431
Automotive air spring, 50
Automotive suspension system, 555
    frequency range, 40
Automotive systems, 9
Automotive transmission, continuously
    variable, 209
Average power, electrical, 468

Back emf, dc motor, 290
Band-pass filter, 182, 572
Bandwidth, 550
    linear amplifier, 317
    PWM amplifier, 318
    SCR amplifier, 321
BASIC computer language, 38
Battery, electrical, 168
    model for, 168
Bearings, magnetic, 51
Belts, modeling, 85–89
Bimetallic actuator, 233
Bioengineering applications 209–213
Biological applications, 22
Biot number, 264, 480
Block diagrams, 6, 7, 32
    multiple inputs and outputs, 348
Blood flow system, human, 212
Bode plots, 438, 633–638
    first-order system, 438
    second-order system, 549
Boiler, frequency-response testing (fluid
    impedance), 249
Boundary conditions, 710
Boundary layer, 258
Brake
    friction, 496
    heating of, 287
    pneumatic, 687, 699
Breakpoint frequency, 439, 443
    first-order system, 439
    second-order system, 549

Brush-type dc motor, 291
Brush-type versus brushless motors, com-
    parison, 293
Brushes, of generator, 273
Brushless dc motor, 291
Buffer amplifier, 527
Bulk modulus
    adiabatic, 218, 681
    effective, 239
    fluid, 218
    isothermal, 218
Bus, computer-controlled hydraulic drive,
    209

Cable dynamics, 452
Cams, modeling, 85–89
Cancellation compensation, 427
Capacitance, how measured, 134
Capacitance element, 131, 135
Capacitance motion sensor, in piezoelectric
    system, 309
Capacitive actuation, 233
Capacitor
    design formulas, 159, 160
    electric absorption, 163
    electrolytic, 160
    energy behavior, 133, 135
    frequency range, 164
    in piezoelectric devices, 279
    laboratory standard, 162
    leakage resistance, 162
    manufacture, 159, 161
    power factor, 162
    realistic models, 162, 163
    temperature coefficient, 163
    thin film, 162
    transfer functions, 133, 134
Capillary tube, 223
Cascade connection, 523
Cascaded subsystems, 523
Cavitation, 327
Chains, modeling, 85–89
Characteristic equation, 339, 350
Charge, electric, 131
Charge pump, 327
Checking methods, 545
    dimensional, 581
    limiting case, 581
    special case, 581
Chemical process control, 213
Classification of system models, 14, 18
Classroom demonstration, 631

Clutch, heating of, 287
Clutch/brake motion control systems, 327–334
Clutch system, torque advantage, 332
Coffee cooling dynamics, 589
Combustion, 316
Command-line simulation language, 33
Commutator, of generator, 273
Commutator/brush action, dc motor, 291–292
Complementary solution, 339, 341
Complex number arithmetic, 60
Compliance, spring, 32
Compressibility, fluid, 218
Computational delay, 126, 186, 190, 197
Computer, analog, 150
Computer aiding, effect on design philosophy, 185
Computer-aided systems, 126, 185–200
Concurrent engineering, 512
Conductance (electrical), definition, 127
Conservation of energy, 15
Constantan, wire material, 154
Constants of integration, 339
Contact resistance, heat transfer, 257
Continuous spectrum, 649, 662, 664
Control matrix, 397
Control vector, 397
Controlled source, 125
Conversion factors, 733, 734
Cooling, electronics, 262
Coordinate measuring machine, 185
Coordinate systems, 406
Copper, wire material, 154
Cornering stiffness, tire, 600
Coulomb, 131
Coulomb friction, 54, 69, 81, 105, 197
  effect on natural frequency, 530
  model for clutch, 330–332
  simulation, 69
Coupling coefficient, piezoelectric, 280
Coupling methods, subsystem, 524
Coupling of subsystems, 91
Critical damping, 529
Critical speed, automobile, 604
Cryogenic systems, heat transfer, 257
CSMP (Continuous System Modeling Program), 33, 370
Current amplifier, 307
Current limit, amplifier, 317
Current node law, 451

Current source, 167
  ideal, 167
  real, 172
Curve fitting, frequency response, 668

D operator, 56
  units of, 582
D/A converter, 188, 190, 191, 195
D'Alembert method, 93, 407
Damper, 54
  adjustable rotary, 67
  capillary tube, 67
  coefficient, 54, 56
  eddy-current, 66
  energy dissipation, 58
  for space structures, 68
  frequency response, 59
  gas, 64
  intentionally nonlinear, 66
  optimum nonlinear, 107
  piston/cylinder, 61
  porous plug, 67
  square-law, 73, 448
  squeeze-film, 67
Dampers, design formulas, 61–67
Damping
  dc motor, 486
  detection of nonviscous, 542
  fluid systems, 586
  hydraulic motor, 490
  hysteresis, 74
  inertial, 685
  parasitic, 68
  structural, 74
Damping fluids, properties, 729–731
Damping methods, temperature sensitivity, 67
Damping ratio, 522, 529
DC motor, 484, 696
DC power supply, characteristics, 171
Dead time, 363, 669–673
Decade, 439
Decibel, 438
Default integrator, 37
Delay theorem, 363
Delayed step function, 364
Delayed-action mechanism, 404
Density, fluid, 219
Derivative, noise accentuation, 37
Describing function, 627

Design, 6
  comparison of viewpoints for mechanical, electrical, and thermo/fluid systems, 206
  conceptual, 504
  constraints, 104
  control systems, role of simulation, 199, 200
  detail, 61, 504
  feedback increases design freedom, 685
  functional concept versus practical implementation, 115
  nonlinear, 431
  of systems, 23
  philosophy, 685, 686
  practical details, 115
  specifications, 104
  substantive, 504
  system versus detail, 61, 102, 107
  team, 504
  use of Routh criterion, 628
  utility of manual Bode plots, 634
Design examples
  accelerometer, 104
  accumulator surge-damping system, 478
  approximate integrator, 459
  comprehensive, 503
  computer-aided motion control system, 192
  electric drive for machine slide, 419
  flywheel, 99
  high-speed scale, 538
  low pass filter, 457
  motor vibration isolation, 550
  op-amp circuit, 578
  optimum decelerator, 107
  optoelectronic sensor, 462
  package cushioning, 535
Design philosophy, effect of computer aiding, 185
Determinants, to solve sets of simultaneous equations, 345, 349
Deterministic inputs, 10
Device size versus wavelength considerations, 124
Diagrams (pictorial, schematic, block), 7
Dielectric, 131
Dielectric coefficient, 159
Dielectric heating, 311
Dielectric loss coefficient, 311
Differential equations
  classical operator method, 338

[Differential equations]
  complementary solution, 339, 341
  forcing function, 341
  Laplace transform method, 350
  particular solution, 339, 341, 343
  simultaneous equations, 344
  solution, 337, 625
  steady-state solution, 344
  transient solution, 344
Differential operator (D operator), 56
Differential/difference equations, 672
Differentiator
  approximate electrical, 500
  mechanical, 431–433
  noise accentuation, 459
Diffusion equation, 720
Digital computer, 125
Digital electronic device, 125–126
Digital simulation, 370
Digital simulation language, 33
Digital-to-analog converter, 188, 190, 191, 195
Discontinuous inputs, 364, 387
Discrete delay, 363
Discrete spectrum, 648, 664
Displacement, of hydraulic pump, 283
Displacement transducer, 75
Distance-velocity lag, 669–675
Distributed model, heat conduction, 719–724
Distributed-parameter models, 14, 16, 707
  of spring, 41
Distributions, theory of, 136
Double-integral control, 506
Drag force, 71
  on cylinder, 71
  on sphere, 71
Driving-point impedance, 98
Dry friction, 73
Dynamic braking, electric motor, 321, 334n
Dynamic compensation, motion control, 427
Dynamically equivalent system, 85

E/P transducer, 2, 330, 686–694, 699
Earthquake, motion input, 93
Earthquake simulation, 93
Eddy-current, induction heating, 311
Eddy-current damper, 66
Edison, Thomas, 295
Effective voltage and current, 468
Eigenvalues, 399

Index

Electo-optics, 123
Electric circuit equations, guidelines for setting up, 564–566
Electric field, capacitor, 133
Electric motor
  algebraic signs, 485
  sizing software, 425
  table of parameters, 424
  thermal time constant, 426
  types, 288
Electric motor control, field and armature, 595
Electric power, 128
  conversion to mechanical, 289
Electric systems, design rule for cascaded circuits, 458
Electric vehicle control, 618
Electrical first-order systems, 450
Electrical impedance methods, 148, 466
Electrical models, lumped versus distributed, 124
Electrical noise, in resistor, 158
Electrical second-order systems, 562
Electrical sources, 167
Electrodynamic shaker, 96
Electrohydraulic control system, 324
Electrohydraulic shaker, 93
Electromagnetic torque, generator, 275
Electromechanical analogies, 148
Electromechanical systems, first-order, 484–489
Electropneumatic transducer, 2, 330, 686–694, 699
Elements of systems, 28
  linear, 30,
  pure and ideal, 29
Emissivity, 260
Energy, kinetic, 9
Energy, potential, 9
Energy converters, 272
Energy dissipation, 34
Energy port, 98
Energy storage
  effect on free oscillations, 266
  capacitor, 134
  fluid compliance, 235
  fluid inertance, 242
  inductor, 142
  mass, 82
  spring, 34
Entrained air, in liquids, 236

Equation setup
  electrical systems, 451–456
  fluid systems, 472
  general case, 621–623
  mechanical systems, 406–410
  thermal systems, 479–480
  using block diagram, 487
Equations, right-hand side/left-hand side, 409
Equivalent damping constant, 88
Equivalent dynamic system, 421
Equivalent inertia constant, 88
Equivalent spring constant, 88
Equivalent system, mechanical, 85
Errors, methods to find, 581
Euler integrator, 37
Expansion by minors, determinant, 346, 349
Expansion, thermal, 313
Experimental data, simulation of, 36
External driving, 8, 9

Failure in engineered systems, 159
Farad, 131
Fast Fourier transform software, 654–667
Fatigue failure, 546, 560
Feedback, benefits of, 488–489
Feedback control systems, 171, 174, 185, 186
  design theory, 193
Feedback resistor, op-amp, 177
Feedback system design, 672
Feedback system examples
  compensated, 610
  dynamically compensated, 428
  electromechanical speed control, 486
  flow control, 254
  piezo motion control, 309
  self-leveling air spring, 679
  temperature source, 267
  tension control, 695
  vibration isolation, 683
Ferromagnetic materials, 138
FFT (fast Fourier transform) software, 654–667
Fictitious mass, aid in equation setup, 410
Field circuit, generator, 273
Field models, 16
Field-oriented control, induction motor, 293
Film coefficient of heat transfer, 258
Filters
  active and passive, 454

[Filters]
active low pass, 457
anti-alias, 192, 667
band-pass, 182, 572
digital, 575
high pass, 461, 499
ideal low-pass, 563
low pass, 452
notch, 573
passive second-order low-pass, 562
Final value theorem, 366
Finite-element method, used to get lumped parameters, 17
Finite-element models, 17
Finite-element software, 622, 726
First harmonic, Fourier series, 648
First-order systems, 51, 403
electrical, 450, 453
electromechanical, 484–489
five percent settling time, 413
fluid, 470
frequency response, 433
generic, 411
hydromechanical, 489–492
impulse response, 444
lab testing, frequency response, 441
lab testing, step input, 417
meaning of K and τ, 414
mechanical, 404
optimum step response, nonlinear, 429
ramp response, 431
random input, 432
simulation, 419
standard form, 412
steady-state gain, 411
step response, 412
thermal, 479
thermomechanical, 492–498
time constant, 411
transfer functions, 412
with numerator dynamics, 498
Fixed-step-size integrator, 37
Flexible automation, 429
Flow rate, fluid, 217
Flow resistance, air spring system, 681
Flow source, 285
Flow velocity, 217
Flowmeter, electromagnetic, 312
Fluid and thermal elements, 206
Fluid compliance (capacitance), 220, 235
elastic tube wall, 235
element, 234-240

[Fluid compliance]
gas bubbles in liquid, 234
Fluid coupling, 406
Fluid energy, conversion to other forms, 311–313
Fluid first-order systems, 470
Fluid friction, 219
Fluid impedance, 248–253
Fluid inertance, 220
element, 240–245
laminar flow, 242
one-dimensional flow, 240
turbulent flow, 243
Fluid inertia effect, immersed solid bodies, 244
Fluid mass density, 219
Fluid power, 217, 222
converting from mechanical, 282
Fluid resistance, 216
adjustable, 233
element, 216, 219, 221
linearized, 222
orifice, 228–231
theoretical formulas, 223–231
Fluid sources, 253–255
ideal flow, 253
pressure, 253
real flow, 253
real pressure, 255
Fluid system models, comparison of lumped and distributed, 245
Fluid systems, equation setup guidelines, 472
Fluid systems, second-order, 579
Flux-vector control, induction motor, 293
Flywheel, 75
design example, 99
used to smooth speed fluctuations, 102
Force source, 92
Force-input transfer function, spring, 32
Forced resonance, 521
Forced response, 417
Forces, fundamental types, 92, 408
Forcing function, 341
FORTRAN, 33
Fourier law of heat conduction, 15, 256, 719
Fourier number, 265
Fourier series, 622, 644–662, 712, 722
definition, 646
numerical calculation, 650–657

Index

743

Fourier transform, 622, 662
  inverse, 252
Free oscillation, 521
Free vibration, second-order system, 530
Free-body diagram, 408
Frequency response, 12, 38
  bandpass filter, 572
  curve-fitting software, 443
  dead-time element, 672
  first-order system, 433
  fluid pipeline, 246
  general applicability, 40
  generalized, 631
  graphs, 39
  high-pass filter, 499
  lag compensator, 503
  lead compensator, 503
  leadlag compensator, 610
  logarithmic plotting, 437, 441
  low-pass filters, 568
  MATLAB plotting, 437, 441
  matrix method, 640
  notch filter, 575
  second-order system, 546–550
  series-resonant circuit, 570
  spring element, 39
  three-mass system, 639, 642
  vibration absorber, tuned, 676
  vibration, distributed rod model, 718
  vibration isolation, force transmissibility,
    554
  vibration, rotating unbalance, 557
  used to define model validity, 85
Frequency
  cyclic, 39
  damped natural, 531
  radian, 39
  peak forced response, 547
  resonant, 547
  undamped natural, 522
Frequency spectrum, continuous, 662
Frequency range of system, 39
Frequency-response curves, fitting mea-
    sured values with analytical func-
    tions, 252
Frequency-response graphing, logarithmic
    method, 633–638
Frequency-response methods, 622
Frequency-response testing, first-order sys-
    tem, 441
Frequency-spectrum analysis, 644

Friction
  Coulomb, 54, 69, 543, 679
  hydraulic motor, 69
  modeling from lab tests, 542–544
  of hydraulic pump, 285
  various types, 542
  viscous, 54
Fuel cell, 316
Fundamental, Fourier series, 648

Gain, closed-loop, 488
Gain margin, control system, 629
Gain (amplitude ratio), 547–549
Gain-bandwidth product, op-amp, 578
Gas flow controller, micromachine type,
    234
Gas system, second-order, 583–587
Gearing
  modeling, 85–89
  effect on motor/load inertia, 89
General linear system dynamics, 621
Generated voltage, of generator, 274, 277
Generator, electrical, dc, 168, 272–277
  modeling, 171
Gibbs phenomenon, 647
Graphical user interface (GUI), 33
Gravity forces, in vibrating systems,
    527–529
GUI (Graphical user interface), 33

Harmonics, Fourier series, 648
Heat pipe, 262
Heat sink, 262
Heat transfer
  conduction, 256
  convection, 258
  radiation, 259
Heat conduction in rod, distributed model,
    719–724
  lumped model, 724–725
Heaters, electrical resistance, 156
Heating, effect on choice of electric motor,
    263
Heating, electrical, 311
Heating system, tank, 483
Helicopter vibration, 675
Helmholtz resonator, 583
Hi-fi sound system, frequency range, 124
Honeywell D-STRUT damper, 68
Human dynamic behavior, 670
Hydraulic actuator, valve-controlled, 324

Hydraulic dynamometer, 489
Hydraulic motor, 283, 311
  pump-controlled, 325
Hydraulic pump, 283
  torque, 284
Hydraulic/pneumatic software, 394
Hydrodynamic mass, 244
Hydromechanical systems
  closed-loop rotary position control, 505
  dynamometer, 490
  first-order, 489–492
  materials testing machine, 593
  motion control 505
  open-loop speed control, 491
Hydrostatic air bearing, 629
Hydrostatic transmission, 326
Hysteresis, piezoelectric systems, 309
Hysteresis, spring, 48
Hysteresis losses, 71

Ideal elements, 29
Ideal source, 9
Ideal voltage source, 10
Impedance
  analyzer, electrical, 151
  driving-point, 98
  fluid, 248–253
  measurements, electrical, 151
  measurements, mechanical, 152
  mechanical, 90
  methods, electrical, 466
  output, 99
  reactive, 152
  resistive, 152
Improper fraction, 356
Impulse function, 136, 444
Impulse response
  approximation, 446
  first-order system, 444
  second-order system, 560
Induced voltage, 138
Inductance
  energy behavior, 142
  frequency response, 142
  incremental, 138
  mutual, 138
  nonlinear, 273
  self, 138
Inductance element, 138, 141
Induction heating, 311
Induction motor, 293–298

Inductor
  core loss, 164
  design formulas, 165
  nonlinear, 143
  presence of resistance, 138
  quality factor, 164
  use of magnetic materials, 164
Inertia
  "electrical", 684
  of hydraulic pump, 285
  negligible with heavy damping, 535
Inertia effect of bodies immersed in fluid, 244
Inertia element, 75, 82
  effect of non-rigidity, 83
  energy behavior, 82, 83
  frequency response, 82, 83
  step response, 82
  transfer function, 82
Inertia force, 93
Inertial damping, 685
Inertial navigation, 95
Inertial properties, 78
  lab measurement, 79
Initial conditions, 339, 347, 350, 351, 354,
      444, 499, 502, 625
  D-operator method, 342
  electric circuit, 609
  integrator, 57
  Laplace transform method, 342
Initial energy storage, 8
Initial value theorem, 366
Ink-jet printer, 310
  bubble type, 315
Input
  almost periodic, 11
  amplitude-modulated, 12
  deterministic, 10
  periodic, 10
  sinusoidal, 12
  transient, 10
  undesired, 8
Input classification, 8, 11
Input impedance, for subsystem coupling,
      524
Input resistor, op-amp, 177
Input/system/output concept, 7
Instability, 685
  automobile, 603
  electropneumatic transducer, 694
  feedback system, 197
Insulator, 131
Integral control, 691

Index

745

Integrated circuit, resistor, 156
Integrating step size, 37
  choice of, 37
Integrator, approximate electrical, 452,
    459
Integrator types, 37
Integro-differential equations, 351
Inverse Fourier transform, 252
Iso-elastic, spring material, 48

Johnson noise, 158

Kinetic energy input, 9
Kirchhoff's laws, 177, 451

Lab testing, 6, 38, 90
  air damper, 65
  air spring area, 681, 682
  automobile, 94
  car inertia, 601
  centrifugal pumps, 285
  control system components, 194
  damper, 63
  earthquake, 93
  electric motor thermal properties, 263
  electrical elements, 152, 154
  electrical generator, 275–278
  electropneumatic transducer, 688, 689,
      693
  first-order frequency response, 441
  first-order step test, 417
  flow coefficient, 630
  fluid impedance, 248–253
  fluid pipelines, 247
  foam packaging, 537
  friction identification, 543
  general utility, 406
  heat transfer, 262
  hydraulic control, 507
  hydraulic pump, 283
  inertial properties, 79, 80
  linearized models, 45
  optimum damper, 113
  orifice, 230, 231
  periodic input, 92
  piezoelectric devices, 280
  pressure sensing, 586
  pulse input, 92
  pump leakage, 285
  random input, 92
  reciprocity theorem, 348, 625

[Lab testing]
  second-order frequency-response tests,
      549–550
  second-order step tests, 539
  spectrum analysis, 667
  springs, 51
  subsystem coupling, 524
  system friction, 594
  thermal actuator, 496
  thermal contact resistance, 257
  tire parameters, 600
  vibrating system parameter
      identification, 551
  vibration, 96
Lagrange energy method, 621
Laminar flow, 222
Laplace transfer function, 56, 355
Laplace transform, 56, 350
  definition, 351
  delay theorem, 363
  differentiation theorem, 351
  initial and final value theorems, 366
  integration theorem, 351
  linearity theorem, 351
  partial-fraction expansion, 355
  repeated roots, 358
  table, 352–353
  transfer functions, 355
Laser interferometer, 185
Laser light, measurement of, 465
Laser, manufacturing processes, 311
Lead lag device, 427
Lead-screw drive
  modeling, 85–89
  stepping motor, 305
Leadlag compensator, 607
Levers, modeling, 85–89
Limit cycle, 197, 199, 627
Linear amplifier, 317
Linear element, 30
Linearization, 43
  accumulator compliance, 239
  air bearing, 630
  average value method, 481
  centrifugal pump, 286
  electric motor system, 598
  fluid resistance, 224
  for large signals, 70
  hydraulic speed control, 491
  justification for control applications,
      474
  multivariable, 45, 286, 491

[Linearization]
orifice flow, 231
resistor, 131
use perturbation method for transfer functions, 474
Linearized model, air spring, 681
Linearized models, validation by simulation, 448
Linearized spring constant, 44
stepping motor, 302
Linkages, modeling, 85–89
Loading effect, 458, 522–527, 564, 578, 579, 598
Logarithmic decrement, 541
Logarithmic frequency-response plotting, 437
Logic operation, electronic, 126
Lookup table, simulation, 112
saving contents, 102
use of spline, 100
used for speed/torque curve, 299
used with pump curves, 285
Loop gain, 488
Loudspeaker, 123, 124, 193
Lumped models, sizing of lumps, 21
Lumped-model choice, based on frequency considerations, 246–248
Lumped-parameter models, 14
Lumped versus distributed models, pipeline, 245
Lumping rule
for wave propagation systems, 246
for rod vibration problem, 714
wavelength/system size rule, 584
LVDT motion sensor, in piezoelectric system, 309

Magnetic bearings, 51
Magnetic field, inductor, 138
Magnetic levitation system, modeling, 45
Magnetic torque effect, generator, 274, 275
Magnification factor, 547
Magnitude scaling, 369
Manganin, wire material, 154
Manometer dynamics, 583
Manufacture, integrated circuit, 678
Manufacturing considerations in design, 512
Manufacturing quality control, 152
Manufacturing tolerances, 63
MAPLE, 346, 625
Mass moment of inertia, 77

Mass units, British system, 194
Materials, piezoelectric, 279
Material-testing machine, resonant, 591
MATHCAD, 340, 346, 625, 724
MATLAB/SIMULINK, 33
MATLAB, 36
bode, 441
bode (num,den,w), 572
fast Fourier transform, 651, 654–657
fcn icon, 75
fft, 656
frequency-response graphing, 638–639
gtext, 383
hold on, 383
invfreqs, 441, 669
m file, 642
matrix frequency response, 641
partial-fraction expansion, 360
plot, 383
poly, 399, 638
repeated poles, 362
RESI2, 362
residue, 361
root finder, 340
roots, 399, 625, 637
sign function, 75
spline, 100, 376, 386, 651, 653
subplot, 384
Maxwell's equations, 124
Measuring instruments
first-order ramp error, 432
frequency-response specifications, 458
optimum damping, 538
time delay, 432, 462
Mechanical impedance, 90
damper, 90
mass, 90
parallel connection, 91
series connection, 91
spring, 90
Mechanical elements, 28
Mechanical engineering, in "electronic" systems, 123
Mechanical power, 34
Mechanisms, modeling, 85
Mechatronics, 185–200
MEMS (microelectromechanical systems), 233
Mho, 128
Micromachine technology, 233
Micromotion control, 306
Microphone, 123, 124

Index

Microstepping, stepping motor, 304
Microtechnology, 306
Microwave, 124
Mixed first-order systems, 484
Mobility, 94
Mode shape, vibration, 712, 713, 716
Modeling, experimental, 668
Models
  choice based on pertinent frequency
    range, 40
  comparison, lumped versus distributed,
    19, 42, 43
  complex versus simple, 454
  complexity, stepwise approach, 43
  complexity needed, judged from fre-
    quency spectrum, 657, 663
  distributed, 14
  finite-element, 17
  hierarchy, 582
  lumped, 14
  lumped versus distributed, 718, 719, 725,
    726
  mathematical, 10
  physical, 10
  simplification, based on frequency
    response, 640
Moment of inertia, 77
  formulas, 79
  lab measurement, 80
Moon vehicle, steering system, 190
Motion control
  clutch/brake systems, 327–334
  design of computer-aided system, 192
  speed control, 484, 486–490, 595–599
  system, 575
Motion source, 92
Motion-input transfer function, spring,
  32
Motion transformers, 85
Motor, electric
  ac induction, 291, 293
  brushless, 291–293
  brush-type, 291, 293
  dc, 274, 277, 289
  stepping, 300–306
  three-phase ac induction, 293
  translatory, 277
Motor, hydraulic, 311, 506
  variable-displacement, 312
Motor, optimum use of, 429
Multivariable linearization, 45, 260, 286,
  491

Mutual inductance, 138
  dot rule, 140

Nanotechnology, 306
Natural frequency, 712
  damped, 531
  piezo systems, 309
  stepping motor, 302
  undamped, 522, 529
Natural oscillations, lack of in some
  systems, 567
Natural response, 417
Neatness, 581
Network, versus field, viewpoint, 124
Network topology, 621
Nichrome, wire material, 154
Noise, current, 158
Noise variance, 386
Noisy signal, use of band-pass filter, 572
Nonlinear models, 661
Nonlinear system
  computer-aided system, 189
  free-convection thermometer, 481
  magnetic levitation, 46
  nonlinear damping, 72
  optimum decelerator, 107
  optimum step response, 430
  response to periodic input, 659
  RL circuit, 145
  square-law damper, 448
  three-mass system, 381
  vibration with nonlinear damping,
    543
Nonlinear versus linearized model, fluid
  system, 472
Nonlinearity, 344
  explored by using various operating
    points, 249
  intentional use in design, 30, 107, 431,
    621
Nonprocedural computer language, 38
Notch filter, 573
Nozzle flapper, 2
Nuclear fission, 316
Numerator dynamics
  automotive suspension, 555
  bandpass filter, 572
  compensated servo system, 610
  first-order systems, 498
  force transmissibility, 553
  notch filter, 573
  power-factor correction, 571

[Numerator dynamics]
  second-order systems, 599
  simulation technique, 612
  sixth-order (three-mass) system, 626
  tank heating system, 589
  tuned vibration absorber, 674
Numerical integration, 37

Octave, 439
Ohm's law, 127
Ohmmeter, 127, 131
One-dimensional flow model, 217
Op-amp circuits, 576
  dynamic compensator, 428
  general second-order, 563
  photodiode, 465
Operational amplifier ("Op-amp"), 125,
    174, 317
  assumptions, 176
  bias current, 182, 183
  circuit with general impedances, 181
  close-loop gain, 179
  coefficient multiplier, 177
  differential input, 176, 180
  errors in simple model, 182–184
  frequency response, 184
  input impedance, 184
  integrator, 177, 180
  inverter, 180
  matched characteristics, 180
  model, 175
  offset voltage, 182
  open-loop gain, 179
  output impedance, 184
  power output, 184
  settling time, 184
  single-ended input, 176
  summer, 177, 180
  summing junction, 177, 180
  trimming (nulling), 182
  used to avoid use of inductor, 163
  virtual ground, 180
Open-circuit output voltage, generator, 277
Open-loop motion control, stepping motor,
    303
Operating point for linearization, 43
Operational transfer function, 6
Optical sensor, design example, 462
Optimum step response, nonlinear first-
    order system, 429
Optimum damping ratio, 538
Optoelectronics, 123

Orifice, 228–231
  discharge coefficient, 230
  inertia and compliance effects, 231n
Output impedance, 99
  current supply, 172
  for subsystem coupling, 524
  voltage supply, 171
Output matrix, 397
Output vector, 397
Overall coefficient of heat transfer, 259
Overdamping, 529
Overheating, generator, 275

Package cushioning, 535
Pade approximant, dead time, 672
Parallel-axis theorem, 88
Partial differential equations, 14, 707
  solution methods, 710
Partial-fraction expansion, 351
Particular solution, 339, 341, 343
Pass band, filter, 182
Passive device, electrical, 125
Passive elements, 29
Perfect gas law, 237
Period of oscillation, 541
Periodic inputs, 10
  nonlinear system response, 661
  practical importance, 646
  system response to, 644–662
  system response using simulation, 657
Period steady-state response, 649
Permanent magnet field, 273
Permanent pressure drop, 229
Permeability, magnetic material, 164
Permittivity, dielectric, 159
Perturbation analysis, 449
  air bearing, 630
  centrifugal pump, 286
  electric motor, 598, 599
  fluid system, 473
  gas system, 585
  hydraulic speed control, 491
Phase angle, degrees or radians, 153
Phase shift, 39
Phase-lag compensator, 502
Phase-lead compensator, 502
Phasor method, 632
Photodiode, 453
Photodiode detector, 462
Pictorial diagram, 7
PID control, 607
Piezoelectric actuation, 233, 686–688

Index

Piezoelectric materials, 279
Piezoelectric motion control, 306–310
Piezoelectric sensors, 278, 453
Piezoelectric system, low-voltage, 309
Piezoelectric systems, 277–282
Pipe flow, distributed model frequency
    response, 227
Pipeline dynamic model
  comparison with measured results, 247
  dynamic model, distributed, 247
  lumped, 243
Pipes, fluid resistance, 217
PM (permanent magnet) field, generator,
    273
Pneumatic actuator, low friction, 66
Pneumatic capacitors, 240n
Pneumatic phase-lag compensator, 500
Pneumatic processes, simplifying assump-
    tions, 501
Pneumatic transmission line, 669
Pneumatic valve positioner, 687
Poles, 355
Polytropic gas process, 240
Positive-displacement pump, 253
Potential energy input, 9
Power
  electrical, 128
  fluid, 222
  interconverting various forms, 272–316
  mechanical, 34, 272
  thermal 261
  use of SI units, 128
Power amplification, 125
Power amplifier, 174, 184, 316–334
  clutch/brake type, 328, 329
  hydraulic, 324
Power factor, 570
Power factor angle, 468
Power factor correction, 570
Power modulators, 316
Power op-amp, 184
Power plant cooling system, 208
Pressure, fluid, 217
Pressure-measuring systems, 470
  gas, 583–587
Pressure sensor, integrated circuit, 686
Pressure source, 285
Printed circuit board, heat transfer, 257
Procedural computer language, 38
Process control, pneumatic, 501
Process control models, 672
Product of inertia, 78

Proof mass, 75
Propagation velocity
  electric waves, 124
  in gas, 124
  in liquid pipeline, 245, 584
Proper fraction, 356
Proportional control, 195
Proportional-plus-integral control, 205
  approximate, 502
Pseudo-inertia, 684
Pulse testing, 92
Pulse-width-modulation (PWM) amplifier,
    317
Pumps, 282
  centrifugal, 282, 285
  electromagnetic, 310
  positive-displacement, 253, 282, 489
  variable-displacement, 254, 325, 505
Pure and ideal elements, 29
PWM (pulse-width-modulation) amplifier,
    291

Q (quality factor), 164, 166
Quality control, manufacturing, 152, 159
Quantization, 126, 186
Quartz, spring material, 48

Rack and pinion system, 405, 419
Radar, 124
Radio receiver, 547, 569
Ramp function, 59
Ramp response
  first-order system, 431
  second-order system, 544
Random input, 10, 432, 697
  loading example, 526
  response to, 12
  stationary, 14
  unstationary, 14
Random number, variance, 386
Random number seed, 386
Random signal, filter for smoothing,
    386
Random signal generator, 174
RCL meter, 151
Reactance
  inductive, 153
  capacitive, 153
Real spring step response, 38
Reciprocity theorem, 348, 623
Redundant systems, 209

Referral of elements, in motion transformers, 85
Referred inertia and friction, induction motor drive, 298
Regenerator, heat transfer device, 589
Regression software, use in modeling, 276
Reliability, electronic components, 158, 163
Relief valve, 254
Repeated roots, 358, 360
Residues, 361
Resistance
  electrical, 127
  fluid, 221
  thermal, 261
Resistance, incremental, 131
Resistance element, electrical, 127
Resistance, heating, 311
Resistivity, electrical, 154
Resistor
  carbon-composition, 156
  heating of, 128
  integrated circuit, 156
  low inductance, 140
  manufacture, 154, 156
  nonlinear, 131
  practical considerations, 154–159
  small temperature effect, 154
  step response, 157
  temperature coefficient, 158
  useful frequency range, 156
  voltage coefficient, 158
  wirewound, 154
Resonance, 42
  electrical, 568
  for periodic inputs, 657
  mechanical, 547
  practical use of, 591
  use of notch filter to suppress, 573
Reynolds number, 222, 223
Rigid body, 75
Rise time, 104
Rocket engine instability, 207
Root finder, 340, 358, 625
Root-mean-square (rms) value, 469
Roots, cubic and quartic, 340
Rotating field, 3-phase induction motor, 295
Rotating unbalance 555–560
Rotor, of electric motor, 291
Routh stability criterion, 627, 685
Runge–Kutta integrator, 37

Sample-and-hold amplifier, 188
Sampling, 126, 186
  choice of frequency, 187
Sampling theorem, 186, 187
Satellite temperature control, 215
Saturation, 199, 691
  amplifier in piezo circuit, 308
  generator magnetic field, 275
  limit on performance, 685
  magnetic, 144
Scale models, 10
  building/earthquake, 93
Schematic diagram, 7
Scotch yoke mechanism, 97
Second-order system, 5
  absence of free oscillations, 567, 583, 587
  critically damped, 529, 533
  damped natural frequency, 531
  damping ratio, 522, 529, 537
  electrical systems, 562
  electromechanical, 595–599
  fluid, 579–587
  free vibration, 527, 530
  frequency response, 546–550
  from cascaded first-order, 522
  hydromechanical, 591–595
  impulse response, 560
  initial condition response, 530–531
  logarithmic plots, 547–549
  mechanical, 527
  mixed, 591
  numerator dynamics, 554, 572, 589, 599
  op-amp circuits, 576
  optimum damping, 538
  overdamped, 529, 534
  peak frequency, 547
  ramp response, 544
  settling time, 538
  significance of K, ζ, ω_n, 537
  standard form, 522
  steady-state gain, 522, 529, 537
  step input lab tests, 539
  step response, 527
  thermal, 587
  undamped, 529, 532
  undamped natural frequency, 522, 529, 537
  underdamped, 529
Self-inductance, 138
Semiconductor diode, 130
Sensors, 186
  acceleration, 75, 104, 459

Index 751

[Sensors]
    position, 193, 695
    pressure, 470, 583–587, 686
    speed, 486–487
    temperature, 480
Series resonant circuit, 568
Servomechanism, 89, 153, 628
Servovalve, 313, 592
    hydraulic, 323
    multistage, 324
Settling time
    first-order system, 413
    second-order system, 538
Shakers, vibration
    electrodynamic, 96
    electrohydraulic, 93
Shannon's sampling theorem, 186, 187
Shock absorber, automotive, 61
Sideslip angle, car, 601
Siemen, 128
Sign conventions, 31, 406
    electrical systems, 451, 452, 565
    mechanical systems, 406–411
Signal frequency content, 187
Signal generator, electronic, 173, 174
Silicon-controlled rectifier (SCR) amplifier, 317
Silicone oil, for dampers, 61, 729
Simplification of models, 8
Simulation examples
    acceleration of rotating unbalance, 559
    automobile handling, 604
    band-pass filter, 573
    coffee cooling, 590
    comparison of linear and PWM amplifier, 319
    compensated servo system, 610
    computer-aided motion control system, 196
    computer-aided system, 189
    damper design and comparison, 112
    dynamic compensation of motion control, 428
    electropneumatic transducer, 689
    engine flywheel, 101
    event-controlled switching, 392
    first-order system, 420
    Fourier series, 651
    friction types, 72, 543
    ideal and real inductors, 145
    induction motor/clutch system, 331
    induction motor drive, 299

[Simulation examples]
    integrator design study, 461
    material test machine, 594
    motor driven machine slide, 422
    multiple parameter values, 390
    nonlinear and linearized model comparison, 450
    notch filter solves resonance problem, 575
    optimum step response, 430
    PWM amplifier, 319
    random signals, 388
    repeated roots, 361
    rotary motion control, 508
    SCR amplifier, 322
    segmented inputs, 387
    spring/damper comparison, 57
    spring energy, 35
    stepping motor, 303
    tank-level control, 393
    tension controller, 697
    thermal actuator, 494
    three-mass system, 643
Simulation
    comparing alternatives, 387
    detailed levels, 419
    digital, 370
    earthquake, 93
    hardware-in-the-loop, 607
    methods, 367
    real-time, 605
    side-by-side, 57
    software, 22, 32
    special-purpose, 177
SIMULINK, 33, 370
    adding notes to graphs, 383
    block orientation, 380
    connecting blocks, 380
    copying icons, 373
    deleting lines and icons, 381
    diagram labeling, 381
    division operation, 378
    event-controlled switching, 391
    fixed step size, 382
    graphing, 379, 381–385
    group, ungroup, 642, 643
    initial conditions, 376
    input signal generation, 385
    integrators, 376, 382
    jagged graphs, 382
    large simulations, 642
    linear blocks, 376

[SIMULINK]
main menu, 372
multiple parameter values, 390
nonlinear modules, 377
printing, 385
random signals, 386
running a simulation, 381
sizing icons, 36, 375
sources, 374
squaring corners, 381
start and stop times, 381
starting new simulation, 374
step size, 382
variable step size, 382
SIMULINK icons
absolute value, 73, 377
A/D converter (quantizer), 188
backlash, 378
chirp, 376
clock, 375
constant, 375
D/A converter (quantizer), 188
dead zone, 378
derivative, 37, 377, 697
function, 100, 378
gain, 34, 37, 376
integrator, 37, 57, 376
lookup table, 36, 378, 385, 651, 658
mux, 652
outport, 389
product, 37, 73, 378, 392
pulse generator, 375
quantizer, 188
random signal, 375
relay, 23, 378, 392, 425
repeating sequence, 319, 376, 590, 658
saturation, 378
signal generator, 144, 375
sine wave, 375
step function, 375
summer, 34, 376
switch, 379, 393, 590
to workspace, 36, 381
transfer function, 377
transport delay, 190, 379, 434
xy graph, 380
zero order hold, 188
Simultaneous differential equations, 344,
    525, 566, 578, 581, 588–590, 594,
    598, 603, 623, 630, 674, 682, 689,
    715–716, 725
Single-integral control, 506

Sinusoidal input, 12, 38
Sinusoidal transfer function, 41, 60
general proof, 631–633
Size versus wavelength considerations, 124
Skin effect
in conductors, 167
induction heating, 311
Skyscraper vibration absorber, 677
Slip, 3-phase induction motor, 296
Slip angle, automobile, 600
Software, used for filtering, 575
Soldered connections, reliability, 158
Sound system, frequency range, 124
Source
electrical, 167
force, 92
ideal, 9
motion, 92
Specific heat, 265
Specific weight, of fluid, 239
Specifications, for design, 104
Spectrum analysis, 644–669
Spectrum analyzer, 667
Speed control
closed-loop electromechanical, 486–489
closed-loop hydromechanical, 486, 492
open-loop electromechanical, 484,
    595–599
open-loop hydromechanical, 490
Speed increase, using phase-lead compen-
    sator, 502
Speed of sound
in air, 124
in pipeline, 245
Speed measuring instrument, 431
Speed/torque curve, ac induction motor,
    295–299
Spline function (smoothing), 36
used in lookup table, 100
Spring
aerodynamic, 51
air, 50
automotive tire, 53
buoyancy, 51
centrifugal, 53
distributed-parameter model, 41
electrostatic, 51
gas, 676
hydraulic, 48
liquid column, 51
magnetic, 51
pendulum, 51

Index

[Spring]
  rubber, 48
Spring element, 29, 31
Springs
  compliance, 32
  constant, 30, 32
  design formulas, 49
  effect of mass, 41–43
  energy losses, 47
  materials, 48
  spring constant, linearized, 44
  stiffness, 30, 32
Squirrel-cage ac induction motor, 291, 293,
    295
Stability
  absolute, 626
  general discussion, 626
Stalled-torque test, of generator, 275
Standard form
  first-order system, 412
  second-order system, 522
State variable, 397
State-variable notation, 396
State vector, 397
Static sensitivity, 5, 411
Stationary random input, 14
Statistical uncertainty analysis, 63
Stator, of electric motor, 291
Steady-state error, 197
Steady-state gain, 5, 411, 547, 549
Steady-state solution, 344
Steer angle, car, 602
Step input, 38
Step-input testing
  first-order system, 417
  second-order system, 539
Step size, integration, 37
  choice of, 37
  fixed, 37
  variable, 37
Stepping motor, 300–306
  magnetic spring constant, 302
  microstepping, 304
  natural frequency, 302
  torque expression, 302
  variable-reluctance, 301
Stereo sound system, 40
Strain, unit, 708
Strain gage motion sensor, in piezoelectric
    system, 309
Structural damping, 74

Substitution and elimination, simultaneous
    equations, 345
Subsystem coupling, 91, 524
Supercapacitor, 134
  as power source, 168
Supercharging, hydraulic system, 327
Superposition principle, 337, 343, 344, 483,
    530, 649, 712
Symbolic processor, 346, 349, 625
Symbols, choice of, 33
Symmetric matrix, 348
Synchronous speed, 3-phase induction
    motor, 295
System analyzer, 668
System design, 6, 23, 61
System dynamics, definition, 1
System engineer, viewpoint, 275
System matrix, 397
System models, classification, 14
Systems with several inputs, 483

Tachometer encoder, 486
Tachometer generator, 486
Tank, fluid compliance, 329–240
Tank heating model, improved, 587
Tank systems, 579
  control, 470
Taylor series, linearization, 44, 45
Temperature control system, 4, 213
Temperature detector, resistance, 158
Temperature sensitivity of damping meth-
    ods, 67
Temperature sensitivity of systems, 8
Temperature sensors, 158, 314, 480
  filter for, 457
Terminated ramp input, 104, 136
Tesla, Nicola, 295
Theory of distributions, 136
Thermal actuation, 233, 314
Thermal capacitance, 263–266
  for phase change, 265
Thermal conductivity, 256
Thermal conductivity integral, 257
Thermal diffusivity, 265, 720
Thermal energy, conversion to other forms,
    313–315
Thermal expansion, 313
Thermal first-order systems, 479
Thermal inductance, 266
Thermal power, conversion from mechani-
    cal, 287

Thermal resistance, 255–263
  conduction, 257
  convection, 258, 259
  electronic device cooling, 262
  overall, 259
  radiation, 260, 261
Thermal sources, 266–268
  heat flow, 267
  temperature, 266
Thermal system properties, 735, 736
Thermal systems
  absence of free oscillations, 587
  equation setup guidelines, 479
Thermistor, 158
Thermocouple, 314
Thermodynamics, second law, 47
Thermoelectric heating/cooling, 311
Thermomechanical systems
  first-order, 492–498
  friction brake, 496
  thermal expansion actuator, 492
Thermometer dynamic model, 480
Time constant, 5, 411
Time scaling, 369
Time response, from frequency response,
    252
Tire, spring effect, 53
Torque, magnetic, in generator, 275
Torque ripple, of brushless motor, 293
Tradeoff, motor armature inductance,
    321
Transconductance amplifier, 290
  with piezo device, 307
Transducer, 173, 272
Transfer function, 6
  Laplace, 355
  multiple inputs and outputs, 347–348
  operational, 56
  sinusoidal, 60
  sinusoidal, proof, 631
  spring, 32
Transformer, motion, 85
Transient duration, relation to frequency
    content, 664
Transient input, 10
Transient resonance, 558, 676
Transient signals, frequency spectrum,
    662
Transient solution, 344
Transistor, 125
Transition flow, 223
Translatory motor, 277

Transmissibility, 550
  force, 553
  motion, 555
Transport delay, 363, 669–675
Transport lag, 363, 669–673
Tuned circuit, 568
Turbine, 311
Turbomachine, 312
Turbulent flow, 222, 243

Unbalance, rotating, 546
Unbalance forces, 95
Uncertainty analysis, 63
Underdamping, 529
Undesired input, 8
Undetermined coefficients, method of,
    339
Units and conversion factors, 733, 734
Unstationary random input, 14

Vacuum diode, 130
Valve
  adjustable fluid resistance, 233
  micro size, 233
  piezoelectric, 310
  relief, 254
Variable multiplier and divider, 367
Variable-displacement pump, 254
Varistor, 130
Vehicle dynamics, 78
Vehicle steering forces, 97
Velocity profiles, various flow conditions,
    241
Velocity-sensing mechanism, 404
Vibrating conveyor, 547
Vibration, 521
  force transmissibility, 553
  forced, 546
  free, 527
  motion transmissibility, 555
  rod longitudinal, 707
  rotating unbalance, 555
  torsional, in ac motor drives, 295
Vibration absorber, tuned, 673
Vibration isolation, 50, 550
  active electromechanical, 685
  improved passive, 680
  self-leveling air spring, 678
Vibration shaker, 318, 668
Virtual ground, op-amp, 180
Viscosimeter, 63, 120

Index

Viscosity, 63
   charts for silicone damping fluids,
      729–731
Viscous friction, 54
Voice-coil actuator, 193, 683
Volt-amperes, 570
Voltage, induced, 138
Voltage loop law, 451
Voltage recorder, use of notch filter, 573
Voltage source, 167
   ideal, 10, 167
   real, 171
Voltmeter, true rms, 469

Wave equation
   pipeline, 245
   rod vibration, 709
Wave propagation, 124
   in gas, 583–584

[Wave propagation in gas]
   wavelength/frequency/velocity relation,
      124, 245
Waveguide, 124
Wavelength/velocity/frequency relation,
      124, 245
Web-tension control system, 695
Wind-force input, 93
Windage torque, electrical machines, 71
Wire, resistance calculations, 154
Wirewound resistor, 154
Wiring, in circuits, 142
WORKING MODEL software, 394
Wound field, 273

Yaw angle, car, 601

Zero-order system, 414
Zeros, 355

HUMBERSIDE UNIVERSITY CAMPUS HULL